The Human Body

An Introduction for the Biomedical and Health Sciences

Gillian Pocock & Christopher D. Richards

OXFORD
UNIVERSITY PRESS

OXFORD

UNIVERSITY PRESS

Great Clarendon Street, Oxford OX2 6DP

Oxford University Press is a department of the University of Oxford.
It furthers the University's objective of excellence in research, scholarship,
and education by publishing worldwide in

Oxford New York

Athens Auckland Bangkok Bogotá Buenos Aires Calcutta
Cape Town Chennai Dar es Salaam Delhi Florence Hong Kong Istanbul

Karachi Kuala Lumpur Madrid Melbourne Mexico City Mumbai

Nairobi Paris São Paulo Singapore Taipei Tokyo Toronto Warsaw

With associated companies in Berlin Ibadan
Oxford is a registered trade mark of Oxford University Press
in the UK and in certain other countries

Published in the United States
by Oxford University Press Inc., New York

© G. Pocock and C. Richards 2009

British Library Cataloguing in Publication Data

Data available

Library of Congress Cataloguing in Publication Data

Data available

ISBN 978–0–19–928907–3

10 9 8 7 6 5 4 3 2 1

Typeset by Macmillan Publishing Solutions
Printed in Italy
on acid-free paper by
Rotolito Lombarda SpA

The Human Body

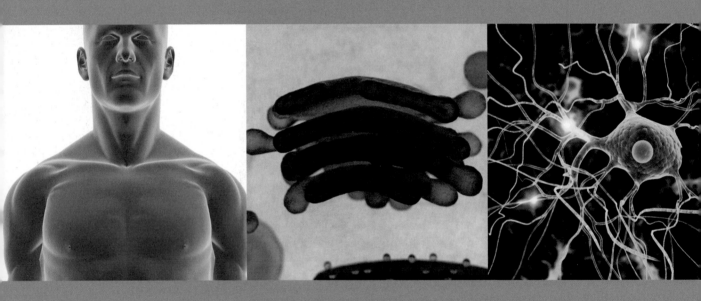

To our long-suffering families

Preface

The idea for this book grew out of our discussions concerning the rapid changes that were occurring in the teaching of the biomedical and health sciences. We felt that an introductory textbook that covered human anatomy, physiology and pharmacology would be a useful learning resource. For this reason, the text is primarily written for students of the health-care sciences so that the clinical implications of the subject matter are emphasized wherever it is appropriate. We have assumed a basic knowledge of chemistry and biology similar to that expected from British students with 'AS' levels in these subjects but we have included a brief discussion of elementary chemistry for those students with no background in that subject.

Our intention has been to provide straightforward accounts of the anatomy and histology of the human body, together with the physiology and pharmacology of the different organ systems. Our treatment of physiology is considerably more detailed than some other texts that cover similar topics. We make no apology for this, as a proper appreciation of the principles of physiology is necessary to understand the modern treatment of many disorders.

The book begins with some elementary chemistry and then proceeds to consider the fundamental organization of cells, their basic properties and their interactions with other cells. The properties of excitable cells (nerve and muscle cells) are then presented in some detail before proceeding with a discussion of the main body systems. A brief discussion of genetics is included after the chapters on human reproduction. The final chapters are mainly concerned with integrative topics: nutrition; energy balance and metabolic rate; the physiology of exercise; the regulation of body temperature; the regulation of body fluid volume and acid–base balance. The book concludes with an introduction to pharmacokinetics. In providing straightforward accounts of complex topics, it has occasionally been necessary to omit some details or alternative explanations. Although this approach occasionally presents a picture that is more clear-cut than the evidence warrants, we believe that this is justified in the interests of clarity.

As far as possible, the illustrations have been presented as simple line drawings or simple images. We have not included extensive accounts of the experimental techniques of modern biomedical research but have tried to make clear the importance of experimental evidence by using original data from the published literature wherever it helps to illustrate a key point. Normal values have been given throughout the text in SI units but important physiological variables have also been given in traditional units (e.g. mmHg for measurements of blood pressure).

To aid student learning, key points are presented in separate feature boxes and we have tried to encourage self-testing by setting some self-assessment questions at the end of most chapters. Annotated answers to the questions are given in all cases. The reading material given at the end of each chapter is intended to provide sources from which more detailed information can be obtained before searching the primary literature.

We are deeply indebted to our colleagues: to Professor Tim Arnett who provided us with access to a large bank of histological material and research images, to Dr Ted Debnam who read through and constructively criticized the chapters on the gastrointestinal tract and to Dr Barrie Higgs of the Department of Anaesthesia, Royal Free Hospital, London, who advised us on clinical matters. We wish to thank our editor Jonathan Crowe who has given us much encouragement and useful advice. Finally, thanks are due to our copy-editor Elizabeth Paul and the staff of Oxford University Press for their work in preparing the manuscript for publication. Any remaining obscurities or errors are entirely our responsibility.

G.P. and C.D.R.
London 2008

Acknowledgements

In this book we have used many original histology images that were kindly provided by Professor Tim Arnett of University College London. He has also allowed us to use his equipment to prepare our own images. We are very grateful for this help. We also wish to thank those who have granted permission to reproduce figures for us to use either unmodified or in a modified form. The original sources are listed here:

Figure 4.1 is adapted from an original figure of J.D. Robertson. Figure 4.5 is based on Figure 2-18 of B. Alberts, D. Bray, J. Lewis, M. Raff, K. Roberts and J.D. Watson (1989) *Molecular Biology of the Cell*, 2nd edn, Garland, New York. Figure 4.7 is based on Figure 12.13 of W.H. Elliott and D.C. Elliott (2005) *Biochemistry and Cell Biology*, 3rd edn, Oxford University Press, Oxford. Figure 4.10 is adapted from Figure 3.24 of L. Weiss and R.O. Greep (1977) *Histology*, 4th edn, McGraw-Hill, New York. Figure 6.6 is reproduced courtesy of Dr R.C. Wagner of the University of Delaware. Figures 6.8, 10.15, 10.17b and 10.20 are reproduced courtesy of Professor Tim Arnett of UCL, London. Figure 6.10 is adapted from Figure 5.3 of P.R. Wheater, H.G. Burkitt and V.G. Daniels (1979) *Functional Histology: A Text and Colour Atlas*, Churchill-Livingstone, Edinburgh. Figures 7.3 and 17.9 are based on Figures 2.3 and 13.6 of H.P. Rang, M.M. Dale and J.M Ritter (1995) *Pharmacology*, 3rd edn, Churchill-Livingstone, Edinburgh. Figure 9.2 is modified after Figure 11.19 of W. Bloom and D.W. Fawcett (1975) *Textbook of Histology*, W.B. Saunders and Co., which was drawn by Sylvia Collard Keene. Figure 9.6 is based on data of A.M. Gordon, A.F. Huxley and F.J. Julian (1966) *Journal of Physiology*, Vol. 184, pp. 170–192. Figures 10.17a and 10.18 are based on Figures 7.2 and 7.20 of A.A. Maximow and W. Bloom (1957) *Textbook of Histology*, W.B. Saunders and Co., Philadelphia. Figure 10.18 is from D.W. Fawcett and R.P Jench *Concise Histology* 2nd edn (Arnold, 2002), reproduced by permission of Edward Arnold. Figures 10.19 and Figure 2 of Box 10.1 were kindly provided by Ruth Denton. Figure 11.1 is a compilation from images of D. Agamanolis and F. Guillemot; and from Figure 3(a) of F. Zhang *et al.* (2008) *Molecular Pain* 4: 15. Figures 11.2, 11.3, 11.4, 11.5, 11.7, 11.9, 11.11, 12.8, 12.9, 12.10, 12.12, 12.14, 14.3, 14.7, 14.8, 14.9, 14.22 and 14.31 have been adapted and redrawn from Figures 2.1, 2.7, 2.26, 2.28, 2.43, 2.38, 2.39, 9.1, 11.2, 11.13, 10.2, 10.3, 4.1, 4.13, 4.14, 13.24, 13.26 and 13.12 of P. Brodal (1992) *The Central Nervous System. Structure and Function*, 2nd edn, Oxford University Press, New York. Figures 14.5, 14.30 and 14.32 are based on Figures 16.6, 16.10 and 17.6 of H.B. Barlow and J.D. Mollon (eds) (1982) *The Senses*, Cambridge University Press, Cambridge. Figures 14.13 and 15.2 are based on Figures 7.18 and 13.4 of R.H.S. Carpenter (1996) *Neurophysiology*, 3rd edn, Edward Arnold, London. Figure 14.12 is based on Figure 20.6 of E.R. Kandel, J.H. Schwartz and T.M Jessell (eds) (1991) *Principles of Neuroscience*, 3rd edn, Elsevier Science, New York. Figure 14.15 is from Figure 4.5 of R.F. Schmidt (ed.) (1986) *Fundamentals of Sensory Physiology*, 3rd edn, Springer-Verlag, Berlin. Figure 14.26 is based on Figure 5 of I.J. Russell and P.M. Sellick (1978) *Journal of Physiology*, Vol. 284, pp. 261–290. Figure 14.27 is based on J.O. Pickles and D.P. Corey (1992) *Trends in Neurosciences*, Vol. 15, p. 255, with the permission of Elsevier Science. Figure 14.30 is adapted from Lindemann (1969) *Ergebnisse der Anatomie*, Vol. 42, pp. 1–113. Figure 13.2 is adapted from Figure 18.2 of G.M. Shepherd (1994) *Neurobiology*, Oxford University Press, New York. Figure 15.3 is modified from Figure 6.3 of S.P. Springer and G. Deutsch (1989) *Left Brain, Right Brain*, 3rd edn, W. H. Freeman & Co., New York. Figure 15.13 is from original data of Dr D.A. Richards. Figures 16.8, 16.9, and 16.21 are adapted from Figures 3.4, 5.2 and 4.12 of C. Brook and N. Marshall (1996) *Essential Endocrinology*, Blackwell Science, Oxford. Figures 16.11, 16.16 and 16.17b are from Plates 4.1, 4.2, 9.1, 9.2 and 9.3 of J. Laycock and P. Wise (1996) *Essential Endocrinology*, 3rd edn, Oxford Medical Publications, Oxford. Figures 16.15, 16.23, 31.2 and 31.4 are printed courtesy of Wellcome

Images. Figure 16.17a is from F.M. Delange (1996) *Endemic Cretinism*, in L.E. Braverman and R.D. Utiger (eds) *Werner and Ingbar's The Thyroid*, 7th edn, Lippincott-Raven, Philadelphia. Figures 17.4 and 19.2 are courtesy of Dr S. Ruehm. Figures 17.1, and 19.6 are based on Figures 1.4 and 1.6 of J.R. Levick (1995) *An Introduction to Cardiovascular Physiology*, 4th edn, Arnold, London. Figures 17.11, 17.7, 22.1 and 22.2 are based on Figures 5.4.10, 5.4.6, 5.4.7, 5.3.3 and 5.3.4 in Vol. 2 of P.C.B McKinnon and J.F. Morris (2005) *Oxford Textbook of Functional Anatomy*, Oxford University Press, Oxford. Figure 17.5 is based on Figures 4 and 10 of O.F. Hutter and W. Trautwein (1956) *Journal of General Physiology*, Vol. 39, pp. 715–733, by permission of the Rockefeller University Press. Figure 19.8 is based on Figure 19.14 of A.C. Guyton (1986) *Textbook of Medical Physiology*, 7th edn, W. B. Saunders & Co., Philadelphia. Figure 19.8 is based on Figure 7 of A.E. Pollack and E.H. Wood (1949) *Journal of Applied Physiology*, Vol. 1, pp. 649–662. Figure 19.13 is based on Figure 5 of L.H. Smaje *et al.* (1970) *Microvascular Research*, Vol. 2, pp. 96–110. Figure 20.4 is from an original figure kindly provided by Dr E.S. Debnam. Figure 20.8 is a compilation of original photomicrographs with Figures 7.2 and 7.4 of A.W. Rogers (1983) *Cells and Tissues*, Academic Press, London. Figures 21.1, 21.2, 21.7 and 21.10 are adapted from Figures 9.1, 9.2, 13.1 and 13.2 of J.H. Playfair (1995) *Infection and Immunity*, Oxford University Press, Oxford. Figures 22.3 is based on Figure 1.5 of J. Widdicombe and A. Davies (1991) *Respiratory Physiology*, Edward Arnold, London. Figure 22.4 is from Plate 8 of P.M. Andrews (1974) *American Journal of Anatomy*, Vol. 139, pp. 399–424. Figures 22.16 and 32.6 are adapted from original figures kindly provided by Professor M. de Burgh Daly. Figure 23.3 is based on Figure 1 of W. Kritz and L. Bankir (1988) *American Journal of Physiology*, Vol. 254, pp. F1–F8. Figure 23.5 is based on Figure 2.3 of B.M. Keoppen and B.A. Stanton (1992) *Renal Physiology*, Mosby, St Louis. Figures 24.19 and 25.1 are based on Figures 6.4.6 and 6.6.2 in Vol. 2 of P.C.B McKinnon and J.F. Morris (2005) *Oxford Textbook of Functional Anatomy*, Oxford University Press, Oxford. Figures 24.31 and 24.32 are based on Figures 6.1 and 6.2 of P.A. Sanford (1992) *Digestive System Physiology*, 2nd edn, Edward Arnold, London. Figures 27.9 and 27.10 were kindly provided by Dr D. Becker, UCL, London. Figure 28.9 is based on Figure 6.10.3 in Vol. 2 of P.C.B. McKinnon and J.F. Morris (2005) *Oxford Textbook of Functional Anatomy*, Oxford University Press, Oxford. Figure 29.4 is adapted from Figure 53.3 of Young and Hobbs (1979), *Life of Mammals*, 2/e, Oxford University Press, Oxford. Figure 29.14 is adapted from Figure 13.1 of D.J. Begley, J.A. Firth and J.R.S. Hoult (1980) *Human Reproduction and Developmental Biology*, Macmillan Press, London. Figures 29.9 and 29.17 are based on Figures 7.2 and 7.3 of N.E. Griffin and S.R. Ojeda (1995) *Textbook of Endocrine Physiology*, 2nd edn, Oxford University Press, Oxford. Figures 30.1 and 30.4 were kindly supplied by David McDonald, Laboratory of Pathology of Seattle, University of Washington. Figures 32.4 and 32.5 are from Figures 7.3 and 7.10 of P.-O. Astrand and K. Rohdal (1986) *Textbook of Work Physiology. Physiological Basis of Exercise*, 3rd edn, McGraw-Hill, New York. Figure 34.5 was kindly provided by Professor R. Levick, St George's Hospital Medical School. Figure 34.7 is from Plates 9 and 10 of R.A. Hope, J.M. Longmore, S.K. McManus and C.A. Wood-Allum (1998) *Oxford Handbook of Clinical Medicine*, 4th edn, Oxford University Press, Oxford. Figures 31.5 and 34.6 are courtesy of WHO. Figure 35.8 is plotted from data of Korsten *et al.* (1975) *New England Journal of Medicine*, Vol. 192, pp. 386–389. Figure 35.9 is plotted from data of O'Reilly *et al.* (1971) *Thrombosis et Diathesis Haemorrhagica*, Vol. 25, pp. 178–186. Figure 35.10 is plotted from data of Levy (1965) *Journal of Pharmaceutical Sciences*, Vol. 54, pp. 959–969. Figure 35.11 is based on data of U. Klotz *et al.* (1975) *Journal of Clinical Investigation*, Vol. 55, pp. 347–359.

Contents

List of abbreviations xvii

Section 1 Introduction 3

1 An introduction to the human body 4

Introduction 4
The hierarchical organization of the body 5
Homeostasis 7
Terms used in anatomical descriptions 8
Self-assessment questions 13

2 Key concepts in chemistry 14

Introduction 14
Molecules are specific combinations
   of atoms 16
Water and solutions 18
Self-assessment questions 24

3 The chemical constitution of the body 26

Introduction 26
Body water 27
The major organic constituents
   of the body 27
Recommended reading 37
Self-assessment questions for
   Chapters 2 and 3 37

Section 2 The organization and basic functions of cells 39

4 Introducing cells 40

Introduction 40
The structure and functions of the cellular
   organelles 41
Cell division 45
Programmed cell death 47
Energy metabolism in cells 47
Cell motility 52
Recommended reading 54
Self-assessment questions 54

5 The functions of the plasma membrane 55

Introduction 55
The permeability of cell membranes to ions and
   uncharged molecules 56
Active transport 59
The resting membrane potential 62
The regulation of ion channel activity 62
Secretion, exocytosis and endocytosis 64
Recommended reading 66
Self-assessment questions 67

6 Cells and tissues 68

Introduction 68
Histological features of the main tissue types 69
Cell–cell adhesion 76
Specialized cell attachments 77
Recommended reading 78
Self-assessment questions 78

7 The principles of cell signalling 80

Introduction 80
Cells use diffusible chemical signals for paracrine,
   endocrine and synaptic signalling 81
Chemical signals are detected by specific receptor
   molecules 86
G protein activation of signalling cascades 89
Some local mediators are synthesized as they are
   needed 91
Steroid and thyroid hormones regulate gene
   transcription 93
Gap junctions permit the exchange of small
   molecules between neighbouring cells 94
Pharmacological aspects of drug–receptor
   interaction 95
Recommended reading 100
Self-assessment questions 100

Section 3 Excitable cells 103

8 Nerve cells and their connections 104

Introduction 104

The structure of axons 105

The primary function of an axon is to transmit information coded as a sequence of action potentials 106

Chemical synapses 112

Recommended reading 116

Self-assessment questions 117

9 Muscle 118

Introduction 118

The structure of skeletal muscle 120

How does a skeletal muscle contract? 121

The activation and mechanical properties of skeletal muscle 124

Neuromuscular transmission 129

Smooth muscle 131

Pharmacology of smooth muscle 136

Recommended reading 136

Self-assessment questions 137

Section 4 The anatomy and physiology of body systems 139

10 The musculoskeletal system 140

Introduction 140

Anatomy of the skeleton 141

The axial skeleton 143

The appendicular skeleton 149

The physiology of bone 154

Disorders of the skeleton 161

Joints 163

The skeletal muscles 167

Disorders of skeletal muscle 180

Recommended reading 181

Self-assessment questions 181

11 The organization of the nervous system 183

Introduction 183

The cells of the CNS 184

The organization of the brain and spinal cord 185

The structure of peripheral nerve trunks 191

The cerebral ventricles and the cerebrospinal fluid 192

Cerebral circulation 194

Recommended reading 195

Self-assessment questions 195

12 The physiology of motor systems 197

Introduction 197

The hierarchical nature of motor control systems 198

Organization of the spinal cord 199

Reflex action and reflex arcs 201

The role of the muscle spindle in voluntary motor activity 204

Effects of damage to the spinal cord 205

Descending pathways involved in motor control 206

Goal-directed movements 208

The role of the cerebellum in motor control 211

The basal ganglia 213

Recommended reading 214

Self-assessment questions 215

13 The autonomic nervous system 216

Introduction 216

Organization of the autonomic nervous system 217

Chemical transmission in the autonomic nervous system 221

Central nervous control of autonomic activity 224

Recommended reading 225

Self-assessment questions 225

14 Sensory systems 226

Introduction 226

The somatosensory system 229

Pain 234

Itch 238

The physiology of the eye and visual pathways 238

The physiology of the ear – hearing and balance 250

The chemical senses – smell and taste 259

Recommended reading 260

Self-assessment questions 261

15 Some aspects of higher nervous function 263

Introduction 263

The specific functions of the left and right hemispheres 263

Speech 268

The EEG can be used to monitor the activity of the brain 269

Sleep 272

Circadian rhythms 275

Learning and memory 277

Recommended reading 281

Self-assessment questions 282

16 The endocrine system 283

Introduction 283
The chemical nature of hormones 285
Measurement of hormone levels in body fluids 285
Patterns of hormone secretion – circadian rhythms and feedback control 286
The pituitary gland and the hypothalamus 286
The role of the hypothalamic hormones in the control of anterior pituitary secretion 288
The posterior pituitary hormones 289
Growth hormone 291
The thyroid gland 296
The adrenal glands 303
Hormonal regulation of plasma calcium and phosphate 310
The hormonal regulation of plasma glucose 315
Recommended reading 324
Self-assessment questions 325

17 The heart 327

Introduction 327
The organization of the circulation 328
The gross anatomy of the heart 329
The structure of cardiac muscle 330
The coronary circulation 331
The origin of the heartbeat 334
The heart as a pump – the cardiac cycle and the heart sounds 337
Cardiac output 341
Heart failure 347
Drugs that act on the heart 350
Recommended reading 352
Self-assessment questions 353

18 The electrocardiogram 354

Introduction to the electrocardiogram 354
Recording the ECG 355
The characteristics of the normal ECG 358
Clinical aspects of electrocardiography 359
Antiarrhythmic drugs 363
Recommended reading 363
Self-assessment questions 364

19 The circulation 365

Introduction 365
The anatomy of the systemic circulation 366
The pulmonary circulation 369
The structure of the blood vessels 370
Arterial blood pressure 371

Pressure and blood flow in the circulation 375
The mechanisms that control the calibre of blood vessels 376
The microcirculation and tissue–fluid exchange 382
Hypertension 384
Recommended reading 386
Self-assessment questions 387

20 The properties of blood 388

Introduction 388
The physical and chemical properties of plasma 389
The cells of the blood 391
Iron metabolism 394
The carriage of oxygen and carbon dioxide by the blood 396
Major disorders of red blood cells 400
White blood cells – leukocytes 401
Disorders of white blood cells 403
Platelets (thrombocytes) 404
Mechanisms of haemostasis 404
Failure of the normal clotting mechanisms 407
Blood transfusions and the ABO system of blood groups 409
Recommended reading 412
Self-assessment questions 413

21 Defence against infection: inflammation and immunity 414

Introduction 414
Passive barriers to infection 415
Self and non-self 416
The natural immune system 416
The adaptive immune system 419
Disorders of the immune system 427
Transplantation and the immune system 428
Anti-inflammatory drugs 429
Recommended reading 429
Self-assessment questions 430

22 Respiration 431

Introduction 431
The composition of expired air 433
The structure of the respiratory tree 434
The structure of the airways 435
The structure of the chest wall 439
Innervation of the respiratory system 440
The mechanics of breathing 441

The intrapleural pressure   443
Surface tension and the elasticity of the lungs   445
Airways resistance   446
Alveolar ventilation and dead space   446
The bronchial and pulmonary circulations   448
The matching of pulmonary blood flow to alveolar
    ventilation   449
Gas exchange and diffusion capacity of the
    lungs   451
Fluid exchange in the lungs   451
The origin of the respiratory rhythm   452
The chemical regulation of respiration   453
Other respiratory reflexes   456
Pulmonary defence mechanisms   457
Tests of ventilatory function   458
Some common disorders of respiration   459
Insufficient oxygen supply to the tissues – hypoxia
    and its causes   461
Drugs used to treat disorders of the respiratory
    system   462
Recommended reading   463
Self-assessment questions   464

23 The kidney and the urinary tract   466

Introduction   466
The anatomy of the kidney and urinary tract   467
The structure of the nephron   467
The organization of the renal circulation   471
Renal blood flow is kept constant by
    autoregulation   472
The kidneys regulate the plasma composition by
    ultrafiltration followed by selective modification
    of the filtrate   473
How can we tell whether a substance is reabsorbed
    or secreted by the kidneys?   475
The functions of the proximal tubule   477
Tubular transport in the loop of Henle   479
Ionic regulation by the distal tubules   480
The kidney regulates the osmolality of the plasma by
    adjusting the amount of water reabsorbed by the
    collecting ducts   483
The collection and voiding of urine   485
Urine composition   487
Drugs acting on the kidneys   489
Recommended reading   492
Self-assessment questions   492

24 The digestive system   493

Introduction   493

Anatomical organization of the gastrointestinal
    tract   494
The principal structural features of the gut wall   495
Peristalsis and segmentation   496
The blood supply of the gastrointestinal tract   497
Regulation of the gastrointestinal tract by nerves
    and hormones   498
Ingestion of food, chewing and salivary
    secretion   501
Swallowing (deglutition)   504
The stomach   506
The small intestine   515
Digestion   519
The absorption of digestion products by the small
    intestine   521
The large intestine   525
Recommended reading   531
Self-assessment questions   531

25 The liver and gall bladder   533

Introduction   533
The structure of the liver   534
The hepatic circulation   536
The role of bile in the digestion and absorption of
    fats   537
The excretory role of the bile   538
Energy metabolism and the liver   540
Endocrine functions of the liver   541
Detoxification mechanisms   542
Liver failure   544
Recommended reading   545
Self-assessment questions   546

26 The pancreas   547

Introduction   547
Anatomy and histology of the pancreas   547
The blood supply and innervation of the
    pancreas   549
The pancreatic juice and its functions   550
The regulation of pancreatic secretion   551
Disorders of pancreatic exocrine secretion and their
    treatment   553
Endocrine functions of the pancreas   554
Recommended reading   555
Self-assessment questions   555

27 Skin structure and function   556

Introduction   556
The main structural features of the skin   557

The epidermis   558

The dermis and hypodermis   560

The accessory structures of the skin – hairs, nails and glands   561

The cutaneous circulation   564

The role of the skin as a sense organ   565

The triple response   567

Wound healing   567

Pharmacological aspects of skin   571

Recommended reading   572

Self-assessment questions   572

Section 5   Reproduction and inheritance   575

28   The physiology of reproduction   576

Introduction   576

The anatomy of the male reproductive system   577

The adult testis makes sperm and male steroid hormones (androgens)   579

Hormone secretion by the testis   581

Spermatogenesis – the production of sperm by the testes   582

The control of spermatogenesis by pituitary and testicular hormones   587

The anatomy of the female reproductive system   589

The physiology of the ovaries   592

Hormonal regulation of the female reproductive tract   599

Why do the plasma concentrations of gonadotrophins and ovarian steroids vary during the ovarian cycle?   602

Puberty and the menopause   604

Recommended reading   605

Self-assessment questions   606

29   Fertilization and pregnancy   607

Introduction   607

The sexual reflexes   608

Fertilization   610

Implantation and the development of the placenta   613

Exchange of substances across the placenta   616

The placenta is an important endocrine organ   618

Delivery of the infant – parturition   620

Important changes to maternal physiology in pregnancy   621

Nutritional requirements of pregnancy   625

Some important aspects of fetal physiology   627

Respiratory and cardiovascular changes at birth   631

Adaptations of the neonatal circulation to pulmonary gas exchange – closure of the fetal shunts   633

Renal function and fluid balance in the fetus and neonate   635

The gastrointestinal tract of the fetus   635

Development of the fetal reproductive organs   636

Lactation – the synthesis and secretion of milk after delivery   638

Recommended reading   644

Self-assessment questions   645

30   The genetic basis of inheritance   646

Introduction   646

Chromosomes, genes and the genetic code   648

DNA and the genetic code   651

Principles of inheritance   654

Simple (Mendelian) inheritance   656

Co-dominance and the inheritance of ABO blood groups   659

Incomplete dominance and polygenic inheritance   661

Sex-linked inheritance   661

Mitochondrial inheritance   664

Genetic testing, genetic counselling and gene therapy   665

Recommended reading   666

Self-assessment questions   666

Section 6   Integrative aspects of physiology and pharmacology   669

31   The nutritional needs of the body   670

Introduction   670

Macronutrients   671

Micronutrients   673

The regulation of food intake – appetite, hunger and satiety   677

Assessment of nutritional status   679

Recommended reading   681

Self-assessment questions   681

32   Energy balance and exercise   682

Introduction   682

The chemical processes of the body produce heat   683

Energy balance   683
How much heat is liberated by metabolism?   685
Basal metabolic rate   686
Physiological factors that affect overall metabolic rate   687
Energy expenditure during exercise   688
Recommended reading   693
Self-assessment questions   693

33   The regulation of body temperature   695

Introduction   695
Mechanisms of heat exchange between the body and the environment   697
The role of the skin in thermoregulation   698
Body temperature is controlled around a 'set-point'   700
Thermoregulatory responses to cold   700
Thermoregulatory response to heat   701
Disorders of thermoregulation   702
Special thermoregulatory problems   703
Pyrexia (fever)   704
Recommended reading   705
Self-assessment questions   706

34   Body fluid and acid–base balance   707

Introduction   707
The distribution of body water between compartments   708
Body fluid osmolality and volume are regulated independently   709
Body water balance   710
Thirst is the physiological mechanism for replacement of lost water   710

Alterations to the effective circulating volume regulate sodium balance   712
Dehydration and disorders of water balance   714
Oral rehydration therapy   715
Oedema   715
Treatment of oedema with diuretics   718
Acid–base balance   718
Primary disturbances in acid–base balance   719
How the body compensates for acid–base balance disorders   722
The clinical evaluation of acid–base status   724
Recommended reading   725
Self-assessment questions   725

35   The uptake, distribution and elimination of drugs   726

Introduction   726
Routes of drug administration   727
Absorption and distribution of drugs   728
Drug metabolism and excretion   733
Effects of age on drug potency   736
Adverse effects of drugs   736
Recommended reading   737
Self-assessment questions   737

Appendix: SI units   739
Glossary of technical terms   741
Answers to self-assessment questions   755
Index   767

List of abbreviations

ACE	angiotensin-converting enzyme		DVT	deep vein thrombosis
ACh	acetylcholine		ECF	extracellular fluid
ACTH	adrenocorticotrophic hormone (corticotrophin)		ECG	electrocardiogram
			ECV	effective circulating volume
ADH	antidiuretic hormone (vasopressin)		EDTA	ethylenediaminetetraacetic acid
ADP	adenosine diphosphate		EDV	end-diastolic volume
AIDS	acquired immunodeficiency syndrome		EEG	electroencephalogram
AMP	adenosine monophosphate		EPP	end-plate potential
ANP	atrial natriuretic peptide		EPSP	excitatory postsynaptic potential
ATP	adenosine triphosphate		ERV	expiratory reserve volume
AV	atrioventricular, arteriovenous		FAD	flavine adenine dinucleotide
BMI	body mass index		$FADH_2$	reduced flavine adenine dinucleotide
BMR	basal metabolic rate		FEV_1	forced expiratory volume at 1 second
BP	blood pressure		FRC	functional residual volume
2,3-BPG	2,3-bisphosphoglycerate		FSH	follicle-stimulating hormone
b.p.m.	beats per minute		FVC	forced vital capacity
CABG	coronary artery bypass grafting		GABA	γ-aminobutyric acid
C	gas content of blood (e.g. CV_{O_2})		GALT	gut-associated lymphoid tissue
CCK	cholecystokinin		GDP	guanosine diphosphate
CJD	Creutzfeld–Jakob disease		GFR	glomerular filtration rate
CN	cranial nerve		GH	growth hormone (somatotrophin)
CNS	central nervous system		GHIH	growth hormone-inhibiting hormone (somatostatin)
CO	cardiac output			
CoA	coenzyme A		GHRH	growth hormone-releasing hormone
COLD	chronic obstructive lung disease		GI	gastrointestinal
COPD	chronic obstructive pulmonary disease		GIP	gastric inhibitory peptide
CRH	corticotrophin-releasing hormone		GLUT	glucose transporter family member
CSF	cerebrospinal fluid		GMP	guanosine monophosphate
CVP	central venous pressure		GnRH	gonadotrophin-releasing hormone
CVS	chorionic villus sampling		G protein	heterotrimeric GTP-binding protein
DAG	diacylglycerol		GTN	glyceryl trinitrate
dB	decibel		GTP	guanosine triphosphate
1,25-DHCC	1,25-dihydroxycholecalciferol		Hb	haemoglobin
DIT	di-iodotyrosine		HbF	fetal haemoglobin
DMT1	divalent metal ion transporter 1		HbS	sickle-cell haemoglobin
DNA	deoxyribonucleic acid		hCG	human chorionic gonadotrophin
DSS	di-octyl sodium sulphosuccinate		HIV	human immunodeficiency virus

HLA	human leukocyte antigen		$PGF_{2\alpha}$	prostaglandin $F_{2\alpha}$
hPL	human placental lactogen		PGI_2	prostacyclin
HR	heart rate		PIH	prolactin inhibitory hormone (dopamine)
HRT	hormone replacement therapy		PKU	phenylketonuria
5-HT	5-hydroxytryptamine (serotonin)		PSA	prostate-specific antigen
ICF	intracellular fluid		PTCA	percutaneous transarterial coronary angioplasty
IgA	immunoglobin A		PTH	parathyroid hormone
IgE	immunoglobin E		RDA	recommended daily amount
IgG	immunoglobin G		REM	rapid eye movement
IgM	immunoglobin M		Rh	rhesus factor (D antigen)
IGF-1	insulin-like growth factor 1		RNA	ribonucleic acid
IGF-2	insulin-like growth factor 2		RPF	renal plasma flow
INR	International Normalized Ratio		RQ	respiratory quotient (also known as the respiratory exchange ratio)
IP_3	inositol trisphosphate			
IPSP	inhibitory postsynaptic potential		RV	residual volume
LH	luteinizing hormone		SA	sinoatrial (node)
LHRH	luteinizing hormone releasing hormone		SCN	suprachiasmatic nucleus
LTP	long-term potentiation		SGLT1	sodium-linked glucose transporter 1
MALT	mucosa-associated lymphoid tissue		SPL	sound pressure level
MAP	mean arterial pressure		STP	standard temperature and pressure (0°C and 101 kPa (760 mmHg))
MHC	major histocompatibility complex (equivalent to human leukocyte antigen or HLA)			
			SV	stroke volume
MI	myocardial infarction		T	absolute temperature
MIH	Müllerian inhibiting hormone		T_3	tri-iodothyronine
MIT	mono-iodotyrosine		T_4	thyroxine
M_r	relative molecular mass		TENS	transcutaneous electrical nerve stimulation
mRNA	messenger RNA			
MSH	melanocyte stimulating hormone		TH	thyroid hormone
MVV	maximum ventilatory volume		Tm	transport maximum
NAD	nicotinamide adenine dinucleotide		TPA	tissue plasminogen activator
NADH	reduced nicotinamide adenine dinucleotide		TPR	total peripheral resistance
			TRH	thyrotrophin-releasing hormone
NSAID	non-steroidal anti-inflammatory drug		tRNA	transfer RNA
P	pressure or partial pressure		TSH	thyroid-stimulating hormone
P_{50}	pressure for half saturation		TXA_2	thromboxane A_2
PAH	p-aminohippurate		UTI	urinary tract infection
Pa_{O_2}	partial pressure of oxygen in the arterial blood		UV	ultraviolet
			V	volume of gas
Pa_{CO_2}	partial pressure of carbon dioxide in arterial blood		$\dot{V}$	flow rate
			$\dot{V}/\dot{Q}$	ventilation/perfusion ratio (in lungs)
PE	pulmonary embolism		VC	vital capacity
PEM	protein–energy malnutrition		VIP	vasoactive intestinal polypeptide
PGE_2	prostaglandin E_2		vWF	von Willebrand factor

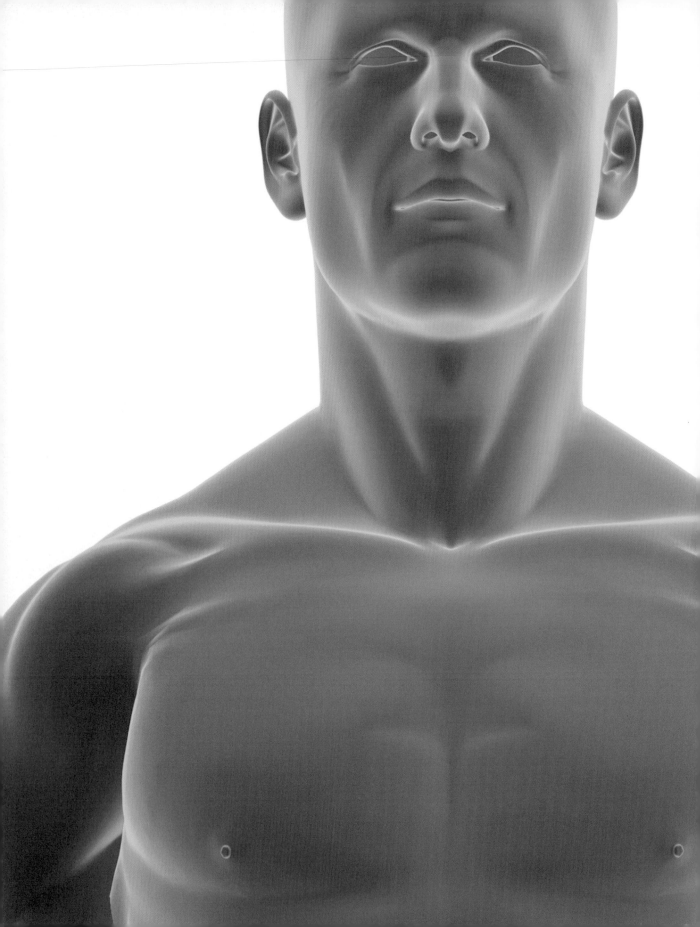

Section 1

Introduction

1 An introduction to the human body   **4**

2 Key concepts in chemistry   **14**

3 The chemical constitution of the body   **26**

1 An introduction to the human body

After reading this chapter you should have gained an understanding of:

- The subject matter of human biology for the health sciences
- The organization of this book
- The hierarchical organization of the body
- The concept of homeostasis
- The important terms used in anatomical descriptions

1.1 Introduction

For anyone concerned with the health sciences or related research, a sound knowledge of the normal structure and function of the body is essential as it provides a foundation on which to build strategies for diagnosis, treatment and prevention of disease. Two main areas of study lie at the heart of an understanding of human health. These are physiology and anatomy. **Physiology** is the study of how an organism works – how it feeds, breathes, eliminates waste products, moves and reproduces. It is impossible, however, to develop a full understanding of function without some knowledge of the structures of the body and their relationships to one another. This is the study of **anatomy**, which embraces not only gross structural anatomy but also the microanatomy of structures too small to be visible to the naked eye – the study of cells and tissues (**histology**).

Anatomy and physiology are inextricably related as the structure of an organ or tissue determines the functions it can perform. For example, the heart possesses a thick layer of muscle in its wall, enabling it to pump blood around the body; the structure of a joint determines the type of movements that it can perform and the structure of the chest determines how the lungs change their volume during breathing. If this knowledge is to be applied to clinical practice, then other aspects of human biology are also important. These include some fundamentals of cellular and molecular biology, genetics, disease processes (**pathology**) and drug treatments (**pharmacology**). A full treatment of all of these areas, even in brief, is beyond the scope of any one textbook. The main focus of this book is the physiology and related anatomy of the human body. To help to clarify normal function, the book will provide descriptions of some of the more common pathologies associated with the body systems. Where appropriate a brief account of the drugs that are currently used to treat these disorders will be given. Introductory chapters on genetics and immunology are also included.

While each chapter can be read on its own, the book has been laid out in six sections, each of which deals with a major aspect of human biology: Section 1 is a broad introduction to the subject, and includes a brief discussion of some basic concepts in chemistry. Section 2 presents basic information on the organization and functions of cells. Section 3 deals with the properties of excitable cells (nerve and muscle), and Section 4 provides an account of the anatomy and physiology of the principal organ systems. Section 5 is concerned with reproductive mechanisms, inheritance and physiology of the fetus and neonate. Section 6 is concerned with the interactions between different organ systems and with some clinical applications of anatomy, physiology and pharmacology.

An appreciation of mathematical principles can provide additional insight into certain aspects of human physiology and pharmacology. Examples include acid–base regulation, some aspects of respiratory function and quantitative aspects of drug–receptor interactions. More advanced topics of this kind are presented within feature boxes. It is not necessary to read these boxes to understand the core material.

Key points are included at the end of each major section and most chapters end with a set of questions that are designed to test your knowledge and to help to lodge the key information in your mind. Most of these questions are in the form of statements that are either true or false. Answers to these questions (with explanations where appropriate) are also provided.

For those students who require a more detailed account of the subjects dealt with here, a list of recommended further reading is included at the end of each chapter. These texts have been chosen for their clarity of exposition, but many other good sources are available.

1.2 The hierarchical organization of the body

The success of modern biological science in explaining how organisms perform their daily tasks is based on the notion that they are intricate and exquisite machines whose operation is governed by the laws of physics and chemistry. Like all matter, that of the body is made of atoms which are combined in many different ways to form more complex structures called molecules. The molecules are the most basic units of living things and are assembled in a vast number of different ways to provide the building blocks of the body: the **cells**.

Cells are composed of a limiting membrane (the plasma membrane) surrounding a fluid medium, the cytoplasm, which houses a number of structures called **organelles** that perform specific functions within the cell. Cells are grouped together to form **tissues**. The principal types of tissue are:

* Blood and lymph
* Connective tissue
* Nervous tissue
* Muscle
* Epithelia and glandular tissue.

Blood and lymph are sometimes classified as connective tissue. Each tissue has its own characteristics, which are described in Chapter 6.

As the fine structure of the tissues requires investigation with a microscope, the tissues are prepared for examination as follows: First they are treated with formaldehyde or another chemical to preserve their constituents. This is known as 'fixation'. The fixed tissue is then cut into thin slices called **histological sections** before being mounted on a microscope slide. The sections are then treated with dyes. This permits the fine details of a tissue to be seen with a light microscope. Detail that cannot be resolved with the light microscope can be revealed by **electron microscopy** in which the tissue is first treated with a heavy metal before being imaged with a beam of electrons.

Organs, such as the brain, the heart, the lungs, the intestines and the liver, are formed by the aggregation of different kinds of tissue. The organs are themselves parts of distinct physiological **systems**. The heart and blood vessels form the cardiovascular system; the lungs, trachea and bronchi together with the chest wall and diaphragm form the respiratory system; and so on. This organization is essentially hierarchical – complexity of function increases from cells through to systems, which function in an integrated fashion within the body as a whole. Each system is adapted to perform a specific set of functions that enable the body to perform the vital activities of living.

The principal organ systems

The cardiovascular system. The cells of large, multicellular animals cannot derive the oxygen and nutrients they need directly from the external environment. These must be transported to the cells. This is one of the principal functions of the blood, which circulates within blood vessels by virtue of the pumping action of the heart. The heart, blood vessels and associated tissues form the cardiovascular system. Fluid exchanged between the blood plasma and the tissues passes into the **lymphatic system**, which eventually drains back into the blood.

The respiratory system. The energy required for performing the various activities of the body is ultimately derived from respiration. This process involves the oxidation of foodstuffs (principally sugars and fats) to release the energy they contain. The oxygen needed for this process is absorbed from the air in the lungs and carried to the tissues by the blood. The carbon dioxide produced by the respiratory activity of the tissues is carried by the blood to the lungs where it is excreted in the expired air.

The digestive system. The nutrients needed by the body are derived from the diet. Food is taken in by the mouth and broken down into its component parts by enzymes in the gastrointestinal tract (or gut). The digestive products are then absorbed into the blood across the wall of the intestine and pass to the liver via the portal vein. The liver makes nutrients available to the tissues both for their growth and repair and for the production of energy.

The kidneys and urinary tract (the renal system). The chief function of the kidneys is to control the composition of the extracellular fluid (the fluid that bathes the cells). In the course of this process, they also eliminate non-volatile waste products from the blood. To perform these functions, the kidneys produce urine of variable composition, which is temporarily stored in the bladder before voiding.

The reproductive system. Reproduction is one of the fundamental characteristics of living organisms. The gonads (the testes in the male and ovaries in the female) produce specialized sex cells known as gametes. At the core of sexual reproduction is the creation and fusion of the male and female gametes, the sperm and ova (eggs), with the result that the genetic characteristics of two separate individuals are mixed to produce offspring that differ genetically from their parents.

The musculoskeletal system. This consists of the bones of the skeleton, skeletal muscles, joints and their associated tissues. Its primary function is to provide a means of movement, which is required for locomotion, for the maintenance of posture and for breathing. It also provides physical support for the internal organs.

The endocrine and nervous systems. The activities of the different organ systems need to be co-ordinated and regulated so that they act together to meet the needs of the body as a whole. Two co-ordinating systems have evolved: the nervous system and the endocrine system. The nervous system uses electrical signals to transmit information very rapidly to specific cells. Thus, the nerves use electrical signals to control the contraction of the skeletal muscles. The endocrine system secretes chemical agents, **hormones**, which travel in the bloodstream to the cells upon which they exert a regulatory effect. Hormones play a major role in the regulation of many different organ systems and are particularly important in the regulation of the menstrual cycle and other aspects of reproduction.

The immune system. This provides the body's defences against infection both by killing invading organisms and by eliminating diseased or damaged cells.

The integumentary system. This includes the skin and its associated structures (hair, nails, sweat glands and sebaceous glands). The skin is important in protection of the internal structures, fluid balance, sensation and the regulation of body temperature (thermoregulation). It also acts as a barrier to invasion by foreign organisms such as bacteria.

Although it is helpful to study how each individual organ performs its functions, it is essential to recognize that the activity of the body as a whole is dependent on the intricate interactions between the various organ systems. If one part fails, the consequences are found in other organ systems throughout the whole body.

KEY POINTS:

- The body has a hierarchical organization of cells, tissues, organs and systems.
- Each system consists of organs that carry out a particular set of functions that contribute to the well-being of the body as a whole.

1.3   Homeostasis

It is common knowledge that the internal temperature of the body of a healthy person is maintained at around 37°C and that it varies very little, despite wide variations in the external environmental temperature. Many other physiological parameters show a similar degree of constancy, for example plasma glucose, blood pressure, plasma concentrations of sodium and potassium, blood gas concentrations and so on. This maintenance of a stable internal environment is essential for the normal healthy function of the body's cells, tissues and organs. It is called **homeostasis** (literally 'staying the same' or 'standing still'). A loss of homeostasis is reflected in ill-health, and much of the work of those involved in the health sciences is concerned with helping the body to maintain or restore homeostasis. Furthermore, many of the clinical observations and diagnostic tools employed in practice are designed to monitor loss and restoration of homeostasis.

How does the body maintain homeostasis?

In a healthy body, individual cells have their own mechanisms for regulating their internal composition, and complex mechanisms work to regulate the volume and composition of the extracellular fluid. Virtually every organ system has a role in homeostasis, but the endocrine and nervous systems are particularly important as they allow communication within the body and the integration of cell and tissue functions.

The maintenance of homeostasis is achieved through balancing intake and output of substances by the body. An obvious example of such a mechanism is the balance achieved between fluid intake and urine output. The thirst mechanism and the production of urine by the kidneys are carefully controlled to ensure that the extracellular fluid volume is kept constant – as everyone knows, the consumption of a large quantity of water or beer is followed very soon by a large increase in the production of very dilute urine!

A similar example is provided by the regulation of the concentration of plasma glucose. In health, plasma glucose is 4–6 mmol l⁻¹ between meals. Shortly after a meal, plasma glucose rises above this level. The increase stimulates the secretion of the hormone insulin by the pancreas, which acts to bring the concentration down. As the concentration of glucose falls, so does the secretion of insulin. In each case, the changes in the circulating level of insulin act (together with other mechanisms) to maintain the plasma glucose at an appropriate level. This type of regulation is known as **negative feedback**.

A **negative feedback loop** is a control system that acts to maintain the level of some variable within a given range following a disturbance. Although the example given above refers to plasma glucose, the basic principle can be applied to other physiological variables such as body temperature, blood pressure and the osmolality of the plasma.

A negative feedback loop requires a **sensor** (or receptor) of some kind that responds to the variable in question but not to other physiological variables. Thus, an osmoreceptor should respond to changes in osmolality of the body fluids but not to changes in body temperature or blood pressure, while a thermoreceptor responds to changes in temperature, and so on. The information from the sensor must be compared in some way with the desired level (known as the 'set point' of the system) by some form of **comparator** (sometimes called an **integrator**). If the two do not match, an error signal is transmitted to an **effector**, a system that can act to restore the variable to its desired level and 'negate' the original change. The basic features of a negative feedback loop are summarized in Figure 1.1.

While it is difficult to over-emphasize the importance of negative feedback control loops in homeostatic mechanisms, they are frequently reset or overridden in stresses of various kinds. For example, arterial blood pressure is monitored by receptors, known as baroreceptors, which are found in the walls of the aortic arch and carotid sinus. These receptors are the sensors for a negative feedback loop that maintains the arterial blood pressure within close limits. If the blood pressure rises, compensatory changes occur that tend to restore it to normal. In exercise, however, this mechanism is reset. Indeed, if it were not, the amount of exercise we could undertake would be very limited.

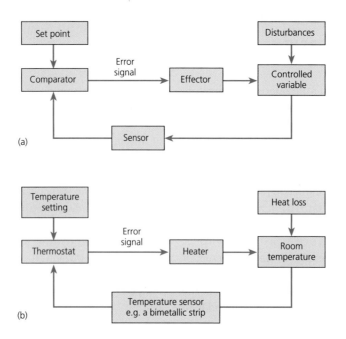

(a)

(b)

Figure 1.1 Schematic drawings of a negative feedback control loop (a) compared with the components of a simple heating system (b).

Negative feedback loops operate to maintain a particular variable within a specific range. Although rare, in some circumstances, **positive feedback** occurs. In this case, the error signal acts to *increase* the initial change rather than to counteract it. An example of such a mechanism is provided by the uterine contractions that bring about delivery of a baby during labour. Small contractions occur at the start of labour but these stimulate the secretion of hormones that increase the excitability of the uterus so that the strength of contractions increases. These contractions stimulate further hormone secretion and so on until the contractions are so powerful that the baby is born. Similar positive feedback effects are seen in the hormonal changes that culminate in ovulation.

KEY POINTS:

- Homeostasis is the maintenance of a stable internal environment despite changes to the external environment. It is essential for health.

- Many physiological parameters are kept within narrow limits by negative feedback mechanisms in which any change is counteracted by appropriate body responses.

- A small number of physiological functions are controlled by positive feedback mechanisms in which a change is enhanced rather than counteracted.

1.4   Terms used in anatomical descriptions

For ease of communication and to describe accurately the location and relationships of body parts, it is conventional to use a prescribed set of terms that relate to:

- The anatomical position

- Anatomical planes
- Movements
- Body cavities
- Anatomical direction.

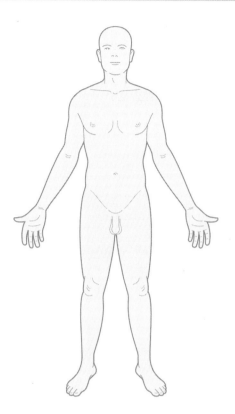

Figure 1.2 An outline of the body in the anatomical position. Note that the palms of the hands face forwards.

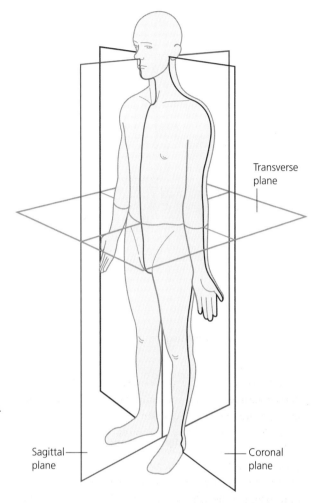

Figure 1.3 The principal anatomical planes.

The anatomical position

Unless it is stated otherwise, for the purposes of description the body is always considered to be erect (upright) and facing ahead, with the feet slightly apart. The arms are by the sides with the palms facing forwards and the fingers extended. This is the so-called anatomical position and is shown in Figure 1.2. Of course, in reality, a person may be in any position including lying face up (**supine**), or lying face down (**prone**).

Anatomical planes

Textbooks (including this one) often include diagrams of structures that have been cut (sectioned) in various ways to reveal their internal organization. Three main types of section are regularly used in this way

and are illustrated in Figure 1.3. They are also called anatomical planes (imaginary lines along which the section has been made) and lie at right angles to one another:

- **Sagittal** section or plane – a vertical (longitudinal) plane that divides the body or internal structure into a right and left portion. This is sometimes called a median section.

- **Coronal** section or plane – a vertical plane at right angles to the sagittal. This is also called a frontal section and divides the body or internal

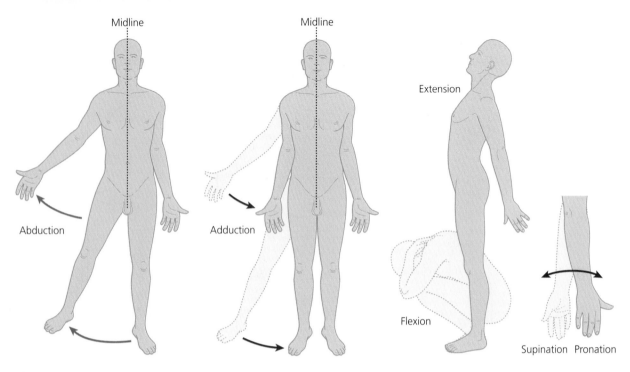

Figure 1.4 The formal description of the movements of the body.

structure into a front (anterior) and back (posterior) portion.

- **Transverse** (or horizontal) section – a horizontal plane at right angles to both coronal and sagittal planes. This divides the body or internal structure into an upper (superior) and lower (inferior) portion. It is also known as a cross-section.

Any section or plane that is not parallel to one of the above is called an **oblique** section.

Movements

The numerous joints of the musculoskeletal system permit movements of many kinds. The principal types of movement are illustrated in Figure 1.4 and are named as follows:

- **Abduction:** movement of a limb or other structure away from the midline or away from its original position

- **Adduction:** movement of a limb or other structure towards the midline or back to its original position

- **Flexion:** a forward or anterior movement of the trunk or of a limb

- **Extension:** a backward or posterior movement of the trunk or of a limb

- **Pronation:** a movement of the arm that allows the palm to face backwards (posteriorly)

- **Supination:** a movement of the arm that allows the palm to face forwards (anteriorly).

Other types of movement include rotation, pivot movements, inversion and eversion.

Body cavities

Many of the internal structures of the body are contained within spaces known as cavities, which offer them some protection and support. There are two

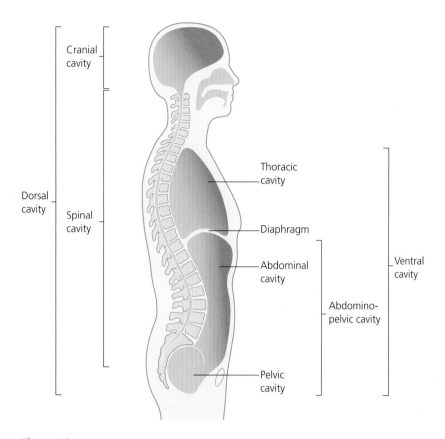

Figure 1.5 The major body cavities and their subdivisions.

major body cavities, the dorsal and ventral cavities. The **dorsal cavity** is made up of the **cranial cavity**, which houses the brain, and the **spinal cavity**, which houses the spinal cord. The **ventral cavity** consists of the **thoracic cavity** and the **abdominopelvic cavity**. These are shown in Figure 1.5.

The thoracic (chest) cavity houses a number of structures, including the heart and lungs. The abdominopelvic cavity is very large and houses many of the structures of the gastrointestinal tract, the kidneys and urinary tract, and the reproductive organs. Because of its size and complexity, the abdominopelvic cavity is often further subdivided in one of two ways:

- Into four quadrants (the right and left upper and right and left lower quadrants) as illustrated in Figure 1.6a
- Into nine regions as illustrated in Figure 1.6b.

Anatomical directional terms

When describing parts of the body, it is often necessary to make reference to their relative positions within the body as a whole or in relation to other structures close by. A number of directional terms have been established to make this easier. These are summarized in Table 1.1.

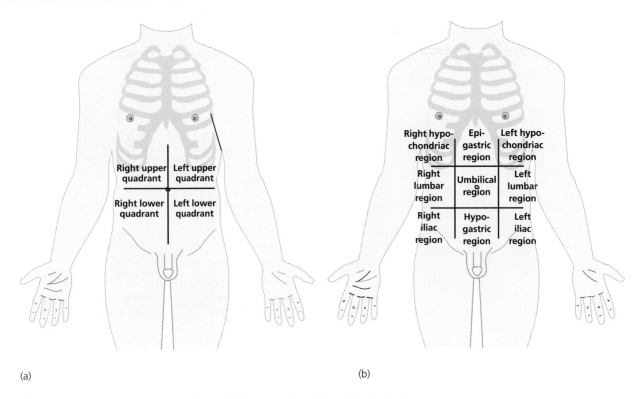

(a)

(b)

Figure 1.6 (a) The four abdominal quadrants. (b) The nine regions of the abdominal cavity.

Table 1.1 Commonly used directional terms and their definitions

Term	Definition	Example
Superior	Above, towards the head or towards the top of a structure	Superior vena cava The nose is superior to the mouth
Inferior	Below, away from the head or towards the lower part of a structure	Inferior vena cava The diaphragm is inferior to the heart
Medial	Towards or at the midline of the body On the inner surface or side of…	The sternum is medial to the arms The medial surface of the lung
Lateral	Away from the midline of the body On the outer surface of…	The kidneys are lateral to the vertebral column The lateral surface of the lung
Dorsal (posterior)	Behind or towards the back of the body or of a structure	The heart is dorsal (posterior) to the sternum
Ventral (anterior)	In front of or towards the front of the body or of a structure	The liver is ventral to the gall bladder
Superficial	Towards or at the body surface	The skin is superficial to the subcutaneous fat
Deep	Away from the body surface	The dermis is deep to the epidermis

Table 1.1 *(Continued)*

Term	Definition	Example
Proximal	Close to the origin of the body part or the point of attachment of a limb to the trunk	Proximal tubule (of the nephron) The knee is proximal to the ankle
Distal	Away from the origin of the body part or the point of attachment of a limb to the trunk	Distal tubule (of the nephron) The wrist is distal to the elbow
Ipsilateral	On the same side of the body	The right eye is ipsilateral to the right ear
Contralateral	On the opposite side of the body	The right cerebral hemisphere is contralateral to the left leg
Afferent	Towards	Afferent fibres of the nervous system (carry impulses towards the CNS) Afferent arteriole (of the nephron)
Efferent	Away from	Efferent fibres of the nervous system (carry impulses away from the CNS) Efferent arteriole (of the nephron)

Self-assessment questions

In Questions 1 to 3, which of the statements are true and which are false? Answers are given at the end of the book (p. 755).

1
 a) Tissues are aggregates of different types of cell.
 b) The endocrine system uses electrical signals to allow communication.
 c) Organelles carry out specific tasks within the body systems.
 d) The intestine is part of the digestive system.
 e) The lungs are part of the cardiovascular system.

2
 a) Most physiological variables are regulated by positive feedback.
 b) Arterial blood pressure is monitored by baroreceptors.
 c) Plasma glucose concentration is controlled by a negative feedback homeostatic loop.
 d) Effectors detect changes in physiological variables.
 e) Delivery of a baby is the end result of a positive feedback control mechanism.

3
 a) A person who is supine is lying face up.
 b) A midline sagittal section of the body will pass through the umbilicus.
 c) The stomach is located in the thoracic cavity.
 d) The spinal cavity is dorsal to the abdominal cavity.
 e) The epigastric cavity is inferior to the hypogastric cavity.

4 Complete the following sentences using the correct anatomical directional term (there may be more than one correct answer):
 a) The ears are ....to the nose.
 b) The big toe is ....to the little toe.
 c) The heart is ....to the bladder.
 d) The ankle is ....to the knee.
 e) The larynx is ....to the oesophagus.

2   Key concepts in chemistry

After reading this chapter you should have gained an understanding of:

- The basic structure of matter: atoms and molecules
- How molecules are bound together: covalent and ionic bonds
- How chemical reactions can transform one kind of molecule into another
- The properties of water as a biological solvent and the nature of polar and non-polar substances
- The ionization of molecules: strong and weak electrolytes
- Weak and strong acids and the pH scale
- The idea of molarity to express concentrations
- The osmotic pressure of aqueous solutions

2.1   Introduction

This chapter presents a brief account of the chemistry necessary to understand the metabolic reactions of the body. It should provide enough information to enable you to understand what is meant by an element and a compound; to differentiate between atoms and molecules; to understand molecular structures; and to calculate the concentration of a substance in solution. A few numerical exercises are given at the end of the chapter to help consolidate some of the key concepts discussed here.

All matter is composed of **chemical elements**, which are substances that cannot be broken down into sim-

pler materials by chemical means. Each element has a specific abbreviation or **chemical symbol** of one or two letters. Thus, carbon is written as C, calcium as Ca, hydrogen as H, oxygen as O and so on. However, most of the material encountered in everyday life is made from combinations of elements called chemical compounds. Thus, chalk is a combination of calcium, carbon and oxygen known as calcium carbonate; common table salt is a combination of the elements sodium and chlorine; and so on. The main elements that make up the human body and their chemical symbols are given in Table 2.1.

Table 2.1 The chemical elements of the body

Element	Chemical symbol	Atomic number	Relative atomic mass	Percentage of body weight
Oxygen	O	8	16	65
Carbon	C	6	12	18
Hydrogen	H	1	1	10
Nitrogen	N	7	14	3.4
Calcium	Ca	20	40.1	1.5
Phosphorous	P	15	31	1.2
Potassium	K	19	39.1	0.28
Sodium	Na	11	23	0.17
Chlorine	Cl	17	35.4	0.16
Magnesium	Mg	12	24.3	0.05
Sulphur	S	16	32.1	0.25
Iron	Fe	26	55.8	0.007
Zinc	Zn	30	65.4	0.002
Iodine	I	53	126.9	0.00004

The body contains trace amounts of other elements in addition to those listed above.

Each element consists of minute particles of the same type known as **atoms**, which are composed of a dense nucleus made up of **protons** and **neutrons**, surrounded by a cloud of **electrons**. The protons are positively charged and the electrons are negatively charged. In an atom of a given element, the number of electrons is equal to the number of protons. Thus, an atom of hydrogen has one proton and one electron; an atom of carbon has six protons and six electrons; and so on. The neutrons have the same mass as the protons but carry no charge.

There are 91 naturally occurring elements, each of which has a particular number of protons (the **atomic number**) that determines its chemical characteristics. The mass of each atom (its **atomic mass**) is determined by the number of protons and neutrons in its nucleus. Atoms with the same number of protons (and therefore the same atomic number) may have different numbers of neutrons. As a result, they have different atomic masses. Such atoms are called **isotopes**. A good example is carbon, which has an atomic number of 6 but has isotopes with mass numbers of 12, 13 and 14. These are known as carbon 12, carbon 13 and carbon 14, written as ^{12}C, ^{13}C and ^{14}C. Carbon 12 has 6 protons and 6 neutrons $(6 + 6 = 12)$, carbon 13 has 6 protons and 7 neutrons $(6 + 7 = 13)$, and carbon 14 has 6 protons and 8 neutrons $(6 + 8 = 14)$.

Carbon 14 is radioactive and breaks down at a constant rate so that half of the ^{14}C in a sample is lost every 5700 years. This property is exploited to date wooden artefacts found in archaeological sites. Many elements have radioactive isotopes and their radioactive decay is used to follow the transformations of chemical compounds that occur in living organisms.

2.2   Molecules are specific combinations of atoms

As indicated above, atoms of different elements are combined in many ways to form the material world around us. A specific combination of two or more atoms is known as a **molecule**. Thus, a water molecule consists of two atoms of hydrogen and one of oxygen. This is commonly written as H_2O. A molecule of common table salt consists of one sodium atom and one chlorine atom and is written as NaCl. The sugar used to sweeten drinks (known as sucrose) consists of 12 atoms of carbon, 22 of hydrogen and 10 of oxygen, and is written as $C_{12}H_{22}O_{10}$. Sometimes two identical atoms combine to form a molecule – for example, the oxygen of the air we breathe is a combination of two oxygen atoms. This is known as molecular oxygen and has the chemical formula O_2. Such formulae give information concerning the number of each type of atom in a molecule and are known as **molecular formulae**. Although useful for many purposes, a molecular formula gives no indication of the way in which the atoms of a particular molecule are arranged. This information is provided by **structural formulae**. Examples of structural formulae for sugars, amino acids and fats are given in the next chapter.

When atoms from different elements are combined to form a molecule, the resulting chemical compound has properties quite distinct from its constituent elements. In this fundamental way, a chemical compound differs from a mixture. A mixture is matter containing two or more elements or chemical compounds that have not undergone any chemical bonding. Unlike the constituents of a chemical compound, the constituents of a mixture can be separated by purely physical means. For example, a mixture of iron filings and sulphur can be separated into its constituents with the aid of a magnet, which will attract the iron filings but not the sulphur. It is not possible to separate iron and sulphur by this means when they have combined together to form ferrous sulphide (FeS).

Relative molecular mass

The relative mass of an atom is measured on a scale in which a single atom of carbon 12 has a mass of 12 exactly. (As mentioned above, carbon 12 is an atom which has 6 protons and 6 neutrons). An **atomic mass unit** is one-twelfth of the mass of an atom of carbon 12. The relative molecular mass of a molecule (M_r) is equal to the sum of the relative atomic masses of the atoms of which it is composed.

Here are some examples of how the relative molecular mass of a molecule is calculated:

The molecular mass of water (H_2O) is 18, which can be calculated as follows:

There are two hydrogen atoms each with a mass of 1, so their combined mass is 2 atomic units (a.u.).

There is one oxygen atom with a mass of 16 a.u.

The molecular mass is therefore 18 a.u. (= $[2 \times 1]$ + $[1 \times 16]$).

As all the calculations of molecular mass are based on the atomic mass unit, this is normally omitted and the molecular mass of water is simply given as 18.

The molecular mass of carbon dioxide (CO_2) may be calculated in the same way:

There is one carbon atom with a mass of 12.

There are two oxygen atoms with a mass of 16; the oxygen atoms thus contribute 32.

The molecular mass is therefore 44 (= $[1 \times 12]$ + $[2 \times 16]$).

The molecular mass of glucose ($C_6H_{12}O_6$) is 180 and is calculated as before:

There are six carbon atoms each with a mass of 12; total = 72.

There are 12 hydrogen atoms each with a mass of 1; total = 12.

There are six oxygen atom with a mass of 16; total = 96.

The molecular mass is therefore 180 (= $[6 \times 12]$ + $[12 \times 1]$ + $[6 \times 16]$).

The molecular mass is sometimes expressed in units called Daltons (abbreviated D or Da). This usage is particularly common in biochemistry.

A mole of any substance has the same number of molecules

In chemistry and biology, it is often useful to have some idea of the number of molecules in a particular volume of a sample. The amount of substance expressed in this way is given in a quantity called the **mole**. One mole of atoms has a mass in grams that is equal to the atomic mass of the atom in atomic mass units. The same applies to compounds – a mole of a compound is equal to its molecular mass in grams. A mole of any element or compound has exactly the same number of atoms or molecules, that is 6×10^{23}. This is known as **Avogadro's number**. For most medical and physiological purposes, the mole is inconveniently large and quantities are given in thousandths of a mole, a quantity known as a **millimole** (1/1000 or 10^{-3} moles).

Covalent bonds and molecular shape

When atoms combine to form molecules, they are held together by chemical bonds, which are of two kinds: covalent bonds and ionic bonds. **Covalent bonds** are formed when two atoms share electrons. When two atoms share one pair of electrons, a single covalent bond is formed. Such bonds are represented on paper by a single line linking the two atoms, as shown in Figure 2.1a. When two atoms share two pairs of electrons, a double bond is formed, which is represented by two lines linking the neighbouring atoms (Figure 2.1b). Note that the atoms of different elements are able to make different numbers of bonds. The number of chemical bonds that can be formed by an element is called its **valence** or **valency**. Many elements have only one valency state, for example carbon always has a valency of four, but some elements – especially metals – have more than one valency state. For example, iron has two valency states: ferrous (Fe^{2+}) and ferric (Fe^{3+}).

Individual molecules have shapes that are determined by the way their bonds are arranged in space. Oxygen atoms make two single bonds that are approximately at an angle of 100°. Carbon atoms make four single bonds that are distributed symmetrically in space in the form of a tetrahedron. Nitrogen atoms make three single bonds distributed in the shape of a three-sided pyramid, and so on. Examples of some simple compounds formed by these elements are shown in Figure 2.1c. Single covalent bonds linking two atoms permit the individual atoms to rotate about the axis of the bond, so molecules can change their shape to some degree without breaking their chemical bonds. A double bond, however, prevents rotation around that axis and gives rigidity to a molecule.

Ions are atoms that possess a charge

When one or more electrons are passed from one atom to another, an **ionic bond** is formed. In this case, it is the attraction between the opposite electrical charges of the constituent atoms that holds the material together. A charged atom is called an ion (hence the term ionic bond). A positively charged atom is known as a **cation**,

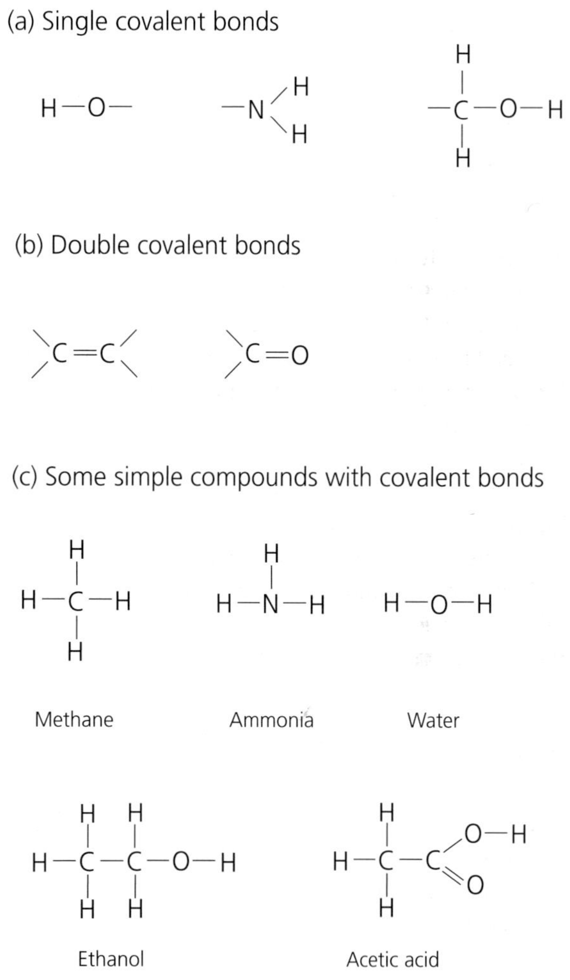

(a) Single covalent bonds

(b) Double covalent bonds

(c) Some simple compounds with covalent bonds

Methane   Ammonia   Water

Ethanol   Acetic acid

Figure 2.1 The conventional representation of the structure of covalent compounds. (a) The representation of single covalent bonds. (b) The representation of double covalent bonds. (c) The structural formulae of some simple chemical compounds.

while a negatively charged atom is called an **anion**. For example, the transfer of an electron from a sodium atom to a chlorine atom that occurs when metallic sodium reacts with chlorine gas to form sodium chloride (common table salt) results in each molecule having a positively charged sodium atom and a negatively charged chlorine atom.

The number of electrons an atom may gain or lose in forming an ionic bond is characteristic of that atom. The resulting ions are represented by their chemical symbols with the charge indicated. Atoms of sodium lose a single electron to become a sodium ion (written as Na^+). Potassium atoms also lose a single electron to form potassium ions (K^+), while atoms of calcium lose two electrons to form calcium ions, which, as they have two positive charges, are written as Ca^{2+}. When chlorine atoms gain an electron they form **chloride ions**, which are written as Cl^-.

Chemical reactions

In the body, one type of molecule is frequently converted into another by way of transformations known as chemical reactions. These reactions constitute the **metabolism** of the body and involve the input or release of energy. In a chemical reaction the total number of atoms remains the same, but their arrangement changes. This is the law of conservation of matter.

Chemical reactions may be represented by chemical equations such as the following:

$$C_6H_{12}O_6 + 6O_2 \rightarrow 6CO_2 + 6H_2O + \text{Heat (Energy)}$$

Note that on each side of the equation there are the same number of carbon atoms (6), hydrogen atoms (12) and oxygen atoms (18). The equation is said to balance. This particular equation shows the oxidation of glucose to carbon dioxide and water with the liberation of heat. To make the reaction go in the opposite direction (i.e. to form glucose from carbon dioxide and water) requires energy. Plants utilize sunlight to do so. This shows that *molecules can act as a store of energy*. Different molecules are able to store different amounts of energy, which can be used by cells to perform their functions.

KEY POINTS:

- When atoms combine to form molecules, they are held together by chemical bonds, which are of two main kinds: covalent bonds and ionic bonds.

- The relative molecular mass of a molecule (M_r) is equal to the sum of the atomic masses of the atoms of which it is composed.

- A mole of a compound is equal to its molecular mass in grams.

- Molecules can undergo chemical reactions in which the atoms are rearranged into other molecules, often with the liberation of energy.

2.3 Water and solutions

Water is the principal constituent of the human body and is essential for life. It is the chief solvent in living cells. *A solvent is a liquid that can dissolve a substance (known as the solute) to form a solution.* The amount of a substance in a given volume of solution is known as its concentration. It may be simply stated in grams per litre of solution or grams per decilitre of solution (gl^{-1} or $g\,dl^{-1}$), but in physiology and medicine it is often much more informative to express the concentration in terms of the number of moles or millimoles per litre of solution. This is known as the **molarity** of the solution, which is often denoted by the letter M. A 0.1 M solution of glucose

will contain 0.1 moles of glucose per litre of solution. This could also be expressed as 100 millimoles per litre of solution ($100\,mmol\,l^{-1}$ or $100\,mM$). Solutions of the same molarity have the same number of molecules of solute per unit volume of solution. Box 2.1 explains how to calculate the molarity of a solution.

Polar and non-polar molecules

Molecules of biological interest can be divided into those that dissolve readily in water and those that do not. Substances that dissolve readily in water are called **polar** or

Box 2.1 How to calculate the molarity of a solution

Very often the concentration of a solution is expressed as so many grams per unit volume. Although this is a proper way to express the concentration of a solution, it gives little information about the number of molecules present in a given volume of the solution. This is important for many purposes – including calculation of the osmolarity of a solution and for calculating electrolyte balance (i.e. the number of cations compared with the number of anions). As different molecules may differ greatly in mass, this should be taken into account when we state the concentration of a solution. A mole of a chemical compound (or element) is equal to its molecular mass in grams (see main text). So to calculate the molarity of a solution we simply need to calculate the number of moles in each litre of solution as follows:

$$\text{Molar concentration} = \frac{\text{weight of substance in one litre}}{\text{relative molecular mass}}$$

Here are some examples of molar calculations:

A solution of 0.9% sodium chloride in water (also known as normal saline) has 0.9 grams (g) of sodium chloride per 100 ml (per decilitre or dl) of solution. This is the same as 9 g per litre. The relative molecular mass (M_r) of sodium chloride is 58.5 so the molarity of this solution is $9 \div 58.5 = 0.154$ moles per litre. This is said to be a 0.154 molar solution of sodium chloride. Rather than express the concentration as a fraction, it is more usual

to express concentrations of this magnitude as so many millimoles per litre – in this case 154 millimoles per litre or 154 mmol l^{-1}. (To convert from moles to millimoles, multiply by 1000; to convert millimoles to moles, divide by 1000).

In normal blood, the glucose concentration is around 85 mg of glucose per 100 ml of blood. The molecular mass (M_r) of glucose ($C_6H_{12}O_6$) is 180 (= [6 × 12] + [1 × 12] + [6 × 16]). To calculate the molarity of glucose in blood, first calculate how many milligrams (mg) there would be per litre of solution: In this case it is 85 × 10 = 850 mg per litre or 0.85 g per litre and the molar concentration is $0.85 \div 180 = 0.00472$ moles per litre or 4.72 millimoles per litre (abbreviated as 4.72 mmol l^{-1}).

To calculate how much potassium chloride (KCl; M_r = 74.6) is required to make 100 ml of a 50 mmol l^{-1} solution, first work out how much potassium chloride is required to make a litre of a 50 millimolar solution:

$$(74.6 \times 50) \div 1000 = 3.730\,\text{g}$$

Then multiply by the fraction of a litre that is required (remember that 1 litre = 1000 millilitres):

$$100 \div 1000 = 0.1$$

The final quantity needed is:

$$3.730 \times 0.1 = 0.373\,\text{g (or 373 mg)}$$

---

hydrophilic while those that are insoluble in water are called **non-polar** or **hydrophobic**. Examples of polar substances are sodium chloride, sucrose (table sugar), ethanol and acetic acid (the pungent ingredient of vinegar). Examples of non-polar materials are fats (e.g. butter), olive oil and waxes. Many molecules of biological interest have mixed properties so that one part is polar while another part is non-polar. These are known as **amphiphilic** or **amphipathic** substances.

Some covalent bonds do not have the shared electrons equally distributed between the two atoms so that the electrons tend to be more associated with one of the two atoms. Such chemical bonds are called **polar bonds**.

Examples of polar bonds are the bonds between hydrogen and oxygen (—O—H) and the bond between nitrogen and hydrogen (—N—H). In these cases, the oxygen and nitrogen atoms tend to attract the electrons of the bond more than the hydrogen atoms.

Hydrogen bonds

When a hydrogen atom of a polar bond is attracted to a neighbouring oxygen or nitrogen atom, a type of bond known as a hydrogen bond is formed. Hydrogen bonds are much weaker than covalent or ionic bonds but are very important in holding large biological molecules

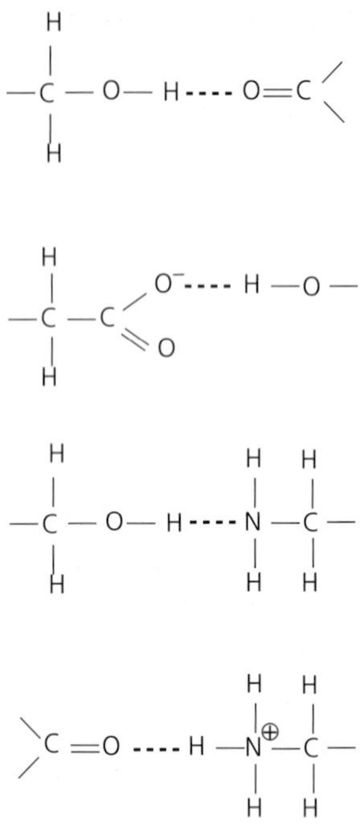

Figure 2.2 Typical examples of hydrogen bond formation.

such as proteins in their correct shape. Common types of hydrogen bond found in biological systems are shown in Figure 2.2.

Ionization

Pure water can dissociate so that one molecule is transformed into one hydrogen ion (H^+) and one hydroxyl ion (OH^-) as shown in the following chemical equation:

$$H_2O \leftrightharpoons H^+ + OH^- \qquad (1)$$

However, as the two ions tend to attract one another to reform water, an equilibrium is established (indicated here by the double arrow) in which the tendency of water molecules to dissociate is balanced by the tendency of the hydrogen and hydroxyl ions to bind together to form a neutral water molecule. In this case, there are vastly more water molecules than hydrogen and hydroxyl ions, so the equilibrium is heavily biased to the left side.

Compounds held together by ionic bonds, such as sodium chloride, potassium chloride and calcium chloride, generally dissolve readily in water and dissociate

completely into their constituent ions. So, when sodium chloride dissolves in water there are no molecules of NaCl in the solution, but a mixture of equal numbers of sodium ions (Na^+) and chloride ions (Cl^-).

$$NaCl \leftrightharpoons Na^+ + Cl^- \qquad (2)$$

A solution of potassium chloride has equal numbers of potassium ions (K^+) and chloride ions.

$$KCl \leftrightharpoons K^+ + Cl^- \qquad (3)$$

However, when calcium chloride ($CaCl_2$) dissolves in water, the resulting solution has twice as many chloride ions as calcium ions (Ca^{2+}), because each calcium ion has two positive charges and so can bind two chloride ions, each of which has a single negative charge.

$$CaCl_2 \leftrightharpoons Ca^{2+} + 2Cl^- \qquad (4)$$

Solutions containing ions conduct electricity. Because of this property, ions in solution are collectively called **electrolytes**. Those substances that dissociate completely into their constituent ions (e.g. NaCl) form solutions that conduct electricity easily and the electrolytes are known as strong electrolytes.

Compounds possessing polar bonds are often able to interact with water and dissociate giving rise to ions. Examples of covalent molecules undergoing ionization are the reaction of ammonia with water to form ammonium hydroxide, which then dissociates into ammonium ions (NH_4^+) and hydroxyl ions:

$$NH_3 + H_2O \leftrightharpoons NH_4OH \leftrightharpoons NH_4^+ + OH^- \qquad (5)$$

and the reaction of carbon dioxide with water in which carbonic acid is formed followed by its dissociation into bicarbonate ions (HCO_3^-) and hydrogen ions:

$$CO_2 + H_2O \leftrightharpoons H_2CO_3 \leftrightharpoons H^+ + HCO_3^- \qquad (6)$$

This process is called **ionization** and occurs very frequently in biochemical reactions. The solutions formed are not very good conductors of electricity and the electrolytes are called weak electrolytes. In physiology, the principal strong electrolytes are sodium (Na^+), potassium (K^+), calcium (Ca^{2+}), magnesium (Mg^{2+}) and chloride (Cl^-). The main weak electrolytes are bicarbonate (HCO_3^-) and phosphate (PO_4^{3-}). Sodium and potassium are the main cations of the body fluids while chloride and bicarbonate are the main anions.

These are routinely measured, together with urea, during many clinical investigations.

Acids and bases

An acid is, in physiological terms, a substance that generates hydrogen ions in solution and a base is a substance that absorbs hydrogen ions. Acids and bases are further classified as weak or strong according to how completely they dissociate in solution. Strong acids, such as hydrochloric acid (HCl), are completely dissociated in solution and there are no neutral hydrogen chloride molecules in the solution, only hydrogen ions (H^+) and chloride ions (Cl^-). Similarly, when sodium hydroxide (NaOH) is added to water, only sodium (Na^+) and hydroxyl (OH^-) ions are present. In contrast, when weak acids such as carbonic acid (H_2CO_3) and weak bases such as ammonium hydroxide (NH_4OH) are in solution they are only partly dissociated – as shown by chemical equations (5) and (6) above. A neutral solution is one in which the concentrations of hydrogen and hydroxyl ions are equal. When pure water dissociates, both hydrogen ions and hydroxyl ions are produced in equal quantities as shown in reaction (1) above.

As weak acids and weak bases can react with hydrogen ions, they can 'mop up' any hydrogen ions that may arise during the chemical reactions that occur in the body. They can thus limit the resulting changes in hydrogen ion concentration. For this reason, solutions of weak acids and bases are called **buffers**.

Hydrogen ion concentration and the pH scale

In the chemical reaction labelled (6) above, a neutral molecule (CO_2) gave rise to a positively charged hydrogen ion (H^+) when it reacted with water. Hydrogen ions are hydrogen atoms that have lost their electron. Another name for a hydrogen ion is a proton, because the hydrogen atom consists of one proton and one electron. Hydrogen ions combine readily with molecules that have polar groups and alter the chemical properties of such molecules. For this reason, the concentration of hydrogen ions in a solution is very important in biology. It may be expressed in moles of hydrogen ions per litre but it is more often expressed on a scale known as the pH scale.

The pH of a solution is the logarithm to the base 10 of the reciprocal of the hydrogen ion concentration (the H^+ in square brackets indicates hydrogen ion concentration in mol l^{-1})

$$pH = \log_{10}\left[\frac{1}{[H^+]}\right] \tag{7}$$

Since, however,

$$\log_{10}\left[\frac{1}{[H^+]}\right] = -\log_{10}[H^+]$$

$$pH = -\log_{10}[H^+] \tag{8}$$

A common alternative definition is, therefore, that the pH of a solution is the negative logarithm of the hydrogen ion concentration. A change of one pH unit corresponds to a tenfold change in hydrogen ion concentration (because $\log_{10}10 = 1$). An explanation of how to calculate the pH of a solution of known hydrogen ion concentration is given in Box 2.2.

While the pH notation is convenient for expressing a wide concentration range, it is potentially confusing as *a decrease in pH reflects an increase in hydrogen ion concentration and vice versa*. Nevertheless, the pH scale is widely used to express how acidic or alkaline a solution is. Pure water has a pH of 7.0 at 25°C. Acidic solutions have a pH value of less than 7, while alkaline solutions have a pH value greater than 7: the lower the pH value, the more acidic the solution; the higher the pH, the more alkaline it would be. Blood pH is normally between 7.35 and 7.4 (i.e. very mildly alkaline) while urine pH can range from 4.7 to 8.2 although it is usually around pH 6 and thus is mildly acidic.

KEY POINT:

- The common logarithm of a number is the power of 10 that will give that number. If our number is x we can write its logarithm (y) as $y = \log_{10}(x)$, which is equivalent to saying that $x = 10^y$. Logarithms can be calculated using any positive number as a base but in biology only logarithms to base 10 and natural logarithms are used. Natural logarithms use the mathematical constant e (2.718 approximately) as their base and are denoted as $\ln(x)$ or $\log_e(x)$. Both natural logarithms and logarithms to the base 10 are given on many calculators.

Box 2.2 How to calculate the pH of a solution

In the age of calculators and computers, this is now a relatively straightforward procedure.

$$pH = -\log_{10}[H^+]$$

First, enter the hydrogen ion concentration $[H^+]$ *expressed as moles per litre* into the calculator. Then press the $\log_{10}$ key. Finally, change the sign (press the $+/-$ key). The resulting number is the pH of the solution.

To convert from pH to free $[H^+]$ we use the following relationship:

$$[H^+] = 10^{-pH}$$

Enter the pH value and change the sign, then press the inverse function key followed by the log key. The resulting number is the hydrogen ion concentration in moles per litre.

Some worked examples:

If a solution has a hydrogen ion concentration of 5×10^{-6} moles per litre, what is its pH?

$$pH = -\log_{10}[5 \times 10^{-6}] = 5.3$$

If the pH of a blood sample is 7.4 (a normal value), what is the hydrogen ion concentration?

$$[H^+] = 10^{-7.4} = 39.8 \times 10^{-9}\,mol\,l^{-1}$$

Using the same method of calculation, show that, if a urine sample has a pH of 5.5, the $[H^+]$ concentration is $3.16 \times 10^{-6}\,mol\,l^{-1}$. (Note that the 1.9 unit difference in pH between these two samples corresponds to an 80-fold difference in hydrogen ion concentration.)

The osmotic pressure of the body fluids

When an aqueous solution is separated from pure water by a membrane that is permeable to water but not to the solute, water moves across the membrane into the solution by a process known as **osmosis**. This is shown in Figure 2.3. The movement of water can be opposed by applying a hydrostatic pressure to the solution. The pressure that is just sufficient to prevent the uptake of water is known as the **osmotic pressure** of the solution. The osmotic pressure of a solution is directly related to the number of molecules or ions present, and is independent of their chemical nature. Rather than measuring osmotic pressure directly, it is more convenient to state the **osmolarity** (moles per litre of solution) or **osmolality** (moles per kg of water). In clinical medicine, the osmotic pressure of a body fluid is generally expressed as its osmolality.

As the osmotic pressure depends on the number of particles present in a given volume of water, solutions that have the same molality will have the same osmolality. This is a direct consequence of the two solutions having the same number of molecules per unit volume. Despite the large difference in their relative molecular mass, the osmotic pressure exerted by a millimole of glucose (M_r 180) is the same as that exerted by a millimole of albumin (M_r 69 000). However, *aqueous salt solutions are an important exception to this rule*: the salts separate into their constituent ions so that a solution of sodium chloride will exert an osmotic pressure double that of its molal concentration. Hence, a 100 mmol kg^{-1} solution of sodium chloride in water will have an osmotic pressure of 200 mOsmol kg^{-1} of which half is due to the sodium ions and half to the chloride ions.

The total osmolality of a solution is the sum of the osmolality due to each of the constituents. The blood plasma has an osmolality of around 0.3 Osmol kg^{-1} (300 mOsmol kg^{-1}). The principal ions (Na$^+$, K$^+$, Cl$^-$, etc.) contribute about 290 mOsmol kg^{-1} (about 96%) while glucose, amino acids and other small non-ionic substances contribute approximately 10 mOsmol kg^{-1}. Proteins contribute only around 0.5% to the total

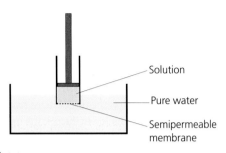

(a) Initial state

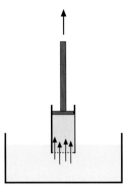

(b) Osmosis is the
uptake of water by
the solution

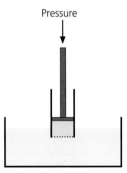

(c) Osmosis can be prevented
by the application of pressure

Figure 2.3 Osmosis and osmotic pressure. The passage of water across a semipermeable membrane into a solution can be opposed by applying pressure. The pressure that just prevents the uptake of water is known as the osmotic pressure of the solution (see text for further details).

osmolality of plasma. This is made clear by the following calculations: Blood plasma has about 6.76 g of sodium chloride and 47.4 g of albumin per kg of plasma water. The osmolality of a solution of 6.76 g of sodium chloride (M_r 58.4) is:

$$(2 \times 6.76) \div 58.4 = 0.231 \, \text{Osmol} \, \text{kg}^{-1} \, \text{or} \, 231 \, \text{mOsmol} \, \text{kg}^{-1}$$

The osmotic pressure exerted by 47.4 g of albumin is:

$$47.4 \div 69\,000 = 6.87 \times 10^{-4} \, \text{Osmol} \, \text{kg}^{-1} \, \text{or}$$
approximately $0.69 \, \text{mOsmol} \, \text{kg}^{-1}$

Thus, the osmotic pressure exerted by 47 g of albumin is only about 0.3% that of 6.76 g of sodium chloride. This makes clear that *the osmotic pressure exerted by proteins is far less than that exerted by the principal ions of the biological fluids.* Nevertheless, the small osmotic pressure that the proteins do exert (known as the **colloid osmotic pressure** or **oncotic pressure**) plays an important role in the exchange of fluids between body compartments (see Chapter 19, p. 383).

The tonicity of solutions

The tonicity of a solution refers to the influence of its osmolality on the volume of cells. For example, red blood cells placed in a solution of 0.9% sodium chloride in water (i.e. 0.9 g sodium chloride in 100 ml of water) neither swell nor shrink. This concentration has an osmolality $\approx 310 \, \text{mOsmol} \, \text{kg}^{-1}$ and is said to be **isotonic** with the cells. (This solution is sometimes referred to as 'normal saline' but would be better called isotonic saline.) If the same cells are added to a solution of sodium chloride with an osmolality of $260 \, \text{mOsmol} \, \text{kg}^{-1}$, they will swell as they take up water to equalize the osmotic pressure across their cell membranes. This concentration of sodium chloride is said to be **hypotonic** with respect to the cells. Conversely, red blood cells placed in a solution of sodium chloride that has an osmolality of $360 \, \text{mOsmol} \, \text{kg}^{-1}$ will shrink as water is drawn from the cells. In this case, the fluid is **hypertonic**. As cells normally maintain a constant volume, it is clear that the osmolality of the fluid inside the cells (the **intracellular fluid**) is the same as that outside the cells (the **extracellular fluid**). The two fluids are said to be iso-osmotic.

Diffusion

When a substance (the **solute**) is dissolved in a **solvent** to form a solution, the individual solute molecules become dispersed within the solvent and are free to move in a random way. Thus, in an aqueous solution, the molecules of both water and solute are in continuous random motion with frequent collisions between them. This process leads to **diffusion**, the random dispersion of molecules in solution. If a drop of ink is added to a volume of pure water, the ink particles slowly disperse throughout the whole volume. If the ink drop had been added to a dilute solution of ink, the same process of dispersion of the ink particles would occur until the whole solution was of a uniform concentration. There is a tendency for any solute to diffuse from a region of high concentration to one of a lower concentration (i.e. down its concentration gradient).

The rate of diffusion of a solute depends on temperature (it is faster at higher temperatures), the magnitude of the concentration gradient and the area over which diffusion can occur. The molecular characteristics of the solute and solvent also affect the rate of diffusion. These characteristics are reflected in a physical constant known as the **diffusion coefficient**. In general, large molecules diffuse more slowly than small ones.

KEY POINTS:

- Water is the chief solvent of the body. Substances that dissolve readily in water are said to be polar (or hydrophilic) while those that are insoluble in water are non-polar (or hydrophobic).

- The concentration of hydrogen ions in a solution determines how acid it is. Acidic solutions have a pH value less than 7; the lower the pH, the more acidic the solution. Alkaline solutions have a pH value greater than 7: the higher the pH, the more alkaline the solution. A neutral solution is one in which the concentrations of hydrogen and hydroxyl ions are equal and at 25°C it would have a pH of 7.

- When a substance dissolves in water it exerts an osmotic pressure that is related to its molal concentration. The osmotic pressure of a solution is expressed as its osmolality, which is related to the number of particles present per kg of water, independent of their chemical nature. The total osmolality of a solution is the sum of the osmolality due to each of the constituents.

Self-assessment questions

Here are some simple calculations that should help to familiarize you with some of the basic chemical concepts discussed above. Answers are given at the end of the book (p. 755).

1 Using the data of Table 2.1, calculate the molecular mass (M_r) of the following compounds:
   a) Urea (CH_4N_2O)
   b) Fructose ($C_6H_{12}O_6$)
   c) Calcium chloride ($CaCl_2$)
   d) Palmitic acid ($C_{16}H_{36}O_2$)
   e) Progesterone ($C_{20}H_{28}O_2$)

2 Calculate the molarity (moles per litre) of the following solutions:
   a) 29.2 g NaCl dissolved in water to make a litre of solution (M_r of NaCl 58.4)

   b) 8.77 g NaCl dissolved in water to make a litre of solution
   c) 8.03 g urea dissolved in water to make a litre of solution (M_r of urea 60.1)
   d) 1 mg ml^{-1} of glucose in water (M_r of glucose 180.2)
   e) 45 g of albumin dissolved to make a litre of solution (M_r of albumin 69 000)

3 Now calculate the osmolarity (in Osmol per litre) of the solutions given in Question 2.

4 Calculate the pH of the solutions containing the following hydrogen ion concentrations:

a) $0.1 M$ ($100 mmol l^{-1}$)
b) $0.005 M$ ($5 mmol l^{-1}$)
c) $0.01 mmol l^{-1}$ ($10 mmol l^{-1}$)
d) $0.00004 mmol^{-1}$ ($40 nmol l^{-1}$)

5 If a small number of red cells were to be isolated from the blood and placed in solutions of the composition given in (a) to (c), would they swell, shrink or stay approximately the same size?

a) a solution of $0.9 g$ NaCl per $100 ml$ (0.9% saline solution or normal saline)
b) a solution of $7.5 g$ NaCl per litre
c) a solution of $10.5 g$ NaCl per litre

The chemical constitution of the body

After reading this chapter you should have gained an understanding of:

- The chemical composition of the body
- The distribution of body water and the distinction between the intracellular and extracellular fluids
- The structure and functions of carbohydrates
- The chemical nature and functions of lipids
- The structure of amino acids and proteins and their functions
- The structure and functions of nucleotides and nucleic acids

3.1 Introduction

The human body consists largely of four elements: oxygen, carbon, hydrogen and nitrogen, which are combined in many different ways to make a huge variety of chemical compounds. The major organic (i.e. carbon-containing) constituents of mammalian cells are the **carbohydrates**, **fats**, **proteins** and **nucleic acids**, which are built from small molecules belonging to four principal groups: the sugars, the fatty acids, the amino acids and the nucleotides.

The major components of the lean body tissue are water (about 65% of total mass), fats (about 17%) and proteins (about 14%). The remainder consists of nucleic acids, carbohydrates (sugars) and minerals of various kinds such as calcium, iron, magnesium, phosphate, potassium and sodium. The chemical composition of the body shown in Figure 3.1 is an approximate average of all the tissues of an adult.

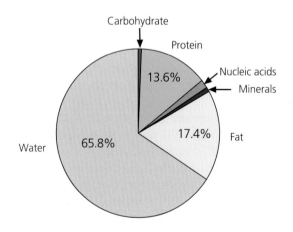

Figure 3.1 The approximate chemical composition of the body tissues. Note the high proportion of water, fat and protein.

3.2   Body water

Water is the principal constituent of the human body and is essential for life. The solutes and water of the space outside the cells constitute the **extracellular fluid** while those inside the cells constitute the **intracellular fluid** (Figure 3.2). The extracellular fluid is further subdivided into the plasma (the liquid portion of the blood), the interstitial fluid, which occupies the space between the blood vessels and the cells of the body, and the transcellular fluid, which is the fluid in serosal spaces such as the joint capsules and the peritoneal cavity. The intracellular fluid is separated from the extracellular fluid by the membrane that surrounds each cell, the plasma membrane.

> **KEY POINTS:**
> - Water accounts for about two-thirds of the lean body tissues of adults.
> - The solutes and water of the fluids inside the cells form the intracellular fluid while those that lie outside the cells form the extracellular fluid.

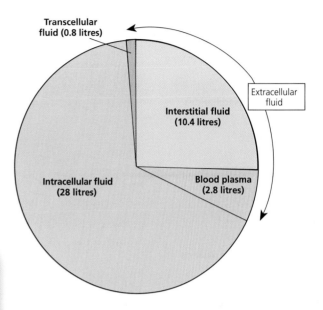

Figure 3.2 The distribution of water between different body compartments. Most of the water is in the cells (the intracellular fluid), only about a third being outside the cells (the extracellular fluid).

3.3   The major organic constituents of the body

The carbohydrates

The carbohydrates or **sugars** are the principal source of energy for cellular reactions. Figure 3.3 shows the structural formulae of some common examples. Those sugars that contain five carbon atoms are **pentoses** while those with six carbon atoms are **hexoses**. Examples are ribose (a pentose), fructose and glucose (both hexoses) all of which are classified as **monosaccharides** (i.e. single sugar molecules). When they are dissolved in water, both pentoses and hexoses adopt a ring structure as shown in Figure 3.3. Pentoses form five-membered rings but hexoses may form either a five-membered ring (e.g. fructose) or a six-membered ring (e.g. glucose). This can be seen in the top section of Figure 3.3.

Many of the carbon atoms in a monosaccharide have four different chemical groups attached. This makes these molecules asymmetric so that two different kinds of each monosaccharide exist, which are mirror images of each other – just as our left and right hands are mirror images. Molecules of this kind are known as **optical isomers**. The two isomers are known as D-isomers and L-isomers. Although they have the same molecular components and chemical properties, they do not have the same biochemical properties. For example, living things can utilize D-glucose but not L-glucose as a source of energy. All naturally occurring carbohydrates are D-isomers. Thus, the glucose we find in nature is more correctly known as D-glucose and, for this reason, it is sometimes called dextrose.

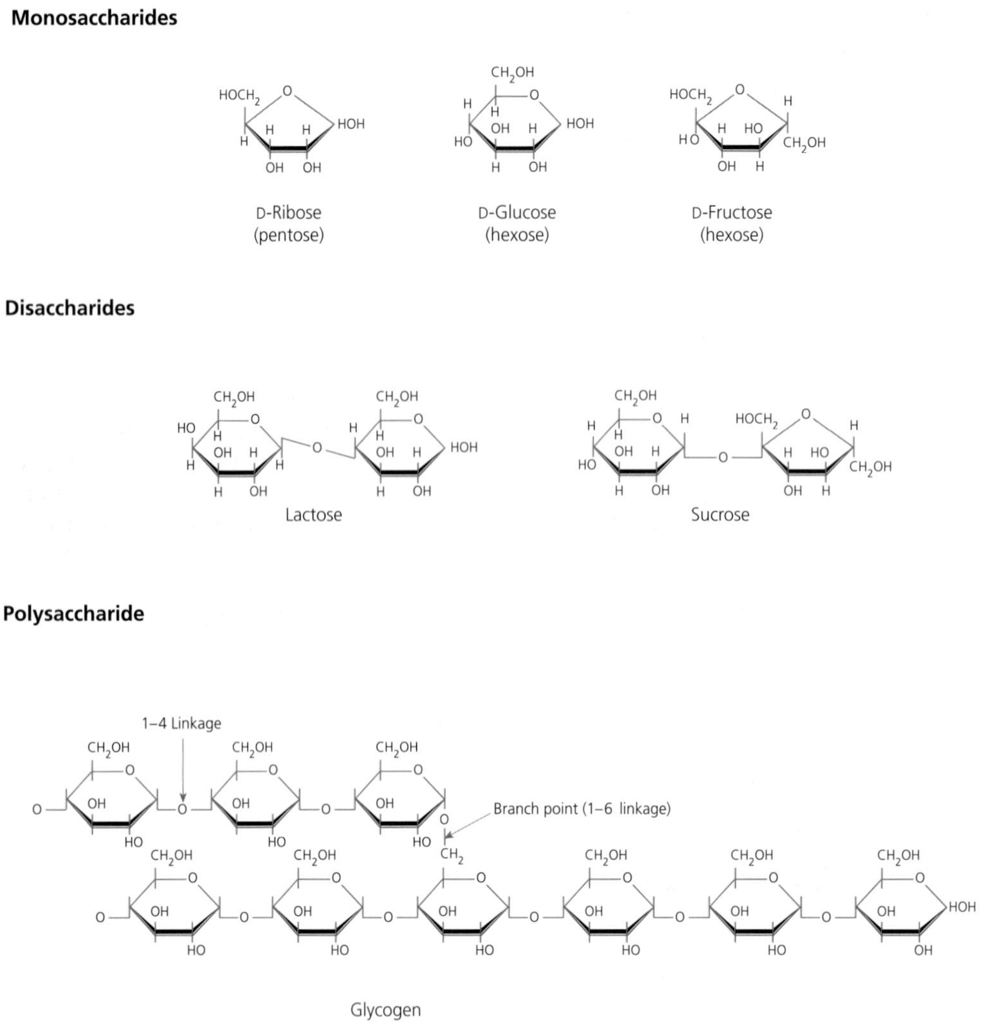

Monosaccharides

D-Ribose
(pentose)

D-Glucose
(hexose)

D-Fructose
(hexose)

Disaccharides

Lactose

Sucrose

Polysaccharide

1–4 Linkage

Branch point (1–6 linkage)

Glycogen

Figure 3.3 The structure of some common carbohydrates. The polysaccharide glycogen consists of many glucose molecules joined together by 1–4 linkages known as glycoside bonds to form a long chain, only part of which is shown in the figure. A number of glucose chains are joined together by 1–6 glycoside linkages to form a single large glycogen molecule.

When two sugar molecules are joined together with the loss of one molecule of water, they form a **disaccharide**. Fructose and glucose combine to form sucrose while glucose and galactose (another hexose) can undergo a similar reaction to form lactose, the principal sugar of milk, as shown in Figure 3.4. Disaccharides can be broken down into their constituent monosaccharides by adding a molecule of water. This process is known as **hydrolysis** and is a common way for large molecules to be broken down to simpler ones.

When many sugar molecules are joined together in this way, they form a **polysaccharide**. Large molecules such as the polysaccharides are known as **macromolecules**. Examples of polysaccharides are **starch**, which is an important constituent of the diet, and **glycogen**, which is the main store of carbohydrate within the muscles and liver. Its structure is shown in the bottom panel of Figure 3.3. By forming glycogen, cells can store large quantities of glucose without making their interior hypertonic (see p. 23). Glycogen is broken down by hydrolysis when glucose is required for energy production (Figure 3.5).

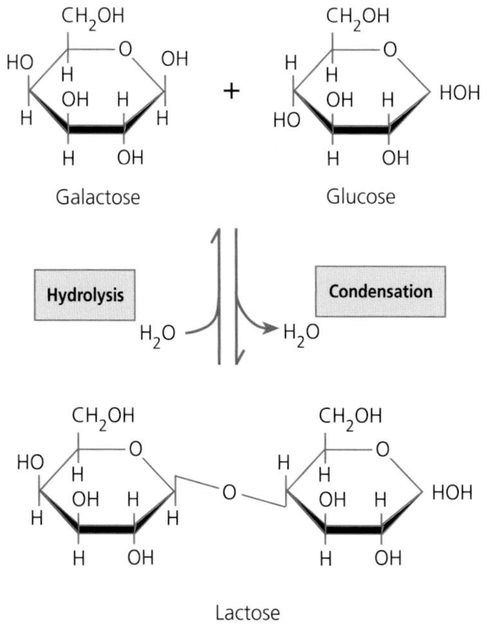

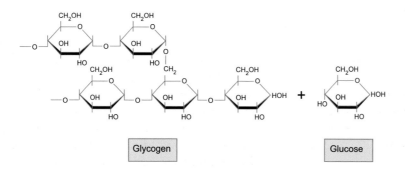

Galactose   Glucose

Hydrolysis   **Condensation**

H_2O H_2O

Lactose

Figure 3.4 The combination of glucose and galactose forms the disaccharide lactose in a condensation reaction in which a molecule of water is eliminated. By adding a molecule of water, lactose can be broken down to glucose and galactose by a process called hydrolysis. Similar reactions occur when the polysaccharide glycogen is synthesized from glucose, a process called glycogenesis and when glycogen is broken down to release glucose for energy production (see also Figure 3.5).

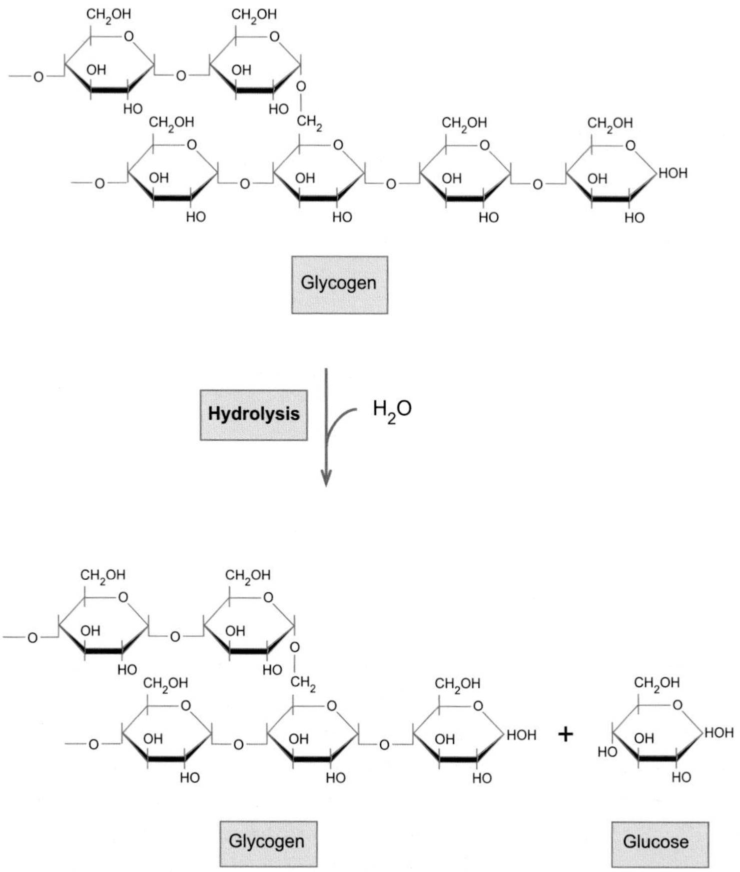

Glycogen

Hydrolysis H_2O

Glycogen   Glucose

Figure 3.5 Glycogen consists of many glucose residues linked together to form a single large molecule. This is how the body stores glucose. When the stored glucose is required for energy production, glycogen is hydrolysed to release individual glucose molecules as shown. This process is called glycogenolysis.

Although sugars are the major source of energy for cells, they are also constituents of a number of molecules that play an important role in metabolism. The nucleic acid DNA contains the pentose sugar 2-deoxyribose. (The structure of DNA and other related molecules is discussed below, see pp. 35–36). Some hexoses have an amino group ($-NH_2$) in place of one of the hydroxyl groups. These are known as the **amino sugars** or **hexosamines**, a well-known example being **glucosamine**. The amino sugars are found in the **glycoproteins** (= sugar + protein) and the **glycolipids** (= sugar + lipid). In the glycoproteins, a polysaccharide chain is linked to a protein by a covalent bond. They are important constituents of bone and connective tissue. The glycolipids are found in the cell membranes – particularly those of the white matter of the brain and spinal cord.

> **KEY POINTS:**
>
> • The carbohydrates, especially glucose, are broken down to provide energy for cellular reactions.
>
> • The body stores glucose as the polysaccharide glycogen, which is hydrolysed back to glucose when required.
>
> • While sugars are the major source of energy for cells, they are also constituents of a large number of molecules that play an important role in biochemical reactions and structural support.

The lipids

The lipids are major constituents of body fat. They are a chemically diverse group of substances that share the property of being insoluble in water but soluble in organic solvents such as ether and chloroform. They serve a wide variety of functions:

• They are an important reserve of energy.

• They provide a layer of heat insulation beneath the skin.

• They provide physical support for many body tissues.

• They are a major structural component of cell membranes.

• Some lipids act as chemical signals (e.g. the steroid hormones and prostaglandins).

Fats consist of fatty acids linked to a glycerol backbone (see below). The fatty acids themselves have a carboxyl group (–COOH) at one end of the molecule, which is attached to a chain of carbon atoms as shown in Figure 3.6. The length of the carbon chain is variable. Typical fatty acids found in animal fats such as butter are **palmitic acid** (which has 16 carbon atoms) and **stearic acid** (which has 18 carbon atoms). These fatty acids have no double bonds in the carbon chain and are known as **saturated fatty acids**.

Many fatty acids have one or more double bonds. Such fatty acids are known as **unsaturated fatty acids**. The examples shown in Figure 3.6 are **oleic acid**

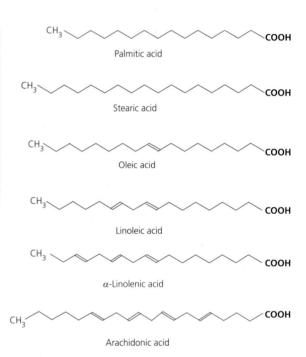

Figure 3.6 The chemical structures of typical fatty acids. The carboxyl group is shown in bold. Note that the carbon chain of long-chain fatty acids is represented by a series of lines thus: ᴧᴧᴧ. Each angle represents a -CH_2- group. Such formulae are known as skeletal structures. Note also that the individual fatty acids have carbon chains of different lengths. Palmitic acid and stearic acid have no double bonds and are known as saturated fatty acids. Oleic acid and other fatty acids that have double bonds between adjacent carbon atoms are unsaturated fatty acids.

(18 carbon atoms with one double bond), **linoleic acid** (18 carbon atoms with two double bonds), *α*-**linolenic** acid (18 carbon atoms with three double bonds) and **arachidonic acid** (20 carbon atoms with four double bonds). Mammals, including man, use the unsaturated fatty acids linoleic acid and *α*-linolenic acid as the starting point for the synthesis of other unsaturated fatty acids that play an important role in cellular metabolism. As linoleic acid and *α*-linolenic acid are required for normal health, they are known as the **essential fatty acids** and must be provided by the diet, as discussed in Chapter 31. In nutritional supplements, these are sometimes called omega-6 and omega-3 fatty acids, respectively.

The **triglycerides** are the body's main store of energy and can be laid down in adipose tissue in virtually unlimited amounts. They consist of three fatty acids joined by ester linkages to a glycerol molecule, as shown in Figure 3.7. Triglycerides generally contain fatty acids with many carbon atoms, for example palmitic and stearic acids. In the digestive system, the triglycerides found in the diet are first hydrolysed to **diglycerides**, which have two fatty acids linked to glycerol, and then to **monoglycerides**, which have only one as shown in Figure 3.7. A similar sequence occurs when the body utilizes its reserves of fats for energy production.

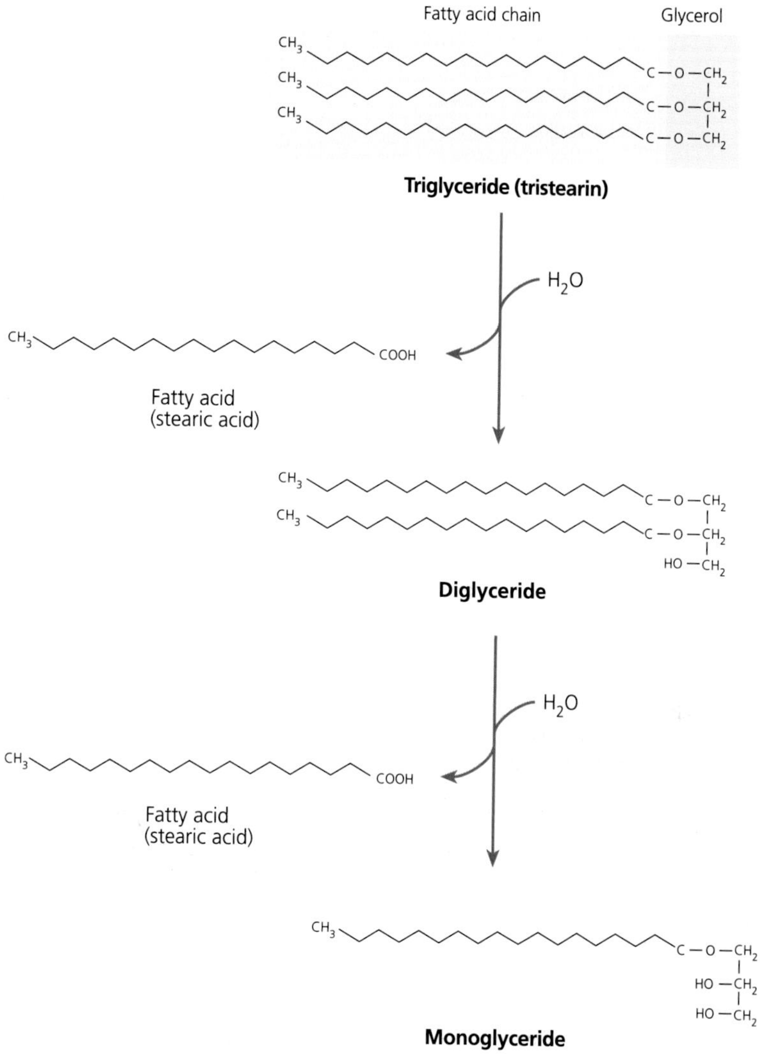

The **structural lipids** are the main component of the cell membranes. They fall into three main groups: **phospholipids**, **glycolipids** and **cholesterol**. The phospholipids fall into two further groups: those based on glycerol and those based on sphingosine. Commonly, one of the fatty acid chains possesses one or more double bonds as shown in Figure 3.8. The **glycolipids** of mammalian membranes are based on sphingosine with one hydroxyl group of glycerol linked to a sugar residue, as shown in Figure 3.8. As with the phospholipids, there is a very large number of glycolipids and the carbohydrate chains may be straight or branched.

The **steroids** are lipids with a structure based on four carbon rings known as the steroid nucleus. The most abundant steroid is cholesterol (Figure 3.9), which is a major constituent of cell membranes and which acts as the precursor for the synthesis of many steroid hormones, for example the **oestrogens**, **progesterone** and **testosterone**.

Figure 3.7 The structure of the glycerides and their interconversion by hydrolysis.

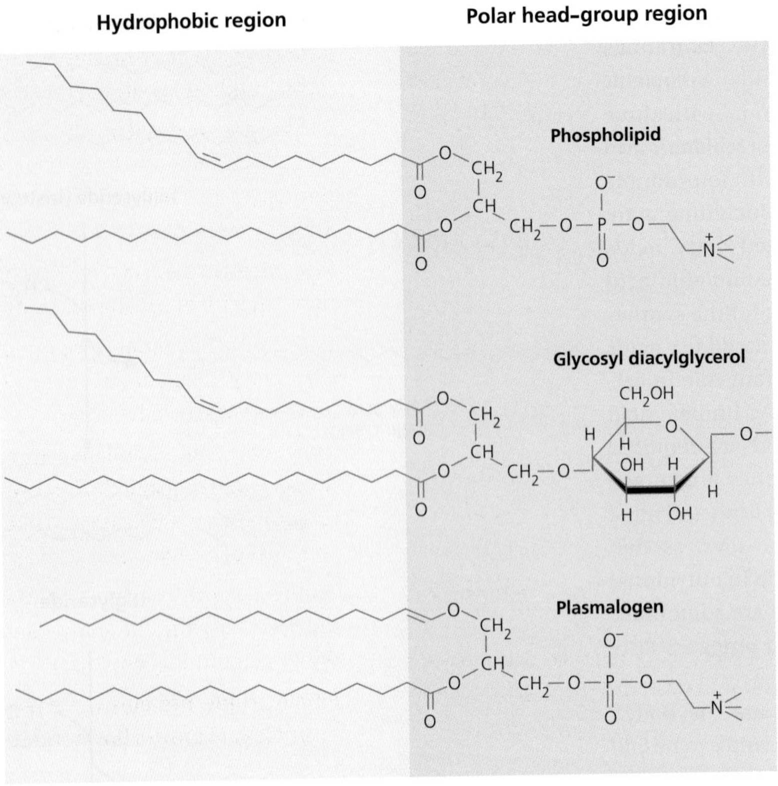

Hydrophobic region   **Polar head-group region**

Figure 3.8 The chemical structures of some of the lipids that form cell membranes. Note that each type has a polar head-group region (highlighted in blue) and a long hydrophobic tail (highlighted in yellow). In reality, the hydrophobic region is more extensive than the polar region.

The structures of three steroid hormones are shown in Figure 3.9.

The long-chain fatty acids and steroids are insoluble in water, but they are transported in the blood and body fluids in association with proteins as **lipoprotein** particles. In cell membranes, the lipids form bilayers, which are arranged so that their polar head groups are oriented towards the aqueous phase while the hydrophobic fatty acid chains face inwards to form a central hydrophobic region. This provides a barrier to the diffusion of polar molecules (e.g. glucose) and ions, but not to small, non-polar molecules such as oxygen and carbon dioxide. The cell membranes divide the cell into discrete compartments that provide the means of storage of various materials and permit the segregation of different

metabolic processes. The way in which lipid membranes divide up cells into different compartments is discussed in greater detail in the next chapter.

KEY POINT:

- The lipids are a chemically diverse group of substances that are insoluble in water but are soluble in organic solvents such as ether and chloroform. They serve a wide variety of functions: the phospholipids form the main structural element of cell membranes; the triglycerides are an important reserve of energy while many steroids act as signalling molecules.

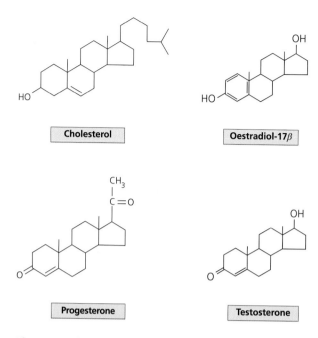

Figure 3.9 The chemical structures of some steroids of physiological importance: cholesterol, oestradiol-17β, progesterone and testosterone.

The amino acids and proteins

The proteins are large molecules that perform many different functions within the body; a few examples are given in the list below. They are assembled from smaller molecules, called the **α-amino acids**, which are carboxylic acids that have an amine group and a side chain attached to the carbon atom next to the carboxyl group (the α-carbon atom), as shown in Figure 3.10. With the exception of the smallest amino acid, glycine, the α-carbon atom of the amino acids is attached to four different groups. As for the carbohydrates, this makes amino acid molecules asymmetric and each has an L- and a D-isomer that are mirror images (i.e. optical isomers). All naturally occurring amino acids belong to the L-series.

Proteins serve an extraordinarily wide variety of functions in the body:

- They form the enzymes that catalyse the chemical reactions of living things.
- They are responsible for the transport of molecules and ions across cell membranes. They thus form the ion channels and carrier proteins of the plasma membrane that are discussed in Chapter 5.

- Some proteins bind ions and small molecules for storage inside cells. An important example is ferritin, which binds excess iron in the gut.
- Others are involved in the transport of molecules and ions around the body. The role of haemoglobin in transporting oxygen and carbon dioxide is a well-known example.
- Proteins form the cytoskeleton that provides the structural strength of cells (actin filaments for example).
- They form the motile components of muscle and of cilia. The proteins actin and myosin form the protein filaments of skeletal muscle.
- They form the connective tissues that bind cells together and transmit the force of muscle contraction to the skeleton. Collagen and elastin are examples of this class of protein.
- Proteins known as the immunoglobulins play an important part in the body's defence against infection, e.g. immunoglobulins A and G (IgA and IgG).
- Finally, some proteins act as signalling molecules. Examples are growth hormone and insulin.

Proteins are assembled from a set of 20 α-amino acids

The α-amino acids of proteins can be grouped into five different classes:

- Acidic amino acids (aspartic acid and glutamic acid)
- Basic amino acids (arginine, histidine and lysine)
- Uncharged hydrophilic amino acids (asparagine, glycine, glutamine, serine and threonine)
- Hydrophobic amino acids (alanine, leucine, isoleucine, phenylalanine, proline, tyrosine, tryptophan and valine)
- Sulphur-containing amino acids (cysteine and methionine).

Amino acids can be combined together by linking the amine group of one with the carboxyl group of another and eliminating water to form a **dipeptide** as shown in Figure 3.11. The linkage between two amino acids joined in this way is known as a **peptide bond**. The addition of a third amino acid would give a tripeptide, a fourth a tetrapeptide and so on. Peptides with large numbers of

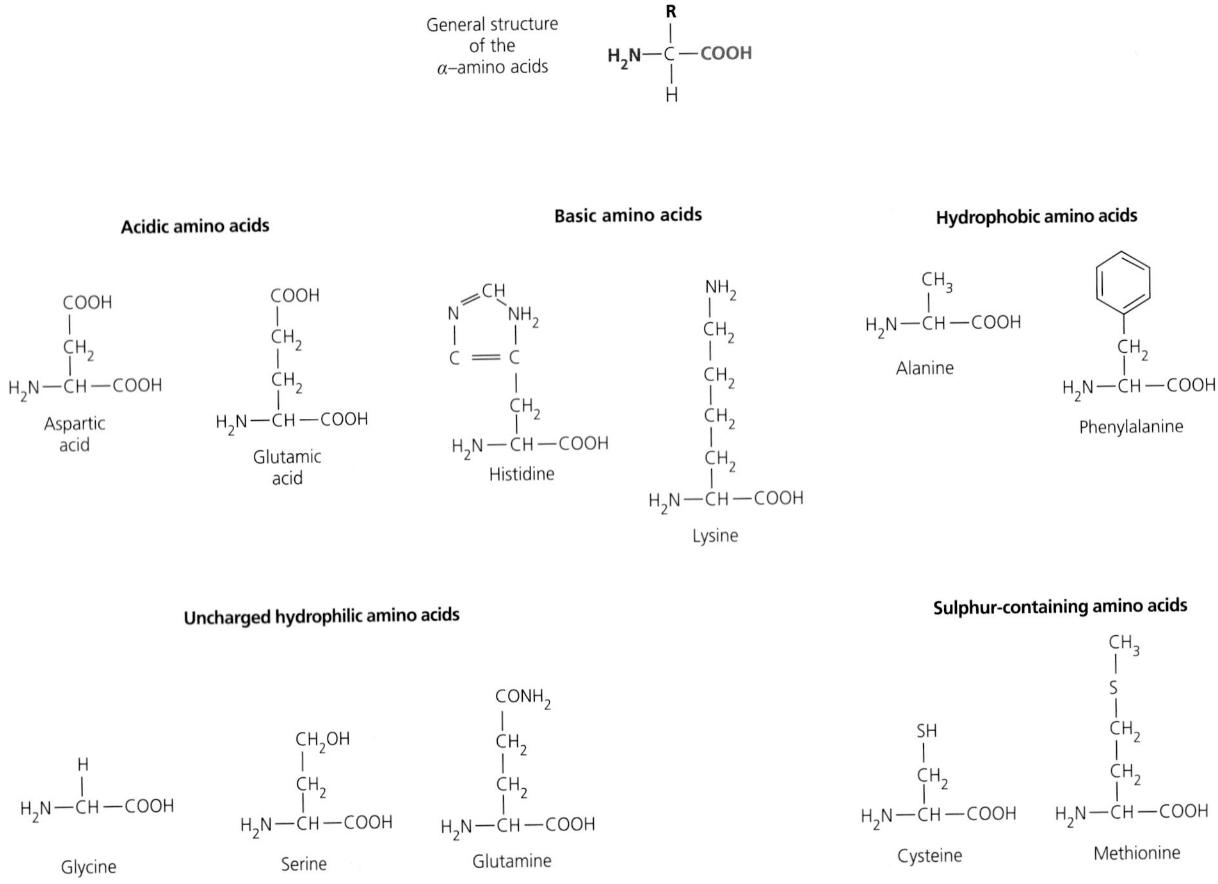

General structure
of the
α–amino acids

$$H_2N-\underset{\underset{H}{|}}{\overset{\overset{R}{|}}{C}}-COOH$$

Acidic amino acids

Aspartic
acid

Glutamic
acid

Basic amino acids

Histidine

Lysine

Hydrophobic amino acids

Alanine

Phenylalanine

Uncharged hydrophilic amino acids

Glycine

Serine

Glutamine

Sulphur-containing amino acids

Cysteine

Methionine

Figure 3.10 The chemical structures of representative α-amino acids. The general structure of the α-amino acids is shown at the top of the figure. R represents the side chains of the different amino acids. The α-amino group is shown in red and the carboxyl group in green.

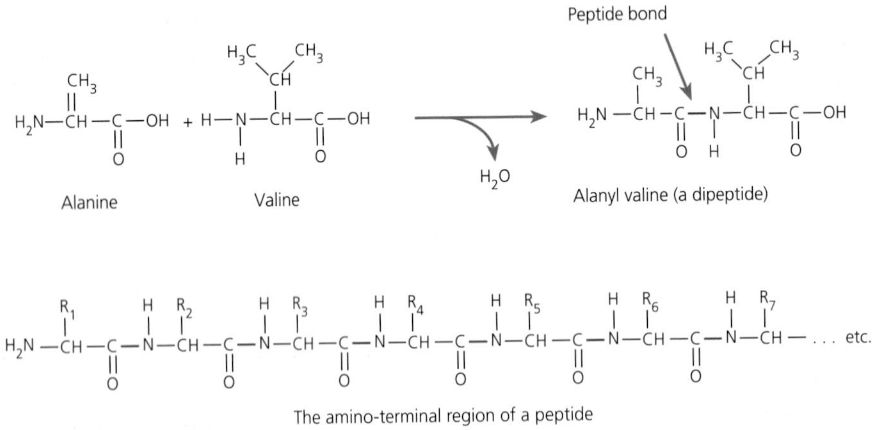

Peptide bond

Alanine   Valine

H_2O

Alanyl valine (a dipeptide)

The amino-terminal region of a peptide

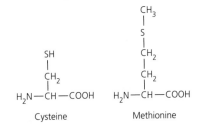

Figure 3.11 The formation of a peptide bond and the structure of the amino-terminal region of a polypeptide showing the peptide bonds in red; R_1, R_2, etc. represent different amino acid side chains.

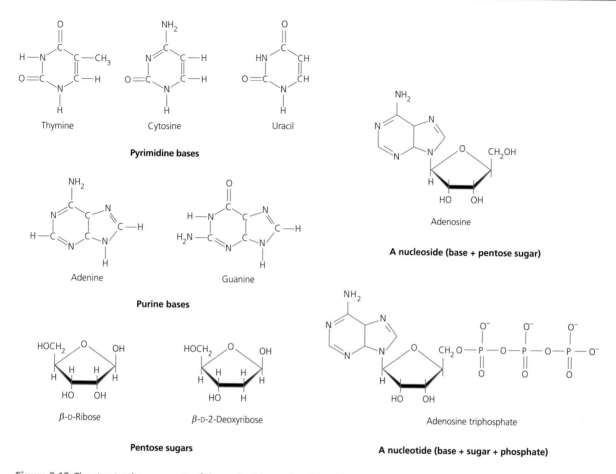

Figure 3.12 The structural components of the nucleotides and nucleic acids.

amino acids linked together are known as **polypeptides**. Proteins are large polypeptides.

As there is no specific order or number of amino acids that can be linked together to form a protein, the number of possible protein structures is essentially infinite. Different proteins have different shapes and different physical properties. Some are soluble in water while others are not. That is why they are such a versatile class of molecules.

Many cellular structures consist of protein assemblies made up of several different kinds of protein. Examples are the filaments of the skeletal muscle fibres, which contain the proteins actin, myosin, troponin and tropomyosin. Actin molecules also assemble together to form the microfibrils that are important components of the cytoskeleton that gives each type of cell its characteristic shape. Enzymes (catalytic proteins) are frequently arranged so that the product of one enzyme can be passed directly to another and so on. These multienzyme assemblies increase the efficiency of cell metabolism.

KEY POINTS:

- Proteins are assembled from amino acids linked together by peptide bonds.

- They serve a wide variety of functions in the body.

(a)

(b)

G

A

T

C

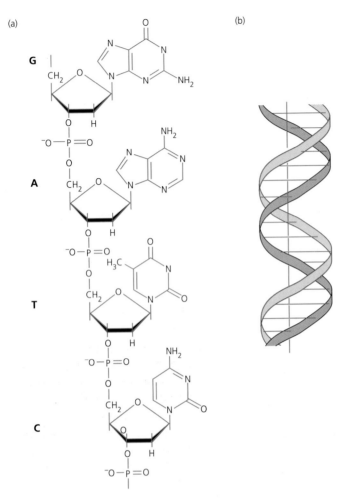

Figure 3.13 The structure of DNA. (a) Diagram of a short length of one of the strands of DNA. The stretch of DNA has the nucleotide sequence guanine (G), adenine (A), thymine (T) and cytosine (C). (b) A diagrammatic representation of the two complementary strands of DNA. The sequence of one strand runs in the opposite sense to the other.

The nucleotides and nucleic acids

The genetic information of the body resides in its DNA (deoxyribonucleic acid), which is stored in the chromosomes of the nucleus. DNA is made by assembling smaller components known as nucleotides into a long chain. Ribonucleic acid (RNA) has a similar primary structure. Each nucleotide consists of a purine or pyrimidine base linked to a pentose sugar, which is in turn linked to a phosphate group, as shown in Figure 3.12. The pyrimi-

dine bases are cytosine, thymine and uracil, and the purine bases are adenine and guanine.

The nucleotide coenzymes. Although nucleotides are the building blocks of DNA and RNA (see below), some are involved in other aspects of metabolism. For example, adenosine triphosphate (ATP) is important in energy metabolism and the nicotinamide and flavine nucleotides play an important role in cellular metabolism as coenzymes.

The nucleic acids. In nature, there are two main types of nucleic acid: DNA and RNA. In DNA, the sugar of the nucleotides is deoxyribose and the bases are adenine, guanine, cytosine and thymine (abbreviated A, G, C and T). In RNA, the sugar is ribose and the bases are adenine, guanine, cytosine and uracil (A, G, C and U). In both DNA and RNA, nucleotides are joined together to form large molecules by phosphate linkages between one nucleotide and the next as shown for DNA in Figure 3.13a.

A molecule of DNA consists of a pair of nucleotide chains linked together by hydrogen bonds in such a way that adenine links with thymine and guanine links with cytosine. The hydrogen bonding between the two chains is so precise that the sequence of bases on one chain automatically determines that of the second. The pair of chains is twisted to form a double helix in which the complementary strands run in opposite directions (Figure 3.13b). The discovery of this base pairing was crucial to the understanding of the three-dimensional structure of DNA and to the subsequent unravelling of the genetic code. Unlike DNA, each RNA molecule has only one polynucleotide chain.

> **KEY POINT:**
>
> • Nucleotides are made of a base, a pentose sugar and a phosphate residue. They can be combined with other molecules to form coenzymes such as the nicotinamide and flavine nucleotides. They can also be assembled into long chains to form DNA and RNA.

Recommended reading

Alberts, B., Johnson, A., Lewis, J., Raff, M., Roberts, K. and Walter, P. (2008). *Molecular Biology of the Cell,* 5th edn. Garland: New York. Chapters 2 (pp. 45–65) and 3 (pp. 125–152).

Elliott, W.H. and Elliott, D.C. (2008). *Biochemistry and Molecular Biology.* Oxford University Press: Oxford. Chapters 1 and 22–25.

Self-assessment questions for Chapters 2 and 3

Which of the following statements are true and which are false? Answers are given at the end of the book (p. 755).

1 a) The osmotic pressure of a solution depends on the number of molecules and ions it contains per litre.

   b) The osmotic pressure of the intracellular fluid is higher than that of the extracellular fluid.

   c) The proteins of the blood plasma account for most of the osmolality of the plasma.

   d) Water accounts for about two-thirds of the body weight of an adult man.

   e) Two solutions that have the same tonicity are iso-osmotic with each other.

   f) Most of the body water is in the extracellular fluid.

2 a) Carbohydrates are broken down to provide energy for cellular reactions.

   b) Starch is the body's main store of carbohydrate.

   c) Polysaccharides are broken down to monosaccharides by hydrolysis.

   d) Sucrose is an example of a monosaccharide.

   e) Glycoproteins are important structural components of connective tissues.

3 a) Lipids are a major reserve of energy.

   b) Lipids are a minor constituent of cell membranes.

   c) Lipids are insoluble in water (hydrophobic).

   d) Palmitic acid is an essential fatty acid.

   e) Cholesterol is a precursor for the synthesis of oestrogens.

4 a) Proteins are made from many amino acids linked together by peptide bonds.

   b) Enzymes are proteins that catalyse the chemical reactions of cells.

   c) Some hormones are proteins.

   d) All proteins are freely soluble in water.

   e) A tripeptide is made from three amino acids linked together by two peptide bonds.

5 a) Nucleotides consist of a pyrimidine base linked to a phosphate group.

   b) RNA contains ribose, which is a monosaccharide with five carbons.

   c) Both DNA and RNA are built from the same bases with different sugars.

   d) Nucleotides can act as coenzymes in cellular metabolism.

   e) RNA only differs from DNA in having a single polynucleotide chain.

Section 2

The organization and basic functions of cells

4   Introducing cells   **40**

5   The functions of the plasma membrane   **55**

6   Cells and tissues   **68**

7   The principles of cell signalling   **80**

4 Introducing cells

After reading this chapter you should have gained an understanding of:

- The structural organization of cells
- The functions of the different cellular organelles
- Cell division by mitosis
- The principles of cellular energy metabolism
- Cell motility

4.1 Introduction

The cells are the building blocks of the body. There are many different types of cell, each with its own characteristic size and shape. Some cells are very large. For example, the cells of skeletal muscle (skeletal muscle fibres) may extend for up to 30 cm along the length of a muscle and be up to 100 μm in diameter. Others are very small – for example the red cells of the blood, which are small, biconcave disks with diameters in the region of 7 μm. Nerve cells (also called neurones) characteristically have a large number of thin extensions – the dendrites and axon (see Chapter 8), while others, such as the lymphocytes of the blood, are round. Skeletal muscle fibres, neurones and red cells represent some of the more striking variations in cell morphology.

The structure of a typical mammalian cell is illustrated in Figure 4.1, which shows it to be bounded by a **cell membrane**, also called the **plasma membrane** or **plasmalemma**. The cell membrane is a continuous sheet that separates the watery phase inside the cell, the **cytoplasm**, from that outside the cell, the extracellular fluid. The shape of an individual cell is maintained by an array of protein filaments known as the **cytoskeleton**.

At some stage of their life cycle, all cells possess a prominent structure called the **nucleus**, which contains the hereditary material, DNA. Most cells have a single nucleus but skeletal muscle fibres have many nuclei reflecting their embryological origin from the fusion of large numbers of progenitor cells known as myoblasts. In contrast, the red cells of the blood lose their nucleus as they mature. Cells possess other structures that perform specific functions such as energy production, protein synthesis and the secretion of various materials. The internal structures of a cell are collectively known as **organelles** and include the nucleus, the mitochondria, the Golgi apparatus, the endoplasmic reticulum and various membrane-bound vesicles (see below).

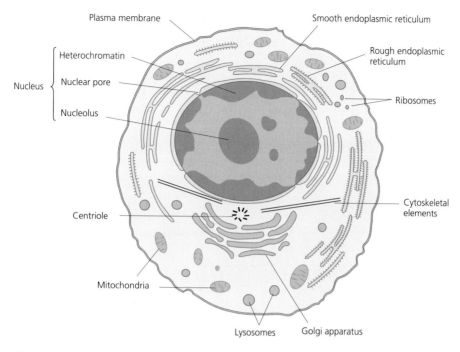

Figure 4.1 A diagram of a typical mammalian cell, showing the major organelles.

4.2 The structure and functions of the cellular organelles

The plasma membrane

The plasma membrane regulates the movement of substances into and out of a cell. It is also responsible for regulating a cell's response to a variety of chemical signals such as hormones. An intact plasma membrane is, therefore, essential for the proper function of a cell. When viewed at high power in an electron microscope, the plasma membrane appears as a sandwich-like structure $5-10\,nm$ thick (i.e. $5-10 \times 10^{-9}\,m$) as shown in Figure 4.2. The membranes of the intracellular organelles (e.g. endoplasmic reticulum, Golgi apparatus, lysosomes and mitochondria) have a three-layered structure, similar to that of the plasma membrane.

The plasma membrane is a lipid bilayer containing proteins

Chemical analysis shows that the plasma membrane is made of lipid and protein in approximately equal amounts by weight. The lipids are arranged so that their

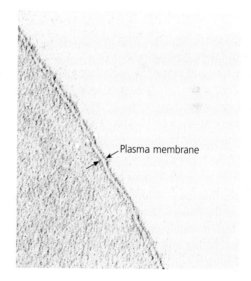

Figure 4.2 An electron micrograph of the plasma membrane of a red cell that shows the characteristic appearance of a lipid bilayer. Note that the two densely staining outer regions are separated by a pale inner zone (indicated by the arrows in the figure).

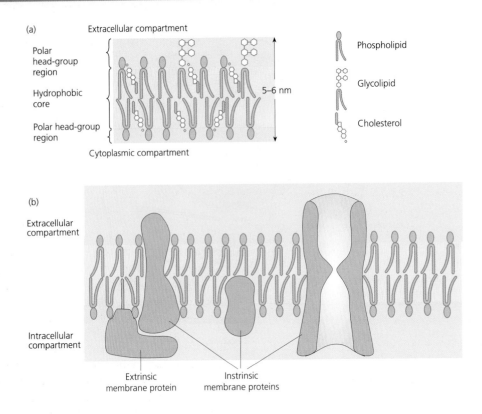

Figure 4.3 The structure of the plasma membrane. (a) The basic arrangement of the lipid bilayer. Note the presence of glycolipid only in the outer leaflet of the bilayer. (b) A simplified model of the plasma membrane showing the arrangement of some of the membrane proteins. Note the hydrophobic chain anchoring the extrinsic membrane protein to the inner face of the membrane.

polar head groups are oriented towards the aqueous phase and the hydrophobic fatty acid chains face inwards to form a central hydrophobic core, as shown in Figure 4.3. The membrane proteins either span the lipid bilayer or are anchored to it in various ways.

The principal lipids of the plasma membrane belong to one of three classes: phospholipids, glycolipids and cholesterol (see Chapter 3 for a discussion of the different kinds of membrane lipid). There is now good evidence that the phospholipids of the outer and inner layers of the plasma membrane differ significantly. The outer leaflet mainly consists of the glycolipids phosphatidylcholine and sphingomyelin, while the inner leaflet is richer in negatively charged phospholipids such as phosphatidylinositol. Cholesterol is present in both leaflets of the bilayer. The presence of phosphatidylinositol in the inner leaflet of the bilayer is important as inositol phosphates play an important role in the

transmission of certain signals from the cell membrane to the interior of the cell (see Chapter 7). While phospholipid molecules are able to diffuse freely in one plane of the bilayer, they rarely flip from one leaf of the bilayer to the other. This indicates that the lipid bilayer is an inherently stable structure.

It is possible to make artificial lipid bilayers in the laboratory and these membranes have been shown to be relatively permeable to carbon dioxide, oxygen and lipid-soluble molecules. However, unlike natural membranes, these artificial bilayers are almost impermeable to polar molecules such as glucose and ions such as sodium (Na$^+$). Moreover, such membranes are relatively impermeable to water. In natural cell membranes, polar molecules, ions and water cross the hydrophobic barrier formed by the lipid bilayer via specific protein molecules – the **carrier proteins**, **ion channels** and **aquaporins**, which are discussed in Chapter 5.

Membrane proteins

The membrane proteins may be divided into two broad groups: **intrinsic membrane proteins** (those that are embedded in the bilayer itself) and **extrinsic membrane proteins** (those that lie outside the lipid bilayer but are linked to it in some way, often by a hydrophobic fatty acid chain). The proteins that facilitate the movement of ions and other polar materials across the plasma membrane are all intrinsic membrane proteins. Some extrinsic proteins link the cell to its surroundings (the extracellular matrix) or to neighbouring cells. Others play a role in the transmission of signals from the plasma membrane to the interior of the cell. The physiological roles of the membrane proteins are discussed in greater detail in the following chapters.

> **KEY POINTS:**
> - The plasma membrane consists of roughly equal amounts by weight of protein and lipid.
> - The lipid is arranged as a bilayer, which forms a barrier to the passage of polar materials.
> - Polar substances and water must enter or leave a cell via specialized transport proteins.
> - The combination of lipid bilayer and transport proteins allows cells to maintain an internal composition that is very different to that of the extracellular fluid.

The nucleus

The nucleus is separated from the rest of the cytoplasm by the **nuclear membrane** (also called the **nuclear envelope**), which consists of two lipid bilayers separated by a narrow space. The nuclear membrane is furnished with small holes known as **nuclear pores**, which provide a means of communication between the nucleus and the cytoplasm. The nucleus contains the genetic material of a cell, DNA, which is associated with proteins called histones in the form of chromatin fibres. During cell division (mitosis), the chromatin becomes distributed into pairs of **chromosomes**, which attach to a structure known as the mitotic spindle before they separate as the cell divides (see below).

Within the nucleus is a strongly staining feature called the **nucleolus**. The nucleolus is concerned with the manufacture of the **ribosomes**, which play an important part in the synthesis of new protein molecules (see Chapter 30, pp. 651–653). Prominent nucleoli are seen in cells that are synthesizing large amounts of protein such as embryonic cells, secretory cells and those of rapidly growing malignant tumours.

The organelles of the cytoplasm

The cytoplasm contains many different organelles that perform specific tasks within the cell. Some of these organelles are separated from the rest of the cytoplasm by membranes so that the cell's internal membranes divide the cytoplasm into various compartments. Examples are the mitochondria, the endoplasmic reticulum, the Golgi apparatus and vesicles such as lysosomes.

Mitochondria. The mitochondria oxidize glucose and fats to provide energy in the form of ATP for other activities such as cell motility and the synthesis of proteins for growth and repair. The mitochondria are rod-shaped structures $2-6\,\mu m$ in length, about $0.2\,\mu m$ in diameter with two distinct membranes: an outer membrane, which is smooth and regular in appearance, and an inner membrane, which is thrown into a number of folds known as **cristae**. It is here, on the inner membrane, that the synthesis of ATP takes place via the **tricarboxylic acid cycle** and **electron transport chain**. Cells that have a high demand for ATP (e.g. the muscle cells of the heart) have many mitochondria. In addition to their role in energy production, mitochondria can also accumulate significant amounts of calcium. They thus play a significant role in regulating the ionized calcium concentration within cells.

Mitochondria are not assembled from scratch from their molecular constituents, but increase in number by the division of existing mitochondria. This division takes place in the interval between cell divisions (the interphase of the cell cycle). When cells divide, the new mitochondria are shared out between the daughter cells.

Endoplasmic reticulum. The endoplasmic reticulum is a system of membranes that extends throughout the cytoplasm of most cells. These membranes are continuous with the nuclear membrane and enclose a significant space within the cell. The endoplasmic reticulum is classified as rough or smooth according to its appearance under the electron microscope. The rough endoplasmic reticulum plays an important role in the synthesis of certain proteins. It is also important for the addition of carbohydrates to membrane proteins (glycosylation), which takes place on its inner surface. The smooth endoplasmic reticulum is involved in the synthesis of lipids and other aspects of metabolism. It also accumulates calcium to provide an intracellular store for calcium signalling, which is regulated in many cells by the inositol lipids of the inner leaflet of the plasma membrane (see Chapter 7). In muscle cells, the endoplasmic reticulum is called the **sarcoplasmic reticulum**. It plays an important role in the initiation of muscle contraction (see pp. 121–122).

The Golgi apparatus (also called the Golgi complex) is a system of flattened, membranous sacs that are involved in modifying and packaging proteins for secretion. The Golgi apparatus has a characteristic appearance in which the face nearest the nucleus appears convex (the *cis* face) while the other (the *trans* face) is concave. Secretory vesicles bud off from the *trans* face before migrating to the plasma membrane prior to being secreted.

Membrane-bound vesicles. Cells contain a variety of membrane-bound vesicles that are integral to their function. Secretory vesicles are formed by the Golgi apparatus. Under the electron microscope, some vesicles appear as simple, round profiles while others have an electron-dense core. After they have discharged their contents, the secretory vesicles are retrieved from the plasma membrane to form endocytic vesicles (see Chapter 5, p. 65). The **lysosomes** are vesicles that contain hydrolytic enzymes that allow cells to digest materials they have taken up during endocytosis or phagocytosis.

The importance of the lysosomes to the economy of a cell is strikingly illustrated by **Tay–Sachs disease** in which gangliosides taken up into the lysosomes are not degraded. The sufferers lack an enzyme known as β-hexosaminidase A, which is responsible for the breakdown of gangliosides derived from recycled plasma membrane. In the absence of this enzyme, the lysosomes accumulate lipid and become swollen. Although other cells are affected, gangliosides are especially abundant in nerve cells, which show severe pathological changes as their lysosomes become swollen with undegraded lipid. This leads to the premature death of neurones, resulting in muscular weakness and retarded development. The condition is fatal and those with the disease usually die before they are 3 years old.

Ribosomes consist of proteins and ribonucleic acid (RNA). They are formed in the nucleolus and migrate to the cytoplasm where they may occur free or in groups called polyribosomes. Ribosomes play an important role in the synthesis of new proteins as outlined in Chapter 30 (pp. 651–653). Some ribosomes become attached to the outer membrane of the endoplasmic reticulum to form the rough endoplasmic reticulum mentioned above, which is the site of synthesis of membrane proteins.

The cytoskeleton. A cell is not just a bag of enzymes and isolated organelles. Different cell types each have a distinctive and stable morphology that is maintained by an internal array of protein filaments known as the cytoskeleton. The protein filaments are of three main kinds: actin filaments, intermediate filaments and microtubules.

- **Actin filaments** play an important role in cell movement, such as the contraction of skeletal muscle. They also help to maintain cell shape in non-motile cells.

- **Intermediate filaments** play an important role in the mechanical stability of cells. Those cells that are subject to a large amount of mechanical stress (such as epithelia and smooth muscle) are particularly rich in intermediate filaments, which link cells together via specialized junctions.

- **Microtubules**, as their name suggests, are hollow tubes, which have an external diameter of about 25 nm and a wall thickness of 5–7 nm. Microtubules are formed from a protein called tubulin. They play an important role in moving organelles (e.g. secretory vesicles) through the cytoplasm. They also play a major role in the movement of cilia and flagella (see p. 53). Microtubules originate from a complex structure known as a **centrosome**.

Between cell divisions, the centrosome is located at the centre of a cell near the nucleus. Embedded in the centrosome are two **centrioles**, which are cylindrical structures, arranged at right angles to each other. At the beginning of cell division, the centrosome divides into two and the daughter centrosomes move to opposite poles of the nucleus to form the mitotic spindle (see below).

Cilia are very small, hair-like projections from certain cells which have a characteristic array of microtubules at their core (see p. 53). In mammals, cilia beat in an orderly, wave-like motion to propel material over the surface of an epithelial layer such as the lining of the upper respiratory tract. **Flagella** are similar in structure to cilia but are much longer. While they are common in single-cell organisms, flagella are only found in mammals as the motile part of the sperm (see Chapter 28).

> **KEY POINTS:**
>
> - Each cell possesses a number of small structures known as organelles that carry out specific cellular functions.
>
> - The nucleus is the most prominent feature of most cells. Within the nucleus is a nucleolus, the site of ribosome assembly.
>
> - Membranes within a cell divide the cytoplasm into a number of separate compartments: the endoplasmic reticulum, the Golgi apparatus, the lysosomes and the mitochondria.
>
> - The mitochondria synthesize most of the ATP required by a cell to carry out its physiological functions.
>
> - Protein filaments form a cytoskeleton that serves to maintain the distinctive shape of each kind of cell.

4.3   Cell division

During life, animals grow by two processes: (1) the addition of new material to pre-existing cells and (2) by increasing the number of cells by division. Cell division occurs by one of two processes:

- **Mitosis**, in which each daughter cell has the same number of genetically identical chromosomes as the parent cell.

- **Meiosis**, in which each daughter cell has half the number of chromosomes as its parent.

Most cells that divide do so by mitosis. Meiosis occurs only in the germinal cells during the formation of the eggs and sperm and is discussed in Chapter 28 (pp. 583–584).

Mitosis

The process of mitosis and DNA replication can be divided into six phases (Figure 4.4):

Prophase. During the early part of prophase, the nuclear chromatin condenses into well-defined chromosomes and the mitotic spindle begins to form outside the nucleus as the centrosomes begin to separate.

Prometaphase begins with the dissolution of the nuclear membrane. It is followed by the movement of the microtubules of the mitotic spindle into the nuclear region. The chromosomes then become attached to the mitotic spindle by a special attachment region called a centromere.

Metaphase. During metaphase, the chromosomes become aligned along the central region of the mitotic spindle.

Anaphase begins with the separation of the two chromatids to form the chromosomes of the daughter cells. The poles of the mitotic spindle move further apart.

Telophase. In telophase, the separated chromosomes reach the poles of the mitotic spindle, which begins to

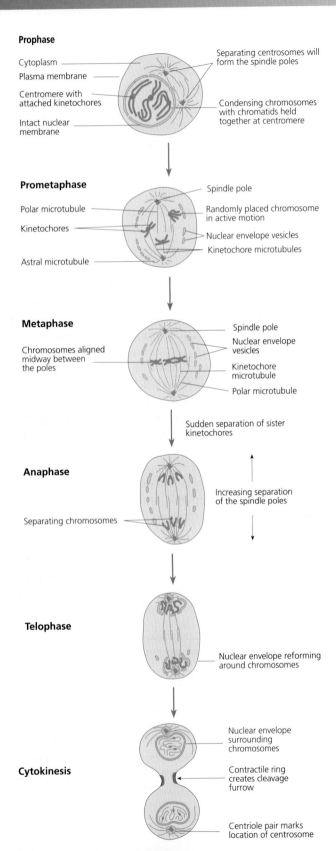

Prophase

Cytoplasm

Plasma membrane

Centromere with
attached kinetochores

Intact nuclear
membrane

Separating centrosomes will
form the spindle poles

Condensing chromosomes
with chromatids held
together at centromere

Prometaphase

Polar microtubule

Kinetochores

Astral microtubule

Spindle pole

Randomly placed chromosome
in active motion

Nuclear envelope vesicles

Kinetochore microtubules

Metaphase

Chromosomes aligned
midway between
the poles

Spindle pole

Nuclear envelope
vesicles

Kinetochore
microtubule

Polar microtubule

Sudden separation of sister
kinetochores

Anaphase

Separating chromosomes

Increasing separation
of the spindle poles

Telophase

Nuclear envelope reforming
around chromosomes

Cytokinesis

Nuclear envelope
surrounding
chromosomes

Contractile ring
creates cleavage
furrow

Centriole pair marks
location of centrosome

Figure 4.4 The principal stages of cell division by mitosis.
Although the different stages are shown as discrete steps,
in reality there is a smooth progression from one stage to
the next.

disappear. A new nuclear membrane is formed around each daughter set of chromosomes and mitosis proper is complete.

Cytokinesis is the division of the cytoplasm between the two daughter cells. It begins during anaphase but is not completed until after the end of telophase. The cell membrane forms a cleavage furrow, which deepens until the narrow neck of cytoplasm joining the two daughter cells finally breaks to give rise to the two daughter cells.

> **KEY POINTS:**
> - Cell division occurs either by mitosis or by meiosis.
> - Most cells divide by mitosis and each daughter cell is genetically identical to its parent.
> - Meiosis occurs only in the germinal cells during the formation of the eggs and sperm.

4.4 Programmed cell death

Each time a cell divides, two daughter cells are produced. If a cell and its daughter cells divide a further ten times, the initial parent cell will have given rise to over a thousand descendants. While this marked increase in numbers is desirable during growth and development, it would become a problem in an adult, where cell numbers must be regulated to maintain a more or less constant body size. Moreover, during development, more cells are produced than are required to form specific tissues. For example, a hand begins as a spade-like structure and the fingers develop by the selective removal of tissue. This process requires a regulated programme of cell death, which is called **apoptosis** (from the Greek for 'falling off' – as in the leaves of trees in autumn). Apoptosis is triggered by specific, local signals that cause the activation of a class of proteases known as caspases. The first set of caspases to be activated within the cell then activates other digestive enzymes, which quickly break it up. The remnants are then taken up by phagocytic cells, which are able to utilize the dead cell's components for energy production and the synthesis of new cell constituents (e.g. proteins).

4.5 Energy metabolism in cells

Animals take in food as carbohydrates, fats and proteins. These complex molecules are broken down in the gut to simple molecules, which are then absorbed. The carbohydrates of the diet mainly consist of starch, which is broken down to glucose. The fats are broken down to fatty acids and glycerol, while the dietary proteins are broken down to their constituent amino acids. These breakdown products can then be used by the cells of the body to make ATP, which provides a convenient way of harnessing chemical energy.

ATP can be synthesized in two ways: Firstly, by the breakdown of glucose to pyruvate via the glycolytic pathway (glycolysis) and secondly by the oxidative metabolism of pyruvate and acetate via the tricarboxylic acid cycle and the electron transport chain. (The tricarboxylic acid cycle is also known as the citric acid cycle or Krebs cycle). The utilization of glucose, fatty acids and amino acids for the synthesis of ATP is summarized in Figure 4.5. In each case, the synthesis of ATP via the tricarboxylic acid cycle requires the uptake of oxygen and results in the production of carbon dioxide and water.

The generation of ATP by glycolysis

Of the simple sugars, glucose is the most important in the synthesis of ATP. It is transported into cells where a phosphate group is added to form glucose-6-phosphate. This is the first stage of the process of glycolysis by which glucose is broken down to form pyruvate. Glycolysis takes place in the cytoplasm of the cell, outside the mitochondria, and does

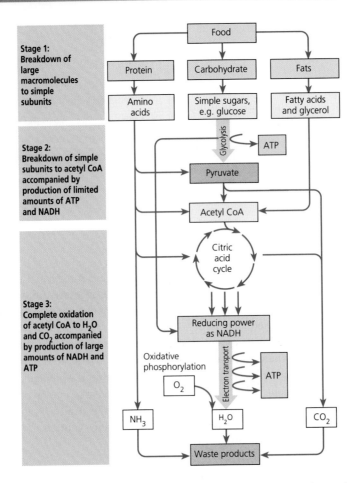

Figure 4.5 An overview of the main steps in the production of energy from carbohydrates, fats and proteins. Most of the ATP is produced by the oxidative metabolism of acetyl CoA via the citric acid cycle (also known as the tricarboxylic acid or TCA cycle) and the electron transport chain.

not require the presence of oxygen. For this reason, the glycolytic breakdown of glucose is said to be **anaerobic**. The glycolytic pathway is summarized in Figure 4.6. Further details can be found in any standard textbook of biochemistry.

The breakdown of a molecule of glucose by glycolysis yields two molecules of pyruvate and two molecules of ATP. In addition, two molecules of reduced nicotinamide adenine dinucleotide (NADH) are produced. When oxygen is present, the NADH generated by glycolysis is oxidized by the mitochondria via the electron transport chain resulting in the synthesis of a further three molecules of ATP and the regeneration of two molecules of nicotinamide adenine dinucleotide (NAD).

Under normal circumstances, the pyruvate that is generated during glycolysis combines with coenzyme A to form acetyl coenzyme A (acetyl CoA), which is oxidized via the tricarboxylic acid cycle to yield ATP. In the absence of sufficient oxygen, however, some of the pyruvate is reduced by NADH in the cytosol to generate lactate. This step regenerates the NAD used in the early stages of glycolysis so that it can participate in the breakdown of a further molecule of glucose. Thus, *glycolysis can generate ATP even in the absence of oxygen* and becomes an important source of ATP for skeletal muscle during heavy exercise (see Chapter 32).

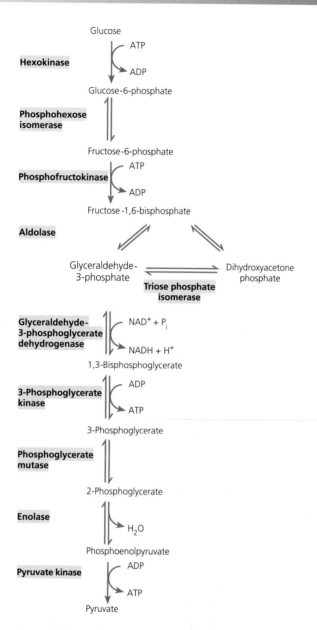

Figure 4.6 A schematic chart showing the principal steps in the glycolytic breakdown of glucose. Note that for each molecule of glucose broken down to pyruvate, two molecules of ATP and two molecules of NADH are produced.

The breakdown of pyruvate by the tricarboxylic acid cycle

The two molecules of pyruvate formed by the glycolytic breakdown of glucose combine with coenzyme A to form acetyl CoA before they enter the tricarboxylic acid cycle. This step occurs in the mitochondria and results in the formation of two molecules of carbon dioxide and two molecules of NADH. Each molecule of acetyl CoA combines with a molecule of oxaloacetate to form the tricarboxylic acid citrate, which then undergoes a series of reactions that results in the complete oxidation of the acetyl CoA to carbon dioxide and water. These reactions are summarized in Figure 4.7.

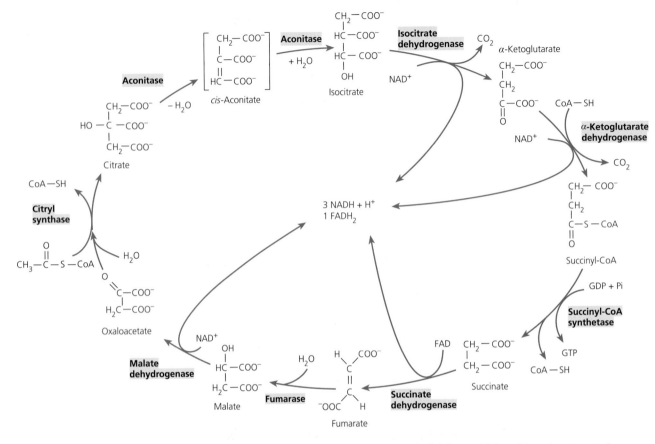

Figure 4.7 The tricarboxylic acid (TCA) cycle. Acetyl CoA enters the TCA cycle on the left by combining with oxaloacetate to form citrate. No ATP is formed directly during the cycle (although one molecule of GTP is formed for each turn of the cycle); the main products are NADH and $FADH_2$, both of which are utilized by the electron transport chain to synthesize ATP. The enzymes involved in each step are highlighted in grey.

During each turn of the tricarboxylic acid cycle, three molecules of NADH, one of reduced flavine adenine dinucleotide ($FADH_2$) and one molecule of GTP are generated. The mitochondria utilize the NADH and $FADH_2$ to generate ATP via an enzyme complex known as the electron transport chain. This process requires molecular oxygen and is called **aerobic metabolism**. The aerobic metabolism is very efficient and yields about 10 molecules of ATP for each molecule of acetyl CoA that enters the tricarboxylic acid cycle. Overall, the complete oxidation of one molecule of glucose to carbon dioxide and water yields about 30 molecules of ATP. This should be compared with the generation of ATP by anaerobic metabolism where only two molecules of ATP are generated for each molecule of glucose used.

Fatty acids are the body's largest store of food energy. They are stored in fat cells (adipocytes) as triglycerides.

Fat cells are found throughout the body, but are most abundant in the adipose tissues. Stored fat is broken down to fatty acids and glycerol by lipases in a process known as lipolysis, which takes place in the cytoplasm of the cell. The glycerol is metabolized by the glycolytic pathway while the fatty acids combine with coenzyme A to form acyl CoA before being broken down by a process known as β-oxidation, which is illustrated in Figure 4.8. The breakdown of fatty acids to provide energy takes place within the mitochondria and, for each two-carbon unit metabolized, 13 molecules of ATP and one molecule of GTP are produced. Overall, the complete oxidation of a molecule of palmitic acid (which has 16 carbon atoms) yields over 100 molecules of ATP.

Although animals can synthesize fats from carbohydrates via acetyl CoA, they cannot synthesize

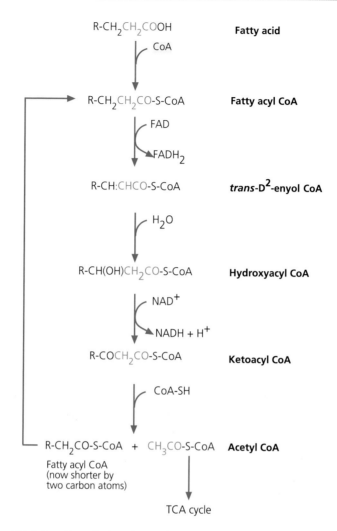

Figure 4.8 The pathway by which fatty acids are broken down to form acetyl CoA. This is known as β-oxidation. No ATP is produced by the β-oxidation pathway. The acetyl CoA formed is subsequently metabolized via the tricarboxylic acid cycle and the electron transport chain to produce ATP. In these reactions, each two-carbon unit (shown in red) generates one molecule of NADH and one molecule of FADH₂ in addition to that generated by the tricarboxylic acid cycle.

carbohydrates from fatty acids. When glucose reserves are low, many tissues preferentially utilize fatty acids liberated from the fat reserves. Under these circumstances, the liver relies on the oxidation of fatty acids for energy production and may produce more acetyl CoA than it can utilize in the tricarboxylic acid cycle. Under such conditions, it synthesizes acetoacetate and D-3-hydroxybutyrate (also called β-hydroxybutyrate). These compounds are known as **ketone bodies** and are produced in large amounts when the utilization of glucose by the tissues is severely restricted as in starvation

or in poorly controlled diabetes mellitus (see Chapter 34). Acetoacetate and β-hydroxybutyrate are not waste products but can be utilized for ATP production by the heart and the kidney. In severe uncontrolled diabetes mellitus, significant amounts of acetone form spontaneously from acetoacetate and this gives the breath a characteristic sweet smell, which can be a useful aid in diagnosis.

Proteins are the main structural components of cells and the principal use for the amino acids derived from dietary protein is the synthesis of new protein.

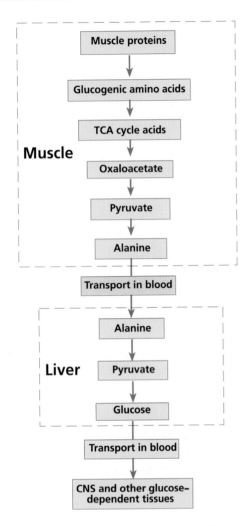

Figure 4.9 An outline of the pathways by which amino acids are broken down for energy (ATP) production and the formation of glucose. This process (gluconeogenesis) is particularly important in starvation.

Those amino acids that are not required for protein synthesis have their amino group ($-NH_2$) replaced by a keto group ($>C=O$), a process called **deamination**. This process results in the liberation of ammonia, which is subsequently metabolized to urea (see Figure 25.9). The carbon skeleton of most amino acids can be used to synthesize glucose (a process known as **gluconeogenesis**) and those that can be so utilized are known as the glucogenic amino acids (Figure 4.9). These steps occur in the liver. Of the 20 amino acids found in proteins, only leucine and lysine cannot be used for gluconeogenesis so they are oxidized via the same pathway as fats. Such amino acids are ketogenic.

KEY POINTS:

- Animals generate the energy they require for movement and growth of body tissues by the progressive breakdown of foodstuffs.

- Carbohydrates are first broken down by the glycolytic pathway to form pyruvate. Fats are broken down by β-oxidation to form acetyl CoA. Proteins are first broken down to amino acids which may then be used to generate pyruvate (glucogenic amino acids) or acetyl CoA (ketogenic amino acids).

- Pyruvate and acetyl CoA are utilized by the tricarboxylic acid (TCA) cycle and electron transport chain to synthesize ATP.

- When glucose reserves are low, many tissues preferentially utilize fatty acids.

- During starvation, the glucose required by the brain is synthesized from amino acids by gluconeogenesis.

4.6 Cell motility

One of the features of cells is their ability to move from place to place. This movement is the result of the activity of specific contractile proteins, notably **actin** and **myosin** (see Chapter 9). The ability of cells to move is not confined to muscle cells. Many cells migrate during embryonic development. White blood cells move from the blood into the tissues during normal immunological surveillance. Fibroblasts invade damaged areas of skin to repair wounds. These cells move by crawling over their neighbours. The swimming motion of spermatozoa results from the whip-like movement of their flagella. Other cells move material over their surface by the beating of cilia – as in the clearing of particles from the airways.

When a cell crawls from one place to another, it does so over a specific surface called the substrate. At first, it extends a process called a **lamellipodium** in the direction of movement. The lamellipodium then attaches itself to the substrate, which provides the traction for the cell to be pulled along. The process is driven by a cycle of actin polymerization and depolymerization in which single actin subunits bind ATP before linking together to form a web of actin filaments. The filaments form at the leading edge of a lamellipodium and become attached to the substrate by other proteins. Once the actin filaments are formed, they slowly hydrolyse their bound ATP and individual actin subunits are released at the trailing edge to start the cycle again. In this way, the cycle of actin polymerization and depolymerization acts as a kind of treadmill within the cell to pull it along.

In mammals, ciliated cells line the upper airways and the Fallopian tubes (oviducts). Cilia have a characteristic internal structure, which is clearly shown in Figure 4.10. The outer surface is an extension of the plasma membrane, which encloses a central core known as the **axoneme**. The axoneme is firmly linked to the main cytoskeleton of the cell via a basal body. The axoneme consists of an array of nine doublet microtubules arranged around a central pair of microtubules. The doublet microtubules are attached to their neighbours by protein links at regular intervals. Flagella have a similar internal structure.

The rapid motion of cilia and flagella is due to a large motor protein called **ciliary dynein**. The ciliary dynein is firmly linked by a polypeptide chain to one side of each doublet microtubule while the free head end is able to interact with the neighbouring microtubule in an ATP-dependent manner. When the dynein head region binds to the neighbouring microtubule and hydrolyses its bound ATP, it is able to bind to the next protein subunit of the microtubule and so on – rather like the interaction of actin and myosin of muscle described in Chapter 9 (p. 122). Unlike the filaments of muscle cells, the microtubules are linked together, so instead of shortening the cilium, the dynein bends it. The bent region progresses along the cilium and results in a whip-like motion that is co-ordinated between adjacent cilia to produce a rhythmic wave-like motion

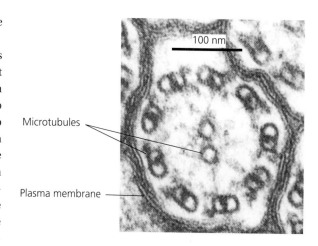

Microtubules

Plasma membrane

100 nm

Figure 4.10 An electron micrograph of a cross-section through a cilium of the respiratory epithelium. Note the characteristic nine-plus-two array of microtubules that forms the core of the cilium.

across the ciliated surface. Furthermore, this wave is co-ordinated between adjacent ciliated epithelial cells so that material can be swept along an epithelial surface. The mechanisms responsible for this co-ordination are still not fully understood.

A defective dynein gene results in **Kartagener's syndrome**. People with this condition suffer from frequent respiratory infections as they cannot clear the mucus and debris from their airways. Moreover, males are infertile as their sperm cannot swim up the Fallopian tubes (oviducts).

> **KEY POINTS:**
> - Many cells are motile. This property is particularly important during embryological development.
> - Even in adults, some cells migrate through the tissues (e.g. white blood cells).
> - Some epithelial cells are specialized to transport material across the epithelial surface by the action of cilia, and sperm progress through the female reproductive tract by means of flagella.
> - Muscle tissue consists mainly of contractile cells that shorten when they are activated.
> - These properties all depend on specific motor proteins.

Recommended reading

Alberts, B., Johnson, A., Lewis, J., Raff, M., Roberts, K. and Walter, P. (2008). *Molecular Biology of the Cell*, 5th edn. Garland: New York. Chapters 1, 10, 16, 17 and 18.

Elliott, W.H. and Elliott, D.C. (2008). *Biochemistry and Molecular Biology*. Oxford University Press: Oxford. Chapters 3, 7 and 11–15.

Junquieira, L.C. and Carneiro, J. (2005). *Basic Histology,* 11th edn. McGraw-Hill. Chapters 2–4.

Young, B. and Heath, J.W. (2000). *Wheater's Functional Histology,* 4th edn. Churchill-Livingstone: Edinburgh. Chapters 1 and 2.

Self-assessment questions

Which of the following statements are true and which are false? Answers are given at the end of the book (p. 755).

1
a) The plasma membrane is approximately equal parts by weight of lipid and protein.
b) Polar molecules cross the plasma membrane freely.
c) The lipid bilayer is an inherently stable part of the plasma membrane.
d) The lipid bilayer forms a barrier to the passage of ions.

2
a) The most prominent features of a cell are its mitochondria.
b) A cell with a large, pale nucleus is about to divide.
c) Proteins are synthesized from RNA located on the rough endoplasmic reticulum.
d) The Golgi apparatus is concerned with packaging proteins prior to their secretion.

3
a) During mitosis, anaphase is associated with the halving of the chromosome number.
b) The separation of the sister chromatids to form the chromosomes occurs during anaphase.
c) The formation of the mitotic spindle occurs during cytokinesis.
d) The nuclear envelope begins to reform during telophase.

4
a) Glucose can be broken down in the absence of oxygen to produce ATP.
b) The anaerobic breakdown of glucose leads to the production of lactate.
c) Proteins can be broken down to their amino acids, which can then be used to synthesize glucose.
d) All amino acids can be used for glucose synthesis.

5
a) Fats provide the richest source of energy on a weight-for-weight basis.
b) Fatty acids can provide energy by anaerobic metabolism.
c) Fatty acids are broken down to pyruvate before entering the TCA cycle.
d) Fatty acids cannot be used by the body to synthesize glucose.

The functions of the plasma membrane

5

After reading this chapter you should have gained an understanding of:

- How polar molecules and ions cross the plasma membrane
- The difference between passive and active transport
- The role of metabolic pumps in generating and maintaining ionic gradients
- How cells exploit the ionic gradients for secondary active transport across epithelial layers
- The origin of the membrane potential
- The nature of ion channels: ligand and voltage-gated channels
- The mechanism of secretion: constitutive and regulated secretion by exocytosis
- Endocytosis and the retrieval of membrane constituents

5.1 Introduction

As explained in Chapter 4, each cell is bounded by a plasma membrane. The region outside the cell (the **extracellular compartment**) is thereby separated from the inside of the cell (the **intracellular compartment**). This physical separation allows each cell to regulate its internal composition independently of the external environment and other cells. Chemical analysis has shown that the composition of the intracellular fluid is very different to that of the extracellular fluid (Table 5.1). It is rich in potassium ions (K^+) but relatively poor in both sodium ions (Na^+) and chloride ions (Cl^-). It is also rich in proteins (enzymes and

structural proteins) and the small organic molecules that are involved in metabolism and signalling (amino acids, ATP, fatty acids, etc.). The first part of this chapter is concerned with the mechanisms responsible for establishing and maintaining the difference in ionic composition between the intracellular and the extracellular compartments. The ways in which cells utilize ionic gradients to perform their essential physiological roles are then discussed and the chapter concludes with a discussion of the mechanisms by which proteins and other large molecules cross the cell membrane – secretion and endocytosis.

Table 5.1 The approximate ionic composition of the intracellular and extracellular fluids of mammalian muscle

Ionic species	Extracellular fluid	Intracellular fluid	Nernst equilibrium potential
Na^+	145	20	$+53\,mV$
K^+	4	150	$-97\,mV$
Ca^{2+}	1.8	c. 2×10^{-4}	$+120\,mV$
Cl^-	114	3	$-97\,mV$
HCO_3^-	31	10	$-30\,mV$

Values are given in $mmol\,l^{-1}$ of cell water and the equilibrium potentials were calculated from the Nernst Equation (see Box 5.1). Note that the resting membrane potential is about $-90\,mV$, close to the equilibrium potentials for potassium (the principal intracellular cation) and Cl^-.

5.2 The permeability of cell membranes to ions and uncharged molecules

The plasma membrane consists of a lipid bilayer in which many different proteins are embedded (see Chapter 4, p. 42). Both natural membranes and artificial lipid bilayers are permeable to gases and lipid-soluble molecules (**hydrophobic molecules**). However, when compared with artificial lipid bilayers, natural membranes are found to have a higher permeability to water and to water-soluble molecules (**hydrophilic** or **polar molecules**) such as glucose, and to ions such as sodium, potassium and chloride. The relatively high permeability of natural membranes to ions and polar molecules can be ascribed to the presence of two classes of integral membrane proteins: the **ion channels** and the **carrier proteins**.

For uncharged molecules such as carbon dioxide, oxygen and urea, the direction of movement across the plasma membrane is simply determined by the prevailing concentration gradient. For charged molecules and ions, however, the situation is more complicated. Measurements have shown that mammalian cells have an electrical potential across their plasma membrane called the **membrane potential** (see p. 62), which affects the diffusion of charged molecules and ions. Overall, the direction in which ions and charged molecules move across the cell membrane is determined by three factors:

1. The concentration gradient
2. The charge of the molecule or ion
3. The membrane potential.

These factors combine to give rise to the **electrochemical gradient**, which can be calculated from the difference between the **equilibrium potential** for the ion in question and the membrane potential. If the intracellular and extracellular concentrations for a particular ion are known, the equilibrium potential can be calculated using the Nernst Equation (Box 5.1).

When molecules and ions diffuse across the plasma membrane down their electrochemical gradients, they do so by **passive transport**. As polar materials cross the plasma membrane much more readily than artificial bilayers, passive transport of these substances is sometimes called **facilitated diffusion**. Cells can also transport

Box 5.1 The Nernst Equation and the resting membrane potential

The flow of any ion across the membrane via an ion channel is governed by its electrochemical gradient, which reflects the charge and the concentration gradient for the ion in question together with the membrane potential. The potential at which the tendency of the ion to move down its concentration gradient is exactly balanced by the membrane potential is known as the equilibrium potential for that ion, so that at the equilibrium potential the rate at which ions enter the cell is exactly balanced by the rate at which they leave. The equilibrium potential can be calculated from the Nernst Equation:

$$E = \frac{RT}{zF} \ln \frac{[C]_o}{[C]_i}$$

where E is the equilibrium potential, ln is the natural logarithm ($\log_e$), and $[C]_o$ and $[C]_i$ are the extracellular (outside) and intracellular (inside) concentrations of the ion in question. R is the gas constant ($8.31\,\mathrm{J\,K^{-1}\,mol^{-1}}$), T the absolute temperature, F the Faraday constant ($96{,}487\,\mathrm{C\,mol^{-1}}$) and z the charge of the ion ($+1$ for Na^+, $+2$ for Ca^{2+}, -1 for Cl^-, etc.). At 37°C, the term RT/F $= 26.7\,\mathrm{mV}$.

The equilibrium potential for K^+ using the data shown in Table 5.1 may be calculated as follows:

$$E_K = \frac{RT}{zF} \ln \frac{[K^+]_o}{[K^+]_i}$$

$$E_K = 26.7 \times \ln \frac{4}{150} = 26.7 \times (-3.62)$$

Hence:

$$E_K = -96.8\,\mathrm{mV}$$

certain molecules and ions against their prevailing electrochemical gradients. This uphill transport is called **active transport** and requires a cell to expend metabolic energy.

Water channels

The relatively high permeability of natural membranes to water can be attributed to the presence of specialized water channels known as **aquaporins**. These are proteins that possess pores, allowing water to pass from one side of the membrane to the other according to the prevailing osmotic gradient. Aquaporins play an important role in tissues that transport large volumes of water in the course of a day, such as the tubules of the kidney and the secretory glands of the gut.

Ion channels

Ions such as sodium and potassium are able to cross the plasma membrane via ion channels, which allow them to diffuse down their electrochemical gradient via a pore (Figure 5.1a). Ion channels exist in one of two states: they are either open and allow the passage of the appropriate ion from one side of the membrane to the other, or they are closed preventing such movements.

An important feature of ion channels is their ability to discriminate between different types of ion. In other words, they show selectivity with respect to the kind of ion they allow through their pore. Each channel type is named after the principal ion to which it is permeable. For example, sodium channels allow sodium ions to cross the membrane but not other cations such as potassium and calcium. Potassium channels are permeable to potassium ions but not to sodium or calcium ions, and so on. The permeability of a cell membrane to a particular ion will depend on how many channels of the appropriate type are open and the number of ions, that can pass through each channel in a given period of time. Many channel proteins have now been identified and their peptide sequence determined with the aid of the techniques of molecular biology.

Carrier proteins

Small organic molecules, such as glucose and some ions, are transported across the plasma membrane via specific carrier proteins (also called **transporters**).

(a) **Ion channel**

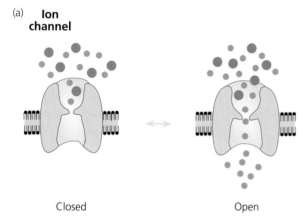

Closed   Open

(b) **Carrier protein**

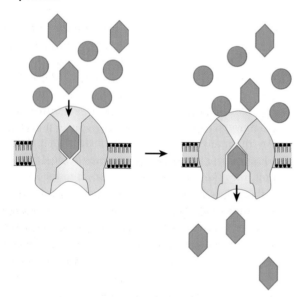

These bind a molecule on one side of the membrane and then undergo a change of shape (known as a conformational change) to move the solute from that side of the membrane to the other, as shown in Figure 5.1b. The capacity of a cell to transport a molecule is limited both by the number of carrier molecules and by the number of molecules each carrier is able to translocate in a given period of time (the 'turnover number'). Carriers are selective for a particular type of molecule and can even discriminate between optical isomers. For example, glucose transporters permit the natural form of glucose (D-glucose) to cross the plasma membrane but the synthetic L- isomer is not transported. Although both isomers have the same chemical constitution, the D- and L-isomers of glucose are mirror images – like our left and right hands. This proves that the carrier can distinguish between optical isomers solely by their shape. This property is known as **stereoselectivity**.

Carrier proteins can be classified into three main groups according to the way in which they permit molecules to cross the plasma membrane. **Uniports** bind a specific molecule or ion on one face of the membrane and then transfer it to the other side, as shown in Figures 5.1b and 5.2. Many substances are, however, transported across the membrane only in association

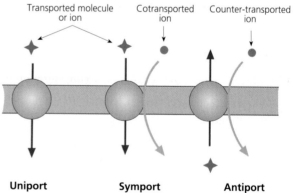

Uniport   **Symport**   **Antiport**

Figure 5.1 Schematic drawings illustrating the differing modes of action of an ion channel (a) and a membrane carrier (b). When an ion channel is activated, a pore is opened and ions are able to diffuse from one side of the membrane to the other in a continuous stream. Molecules that are transported across biological membranes by carrier proteins first bind to a specific site. The carrier then undergoes a conformational change and the bound molecule is able to leave the carrier on the other side of the membrane. Both ions channels and carriers show selectivity for a particular ion or molecule.

Figure 5.2 A diagrammatic representation of the main types of carrier proteins employed by mammalian cells. Unidirectional transport and counter transport may be linked to hydrolysis of ATP (and thus play a role in active transport). Symports exploit existing ionic gradients for secondary active transport.

with a second molecule or ion. Carriers of this type are called **cotransporters** and the transport itself is called **cotransport** or coupled transport. When both molecular entities move in the same direction across the membrane, the carrier is called a **symport**, and when the movement in one direction is coupled to the movement of a second ion or molecule in the opposite direction, the carrier is called an **antiport**. Figure 5.2 shows schematic representations of the different kinds of transport proteins.

> **KEY POINTS:**
>
> - Lipid-soluble molecules cross pure lipid membranes relatively easily while water-soluble molecules cross only with difficulty.
>
> - In natural membranes, water-soluble molecules move from one side of the membrane to the other via ion channels and carrier proteins.

5.3 Active transport

Active transport requires a cell to expend metabolic energy either directly or indirectly and involves carrier proteins. In many cases, the activity of a carrier protein is directly dependent on metabolic energy derived from the hydrolysis of ATP (**primary active transport**, e.g. the sodium pump discussed below). In other cases, the transport of a substance (e.g. glucose) can occur against its concentration gradient by coupling its 'uphill' movement to the 'downhill' movement of sodium ions into the cell. This type of active transport is known as **secondary active transport**, and depends on the ability of the sodium pump to keep the intracellular concentration of sodium significantly lower than that of the extracellular fluid.

The sodium pump

As Table 5.1 shows, the intracellular concentration of sodium is much lower than the extracellular concentration while the intracellular concentration of potassium is much greater than that of the extracellular fluid. These differences in composition are common to all healthy mammalian cells. What mechanisms can account for these differences in ionic composition?

The first clues about the mechanisms by which cells regulate their intracellular sodium came from the problems associated with the storage of blood for transfusion. Like other cells, the red cells of human blood have a high intracellular potassium concentration and a low intracellular sodium. When blood is stored at a low temperature in a blood bank, the red cells lose potassium and gain sodium over a period of time – a trend that can be reversed by warming the blood to body temperature (37°C). If red cells that have lost their potassium during storage are incubated at 37°C in an artificial solution similar in ionic composition to that of plasma, they only reaccumulate potassium if glucose is present. This glucose-dependent uptake of potassium and extrusion of sodium by the red cells occurs against the concentration gradients for these ions. It is therefore clear that the movement of these ions is dependent on the activity of a membrane pump driven by the energy liberated by the metabolic breakdown of glucose – in this case the sodium pump.

The sodium pump is found in all mammalian cells and relies on the hydrolysis of ATP to pump sodium ions out of the cell in exchange for potassium ions. For each ATP molecule hydrolysed, a cell pumps out three sodium ions in exchange for two potassium ions. Figure 5.3 shows a schematic diagram of the operation of the sodium pump, which is sometimes called the Na^+, K^+ ATPase.

The calcium pump

The intracellular fluid of mammalian cells has a very low concentration of calcium ions – typical values for a resting cell are about $10^{-7}\,mol\,l^{-1}$ while that of the extracellular fluid is $1–2 \times 10^{-3}\,mol\,l^{-1}$. There is, therefore, a very steep concentration gradient for calcium ions across the plasma membrane, which is exploited by a wide variety

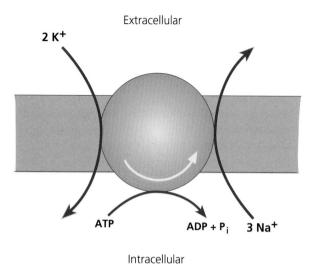

Extracellular

2 K⁺

ATP   ADP + P$_i$ 3 Na⁺

Intracellular

Figure 5.3 A schematic diagram of the sodium pump. For every molecule of ATP hydrolysed, three sodium ions are pumped out of the cell and two potassium ions are pumped into the cell.

of cells to provide a means of transmitting signals to the cell interior. For example, a rise in intracellular Ca^{2+} triggers the contraction of skeletal muscle fibres. It is therefore essential that cells at rest maintain a low resting level of intracellular Ca^{2+}. As Figure 5.4 shows, this is accomplished in a rather interesting and intricate way:

1. Calcium ions can be pumped out of a cell via a **calcium pump**, which, like the sodium pump, uses energy derived from ATP to pump calcium against its concentration gradient.

2. Another type of calcium pump is used to pump calcium into the endoplasmic reticulum to provide a store of calcium within the cell itself while keeping the calcium concentration of the cytosol very low. This store of calcium can be released in response to signals from the plasma membrane (see Chapter 7 for further details).

3. The mitochondria can also take up and store large amounts of calcium from the cytosol.

4. Intracellular calcium can be exchanged for extracellular sodium via an antiport (the Na^+–Ca^{2+} exchanger). In this case the inward movement of sodium ions down their electrochemical gradient provides the energy for the uphill movement of calcium ions from the inside to the outside of the cell against their electrochemical gradient.

Secondary active transport across epithelial membranes

Glucose and amino acids are required to generate energy and for cell growth. Both are obtained by the digestion of food so the cells that line the small intestine (the enterocytes) must be able to transport glucose and amino acids from the central cavity of the gut (the lumen) to the blood. To take full advantage of all the available food, the enterocytes must be able to continue this transport even when the concentration of these substances in the lumen has fallen below that of the blood. This requires active transport across the intestinal epithelium.

How is this transepithelial transport achieved? Like all epithelial cells, those of the small intestine have an apical surface, which faces the gut lumen, and a basolateral surface, which faces towards the bloodstream. These regions are separated by a junctional complex (see Chapter 6, p. 77) that prevents glucose and other small solutes passing between the cells. Consequently, these substances must pass through the cells

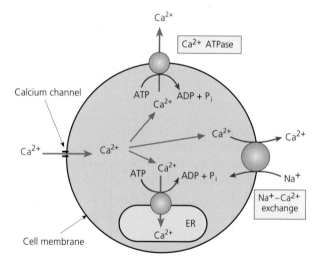

Figure 5.4 The mechanisms employed to regulate intracellular Ca^{2+}. Calcium ions enter cells by way of calcium channels. This calcium can be stored within the cell either in the endoplasmic reticulum (sarcoplasmic reticulum in muscle) or in the mitochondria (not shown). It can be removed from the cell either by the calcium pump (the Ca^{2+} ATPase) or by Na^+–Ca^{2+} exchange.

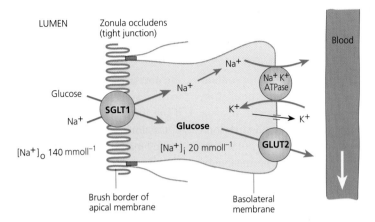

Figure 5.5 The transport of glucose across the epithelium of the small intestine occurs in two stages. First, glucose entry into the enterocyte is coupled to the movement of sodium ions down their concentration gradient. This coupled transport allows the cell to accumulate glucose until its concentration inside the cell exceeds that bathing the basolateral surface of the cell. Glucose crosses the basolateral membrane down its concentration gradient via another carrier protein that is not dependent on extracellular sodium. The sodium pump removes the sodium ions accumulated during glucose uptake. $[Na^+]_o$, extracellular concentration of sodium; $[Na^+]_i$, intracellular concentrations of sodium.

(**transcellular transport**) if they are to be absorbed into the blood.

As Figure 5.5 shows, the plasma membranes of the apical and basolateral regions of the enterocytes possess different carriers: the apical membrane contains a symport called SGLT1 (for sodium-linked glucose transporter 1) that binds both sodium and glucose. As the concentration of sodium in the enterocytes is about $20\,mmol\,l^{-1}$, the movement of sodium from the lumen (where the sodium concentration is about $140\,mmol\,l^{-1}$) into these cells is favoured by its electrochemical gradient. The carrier links the inward movement of sodium down its electrochemical gradient to the uptake of glucose against its concentration gradient and this enables the enterocytes to accumulate glucose. The basolateral membrane has a different type of glucose carrier – a uniport called GLUT2 (for glucose transporter type 2) – that permits the movement of glucose from the cell interior down its concentration gradient into the space surrounding the basolateral surface. Finally, the sodium absorbed with the glucose is removed from the enterocyte by the sodium pump of the basolateral membrane. In this way, a sodium gradient established by the sodium pump provides the energy

for the uptake of glucose. This is called **secondary active transport**. Similar mechanisms exist for the transport of the amino acids (see Chapter 24).

> **KEY POINTS:**
>
> - Molecules and ions cross the plasma membrane via ion channels or carrier molecules.
>
> - Ions that cross the membrane down their electrochemical gradient via ion channels do so by passive transport.
>
> - Transport of an ion or molecule against an electrochemical gradient requires the expenditure of metabolic energy (active transport).
>
> - The sodium pump is the prime example of active transport.
>
> - Certain carriers use the ionic gradient established by the sodium pump to provide the energy to move another molecule (e.g. glucose) against its concentration gradient. This process is known as secondary active transport.

5.4 The resting membrane potential

The activity of the sodium pump not only pumps sodium ions out of the cells, but also leads to an accumulation of potassium ions inside cells. However, the plasma membrane is not totally impermeable to potassium ions so some are able to diffuse out of the cell via potassium channels. As the membrane is much less permeable to sodium ions (the membrane at rest is between 10 and a 100 times more permeable to potassium ions than it is to sodium ions), the lost potassium ions cannot be readily replaced by sodium ions and this leads to the build up of negative charge on the inside of the membrane. This negative charge gives rise to a potential difference across the membrane known as the **membrane potential**.

The membrane potential is a physiological variable that is used by many cells to control various aspects of their activity. The **resting membrane potential** is the membrane potential of cells that are not engaged in a major physiological response. When the membrane potential becomes less negative, the cell is said to undergo a **depolarization**. If the membrane potential becomes more negative, the change is called a **hyperpolarization**.

The negative value of the membrane potential tends to attract positively charged potassium ions into the cell. Thus, on one hand, potassium ions tends to diffuse out of the cell down their concentration gradient and, on the other, the negative charge on the inside of the membrane tends to attract potassium ions from the external medium into the cell. The potential at which these two opposing tendencies are exactly balanced is known as the **potassium equilibrium potential**, which is very close to the resting membrane potential of many cells. Indeed, if the intracellular and extracellular concentrations of potassium ions are known, the approximate value of the resting membrane potential can be calculated using the **Nernst Equation** (Box 5.1).

The membrane potential of a cell can be measured with fine glass electrodes (microelectrodes) that can puncture the cell membrane without destroying the cell. The magnitude of the resting membrane potential varies from one type of cell to another but is a few tens of millivolts (1 mV = 1/1000 volt). It is greatest in nerve and muscle cells (excitable cells) where it is generally −70 to −90 mV (the minus sign indicates that the inside of the cell is negative with respect to the outside). In non-excitable cells, the membrane potential may be significantly lower. For example, the membrane potential of the hepatocytes of the liver is about −35 mV.

> **KEY POINT:**
>
> - As potassium ions are able to diffuse out of the cell via potassium channels, the potassium gradient generated by the activity of the sodium pump gives rise to a membrane potential in which the cell interior is negative with respect to the outside.

5.5 The regulation of ion channel activity

Some ion channels open when they bind a specific chemical agent known as an **agonist** or **ligand**, as shown in Figure 5.6. This kind of channel is known as a **ligand-gated ion channel**. (A ligand is a molecule that binds to another; *an agonist is a molecule that both binds to and activates a physiological system.*) Other ion channels open when the membrane potential becomes depolarized. These are known as **voltage-gated ion channels**. Both kinds of channel are widely distributed throughout the cells of the body but a particular type of cell will possess a specific set of channels that are appropriate to its function. Some ion channels are regulated both by a ligand and by changes in membrane potential.

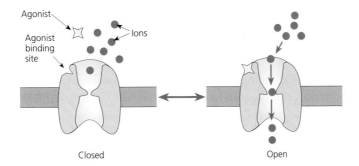

Figure 5.6 Agonists bind to specific sites on the channels they activate. Once they are bound, they cause the channel to open and ions can pass through the pore. When the agonist dissociates from its binding site, the channel is able to revert to the closed state. This type of channel is called a ligand-gated channel.

Ligand-gated channels are widely employed by cells to detect and respond to various chemical signals. As each type of channel has its own unique set of properties, cells are able to respond to different chemical signals (agonists) in a variety of ways. To take two examples: acetylcholine elicits a contraction in skeletal muscle fibres but causes a slowing of the heart rate. In skeletal muscle, the acetylcholine elicits a depolarization of the motor end plate (see Chapter 9). However, it elicits a hyperpolarization of the pacemaker cells of the heart (see Chapter 17). The different physiological responses are a reflection of the specific ion channels present in each case.

Voltage-gated ion channels open in response to depolarization of the membrane (i.e. they open when the membrane potential becomes less negative). When they are open, they allow ions to cross the membrane. Quite commonly, voltage-gated channels spontaneously close after a brief period of time even though the membrane remains depolarized – a property known as **inactivation**. Thus, this type of channel can exist in three distinct states: it can be primed ready to open (the closed or resting state); it can be open and allow ions to cross the membrane; or it can be in an inactivated state from which it must first return to the closed state before it can reopen (Figure 5.7). Examples are voltage-gated sodium channels, which are employed by nerve and muscle cells to generate action potentials, and voltage-gated calcium channels, which are utilized by cells to control a variety of cell functions including secretion.

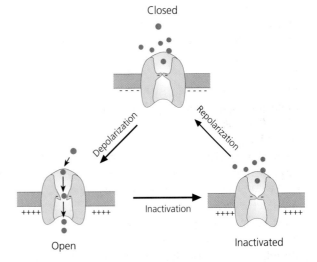

Figure 5.7 Voltage-gated ion channels are opened by a change in membrane potential (usually a depolarization) and ions are then able to pass through the channel pore down their concentration gradient. The open state of most voltage-gated channels is unstable and the channels close spontaneously even though the membrane remains depolarized. This is known as channel inactivation. When the membrane is repolarized, the channel returns to its normal closed state from which it can again be opened by membrane depolarization.

KEY POINT:

- Ion channels may be opened as a result of binding a chemical (ligand-gated channels) or as a result of depolarization of the cell membrane (voltage-gated channels).

5.6   Secretion, exocytosis and endocytosis

Many cells release molecules that they have synthesized into the extracellular environment. This is known as **secretion**. In some cases, secretion occurs by simple diffusion through the plasma membrane – the secretion of steroid hormones by the cells of the adrenal cortex occurs in this way. (A hormone is a signalling molecule carried in the blood – see Chapter 7). This method of secretion is, however, limited to those molecules that can penetrate the lipid barrier of the cell membrane. Polar molecules (e.g. digestive enzymes) are packaged in membrane-bound **vesicles**, which can fuse with the plasma membrane to release their contents into the extracellular space in a process called **exocytosis**.

Exocytosis occurs by two pathways: constitutive and regulated

Secretion occurs via two pathways: constitutive secretion, which is continuous, and regulated secretion, which, as the name implies, is secretion in response to a specific signal. Constitutive secretion is common to all cells and is the mechanism by which cells are able to insert newly synthesized lipids and proteins, such as carriers and ion channels, into their cell membranes (Figure 5.8).

Regulated secretion is used by exocrine and endocrine cells to control the timing and rate of release of their vesicles into the extracellular space. This is exemplified by the acinar cells of the salivary glands. These cells secrete the enzyme amylase in response to nerve impulses arising from the salivary nerves. In other cells, secretion is stimulated by the release of a hormone into the blood (e.g. the acinar cells of the pancreas). Many hormonal secretions are regulated by the concentration of a substance that is continuously circulating in the plasma. This may be another hormone or some other chemical constituent of the blood. For example, the rate at which the β-cells of the pancreatic islets of Langerhans secrete the hormone insulin is regulated by the glucose concentration in the plasma.

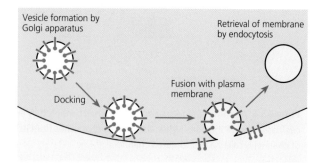

Figure 5.8 The principal phases of constitutive secretion. The vesicles are formed by the Golgi apparatus. They then move towards the plasma membrane to which they become closely apposed ('docking') before releasing their contents ('fusion'). After the two membranes have fused, the proteins and lipids of the vesicular membrane are able to diffuse into the plasma membrane itself. Note that the orientation of the proteins is preserved so that the protein on the inside of the vesicle becomes exposed to the extracellular space.

In many types of cell, regulated exocytosis is triggered by an increase in the concentration of ionized calcium within the cytosol. This occurs via two processes:

1. Entry of calcium through calcium channels in the plasma membrane

2. Release of calcium from intracellular stores (mainly the endoplasmic reticulum).

The entry of calcium ions into the cell usually occurs through voltage-gated calcium channels that open following depolarization of the plasma membrane in response to a specific chemical signal. Calcium diffuses into the cell down its electrochemical gradient. It may also be mobilized from internal stores following activation of G-protein-coupled receptor systems (see Chapter 7, pp. 89–90). In both cases, the intracellular free calcium increases and triggers the fusion of docked secretory vesicles with the plasma membrane. As fusion proceeds, a pore is formed that connects the extracellular space with the interior of the vesicle, and this pore provides a pathway for the contents of the vesicle to diffuse into the extracellular space (Figure 5.9).

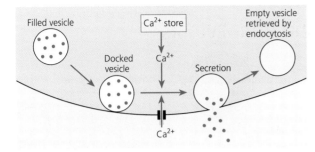

Figure 5.9 A rise in intracellular Ca^{2+} can trigger regulated secretion. In exocrine cells and many endocrine cells, secretion is dependent on extracellular Ca^{2+}. Following stimulation, the cell depolarizes and this opens voltage-gated Ca^{2+} channels. The Ca^{2+} ions enter and cause docked vesicles to fuse with the plasma membrane and release their contents. Calcium ions can also be released into the cytoplasm from intracellular stores to initiate secretion – this calcium is derived mainly from the endoplasmic reticulum.

KEY POINTS:

- Cells release materials that they have synthesized into the extracellular space by means of secretion.
- Secretion of lipid-soluble materials, such as steroid hormones, is by diffusion across the cell membrane.
- Small water-soluble molecules and macromolecules are secreted by exocytosis.

Endocytosis

When cells undergo exocytosis, their cell membrane increases in area as the vesicular membrane fuses with the plasma membrane. This increase in area is offset by membrane retrieval, known as endocytosis, in which small areas of the plasma membrane are pinched off to form endocytic vesicles, which are generally less than 150 nm in diameter. Almost all the cells of the body continuously undergo membrane retrieval via endocytosis. As the formation of the vesicles traps some

of the extracellular fluid, this process is also known as **pinocytosis** ('cell drinking'). Some proteins and other macromolecules are absorbed with the extracellular fluid (**fluid phase endocytosis**). In other cases, proteins bind to specific surface receptors. For example, the cholesterol required for the formation of new membranes is absorbed by cells via the binding of specific carrier protein–cholesterol complexes (low-density lipoproteins) to a surface receptor. This is called **receptor-mediated endocytosis**.

The endocytic vesicles fuse with larger vesicles known as **endosomes**, which eventually become **lysosomes**. The interior of the lysosomes is acidified by the activity of an ATP-driven proton pump. This acidic environment allows the lysosomal enzymes to break down the macromolecules trapped during endocytosis into their constituent amino acids and sugars. These small molecules are then transported out of the lysosomes to the cytosol where they may be reincorporated into newly synthesized proteins.

Not all endocytosed molecules are broken down by the lysosomes. Some elements of the plasma membrane that are internalized during endocytosis are eventually returned to the plasma membrane by transport vesicles. This is often the case with membrane lipids, carrier proteins and receptors. Indeed, in many cases, cells regulate the number of active carriers or receptors by adjustment of the rate at which these proteins are reinserted into the plasma membrane. For example, when the body needs to conserve water, the cells of the collecting duct of the kidney increase their permeability to water by increasing the rate at which water channels (aquaporins) are inserted into the apical membrane by transport vesicles (see Chapter 23, p. 484).

KEY POINT:

- The membrane of secretory vesicles that became incorporated into the plasma membrane during exocytosis is later retrieved by endocytosis.

Phagocytosis

Phagocytosis ('cell eating') is a process in which large particles (e.g. bacteria or cell debris) are ingested by cells. Within the body, this activity is confined to specialized cells called **phagocytes**. The cells that perform this function are the **neutrophils** of the blood and the **macrophages**, which are widely distributed throughout the body (see Chapters 20 and 21). Unlike endocytosis, phagocytosis is triggered only when receptors on the surface of the cell bind to the particle to be engulfed. When the particle has been engulfed, a membrane-bound **phagosome** is formed around it. The phagosomes fuse with lysosomes to digest the ingested material (Figure 5.10). The material that cannot be digested remains in the phagocyte as a membrane-bound particle called a residual body.

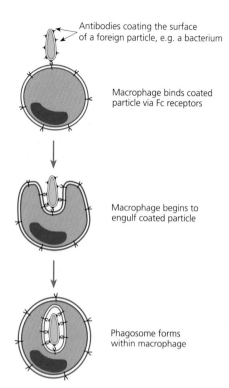

Antibodies coating the surface of a foreign particle, e.g. a bacterium

Macrophage binds coated particle via Fc receptors

Macrophage begins to engulf coated particle

Phagosome forms within macrophage

KEY POINTS:

- Specialized cells engulf foreign particles and cell debris by a process called phagocytosis, which is triggered when receptors on the cell surface recognize specific proteins on the surface of a foreign particle.

- The ingested material is digested in vacuoles called phagosomes.

Figure 5.10 The main stages of phagocytosis. A bacterium or other particle must first adhere to the surface of a phagocyte before it can be engulfed. This is achieved by linking surface receptors on the phagocyte with complementary binding sites on the microbe (e.g. the Fc region of antibodies manufactured by the host). The microbe is then engulfed by a zipper-like action until it becomes fully enclosed in a vacuole (a phagosome).

Recommended reading

Biochemistry and cell biology

Alberts, B., Johnson, A., Lewis, J., Raff, M., Roberts, K. and Walter, P. (2008). *Molecular Biology of the Cell*, 5th edn. Garland: New York. Chapters 11 and 13.

Berg, J.M., Tymoczko, J.L. and Stryer, L. (2002). *Biochemistry*, 5th edn. Freeman: New York. Chapter 12.

Biophysics

Hille, B. (2001). *Ionic Channels of Excitable Membranes*, 3rd edn. Sinauer Associates: Sunderland, Massachusetts.

Aidley, D.J. and Stanfield, P.R. (1996). *Ion Channels: Molecules in Action*. Cambridge University Press: Cambridge.

Self-assessment questions

In the following questions, a number of different options are given to complete each statement. Each option generates a statement that is either true or false. Which are true and which are false? Answers are given at the end of the book (p. 755).

1 Gases such as oxygen and carbon dioxide cross the plasma membrane by:
   a) active transport.
   b) passive diffusion through the lipid bilayer.
   c) a specific carrier protein.

2 Ions can cross the plasma membrane by:
   a) diffusion through the lipid bilayer.
   b) diffusion through the central pore of channel proteins.
   c) binding to specific carrier proteins.

3 A substance can be accumulated against its electrochemical gradient by:
   a) active transport.
   b) facilitated diffusion.
   c) ion channels.
   d) a symport.

4 The sodium pump:
   a) exchanges intracellular Na^+ for extracellular K^+.
   b) requires ATP.
   c) directly links Na^+ efflux with K^+ influx.
   d) is an ion channel.

5 The resting membrane potential of a muscle fibre is close to:
   a) $0\,mV$.
   b) $-90\,mV$.
   c) $+50\,mV$.
   d) the K^+ equilibrium potential.

6 Secretion:
   a) always involves membrane vesicles.
   b) may be triggered by a rise in intracellular Ca^{2+}.
   c) provides a means of inserting proteins into the plasma membrane.

7 Endocytosis is used by cells to:
   a) ingest bacteria and cell debris.
   b) retrieve elements of the plasma membrane after exocytosis.
   c) take up large molecules from the extracellular space.

6 Cells and tissues

After reading this chapter you should have gained an understanding of:

- The principal tissue types:
  - Blood and lymph
  - Connective tissue
  - Nervous tissue
  - Muscle
  - Epithelia and glandular tissue
- Cell–cell adhesion and the extracellular matrix
- Cell–cell recognition

6.1 Introduction

The basic features of mammalian cells were discussed in Chapter 4. This chapter is concerned with the arrangement of cells into functional aggregates known as **tissues**. There are five main types of tissue:

1. Blood and lymph
2. Connective tissue
3. Nervous tissue
4. Muscle tissue
5. Epithelial tissue.

Each type of tissue has its own characteristics and the organs of the body, such as the brain, the heart, the lungs, the intestines and the liver, are built from these basic types of tissue. For example, the stomach is lined by epithelial tissue; its wall consists mainly of connective tissue and smooth muscle, which are supplied with blood vessels and nerve fibres. The organs themselves are parts of distinct physiological systems: the heart and blood vessels form the cardiovascular system; the lungs, trachea and bronchi together with the chest wall and diaphragm form the respiratory system; the skeleton and skeletal muscles form the musculoskeletal system; the brain, spinal cord, autonomic nerves and ganglia, and peripheral somatic nerves form the nervous system, and so on.

6.2 Histological features of the main tissue types

Blood and lymph

Blood and lymph are characterized by having cells dispersed within a fluid. As the cells of blood and lymph do not form a continuous mass but are dispersed within the fluid phase, these tissues are sometimes classed as connective tissue (see below). In the case of blood, the fluid phase is known as **plasma**. The vast majority of blood cells are **red cells** or **erythrocytes**, which have no nucleus (Figure 6.1). The red cells contain the oxygen-binding protein haemoglobin and their chief function is to transport oxygen from the lungs to the tissues and return the carbon dioxide generated by the cells to the lungs. There are five different kinds of white blood cell (**leukocyte**). These are the granulocytes (basophils, eosinophils and neutrophils), the monocytes and the lymphocytes. Examples of some of these can be seen in Figure 6.1. The leukocytes play a major role in the defence against infection. In addition, the blood contains small cell fragments known as **platelets**, which play a crucial role in blood clotting. The detailed properties of blood are discussed in Chapter 20.

Lymph is a clear fluid formed in the tissues. It contains many fewer cells than blood and its main cells are lymphocytes, although a small number of red blood cells may also be present. The cell count is significantly increased after the lymph has passed through a lymph node from which it acquires many lymphocytes (see p. 383).

> **KEY POINT:**
> - Blood and lymph are specialized tissues that are characterized by having cells dispersed within a fluid medium.

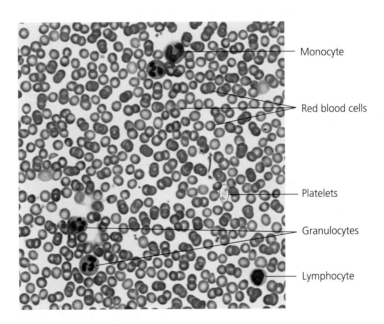

Monocyte

Red blood cells

Platelets

Granulocytes

Lymphocyte

Figure 6.1 A blood film showing the main classes of cell. Note the large number of red blood cells relative to the number of white blood cells. The granulocytes have lobed nuclei – for that reason they are also called polymorphs. The small densely staining fragments are platelets.

Connective tissue

The connective tissues provide structural support to the various organs of the body. For this reason, they are sometimes called **structural tissues**. Connective tissues are very variable in appearance as each type is adapted to perform a specific function, but all are characterized by having cells that are dispersed within an extensive extracellular matrix. The main cells of connective tissues are **fibroblasts**, which synthesize and secrete the extracellular matrix (often called ground substance, and the fibres that provide the mechanical strength of the connective tissues. The extracellular matrix consists of polysaccharides known as **glycosaminoglycans**. Most connective tissues contain fat cells (**adipocytes**) and are well supplied with blood vessels (the exception being cartilage).

Most connective tissue fibres are formed of **collagen**, which exists in a variety of forms. Examples are the dense type I collagen fibres found in the ligaments and tendons (Figure 6.2) and the delicate fibrils found interspersed between the cells of organs such as the liver (known as type III collagen). In many tissues, such as the kidney, the connective tissue acts as a loose packing material with little mechanical strength (Figure 6.3). In others, for example the dermis of the skin, the connective tissue provides a tough, protective layer. The tough outer capsule of the liver and kidneys is formed of connective tissue, as is the mesentery of the gastrointestinal tract. Bone and

cartilage are highly specialized forms of supporting tissue and will be discussed in Chapter 10 (pp. 154–160). Many tissues have fibres of **elastin**, which provide the elasticity that allows tissues to resume their original shape after they have been stretched. Elastin fibres are found in the skin, lungs and in the walls of elastic arteries (Figure 6.4).

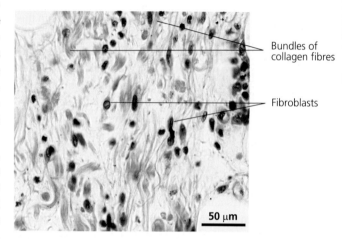

Figure 6.3 Loose connective tissue stained to show the fibroblasts and the bundles of collagen. Note the random arrangement of the collagen bundles.

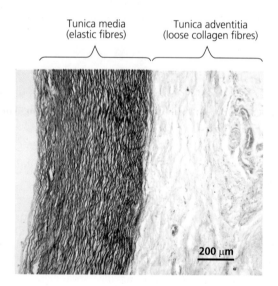

Figure 6.4 A section of the wall of an elastic artery stained to show the elastic fibres of the tunica media. The section also shows the loose arrangement of the collagen bundles of the tunica adventitia.

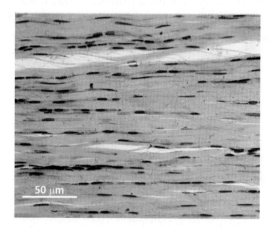

Figure 6.2 A section of tendon stained to reveal the nuclei of the fibroblasts. Note the alignment of the fibroblast nuclei with the bundles of collagen fibres, which are less intensely stained.

There are two main forms of fatty connective tissue: **white adipose tissue** and **brown adipose tissue**, more commonly called **brown fat**. The white adipose tissue is widely dispersed throughout the body and forms its main energy reserve. The fat is stored within adipocytes as a single large droplet that forces the cytoplasm into a thin layer around the cell. The nucleus is compressed and forced to one side of the cell, as shown in Figure 6.5. In contrast, the cells of brown adipose tissue store their fat in a number of small vesicles surrounded by a large number of mitochondria, which give the tissue its characteristic brown colouration. The distribution of brown fat is more restricted than that of white adipose tissue. It is especially important in neonates, where it plays an important role in the maintenance of body temperature (see Chapter 33).

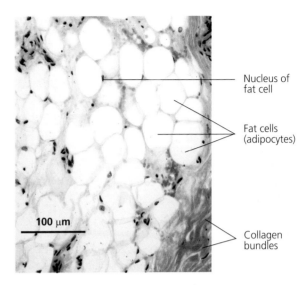

Figure 6.5 The appearance of white adipose tissue in a histological section. The nuclei are stained blue and appear on the edges of large spaces, which contained fat droplets that have been removed in the processing of the section.

> **KEY POINTS:**
>
> - The connective tissues provide structural support to the various organs of the body. They are also called structural tissues.
> - They are very variable in appearance, but all are characterized by having cells that are dispersed within an extensive extracellular matrix.

Nervous tissue

The brain and spinal cord constitute the **central nervous system** (CNS) while the nervous tissues that lie outside the CNS make up the **peripheral nervous system**. The CNS is made up of two main types of cell, **nerve cells** (also called **neurones**) and **glial cells**. The nerve cells provide rapid and discrete signalling over long distances (from millimetres to a metre or more) while the glial cells have a complex supporting role. The nerve cell bodies are very varied in both size and shape, but all stain strongly with basic dyes, as shown in Figure 6.6. The material is called **Nissl substance** and corresponds to the RNA of the rough endoplasmic reticulum described in Chapter 4 (p. 44). Figure 6.6 also shows the nuclei of glial cells, which are more widespread and plentiful than the nerve cells.

Each nerve cell has an extensive set of fine branches called **dendrites** that receive information from other neurones, integrate it and transmit it to its target cells via a delicate extension of the cell body called an **axon** (Figure 6.7). Axons and dendrites are collectively called neuronal processes. The space between the nerve cell bodies is known as the **neuropil**. It contains the cytoplasmic extensions of both neurones and glia.

The peripheral nervous system consists of nerve trunks and ganglia. The nerve trunks contain large numbers of axons while the ganglia contain nerve cell bodies, as in the example shown in Figure 6.8. Both the peripheral and central nervous system are well provided with blood vessels and both are protected with layers of connective tissue, which will be described in detail in Chapters 8 and 11.

> **KEY POINTS:**
>
> - Nervous tissue consists mainly of neurones and glial cells.
> - It is found mainly in the brain and spinal cord, but small ganglia are scattered throughout the body.
> - Nerve trunks connect the CNS to the various organ systems.

Labels on Figure 6.5:
- Nucleus of fat cell
- Fat cells (adipocytes)
- Collagen bundles
- 100 µm

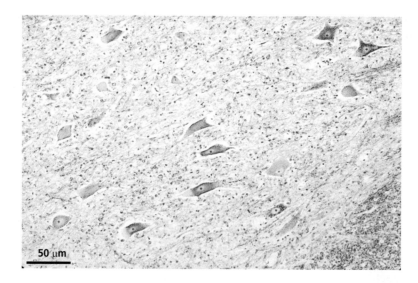

Figure 6.6 A section of the grey matter of the spinal cord stained to show the motoneurones (violet) and the glial cells of the neuropil (blue). The cell bodies of the neurones are well defined and are much larger than those of the glial cells. However, glial cells vastly outnumber the neurones.

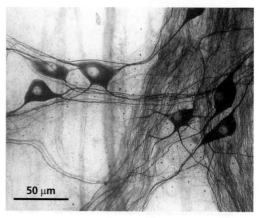

Figure 6.8 This image shows neurones in a ganglion from the wall of the gastrointestinal tract. The main processes of the individual cells are clearly visible.

Muscle tissue

Muscle tissue is characterized by its ability to contract in response to an appropriate stimulus. Muscle cells are also known as **myocytes**. There are three main types of muscle: skeletal muscle, cardiac muscle and smooth muscle. **Skeletal muscle** is the muscle directly attached to the bones of the skeleton, **cardiac muscle** is the muscle of the heart and **smooth muscle** is the muscle that lines the blood vessels and the hollow organs of the body.

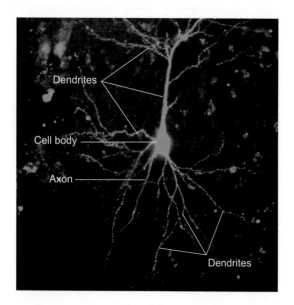

Figure 6.7 A pyramidal cell (a type of neurone) from the CNS stained with a green fluorescent dye. Note the extensive dendrites, which have a roughened appearance and arise from both the apex and base of the cell. A single axon emerges from the base of the cell.

The appearance of the different kinds of muscle tissue is shown in Figure 6.9. Their detailed structure and function will be discussed in Chapter 9.

When the cells of skeletal and cardiac muscle are viewed down a microscope, they are seen to have characteristic striations – small regular stripes running across the individual muscle cells (Figure 6.9a and b). For this reason, skeletal and cardiac muscles are sometimes called **striated muscles**. Smooth muscle lacks striations and consists of sheets of spindle-shaped cells. Between the muscle cells are layers of delicate connective tissue that bind them together into bundles called fascicles. The individual fascicles are bound together in a denser wrapping of connective tissue that provides support for the blood vessels and nerve fibres as they course through the tissue. The connective tissues of muscle will be described in more detail in Chapter 9. In skeletal and cardiac muscle, the smaller blood vessels run parallel to the main course of the bundles of muscle fibres, as shown in Figure 6.10.

(a)

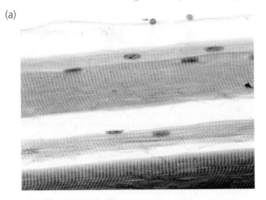

(b)

(c)

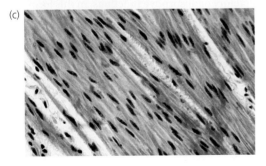

KEY POINTS:

- There are three kinds of muscle tissue: skeletal, cardiac and smooth muscle.

- Skeletal and cardiac muscle fibres are striated while smooth muscle cells are not.

- Muscle fibres are bound together with layers of connective tissue and are well supplied with blood vessels.

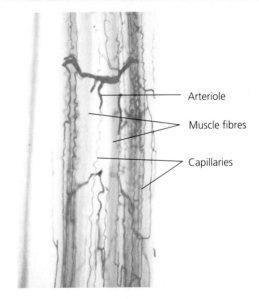

Arteriole

Muscle fibres

Capillaries

Figure 6.9 The appearance of the three principal types of muscle. (a) This shows the appearance of skeletal muscle with its characteristic striations. Note the large size of the individual fibres and the distribution of the nuclei on the outside of each fibre. (b) The appearance of cardiac muscle. Note the prominent intercalated disks and the centrally place nuclei. (c) Smooth muscle from the gastrointestinal tract. Note the absence of striations and the elongated nuclei within the spindle-shaped cells.

Figure 6.10 The blood supply of skeletal muscle revealed by perfusing the blood vessels with a red dye. The larger vessels enter the body of the muscle and branch to give rise to a series of capillaries that run alongside the bundles of muscle fibres.

Epithelia

Continuous sheets of cells are known as **epithelia.** They form the epidermis of the skin and line the hollow organs of the body, such as the gut, lungs and urinary tract, as well as the fluid-filled spaces such as the peritoneal cavity. However, the cell layer that lines the blood vessels is called the vascular **endothelium** and the epithelial coverings of the pericardium, pleura and peritoneal cavity are known as **mesothelium**, reflecting their origins from the embryonic mesoderm. Nevertheless, all three separate one compartment of the body from another. Their detailed structure reflects their differing functional requirements. For example, the epithelium of the skin is thick and tough to resist abrasion, and to prevent the loss of water from the body. In contrast, the epithelial lining of the alveoli of the lungs is very delicate and thin to permit free exchange of the respiratory gases.

Despite their differences in form and function, all epithelia share certain features:

- Firstly, they are composed of cells that are tightly joined together via specialized cell–cell junctions to form a continuous sheet (see p. 77).

- Secondly, epithelial cells lie on a matrix of connective tissue fibres called the **basement membrane**. The basement membrane provides physical support and separates the epithelium from the underlying vascular connective tissue, which is known as the **lamina propria**.

- Thirdly, to replace damaged and dead cells, all epithelia undergo continuous cell replacement. The rate of replacement depends on the physiological role of the epithelium and is highest in the skin and gut, both of which are continually subjected to abrasive forces.

- Finally, unlike cells scattered throughout a tissue, the arrangement of cells into epithelial sheets permits the directional transport of materials either into or out of a compartment. In the gut, kidney and many glandular tissues, this feature of epithelia is of great functional significance. The surface of an epithelial layer that is oriented towards the central space of a gland or hollow organ is known as the **apical surface**. The surface that is oriented towards the basement membrane and the interior of the body is called the **basolateral surface**.

The classification of epithelia

The main types are known as **simple**, **stratified** and **pseudostratified**. Simple epithelia consist of a single cell layer and are classified according to the shape of the constituent cell type. Stratified epithelia are also classified according to the appearance of their constituent cells. They are characterized by having more than one cell layer. Pseudostratified epithelia consist of a single layer of cells in contact with the basement membrane, but the varying height and shape of the constituent cells gives the appearance of more than one cell layer.

The morphological characteristics of the main types of epithelium are shown in Figure 6.11 and are summarized below:

- Simple squamous epithelium (squamous = flattened) consists of thin and flattened cells, as shown in Figure 6.11 (i) and (ii). These epithelia are adapted for the exchange of small molecules between the separated compartments. The walls of the alveoli of the lungs and the endothelium of the blood vessels are squamous epithelia.

- Simple cuboidal epithelium, as the name implies, consists of a single layer of cuboidal cells whose width is approximately equal to their height (Figure 6.11 (iii)). A simple cuboidal epithelium forms the walls of the small collecting ducts of the kidneys.

- Simple columnar epithelium is adapted to perform secretory or absorptive functions. In this form of epithelium, the height of the cells is much greater than their width, as shown in Figure 6.11 (iv). It occurs in the large-diameter collecting ducts of the kidneys. It is also found lining the small intestine.

- Ciliated epithelium consists of cells that have cilia on their apical surface. Non-ciliated cells are also interspersed between the ciliated cells, as illustrated in Figure 6.11 (v). Ciliated epithelia line the Fallopian tubes.

- Pseudostratified columnar ciliated epithelium consists of cells of differing shapes and height, as shown in Figure 6.11 (vi). This type of epithelium predominates in the upper airways (trachea and bronchi).

- Stratified squamous epithelium is adapted to withstand chemical and physical stresses. The best-known stratified epithelium is the epidermis of the skin. In

(a) Simple epithelia

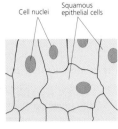

(i) Simple squamous epithelium viewed from above

(ii) Sectional view of squamous epithelium

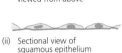

(iii) Cuboidal epithelium

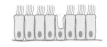

(iv) Columnar epithelium

(v) Ciliated columnar epithelium

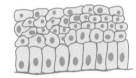

(vi) Pseudostratified epithelium

(b) Stratified epithelia

(vii) Stratified squamous epithelium

(viii) Transitional epithelium

(c) Glandular epithelia

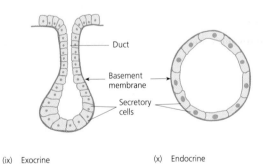

(ix) Exocrine

(x) Endocrine

Figure 6.11 A diagrammatic representation of the principal types of epithelium. Panel (a) shows the general structure of the common types of simple epithelia. Note that when viewed from the apical surface there are no gaps between the cells (i). Pseudostratified epithelia (vi) have cells of differing shape and size, so giving a false appearance of multiple cell layers. The nuclei are found at many levels within the epithelium. Panels (b) and (c) show the general structure of stratified, transitional and glandular epithelia. See text for further details.

this case, the flattened epithelial cells form many layers, only the lowest layer being in direct contact with the basement membrane (Figure 6.11 (vii)). The more superficial cells are filled with a special protein called keratin, which renders the skin almost impervious to water and provides an effective barrier against invading organisms such as bacteria.

- Transitional epithelium is found in the bladder and ureters. It is similar in structure to stratified squamous epithelium except that the superficial cells are larger and rounded (Figure 6.11 (viii)). This adaptation allows stretching of the epithelial layer as the bladder fills.

KEY POINTS:

- Epithelia are formed entirely from sheets of cells and consist of one or more cell layers.
- They separate one compartment of the body from another.
- An epithelium consisting of a single cell layer is known as a simple epithelium while those with more than one layer are called stratified epithelia.

Glandular epithelia

Glandular epithelia are specialized for secretion. The epithelia that line the airways and part of the gut are covered with a thick layer of mucus that is secreted by specialized secretory cells, known as goblet cells, which discharge their contents directly onto an epithelial surface. Glands, known as **exocrine glands**, secrete material via a specialized duct onto an epithelial surface, as shown in Figure 6.11 (ix). Examples are the salivary glands and sweat glands. Other glandular epithelia lack a duct and secrete material across their basolateral surfaces from where it passes into the blood. These are the **endocrine glands**. Examples include the thyroid gland and the irregular clusters of epithelial cells that constitute the islets of Langerhans of the pancreas.

The exocrine glands may be grouped into various classes according to the arrangement of their duct system. Some examples are shown in Figure 6.12. The gastric glands in the body of the stomach are simple tubular glands. The sweat glands are simple, coiled, tubular glands. Other exocrine glands have a system of branching ducts that link groups of secretory cells called **acini** (singular **acinus**). The acinar cells of an individual acinus are held in a tight ball by a capsule of connective tissue. This type of organization is found in the pancreas and salivary glands.

Serous membranes enclose a fluid-filled space such as the peritoneal cavity. The peritoneal membranes consist of a layer of epithelial cells (called the mesothelium) overlying a thin basement membrane, beneath which is a narrow band of connective tissue. The pleural membranes that separate the lungs from the wall of the thoracic cavity are another example of serous epithelia. Not all fluid-filled spaces are lined by an epithelium: while the articular surfaces of joints such as the knee are lubricated by synovial fluid, they are not lined by a continuous layer of cells. Instead, the synovial cells form a discontinuous layer up to four cells deep, which does not constitute a true epithelium.

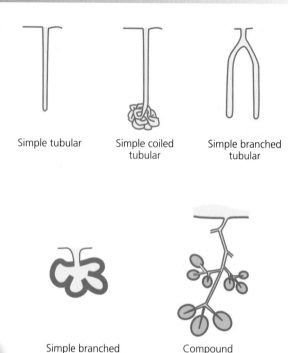

Simple tubular   Simple coiled tubular   Simple branched tubular

Simple branched acinar   Compound acinar

Figure 6.12 This figure shows the duct arrangement of several types of exocrine gland. The compound glands all have a set of branching ducts.

KEY POINTS:

- Glandular epithelia are specialized for secretion.
- If their secretion is via a duct, they form part of an exocrine gland.
- If their secretion passes directly into the blood, they form part of an endocrine gland.

6.3   Cell–cell adhesion

To form complex tissues, different cell types must aggregate together. As all cells have their origin in the fertilized egg, some cells must migrate from their point of origin to another part of the body during development. When they arrive at the appropriate region, they must recognize their target cells and participate in the differentiation of the tissue. To do so, they must attach to other cells and to the extracellular matrix. What signals are employed by developing cells to establish their correct positions and why do they cease their migrations when they have found their correct target?

Unlike adult cells, embryonic cells do not form strong attachments to each other. Instead, when they interact, their cell membranes become closely apposed to each other leaving a very small gap of only 10–20 nm. Exactly how cells are able to recognize their correct associations is not fully understood, but each type of cell has a specific marker on its surface. When cell membranes touch each other, these marker proteins can interact and allow the cells to adhere. This must be an early step in tissue formation. It has been shown that cells will associate only if they recognize the correct surface markers. Thus, if differentiated embryonic liver cells are dispersed by treatment with enzymes and grown in culture with cells from the retina, the two cell types aggregate with others of the same kind. Thus, liver cells aggregate together and exclude the retinal cells and vice versa.

> **KEY POINTS:**
>
> - For cells to assemble into tissues, they need to adhere to other cells of the correct type. This recognition requires tissue-specific cell-surface marker molecules.
>
> - Cell–cell adhesion and cell–matrix interactions play an important role in tissue maintenance and development.

6.4 Specialized cell attachments

Epithelial cells are joined together by a characteristic structure known as the **junctional complex** (Figure 6.13), which consists of three structural components: the **tight junction** (also known as the zonula occludens), the **adherens junction** (or zonula adherens) and the **desmosomes**. Within the junctional complexes, specialized regions of contact, called **gap junctions**, are found. Gap junctions allow small molecules to diffuse between adjacent cells. In this way, they play a role in communication between neighbouring cells (see Chapter 7, p. 94).

The tight junction forms a continuous band in which the membranes of adjacent cells are fused together. Indeed, the adjacent cells are held so closely apposed that the extracellular space is eliminated. As a result, the tight junction separates the space above the apical surface of an epithelial cell from that surrounding the basolateral surface and prevents ions and water-soluble molecules from leaking between the cells. The proteins responsible for linking the epithelial cells so closely together are transmembrane proteins called **claudins**.

As the tight junctions have little structural rigidity, they are supported by protein filaments in the cytoskeleton. These filaments form the adherens junctions and desmosomes. The adherens junctions form a continuous band around each epithelial cell. On the cytoplasmic side of the membrane, anchor proteins link to the actin and intermediate filaments of the cytoskeleton. This is known as an attachment plaque and appears as a densely staining band in electron micrographs. The anchor proteins are connected to transmembrane adhesion proteins called **cadherins**, which bind neighbouring cells together.

The desmosomes are points of contact between the adjacent cells. They consist of various anchoring proteins that link intermediate filaments of the cytoskeleton to the cadherins. The cadherins of one cell bind to those of its neighbours to form focal attachments of great mechanical strength, rather as spot welds can be used to secure sheets of metal to one another.

Hemidesmosomes, as their name implies, have a similar appearance to half a desomosome as seen in an electron micrograph. However, they are formed by different anchoring proteins that bind cytoskeletal intermediate filaments to transmembrane adhesion proteins known as **integrins**. The integrins fix the epithelial cells to the basal lamina, so linking the cell layer to the underlying connective tissue. The integrins also play an important role in development and wound repair. The cadherins, integrins and immunoglobulin-like cell adhesion molecules (e.g. **N-CAM**) are also involved in non-junctional cell–cell adhesion, which must play an important role in the formation of integrated tissues. Other proteins that promote cell adhesion are the **selectins**, which are a family of adhesion molecules that mediates the initial attachment of a white blood cell to the wall of a blood vessel before it can migrate to a site of tissue injury.

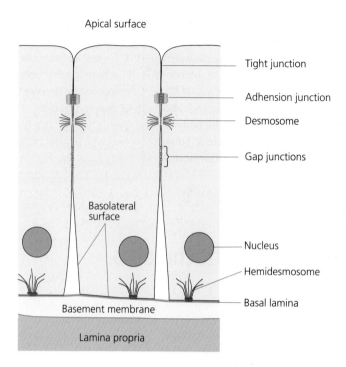

Apical surface

Tight junction

Adhension junction

Desmosome

Gap junctions

Basolateral surface

Nucleus

Hemidesmosome

Basal lamina

Basement membrane

Lamina propria

Figure 6.13 A diagram that shows the main structural features of the junctional complexes of epithelia. The tight junctions and adhesion junction form continuous bands around each cell, linking it with its neighbours. The desomosomes give great mechanical strength and the hemidesmosomes link the cells firmly to the underlying tissues.

Recommended reading

Alberts, B., Johnson, A., Lewis, J., Raff, M., Roberts, K. and Walter, P. (2008). *Molecular Biology of the Cell*, 5th edn. Garland: New York. Chapter 19.

Junquieira, L.C. and Carneiro, J. (2005). *Basic Histology*, 11th edn. McGraw-Hill. Chapters 2–4.

Young, B. and Heath, J.W. (2000). *Wheater's Functional Histology*, 4th edn. Churchill-Livingstone: Edinburgh. Chapters 3–7.

Self-assessment questions

Which of the following statements are true and which are false? Answers are given at the end of the book (p. 755).

1 a) An organ usually contains a single tissue type.
   b) Connective tissues are characterized by an extensive extracellular matrix.
   c) The principal cells of connective tissue are called fibroblasts.
   d) Epithelia are connective tissues.
   e) The CNS contains a large number of collagen fibres.

2 a) Connective tissues contain either collagen or elastic fibres.
   b) Fibroblasts secrete the proteins that make up the fibres of the extracellular matrix.

c) Elastic fibres provide the supporting framework for the cells of the liver.
d) Bone is a type of connective tissue.
e) Adipocytes (fat cells) are commonly found in loose connective tissue.

3 a) Epithelia consist of cells that lie on a basement membrane.
b) The lung alveoli are lined with ciliated columnar epithelium.
c) The epidermis of the skin is a stratified squamous epithelium.
d) Epithelia separate body compartments from each other.
e) The joints are lined with an epithelium.

4 a) The cells of an epithelium are linked together by hemidesmosomes.
b) Epithelial cells communicate with their immediate neighbours via gap junctions.
c) The tight junctions between epithelial cells seal off one body compartment from another.
d) The cadherins are membrane proteins that bind neighbouring cells together.
e) The intermediate filaments of the cytoskeleton are linked to the basal lamina by integrins.

The principles of cell signalling

After reading this chapter you should have gained an understanding of:

- The need for cell signalling

- The differing roles of paracrine, endocrine and synaptic signalling

- How receptors in the plasma membrane regulate the activity of target cells

- The roles of cyclic AMP and inositol trisphosphate as second messengers

- How steroid and thyroid hormones control gene expression via intracellular receptors

- The functions of gap junctions between cells

7.1 Introduction

Individual cells are specialized to carry out a specific physiological role, such as secretion or contraction. These processes need to be well co-ordinated if an organism is to deal appropriately with the various challenges it encounters throughout life. For example, to enable it to eat and digest food, an animal must first be able to take the food into the mouth and pass it along the alimentary tract. This requires the co-ordination of the contractions of different sets of muscles. It must then secrete digestive enzymes in the correct amounts and in the correct order. The digested food must then be transferred to the blood before it can be used by the body as a whole. To do all this, the different types of tissue need to receive or transmit signals of various kinds. The principal ways in

which cells make use of such signals form the substance of this chapter.

In essence, cells communicate with each other in three different ways:

- By diffusible chemical signals
- By direct contact between the plasma membranes of adjacent cells
- By direct cytoplasmic contact via gap junctions.

Diffusible chemical signals allow cells to communicate at a distance, while direct contact between cells is particularly important in cell–cell recognition during development and during the passage of lymphocytes through the tissues where they scan cells for the presence of foreign antigens (see Chapter 21). Direct cytoplasmic contact between neighbouring cells via gap junctions permits the electrical coupling of cells and plays an important role in the spread of excitation between adjacent cardiac muscle cells. It also allows the direct exchange of chemical signals between adjacent cells.

> **KEY POINTS:**
> - Cells send and receive signals of various kinds.
> - Many do so by secreting specific chemical signals (diffusible signals) or by sharing their metabolic products via gap junctions.
> - In some cases, cells communicate by direct cell–cell contact.

7.2 Cells use diffusible chemical signals for paracrine, endocrine and synaptic signalling

Cells release a variety of chemical signals. Some are local mediators that act on neighbouring cells, reaching their targets by diffusion over relatively short distances (up to a few millimetres). This is known as **paracrine signalling**. When the secreted chemical also acts on the cells that secreted it, the signal is said to be an **autocrine signal**. Frequently, substances are secreted into the blood by specialized glands – the **endocrine glands** – to act on various tissues around the body. The secreted chemicals themselves are called **hormones**. Finally, nerve cells release chemicals at their endings to affect the cells they contact. This is known as **synaptic signalling**.

Local chemical signals – paracrine and autocrine secretions

Paracrine secretions are derived from individual cells rather than from a collection of similar cells or a specific endocrine gland. Cells engage in paracrine signalling in response to local stimuli of various kinds and the secreted chemicals have a local effect (Figure 7.1a) before being rapidly destroyed by extracellular enzymes or by uptake into the target cells. Consequently, very little of the secreted material enters the blood.

One example of paracrine signalling is provided by histamine-containing cells called mast cells. These cells are found in connective tissues all over the body and secrete histamine in response to local injury or infection.

The secreted histamine causes the local arterioles to dilate and increase the local blood flow. In addition, it increases the permeability of the nearby capillaries to proteins such as immunoglobulins to counter any foreign antigens. However, this increase in capillary permeability must be local and not widespread, otherwise there would be a significant loss of protein from the plasma to the interstitial fluid, which could result in circulatory failure. The mast cells also secrete small peptides that are released with the histamine to stimulate the invasion of the affected tissue by white blood cells. These actions form part of the inflammatory response and play an important role in halting the spread of infection (see Chapter 21, p. 418).

The inflammatory response is also associated with an increase in the synthesis and secretion of a group of local chemical mediators called the prostaglandins (see below, p. 91). The prostaglandins secreted by a cell act on neighbouring cells to stimulate them to produce more prostaglandins. This is another example of paracrine signalling. In addition, the secreted prostaglandin stimulates further prostaglandin production by the cell that initiated the response. Here, the prostaglandin is acting as an **autocrine signal**. This autocrine action amplifies the initial signal and may help it to spread throughout a population of cells to ensure a rapid mobilization of the body's defences in response to injury or infection.

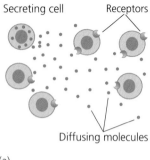

(a)

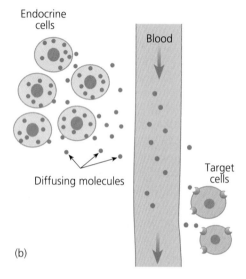

(b)

Figure 7.1 A comparison between paracrine and endocrine signalling. (a) Schematic diagram illustrating paracrine secretion. The secreting cell releases a signalling molecule (e.g. a prostaglandin) into the extracellular fluid from where it reaches the receptors of neighbouring cells by diffusion. (b) The chief features of endocrine signalling. Endocrine cells secrete a hormone into the extracellular space in sufficient quantities for it to enter the bloodstream, from where it can be distributed to other tissues. The target tissues may be at a considerable distance from the secreting cells.

Hormonal secretions provide a means of diffuse, long-distance signalling to regulate the activity of distant tissues

Hormones play an extensive and vital role in regulating many physiological processes and their physiology will be discussed at length in Chapter 16 and in subsequent chapters. Here, their role as cell signals will be discussed only in general terms.

Endocrine cells synthesize hormones and secrete them into the extracellular space from where they are able to diffuse into the blood. Once in the bloodstream, a hormone will be distributed throughout the body. It is then able to influence the activity of tissues a long way from the gland that secreted it. In this important respect, hormones differ from local chemical mediators. The secretion and distribution of hormones is shown schematically in Figure 7.1b.

The endocrine glands secrete hormones in response to a variety of signals:

• They may respond to the level of some constituent of the blood. For example, insulin secretion by the β-cells of the islets of Langerhans is regulated by the blood glucose concentration.

• The circulating levels of other hormones may closely regulate their activity. This is the case for the secretion of the sex hormones from the ovaries (oestrogens) and testes (testosterone), which are secreted in response to hormonal signals from the anterior pituitary gland.

• They may be directly regulated by the activity of nerves. This is how oxytocin secretion from the posterior pituitary gland is controlled during lactation (a neuroendocrine reflex).

As hormones are distributed throughout the body via the circulation, they are usually capable of affecting widely dispersed populations of cells. The releasing hormones of the hypothalamus (a small region in the base of the brain) are an important exception to this rule. These hormones are secreted into the hypophyseal portal blood vessels in minute quantities and travel only a few millimetres to the anterior pituitary where they control the secretion of the anterior pituitary hormones (see Chapter 16). While cells are exposed to almost all the hormones secreted into the bloodstream, a particular cell will only respond to a hormone if it possesses receptors of the appropriate type (see Section 7.3 below). Thus, the ability of a cell to respond to a particular hormone depends on whether it has the right kind of receptor. This diffuse signalling is beautifully adapted to regulate a wide variety of cellular activities in different tissues.

Compared with the signals mediated by nerve cells (synaptic signalling, see below), the effects of hormones are usually relatively slow in onset and long lasting. However, some hormones have a rapid and brief action – oxytocin is an example. This hormone is the efferent arm of a neuroendocrine reflex that causes the ejection of milk from the lactating mammary gland in response to suckling. The steroid hormone aldosterone acts on the renal tubules over a period of hours to regulate the amount of sodium reabsorbed by the kidneys. The secretion of other hormones may determine the pattern of life for an individual: the development of the male reproductive system is critically dependent on the secretion of testosterone by the fetus and thyroid hormone must be secreted in appropriate quantities during early life to permit the normal development of the nervous system.

The difference between endocrine and paracrine signalling fundamentally depends on the quantity of the chemical signal secreted. If sufficient quantities of a signalling molecule are secreted for it to enter the blood, it is being employed as a hormone. If, however, the amounts secreted are sufficient only to affect neighbouring cells, the chemical signal is acting as a paracrine signal. Consequently, a substance can be a hormone in one situation and a paracrine signal in another. For example, the peptide somatostatin is found in the hypothalamus. It is secreted into the hypophyseal portal blood vessels, which carry it to the anterior pituitary gland where it acts to inhibit the release of growth hormone (see Chapter 16). As somatostatin has entered the blood to be carried to its target tissue, it is acting as a hormone. Somatostatin is also found in the D cells of the gastric mucosa. It is secreted when the acidity in the stomach rises and it inhibits the secretion of the hormone gastrin, which stimulates acid secretion. In this situation, somatostatin acts on neighbouring cells as a paracrine signal rather than as a hormone.

Fast signalling over long distances is accomplished by nerve cells

Chemical signals that are released into the extracellular space and need to diffuse some distance to reach their target cells have two disadvantages:

- Their effect cannot be restricted to an individual cell.

- Their speed of signalling is relatively slow – particularly if there is any great distance between the secreting cell and its target.

For many purposes these factors are not important, but there are circumstances where the signalling needs to be both rapid (occurring within a few milliseconds) and precise. For example, during walking, different sets of muscles are called into play at different times to provide co-ordinated movement of the limbs. This type of rapid signalling is performed by nerve cells. The contact between the nerve ending and the target cell is called a **synapse** and the overall process is known as synaptic signalling.

To perform their role, nerve cells need to make direct contact with their target cells. They do this via long, thin, hair-like extensions of the cell called **axons**. Each nerve cell gives rise to a single axon, which usually branches to contact a number of different target cells. As axons may extend over considerable distances (in some cases up to a metre), nerve cells need to be able to transmit their signals at relatively high speeds. This is achieved by means of an electrical signal (an **action potential**) that passes along the length of the axon from the cell body to its terminal (the **nerve terminal**). When an action potential reaches a nerve terminal, it triggers the release of a small quantity of a chemical (a **neurotransmitter**), which then acts on the target cell. The gap between the nerve terminal and its target is very small so that the neurotransmitter only has to diffuse a very short distance (about 20 nm) to reach its point of action. For this reason, very small quantities of neurotransmitter are required to activate each target cell and neighbouring cells are not affected. The combination of electrical signalling and a very short diffusion time therefore permits both rapid and discrete activation of the target. The essential features of synaptic signalling are summarized in Figure 7.2. The organization and properties of nerve cells and synapses will be discussed more fully in the following chapter (Chapter 8).

Signalling molecules are very diverse in structure

The chemical signals employed by cells are very diverse. Cells use both water-soluble molecules and hydrophobic molecules for signalling. Some, such as nitric oxide

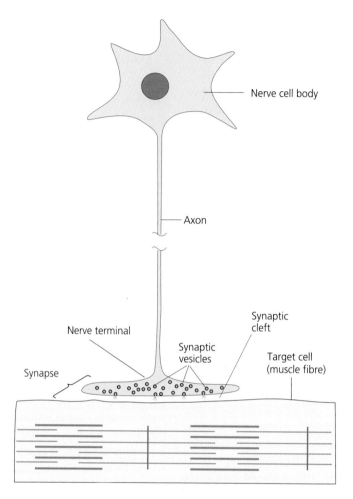

Figure 7.2 Synaptic signalling is performed by nerve cells. Electrical signals (action potentials) originating in the cell body pass along the axon and trigger the secretion of a signalling molecule by the nerve terminal. As the nerve terminal is closely apposed to the target cell (in this case a skeletal muscle fibre), the signal is highly localized. The junction between the nerve cell and its target is called a synapse.

and glycine, have a small molecular mass (<100) while others are very large molecules, such as growth hormone (which has 191 amino acids linked together and a molecular mass of 21 500). Many of the water-soluble signalling molecules are derived from amino acids – examples are peptides, proteins and the biological amines. Hydrophobic signalling molecules include the prostaglandins (see below), and the steroid hormones such as cortisol and testosterone.

Many chemical signalling molecules are stored in membrane-bound vesicles prior to secretion by exocy-tosis (see Chapter 5, p. 64). This is the case with many hormones (e.g. adrenaline) and neurotransmitters (e.g. acetylcholine). Other signalling molecules are so lipid soluble that they cannot be stored in vesicles on their own but must be stored bound to a specific storage protein. This is the case with thyroid hormone. Finally, some chemical mediators are secreted as they are formed. This happens with the steroid hormones and the prostaglandins. Table 7.1 lists the principal families of signalling molecule and their chief modes of action.

Table 7.1 Examples of signalling molecules used in cell–cell communication

Class of molecule	Specific example	Physiological role and mode of action
Signalling molecules secreted by fusion of membrane-bound vesicles		
Ester	Acetylcholine	Synaptic signalling molecule; opens ligand-gated ion channels; also activates G protein-linked receptors
Amino acid	Glycine	Synaptic signalling molecule; opens a specific type of ligand-gated ion channel
	Glutamate	Synaptic signalling molecule; affects specific ligand-gated ion channels; also activates G protein-linked receptors
Amine (bioamine)	Adrenaline (epinephrine)	Hormone; wide variety of effects, acts via G protein-linked receptors
	5-Hydroxytryptamine (5-HT, serotonin)	Local mediator and synaptic signalling molecule; acts via G protein-linked receptors
	Histamine	Local mediator; acts via G protein-linked receptors
Peptide	Somatostatin	Hormone and local mediator; inhibits secretion of growth hormone by anterior pituitary via G protein-linked receptor
	Vasopressin (antidiuretic hormone (ADH))	Hormone; increases water reabsorption by collecting tubule of the kidney via G protein-linked receptor
Protein	Insulin	Hormone; activates a catalytic receptor in plasma membrane; increases uptake of glucose by liver, fat and muscle cells
	Growth hormone	Hormone; activates a non-receptor tyrosine kinase in target cell
Signalling molecules that can diffuse through the plasma membrane		
Steroid	Oestradiol-17β	Hormone; binds to nuclear receptor; hormone–receptor complex regulates gene expression
Thyroid hormone	Tri-iodothyronine (T3)	Hormone; binds to nuclear receptor; hormone–receptor complex regulates gene expression
Eicosanoid	Prostaglandin E_2 (PGE$_2$)	Local mediator; diverse actions on many tissues; activates G protein-linked receptors in plasma membrane
Inorganic gas	Nitric oxide	Local mediator; acts by binding to guanylyl cyclase in target cell

KEY POINTS:

- Cells use diffusible chemical signals in three ways: as local signals (paracrine signalling), as diffuse signals that reach their target tissues via the bloodstream (endocrine signalling) and as rapid, discrete signals (synaptic signalling).
- A wide variety of molecules are employed as chemical signals.
- Signalling molecules may be secreted by exocytosis or they may diffuse across the plasma membrane as they are synthesized.

7.3   Chemical signals are detected by specific receptor molecules

As cells are exposed to a wide variety of chemicals in the extracellular space, they need to have a means of detecting those signals that are intended for them. They do so by means of molecules known as **receptors**, which are specific for particular chemical signals. For example, an acetylcholine receptor binds acetylcholine but does not bind adrenaline. When a receptor has detected a chemical signal, it initiates an appropriate cellular response. The link between the detection of the signal and the response is called **signal transduction**. Substances that bind to and activate a particular receptor are called **agonists**, while those drugs that block the effect of an agonist are called **antagonists** (see also p. 96).

All receptors are proteins and many are located in the plasma membrane where they are able to bind the water-soluble signalling molecules that are present in the extracellular fluid. Hydrophobic signalling molecules, such as the steroid hormones, cross the plasma membrane and bind to receptors within the cell. Finally, some intracellular organelles possess receptors for molecules that are generated within the cell (second messengers – see below).

In recent years, it has become clear that an individual cell may possess many different types of receptor so that it is able to respond to a variety of extracellular signals. The response of a cell to a specific signal depends on which receptors are activated. Consequently, a particular chemical mediator can produce different responses in different cell types. For example, acetylcholine released from the nerve terminals of motor nerve fibres onto skeletal muscle causes the muscle to contract. When it is released from the endings of the vagus nerve, it slows the rate at which the heart beats (the heart rate). The same mediator has different effects in these two tissues because it acts on different receptors. The acetylcholine receptors of skeletal muscle are known as **nicotinic receptors** because the alkaloid nicotine can also activate them. Those of the heart have a different structure and are called **muscarinic receptors** as they can be activated by another chemical – muscarine.

How do receptors control the activity of the target cells?

There are four basic ways in which activation of a receptor can alter the activity of a cell (Figure 7.3):

- Firstly, it may open an ion channel and so modulate the membrane potential.

- Secondly, it may directly activate a membrane-bound enzyme to activate a metabolic sequence.

- Thirdly, it may activate a receptor that activates a membrane protein called a G protein (see below). Activated G proteins cause changes in the intracellular concentration of **second messengers** – chemical signals that regulate metabolic activity within a cell. (The original signalling molecule is the first messenger.) Certain G protein-linked receptors are also known to modulate the activity of some ion channels.

- Finally, the signal may act on an intracellular receptor to modulate the transcription of specific genes.

Ion channel modulation

Many receptors are directly coupled to ion channels. These receptor–channel complexes are called **ligand-gated ion channels** (see Chapter 5, p. 62) and are employed by cells to regulate a variety of functions. In general, ligand-gated channels open for a short period of time following the binding of their specific agonist; this transiently alters the membrane potential of the target cell and thereby modulates its physiological activity. In some instances, control of the membrane potential is the final response. This is the situation when one nerve cell inhibits the activity of another (see Chapter 8, p. 114). More often, the change in membrane potential triggers some further event. Thus, in many cases, activation of ligand-gated channels causes the target cell to depolarize. This depolarization then activates voltage-gated ion channels that trigger the appropriate cellular response. In this way, the activation of a ligand-gated ion channel can be used to control the activity of even the largest of cells.

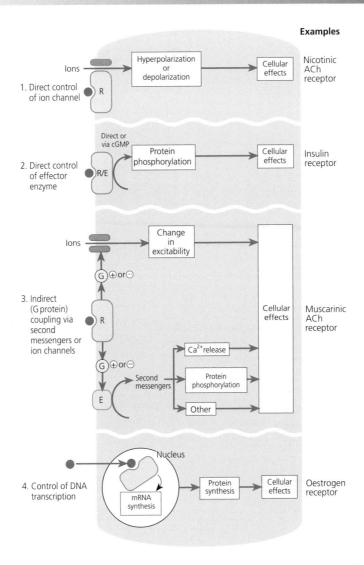

Figure 7.3 Schematic drawing to show the principal ways in which chemical signals affect their target cells. Examples of each type of coupling are shown on the right of the figure. R, receptor; E, enzyme; G, G protein; ACh, acetylcholine; + indicates increased activity, – decreased activity.

This pattern of events is illustrated by the stimulatory effect of acetylcholine on adrenaline secretion by the chromaffin cells of the adrenal medulla. Acetylcholine released by the terminals of the splanchnic nerve binds to nicotinic receptors on the plasma membrane and this increases the permeability of the membrane to Na^+ so that the membrane depolarizes. The depolarization results in the opening of voltage-gated calcium channels and calcium ions flow down their concentration gradient into the cell via these channels. The intracellular Ca^{2+} concentration rises and triggers the secretion of adrenaline. This complex sequence of events is summarized in Figure 7.4.

Catalytic receptors

Catalytic receptors are membrane-bound **protein kinases** that become activated when they bind their specific ligand. (A kinase is an enzyme that adds a phosphate group to its substrate – which can be another enzyme). A typical

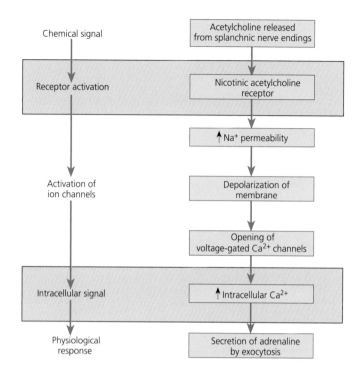

Figure 7.4 The sequence of events that links activation of a nicotinic receptor to the secretion of adrenaline by adrenal chromaffin cells. Acetylcholine binds to a nicotinic receptor on the plasma membrane of the chromaffin cell and opens an ion channel. This increases the permeability of the membrane to Na^+ and the membrane depolarizes. The depolarization activates voltage-gated Ca^{2+} channels and Ca^{2+} ions enter the cell to trigger the secretion of adrenaline by exocytosis.

example of a catalytic receptor is the insulin receptor that is found in liver, muscle and fat cells. This receptor is activated when it binds insulin and in turn it activates other enzymes by adding a phosphate group to tyrosine residues. This results in an increase in the activity of the affected enzymes, which culminates in an increase in the rate of glucose uptake. Many peptide hormone and growth factor receptors are tyrosine-specific kinases.

G protein-linked receptors

G proteins (GTP-binding regulatory proteins) are a specific class of membrane-bound regulatory protein that are activated when a receptor binds its specific ligand. When a G protein-linked receptor is activated, the G protein exchanges bound GDP for GTP. The G protein is then able to modulate the activity of an ion channel or membrane-bound enzyme (Figure 7.5) to alter the

rate of production of a **second messenger** – for example, cyclic AMP or IP_3 (the initial chemical signal is the first messenger). In their turn, the second messengers regulate a variety of intracellular events. Changes in the level of cyclic AMP alter the activity of a variety of enzymes while IP_3 mainly acts by releasing Ca^{2+} from intracellular stores. The series of events linking the change in the level of the second messenger to the final response is called a **signalling cascade**.

> **KEY POINT:**
>
> • Chemical signals are detected by specific receptor molecules that alter the behaviour of the target cell either by opening an ion channel or by activating a membrane-bound enzyme.

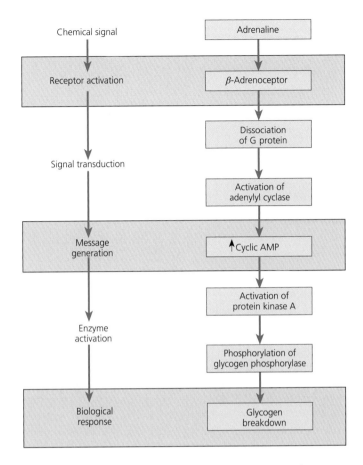

Figure 7.5 A simplified diagram to show the transduction pathway for the action of adrenaline on the glycogen stores of skeletal muscle. On the left are the main stages in the signalling pathway. The details of the individual steps are shown on the right. Adrenaline binds to β-adrenoceptors, which are linked to a specific type of G protein (G_s) that can activate adenylyl cyclase. This enzyme increases the intracellular concentration of the second messenger, cyclic AMP, which in turn leads to the activation of enzymes that break glycogen down to glucose.

7.4 G protein activation of signalling cascades

Cyclic AMP (cAMP) is generated when adenylyl cyclase is activated by a G protein called G_s. The cAMP formed as a result of receptor activation then binds to other proteins (enzymes and ion channels) within the cell to alter their activity. The exact response elicited by cAMP in a particular type of cell will depend on which enzymes that cell possesses. As only one molecule of the appropriate chemical signal is required to activate the membrane receptor, and as adenylyl cyclase can produce many mol-

ecules of cAMP, the initial signal is amplified many times. An example of metabolic control via cAMP production is summarized in Figure 7.5. In this example, the hormone adrenaline is shown acting on skeletal muscle to trigger the breakdown of its energy store (glycogen) to glucose. This glucose is used to generate the increased amounts of ATP the muscle requires for contraction.

Certain G protein-linked receptors, such as the muscarinic receptor (see p. 222) activate an enzyme known

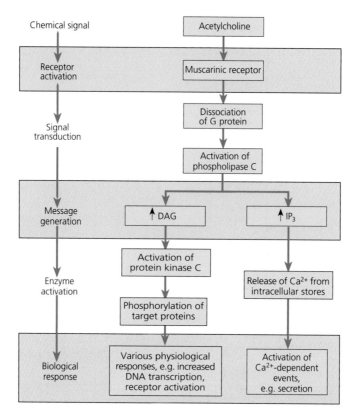

Figure 7.6 A simplified diagram to show the transduction pathway for the formation of inositol trisphosphate (IP_3) and diacylglycerol (DAG). Both IP_3 and DAG act as second messengers. The main stages in the signal transduction pathway are shown on the left, while the detailed steps are shown on the right. In this example, acetylcholine acts on muscarinic receptors that are linked to a G protein that can activate phospholipase C. This enzyme breaks down membrane phosphoinositides to form DAG and IP_3.

as phospholipase C. This enzyme produces diacylglycerol and inositol 1,4,5-trisphosphate (IP_3) from certain membrane lipids, both of which act as intracellular mediators.

IP_3 is a water-soluble molecule that binds to a specific receptor (the IP_3 receptor) to mobilize Ca^{2+} stored within the endoplasmic reticulum. IP_3 generation is therefore able to couple activation of a receptor in the plasma membrane to the release of Ca^{2+} from an intracellular store (Figure 7.6). Many cellular responses depend on this pathway. Examples are enzyme secretion by the pancreatic acinar cells and smooth muscle contraction.

Diacylglycerol is a hydrophobic molecule that is retained in the membrane when IP_3 is formed. It interacts with and activates an enzyme called protein kinase C, which regulates a variety of cellular responses (Figure 7.6) including DNA transcription. Diacylglycerol can also be metabolized to form prostaglandins (see below).

KEY POINTS:

- G protein-linked receptors alter the level of second messengers within a cell (e.g. cyclic AMP).

- The second messengers then modulate ion channels or activate intracellular enzymes.

7.5 Some local mediators are synthesized as they are needed

Certain chemical signals are very lipid soluble and, unlike peptides and amino acids, cannot be stored in vesicles. Instead, the cells synthesize them as they are required. Important examples are the prostaglandins, leukotrienes and the inorganic gas nitric oxide. The secretion of these substances is continuously regulated by increasing or decreasing their rate of synthesis from membrane phospholipids. Once formed, they are rapidly degraded by enzyme activity.

The specific effect exerted by a particular prostaglandin depends on the individual tissue (Table 7.2). For example, prostaglandins PGE_1 and PGE_2 are powerful vasodilators (they relax the smooth muscle in the vessel walls). In the gut and uterus, however, they cause contraction of the smooth muscle. The diversity of effects in response to a particular prostaglandin is explained by the presence of different prostaglandin receptors in different tissues.

Thromboxane A_2 (TXA_2) plays an important role in haemostasis (blood clotting, see p. 404) by causing platelets to aggregate (i.e. to stick together). It is produced by platelets in response to the blood clotting factor thrombin. Thrombin acts on a receptor in the cell membrane that activates phospholipase C. In turn, the phospholipase C liberates diacylglycerol from which TXA_2 is synthesized. It diffuses to neighbouring platelets inducing them to generate TXA_2 (a paracrine action) and it activates a protein that enables the platelets to stick to each other. In this way, tissue damage leads to the formation of a platelet plug. This process is normally held in check by prostacyclin (PGI_2), which is secreted by the endothelial cells that line the blood vessels.

Nitric oxide dilates blood vessels by increasing the production of cyclic GMP in smooth muscle. It is derived from the amino acid arginine by an enzyme called nitric oxide synthase, which is activated when the intracellular free Ca^{2+} concentration in the endothelial cells is increased by acetylcholine, and other signalling molecules. It is also activated by ion channels that open in response to stretching of the plasma membrane. As it is a gas, nitric oxide readily diffuses across the plasma membrane of the endothelial cell and into the neighbouring smooth muscle cells where it activates an enzyme called guanylyl

Table 7.2 The actions of prostaglandins and related compounds

Eicosanoid	Effect on blood vessels	Effect on platelets	Effects on the lung
Prostaglandin E_1 (PGE_1)	Vasodilatation	Inhibition of aggregation	Bronchodilatation
Prostaglandin E_2 (PGE_2)	Vasodilatation	Variable effects	Bronchodilatation
Prostacyclin (PGI_2)	Vasodilatation	Inhibition of both aggregation and adhesion	Bronchodilatation
Thromboxane A_2 (TXA_2)	Vasoconstriction	Aggregation	Bronchoconstriction
Leukotriene C_4 (LTC_4)	Vasoconstriction	–	Bronchoconstriction, increased secretion of mucus

Note: All these molecules are eicosanoids, which are long-chain (20-carbon) molecules derived from certain essential fatty acids such as linolenic and linoleic acid. These fatty acids are popularly known as omega-3 and omega-6 fatty acids, respectively. All four families of eicosanoids are represented in this table: the prostaglandins, the prostacylins, the thromboxanes and the leukotrienes.

cyclase that converts GTP into cyclic GMP. The increase in cyclic GMP within the smooth muscle brings about muscle relaxation and vasodilatation. This sequence of events is summarized in Figure 7.7.

Nitric oxide synthase is not normally present in macrophages but when these cells are exposed to bacterial toxins, the gene controlling the synthesis of this enzyme is switched on and the cells begin to make nitric oxide. In this case, the nitric oxide is not used as a signalling molecule but as a lethal agent to kill invading organisms.

Organic nitrites and nitrates, such as **amyl nitrite** and **nitroglycerine**, have been used for over a hundred years to treat the pain of **angina pectoris** (angina). These compounds promote the relaxation of the smooth muscle in the walls of blood vessels. It is now known that this effect is due to the formation of nitric oxide. This exogenous nitric oxide then acts in a similar way to that derived from normal metabolism to bring about vasodilatation and an increase in blood flow through the blood vessels of the heart.

KEY POINTS:

- Prostaglandins and related substances act as autocrine and paracrine signalling molecules. They are synthesized as required from membrane phospholipids.

- Nitric oxide is a powerful vasodilator that is synthesized on demand by the endothelial cells of blood vessels. It relaxes the smooth muscle of the blood vessels and this, in turn, leads to vasodilatation and increased blood flow.

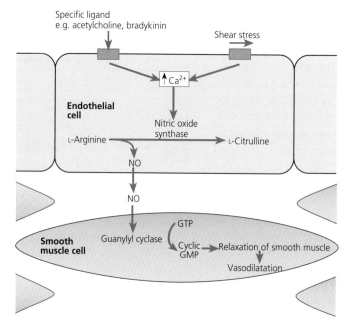

Figure 7.7 The synthesis of nitric oxide (NO) by endothelial cells and its action on vascular smooth muscle. The trigger for increased synthesis of NO is a rise in Ca^{2+} in the endothelial cell. This can occur as a result of stimulation by chemical signals (acetylcholine, bradykinin, ADP, etc.) acting on receptors in the plasma membrane or as a result of the opening of ion channels by stretching of the plasma membrane (shear stress). The NO diffuses across the plasma membrane of the endothelial cell into the neighbouring smooth muscle cells and converts guanylyl cyclase into its active form. The increased production of cyclic GMP leads to relaxation of the smooth muscle.

7.6 Steroid and thyroid hormones regulate gene transcription

The steroid hormones are themselves lipids, so they are able to pass through the plasma membrane freely – unlike more polar, water-soluble signalling molecules such as peptide hormones. Thus, steroid hormones are not only able to bind to specific receptors in the plasma membrane of the target cell but they can also bind to receptors within its cytoplasm and nucleus. The existence of cytoplasmic receptors for steroid hormones was first shown for oestradiol. This hormone is accumulated by its specific target tissues (the uterus and vagina) but not by other tissues. It was found that the target tissues possess a cytoplasmic receptor protein for oestradiol, which, when it has bound the hormone, increases the synthesis of specific proteins.

The full sequence of events for the action of oestradiol can be summarized as follows: the hormone first diffuses through the lipid bilayer and binds to its receptor inside the target cell. The receptor–hormone complex then increases the transcription of DNA into the appropriate mRNA, which is then used as a template for protein synthesis (see Chapter 30). Other steroid hormones, such as the glucocorticoids and aldosterone, are now known to act in a similar way. This scheme is outlined in Figure 7.8. Thyroid hormones regulate gene expression in a similar manner to steroid hormones but gain entry to their target tissues via specific transport proteins.

The receptors for the steroid and thyroid hormones are members of a large group of proteins involved in regulating gene expression, known as the nuclear receptor superfamily. However, not all nuclear receptors are located in the cytoplasm prior to binding their hormone. Some are bound to DNA in the nucleus even in the absence of their normal ligand. For example, the thyroid hormone receptor is bound to DNA in the nucleus even in the absence of thyroid hormone.

The response to thyroid hormone, the steroid hormones and other ligands that bind to members of the nuclear receptor superfamily is determined by the nature of the target cells themselves. It depends on the presence of the appropriate intracellular receptor and the particular set of regulatory proteins present, both of which are specific to certain cell populations. The final physiological effects are slow in onset and long lasting. For example, sodium retention by the kidney occurs after a delay of 2–4 hours following the administration of aldosterone and may last for many hours.

> **KEY POINTS:**
>
> - Steroid and thyroid hormones are very hydrophobic and are carried in the blood bound to specific carrier proteins.
> - They enter cells by diffusing across the plasma membrane and bind to cytoplasmic and nuclear receptors to modulate the transcription of specific genes.

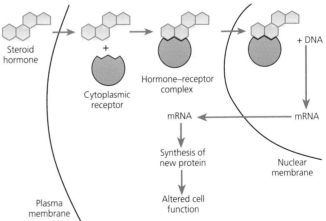

Figure 7.8 A simplified diagram to show how steroid hormones regulate gene transcription in their target cells. Steroid hormones are lipophilic and are able to pass through the plasma membrane to bind to specific receptor proteins in the cytoplasm of the target cells. The hormone–receptor complex diffuses to the cell nucleus where it binds to a specific region of DNA to regulate gene transcription.

7.7 Gap junctions permit the exchange of small molecules between neighbouring cells

Some cells are connected together by gap junctions, which are built from membrane proteins that form doughnut-shaped structures known as **connexons**. When the connexons of two adjacent cells are aligned, the cells become joined by a water-filled pore. As the connexons jut out above the surface of the plasma membrane, the cell membranes of the two cells forming the junction are separated by a small gap – hence the name gap junction (see Chapter 6, p. 77).

Unlike ion channels, the pores of gap junctions are kept open most of the time so that small molecules and inorganic ions can readily pass from one cell to another (Figure 7.9). Consequently, gap junctions form a low-resistance pathway between the cells, and electrical current can spread from one cell to another. The cells are thus electrically coupled. This property is exploited by the myocytes of the heart, which are connected by gap junctions. As the cells are electrically coupled, depolarization of one myocyte causes current to pass between it and its immediate neighbours, which become depolarized. In their turn, these cells cause the depolarization of their neighbours and so on. As a result, current from a single point of excitation spreads across the whole of the heart via the gap junctions and the heart muscle behaves as a **syncytium** (a collection of cells fused together). By this means, the electrical and contractile activity of individual myocytes is co-ordinated, allowing a wave of contraction to travel across the heart and propel the blood around the body (see Chapter 17, p. 335).

In the liver, the role of gap junctions between adjacent liver cells (hepatocytes) is quite different. Here, the gap junctions allow the exchange of second messengers between cells. For example, the hormone glucagon stimulates the breakdown of glycogen to glucose by increasing the level of cyclic AMP. The cyclic AMP produced by one cell can diffuse through the water-filled pores of the gap junctions and pass to the neighbouring cells. Thus, cells not directly activated by a molecule of glucagon can be stimulated to initiate glycogen breakdown as the gap junctions provide a means of spreading the initial stimulus from one cell to another.

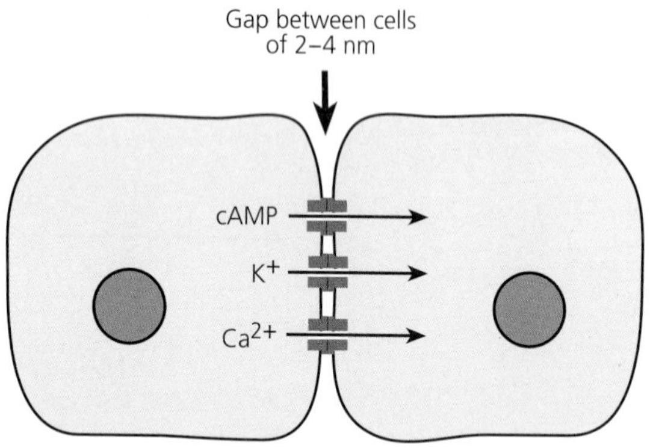

Figure 7.9 A simple diagram showing two cells linked by gap junctions. Ions can pass between the neighbouring cells and so the cells are electrically coupled. In addition, small organic molecules (M_r <1500) can pass through gap junctions, allowing the spread of second-messenger molecules (e.g. cyclic AMP) between adjacent cells.

KEY POINTS:

- Gap junctions between adjacent cells permit small molecules and ions to diffuse from the cytoplasm of one cell to that of its neighbour.

- Gap junctions allow electrical current to flow from one cell to another. Gap junctions, therefore, allow the electrical signals to spread between neighbouring cells.

7.8 Pharmacological aspects of drug–receptor interaction

Many chemical substances alter the normal function of biological systems. When a chemical is poisonous, it is called a **toxin** but when it is of therapeutic interest it is usually called a **drug**, a term that is popularly applied to substances of abuse. The term **pharmacology** applies to all aspects of the study of drugs, including their mode of action and therapeutic uses. The study of the actions of drugs on the physiological systems of the body is called **pharmacodynamics**. The uptake, distribution, metabolism and excretion of drugs within the body form part of the discipline of pharmacology known as **pharmacokinetics**, which will be discussed in Chapter 35.

Although some drugs produce their effects by simple chemical processes (e.g. the neutralizing effect of antac-ids such as magnesium hydroxide or 'milk of magnesia'), most therapeutically useful drugs work by influencing the properties of a specific protein or group of proteins. The proteins that bind drugs are generally called **receptors**, although not all such drug targets are receptors in the strict physiological sense described earlier in this chapter. In fact, while many drug targets are receptors for normal physiological signals, others are enzymes, transport molecules or ion channels. Some drug targets are nucleic acids. This is the mechanism by which certain anticancer drugs work. An example is cyclophosphamide (Cytoxan), which chemically modifies the DNA to prevent the proliferation of cancer cells. Examples of same types of drug target are given in Table 7.3 and many others will be found throughout this book.

Table 7.3 Some typical examples of drug targets

Drug target	Physiological function	Drugs affecting target
Membrane carrier protein: $Na^+/K^+/Cl^-$ carrier	Transport of ions in the kidney	Loop diuretic such as furosemide (antagonist)
Ion channel: Muscle nicotinic receptor	Excitation of muscle	Acetylcholine (normal agonist) Curare (antagonist)
Effector enzyme: Cholinesterase	Breakdown of acetylcholine	Eserine (antagonist)
G protein-linked receptors: β-Adrenoceptors	Activate adenylyl cyclase	Propranolol (antagonist)
Muscarinic receptors	Activates phospholipase C	Atropine (antagonist)
Nuclear receptors: Steroid receptors	Activates gene transcription	Spironolactone (antagonist at the aldosterone receptor)
DNA	Encodes the entire genome	Cyclophosphamide – chemically modifies DNA to interfere with gene transciption

The way many drugs work can be described by a simple scheme that can be represented as follows:

$$\text{Drug} + \text{Receptor} \rightarrow \text{Drug–receptor complex} \rightarrow \text{Effect} \quad (1)$$

According to this scheme the response of a biological system to a particular drug depends on the number of receptors occupied by the drug.

The first step in this sequence can be represented by the following simple chemical reaction scheme:

$$A + R \rightleftharpoons AR \quad (2)$$

where A represents the drug, R the receptor and AR the drug–receptor complex.

Implicit in this description is the idea that drug–receptor interactions obey the Law of Mass Action. This requires that the amount of drug bound will depend both on the concentration of the drug and on the number of available receptors. The forward and backward pointing arrows indicate that the reaction is reversible. At equilibrium, the rate at which the drug binds to the receptor is equal to the rate at which the drug–receptor complex dissociates; in other words, the rate at which the forward reaction proceeds is the same as that of the backward reaction. From this basic assumption, it can be shown that the proportion of receptors (p_A) that have bound the drug is given by the following equation:

$$p_A = \frac{x_A}{x_A + K_A} \quad (3)$$

where x_A is the concentration of the drug and K_A is the ratio of the backward reaction rate divided by that of the forward reaction rate. K_A is called the **equilibrium constant**. It is numerically equal to the drug concentration at which half the receptors are occupied. Figure 7.10 shows the theoretical relationship between drug concentration and the proportion of receptors that have bound the drug. When the drug concentration is plotted on a normal linear scale, the curve described is a rectangular hyperbola (Figure 7.10a); when it is plotted on a logarithmic scale, the curve is a sigmoid curve as shown in Figure 7.10b. The relationships described by these curves are called **dose–response curves**. As drugs acting on the same receptor often have very different potencies, it is usual to plot dose–response curves with the concentration axis on a logarithmic scale.

A drug that interacts with a receptor and activates it is called an **agonist**. Agonists that achieve the maximum response of which a biological system is capable are sometimes called **full agonists**. Those agonists that are unable to evoke a maximal response are called **partial agonists**. Partial agonists may have a similar potency to certain full agonists on a biological system, but they are less effective at activating the system (i.e. they have a lower efficacy, see below). Some drugs that interact with a receptor do not produce a biological effect but prevent its activation by an agonist. These drugs are called **antagonists**.

The **potency** of a drug (an agonist or antagonist) is determined by the concentration required to produce a given effect and is indicated by the location of the dose–response curve along the concentration axis. The **efficacy** of a drug is the magnitude of the effect the drug can produce. The maximal efficacy is given by the plateau at the top of the dose–response curve. The concentration at which a drug achieves half its maximum effect is called its EC_{50}. Figure 7.11 illustrates the use of these terms.

While a dose–response curve provides a useful way of summarizing key information about the effect of a particular drug on a biological system, it cannot be used to determine the affinity of a drug for its receptor. This is usually determined by **ligand binding studies**, which directly measure the amount of drug bound to a purified receptor.

The drug–receptor interaction scheme in Equation 1 above indicates that a drug will elicit a biological response whose magnitude depends on the number of its receptors that are occupied. However, the magnitude of the response is not normally linearly related to the drug concentration. There are several reasons for this state of affairs: in some tissues, a maximum effect can be produced when only a small proportion of the available receptors have been occupied. This is often the case when an agonist is acting on a G protein-linked receptor. Not all responses are smoothly graded: the measured response may depend on a threshold event that must first be reached. An example is the loss of an animal's ability to

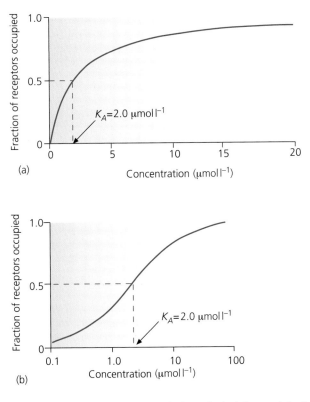

Figure 7.10 Theoretical relationship between the concentration of a hypothetical drug and the fraction of receptors occupied. In (a), the fraction of receptors occupied is plotted as a linear function of drug concentration. Note that the curve is initially very steep (so a small increase in drug concentration will result in a greater proportion of receptors being occupied). As the drug concentration increases, the increase in the proportion of occupied receptors becomes progressively smaller as the response reaches its maximum – the curve reaches an asymptote. In (b), the same data plotted on a semilogarithmic scale. In this case, the curve is S-shaped (sigmoid) and it is clear that the steepest part of the curve occurs when the proportion of occupied receptors lies between about 25 and 75%.

right itself after being given an anaesthetic. This is an all-or-none response and many steps are involved between the drug–receptor interaction and the final response. As partial agonists are less effective in eliciting a response than full agonists, they must occupy a higher proportion of the available receptors for a given magnitude of response.

Competitive and non-competitive antagonism

Many antagonists bind reversibly to a receptor and their effect can be overcome by increasing the con-centration of an appropriate agonist, as shown in Figure 7.12. This kind of antagonism is called **competitive antagonism**. The antihypertensive drug propranolol is a competitive antagonist of adrenaline and noradren-aline at β-adrenoceptors (see p. 223). Sometimes an antagonist binds irreversibly to its receptor and this reduces the maximum effect that can be produced by an agonist because the number of available recep-tors is reduced (Figure 7.12). A third type of antago-nism occurs when the antagonist blocks one of the steps between receptor activation and the biological response being measured. This type of antagonism is called **non-competitive antagonism**. An example is

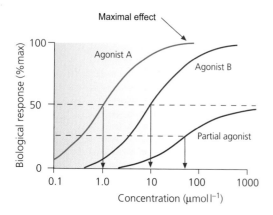

Figure 7.11 Dose–response curve for three different agonists. Agonist A is a full agonist and has the highest potency with an EC_{50} value of $1.0\,\mu mol\,l^{-1}$. Agonist B is also a full agonist but has a lower potency than agonist A ($EC_{50} = 10\,\mu mol\,l^{-1}$). The partial agonist has a much lower efficacy (if can only elicit half the response that the full agonists can achieve). It also has a lower potency ($EC_{50} \approx 75\,\mu mol\,l^{-1}$). The EC_{50} is the concentration at which the response is half its maximum value and is indicated by the downward-pointing arrows for each of the drugs shown.

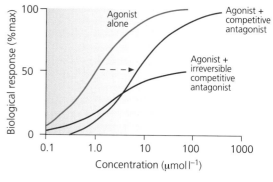

Figure 7.12 A set of dose–response curves that show competitive antagonism. The green curve shows the dose–response relationship for the agonist alone. In the presence of a reversible competitive antagonist, the curve is shifted to the right, as indicated by the blue curve. This parallel shift indicates that the effect of the antagonist can be overcome by increasing the concentration of agonist as both drugs are competing for the same site. If the competitive antagonist binds irreversibly to the receptor, the number of available receptors is reduced. As a result, the maximum response is also reduced (the red curve). The resulting dose–response curve may also be displaced to the right.

the effect of calcium channel blockers on the contraction of smooth muscle stimulated by acetylcholine. The slope and maximum of the dose–response curve are reduced, although there may also be a rightward shift.

Desensitization and drug tolerance

The effect of a drug may progressively diminish if it is given repeatedly over a short period of time, that is, over a few minutes. This phenomenon is called **desensitization** or **tachyphylaxis**. Often the effect of a drug will slowly decline over days or weeks; this slow change in the effective of a drug is generally called **tolerance**. In the case of antibacterial drugs, the loss of effectiveness is known as **drug resistance**.

KEY POINTS:

- A drug is a chemical that changes a biological process in a particular way.

- The effect of a drug on a biological system can be summarized by a dose–response curve.

- The potency of a drug reflects its ability to bind to its receptor.

- Drugs that activate a biological system are called agonists.

- Substances that prevent agonists having their normal pharmacological effects are called antagonists.

- The efficacy of a drug is a measure of the maximum effect it can elicit compared with the maximum effect that can be elicited.

Box 7.1 Competitive antagonism, Schild plots and pA_2 values of drug antagonists

The number of receptors occupied by a drug is determined by its concentration and by its affinity for the binding site. The proportion of receptors occupied by drug A is given by Equation (3) (see main text) namely:

$$p_A = \frac{x_A}{x_A + K_A} \tag{B1}$$

Dividing by K_A we arrive at the following expression:

$$p_A = \frac{x_A/K_A}{x_A/K_A + 1} \tag{B2}$$

When two drugs bind to the same receptor site, one drug competes with the other for the available receptors. By applying the Law of Mass Action to this situation, it can be shown that the proportion of receptors occupied by drug A is:

$$p_A = \frac{x_A/K_A}{x_A/K_A + x_B/K_B + 1} \tag{B3}$$

where x_B and K_B are the concentration and equilibrium constant for drug B; a similar expression applies to the proportion of receptors occupied by drug B.

For the purposes of argument, assume that drug A is an agonist and that drug B is an antagonist. Equation (B3) shows that adding a competitive antagonist (drug B) results in a reduction in the number of receptors occupied

by the agonist A for any given concentration of A. In effect, the concentration–response curve is displaced to the right along the concentration axis, as shown by the blue curve in Figure 7.12. The effect of the antagonist on the response of the system can be overcome by increasing the concentration of the agonist A. The concentration of agonist that must be used to restore a particular level of receptor occupancy (and therefore a given biological response) will depend on the affinity of the antagonist for the receptor. The ratio of the concentrations that give the same receptor occupancy is $r = x'_A/x_A$ where x'_A is the concentration in the presence of antagonist and x_A is the occupancy in the absence of antagonist. It can be shown that:

$$r = \frac{x_B}{K_B} + 1 \tag{B4}$$

The relationship given by Equation (B4) is known as the **Schild equation**, and can be expressed logarithmically as:

$$\log(r-1) = \log x_B - \log K_B \tag{B5}$$

A plot of $\log(r-1)$ against $\log x_B$ is called a **Schild plot**, which should be a straight line with a slope of 1. The intercept with the x axis (the abscissa) is equal to $\log K_B$, as shown in Figure 1.

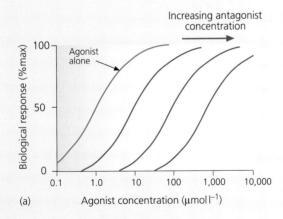

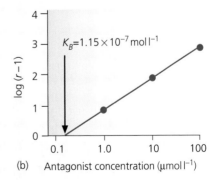

(a)

(b)

Figure 1 (a) The concentration–response curve for a hypothetical agonist and its rightward displacement in the presence of increasing concentrations of an antagonist. (b) A Schild plot of this hypothetical data. The slope of the line is 1 and the intercept with the abscissa gives the equilibrium constant (in this case $1.15 \times 10^{-7}\,\mathrm{mol\,l^{-1}}$). Note that the abscissa is plotted on a logarithmic scale.

The potency of antagonists is often expressed by their pA_2 **values**, where $pA_2 = -\log K_B$. Thus, the pA_2 of an antagonist is the negative logarithm of the equilibrium binding constant. Thus, an antagonist with a K_B of 2×10^{-9} mol l^{-1} has a pA_2 of 8.7 and one of 2.5×10^{-7} mol l^{-1} has a pA_2 of 6.6: the higher the pA_2 value, the greater the affinity of the antagonist for its receptor. The advantage of the pA_2 notation is that it allows the potency of an antagonist to be expressed simply, rather like the pH notation for the hydrogen ion concentration of solutions (see Chapter 2, p. 21).

Recommended reading

Biochemistry and cell biology

Alberts, B., Johnson, A., Lewis, J., Raff, M., Roberts, K. and Walter, P. (2008). *The Molecular Biology of the Cell,* 5th edn. Garland: New York. Chapters 15 and 19.

Berg, J.M., Tymoczko, J.L. and Stryer, L. (2002). *Biochemistry,* 5th edn. Freeman: New York. Chapter 15.

Elliott, W.H. and Elliott, D.C. (2008). *Biochemistry and Molecular Biology,* 4th edn. Oxford University Press: Oxford. Chapter 27.

Gomperts, B.D., Tatham, P.E.R. and Kramer, I.M. (2002). *Signal Transduction*. Academic Press: San Diego.

Pharmacology

Rang H.P., Dale, M.M. and Ritter, J.M. (2007). *Pharmacology,* 6th edn. Churchill-Livingstone: Edinburgh. Chapters 1–3.

Self-assessment questions

Which of the following statements are true and which are false? Answers are given at the end of the book (p. 755).

1 Hormones:
  a) are chemical signals that are secreted into the blood.
  b) can influence the behaviour of many different cell types.
  c) act only on neighbouring cells.
  d) are secreted by specialized glands.

2 The following are secreted by one type of cell specifically to regulate the activity of others:
  a) nitric oxide
  b) prostaglandins
  c) insulin
  d) adrenaline
  e) Ca^{2+}
  f) glucose
  g) cyclic AMP.

3 Physiological receptors:
  a) are always proteins.
  b) are always located in the plasma membrane.
  c) may be membrane-bound enzymes.
  d) can activate second-messenger cascades via G proteins.

4 Prostaglandins:

   a) are secreted by exocytosis.

   b) are synthesized from phospholipids.

   c) are hormones.

   d) act on G protein-linked receptors.

5 Steroid hormones:

   a) are lipid soluble.

   b) directly activate ion channels.

   c) can alter gene transcription.

   d) are secreted as they are synthesized.

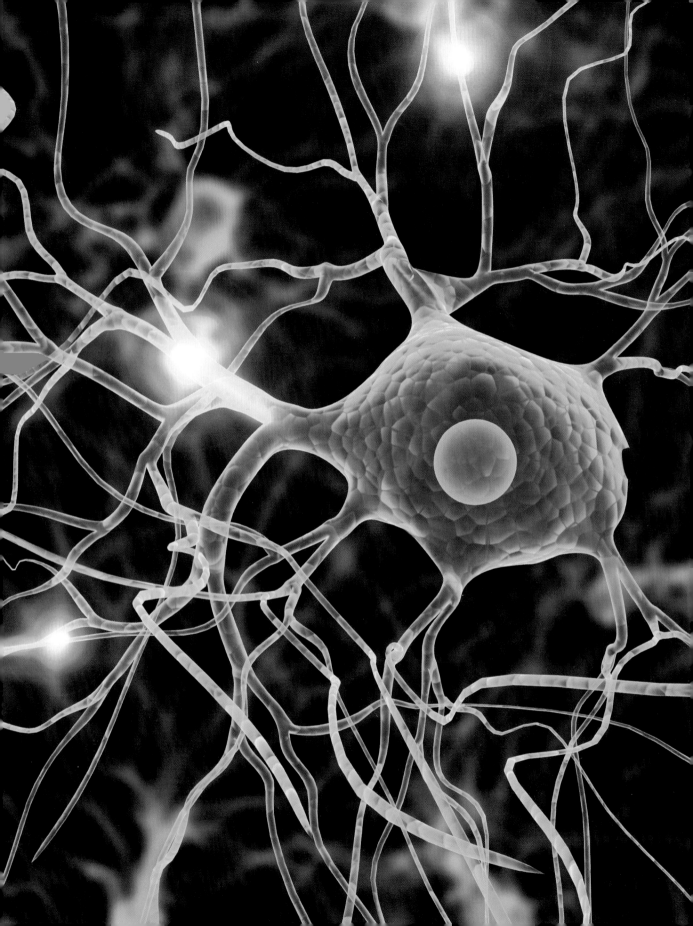

Section 3

Excitable cells

8 Nerve cells and their connections 104

9 Muscle 118

3

8 Nerve cells and their connections

After reading this chapter you should have gained an understanding of:

- The structure of nerve cells
- The ionic basis of the action potential in neurones and axons
- The principal features of synaptic transmission between nerve cells – the synaptic basis of excitation and inhibition

8.1 Introduction

The nervous system is adapted to provide rapid and discrete signalling over long distances (from millimetres to a metre or more) and this chapter is chiefly concerned with the cellular mechanisms that enable it to do so. The principal signalling unit of the nervous system is the **nerve cell** or **neurone**, which was briefly mentioned in Chapter 6. Nerve cells are associated with various other cell types within the brain and spinal cord, which are called **glial cells** or **neuroglia**. Their properties will be discussed briefly in Chapter 11. The cell bodies of neurones are found throughout the grey matter of the brain and spinal cord. They are very varied in both size and shape but all are rich in ribosomes that stain strongly with basic dyes. The space between the cell bodies is known as the **neuropil**.

Each neurone has an extensive set of fine branches called **dendrites** that receive information from other neurones, integrate it and transmit the result to its target cells (which may be other neurones) via a single, thread-like extension of the cell body called an **axon** (Figure 8.1). Axons and dendrites are collectively called neuronal processes. The axon branches as it passes through the tissues and each branch ends in a small swelling as it contacts its target cell. The final segment of the axon is called the **axon terminal** or **nerve terminal** and the contact between an axon terminal and its target is called a **synapse**. Synapses may be made between nerve cells or between an axon and a non-neuronal cell, such as a muscle fibre.

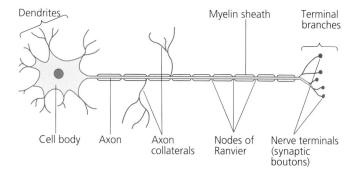

Figure 8.1 The principal features of a typical spinal motoneurone.

Those nerve cells that transmit information from the central nervous system (CNS) to the muscles and secretory glands are called **motoneurones** and their axons reach their target tissues via **efferent nerves**. Those nerve cells that transmit signals from specific sensory end organs to the CNS are **sensory neurones**. Their axons travel to the CNS via **afferent nerves**. The structure of the peripheral nerves will be considered in Chapter 11 (p. 191). Many nerve cells do not connect directly with the peripheral tissues but are concerned with processing information within the brain and spinal cord. These are known as **interneurones**.

> **KEY POINTS:**
>
> • The neurone is the principal functional unit of the nervous system.
>
> • The cell bodies of neurones give rise to two types of process: dendrites and axons.
>
> • The dendrites are highly branched and receive information from many other nerve cells.
>
> • Each cell body gives rise to a single axon, which subsequently branches to make contact with many cells.

8.2 The structure of axons

Axons are delicate structures and outside the brain and spinal cord they run in bundles called peripheral nerves, which are protected by layers of connective tissue (see Chapter 11, p. 191). Individual axons may be either myelinated or unmyelinated. **Myelinated axons** are covered by a thick layer of fatty material called **myelin**. Although the myelin sheath extends along the length of an axon, it is interrupted at regular intervals by gaps known as the **nodes of Ranvier**. At the nodes of Ranvier, the axon membrane is not covered by myelin but is in direct contact with the extracellular fluid. The distance between adjacent nodes varies with the axon diameter, larger fibres having a greater distance between each node.

In peripheral nerves, the myelin is derived from a type of cell called a **Schwann cell** (Figure 8.2) while in the CNS the myelin is formed by the **oligodendrocytes** (see Chapter 11). In peripheral nerves, **unmyelinated axons** are also covered by Schwann cells, but, in this case, there is no layer of myelin and a number of nerve fibres are covered by a single Schwann cell. Unmyelinated axons are in direct communication with the extracellular fluid via a longitudinal cleft in the Schwann cell called a **mesaxon** (Figure 8.2d). Within the CNS, unmyelinated axons have no protective covering as they course through the neuropil.

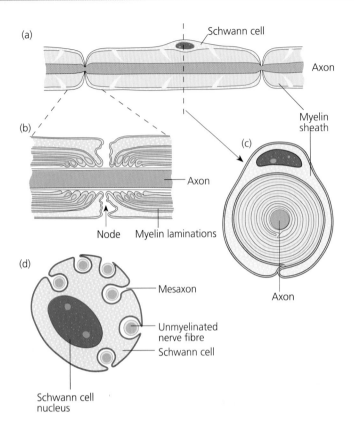

Figure 8.2 The main structural differences between myelinated and unmyelinated nerve fibres: (a) shows a short length of myelinated axon with an enlarged image illustrating the detailed structure of a single node of Ranvier below in (b); (c) is a diagrammatic view of a cross-section of the internodal region of a myelinated nerve fibre to show how the Schwann cell forms the myelin layers; (d) illustrates how unmyelinated nerve fibres are supported by the Schwann cells. The cleft labelled mesaxon communicates directly with the extracellular space. Each unmyelinated fibre has its own mesaxon.

KEY POINTS:

- Individual axons may be either myelinated or unmyelinated. Myelin is formed by oligodendrocytes in the CNS and by Schwann cells in the periphery.
- The axon of a neurone transmits information to other neurones or to non-neuronal cells such as muscles.
- After leaving the CNS, axons run in peripheral nerve trunks, which contain layers of connective tissue to provide structural support.

8.3 The primary function of an axon is to transmit information coded as a sequence of action potentials

In the late eighteenth century, Galvani showed that electrical stimulation of the nerve in a frog's leg caused the muscles to twitch. This key observation led to the discovery that the excitation of nerves was accompanied by an electrical wave that passed along the nerve. This wave of excitation is now called the **nerve impulse** or **action potential**.

To generate an action potential, an axon requires a stimulus of a certain minimum strength, known as the threshold stimulus or **threshold**. An electrical stimulus that is below the threshold (a subthreshold stimulus) will not elicit an action potential while a stimulus that is above threshold (a suprathreshold stimulus) will do so (see

Figure 8.5 on p. 110). With stimuli above threshold, the action potential has approximately the same magnitude and duration irrespective of the strength of the stimulus. This is known as the 'all-or-none' law of action potential transmission.

In a mammalian axon, each action potential lasts for about 0.5–1 ms. If a stimulus is given immediately after an action potential has been elicited, a second action potential is not generated until a certain minimum time has elapsed. The interval during which it is impossible to elicit a second action potential is known as the **absolute refractory period**. This determines the maximum number of action potentials a particular axon can transmit in a given period of time. Following the absolute refractory period (which is usually 0.5–1 ms in duration), there is a phase of reduced excitability during which a stronger stimulus than normal is required to elicit an action potential. This phase lasts for about 5 ms and is called the **relative refractory period**.

What mechanisms are responsible for the generation of the action potential? During the action potential, the membrane potential briefly reverses in polarity reaching a peak value of +40 to +50 mV before falling back to its resting level of about −70 mV (Figure 8.3). At the peak of the action potential, the membrane potential is close to the equilibrium potential for sodium ions (E_{Na^+}), whereas its resting membrane potential lies close to the equilibrium potential for potassium ions (E_{K^+} – see Chapter 5 for an explanation of the resting potential and equilibrium

potentials). From this key observation, it is evident that the action potential is caused by a large and short-lived increase in the permeability of the axon membrane to sodium ions.

What underlies the change in the permeability of the axon membrane to sodium ions during an action potential? The axonal membrane contains ion channels of a specific type called **voltage-gated sodium channels**. When an axon is excited, the membrane depolarizes and this small change in membrane potential causes some of the voltage-gated sodium channels to open, permitting sodium ions to move into the axon down their electrochemical gradient. This inward movement of sodium depolarizes the membrane further, causing more sodium channels to open. This results in a positive feedback in which a greater influx of sodium causes a greater degree of depolarization until, at the peak of the action potential, all the voltage-gated sodium channels are open and the membrane is, for a brief time, highly permeable to sodium ions. Indeed, at the peak of the action potential, the membrane is more permeable to sodium ions than to potassium ions, and this explains the fact that, at the peak of the action potential, the membrane potential is positive and close to E_{Na^+}.

Why does the membrane repolarize after the action potential? The open state of the sodium channels is unstable and the sodium channels begin to inactivate (see Chapter 5, p. 63) and the permeability of the axon membrane to sodium begins to fall. At the same time, voltage-

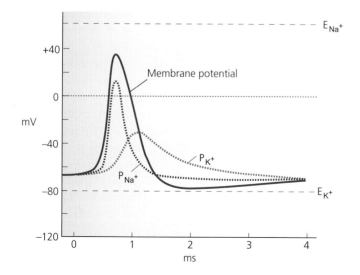

Figure 8.3 The changes in membrane permeability that occur during the action potential. The time course of an action potential is shown by the blue line. The red line shows the changes in membrane permeability to sodium ions (P_{Na^+}) that occur during an action potential. The initial increase in P_{Na^+} is responsible for the upstroke of the action potential. As P_{Na^+} decreases, the membrane potential declines towards its resting level. The restoration of the resting membrane potential is further aided by the delayed increase in the membrane permeability to potassium (P_{K^+}), as shown by the green line. E_{Na^+} and E_{K^+} are the equilibrium potentials of sodium and potassium ions, respectively.

activated potassium channels begin to open in response to the depolarization and potassium ions leave the axon down their electrochemical gradient. This increase in the permeability of the membrane to potassium occurs at the same time as the permeability of the membrane to sodium begins to fall. The decrease in sodium permeability and the increase in potassium permeability combine to drive the membrane potential from its positive value at the peak of the action potential towards the equilibrium potential for potassium ions. As the membrane potential approaches its resting level, the voltage-activated potassium channels begin to close and the membrane potential assumes its resting level. At the resting membrane potential, the inactivated sodium channels revert to their closed state and the axon is primed to generate a fresh action potential.

This sequence of events explains both the threshold and the refractory period. When a weak stimulus is given, the axon does not depolarize sufficiently to allow enough voltage-gated sodium channels to open to depolarize the membrane further. Such a stimulus is subthreshold. At the threshold, sufficient sodium channels open to permit further depolarization of the membrane as a result of the increased permeability to sodium. The resulting depolarization causes the opening of more sodium channels and the process continues until all available sodium channels are open.

Immediately after the action potential, most sodium channels are in their inactivated state and cannot reopen until they have returned to their closed state. To do so, they must spend a brief period at the resting membrane potential. This accounts for the absolute refractory period. One plausible explanation of the relative refractory period of a single nerve fibre is that an action potential can be supported when enough sodium channels have returned to the closed state provided that they are all activated together. This will require a stronger stimulus than normal to ensure that all the sodium channels open at the same time.

The effect of the passage of an action potential along an axon is to leave it with a little more sodium and a little less potassium than it had before. Nevertheless, *the quantities of ions exchanged during a single action potential are very small and are not sufficient to alter the ionic gradients across the membrane.* In a healthy axon, the ionic gradients are maintained by the continuous activity of the sodium pump (p. 59). If the sodium pump is blocked by a metabolic poison, a nerve fibre is able to conduct impulses only until its ionic gradients are dissipated. Thus, the sodium gradient established by the sodium pump indirectly powers the action potential.

> **KEY POINTS:**
> - Nerve cells transmit information along their axons by means of action potentials. This enables them to transmit signals rapidly over considerable distances.
> - Action potentials are excited by a stimulus of a certain minimum strength known as the threshold. With stimuli above threshold, each action potential has approximately the same magnitude and duration. This is known as the 'all-or-none' law.
> - The action potential is caused by a large, short-lived increase in the permeability of the plasma membrane to sodium ions.

How is the action potential propagated along an axon?

When an action potential has been initiated, it is transmitted rapidly along the entire length of an axon. How does this happen? The explanation is as follows. At rest the membrane potential is about $-70\,mV$, while during the peak of the action potential, the membrane potential is positive (about $+50\,mV$). So, when an action potential is being propagated along an axon, the active zone (point A in Figure 8.4) and the resting membrane (point B in Figure 8.4) will be at different potentials. A small electrical current will flow between the two regions, rather like the current flowing between the two terminals of a battery when the circuit is completed. The active zone and the neighbouring resting membrane form a local circuit in which the electrical current is able to depolarize the resting membrane (Figure 8.4). This depolarization causes voltage-gated sodium channels in the membrane ahead of the active zone to open and, when sufficient channels have opened, the action potential invades this part of the membrane. This in turn spreads the excitation farther along the axon where the

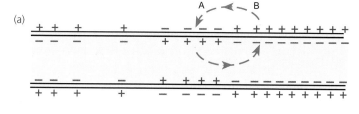

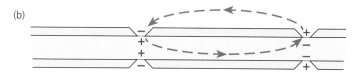

Figure 8.4 A simplified diagram illustrating the local circuits responsible for action potential propagation in unmyelinated and myelinated axons. In unmyelinated axons (a), the local circuit allows current to pass from the active region (point A) to the neighbouring resting membrane (point B) and depolarize it. The action potential passes along the axon as a continuous wave. In myelinated fibres (b), the current can only cross the axon membrane at the nodes of Ranvier where there are breaks in the insulating layer of myelin. As a result, the action potential is propagated along the axon by a series of jumps. This is known as saltatory conduction.

process is repeated until the action potential has traversed the length of the axon.

When an action potential has invaded one part of the membrane and has moved on, it leaves the sodium channels in an inactivated state from which they cannot be reactivated until they have passed through their closed (or resting) state. The time taken for this transition is the absolute refractory period of the axon during which the membrane cannot support another action potential (see above, p. 107). By the time the sodium channels have reverted to their closed state, the peak depolarization has passed farther along the axon. It is then no longer capable of depolarizing the regions through which it has passed. In the intact nervous system, an action potential is propagated by the axon from its point of origin to the axon terminals (**orthodromic propagation**). However, if an axon is artificially stimulated near its terminus, an action potential will be propagated towards the cell body, that is, in the direction opposite to its normal mode of propagation. This is called **antidromic propagation.**

The precise way in which an action potential is propagated depends on whether the axon is myelinated or unmyelinated. In unmyelinated axons, the axonal membrane is in direct electrical contact with the extracellular fluid for the whole of its length via the mesaxon (see above, p. 105). Each region depolarized by the action potential results in the spread of current to the immediately adjacent membrane, which then becomes depolarized. This ensures that the action potential propagates smoothly along the axon in a continuous

wave. In myelinated axons, the axon membrane is insulated from the extracellular fluid by the layers of myelin except at the nodes of Ranvier. In this case, an action potential at one node of Ranvier completes its local circuit via the next node. This enables the action potential to jump from one node to the next – a process called **saltatory conduction**. This adaptation serves to increase the speed at which the action potential is propagated so that, size for size, myelinated nerve fibres have a much higher conduction velocity than unmyelinated fibres. The speed with which axons conduct action potentials (i.e. their **conduction velocity**) varies from less than $0.5\,\mathrm{m\,s^{-1}}$ for small unmyelinated fibres to over $100\,\mathrm{m\,s^{-1}}$ for large myelinated fibres: for both myelinated and unmyelinated axons, the greater the diameter, the faster the conduction velocity.

KEY POINTS:

- After the passage of an action potential, the axon cannot propagate another until sufficient sodium channels have returned to their resting state. This period of inexcitability is known as the absolute refractory period.

- Large-diameter axons conduct impulses faster than small-diameter axons and myelinated axons conduct impulses faster than unmyelinated axons. Myelinated axons conduct impulses by saltatory conduction.

Action potentials can be recorded from intact nerves

While it is possible to record action potentials from individual axons, for diagnostic purposes it is more useful to stimulate a nerve trunk through the skin and record the summed action potentials of all the different fibres present. This summed signal is called a **compound action potential**. Unlike individual nerve fibres, which show an all-or-none response, the action potential of a nerve trunk is graded with stimulus strength (Figure 8.5b). For weak stimuli below threshold, no action potential is elicited. Above threshold, those axons with the lowest threshold are excited first. As the stimulus intensity is increased, more axons are excited and the action potential becomes larger. In this way, the compound action potential grows with increasing stimulus strength until all the available nerve fibres have been excited.

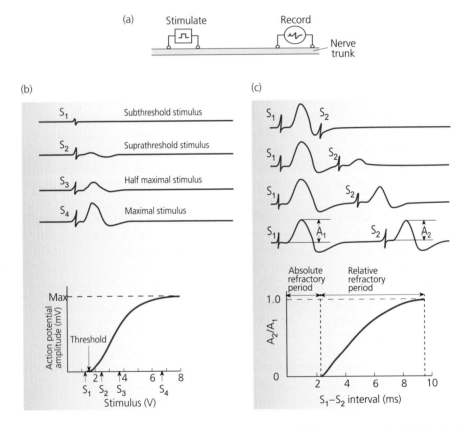

Figure 8.5 Some characteristic properties of compound action potentials recorded from nerve trunks. (a) A simple diagram of the arrangement of the stimulating and recording electrodes. Panel (b) shows how increasing the strength of the electrical stimulus given to the nerve gives rise to an action potential that becomes progressively larger in amplitude until it reaches a maximum value. S_1 is a very weak shock that does not elicit an action potential (a **subthreshold stimulus**). S_2–S_4 elicit action potentials of progressively larger amplitude until a maximum is reached (S_4 is called a **maximal stimulus**). The progressive increase in the amplitude of the action potential with increasing stimulus strength is shown at the bottom of the panel and reflects the ability of stronger stimuli to activate more and more of the axons present within the nerve (recruitment). Panel (c) shows the change in excitability that follows the passage of an action potential. Two shocks (S_1 and S_2) of equal intensity are given at various time intervals. If S_2 follows S_1 within 2 ms, no fibres are excited (top record – this is the absolute refractory period). As the interval between S_1 and S_2 is increased, more and more fibres are excited by S_2 (middle two records) until all are excited (bottom record). The relative change in the amplitude of the compound action potential with the interval between the stimuli is shown at the bottom of the panel.

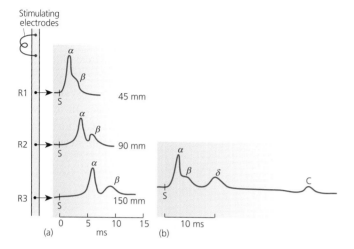

Figure 8.6 Panel (a) shows a compound action potential recorded at different points along an intact nerve (shown as R_1 to R_3 in the figure) following an electrical stimulus. Note that as the action potential travels along the nerve, it progressively breaks up into distinct components reflecting the different conduction velocities of the individual nerve fibres. The two fast components are marked α and β – the slower components are shown in (b). Each wave reflects the activity of a group of fibres with a similar conduction velocity. The fastest are called α fibres. They have a maximum conduction velocity of about $100\,m\,s^{-1}$. The slowest are unmyelinated fibres called C-fibres, which have a conduction velocity of about $1.3\,m\,s^{-1}$.

Compound action potentials exhibit both absolute and relative refractory periods. Just as with a single axon, a second stimulus applied within 1–2 ms of the first will fail to elicit an action potential (absolute refractory period). As the interval between successive stimuli is increased, a second action potential is generated, which progressively grows in amplitude until it reaches the amplitude of the first (Figure 8.5c). During the relative refractory period, more and more of the fibres in the nerve trunk recover their excitability as the interval between the two stimuli increases.

When a long length of nerve is stimulated and the compound action potential is recorded well away from the stimulating electrode, a number of peaks become visible in the record (Figure 8.6). These different peaks reflect differences in the conduction velocity of the different axons present within the nerve trunk. Axons can be classified on the basis of their conduction velocity and physiological role. The axons with the highest conduction velocity belong to neurones involved in motor control (see Chapter 9). Unmyelinated fibres (sometimes called C-fibres) have the slowest conduction velocity of all and serve to transmit certain kinds of sensory information (e.g. pain and temperature) to the CNS.

Some diseases, such as multiple sclerosis, are characterized by a loss of myelin from axons. These diseases, which are known as **demyelinating diseases**, may be diagnosed by measuring the conduction velocity of peripheral nerves. Affected nerves have an abnormally slow conduction velocity. The loss of myelin affects conduction velocity because the normal pattern of current spread is disrupted. Action potentials are propagated through the affected region by continuous conduction similar to that seen in small unmyelinated fibres rather than by saltatory conduction as seen in healthy fibres. In severe cases, there may even be a total failure of conduction. In either eventuality, the function of the affected pathways is severely impaired.

Local anaesthetics block conduction in nerve trunks

Local anaesthetics are used to block the conduction of action potentials in peripheral nerve trunks. The area supplied by the nerve in question is rendered numb and insensitive to pain. This permits a dentist or doctor to carry out a painful procedure, such as the extraction of a tooth or removal of a skin growth, without the necessity of a general anaesthetic. Local anaesthetics act by blocking the voltage-gated sodium channels that are responsible for generating and propagating the action potential. Although all nerve fibres are affected, the small myelinated and unmyelinated nerve fibres, which transmit pain and other sensations, are the first to be blocked.

Local anaesthetics are given topically (i.e. applied directly to the skin) or by direct injection at an appropriate site. As they must penetrate the layers of connective tissue surrounding the nerve fibres themselves, they need to be moderately lipid soluble to reach their site of action. The degree of lipid solubility determines how quickly they act and their duration of action. The best-known, clinically useful local anaesthetic is **lignocaine** (called **lidocaine** in North America), which has a rapid onset and an intermediate duration of action. Local anaesthetics with a slower onset but longer duration of action are **bupivacaine** and **ropivacaine**. **Benzocaine** is used as a topical agent in many creams, gels and ointments. Local anaesthetics, particularly bupivacaine, are also used for epidural and spinal anaesthesia.

KEY POINTS:

- An electrical stimulus applied to a nerve trunk elicits a compound action potential, which is the summed activity of all the nerve fibres present in the nerve trunk.
- Compound action potentials grow in amplitude as the stimulus strength is increased above threshold until all the axons in the nerve are recruited.
- Local anaesthetics act by blocking the propagation of action potentials.

8.4 Chemical synapses

A synapse is formed when an axon makes contact with its target cell (see above). The nerve cell that gave rise to the axon is called the **presynaptic neurone** and the target cell is called the **postsynaptic cell**, which may be another neurone, a muscle cell or a gland cell. A synapse has the function of transmitting the information coded in a sequence of action potentials to the postsynaptic cell so that it responds in an appropriate way. This section is concerned with the mechanisms by which nerve cells transmit information to other nerve cells.

Synaptic transmission is a one-way signalling mechanism – information flows from the presynaptic to the postsynaptic cell and not in the other direction. When the activity of the postsynaptic cell is increased, the synapse is called an **excitatory synapse**. Conversely, when activity in the presynaptic neurone leads to a fall in activity of the postsynaptic cell, the synapse is called an **inhibitory synapse**. In mammals, including man, most synapses operate through the secretion of a small quantity of a chemical (a neurotransmitter) from the nerve terminal. This type of synapse is called a **chemical synapse**.

The structure of chemical synapses

When an axon reaches its target cell, it loses its myelin sheath and ends in a small swelling known as a **nerve terminal** or **synaptic bouton** (see Figure 8.7). A nerve terminal together with the underlying membrane on the target cell constitutes a synapse. The nerve terminals contain mitochondria and a large number of small vesicles known as **synaptic vesicles**. The membrane immediately under the nerve terminal is called the **postsynaptic membrane**. It contains electron-dense material that makes it appear thicker than the plasma membrane outside the synaptic region. This is known as the postsynaptic thickening. The postsynaptic membrane contains specific receptor molecules for the neurotransmitter released by the nerve terminal. Between a nerve terminal and the postsynaptic membrane, there is a small gap of about 20 nm, which is known as the **synaptic cleft**.

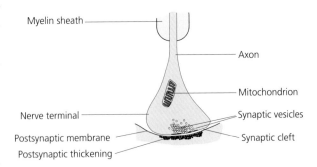

Myelin sheath
Axon
Mitochondrion
Synaptic vesicles
Nerve terminal
Synaptic cleft
Postsynaptic membrane
Postsynaptic thickening

Figure 8.7 A simple diagram to illustrate the main structural elements of a typical chemical synapse.

How does a chemical synapse work?

The transmission of information across a synapse occurs when an action potential reaches the presynaptic nerve terminal. The nerve terminal becomes depolarized and this depolarization causes voltage-gated calcium channels in the presynaptic membrane to open. Calcium ions flow into the nerve terminal down their electrochemical gradient and the resulting rise in free calcium triggers the fusion of one or more synaptic vesicles with the presynaptic membrane, resulting in the secretion of neurotransmitter into the synaptic cleft. Synaptic transmission is therefore an example of regulated secretion (see p. 64). In nerve terminals, this secretory process is extremely rapid and occurs within 0.5 ms of the arrival of the action potential. The secreted transmitter diffuses across the synaptic cleft and binds to receptors on the postsynaptic membrane. Subsequent events depend on the kind of receptor present.

Both excitatory and inhibitory synaptic transmission may be fast and short lasting, or slow and long lasting. If the transmitter activates a ligand-gated ion channel, synaptic transmission is usually both rapid and short-lived. This type of transmission is called **fast synaptic transmission** and is typified by the action of acetylcholine on the neuromuscular junction (see p. 129). If the transmitter activates a G protein-linked receptor, the change in the postsynaptic cell is much slower in onset and lasts much longer. This type of synaptic transmission is called **slow synaptic transmission**. A typical example is the excitatory action of noradrenaline (norepinephrine) on α_1-adrenoceptors in the peripheral blood vessels.

KEY POINTS:

- In response to an action potential, a nerve terminal secretes a neurotransmitter into the synaptic cleft.
- The secretion of neurotransmitter occurs by calcium-dependent exocytosis of synaptic vesicles.
- The neurotransmitter rapidly diffuses across the synaptic cleft and binds to specific receptors to excite the postsynaptic cell or inhibit its activity.

Nerve cells employ a wide variety of signalling molecules as neurotransmitters including acetylcholine, noradrenaline (norepinephrine), glutamate and γ-amino butyric acid (GABA). Chemical analysis of nerve terminals isolated from the brain has shown that they contain high concentrations of neurotransmitters, almost all of which is contained within synaptic vesicles.

Fast excitatory synaptic transmission occurs when a neurotransmitter (e.g. acetylcholine or glutamate) is released from the presynaptic nerve ending and is able to bind to and open non-selective cation channels. The opening of these channels causes a brief **depolarization** of the postsynaptic cell. This shifts the membrane potential closer to the threshold for action potential generation and so renders the postsynaptic cell more excitable. When the postsynaptic cell is a neurone, the depolarization is called an **excitatory postsynaptic potential** or **EPSP**. A single EPSP occurring at a fast synapse reaches its peak value within 1–5 ms of the arrival of the action potential in the nerve terminal and decays to nothing over the ensuing 20–50 ms (Figure 8.8).

How does activation of a non-selective cation channel lead to depolarization of the postsynaptic membrane? The explanation is as follows: the membrane potential is determined by the distribution of ions across the plasma membrane and the permeability of the membrane to those ions. At rest, the membrane is much more permeable to potassium ions than it is to sodium ions. The membrane potential (about -70 mV) is therefore close to the equilibrium potential for potassium (about -80 mV). If, however, the membrane was equally permeable to sodium and potassium ions, the membrane potential would be close to zero (i.e. the membrane would be depolarized) because the sodium equilibrium potential is around $+50$ mV. Consequently, when a neurotransmitter such as acetylcholine opens a non-selective cation channel in the postsynaptic membrane, the membrane depolarizes at the point of excitation. The exact value of the depolarization will depend on how many channels have been opened, as this will determine how far the membrane's permeability to sodium ions has increased relative to that of potassium. In neurones, single EPSPs rarely exceed a few mV, while at the neuromuscular junction the synaptic potential (known as the **end-plate potential**, see p. 130) has

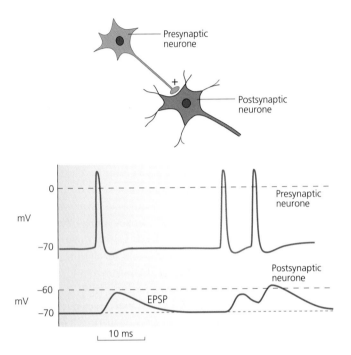

Figure 8.8 This diagram shows the relationship between an action potential in a presynaptic neurone and the generation of an excitatory postsynaptic potential (EPSP) in the postsynaptic neurone. Note that if two EPSPs occur in rapid succession, they summate and the final degree of depolarization is greater than either EPSP could achieve on its own (temporal summation). If the depolarization reaches threshold, an action potential will occur in the postsynaptic neurone.

an amplitude of about 40 mV. However, if an EPSP or sequence of EPSPs depolarizes the postsynaptic cell beyond the threshold, an action potential will be generated by the postsynaptic cell.

Fast inhibitory synaptic transmission occurs when a neurotransmitter such as GABA or glycine is released from a presynaptic nerve ending and is able to activate chloride channels in the postsynaptic membrane. The opening of these channels causes the postsynaptic cell to become **hyperpolarized** for a brief period. This negative shift in membrane potential is called an **inhibitory postsynaptic potential** or **IPSP** as the membrane potential is moved further away from threshold, making it harder for the cell to generate an action potential. A single IPSP occurring at a fast synapse reaches its peak value within 1–5 ms of the arrival of the action potential in the nerve terminal and decays to nothing within a few tens of milliseconds (Figure 8.9).

Why does the membrane hyperpolarize during an IPSP? The resting membrane potential is less (at about –70 mV) than the equilibrium potential for potassium ions (which is about –80 mV) as the membrane has a low permeability to sodium ions. The distribution of chloride ions across the plasma membrane, however, mirrors that of potassium so that the chloride equilibrium potential lies close to that of potassium. When an inhibitory neurotransmitter such as GABA opens chloride channels, the membrane permeability to chloride ions is increased and chloride ions can pass into the cell down their electrochemical gradient. This increase in chloride permeability shifts the membrane potential towards a more negative value that is closer to the chloride equilibrium potential (E_{Cl^-}). The extent of the hyperpolarization will depend on how much the chloride permeability has increased relative to that of both potassium and sodium. IPSPs are generally small in amplitude, about 1–5 mV. Nevertheless, they tend to last for tens of milliseconds and play an important role in determining the excitability of neurones.

EPSPs and IPSPs are not all-or-none events but are graded with the intensity of activation. Consequently, they can be superimposed on each other (a process called **summation**). If a synapse is activated repeatedly, the resulting synaptic potentials become superimposed (see Figure 8.8), a phenomenon called **temporal summation**. As individual neurones receive very many contacts, it is possible for two synapses on different parts of the

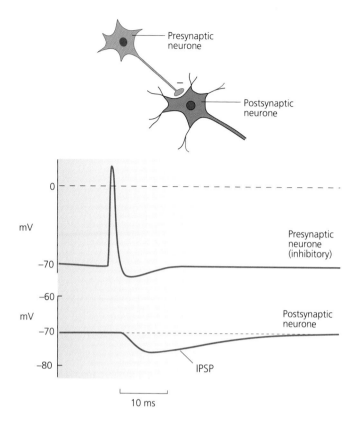

Figure 8.9 The relationship between an action potential in a presynaptic neurone and the generation of an inhibitory postsynaptic potential (IPSP) in the postsynaptic neurone. Note that in this case the membrane potential of the postsynaptic neurone becomes more negative (it hyperpolarizes) and thus moves further away from the threshold for action potential generation. (There is a decrease in the excitability of the postsynaptic neurone.)

cell to be activated at the same time. The resulting synaptic potentials also summate, but, because they were activated at different points on the cell, this type of summation is called **spatial summation**.

KEY POINTS:

- Activation of an excitatory synapse causes depolarization of the membrane potential, which is called an excitatory postsynaptic potential or EPSP. Activation of an inhibitory synapse leads to hyperpolarization of the membrane potential, which is called an inhibitory postsynaptic potential, or IPSP.

- EPSPs and IPSPs are graded in intensity. They greatly outlast the action potentials that initiated them. As a result, synaptic potentials can be superimposed on each other, leading to temporal and spatial summation.

Neurotransmitters are chemically diverse

As with other kinds of chemical signalling, synaptic signalling is mediated by a wide variety of substances. They can be grouped into six main classes:

1. Esters – acetylcholine (ACh)

2. Monoamines – such as noradrenaline, dopamine and serotonin

3. Amino acids – such as glutamate, GABA and glycine

4. Purines – such as adenosine and ATP

5. Peptides – such as the enkephalins, substance P and vasoactive intestinal polypeptide (VIP)

6. Inorganic gases – nitric oxide (NO).

Most of these transmitters activate ion channels or G-protein-linked receptors, both of which are targets for many drugs. Those acting on the nicotinic receptors of the neuromuscular junction are discussed in Chapter 9 and those affecting the autonomic nervous system are discussed in Chapter 13. Other drugs that act in a similar manner are discussed in the chapters relating to specific body systems.

What limits the duration of action of a neurotransmitter?

Neurotransmitters are highly potent chemical signals that are secreted in response to very specific stimuli. If the effect of a particular neurotransmitter is to be restricted to a particular synapse at a given time, there needs to be some means of terminating its action. This can be achieved in three different ways:

1. By rapid enzymatic destruction
2. By uptake either into the secreting nerve terminals or into neighbouring cells
3. By diffusion away from the synapse followed by enzymatic destruction, uptake, or both.

Of the currently known neurotransmitters, only acetylcholine is inactivated by rapid enzymatic destruction. It is hydrolysed to acetate and choline by the enzyme **acetylcholinesterase**. The acetate and choline are not effective in stimulating the cholinergic receptors but they may be taken up by the nerve terminal and resynthesized into new acetylcholine. The role of acetylcholine as a fast neurotransmitter at the neuromuscular junction is discussed in more detail on pp. 129–131.

The monoamines, such as noradrenaline (norepinephrine), are inactivated by uptake into the nerve terminals where they may be reincorporated into synaptic vesicles for subsequent release. Any monoamine that is not removed by uptake into a nerve terminal is metabolized either by **monoamine oxidase** or by **catechol-*O*-methyl transferase**. These enzymes are present in nerve terminals. They are also found in other tissues such as the liver. Peptide neurotransmitters become diluted in the extracellular fluid as they diffuse away from their site of action. They are subsequently destroyed by extracellular peptidases. The amino acids released in the process are taken up by the surrounding cells where they enter the normal metabolic pathways.

> **KEY POINTS:**
>
> - Many different kinds of chemical can serve as neurotransmitters.
> - Some nerve terminals secrete more than one kind of neurotransmitter and some neurotransmitters activate more than one kind of receptor at the same synapse.

Recommended reading

Pharmacology of synaptic transmission

Rang, H.P., Dale, M.M., Ritter, J.M. and Flower, R. (2007). *Pharmacology,* 6th edn. Churchill-Livingstone: Edinburgh. Chapters 31–33.

Physiology

Alberts, B., Johnson, A., Lewis, J., Raff, M., Roberts, K. and Walters, P. (2008). *Molecular Biology of the Cell*, 5th edn. Garland: New York and London. pp. 667–687.

Nicholls, J.G., Martin, A.R., Wallace B.G. and Fuchs, P.A. (2001). *From Neuron to Brain*, 4th edn. Sinauer: Sunderland, Massachusetts. Chapters 6–14.

Self-assessment questions

In Questions 1, 2 and 4, which of the statements are true and which are false? Answers are given at the end of the book (p. 755).

1 The axons of peripheral nerves:
 a) are protected by layers of connective tissue.
 b) are always associated with Schwann cells.
 c) are always covered by a layer of myelin.
 d) can only conduct action potentials in one direction.

2 The action potential of a single nerve fibre:
 a) is caused by a large change in the permeability of the membrane to sodium.
 b) is terminated when the sodium channels have inactivated.
 c) can summate with an earlier action potential.
 d) becomes larger as the stimulus is increased above threshold.

3 If a motor axon has a conduction velocity of $50\,\mathrm{m\,s^{-1}}$, how long will it take an action potential to reach a muscle that is $0.5\,\mathrm{m}$ from the cell body?
 a) $5\,\mathrm{ms}$
 b) $10\,\mathrm{ms}$
 c) $2\,\mathrm{ms}$
 d) $20\,\mathrm{ms}$

4 The following statements relate to excitatory and inhibitory synaptic transmission:
 a) During an EPSP, the membrane potential of the postsynaptic neurone always depolarizes.
 b) Unlike action potentials, EPSPs can summate to produce larger depolarizations of the postsynaptic membrane.
 c) The effects of a synaptic transmitter are always terminated by enzymatic destruction.
 d) Both EPSPs and IPSPs can result from activation of second messenger systems.
 e) During an IPSP, the postsynaptic membrane depolarizes.
 f) Synaptic transmission is always triggered by an influx of calcium into the presynaptic nerve terminal.

9 Muscle

After reading this chapter you should have gained an understanding of:

- The histological appearance of different muscle types
- The mechanism of muscle contraction
- The mechanical properties of skeletal muscle:
  - Twitch tension, summation, tetanus
  - Force–velocity relationships
  - Fast and slow muscles
  - Fatigue
- Neuromuscular transmission
- Drugs affecting the neuromuscular junction:
  - Neuromuscular blockers and anticholinesterases
- Smooth muscle contraction
- Drugs acting on smooth muscle:
  - Cholinergic and adrenergic drugs, eicosanoids, nitric oxide

9.1 Introduction

One of the distinguishing characteristics of animals is their ability to use co-ordinated movement to explore their environment. For large, multicellular animals, this movement is achieved by the use of muscles, which consist of cells that can change their length by a specific contractile process. In vertebrates, including man, three types of muscle can be identified on the basis of their structure and function. They are:

- Skeletal muscle
- Cardiac muscle
- Smooth muscle.

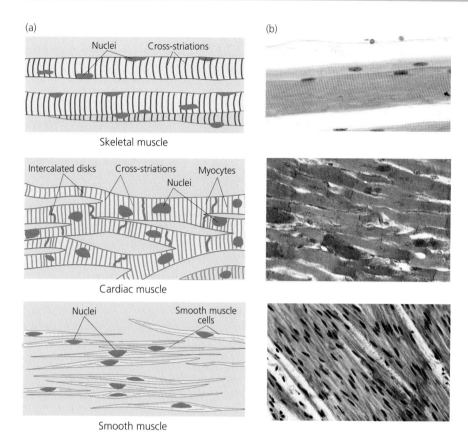

Figure 9.1 (a) Shows diagrammatic representations of the microscopic appearance of skeletal, cardiac and smooth muscle at similar magnification. These can be compared to the histological appearance of the same types of muscle shown in (b).

As its name implies, **skeletal muscle** is the muscle directly attached to the bones of the skeleton and its role is both to maintain posture and to move the limbs by contracting. **Cardiac muscle** is the muscle of the heart and **smooth muscle** is the muscle that lines the blood vessels and the hollow organs of the body. Together, the three kinds of muscle account for nearly half of body weight, the bulk of which is contributed by skeletal muscle (about 40% of total body weight).

Under normal circumstances, a skeletal muscle will only contract when the motor nerve to the muscle is activated. The signal to cause contraction of a skeletal muscle therefore originates in the CNS and the resulting contraction is said to be **neurogenic** in origin. In contrast, provided it is placed in a suitable nutrient medium, the heart will continue to beat spontaneously even when it is isolated from the body. Some smooth muscles behave in the same way. Contractions that arise from activity within the muscle itself are said to be **myogenic** in origin.

The cells of skeletal, cardiac and smooth muscle are known as **myocytes**. When the cells of skeletal and cardiac muscle are viewed down a microscope, they are seen to have characteristic striations – small, regular stripes running across the individual muscle cells. For this reason, skeletal and cardiac muscles are sometimes called striated muscles. Smooth muscle lacks striations and consists of sheets of spindle-shaped cells. The microscopic appearance of the different kinds of muscle is shown in Figure 9.1. Despite these differences in structure, the molecular basis of the contractile process is very similar for all types of muscle.

This chapter is chiefly concerned with the physiological properties of skeletal muscle and neuromuscular transmission. This is followed by a brief account of the properties of smooth muscle. The properties of cardiac muscle are discussed in Chapter 17 along with the general physiology of the heart.

9.2   The structure of skeletal muscle

Each skeletal muscle is made up of a large number of skeletal **muscle fibres**, which are long, thin, cylindrical cells that contain many nuclei. The length of individual muscle fibres varies according to the length of the muscle and ranges from a few millimetres to 10 cm or more. Their diameter also depends on the size of the individual muscle and ranges from about 50 to 100 μm. Despite their great length, few muscle fibres extend for the full length of a muscle. Individual muscle fibres are embedded in connective tissue called the **endomysium** and groups of muscle fibres are bound together by connective tissue called the **perimysium** to form bundles called muscle **fascicles**. Surrounding the whole muscle is a coat of connective tissue called the **epimysium** or fascia that binds the individual fascicles together. The connective tissue matrix of the muscle is secreted by fibroblasts that lie between the individual muscle fibres. It contains collagen and elastic fibres that merge with the connective tissue of the tendons which transmit the mechanical force generated by the muscle to the skeleton. Finally, within the body of a skeletal muscle there are specialized sense organs, known as muscle spindles, that play an important role in the regulation of muscle length (see Chapter 12, p. 204).

The individual muscle fibres are made up of filamentous bundles that run along the length of the fibre. These bundles are called **myofibrils** and have a diameter of about 1 μm. Each myofibril consists of a repeating unit known as a **sarcomere**. The alignment of the sarcomeres between adjacent myofibrils gives rise to the characteristic striations of skeletal muscle. The sarcomere is the fundamental contractile unit within skeletal and cardiac muscle. Each sarcomere is only about 2 μm in length so that each myofibril is made up of many sarcomeres placed end to end.

When a muscle fibre is viewed by polarized light, the sarcomeres are seen as alternating dark and light zones. The regions that appear dark do so because they refract the polarized light. This property is called anisotropy and the corresponding band is known as an **A band**. The light regions do not refract polarized light and are said to be isotropic. These regions are called **I bands**. Each I band is divided by a characteristic line known as a Z line and the unit between successive Z lines is a sarcomere. At high magnification in the electron microscope, the A bands are seen to be composed of thick filaments arranged in a regular order. The I bands consist of thin filaments. At the normal resting length of a muscle, a pale area can be seen in the centre of the A band. This is known as the **H zone**, which corresponds to the region where the thick and thin filaments do not overlap. In the centre of each H zone is the M line at which links are formed between adjacent thick filaments. The principal protein of the A bands is **myosin** while that of the I bands is **actin**. The interaction between these proteins is fundamental to the contractile process (see below). The main features of the structure of skeletal muscle are summarized in Figure 9.2.

Like all cells, skeletal muscle fibres are bounded by a cell membrane, which, in the case of muscle, is known as the **sarcolemma**. Beneath the sarcolemma lie the nuclei and many mitochondria. In mammalian muscle, narrow tubules run from the sarcolemma transversely across the fibre at the junction of the A and I bands. These are known as **T-tubules**. The lumen of each tubule is continuous with the extracellular space. Each myofibril is surrounded by the **sarcoplasmic reticulum**, which is a membranous structure homologous with the endoplasmic reticulum of other cell types. Where the T-tubules and the sarcoplasmic reticulum come into contact, the sarcoplasmic reticulum is enlarged to form the **terminal cisternae**. Each T-tubule is in close contact with the cisternae of two regions of sarcoplasmic reticulum and the whole complex is called a **triad** (see Figure 9.3). The T-tubules and triads play an important role in excitation–contraction coupling.

KEY POINTS:

- Skeletal muscle is made up of long, cylindrical, multinucleated cells called muscle fibres.

- Muscle fibres contain myofibrils, which are made up of repeating units called sarcomeres.

- Each sarcomere is separated from its neighbours by Z lines and consists of two half I bands, one at each end, separated by a central A band.

- The principal contractile proteins of the sarcomeres are actin and myosin.

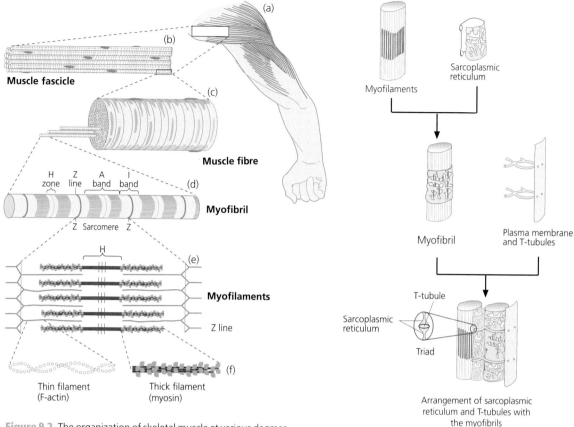

Figure 9.2 The organization of skeletal muscle at various degrees of magnification: (a) the appearance of the whole muscle; (b) the appearance of a bundle of muscle fibres (a muscle fascicle), with the nuclei of the muscle fibres shown in blue; (c) the appearance of a single muscle fibre; (d) the structure of an individual myofibril; (e) the appearance of the protein filaments that make up the individual sarcomeres; (f) the organization of actin and myosin in the thin and thick filaments. In (e) and (f), the head groups of the myosin molecules are shown in green.

Figure 9.3 The detailed organization of the T-system and sarcoplasmic reticulum of skeletal muscle. The sarcoplasmic reticulum is a membranous sac that envelops the myofibrils, and stores and releases calcium to regulate muscle contraction. The T-tubules have their origin at the sarcolemma (the plasma membrane of the individual muscle fibres) and form a network that crosses the muscle fibre. At the junction of the A and I bands, the T-tubules come into close apposition with enlargements of the sarcoplasmic reticulum (the terminal cisternae) to form the triads.

9.3 How does a skeletal muscle contract?

Skeletal muscle, like nerve, is an excitable tissue and stimulation of a muscle fibre at one point will rapidly lead to excitation of the whole cell. In the body, each skeletal muscle fibre is innervated by one motoneurone. As it enters a muscle, the axon branches and each branch supplies a single muscle fibre. An action potential in the motoneurone excites each of the muscle fibres to which it is connected and triggers an action potential that prop-agates along the whole length of the muscle fibre. This action potential is followed by contraction of the muscle fibre. The process by which a muscle action potential triggers a contraction is known as **excitation–contraction coupling**.

What steps link the muscle action potential to the contractile response? Significant amounts of Ca^{2+} are stored in the sarcoplasmic reticulum and it is now thought that

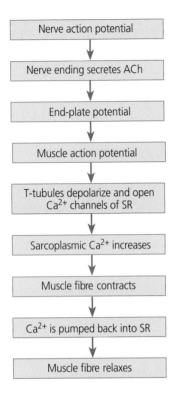

Figure 9.4 A flow diagram to illustrate the sequence of events leading to the contraction and subsequent relaxation of a skeletal muscle fibre. SR, sarcoplasmic reticulum; ACh, acetylcholine.

the depolarization of the plasma membrane during the muscle action potential spreads along the T-tubules where it causes Ca^{2+} channels in the sarcoplasmic reticulum to open. As a result, the stored Ca^{2+} is released and the level of Ca^{2+} in the sarcoplasm rises. This rise in Ca^{2+} triggers the contraction of the muscle fibre. Relaxation of the muscle occurs as the Ca^{2+} in the sarcoplasm is pumped back into the sarcoplasmic reticulum by a Ca^{2+} pump of the kind described in Chapter 5 (p. 59). These events are summarized in Figure 9.4.

How is tension generated and how does a rise in Ca^{2+} lead to the contractile response? All muscles contain two proteins – **actin** and **myosin**. In skeletal and cardiac muscle, the thick filaments are chiefly composed of myosin while the thin filaments contain actin (the principal protein) and lesser quantities of two other proteins known as troponin and tropomyosin that only allow actin and myosin to interact if the sarcoplasmic Ca^{2+} is increased above its resting level.

Under these conditions, each actin molecule in the chain is able to bind one myosin head region. Actin and myosin molecules dissociate when a molecule of ATP is bound by the myosin. The breakdown of ATP and the subsequent release of inorganic phosphate cause a change in the angle of the head region of the myosin molecule that enables it to move relative to the thin filament (Figure 9.5). Once again, ATP causes the dissociation between the actin and myosin and the cycle is repeated. This process is known as cross-bridge cycling and results in the thick and thin filaments sliding past each other, so shortening the fibre. This is known as the **sliding filament theory** of muscle contraction. The formation of cross-bridges between the actin and myosin head groups is not synchronized across the myofibril. While some myosin head groups are dissociating from the actin, others are binding or developing their power stroke. Consequently, tension is developed smoothly.

The sliding filament theory of muscle contraction explains why the tension developed by a muscle depends on its resting length (Figure 9.6). When the muscle is at its natural resting length, the thin and thick filaments overlap optimally and form the maximum number of cross-bridges. When the muscle is stretched, the degree of overlap between the thin and the thick filaments is reduced and the number of cross-bridges falls. This leads to a decline in the ability of the muscle to generate tension. When the muscle is shorter than its natural resting length, the thin filaments already fully overlap the thick filaments, but the filaments from each end of the sarcomere touch in the centre of the A band and each interferes with the motion of the other. As a result, tension development declines. When the thin and thick filaments fully overlap, the A bands abut the Z lines and tension development is no longer possible.

The role of ATP and creatine phosphate

At rest, a skeletal muscle is plastic and can readily be stretched. In this state, although the myosin head groups have bound ATP, there are no cross-bridges being formed between the thick and thin filaments because the troponin complex prevents the interaction between actin and myosin. When ATP levels fall to zero after death, the cross-bridges between the actin and myosin do not

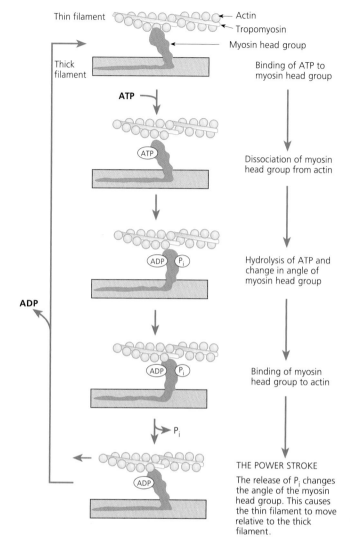

Figure 9.5 A schematic representation of the molecular events responsible for the relative movement of the thin and thick filaments of a striated muscle. Tropomyosin is a regulatory protein present on the thin filaments that regulates the binding of actin and myosin.

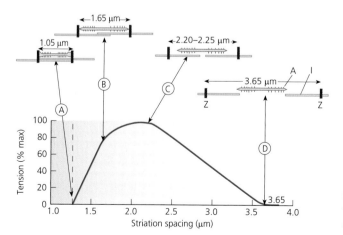

Figure 9.6 The sliding filament theory of muscle contraction and the length–tension relationship of a single skeletal muscle fibre. The sequence A–D represents the overlap between the thin and thick filaments as the resting length of a single skeletal muscle fibre is increased. When the thin and thick filaments fully overlap and the A band is compressed against the Z line (point A), the muscle cannot shorten further and is unable to develop tension. As the fibre is stretched so that the thin and thick filaments overlap without compressing the A band, active tension is generated when the muscle is stimulated (point B). Further stretching provides optimal overlap between the thin and thick filaments leading to maximal development of tension (point C). If the fibre is stretched to such a degree that the thin and thick filaments no longer overlap, there is no tension developed (point D).

dissociate. The muscles lose their plasticity and become stiff – a state known as **rigor mortis**.

The energy for contraction is derived from the hydrolysis of ATP. As with other tissues, the ATP is derived from the oxidative metabolism of glucose and fats (see Chapter 4, pp. 47–50). During the contractile cycle, however, it is important for the levels of ATP to be maintained. This need is met by **creatine phosphate** (also known as **phosphocreatine**), which is present in muscle at high concentrations (about 15–20 mM) and has a phosphate group that is readily transferred to ATP. This reaction is catalysed by the enzyme **creatine kinase**:

Creatine phosphate + ADP + H$^+$ ⇌ creatine + ATP

During heavy exercise, there may be insufficient oxygen delivered to the exercising muscles for oxidative metabolism. In this situation, the generation of ATP from glucose and fats via the tricarboxylic acid cycle is compromised and ATP is generated from glucose via the glycolytic pathway instead. This anaerobic phase of muscle contraction is much less efficient in generating ATP (see pp. 47–48) and hydrogen ions, lactate and phosphate ions are produced in increasing quantities. As these ions accumulate, muscle pH falls, the muscular effort becomes progressively weaker and the muscle relaxes more slowly. This is known as **fatigue**. During muscle fatigue, the muscle fibres remain able to propagate action potentials but their ability to develop tension is impaired.

KEY POINTS:

- The link between the electrical activity of a muscle and the contractile response is called excitation–contraction coupling in which the action potential triggers an increase in Ca^{2+} around the myofibrils.

- The A band of each sarcomere consists mainly of myosin molecules arranged in thick filaments and the I band consists mainly of actin. During contraction, the thick and thin filaments slide past each other.

- The energy for muscle contraction is provided by the hydrolysis of ATP.

9.4   The activation and mechanical properties of skeletal muscle

The innervation of skeletal muscles

The neurones that are directly responsible for controlling the activity of muscle fibres are known as **motoneurones** and the nerves that transmit signals from the CNS to the skeletal muscles are known as **motor nerves**. As a motor nerve enters the muscle, it branches to supply different groups of muscle fibres. Within the motor nerve bundle, individual axons also branch, so that one motor fibre makes contact with a number of muscle fibres. When a motoneurone is activated, all the muscle fibres it supplies contract in an all-or-none fashion so they act as a single unit – a **motor unit**. The size of motor units varies from muscle to muscle according to the degree of control required. Where fine control of a movement is not required, the motor units are large. Thus, in the gastrocnemius muscle of the calf, a motor unit may contain up to 2000 muscle fibres. In contrast, the motor units of the extraocular muscles (which control the direction of the gaze) are much smaller (as few as 6–10 muscle fibres being supplied by a single motoneurone). The process by which a motoneurone activates a skeletal muscle (neuromuscular transmission) is considered later in this chapter (p. 129).

The mechanical properties of skeletal muscle

Like nerve, skeletal muscle is an excitable tissue that can be activated by direct electrical stimulation. When a muscle is activated, it shortens and, in doing so, exerts a force on the tendons to which it is attached. The amount of force exerted depends on many factors:

1. The number of active muscle fibres
2. The frequency of stimulation

3. The rate at which the muscle shortens

4. The initial resting length of the muscle

5. The cross-sectional area of the muscle.

Consider first the situation where the muscle is being artificially activated by a single electrical shock to its motor nerve. The response of a muscle to a single stimulus is called a **muscle twitch** and the force developed is called the **twitch tension**. If the electrical shock is weak, only a small proportion of the nerve fibres will be activated. Consequently, only a small number of the muscle fibres will contract and the amount of force generated will be weak. If the intensity of the stimulus is increased, more motor fibres will be recruited and there will be a corresponding increase both in the number of active muscle fibres and in the total tension developed. When all the muscle fibres are activated, the total tension developed will reach its maximum. A stimulus that is sufficient to activate all the motor fibres supplying a muscle is called a **maximal stimulus**.

Muscles do not elongate when they relax unless they are stretched. In the body, this stretching is achieved for skeletal muscles by the arrangement of muscle pairs acting on particular joints. These pairs of muscles are known as **antagonists**.

The force developed by a muscle increases during repetitive activation

When a muscle fibre is activated, the process of contraction begins with an action potential passing along its length. This is followed by the contractile response, which consists of an initial phase of acceleration during which the tension in the muscle increases until it equals that of the load to which it is attached. The fibre then begins to shorten. At first, the fibre shortens at a relatively constant rate, but, as the muscle continues to contract, the rate of shortening progressively falls – the phase of deceleration. Finally, the muscle relaxes and the tension it exerts declines to zero. The whole cycle of shortening and relaxation takes several tens of milliseconds while the muscle action potential is over in a matter of two or three milliseconds. Consequently, the mechanical response greatly outlasts the electrical signal that initiated it (Figure 9.7).

If a muscle is activated by a pair of stimuli that are so close together that the second stimulus arrives before the muscle has fully relaxed after the first, the total tension developed following the second stimulus is greater than that developed in response to a single stimulus. This increase in developed tension is called **summation** and is illustrated in Figure 9.8b. The degree of summation is at its maximum with short intervals and declines as the interval between the stimuli increases. If a number of stimuli are given in quick succession, the tension developed progressively summates and the response is called a **tetanus**. At low frequencies, the tension oscillates at the frequency of stimulation as shown in Figure 9.8c, but, as the frequency of stimulation rises, the development of tension proceeds more and more smoothly. When tension develops smoothly (Figure 9.8d), the contraction is known as a **fused tetanus**.

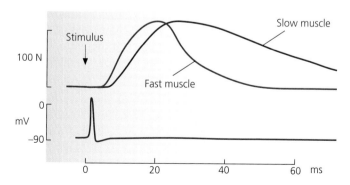

Figure 9.7 The time course of the contractile response (a twitch) in response to a single action potential for a fast and a slow skeletal muscle. Note that the contractile response begins after the action potential and lasts much longer. Force is expressed as newtons (N).

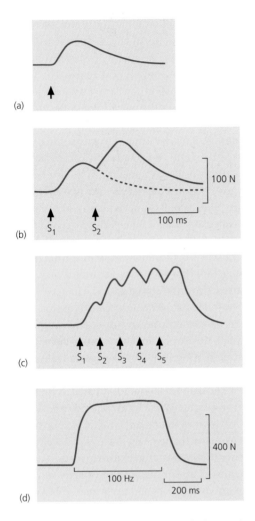

Figure 9.8 The summation and fusion of the mechanical response of a muscle in response to repetitive stimulation. (a) The response to a single electrical stimulus (arrow). (b) The summation of tension when a second stimulus (S_2) is given before the mechanical response to the first has decayed to zero. (c) The summation of tension during a rapid sequence of stimuli. (d) The smooth development of tension during a high-frequency train of stimuli (giving a fused tetanus). Note the different time and tension calibrations for (a) and (b) compared with (c) and (d).

Why is a muscle able to develop more tension when two or more stimuli are given in rapid succession? Consider first how tension develops during a single twitch. When the muscle is activated, it must transmit the tension developed to the load. To do this, the tension of the tendons and of the connective tissue of the muscle itself must be raised to that of the load. As the tendons are to some degree elastic, they need to be stretched a little until the tension they exert on the load is equal to that of the muscle. This takes a short, but finite, amount of time during which the response of the tension-generating machinery begins to decline. The transmission of force to the load is accordingly reduced in efficiency. This also accounts for the initial acceleration in the rate of contraction described above. During a train of impulses, however, the contractile machinery is activated repeatedly and the tension in the tendons has little time to decay between successive contractions. Transmission of tension to the load is more efficient and a greater tension is developed.

Fast- and slow-twitch muscle fibres

Different types of muscle show different rates of contraction and susceptibility to fatigue. Broadly speaking, skeletal muscle fibres can be classified as slow (type 1) or fast (type 2) according to their rate of contraction. Most muscles contain a mixture of both kinds of fibre but some muscles have a predominance of one type.

Slow muscles contract at about $15\,\mathrm{mm\,s^{-1}}$ and relax relatively slowly. They are composed of thin fibres, rich in both mitochondria and the oxygen-binding protein myoglobin. This gives them a reddish appearance. They rely mainly on oxidative metabolism of fats for their energy supply and have a rich blood supply, which makes them very resistant to fatigue. Slow muscles play an important role in the maintenance of posture. They are activated by their motoneurones at a continuous, steady rate, which enables them to maintain a steady muscle tone. Fast muscles shorten at about $40–45\,\mathrm{mm\,s^{-1}}$ and relax relatively quickly. However, as they have a limited blood supply, few mitochondria and little myoglobin, they are easily fatigued. They are therefore well adapted to provide short periods of high tension development suitable for short periods of intense muscular activity (e.g. sprinting). Their relative lack of mitochondria and myoglobin gives them a pale appearance.

Asynchronous activation of motor units allows muscle tension to develop smoothly

In the absence of organic disease, the development of tension during normal movement is smooth and progressive. This arises because the CNS recruits motoneurones progressively. Consequently, the motor units comprising a muscle are activated at different times – very unlike the experimental situations described in the previous sections. Individual motor units may be activated by relatively low frequencies of stimulation but the maintenance of a steady tension ensures the efficient transmission of force to the load. The ability of a muscle to maintain a smooth contraction is further enhanced by the presence of both fast and slow muscle fibres in most muscles.

The power of a muscle depends on the rate at which it shortens

The rate at which a muscle can shorten depends on the load against which it acts. If there is no external load, a muscle shortens at its maximum rate. With progressively greater loads, the rate of shortening decreases until the load is too great for the muscle to move. The relationship between the load imposed on a muscle and its rate of shortening is known as the **force–velocity curve** (Figure 9.9). If a muscle contracts against a load that prevents shortening, the muscle is said to undergo an **isometric contraction** while if it shortens against a constant load it is said to undergo an **isotonic contraction**. Thus, isometric contraction and isotonic contraction with no external load represent the extreme positions of the relationship between the force developed by a muscle and the rate at which it shortens.

The work of a muscle is determined by the distance it is able to move a given load and the power of a muscle is the rate at which it performs work (Box 9.1). Thus:

$$\text{Power} = \text{Force} \times \text{Velocity}$$

From the curve relating the velocity of shortening to the power developed (Figure 9.9), it is clear that the power developed by a muscle passes through a definite maximum. When a muscle shortens isometrically, it does no work, as the load is not moved through a distance. Consequently, no power is developed. Equally, if the muscle contracts while it is not acting on an external load, no work is done and no power is developed. In between these two extremes, the muscle performs useful work and develops power. In general, the greatest power is developed when the muscle is shortening at about one-third of its maximum rate.

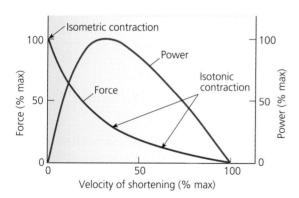

Figure 9.9 The force–velocity relation for a skeletal muscle. Note that maximum force is developed during an isometric contraction but maximum speed of shortening occurs in an unloaded muscle. Maximum power is developed when the muscle shortens at about one-third of maximum velocity.

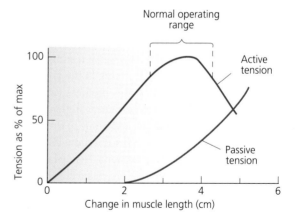

Figure 9.10 Isometric force–tension relationship at different muscle lengths. The data are for a human triceps muscle, which is about 20 cm in length. As the muscle is stretched, passive tension increases. Active tension increases from zero to a maximum and then declines with further stretching. The total tension developed during a contraction is the sum of the active and passive tensions.

Box 9.1 The efficiency and power of muscles

It is a matter of common experience that it is more difficult to move heavy objects than light ones, but how efficient are muscles in converting chemical energy into useful work?

The force exerted on a given load is defined as:

$$\text{Force} = \text{Mass} \times \text{Acceleration} \qquad (B1)$$

and is given in **newtons** (N). One newton is the force that will give a mass of 1 kg an acceleration of $1\,\mathrm{m\,s^{-1}}$.

The work performed on a load is the product of the load and the distance through which the load is moved. Thus:

$$\text{Work} = \text{Force} \times \text{Distance} \qquad (B2)$$

The unit for work is therefore N.m and one N.m is a **joule** (J).

Power is defined as the capacity to do work or the work per unit time and is expressed in $\text{joules\,s}^{-1}$ or **watts** (W).

$$\text{Power} = \text{Work/Time}$$

$$= \text{Force} \times \text{Distance/Time}$$

$$= \text{Force} \times \text{Velocity} \qquad (B3)$$

The key to understanding the power and efficiency of a muscle is its force–velocity curve. For an isometric contraction, maximum force is exerted but the load is not moved so that no work is done and the power is also zero. When the muscle shortens without a load, no useful work is done. Between these two extremes, the work is given by Equation B2 above and the power by Equation B3. The power is usually at a maximum when the muscle is shortening at about one-third of the maximum possible rate (Figure 9.9).

The mechanical efficiency of muscular activity or work is expressed as a percentage of work done relative to the increase in metabolic rate attributable to the activity of the muscles employed in the task.

$$\text{Efficiency} = (\text{Work done/Energy expended in task}) \times 100 \qquad (B4)$$

For our examples above, both isometric contraction and contraction with no external load have zero efficiency. When a muscle does external work, for example walking up stairs or cycling, its efficiency is about 20–25%.

The effect of muscle length on the development of tension

If the force generated by a muscle during isometric contraction is measured for different initial resting lengths, a characteristic relationship is found: in the absence of stimulation, the tension increases progressively as the muscle is stretched beyond its normal resting length. This is known as the **passive tension** and is due to the stretching of the muscle fibres themselves, the connective tissue of the muscle and that of the tendons. The extra tension developed as a result of stimulation (called the **active tension**) is at its maximum when the muscle is close to its resting length (i.e. the length it would have had in its resting state in the body). If the muscle is stimulated when it is shorter than normal, it develops less tension and, if it is stretched beyond its normal resting length, the tension developed during contraction is also less than normal. Overall, the relationship between the initial length of a muscle and the active tension is described by a bell-shaped curve, as shown in Figure 9.10, that is very simi-

lar to the length–tension relationship seen for individual skeletal muscle fibres (Figure 9.6). In the body, the range over which a muscle can shorten is determined by the anatomical arrangement of the joint on which it acts. The length of muscles attached to the skeleton range from 0.7 to 1.2 times their normal resting length.

Effect of cross-sectional area on the power of muscles

The force generated by a single skeletal muscle fibre does not depend on its length but on its cross-sectional area (which determines the number of myofibrils that can act in parallel). This can be seen if we consider the force generated by a myofibril consisting of two sarcomeres. The force generated by the two central half sarcomeres cancel out as one pulls to the left and the other to the right. Consequently, it is the two end half sarcomeres that generate useful force. This remains true whether the myofibril has a hundred or a thousand sarcomeres arranged in series.

Thicker fibres have more myofibrils arranged in parallel. They therefore develop more tension. When muscles hypertrophy (i.e. enlarge) in response to training, the number of muscle fibres does not increase. Rather, there is an increase in the number of myofibrils in the individual fibres and this leads to an increase in their cross-sectional area. The difference in strength between individuals is due to the difference in cross-sectional area of the individual muscles.

The power of a muscle is equal to the force generated multiplied by the rate of contraction (see above). As the force of contraction depends on the cross-sectional area of the muscle and the rate of contraction depends on the length of the muscle, the power of a muscle is proportional to its volume. A short, thick muscle will therefore develop the same power as a long, thin one of the same volume. The thick muscle will develop more force but will shorten more slowly than the long thin one.

KEY POINTS:

- The nerves supplying a skeletal muscle are known as motor nerves. They are myelinated and individual axons branch to make contact with a number of muscle fibres. A motoneurone and its associated muscle fibres are called a **motor unit**.

- A skeletal muscle contracts in response to an action potential in its motor nerve. A single action potential gives rise to a contractile response called a twitch. During repeated activation of a muscle, the tension summates and the muscle is said to undergo tetanic contraction.

- The force developed by a muscle depends on the number of active motor units, its cross-sectional area and the frequency of stimulation.

9.5 Neuromuscular transmission

The region of contact between a motor axon and a muscle fibre is called the **neuromuscular junction** or **motor end plate** and the process of transmitting a signal from a motor nerve to a skeletal muscle to cause it to contract is called **neuromuscular transmission.**

Motor axons are myelinated but as they enter a muscle they lose the myelin sheath and send branches to separate muscle fibres. Each axon terminal has a large number of synaptic vesicles containing the neurotransmitter **acetylcholine**. Beneath the axon terminal, the muscle membrane is thrown into elaborate folds known as junctional folds. This is the postsynaptic region of the muscle fibre membrane and it contains the nicotinic receptors that bind the acetylcholine. Finally, as with other chemical synapses, there is a small gap of about 20 nm, the synaptic cleft, separating the nerve membrane from that of the muscle fibre. The synaptic cleft contains acetylcholinesterase, which rapidly inactivates acetylcholine by breaking it down to acetate and choline. A diagram of the neuromuscular junction is shown in Figure 9.11.

The neuromuscular junction operates in a similar way to other chemical synapses:

- First, an action potential in the motor axon invades the nerve terminal and depolarizes it.

- This depolarization opens voltage-gated calcium channels in the membrane of the axon terminal, allowing calcium ions to flow into the nerve terminal down their electrochemical gradient.

- This leads to a local rise in free calcium within the terminal, which triggers the fusion of synaptic vesicles with the plasma membrane.

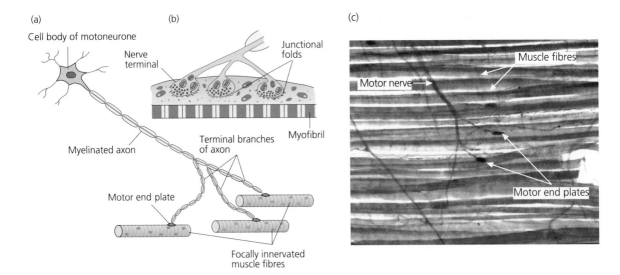

Figure 9.11 This diagram shows how motor nerves innervate skeletal muscle (a) and a simplified drawing of the neuromuscular junction (b). A motoneurone in the CNS gives rise to a motor axon, which branches to supply a number of muscle fibres. The motoneurone and the muscle fibres it supplies form a single motor unit. (c) A photomicrograph of a mammalian skeletal muscle stained to show the motor nerve and its endings on the muscle fibres (motor end plates).

- The acetylcholine contained within the synaptic vesicles is released into the synaptic cleft and diffuses across the cleft.

- When the receptors bind acetylcholine, non-selective cation channels open and this depolarizes the muscle membrane in the end-plate region to give rise to the **end-plate potential** or EPP, which is analogous to the **EPSP** of nerve cells.

- Finally, when the EPP has reached threshold, the muscle membrane generates an action potential that propagates along the length of the fibre. This action potential triggers the contraction of the muscle fibre.

After the acetylcholine has activated the nicotinic receptors, it is rapidly broken down by **acetylcholinesterase** to acetate and choline. This enzymatic breakdown limits the action of acetylcholine to a few milliseconds. The acetate and choline can be taken up by the nerve terminal and resynthesized into acetylcholine for recycling as a neurotransmitter. If acetylcholinesterase is inhibited by a specific blocker such as **eserine**, activation of the motor nerve leads to a maintained depolarization of the muscle and neuromuscular transmission is blocked.

The muscle nicotinic receptors can be blocked by a number of specific drugs and poisons (called **neuromuscular blockers**), the best known of which is **curare** – an arrow poison used by South American Indians for hunting. Neuromuscular blockers such as **d-tubocurarine** and **succinylcholine** are now routinely used in conjunction with general anaesthetics to prevent muscle contractions and so provide complete muscular relaxation during complex surgery. As the respiratory muscles also rely on acetylcholine to trigger their contractions, this way of providing muscle relaxation is only possible while the patient's breathing is being supported by artificial respiration.

Paralysis of a skeletal muscle leads to its atrophy

If the motor nerve to a muscle is cut or crushed, the muscle loses its nerve supply (it is said to be denervated) and becomes paralysed. After a time, the nerve terminals degenerate and disappear. The muscle becomes weak and atrophies. Functional denervation occurs in a disease of the neuromuscular junction called **myasthenia gravis**. Patients with this disease make antibodies against their own nicotinic receptors. The circulating antibodies bind to the nicotinic receptors of the neuromuscular junction and so reduce the number available for neuromuscular transmission. The decline in the number of active receptors leads to a progressive failure of neuromuscular transmission and there is progressive paralysis of the affected muscles, particularly those supplied by the cranial nerves – drooping of the eyelids is a characteristic early sign. The symptoms of the disease can be ameliorated by giving the sufferer a low dose of an inhibitor of acetylcholinesterase (an anticholinesterase, e.g. **neostigmine**) to prolong the activity of the acetylcholine.

KEY POINTS:

- During neuromuscular transmission, the motor nerve endings release acetylcholine.
- Acetylcholine activates nicotinic receptors in the junctional region of the muscle membrane and causes a depolarization known as an end-plate potential.
- The end-plate potential triggers an action potential in the muscle membrane that leads to contraction of the muscle.
- The effect of acetylcholine is terminated by the enzyme acetylcholinesterase.
- Neuromuscular transmission can be blocked by drugs such as curare that compete with acetylcholine for binding sites on the nicotinic receptors.

9.6 Smooth muscle

Smooth muscle is the muscle of the internal organs such as the gut, blood vessels, bladder and uterus. It forms a heterogeneous group with a range of physiological properties. Each type of smooth muscle is adapted to serve a particular function. In some cases, the muscle must maintain a steady contraction for long periods of time and then rapidly relax (as in the case of the sphincter muscles that control the emptying of the bladder and rectum). In others, the muscles are constantly active (such as those of the stomach and small intestine).

A particular type of smooth muscle may express different properties at different times, as in the case of uterine muscle, which must be quiescent during pregnancy but contract forcefully during labour. It is therefore not surprising that the smooth muscle serving a specific function will have distinct properties. For this reason, rather than classifying smooth muscles into particular types, it is more useful to determine how the properties of a specific muscle are adapted to serve its particular function.

Each smooth muscle consists of sheets of many small, spindle-shaped cells (Figure 9.1) linked together by two types of junctional contact, as shown in Figure 9.12. These are mechanical attachments between neighbouring cells and gap junctions, which provide electrical continuity between cells and thus provide a pathway for the passage of electrical signals between cells. Each smooth muscle cell has a single nucleus, is about $2–5\,\mu m$ in diameter at its widest point and about $50–200\,\mu m$ in length. In some tissues – such as the alveoli of the mammary gland and in some small blood vessels – the smooth muscle cells are arranged in a single layer known as **myoepithelium**.

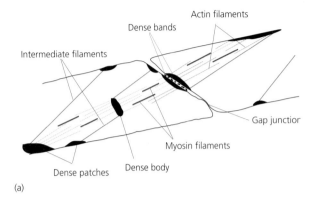

(a)

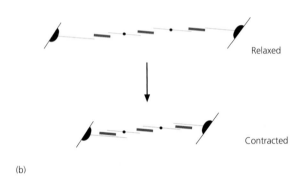

Relaxed

Contracted

(b)

Figure 9.12 The organization of the contractile elements of smooth muscle fibres. (a) Note the points of close contact for mechanical coupling (dense bands) and the gap junction for electrical signalling between cells. (b) A simple model of the contraction of smooth muscle. As the obliquely running contractile elements contract, the muscle shortens.

Myoepithelial cells have broadly similar physiological properties to other smooth muscle cells.

In smooth muscle, no cross-striations are visible under the microscope (hence its name), but, like skeletal and cardiac muscle, smooth muscle contains actin and myosin filaments. These are not arranged in a regular manner like those of skeletal muscle, but are arranged in a loose lattice with the filaments running obliquely across the smooth muscle cells, as shown in Figure 9.12a.

In smooth muscle, there are many more of the thin actin filaments than thick myosin filaments. In addition to actin and myosin filaments, smooth muscle cells also contain **intermediate filaments** that assist in transmission of the force generated during contraction to the neighbouring smooth muscle cells and connective tissue. While there are no Z lines in smooth muscle, they have a functional counterpart in dense bodies that are distributed throughout the cytoplasm and which serve as attachments for both the thin and intermediate filaments. The filaments of contractile proteins are attached to the plasma membrane at the dense bands that link neighbouring cells, as shown in Figure 9.12. Unlike skeletal muscle, smooth muscle cells do not have a T-system and the sarcoplasmic reticulum is not as extensive. A comparison of some of the properties of smooth, cardiac and skeletal muscle is given in Table 9.1.

Smooth muscle is innervated by fibres of the autonomic nervous system, which have swellings called varicosities along their length. These varicosities correspond to the nerve endings of the motor axons of the neuromuscular junction (see above). In some tissues, each varicosity is closely associated with an individual muscle cell (e.g. the piloerector muscles of the hairs) while in others, the axon varicosities remain in small bundles within the bulk of the muscle and are not closely associated with individual fibres (e.g. the smooth muscle of the gut). The varicosities release their neurotransmitter into the space surrounding the muscle fibres rather than onto a clearly defined synaptic region as is the case at the neuromuscular junction of skeletal muscle. The neurotransmitter receptors are distributed over the surface of the cells instead of being concentrated at one region of the membrane as they are at the motor end-plate.

In many tissues, particularly those of the viscera, the individual smooth muscle cells are grouped loosely into clusters that extend in three dimensions. Gap junctions connect the cells so that the whole muscle behaves as a functional syncytium. In this type of muscle, activity originating in one part spreads throughout the rest of the muscle. This is known as **single-unit smooth muscle**. The smooth muscle of the gut, uterus (Figure 9.13) and bladder are good examples of single-unit smooth muscle. In some tissues, such as the gut, there are regular spontaneous contractions (**myogenic contractures**) that originate in specific pacemaker areas.

The activity of many single-unit muscles is strongly influenced by hormones circulating in the bloodstream

Table 9.1 A comparison of the properties of cardiac, skeletal and smooth muscle

Property	Skeletal muscle	Cardiac muscle	Smooth muscle
Cell characteristics	Very long cylindrical cells with many nuclei	Irregular rod-shaped cells, usually with a single nucleus	Spindle-shaped cells with a single nucleus
Maximum cell size (length × diameter)	30 cm × 100 μm	100 μm × 15 μm	200 μm × 5 μm
Visible striations	Yes	Yes	No
Myogenic activity	No	Yes	Yes
Motor innervation	Somatic	Autonomic (sympathetic and parasympathetic)	Autonomic (sympathetic and parasympathetic)
Type of contracture	Phasic	Rhythmic	Mostly tonic, some phasic
Basis of muscle tone	Neural activity	None	Intrinsic and extrinsic factors
Cells electrically coupled	No	Yes	Yes
T-system	Yes	Only in ventricular muscle	No
Mechanism of excitation–contraction coupling	Action potential and T-system	Action potential and T-system	Action potential, Ca^{2+} channels and second messengers
Force of contraction regulated by hormones	No	Yes	Yes

Note: The properties of cardiac muscle are discussed in Chapter 17.

as well as by the activity of autonomic nerves. For example, during pregnancy the motor activity of the uterine muscle (the myometrium) is much reduced due to the presence of high circulating levels of the hormone progesterone.

Certain smooth muscles do not contract spontaneously and are normally activated by motor nerves. The muscles themselves are organized into motor units similar to those of skeletal muscle except that the motor units are more diffuse. These muscles are known as **multiunit smooth muscle**. The smooth muscle of the iris, the piloerector muscles of the skin and the smooth muscle of the larger blood vessels are all examples of the multiunit type. Nevertheless, the distinction between the two types of muscle is not

rigid as, for example, the smooth muscle of certain arteries and veins shows spontaneous activity but also responds to stimulation of the appropriate sympathetic nerves.

Excitation–contraction coupling in smooth muscle

The membrane potential of smooth muscle is often quite low – typically about −50 to −60 mV, which is some 30 mV more positive than the potassium equilibrium potential. This low value of the resting membrane potential arises because the sodium ion permeability of the cell membrane is about one-fifth that of potassium (compared with a $Na^+ : K^+$ permeability ratio

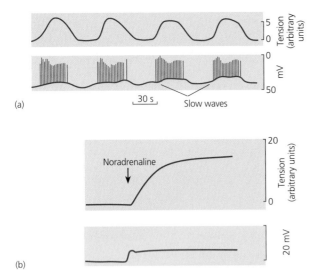

(a)

30 s   Slow waves

Noradrenaline
↓

(b)

Figure 9.13 Patterns of electrical activity recorded from different kinds of smooth muscle. (a) The electrical and mechanical response of uterine smooth muscle isolated from a rat giving birth. Note the slow waves leading to bursts of action potentials and the slow and sustained development of tension. (b) The development of tension in a sheep carotid artery following the application of noradrenaline. Note that in this case the development of tension is not preceded by an action potential.

of about 1:100 for skeletal muscle). When a smooth muscle fibre generates an action potential, the depolarization depends on an influx of both sodium and calcium ions, although the exact contribution of each ion depends on the individual muscle. For example, in the smooth muscle of the vas deferens and the gut, the action potential appears to be mainly dependent on an influx of calcium ions. In contrast, the action potentials of the smooth muscle of the bladder and ureters depends on an influx of sodium ions in just the same way as the action potential of skeletal muscle. Unlike skeletal muscle, however, the action potential of this smooth muscle lasts 10–50 ms (i.e. five to ten times as long). In addition, in some smooth muscles the action potential may develop a prolonged plateau phase similar in appearance to that seen in cardiac muscle (see p. 336).

In single-unit smooth muscle, certain cells act as pacemaker cells and these show spontaneous fluctuations of the membrane potential known as slow waves. During an excitatory phase, slow-wave activity builds up progressively until the membrane potential falls below about –35 mV when a series of action potentials is generated. These are propagated through many cells via gap junctions and the muscle slowly contracts. This pattern of electrical activity and force generation is seen in the gut during peristalsis and in uterine muscle during parturition (Figure 9.13a).

In many smooth muscles, the pacemaker activity is regulated by the activity of the sympathetic and parasympathetic nerves. In the intestine, the release of acetylcholine from the parasympathetic nerve varicosities makes the muscle more active. Conversely, noradrenaline secreted by the sympathetic nerves inhibits contractile activity. Neither acetylcholine nor noradrenaline appear to have a direct action on the pacemaker activity, which is intrinsic to the muscle itself. In other smooth muscles, the role of the sympathetic and parasympathetic innervation is reversed so that sympathetic activity causes excitation while parasympathetic activity inhibits muscle activity (see Chapter 13 for further details).

Like skeletal and cardiac muscle, smooth muscle contracts when intracellular Ca^{2+} rises. As smooth muscle does not possess a T-system, the rise in intracellular Ca^{2+} can either occur as a result of calcium influx through calcium channels in the plasma membrane or by the release of calcium from the sarcoplasmic reticulum following activation of receptors that increase the formation of IP_3 (see Chapter 7). The contractile response is slower and much longer lasting than that of skeletal and cardiac muscle (Figure 9.13). Furthermore, not all smooth muscles require an action potential to occur before they contract. In some large blood vessels, the carotid and pulmonary arteries for example, noradrenaline causes a strong contraction but only a small change in membrane potential (Figure 9.13b). In this case, the contractile response is initiated by a rise in intracellular Ca^{2+} in response to the generation of IP_3 following the activation of α_1-adrenoceptors, as described in Chapter 13 (p. 223).

Smooth muscle is able to maintain a steady level of tension called tone

The smooth muscle of the hollow organs maintains a steady level of contraction that is known as **tone** or tonus. Tone is important in maintaining the capacity of the hollow organs. For example, the flow of blood through a particular tissue depends on the calibre of the arterioles and this, in turn, is determined by the tone in the smooth muscle of the vessel wall (i.e. by the degree of contraction of the smooth muscle).

Smooth muscle tone depends on many factors, which may be either extrinsic or intrinsic to the muscle. Extrinsic factors include activity in the autonomic nerves and circulating hormones, while intrinsic factors include local metabolites, locally secreted chemical agents (e.g. nitric oxide in the blood vessels) and temperature. Thus, smooth muscle tone does not depend solely on activity in the autonomic nerves or on circulating hormones.

Length–tension relationships in single-unit smooth muscle

If a smooth muscle is stretched, there is a corresponding increase in tension immediately following the stretch. This is followed by a progressive relaxation of the tension towards its initial value. This property is unique to smooth muscle and is called stress relaxation or **plasticity**. The converse happens if the tension on smooth muscle is decreased (e.g. by voiding the contents of a hollow organ such as the gut or bladder). In this case, the tension initially falls but returns to its original level in a short period of time. This is called reverse stress relaxation and permits the internal diameter of a hollow organ to be altered to suit the volume of material it contains.

Compared with skeletal or cardiac muscle, smooth muscle can shorten to a far greater degree. A stretched striated muscle can shorten by as much as a third of its resting length while a normal resting muscle would shorten by perhaps a fifth – perfectly adequate for it to perform its normal physiological role. In contrast, a smooth muscle may be able to shorten by more than two-thirds of its initial length. This unusual property is conferred by the loose arrangement of the thick and thin myofilaments in smooth muscle cells. This ability of smooth muscle permits hollow organs to adjust to much wider variations in the volume of their contents than would be possible for skeletal muscle.

A simple calculation shows the advantage conferred by this property of smooth muscle: when the bladder is full it contains about 400 ml of urine. Almost all the urine is expelled when the smooth muscle in the bladder wall contracts. If the bladder were a simple sphere, its circumference would be about 30 cm when it contained 400 ml of urine but would be only about 6 cm if 4 ml of urine were left after it had emptied (i.e. if it contained only 1% of the original volume). This corresponds to a change in muscle length of about 80%. If the bladder were made of skeletal muscle, however, the maximum length change would be only about 30% and the bladder would only be able to void about 70% of its contents, leaving behind about 120 ml of urine.

KEY POINTS:

- Smooth muscle consists of small, spindle-shaped cells linked together at specific junctions. They contain actin and myosin filaments, but these proteins are not arranged in regular sarcomeres.

- Smooth muscle is of two types: single-unit smooth muscle, which shows myogenic activity and behaves as a syncytium, and multiunit smooth muscle, which has little spontaneous activity and is activated by impulses in specific motor nerves.

- Smooth muscle maintains a steady level of tension, known as tone, which can be regulated by circulating hormones, by local factors or by the activity of autonomic nerves.

- Smooth muscle is much more plastic in its properties than other types of muscle and is able to adjust its length over a much wider range than skeletal or cardiac muscle.

9.7 Pharmacology of smooth muscle

The tone of smooth muscle is altered by a wide range of substances, but different smooth muscles react to specific drugs in ways that reflect their varied physiological roles. The bioassay of drugs using a gut bath is well established. A strip of gut is arranged in a thermostatically controlled bath so that the tension developed by the muscle can be recorded continuously. The drugs of choice are added and the response of the smooth muscle recorded. The system can readily be adapted to study the smooth muscle of other tissues.

As smooth muscle is innervated by the autonomic nervous system, it responds to acetylcholine, adrenaline and noradrenaline. These actions will be discussed in Chapter 13 along with other aspects of the autonomic nervous system. This section will briefly outline the effects of some hormones and local mediators that act on smooth muscle.

Histamine, which is released by mast cells is a potent vasodilator, relaxing the smooth muscle of the arterioles. This action is an important part of the inflammatory response (see p. 418). This action can be antagonized by drugs that act on histamine H_1 receptors such as **mepyramine** and **triprolidine**. **Serotonin** (5-hydroxytryptamine or 5-HT) released from platelets is a powerful vasoconstrictor. This effect is most pronounced in the lungs and kidneys. It also causes the smooth muscle of the gut to contract.

The prostaglandins also exert powerful effects on smooth muscle. **Prostaglandin E_2** (PGE$_2$) and **prostaglandin I_2** (PGI$_2$) relax vascular smooth muscle and cause vasodilatation. While they also relax bronchial smooth muscle, they contact the longitudinal smooth muscle of the gastrointestinal tract. Thromboxane A_2 (TXA$_2$) and prostaglandin $F_{2\alpha}$ (PGF$_{2\alpha}$) cause smooth muscle to contract and are powerful vasoconstrictors. The powerful spasmogenic action of PGF$_{2\alpha}$ appears to play a role in parturition.

A number of peptides have powerful effects on smooth muscle. The hormone **vasopressin** is, as its name implies, a powerful vasoconstrictor, and the closely related hormone **oxytocin** causes the ejection of milk by stimulating myoepithelial cells in the breast to contract. It also causes powerful uterine contractions during labour (see Chapter 16). A number of other peptides are also powerful vasoconstrictors, including **angiotensin II** and **endothelin**. Others are vasodilators and relax vascular smooth muscle, for example **bradykinin**, neurotensin and substance P.

 ## Recommended reading

Histology

Junquiera, L.C. and Carneiro, J. (2005). *Basic Histology,* 11th edn. McGraw-Hill. Chapter 10.

Biochemistry of muscle contraction

Alberts, B., Johnson, A., Lewis, J., Raff, M., Roberts, K. and Walter, P. (2008). *Molecular Biology of the Cell,* 5th edn. Garland: New York and London. pp. 1010–1031.

Berg, J.M., Tymoczko, J.L. and Stryer, L. (2002). *Biochemistry,* 5th edn. Freeman: New York. Chapter 15.

Physiology

Jones, D.A., Round, J.M. and De Haan, A. (2004). *Skeletal Muscle from Molecules to Movement.* Churchill-Livingstone: London.

Levick, J.R. (2002). *An Introduction to Cardiovascular Physiology,* 5th edn. Hodder Arnold: London. Chapter 11.

Pharmacology

Trevor, A.J., Katzung, B.G. and Masters, S.B. (2004). *Basic and Clinical Pharmacology*, 9th edn. Appleton & Lange: Norwalk, Conneticut. Chapters 16–19.

 ## Self-assessment questions

Which of the following statements are true and which are false? Answers are given at the end of the book (p. 755).

1 The following statements relate to the structure of muscle tissue:
   a) All muscle cells contain actin and myosin.
   b) Skeletal muscle has the same structure as smooth muscle.
   c) In skeletal muscle, actin is the major protein of the thin filaments.
   d) The myofibrils of skeletal muscle are surrounded by the sarcoplasmic reticulum.
   e) Skeletal muscle fibres are electrically coupled.

2 The following statements relate to the role of Ca^{2+} in muscle contraction:
   a) The sarcoplasmic reticulum acts as a store of Ca^{2+} for the contractile process.
   b) Ca^{2+} entry across the plasma membrane is important in sustaining the contraction of smooth muscle.
   c) A muscle will relax when intracellular Ca^{2+} is raised.
   d) A rise in intracellular Ca^{2+} prevents actin interacting with myosin.

3 In skeletal muscle:
   a) a motor unit consists of a single motoneurone and the muscle fibres it innervates.
   b) the action potential propagates from the neuromuscular junction to both ends of the muscle fibre.
   c) the muscle action potential is an essential step in excitation–contraction coupling.
   d) the muscle fibres are electrically coupled so that one nerve fibre can control the activity of several muscle fibres.
   e) the energy for muscle contraction comes from the hydrolysis of acetylcholine.

4 The following statements relate to neuromuscular transmission at the motor end plate:
   a) Excitation of a motor nerve fibre leads to the contraction of all the muscle fibres innervated by its branches.
   b) The end-plate potential (EPP) is the result of an increase in the permeability of the junctional membrane to chloride ions.
   c) At the neuromuscular junction, the cholinergic receptors are muscarinic.
   d) Neuromuscular transmission can be blocked by curare.
   e) End-plate potentials can be prolonged by drugs that inhibit acetylcholinesterase.

5 The following statements relate to smooth muscle:
   a) Autonomic nerves never innervate individual muscle fibres.
   b) In some muscles, the action potential is due to pacemaker activity.
   c) All smooth muscles behave as a single motor unit.
   d) In many types of smooth muscle, the action potential results from an increase in the permeability of the sarcolemma to both Ca^{2+} and Na^+.
   e) Some smooth muscles contract without an action potential.

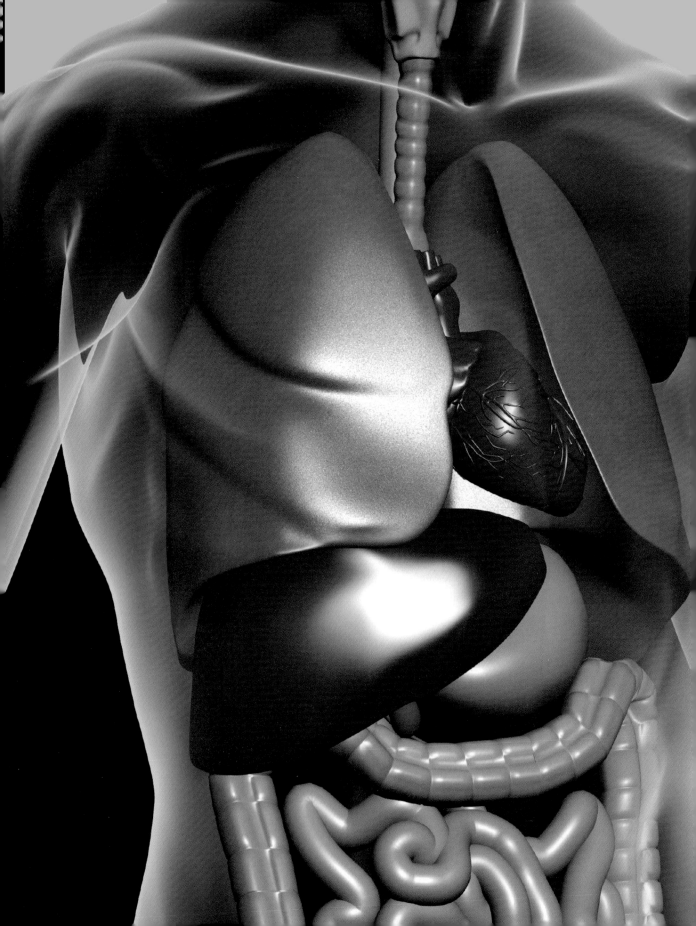

Section 4

The anatomy and physiology of body systems

10 The musculoskeletal system   **140**

11 The organization of the nervous system   **183**

12 The physiology of motor systems   **197**

13 The autonomic nervous system   **216**

14 Sensory systems   **226**

15 Some aspects of higher nervous function   **263**

16 The endocrine system   **283**

17 The heart   **327**

18 The electrocardiogram   **354**

19 The circulation   **365**

20 The properties of blood   **388**

21 Defence against infection: inflammation and immunity   **414**

22 Respiration   **431**

23 The kidney and the urinary tract   **466**

24 The digestive system   **493**

25 The liver and gall bladder   **533**

26 The pancreas   **547**

27 Skin structure and function   **556**

10 The musculoskeletal system

After reading this chapter you should have gained an understanding of:

- The anatomy of the skeleton
- The structure and physiology of bone
- Bone growth
- Bone fractures and fracture healing
- The anatomy of the principal muscle groups and their attachments
- Fixed, cartilaginous and synovial joints and their movements
- Drugs acting on the musculoskeletal system

10.1 Introduction

The skeleton and the skeletal muscles are concerned primarily with mobility and support of the body. Articulations, or joints between bones, allow movement of the whole body from one place to another (walking, running, swimming and so on) and of parts of the body relative to one another, as in breathing. Skeletal muscles also permit movements of tissues other than bones. The eyes, for example, are moved by the extraocular muscles, while the muscles that control movements of the tongue, palate, pharynx and larynx have a vital role in the processes of swallowing and speech.

In addition to its role in support and movement, the skeleton also affords protection to many of the vital organs of the body. It provides a store of calcium and phosphate and within the bone marrow, it houses the haematopoietic tissue from which all the cellular components of blood are derived.

The role of the bone marrow stem cells in blood cell production is discussed in Chapter 20 and the physiology of skeletal muscle is described in Chapter 9. This chapter will first describe the main anatomical features of the skeleton. This will be followed by a discussion of the normal physiology of bone, including its growth and development. The chapter concludes with a description of the principal types of joints and the major muscle groups.

10.2   Anatomy of the skeleton

The skeleton is made up of 206 named bones together with tendons, ligaments and cartilages. It accounts for about 20% of body weight. For the purposes of description and discussion, the skeleton is often subdivided into two parts, the **axial** skeleton and the **appendicular** skeleton. The axial skeleton forms the central longitudinal axis of the body and consists of the skull, vertebral column, sternum and ribcage. The appendicular skeleton is made up of the bones of the pectoral and pelvic girdles and of the upper and lower limbs. The axial skeleton is concerned in particular with support and protection, while the appendicular skeleton allows movements. Figure 10.1 shows anterior and posterior views of the skeleton with the names of the principal bones.

As Figure 10.1 shows, the bones of the human skeleton are of various shapes and sizes. Although not every bone fits easily into a category, it is customary to classify bones into five types. These are:

- Long bones, which are longer than they are wide, e.g. femur and humerus

- Short bones, which are roughly as wide as they are long, e.g. the carpals of the wrist and the tarsals of the foot

- Flat bones, which are thin, flattened and often curved, e.g. skull plates, ribs and scapula

- Sesamoid bones, which are small and rounded, e.g. patella

- Irregular bones, such as the vertebrae and ossicles of the middle ear (malleus, incus and stapes).

Finally, bones possess characteristic depressions, articular surfaces (condyles) and ridges (tuberosities) that distinguish them from one another. A femur, for example, is instantly recognizable as such from its characteristic head, neck and shaft, its ridges and its articular surfaces.

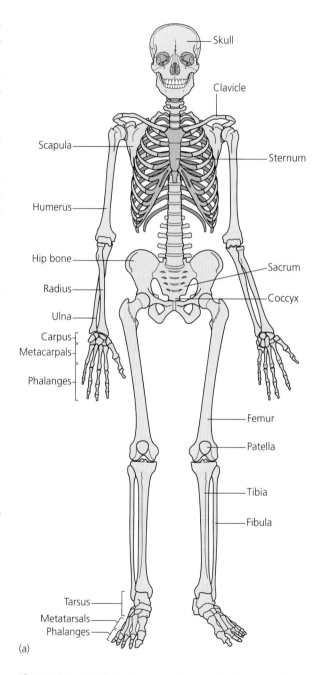

(a)

Figure 10.1 Simplified diagrams to illustrate the location of the principal bones of the skeleton. An anterior (front) view is seen in (a) and a posterior (back) view is seen in (b). *(Continued)*

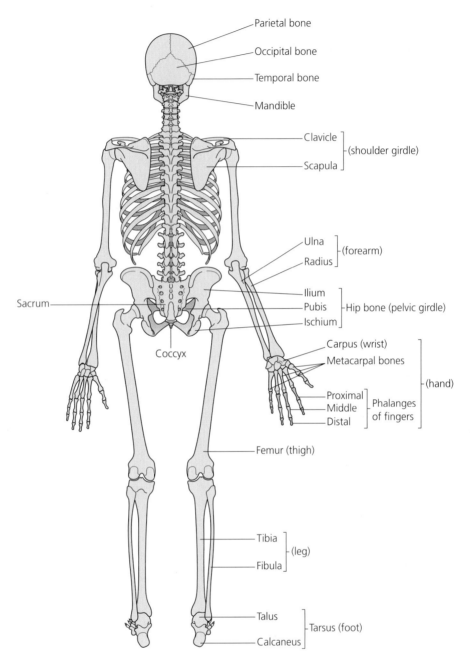

Parietal bone
Occipital bone
Temporal bone
Mandible
Clavicle
(shoulder girdle)
Scapula
Ulna
(forearm)
Radius
Ilium
Sacrum
Pubis — Hip bone (pelvic girdle)
Ischium
Coccyx
Carpus (wrist)
Metacarpal bones
(hand)
Proximal
Middle — Phalanges
Distal — of fingers
Femur (thigh)
Tibia
(leg)
Fibula
Talus
Tarsus (foot)
Calcaneus

(b)

Figure 10.1 *(Continued)*

KEY POINTS:

- The skeleton provides protection, support and mobility. It manufactures blood cells and stores minerals (calcium and phosphates).
- The skeleton is divided into two parts, the axial skeleton (skull, vertebral column and thoracic cage) and the appendicular skeleton (girdles and limbs).
- Bones are classified according to their shape into long, short, flat, irregular and sesamoid bones.

10.3 The axial skeleton

The three main components of the axial skeleton are the skull, the vertebral column and the thoracic cage (sternum and ribs).

The skull (cranium)

The skull consists of cranial bones (the neurocranium) and facial bones (the viscerocranium), which can be seen in Figures 10.2 and 10.3. Figure 10.2 illustrates the bones of the skull viewed from the front (anterior aspect) and from the side (lateral aspect), while Figure 10.3 provides a view of the inside of the base of the skull.

The **cranial bones** form a protective case around the brain. They include:

- The dome-shaped **frontal bone**, which forms the anterior aspect of the skull (the forehead) and the roof of the orbits (eye sockets) (Figure 10.2). The frontal air sinuses lie immediately above and medial to the upper margins of the orbits.

- The left and right **parietal bones**, which form part of the superior and lateral aspects of the skull. There are no cavities in the parietal bones but numerous protrusions and depressions that house blood vessels supplying the dura mater.

- The left and right **temporal bones**, which form the lateral walls of the skull and house the left and right auditory tubes within their petrous regions (Figure 10.2a). On each side of the head, there is a rounded extension of the temporal bone posterior to the auditory tube. This is called the mastoid process, which serves as a point of attachment for several neck muscles. The zygomatic process of the temporal bone joins the temporal process of the zygomatic bone to form the cheekbone (zygomatic arch). Another pointed process extends downwards from the base of the petrous portion of the temporal bone. This is called the styloid process and it provides an attachment point for the ligaments that suspend the hyoid bone (see below).

- The **occipital bone**, which forms the posterior aspect of the skull. It contains a cavity called the foramen magnum through which the spinal cord passes as it leaves the brainstem (Figure 10.3). Vertebral and spinal arteries also pass through the foramen magnum.

- The **sphenoid bone**, which forms the floor of the skull and part of the orbit on each side. It articulates with all the other cranial bones and holds them together, as shown in Figure 10.3. It contains the sphenoidal paranasal sinuses, which drain into the nasal cavity, and a number of other small cavities (foramina) through which cranial nerves and blood vessels pass. It also has a depression on its superior surface, called the sella turcica, which houses the pituitary gland.

- The **ethmoid bone**, which forms the roof of the nasal cavity and the medial regions of the orbits (Figure 10.2b). The ethmoid bone houses the ethmoid sinuses, a series of small, intercommunicating cavities (sometimes called ethmoid air cells) between the nose and the medial wall of the orbit. Surgical access to the pituitary gland may be gained via the ethmoid bone.

The butterfly-shaped sphenoid bone and the **foramen magnum** can be clearly seen in Figure 10.3, which also reveals the orbital plates of the frontal bone and the cribriform plate of the ethmoid bone, forming the roof of the nose. The interior of the cranial cavity consists of three concave areas: the anterior, middle and posterior fossae (Figure 10.3). Each of these supports particular regions of the brain. The anterior fossa houses the frontal lobes, the middle fossa houses the temporal lobes, hypothalamus and pituitary gland, and the posterior fossa supports the cerebellum and the brainstem. The shape of the skull gives it great strength for its weight (rather like the shell of an egg) and allows the bones to be thin and therefore light. Nevertheless, a heavy blow to the head can result in a fracture of the skull, which may be depressed and localized, or may involve large areas of bone with damage to underlying tissues. A blow to the side of the head may cause the brain to move and strike the inside of the cranial cavity on the opposite side – a 'contre-coup' injury.

Within the skull there is also a series of air-filled spaces, the **paranasal sinuses**, which are lined by respiratory epithelium. The cells of this epithelium secrete mucus,

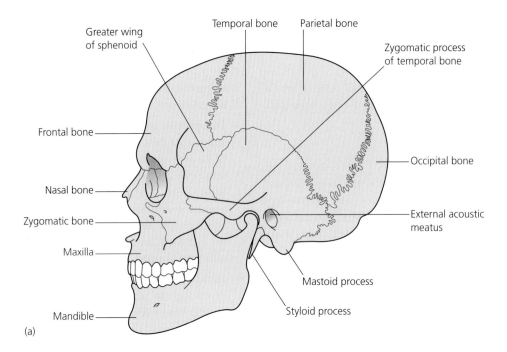

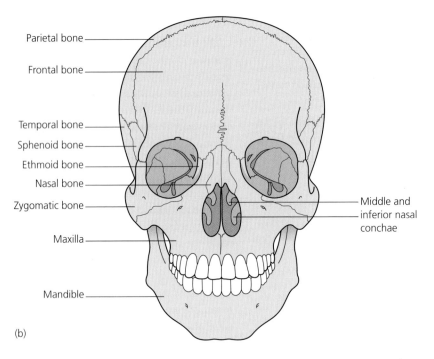

Figure 10.2 Views of the skull showing the principal bones of the face and cranium; (a) shows a lateral view and (b) an anterior view.

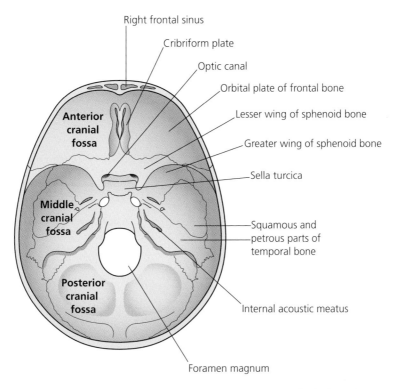

Right frontal sinus

Cribriform plate

Optic canal

Orbital plate of frontal bone

Lesser wing of sphenoid bone

Greater wing of sphenoid bone

Sella turcica

Squamous and petrous parts of temporal bone

Internal acoustic meatus

Anterior cranial fossa

Middle cranial fossa

Posterior cranial fossa

Foramen magnum

Figure 10.3 The internal surface of the skull. The sphenoid bone, foramen magnum and cribriform plate are visible in this view. It is also possible to see the fossae (hollows) of the skull that house and cushion the various parts of the brain.

which drains into the nasal cavities and helps to prevent dehydration of the nasal mucosa. The paranasal sinuses also act as resonating chambers for the voice and help to make the skull lighter. Their location is illustrated in the anterior and medial views of Figure 10.4.

Inspection of the skull reveals that the cranial bones are connected by interlocking joints that appear serrated. These are fibrous, fixed joints known as **sutures**, which in adults prevent movement between the bones of the skull. In newborn babies, the sutures have not yet fused and there are extensive fibrous areas at the junctions between the frontal and parietal bones and at the junction between the parietal and occipital bones. These are the soft spots, or **fontanelles**, which can be felt on a baby's skull. They allow for moulding of the skull during delivery and for growth of the skull during early childhood. The fontanelles normally close during the first year of life and certainly by the time a child reaches the age of 2. Figure 10.5a illustrates the skull of a neonate and shows the location of the fontanelles. This should be compared with Figure 10.5b, which clearly shows the sutures between the fused cranial bones of the skull of an adult.

The **facial bones** of the viscerocranium make up the rest of the skull. All except the mandible are connected by immoveable fibrous joints. The facial bones are illustrated in Figure 10.2 and they include:

- The right and left **nasal bones**, which join to form the bridge of the nose.

- The right and left **maxilla bones** (maxillae), which form the upper jaw and are the largest of the facial bones. They support the upper teeth and form the lower (inferior) borders of the orbits, the lateral margins of the nostrils (nares) and most of the hard palate. Lateral to the nose on each side is a large space, the maxillary sinus, whose epithelial cells secrete mucus, which bathes the inferior surfaces of the nasal cavities. A projection from the maxilla, the zygomatic process, articulates with the zygomatic bone.

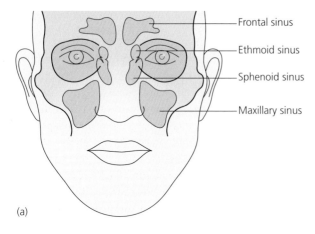

(a)

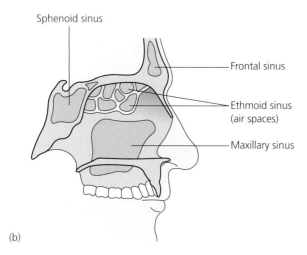

(b)

Figure 10.4 The location of the paranasal sinuses seen in anterior (a) and medial (b) view. Note that the ethmoid sinus is not a single space but is made up of a number of 'air cells'.

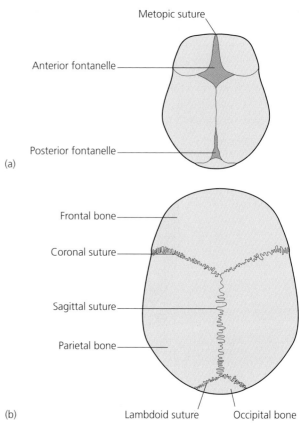

(a)

(b)

Figure 10.5 Comparison of the skulls of a newborn infant (a) and an adult (b). In the newborn skull, the skull plates are unfused and separated by areas of fibrous tissue (the fontanelles). In the adult skull, the plates have fused to form fixed (fibrous) joints called sutures.

- The right and left **palatine bones** form the posterior portion of the roof of the mouth (hard palate) and part of each nasal cavity.

- The right and left **zygomatic bones** (also called the malar bones), which rest on the maxilla bones and form the prominences of the cheeks. As described earlier, a process of the zygomatic bone (the temporal process) joins the zygomatic process of the temporal bone to form the zygomatic arch or cheek bone.

- The right and left **lacrimal bones** form part of the medial borders of the orbits. They are the smallest of the facial bones. Each lacrimal bone possesses on its lateral surface a small depression that houses the lacrimal sac from which tear fluid drains via the lacrimal duct into the nasal cavity.

- The right and left **turbinates** (or inferior nasal conchae), which are thin, curved bones located on the lateral side of each nostril. These may be seen in Figure 10.2b.

- The single **vomer bone**, which articulates with the sphenoid bone at the base of the skull and forms the posterior part of the nasal septum (the vertical partition that separates the right and left nasal cavities – it is mostly made of cartilage).

- The single **mandible bone**, which forms the lower jaw. This bone houses the lower teeth and articulates with the temporal bones to form the freely moveable temporomandibular joints that allow the wide range of movements required for chewing, speaking, etc. The shape of the mandible changes significantly during life, as teeth erupt during childhood and adolescence and are lost in old age.

Suspended from the temporal bones by ligaments attached to the styloid processes is the **hyoid bone**, which is located in the neck just above the larynx. This bone forms part of the axial skeleton but is unique in that it does not articulate with any other bones. It is shaped like a horseshoe and supports the tongue and its associated muscles. It also helps to elevate the larynx and pharynx during swallowing and speech.

The vertebral column

The vertebral column or spine is illustrated in Figure 10.6. It consists of 32 to 34 irregularly shaped vertebrae, which are separated by disks of cartilage (the **intervertebral disks**). The vertebral column provides support for the body and protection for the spinal cord. The cartilaginous intervertebral disks cushion the vertebrae, permit flexion of the spine and enable the vertebral column to withstand considerable compression forces such as those caused by jumping and running. The disks tend to compress and shrink as a person ages and this can cause as much as 2 cm of height to be lost. The vertebrae are

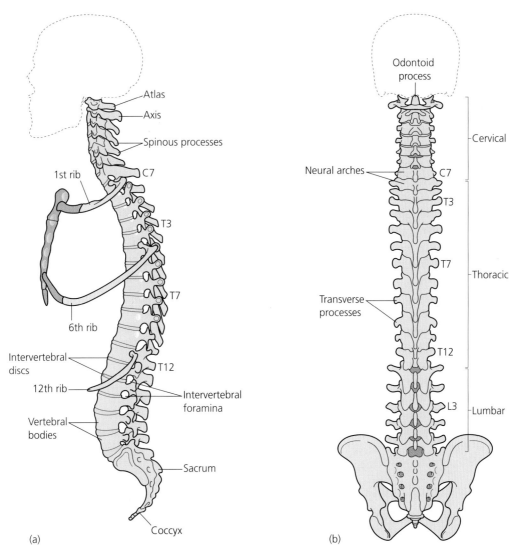

Figure 10.6 The vertebral column seen from the side (a) and from the back (b).

numbered and named according to their relative longitudinal position within the neck, thorax and lower trunk. There are:

- 7 **cervical vertebrae** (numbered C1–C7). C1 is also called the atlas and C2 is the axis
- 12 **thoracic vertebrae** (numbered T1–T12)
- 5 **lumbar vertebrae** (numbered L1–L5)
- 5 **sacral vertebrae**, which are fused in adults to form a roughly triangular bone, the sacrum
- 3–5 tiny **coccygeal vertebrae**, which are fused in adults to form the coccyx.

Figure 10.6a shows that the vertebral column is not straight but has four curves: the cervical, thoracic, lumbar and sacral curves. These curves distribute the weight of the body in line with the body axis and thus allow an upright stance and two-legged locomotion. The thoracic and sacral curves develop during fetal life, but the cervical curve does not develop until the child starts to hold its head up at a few months of age. The lumbar curve is not established until the child starts to walk. Occasionally, abnormal spinal curvatures are seen. These include **scoliosis**, in which the spine is curved laterally, **kyphosis**, in which the thoracic curve is exaggerated (giving a round-backed appearance), and **lordosis**, in which the lumbar curve is exaggerated anteriorly.

Each vertebra has its own characteristic features but all share the following general characteristics. A typical vertebra is made up of a vertebral body, a vertebral arch and a number of processes. These can be seen in Figure 10.7. Between the vertebral arch and the vertebral body is a space, the vertebral canal, through which the spinal cord passes. Each vertebra has two transverse processes and a larger spinous process to which muscles are attached, and superior and inferior articular processes, which allow the vertebra to articulate with the bones above and below it in the column. The first cervical vertebra (the atlas) articulates with the occipital condyles of the skull.

Sternum, ribs and associated cartilages

The rest of the axial skeleton is made up of the sternum, 12 pairs of ribs and their associated cartilages. Together with the thoracic vertebrae, these bones form

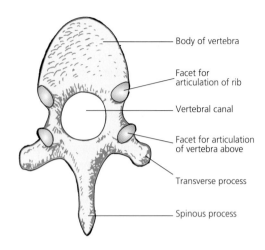

Body of vertebra

Facet for articulation of rib

Vertebral canal

Facet for articulation of vertebra above

Transverse process

Spinous process

Figure 10.7 Diagram of a thoracic vertebra to show its main features. Other vertebrae have similar features but differ in the size and the number of processes they possess.

the thoracic cage, which protects the internal organs of the thoracic cavity including the heart and lungs. The **sternum** (breastbone) is in reality three fused bones: the upper manubrium, the middle body and the lower xiphoid process. Its structure is illustrated in Figure 10.8, which also shows the main features of the thoracic cage.

The **ribs** are curved, flat bones that articulate posteriorly with the thoracic vertebrae. The first seven pairs of ribs (called the 'true ribs') articulate anteriorly with the sternum via **costal cartilages**. Pairs 8–10 attach indirectly to the sternum (Figure 10.8) and pairs 11 and 12 have no attachment (they are sometimes called the 'floating ribs'). The spaces between adjacent ribs (intercostal spaces) contain the intercostal muscles, which are important in breathing (see Chapter 22).

KEY POINTS:

- The skull consists of cranial bones and facial bones.
- The vertebral column consists of 32–34 irregularly shaped vertebrae cushioned by cartilaginous intervertebral discs.
- The thoracic cage consists of the sternum, ribs and costal cartilages.

(a)　　　　　　　　　　　　　　　　　　(b)

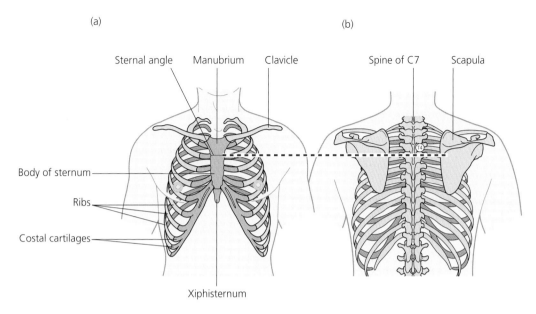

Sternal angle　Manubrium　Clavicle　　Spine of C7　Scapula

Body of sternum

Ribs

Costal cartilages

Xiphisternum

Figure 10.8 The bones of the thoracic cage seen in anterior (a) and posterior (b) view. The red line indicates the relationship between the sternal angle and the junction between the fourth and fifth thoracic vertebrae.

10.4　The appendicular skeleton

The appendicular skeleton is concerned primarily with movement, although the pelvic girdle provides protection for the pelvic organs and both the girdles help to support the body in the upright position. This part of the skeleton consists of:

- The pectoral (shoulder) girdle
- The bones of the upper limbs
- The pelvic (hip) girdle
- The bones of the lower limbs.

The pectoral girdle and the upper limbs

The **pectoral girdle** is made up of two scapulae (shoulder blades) and two clavicles (collar bones). These are illustrated in Figure 10.9. The clavicles are slender, rod-like bones that are sigmoid in shape. They lie horizontally just above the first pair of ribs and articulate medially with the manubrium of the sternum just above the first rib and laterally with the scapula. The clavicles support the

scapulae and hold them back so that the arms can hang freely at the sides of the body. The clavicle is the most frequently fractured bone in the body and when its bracing action is lost, the shoulder drops and the upper limb loses its support.

Each scapula is a flat, roughly triangular bone that covers the area between the second and seventh ribs. On the posterior surface of the bone is a spiny ridge that projects to form the acromion, a protuberance that can easily be felt and forms the high point of the shoulder. The clavicle articulates with the scapula at the acromion. Just below the acromion is a shallow depression, the glenoid cavity, which receives the head of the humerus to form the shoulder joint. Another bony projection of the scapula, the coracoid process, provides attachment sites for muscles. This process and the relationships between the humerus, scapula and clavicle can be seen in Figure 10.9.

The **upper limb** may be divided into the upper arm, lower arm, wrist, palm and fingers. A single long bone, the **humerus**, forms the skeleton of the upper arm

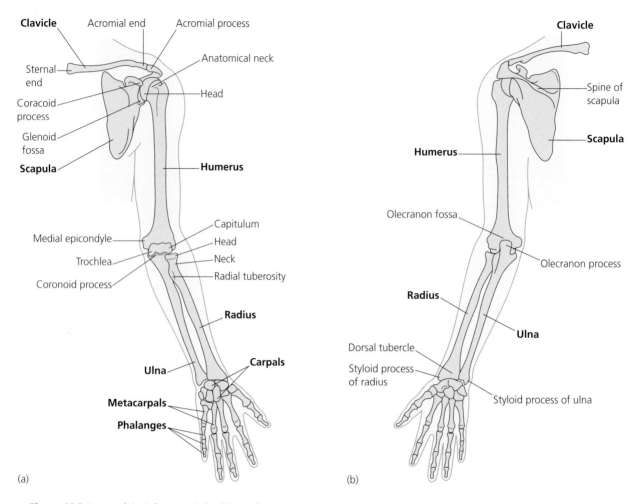

Figure 10.9 Bones of the left arm and shoulder girdle seen in anterior (a) and posterior (b) view.

(Figure 10.9.). This bone has a rounded head which fits into the glenoid cavity of the scapula, a neck region and a long shaft (diaphysis). At its distal end, the humerus articulates with the bones of the lower arm.

The lower arm has two long bones: the **radius**, located on the thumb side of the lower arm (the lateral side), and the slightly longer **ulna** on the side of the little finger (the medial side). The radius and ulna articulate with each other at both ends of the bones and, when the lower arm is rotated, the distal end of the radius moves over the ulna (Figure 10.9). The head of the radius articulates with the capitulum of the humerus, while the upper end of the ulna (the trochlear notch) articulates with the lower end of the humerus (the trochlea) to form a strong hinge

joint – the elbow. The prominent point of the elbow is formed by a projection of the ulna called the olecranon (Figure 10.9b). The lower part of the radius is wider than the ulna and forms a large part of the wrist joint where it articulates with the carpals.

The wrist is made up of eight short bones called **carpals**, arranged in two rows of four and bound together by ligaments. Although movements of the carpals are restricted, they can slide over one another to a limited extent. A dorsal view of the carpals is seen in Figure 10.10. The palm of the hand is formed by five small long bones, the **metacarpals**. These are numbered 1–5 starting with the thumb as number 1. They articulate proximally with the carpals and distally with the bones

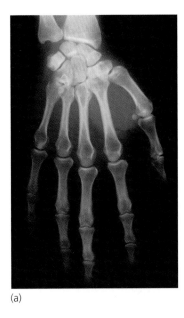

(a)

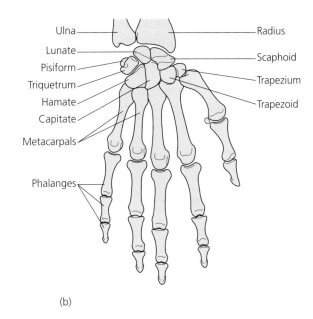

Ulna — Radius
Lunate — Scaphoid
Pisiform — Trapezium
Triquetrum — Trapezoid
Hamate —
Capitate —
Metacarpals —

Phalanges —

(b)

Figure 10.10 The skeleton of the hand: (a) shows an X-ray image of a right hand in which the main bones are clearly visible; (b) is a drawing of the bones of the lower arm, wrist and hand to illustrate the relationship between them.

of the fingers, the **phalanges**. Digits 2–5 (index, middle, ring and little fingers) each consist of three phalanges (proximal, middle and distal) while the thumbs (digit 1) have two phalanges (proximal and distal).

The pelvic girdle and the lower limbs

The bony pelvis supports the trunk and distributes the weight of the upper body to the lower limbs. It protects some of the organs of the lower abdomen and provides strong attachment sites for the muscles of the back, legs and buttocks. It consists of two coxal (hip) bones that form the pelvic girdle, the sacrum (made up of five fused sacral vertebrae) and the coccyx.

The **hip bone** is formed by the fusion of three bones, the ilium, the ischium and the pubis. At the point of fusion lies a deep hollow (socket) called the acetabulum into which fits the head (ball) of the femur. These structures are illustrated in Figure 10.11. The **ilium** is the largest of the three and forms the upper part of the hip bone. It is wider and broader in females and smaller and narrower in males. The upper edge of the ilium is called the iliac crest and can easily be felt. The ilium articu-

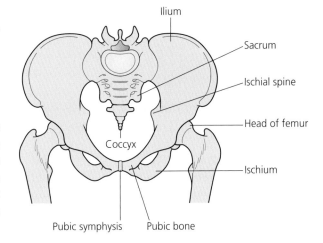

Ilium
Sacrum
Ischial spine
Head of femur
Coccyx
Ischium
Pubic symphysis   Pubic bone

Figure 10.11 An anterior view of the hip to show the bones that make up the pelvic girdle.

lates posteriorly with the sacrum to form the sacroiliac joint. The **ischium** forms the inferior (lower) part of the hip bone and is the part of the pelvis that bears much of the weight of the body when sitting. The **pubis** forms the anterior portion of the hip bone. The two pubic bones are fused anteriorly to form a cartilaginous joint (see below) called the symphysis pubis. As a result, the two hip bones together with the sacrum form a complete ring of bone.

The skeleton of the lower limb may be divided into the upper and lower leg, the ankle, the arches of the foot (or sole) and the toes. The bones that form these structures are shown in Figures 10.12 and 10.13.

The **femur** is the long bone of the upper leg or thigh. The femur has a large, rounded head that fits into the acetabulum of the pelvic girdle to form the freely moveable hip joint. Between the head and the long shaft of

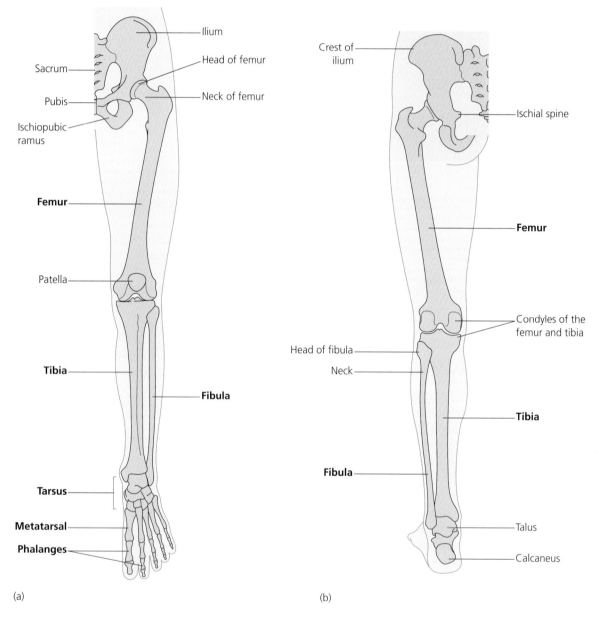

(a)

(b)

Figure 10.12 Bones of the left leg and pelvic girdle seen in anterior (a) and posterior (b) view. Note the ball-and-socket joint between the head of the femur and the pelvis.

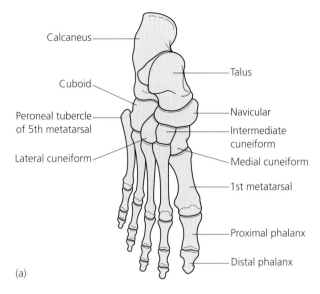

(a)

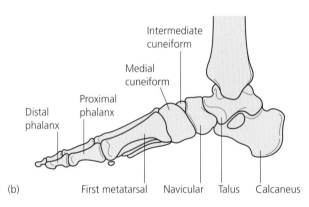

(b)

Figure 10.13 Diagrams to illustrate the bones of the right foot from above (a) and from the medial side (b).

The lower leg consists of the tibia (shin) and the slender fibula, which lies behind it. The **tibia** possesses a flat condyle at its proximal end, which articulates with the femur to form the knee. The **fibula** does not articulate with the femur but is attached to the proximal end of the tibia. Its distal end forms the outer part of the ankle joint.

The foot consists of tarsals, metatarsals and phalanges, which are illustrated in Figure 10.13. Seven **tarsals** form the ankle. The largest of them is called the **calcaneus** or heel bone and this bears the weight of the body while walking. It also forms the point of attachment for the calf muscles (via the Achilles tendon), which lift the heel during locomotion. The prominent ankle bone (the **talus**) lies between the tibia and the calcaneus. This transmits the body weight from the tibia towards the toes.

Five **metatarsals** form the sole of the foot. They are numbered 1–5 across the foot (see Figure 10.13), with number 1 being the most medial (on the side of the big toe). The metatarsals articulate proximally with the tarsals and distally with the **phalanges**. The big toe possesses two phalanges while the other four toes each have three.

The bones of the foot are arranged to form three arches. This shape increases the weight-bearing capacity of the foot and is maintained by ligaments that attach the calcaneus to the distal portions of the metatarsals. Occasionally these arches may fail to develop normally during the fetal period (as in congenital clubfoot) or may be lost during later life (fallen arches) leading to the condition known as 'flat feet'.

the femur is the region known as the neck. This portion of the femur is susceptible to fracture, particularly in elderly women with osteoporosis (see below). At its distal end, the femur widens to form the lateral and medial condyles, which articulate with the tibia of the lower leg to form the knee joint. The **patella** (kneecap) articulates with the anterior surface of these condyles and is attached to the tibia at its apex by the patellar ligament.

KEY POINTS:

- The pectoral girdle consists of two scapulae and two clavicles.

- The skeleton of the upper limb is formed by the humerus (upper arm) and the radius and ulna (lower arm).

- The pelvic girdle consists of the ilium, ischium, pubis and sacrum.

- The skeleton of the lower limb is formed by the femur (upper leg) and the tibia and fibula (lower leg).

10.5   The physiology of bone

In an adult, bone forms one of the largest masses of tissue, weighing 10–12 kg. The adult skeleton contains between 1 and 2 kg of calcium (about 99% of the body total) and between 0.5 and 0.75 kg of phosphorus (about 88% of the body total). Far from being the inert supporting structure its outward appearance might suggest, bone is a dynamic tissue with a high rate of metabolic activity, which is continuously undergoing complex structural alterations under the influence of mechanical stresses and a variety of hormones.

Bone is a specialized form of connective tissue that is made durable by the deposition of mineral within its infrastructure. About 30% of the total skeletal mass is made up of **osteoid**, an organic matrix consisting largely of collagen

together with hyaluronic acid, chondroitin sulphate and a protein called osteocalcin, which is an important calcium-binding molecule. The remainder consists of a **mineral matrix** of calcium phosphate (hydroxyapatite) crystals and **bone cells** including osteoblasts (bone-forming cells), osteoclasts (bone-resorbing cells), osteocytes (mature bone cells) and fibroblasts.

The anatomical features of a typical long bone are illustrated in Figure 10.14. The central shaft is called the **diaphysis** while the regions at either end of the bone are the **epiphyses**. Between the diaphysis and epiphyses is a region of bone known as the **epiphyseal plate** or growth plate. Adjacent to this is the growing end of the diaphysis, known as the **metaphysis**. Growth in length occurs

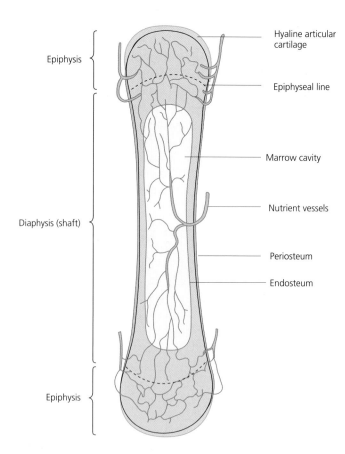

Epiphysis

Diaphysis (shaft)

Epiphysis

Hyaline articular cartilage

Epiphyseal line

Marrow cavity

Nutrient vessels

Periosteum

Endosteum

Figure 10.14 The structural components of a typical long bone.

by deposition of new cartilage at the metaphysis and its subsequent mineralization with calcium phosphate crystals. Once growth is completed following puberty, the epiphyseal plate becomes fully calcified and remains as the **epiphyseal line**, which is indicated in Figure 10.14.

Except at articulations, the surfaces of the bones are covered by **periosteum**, which consists of an outer layer of tough, fibrous connective tissue and an inner layer of osteogenic ('bone-forming') tissue. The articular surfaces of bones are covered by smooth, glassy hyaline cartilage, which helps to reduce the friction associated with movement at joints. A space runs through the centre of bones. This is the **marrow cavity** (or medullary space), which is lined with osteogenic tissue (the **endosteum**). The marrow cavities of the long bones contain mainly fatty, yellow marrow, which is not involved in haematopoiesis (the formation of blood cells) under normal circumstances. Red marrow containing haematopoietic tissue is found within the small, flat and irregular bones of the skeleton such as the sternum, ilium and vertebrae. It is here that blood cells are produced (see Chapter 20).

Long bones are supplied by the nutrient artery, the periosteal arteries and the metaphyseal and epiphyseal arteries. The nutrient artery branches from a systemic artery and pierces the diaphysis before giving rise to ascending and descending medullary arteries within the marrow cavity. In turn, these give rise to arteries supplying the endosteum and diaphysis. The periosteal blood supply takes the form of a capillary network, while the metaphyseal and epiphyseal vessels branch off from the nutrient artery. At rest, the arterial flow rate to the skeleton is around 12% of the total cardiac output (or about 2–3 ml per 100 mg tissue min^{-1}). The mechanisms that control the blood flow to the skeleton are poorly understood, but it is known that blood flow is significantly increased during inflammation and infection and following fracture.

Bone is not uniformly solid but contains spaces that provide channels for blood vessels and reduce the weight of the skeleton. Bone may be classified as either compact (dense) or spongy (trabecular, cancellous) according to the size and distribution of the spaces. **Compact bone** forms the outer regions of all bones, the diaphysis of long bones and the outer and inner regions of flat bones. It contains few spaces and provides protection and support, especially for the long bones in which it helps to reduce the stress of weight bearing.

The functional units of compact bone are the **Haversian systems** or **osteons**. As Figure 10.15 shows,

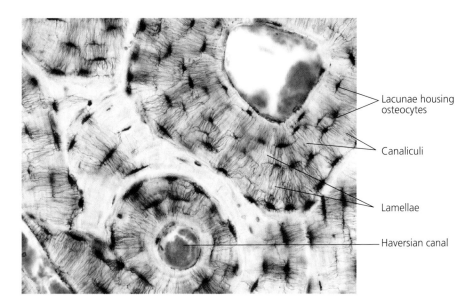

Lacunae housing osteocytes

Canaliculi

Lamellae

Haversian canal

Figure 10.15 The structure of compact bone. Note the lamellar organization of the Haversian systems and the lacunae that imprison the osteocytes.

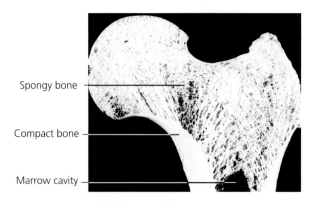

Spongy bone

Compact bone

Marrow cavity

Figure 10.16 A section through the head of the femur showing regions of dense (compact) and spongy bone.

The bone cells

Three major cell types are recognized in histological sections of bone. These are:

- Osteoblasts
- Osteocytes
- Osteoclasts.

Osteoblasts are present on the surfaces of all bones and line the internal marrow cavities such as that shown in Figure 10.17a. They contain numerous mitochondria and an extensive Golgi apparatus associated with rapid

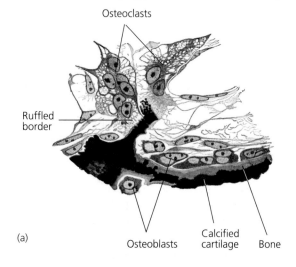

Osteoclasts

Ruffled border

(a)

Osteoblasts

Calcified cartilage

Bone

each osteon consists of a central canal that contains blood vessels, lymphatics and nerves, surrounded by concentric rings of hard intercellular substance (lamellae) between which are spaces (lacunae) containing osteocytes (mature bone cells). Radiating from the lacunae are tiny canals (canaliculi) that connect with adjacent lacunae to form a branching network by which nutrients and waste products may be transported to and from the osteocytes.

By contrast, **spongy bone** contains no true osteons but consists of an irregular lattice of thin plates or spicules of bone (trabeculae) between which are large spaces filled with bone marrow. Lacunae containing osteocytes lie within the trabeculae. Spongy bone makes up most of the mass of short, flat and irregular bones and is present within the epiphyses of long bones and at the growth plates. Figure 10.16 illustrates the different appearance of dense and spongy bone.

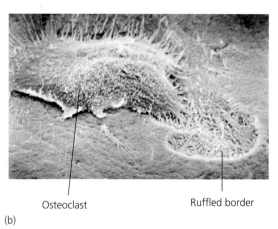

Osteoclast

Ruffled border

(b)

Figure 10.17 (a) A section through the marrow cavity of bone showing calcified cartilage (here stained dark blue), osteoblasts and two large, multinucleate osteoclasts. Note the ruffled border next to the calcified cartilage. (b) A scanning electron micrograph of an osteoclast. Note the prominent ruffled border.

KEY POINTS:

- Bone is a specialized connective tissue made up of an inorganic matrix (osteoid) strengthened by mineral and bone cells.
- Bone is of two types, compact or spongy.

protein synthesis. They secrete the constituents of osteoid, the organic matrix of bone, and are also important in the process of mineralization (calcification) of this matrix. Osteoblasts possess specific receptors for parathyroid hormone and calcitriol (see below).

Osteocytes are mature bone cells derived from osteoblasts that have become trapped in lacunae (small spaces) within the matrix that they have secreted. As described above, adjacent osteocytes are linked by fine cytoplasmic processes that pass through tiny canals (canaliculi) between lacunae (Figure 10.15). This arrangement permits the exchange of calcium from the interior to the exterior of bones and thence into the extracellular fluid should plasma calcium levels fall.

Osteoclasts are giant, multinucleated cells that contain numerous mitochondria and lysosomes. They are highly mobile cells that are responsible for the resorption of bone during growth and skeletal remodelling. They are abundant at or near the surfaces of bone undergoing erosion. At their site of contact with the bone is a highly folded 'ruffled border' of microvilli that infiltrates the disintegrating bone surface (Figure 10.17b). Bone dissolution is brought about by the actions of collagenase, lysosomal enzymes and acid phosphatase. Calcium, phosphate and the constituents of the bone matrix are released into the extracellular fluid as bone mass is reduced. The activity of the osteoclasts appears to be controlled by a number of hormones, notably parathyroid hormone, calcitonin, thyroxine, oestrogens and the metabolites of vitamin D (see p. 312).

Bone development and growth (osteogenesis)

At 6 weeks of gestation, the fetal skeleton is constructed entirely of fibrous membranes and hyaline cartilage. From this time, bone tissue begins to develop and eventually replaces most of the existing structures. Although this process of ossification begins early in fetal life, it is not complete until the third decade of adult life. The bones of the cranium, lower jaw, scapula, pelvis and the clavicles develop from fibrous membranes by a process called **intramembranous ossification**. In this process, new bone is formed on the surface of existing bone. The bones of the rest of the skeleton grow in length as hyaline cartilage templates are replaced by bone (a process known as **endochondral ossification**).

Growth of bone length

A long bone such as the radius in the forearm is laid down first as a cartilage model. At the centre of this model, the so-called primary centre of ossification, the cartilage cells break down and bone appears. This process begins early in fetal life and, shortly before birth, secondary centres of ossification have also developed, predominantly at the ends of the bone or epiphyses. Smaller bones such as the carpals and tarsals of the hands and feet develop from a single ossification centre. The areas of cartilage between the diaphysis and the epiphyses are known as the **growth plates**. In the part of the growth plate immediately under the epiphysis is a layer of stem cells or chondroblasts. These give rise to clones of cells (chondrocytes) arranged in columns extending inwards from the epiphysis towards the diaphysis.

Several zones may be distinguished within the columns of chondrocytes. The outer zone is one of proliferation in which the cells are dividing rapidly. Beneath this are layers in which the cells mature, enlarge and eventually degenerate, as shown in Figure 10.18. The innermost layer of cells is the region of calcification.

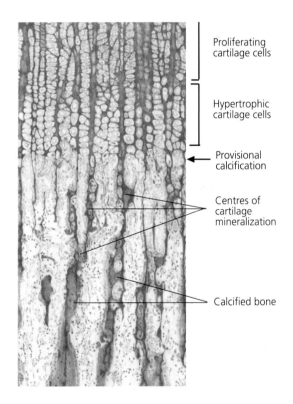

Proliferating cartilage cells

Hypertrophic cartilage cells

Provisional calcification

Centres of cartilage mineralization

Calcified bone

Figure 10.18 A section of a typical long bone prepared to show the process of bone growth and ossification. New cells formed in the proliferative region move down to the hypertrophic region to add to bone accumulating on the top of the diaphysis. Calcified bone is stained red.

Here, the osteogenic cells differentiate into osteoblasts and lay down bone. Thus, at one end of the epiphyseal plate, cartilage is produced while at the other end it is degenerating. Growth in length is therefore dependent upon the proliferation of new cartilage cells. In man, it takes around 20 days for a cartilage cell to complete the journey from the start of proliferation to degeneration. Clearly, the bone marrow cavity must also increase in size as the bone grows, and, to ensure this, osteoclasts erode bone within the diaphysis. In the radiographs of growing hands illustrated in Figure 10.19, the regions of calcification are visible as areas of high density.

At the end of the growth period, following the growth spurt of puberty, the growth plate thins as it is gradually replaced by bone until it is eliminated altogether and the epiphysis and diaphysis are fused. No further increase in bone length is possible once fusion has occurred. Although growth in length of most bones is complete by the age of 20, in males the clavicles do not ossify completely until the third decade of life. The dates of ossification are fairly constant among individuals but different between bones. This fact is exploited in forensic science to determine the age of a body according to which bones have ossified and which have not.

Growth of bone diameter

The growth in width of long bones is achieved by **appositional** bone growth in which osteoblasts beneath the periosteum of the bone form new osteons on the external surface of the bone. The bone thus becomes thicker and stronger. Rapid ossification of this new tissue takes place to keep pace with the growth in length of the bone. This process is similar to the mechanism by which the flat bones grow.

Remodelling of bone

Even after growth has ended, the skeleton is in a continuous state of remodelling as it is renewed and revitalized at the tissue level and remodelled in response to the physical and metabolic demands placed upon it. Furthermore, following a break to a bone, self-repair takes place remarkably quickly (Box 10.1).

To bring about remodelling of the skeleton, old bone is eroded by the enzymes secreted by osteoclasts while new

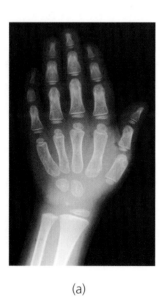

(a)

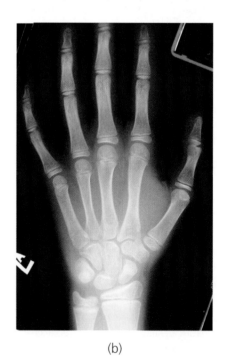

(b)

Figure 10.19 Radiographs of the left hands of two children aged (a) 2 years and (b) 11 years. Note the increase in the number of ossified carpal bones in the wrist of the 11-year-old.

Box 10.1 Bone healing following fracture

When bone is fractured, its original structure and strength is restored quite rapidly through the formation of new bone tissue. Provided that the edges of the fractured bone are repositioned and that the bone is immobilized by splinting, repair will normally occur with no deformity of the skeleton. There are three stages in the repair of a fractured bone. The first stage occurs during the first 4 or 5 days after injury and involves the removal of debris resulting from the tissue damage. This includes bone and other tissue fragments as well as blood clots formed by bleeding between the bone ends and into surrounding muscle when the periosteum is damaged. Phagocytic cells such as macrophages clear the area and granulation tissue forms.

Granulation tissue is formed from a loosely gelled, protein-rich exudate that forms at any site of tissue damage and which later becomes organized into scar tissue. As blood vessels from undamaged capillaries in adjacent tissue restore the blood supply, the granulation tissue takes on a pink, granular appearance. Osteoblasts within the endosteum and periosteum migrate to the site of damage to initiate the second stage of healing (Figure 1a). During this stage, which normally lasts for the next 3 weeks or so,

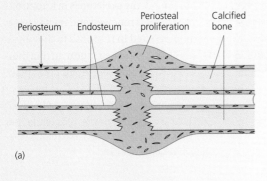

(a)

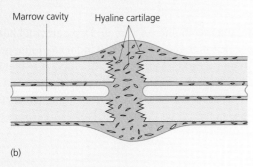

(b)

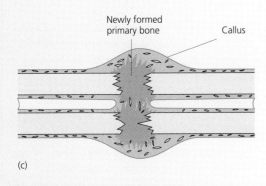

(c)

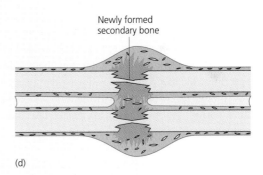

(d)

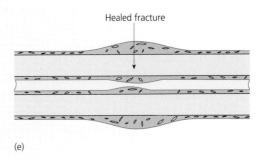

(e)

Figure 1 The principal stages of bone repair following a fracture.

osteoid is secreted by the osteoblasts into the granulation tissue to form a mass between the fractured pieces of bone to bridge the gap. This tissue mass is also known as **soft callus** (Figure 1b). The soft callus gradually becomes ossified to form a region of woven bone (similar to cancellous bone), which is called **hard callus** (Figure 1c). At this stage of healing, there is normally some degree of local swelling at the site of the fracture caused by the hard callus deposit. During the final stage in the process of healing, the mass of hard callus is restructured to restore the original architecture of the bone. This stage may take place over many months and involves the actions of both osteoblasts and osteoclasts. During this time, the periosteum also reforms and the bone is able to tolerate normal loads and stresses. The radiographs of Figure 2 show key stages in the healing process following a fracture.

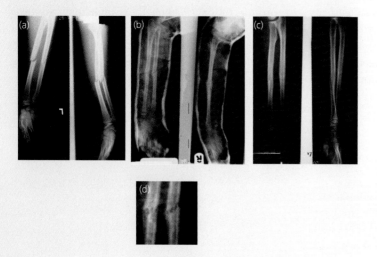

Figure 2 Radiographs showing the stages of repair following a fracture. (a) Simple fracture of humerus and radius of forearm. (b) Development of soft callus. (c) Final healing of fracture. (d) The soft callus in more detail. Note that the radiograph shown in (b) was taken through the immobilizing plaster cast.

bone is laid down by osteoblasts. Indeed, as much as 5–7% of bone mass is recycled each week. In general terms, bone is deposited in proportion to the load it must bear. It follows, therefore, that in an immobilized person bone mass is rapidly (though reversibly) lost. Astronauts experiencing prolonged periods of weightlessness in space have been shown to lose up to 20% of their bone mass in the absence of properly planned exercise programmes. Similarly, appropriate exercise during childhood and adolescence is thought to enhance the development of bone and result in a stronger, healthier skeleton in adult life, a factor that may be particularly important in women (see later).

The role of hormones in the process of growth

The role of growth hormone in growth of the skeleton is discussed in Chapter 16. However, although growth hormone undoubtedly plays a pivotal role in the process of physical growth, many other hormones are also important. Hormones of particular significance include thyroxine and the sex steroids, but other hormones including insulin, the metabolites of vitamin D, parathyroid hormone, calcitonin and cortisol may indirectly influence growth and development through their general metabolic actions or their actions on the physiology of bone. Important aspects of the growth-promoting effects of these hormones are described in the relevant sections of Chapters 16 and 31.

KEY POINTS:

- There are three major types of bone cells: osteoblasts, which secrete osteoid, osteocytes (mature bone cells) and osteoclasts, which are responsible for dissolving old bone during bone remodelling.

- Bones grow longer by the addition of cartilage at the epiphyseal growth plates. This process is stimulated by growth hormone.

- The skeleton is in a continuous state of remodelling as it responds to the demands of the body.

10.6 Disorders of the skeleton

Aside from fractures caused by trauma, a number of disorders affect the skeleton. These include congenital malformations of the bones, genetic developmental disorders such as Marfan's syndrome and achondroplastic dwarfism, and a number of endocrine abnormalities, for example acromegaly, pituitary gigantism and dwarfism, and thyroid hormone deficiency. Furthermore, certain nutritional deficiencies can influence bone formation. For example, rickets (in children) and osteomalacia (in adults) are caused by a diet lacking sufficient calcium and/or vitamin D. Other diseases affecting the skeleton include osteoporosis, osteomyelitis, Paget's disease of bone and bone tumours.

Osteoporosis

The skeleton changes throughout life as the body ages. Peak bone density is usually achieved around the age of 20–25 years. It then remains stable for around 10 years as bone resorption is roughly matched by bone accretion. After this time, the rate of bone resorption begins to outstrip new bone production and bone density gradually declines throughout middle and old age. This progressive reduction in bone density is a normal consequence of ageing. In some people, however, bone resorption greatly exceeds production so that excessive amounts of bone mineral are lost and the architecture of the bone deteriorates significantly. This metabolic bone disease is called osteoporosis. The electron micrographs in Figure 10.20 illustrate the difference in internal structure of normal and osteoporotic bone.

Although anyone, male or female, may suffer from osteoporosis, the condition is most often seen in post-menopausal women. Indeed, around a third of women between the ages of 45 and 80 suffer from some degree of osteoporosis. It is believed that the fall in oestrogen secretion that occurs after the menopause alters the relative activity of osteoblasts and osteoclasts in favour of accelerated bone resorption. As bone density falls, the fragility of the skeleton increases and there is an increased susceptibility to fractures. Common fracture sites include the distal radius, the neck of the femur and the pelvis. Compression fractures of the vertebral column (particu-

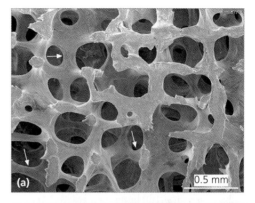
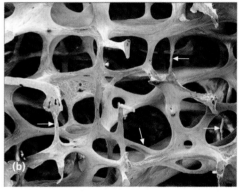

Figure 10.20 Scanning electron micrographs that show the difference in appearance between normal spongy bone (a) and bone from a person suffering from osteoporosis (b). Note the pronounced difference in the thickness of the trabeculi (some of which are indicated by the arrows).

larly those of the weight-bearing regions below T6) are also common in people with osteoporosis.

Osteomyelitis

This is a painful bacterial, or more rarely fungal, infection that causes significant destruction of bone and bone marrow. It most frequently develops as a result of an infection originating in a skin wound or bone fracture site. It causes pain in the affected bone and fever. Rapid treatment (with immobilization and antibacterial drugs) is important to halt the progress of the disease and to prevent further necrosis (death) of bone tissue.

Paget's disease of bone

After osteoporosis, Paget's disease of bone is the most common metabolic bone disorder. Indeed, its prevalence in the UK is the highest in the world. It is a chronic condition characterized by localized regions of accelerated bone turnover in which the activity of both osteoblasts and osteoclasts is increased. It may involve any part of the skeleton and, in the affected areas, normal bone matrix is replaced by patches of softened and enlarged bone. The disease often causes pain and gradual alterations of bone structure that lead to deformity.

Bone tumours

Most bone tumours arise as metastases from other sites (secondary tumours), but benign and malignant primary bone tumours do occur. These are usually described according to their behaviour and whether they form bone or cartilage. For example, an osteoid osteoma is a benign bone-forming tumour (occurring most frequently in children or young adults) while osteosarcomas and chondrosarcomas are malignant tumours that form bone and cartilage, respectively.

Drugs acting on the skeleton

Some of the drug treatments that may contribute to the management of bone disorders are summarized in Table 10.1. It is important to remember that diet, exercise and minimization of risk factors also play an important part in treating conditions such as osteoporosis.

Table 10.1 Drugs used in the treatment of bone disorders

Disorder	Treatments	Comments
Osteoporosis (reduction in bone density, most characteristically seen in postmenopausal women)	Calcium (1200–1500 mg day^{-1}), as Ca carbonate or Ca citrate Vitamin D (800–2000 U day^{-1})	
	Bisphosphonates (alendronate and risedronate)	Inhibit bone resorption by suppressing activity of osteoclasts to preserve bone mass
	Oestrogen therapy	Most effective if started within 4–6 years of menopause
	PTH (parathyroid hormone) is used in patients who fail to respond to other treatments	
Osteomyelitis (infection of bone tissue)	Antimicrobial drugs (antibiotics or antifungals depending on the source of infection)	In chronic or severe cases, surgical removal of dead bone may be necessary to allow antimicrobial drugs to work more efficiently
Paget's disease of bone (characterized by regions of increased bone turnover)	NSAIDs or stronger analgesics for pain Bisphosphonates (e.g. alendronate, etidronate, risedronate, tiludronate)	Inhibit bone resorption by suppressing activity of osteoclasts
		Joint (hip or knee) replacement may be necessary in some cases

KEY POINT:

• Principal disorders of bone include osteoporosis (especially in postmenopausal women), osteomyelitis, Paget's disease of bone and bone tumours.

10.7   Joints

The sites where bones meet are called joints (arthroses). They are the points at which the bones of the skeleton articulate to allow movements and they are stabilized by ligaments made of dense, fibrous connective tissue arranged in bands. **Arthrology** is the study of joints and their associated connective tissues.

Joints vary widely in their structure and in the range of movements that they permit. Some joints are virtually immobile while others permit a wide range of movements in several planes. In general, the most freely moveable joints tend to be the least stable. Bones are moved around joints by the actions of skeletal muscles to which they are attached by tendons. The major muscle groups will be considered in a later section.

The classification of joints

Joints may be classified according to their function or according to their structure and the nature of the tissue that binds the bones together. The three major structural types of joint are:

• Fibrous joints in which the bones are held together by fibrous connective tissue

• Cartilaginous joints in which bones are held together by cartilage

• Synovial joints in which the bones are held together by membranes that form a joint capsule enclosing a joint cavity containing synovial fluid.

Fibrous joints are also known as synarthroses. They provide great stability but very little, if any, movement. The degree of movement permitted by a fibrous joint depends on the width of fibrous tissue that separates the bones. The bones of the adult skull, for example, interlock and are bound together very tightly so that no movement is possible (Figure 10.5). These joints are fixed and

are known as **sutures**. The bones of the neonatal skull, by contrast, are separated by relatively long fibres of connective tissue (at the fontanelles), so that some movement is allowed. A similar type of fibrous joint is found between the distal ends of the tibia and fibula. Fibrous joints of this type are also called **syndesmoses** (singular syndesmosis). A third type of fibrous joint is the peg and socket joint between a tooth and the jaw bone into which it is inserted. This type of joint is called a **gomphosis**.

Cartilaginous joints allow a greater degree of movement than fibrous joints but less stability. There are two types of cartilaginous joint, synchondroses (singular synchondrosis) and symphyses (singular symphysis). **Synchondroses** are held together by hyaline cartilage and, like fibrous joints, allow very little movement. For this reason, they are classified functionally as synarthroses. Examples of such joints include the epiphyseal plates at the growing ends of the long bones and the joint between the first rib and the sternum. Both of these undergo ossification in adulthood and become immoveable. **Symphyses** are held together by a disc of fibrocartilage (although their articulating surfaces are covered by hyaline cartilage). Joints of this type include the symphysis pubis (see above) and the intervertebral discs. Symphyses are slightly moveable joints and are also called amphiarthroses.

Synovial joints (also known as diarthroses) make up the majority of the joints in the skeleton. They are the least stable joints but their structure allows the greatest degree of movement. Figure 10.21 illustrates the principal characteristics of a synovial joint. The articular surfaces of the bones forming the joint are covered by hyaline cartilage, which reduces the friction associated with movement. The space between the bones (the joint space) is enclosed by a tough fibrous capsule, the joint capsule. This capsule is very flexible and allows considerable movement together with great strength, which helps to

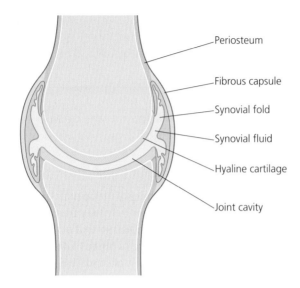

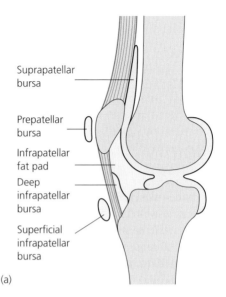

(a)

Figure 10.21 A simple diagram to illustrate the principal features of a typical synovial joint. The articular (hyaline) cartilage is shown in green. The joint cavity is filled with synovial fluid.

prevent dislocation of the joint. The capsule possesses a well-developed sensory nerve supply, which conveys information to the CNS regarding the direction, rate and acceleration of joint movement and signals excessive movement of the joint via pain fibres.

The joint space is lined by the **synovial membrane**, whose cells secrete synovial fluid, which fills the space and provides lubrication and nourishment to the articular (hyaline) cartilage. Synovial fluid also contains phagocytic cells, which remove microbes and any debris that accumulates as a result of wear and tear at the joint. The joint space is often called the synovial cavity.

Certain joints, including the knee and shoulder, in which skin rubs over bone, contain small sac-like structures called **bursae** (singular bursa, the Latin word for 'purse'). These are filled with a fluid similar to synovial fluid, which provides a cushion between the skin and bone and reduces the frictional forces associated with movement of the joint. Inflammation of a bursa (a painful condition known as **bursitis**) may be caused by repeated or excessive use of a joint, a physical trauma to the joint, by a chronic infection such as tuberculosis, or by rheumatoid arthritis. The diagram in Figure 10.22a illustrates the location of bursae within the knee joint.

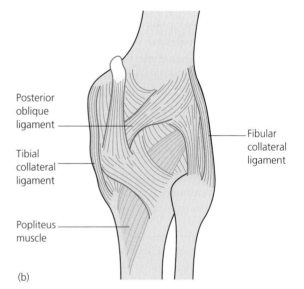

(b)

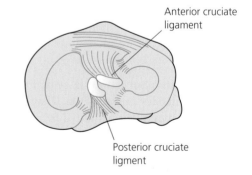

(c)

Figure 10.22 The structure of the knee joint. (a) The bursae that help to reduce the friction of movement within the joint. The large suprapatellar bursa is continuous with the capsule of the knee joint. The other bursae are separate sacs located at points of greatest friction. (b) The extracapsular ligaments that help to stabilize the knee joint. (c) The arrangement of the intracapsular (cruciate) ligaments.

The stability of joints

Although the principal role of synovial joints is to allow movement, it is also important that joints are as stable as possible. The degree of stability of a joint is determined by the shape of its articulating surfaces, the ligaments associated with the joint, and the muscles that provide movement of the joint. Many synovial joints are strengthened and stabilized by ligaments situated both within the joint capsule (intracapsular ligaments) and outside it (extracapsular ligaments). The knee joint, for example, possesses both intracapsular ligaments (the anterior and posterior cruciate ligaments) and extracapsular ligaments (the fibular and tibial collateral ligaments), which are illustrated in Figure 10.22.

The shoulder joint is the most freely moveable of all the joints. It is also one of the least stable, because of the shape of its articulating surfaces (see below). It therefore dislocates relatively easily. To minimize the risk of dislocation, the shoulder joint is strengthened and stabilized by four deep muscles, which, together with their tendons, form the so-called **rotator cuff**. These muscles are the subscapularis, the supraspinatus, the infraspinatus and the teres minor. Their tendons form an almost complete ring (or cuff) around the shoulder joint, as shown in Figure 10.23. The muscles of the rotator cuff, in particular

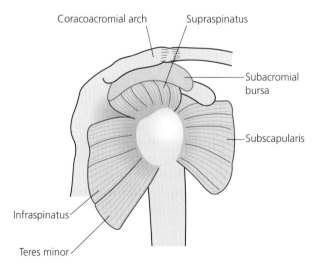

Figure 10.23 The arrangement of muscles of the rotator cuff that help to stabilize the shoulder joint.

Coracoacromial arch
Supraspinatus
Subacromial bursa
Subscapularis
Infraspinatus
Teres minor

the supraspinatus, are highly susceptible to damage through wear and tear, and rotator cuff lesions are a common cause of shoulder pain.

Types of synovial joints

Although all synovial joints share the characteristics illustrated in Figure 10.21, there are many variations between individual joints. These structural differences determine the exact degree and plane of movements that the joint can perform. Six subtypes of synovial joints are recognized: hinge, ball-and-socket, planar (or gliding), pivot, condyloid and saddle joints.

- **Hinge joints** permit movements in one plane only, rather like the opening and closing of a door. They are, therefore, described as monaxial joints. The convex end of one bone fits into a concave or trough-like depression on the other bone. Typically, at hinge joints, one bone remains in a fixed position while the other bone moves around it to bring about flexion (bending) and extension (straightening) of the joint. Examples of hinge joints include the knee, the elbow and the joints of the fingers and toes.

- **Ball-and-socket joints** are multiaxial, i.e. they permit movement in all planes. The rounded, ball-like head of one bone fits into a spherical socket of the other bone and can rotate within it. The most familiar example of a ball-and-socket joint is the hip in which the head of the femur fits into the acetabulum of the hip bone. The shoulder is also a ball-and-socket joint although here the socket (the glenoid cavity of the scapula) is very shallow. As explained above, this gives the shoulder great freedom of movement but also means that it is relatively unstable.

- **Planar or gliding joints** are described as non-axial because the short slipping or gliding movements they allow do not occur around an axis. The articulating surfaces of the bones at planar joints are flat or slightly curved. Examples include the joints between the carpals of the wrist and between the tarsals of the foot.

- **Pivot joints** are monaxial, allowing movement only around their longitudinal axis. An example of a pivot joint is provided by the joint in the neck between the atlas and the axis (the first and second cervical vertebrae) in which the anterior arch of the atlas rotates around the axis to allow side-to-side movements of the head (as in shaking the head to indicate 'no').

- **Condyloid joints** are biaxial. They allow movement in two planes or axes. Here, an oval projection of one bone articulates with an oval depression in another. An example is the wrist (at the radiocarpal joint).

- **Saddle joints** are biaxial and similar to condyloid joints except that the projection of one bone fits into a saddle-shaped depression in the other. An example of a saddle joint is the joint at the thumb between the carpal and metacarpal.

KEY POINTS:

- Joints are the sites at which bones meet. They are moved by the action of skeletal muscles and are stabilized by ligaments and tendons.

- They are classified according to their structure or the degree of mobility they allow.

- Fibrous joints are immovable, cartilaginous joints allow limited movement and synovial joints are freely moveable.

- Synovial joints are classified according to their degree of freedom of movement as hinge, ball-and-socket, planar, pivot, condyloid or saddle joints.

Disorders of joints

Throughout life, the joints are subjected to continual wear and tear, which, over time, can result in degenerative and other changes. Furthermore, if a person regularly engages in sport, particularly of the kind that demands explosive or repetitive movements, the risk of damage to joint structures is increased. Dislocations and fractures are usually the result of physical trauma. These are common, but a number of other diseases can

also affect the joints. The following are among the most important:

- **Osteoarthritis** (also known as degenerative arthritis or degenerative joint disease) is a chronic condition that affects mainly older people and seems to result from wear and tear on the joint surfaces. It is estimated that around 25% of women and 15% of men over the age of 60 years show signs of this disease. It is characterized by thinning and softening of the articular cartilage with subsequent disintegration. It causes pain, stiffness and sometimes swelling of the affected joints, particularly after inactivity.

- **Rheumatoid arthritis** is an inflammatory condition in which joints (most often those of the hands and feet) become swollen and painful. The most severe cases may lead to destruction of the affected joints. The condition affects around 1% of the population worldwide, with women affected approximately two to three times as often as men. It usually develops between the ages of 35 and 50 years, but it can occur at any age, including childhood. Although the exact cause of rheumatoid arthritis is unknown, it is considered to be an autoimmune disease in which the immune system causes the inflammatory changes that damage the joints and connective tissues. Fever and damage to other tissues such as the blood vessels and lungs may also occur. A range of tests is used to diagnose rheumatoid arthritis, including X-rays, MRI scans, examination of synovial fluid and a variety of blood tests to detect the presence of the particular antibodies that are characteristic of the condition.

- **Gout** is really a form of arthritis that is caused by sustained high levels of uric acid in the blood (hyperuricaemia). Uric acid crystals form and accumulate in the synovial cavities of joints giving rise to an inflammatory reaction. Men are about ten times more likely than women to suffer from gout, which causes severe pain, redness and swelling of the affected joint. The big toe is the most commonly affected joint, but gout can occur in any synovial joint.

Table 10.2 summarizes some of the drug treatments that can be used for the management of joint disorders.

Table 10.2 Drugs used in the treatment of joint disorders

Disorder	Treatments	Comments
Osteoarthritis (degenerative disease affecting the articular cartilage of joints)	Analgesics (paracetamol and NSAIDs) are used to alleviate pain	Life-style changes, weight loss and appropriate exercise programmes are all important in treating the condition
Rheumatoid arthritis (inflammatory disease affecting joints and many other body tissues)	NSAIDs relieve pain and stiffness	Joint replacement is an option for severe cases
	Prednisolone (a corticosteroid anti-inflammatory drug)	Steroid drugs have numerous side-effects including thinning of bones and skin, weight gain and hypertension
	Combinations of disease-modifying antirheumatic drugs (DMARDs) help to slow the progress of the disease	
	Examples include methotrexate, sulphasalazine and a variety of newer immunosuppressive agents (e.g. rituximab, etanercept, infliximab, anakinra, abatacept)	
Gout (a form of arthritis caused by chronic elevation of plasma uric acid levels)	NSAIDs to relieve pain	
	Oral colchicine may treat or prevent attacks	
	Allopurinol	Inhibits urate synthesis and lowers plasma uric acid
	Probenecid and sulphinpyrazone	Uricosuric drugs that increase the rate of excretion of urate

KEY POINTS:

- Principal disorders of joints include osteoarthritis, rheumatoid arthritis and gout.
- Analgesic, anti-inflammatory and immunosuppressant drugs are used to treat joint disorders.

10.8 The skeletal muscles

More than 600 skeletal muscles form the muscular system. The locations of some of the more important of these may be seen in Figure 10.24. They contract in response to electrical activity in the motor neurones that innervate them and act as levers to pull bones into new positions around joints. At most joints, one bone remains stationary (or almost so) while the other moves in relation to it. In this way, they are responsible for the voluntary movements that are required for fine manipulation, locomotion, the maintenance of posture and balance, the expression of feelings by the face and for the intricate movements associated with activities such as speech.

Skeletal muscles are attached to the periosteum of bones by strong cords of dense connective tissue called **tendons**. A familiar example is the Achilles (calcaneal) tendon, which connects the calf muscle (gastrocnemius) to the heel (calcaneus). Many tendons, including those of the ankle and wrist, are enclosed by sheaths of fibrous connective tissue containing synovial fluid. These are called **tendon (or synovial) sheaths** and they are similar in structure and function to the bursae described earlier (p. 164). Their function is to reduce the friction associated with the sliding movements of the tendons during muscle contraction. Trauma or repetitive use may lead to inflammation of a tendon, tendon sheath or synovial

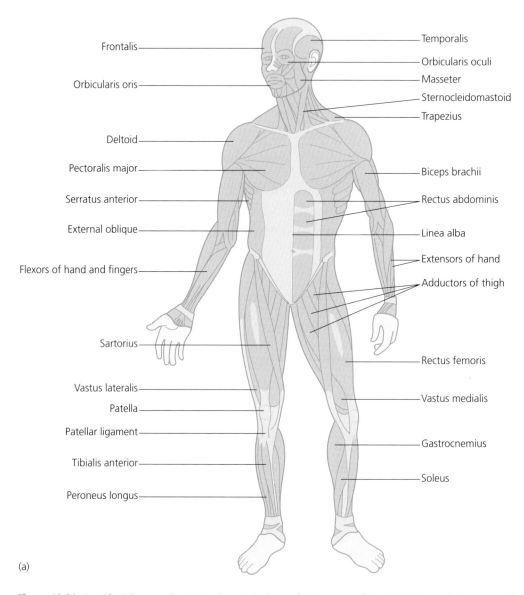

Frontalis

Orbicularis oris

Deltoid

Pectoralis major

Serratus anterior

External oblique

Flexors of hand and fingers

Sartorius

Vastus lateralis

Patella

Patellar ligament

Tibialis anterior

Peroneus longus

Temporalis

Orbicularis oculi

Masseter

Sternocleidomastoid

Trapezius

Biceps brachii

Rectus abdominis

Linea alba

Extensors of hand

Adductors of thigh

Rectus femoris

Vastus medialis

Gastrocnemius

Soleus

(a)

Figure 10.24 Simplified diagrams illustrating the principal superficial muscles of the body; (a) an anterior view and (b) a posterior view. *(Continued)*

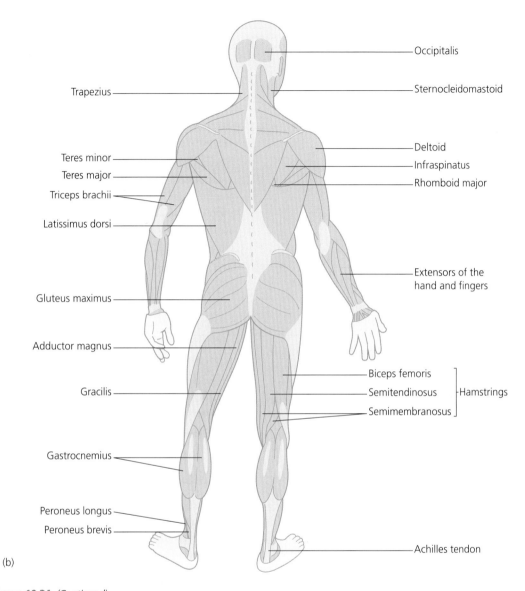

Occipitalis

Sternocleidomastoid

Trapezius

Deltoid

Teres minor

Infraspinatus

Teres major

Rhomboid major

Triceps brachii

Latissimus dorsi

Extensors of the
hand and fingers

Gluteus maximus

Adductor magnus

Biceps femoris

Gracilis

Semitendinosus

Hamstrings

Semimembranosus

Gastrocnemius

Peroneus longus

Peroneus brevis

Achilles tendon

(b)

Figure 10.24 *(Continued)*

membrane – a condition known as **tenosynovitis**. Commonly affected joints are the elbow (where it is known colloquially as 'tennis elbow'), fingers, feet and wrists. Although it may affect anybody, sportsmen and women (especially dancers and gymnasts) who perform repetitive or excessive joint movements are particularly prone to this condition.

The principal muscle groups

The anatomy, histological appearance and physiological properties of skeletal muscle were described in Chapter 9, along with neuromuscular transmission. This section presents a brief account of the principal muscle groups, together with a description of the movements that they bring about as a result of their contraction.

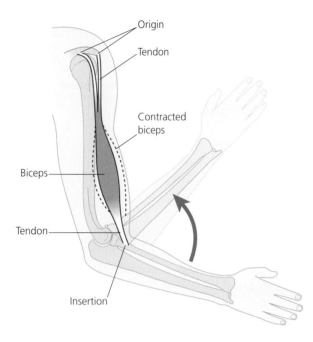

Origin

Tendon

Contracted biceps

Biceps

Tendon

Insertion

Figure 10.25 A diagram to illustrate the lever action of the biceps brachii muscle at the elbow joint. When the muscle contracts, its insertion moves towards its origin. In this case, the forearm is pulled towards the upper arm and the elbow is flexed. The biceps brachialis is classified as a flexor muscle of the elbow joint.

- In the muscles of the limbs, the origin is usually proximal and the insertion distal. Furthermore, the insertion is usually pulled towards the origin when the muscle contracts. Figure 10.25 illustrates the origin and insertion of the biceps brachii muscle and shows how contraction of the muscle moves the insertion towards the origin and thus moves the lower arm in relation to the stationary upper arm to flex the elbow.

- Muscles act in pairs or groups around a joint and the members of a pair or group often have antagonistic actions, e.g. at the elbow joint the biceps flexes the elbow while the triceps extends it. These muscles form an antagonistic pair.

KEY POINTS:

- Skeletal muscles contract in response to action potentials passing along motor nerves. They act in pairs or groups at joints to pull bones into new positions, thus allowing movement.

- Muscles are attached to bones by tendons.

- Muscles are named according to their size, location, shape, action or the orientation of their fibres.

The naming of muscles and their locations is highly complex. It will be helpful, therefore, to bear the following general rules in mind:

- Muscles may be named according to their relative size (maximus, minimus, longus, etc.), the direction of their muscle fibres (rectus, oblique, etc.), their location (temporalis, frontalis, etc.), their shape (e.g. deltoid muscle) or their action (flexor, adductor, etc.).

- The bulging middle region of a muscle is called the belly (or gaster).

- Every muscle has both an origin and an insertion. Its **origin** is its point of attachment to the stationary bone of the joint. Its **insertion** is its point of attachment to the moving bone.

- Some muscles have more than one origin, e.g. the biceps has two origins, the triceps has three and the quadriceps has four.

The muscles of the head and neck

Figure 10.26 illustrates the superficial muscles of the face viewed from one side. These muscles permit a large variety of different facial expressions and are important in speaking and in chewing. Table 10.3 lists the principal muscles of the head and neck, describes their origins and insertions, and indicates their principal actions.

The muscles of the trunk

Figure 10.27 shows the principal muscles of the trunk, which can be divided into anterior and posterior groups. The anterior muscles form the anterior thorax (chest) and the muscles of the abdominal wall. The anterior muscles of the chest move the ribs, head and arms. The posterior muscles support and move the vertebral column while the muscles of the abdominal wall form a strong 'girdle' that reinforces the trunk and contains and protects the

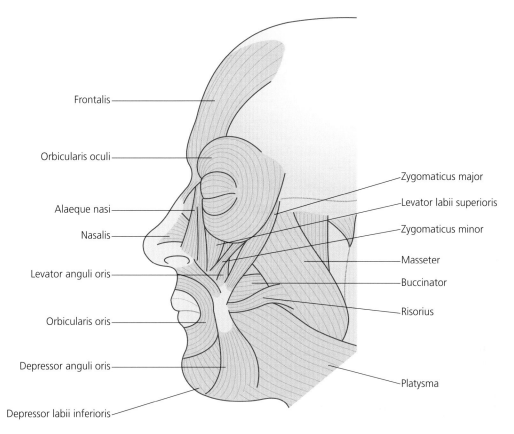

Frontalis

Orbicularis oculi

Alaeque nasi

Nasalis

Levator anguli oris

Orbicularis oris

Depressor anguli oris

Depressor labii inferioris

Zygomaticus major

Levator labii superioris

Zygomaticus minor

Masseter

Buccinator

Risorius

Platysma

Figure 10.26 A left lateral view of the superficial muscles of the face that are involved in chewing, speaking and facial expression.

abdominal organs. Table 10.4 lists the names, locations and principal actions of these muscles.

The muscles of the upper limbs

Three groups of muscles form the skeletal system of the arm. These are the muscles that move the entire arm at the shoulder, the muscles that move the elbow and the muscles that move the bones of the hand.

Muscles that move the arm from the shoulder

These include the pectoralis major, latissimus dorsi and the deltoids, which are described above and shown in Figure 10.27. A number of scapular muscles also contribute to shoulder movements. These include the muscles of the rotator cuff (see p. 165), the teres major and the coracobrachialis (see Figure 10.29).

Muscles that move the elbow

These are located within the upper arm and are illustrated in Figures 10.28 and 10.29. The anterior upper arm muscles are all flexor muscles i.e. they flex or bend the elbow. The most significant of these are the **biceps brachii**, and the **brachialis** muscles. The biceps brachii bulges when the elbow is bent. The biceps brachii has (as its name suggests) two origins from the scapula and it inserts into the radius. Contraction of this muscle brings about flexion of the elbow and supination of the forearm. The brachialis muscle acts in a similar way to the biceps, causing flexion of the ulna when the arm is bent.

The **triceps brachii** forms the bulk of the posterior portion of the upper arm (Figure 10.29) and its contraction brings about extension of the elbow (straightening of the arm). It is therefore described as an

Table 10.3 Principal muscles of the head and neck

Name of muscle	Location within head or neck	Principal action
Frontalis	Covers the frontal bone of the skull It originates in the cranial aponeurosis (a flattened tendon that covers the dome of the skull) and inserts into the skin of the eyebrow and nose	Moves the scalp to raise the eyebrows and wrinkle the forehead
Orbicularis oris	Circular muscle around the lips Origins are the mandible and maxilla Insertion is into the skin and muscle around the mouth	Closes the mouth and protrudes the lips
Orbicularis oculi	Concentric fibres encircling the orbit Origins are in the frontal bone and maxilla Insertion is into tissue around the eyes	Closes the eyes, squints and blinks
Zygomaticus	Originates in the zygomatic arch and inserts into the muscles at the angle (corner) of the mouth	Raises the corners of the mouth, as in smiling
Buccinator	Lies horizontally across the cheek It originates in the maxilla and mandible and inserts into the orbicularis oris	Flattens the cheek, as in blowing, and contracts during chewing
Masseter	Originates in the zygomatic arch and is inserted in the mandible	Elevates the mandible to close the jaw while chewing or speaking
Temporalis	A fan-shaped muscle that covers part of the temporal bone of the skull Originates in the temporal fossa and inserts into the coranoid process of the mandible	Closes the mouth and assists with chewing
Occipitalis	Covers the occipital bone of the skull It originates in the occipital bone and inserts into the cranial aponeurosis	Moves the scalp
Platysma	Covers anterior and lateral neck It originates from chest muscles and inserts into the lower border of the mandible	Pulls the corners of the mouth downwards
Sternocleidomastoid	Located on either side of the neck Origins are in the sternum and clavicle and insertion is into the mastoid process of the temporal bone	Flexes the neck and tilts the head towards the shoulder

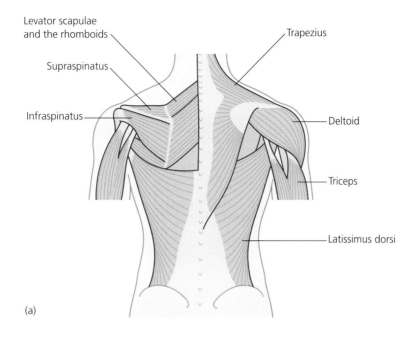

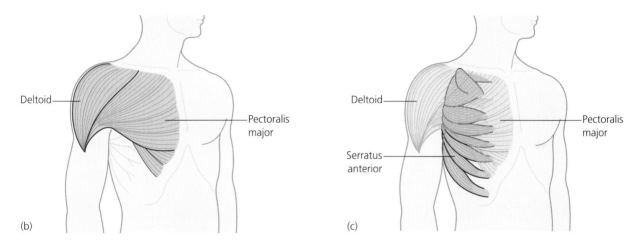

Figure 10.27 Diagrams to illustrate some of the superficial muscles of the trunk and shoulder; (a) the muscles of the back that act to move the scapula and humerus; (b) some of the muscles of the shoulder and (c) the serratus muscles of the medial wall of the axilla, most of which are obscured by the pectoralis major when viewed from the front.

extensor muscle and its action is antagonistic to that of the biceps. The triceps has three origins in the scapula and proximal humerus and inserts into the ulna at the olecranon process.

Muscles that move the wrist and hand

The muscles of the lower arm move the bones of the wrists and hands. This group includes many thin muscles that flex and extend the wrists and fingers such as the

Table 10.4 The principal muscles of the trunk

Name of muscle	Location within the trunk	Principal action
Muscles of the anterior thorax (chest)		
Pectoralis major	A large fan-shaped muscle covering the upper chest	Flexes the humerus and pulls it towards the body (adducts)
	Its origins are in the sternum, clavicle and ribs 1–6 and its insertion is in the humerus	
Intercostal muscles	Deep muscles located between the ribs (in the intercostal spaces)	External intercostals raise the ribcage during inspiration Internal intercostals lower the ribcage during forced expiration
Muscles of the abdominal wall		
Paired rectus abdominis muscles	Extend along the length of both sides of the abdomen	Flex the vertebral column, tense the anterior abdominal wall and help to contain the abdominal contents
	Their origin is the pubis and they insert in the sternum and ribs 5–7	
Paired external oblique muscles	Form the lateral walls of the abdomen	Flex the vertebral column, rotate the trunk and bend it laterally
	Originate in ribs 5–12 and insert into the iliac crest	
Paired internal oblique muscles	Deep muscles whose fibres run at right angles to those of the external oblique muscles	As for the external oblique muscles
Transversus abdominis	Runs horizontally across the abdomen	Compresses the abdominal contents
	Originates in ribs 7–12 and the iliac crest and inserts into the pubis	
Posterior (back) muscles		
Paired trapezius muscles	The most superficial muscles of the neck and posterior trunk forming a distinctive diamond shape	Extend the head and bring about movements of the scapula
	Origins are very broad	
	Each muscle extends from the occipital bone to the last thoracic vertebra	
Paired latissimus dorsi	Cover the lower back	Adduct and extend the arms and rotate the arms medially
	Origins are in the lower thoracic and lumbar vertebrae, sacrum and iliac crest	

(Continued)

Table 10.4 *(Continued)*

Name of muscle	Location within the trunk	Principal action
	Insertion is into the humerus	
Paired deltoid muscles	Form the rounded part of the shoulders	Abduct, extend, flex and rotate the arms (often used for intramuscular injections)
	Origins are in the clavicle and scapula and insertion is into the lateral shaft of the humerus	
Paired quadratus lumborum muscles	Form part of the posterior abdominal wall	Flex and extend the lumbar spine
	They originate in the iliac crests and insert into the 12th ribs and upper lumbar vertebrae	
Paired erector spinae muscles	Originate from the sacrum, ilium and lumbar vertebrae	Extend the back to maintain an erect posture
	Muscle divides to form columns that, together, span the length of the vertebral column	

flexor digitorum superficialis, flexor carpi and extensor digitorum. Figures 10.28 and 10.29 illustrate some of the superficial muscles of the arm.

The muscles of the lower limbs

The muscles of the legs are specialized for walking and for balancing the body in an upright position. Their actions allow movement of the hips, knees and feet.

Muscles that move the hip

The muscles responsible for moving the hip joint are illustrated in Figures 10.30 and 10.31. They include the gluteus maximus, gluteus medius, iliopsoas and the adductor group. The powerful **gluteus maximus** forms most of the flesh of the buttock. Its origin is in the ilium, sacrum and coccyx, and its insertion is in the gluteal ridge of the proximal femur. It extends the hip and is very important in movements such as those required for jumping or climbing. The **gluteus medius** runs mostly behind the gluteus maximus. It originates in the ilium and inserts into the proximal femur. Its principal action is to rotate the thigh and abduct the hip to stabilize the pelvis during walking. It is also an important site for giving intramuscular

injections. The **iliopsoas** is an important flexor muscle of the hip formed from the iliacus and psoas major muscles. It helps to stop the body from falling over backwards when standing upright. It originates at lumbar vertebrae 1–5 and inserts into the femur. The **adductor muscles** are a group of muscles of the inner (medial) thigh. They adduct the thighs (compress them together). They originate in the pubic region of the pelvis and insert into the proximal femur.

Muscles that move the knee

The principal muscles that move the knee include the hamstring group, the sartorius muscle and the quadriceps group. All of these can be seen in Figures 10.30 and 10.31. The **hamstring** muscles actually consist of three separate muscles that form the bulk of the posterior thigh. They are the biceps femoris, the semitendinosus and the semimembranosus muscles as shown in Figure 10.31. The hamstrings span the hip and knee joints and are thus responsible for flexion of the hinge joint at the knee and extension of the upper leg. They have their origins in the ischium and insert into the proximal tibia. Injuries to these muscles (especially at their proximal regions) are common in sportsmen and -women who

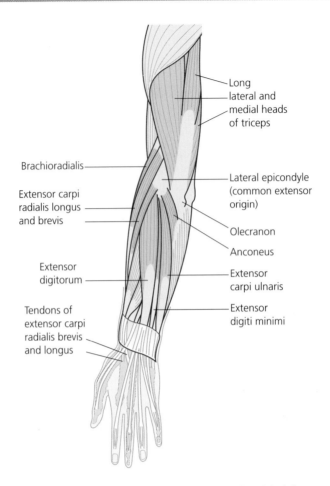

Long
lateral and
medial heads
of triceps

Brachioradialis

Extensor carpi
radialis longus
and brevis

Lateral epicondyle
(common extensor
origin)

Olecranon

Anconeus

Extensor
digitorum

Extensor
carpi ulnaris

Tendons of
extensor carpi
radialis brevis
and longus

Extensor
digiti minimi

Figure 10.28 A posterior view of the superficial muscles of the left arm.

carry out violent muscular exertion, particularly sudden starts and stops.

The gracilis originates from the pubis and passes along the medial side of the thigh before inserting in the tibia. The muscle adducts the thigh (moves it towards the midline) and flexes the knee. Its chief importance is as a balancing muscle.

The **sartorius** muscle is the longest muscle in the body. It originates in the iliac crest and inserts into the medial side of the tibia. Its contraction brings about flexion of the knee joint and lateral rotation of the thigh at the hip joint. The **quadriceps** muscle group comprises four muscles, the rectus femoris, vastus lateralis, vastus medialis and vastus intermedius (which lies deeper and is not shown in Figure 10.30). These form the flesh of the front of the thigh and

their contraction brings about powerful extension of the knee (as when kicking a ball). They also contribute to flexion of the hip. The rectus femoris originates in the pelvis and the three vastus muscles originate from the femur. All of them insert into the patella (via the quadriceps tendon) and then the tibia (via the patellar ligament).

Muscles that move the foot and ankle

The superficial anterior muscles of the lower leg include the **tibialis anterior**, the **extensor digitorum longus** and the **peroneus** muscles. All of these bring about dorsiflexion of the foot (a movement in which the sole is moved upwards to reduce the angle between the foot and the leg). The **tibialis anterior** originates from the proximal tibia and inserts into the tarsal bones. The **extensor digitorum longus** runs

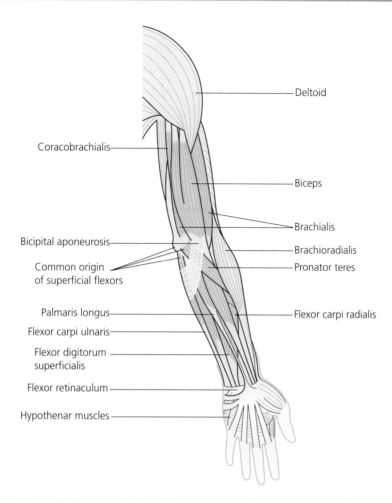

Deltoid

Coracobrachialis

Biceps

Brachialis

Bicipital aponeurosis

Brachioradialis

Common origin
of superficial flexors

Pronator teres

Palmaris longus

Flexor carpi radialis

Flexor carpi ulnaris

Flexor digitorum
superficialis

Flexor retinaculum

Hypothenar muscles

Figure 10.29 An anterior view of the principal superficial muscles of the left arm.

laterally to the tibialis anterior. It originates at the lateral tibia and anterior fibula and inserts into the phalanges of toes 2–5. It dorsiflexes the foot and extends the toes. The peroneus muscles are located on the lateral part of the lower leg. The peroneus longus has its origin on the upper lateral surface of the fibula while the lower region of the lateral surface of the fibula gives rise to the peroneus brevis, which lies beneath the peroneus longus. These muscles insert into the metatarsals of the foot.

A number of muscles contribute to movements of the ankle and foot. The principal superficial muscles on the posterior surface of the lower leg (the calf muscles) are the gastrocnemius, soleus and plantaris muscles. These may be seen in Figure 10.31. All three insert into the calcaneus (heel) by way of the Achilles (calcaneal) tendon and all bring about plantar flexion of the foot at the ankle joint (plantar flexion is movement of the sole of the foot downwards so that the angle between the foot and the leg is increased, as in pointing the toes). The **gastrocnemius** is the most superficial of the calf muscles. It originates at the distal femur and forms the bulge of the calf. The **soleus** muscle originates at the head of the fibula and the medial border of the tibia and the small **plantaris** muscle originates at the femur.

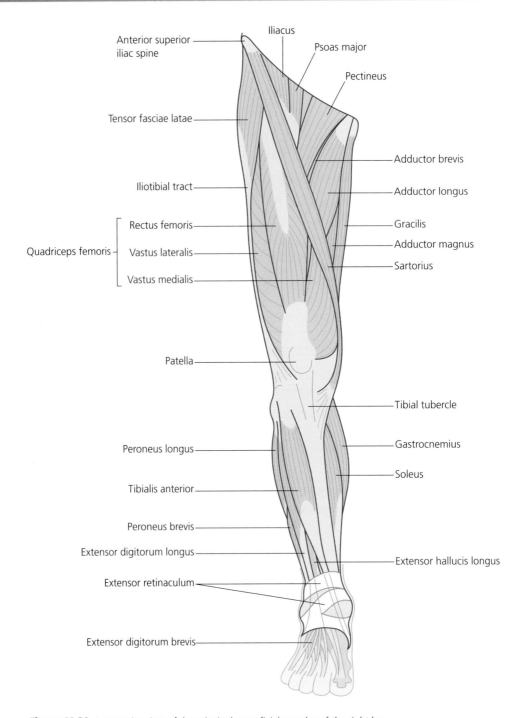

Figure 10.30 An anterior view of the principal superficial muscles of the right leg.

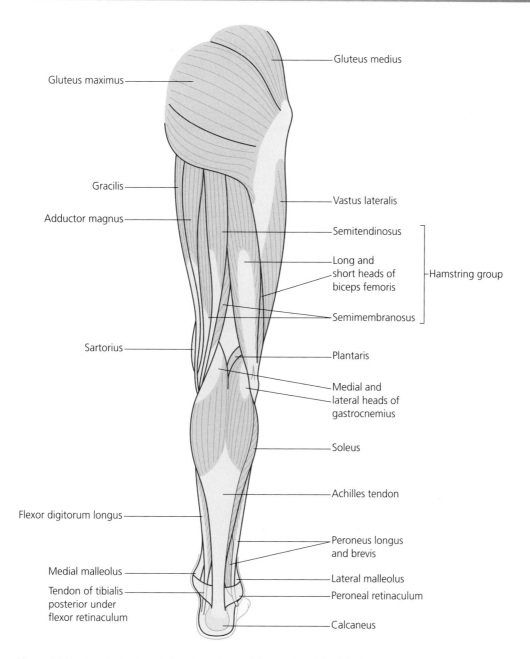

Figure 10.31 A posterior view of the principal superficial muscles of the right leg.

10.9   Disorders of skeletal muscle

A number of disorders have metabolic effects that can result in muscular degeneration and weakness. These include Cushing's syndrome (in which there is excessive secretion of cortisol), hyperthyroid disease and osteomalacia. The principal primary disorders of muscle are polymyositis, muscular dystrophy and myasthenia gravis.

- **Polymyositis** is an inflammatory autoimmune condition in which T lymphocytes (see Chapter 21) destroy muscle fibres, causing muscle weakness. Its cause is unknown.
- **Muscular dystrophy** is not a single condition but rather a group of inherited diseases (the muscular dystrophies) characterized by a progressive loss of muscle fibres. The most severe forms of the disease develop in childhood (e.g. Duchenne muscular dystrophy) but milder forms may not appear until adulthood.
- **Myasthenia gravis** is an autoimmune disease that affects the acetylcholine receptors at the neuromuscular junction (see Chapter 9).

The principal drugs that are used in the treatment and management of muscular disorders are summarized in Table 10.5.

> **KEY POINTS:**
>
> - Principal disorders of skeletal muscle include polymyositis, myasthenia gravis and the muscular dystrophies. All result in muscle weakness.
> - Anti-inflammatory and immunosuppressant drugs are helpful in the treatment of polymyositis. Anticholinesterase drugs are used in the treatment of myasthenia gravis.

Table 10.5 Drugs used in the treatment of muscular disorders

Disorder	Drug	Comments
Polymyositis	Anti-inflammatory corticosteroids, e.g. prednisolone	
	Immunosuppressants such as methotrexate, zathioprine and cyclosporine	Emerging therapies include rituximab and antitumour necrosis factor (anti-TNF)
Muscular dystrophies	No specific drug therapies are available but gene therapy research is ongoing	
Myasthenia gravis	Anticholinesterase drugs, e.g. pyridostigmine, neostigmine	Inhibit acetylcholinesterase (the enzyme that breaks down acetylcholine at the neuromuscular junction)
	Immunosuppressants, e.g. corticosteroids, azathioprine, cyclosporine	

Recommended reading

Anatomy

Rohen, J.W., Yokochi, C. and Lutjen-Drecoll, E. (2006). *Colour Atlas of Anatomy*, 6th edn. Lippincott, Williams and Wilkins: Baltimore.

Dean, C. and Pegington, J. (2005). *Core Anatomy for Students*, Vols 1, 2 and 3. Saunders: London.

MacKinnon, P.C.B. and Morris, J.F. (2005). *Oxford Textbook of Functional Anatomy*, Vol. 1. Musculo-skeletal system. Oxford University Press: Oxford.

Physiology

Pocock, G. and Richards, C.D. (2006). *Human Physiology, the Basis of Medicine*, 3rd edn. Oxford University Press: Oxford. Chapter 23.

Self-assessment questions

Which of the following statements are true and which are false? Answers are given at the end of the book (p. 755).

The following statements relate to the skeleton:

1. a) The ilium is part of the appendicular skeleton.
   b) The sternum is a sesamoid bone.
   c) The sphenoid bone supports the pituitary gland.
   d) The auditory tube is located in the parietal bone of the skull.
   e) The metatarsals are distal to the tarsals.

2. a) The acetabulum is found in the scapula.
   b) The fibula is the main weight-bearing bone of the lower leg.
   c) The talus articulates with the tibia and fibula.
   d) The olecranon is part of the ulna.
   e) There are five lumbar vertebrae.

The following statements relate to the physiology of bone:

3. a) Bone is a specialized connective tissue.
   b) Osteoid is secreted by osteocytes.
   c) Long bones increase in length as new cartilage is deposited at the epiphyseal plate.
   d) The marrow cavity of the sternum houses haematopoietic tissue.
   e) The shaft of a long bone is called the epiphysis.

4. a) Osteoporosis is more common in men than women.
   b) Vitamin D deficiency causes rickets in children.
   c) Trabecular bone is lighter than compact bone.
   d) Calcium and potassium are the principal inorganic components of bone.
   e) Bone mass starts to fall after puberty.

The following statements relate to joints:

5. a) All joints are freely moveable.
   b) The elbow is an example of a hinge joint.
   c) Fibrous joints connect the bones of the cranium.
   d) Synovial fluid is found in cartilaginous joints.
   e) The rotator cuff is found in the knee joint.

6 a) Pivot joints are synovial joints.
   b) Intervertebral discs are made of bone.
   c) Joints are stabilized by ligaments.
   d) The temporomandibular joint is the only moveable joint found in the skull.
   e) Gout is caused by too little uric acid in the blood.

The following statements relate to skeletal muscle:

7 a) Skeletal muscles are innervated by motor nerves.
   b) Muscles are attached to bones by tendons.
   c) A muscle is inserted into the moving bone near a joint.
   d) The biceps brachii is the extensor muscle of the elbow joint.
   e) The masseter muscle closes the jaw during chewing.

8 a) The pectoralis major muscle covers the upper back.
   b) The internal intercostal muscles lower the ribcage during forced expiration.
   c) The quadriceps muscle is found in the lower leg.
   d) The biceps femoris muscle is one of the hamstrings.
   e) The gastrocnemius muscle forms the bulge of the calf.

The organization of the nervous system

11

After reading this chapter you should have gained an understanding of:

- The main divisions of the nervous system
- Its principal cell types
- The principal structural features of the brain and spinal cord
- The structure of the peripheral nerve trunks
- The role of the cerebrospinal fluid
- The cerebral circulation

11.1   Introduction

The nervous system may be divided into five main parts:

- The brain
- The spinal cord
- The peripheral nerves
- The autonomic nervous system
- The enteric nervous system.

The brain and spinal cord constitute the **central nervous system** (CNS) while the peripheral nerves, auto-nomic nervous system and enteric nervous system of the gastrointestinal tract make up the **peripheral nervous system**. The **autonomic nervous system** is concerned with the innervation of the blood vessels and the internal organs and includes the paired ganglia that run parallel to the spinal column (the paravertebral ganglia) and their associated nerves. It will be discussed in Chapter 13. The **enteric nervous system** controls the activity of the gut and will be discussed in Chapter 24.

11.2 The cells of the CNS

Information processing by the nervous system is carried out by the neurones, whose properties were discussed in Chapter 8 (p. 104). The cell bodies of neurones are found throughout the grey matter of the brain and spinal cord (see below) and in various ganglia scattered throughout the body. Neurones are very varied in both size and shape, but they all receive information from other neurones via their dendrites and give rise to a single axon that sends branches to other neurones or to innervate the peripheral organs (see Figure 8.1).

The **glial cells** or **neuroglia** play a supporting role in the CNS and occupy the spaces between the neuronal processes. As Figure 11.1 shows, there are three different classes of glial cell in the brain and spinal cord:

1. **Astroglia** or **astrocytes.** These are star-shaped cells that have long processes (Figure 11.1a). They are found throughout the CNS and their processes end on small blood vessels, as shown in Figure 11.1c, to form an additional barrier between the blood and the extracellular

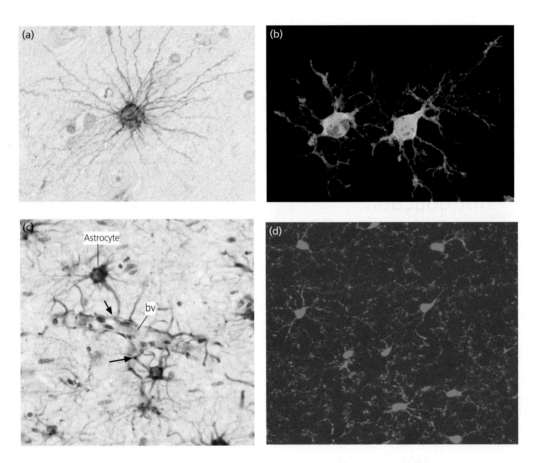

Figure 11.1 This figure shows examples of the three classes of glial cell that are found in the CNS. (a) An astrocyte in the cerebral cortex stained to show a specific protein known as glial fibrillary acidic protein (GFAP). (b) Two oligodendrocyte precursors tagged with two fluorescent proteins. (c) The close association of astrocyte processes with small blood vessels (bv). The cell bodies of several astrocytes are visible in this image. (d) Microglia from the cerebral cortex labelled with green fluorescent protein.

fluid of the brain and spinal cord. This barrier is known as the **blood–brain barrier** and it serves to prevent changes in the composition of the blood influencing the activity of the nerve cells within the CNS.

2. **Oligodendroglia** or **oligodendrocytes**. Oligodendrocytes account for about 75% of all glial cells in white matter, where they form the myelin sheaths of axons. In the peripheral nervous system, the myelin sheaths are formed by Schwann cells (see p. 105). Oligodendrocyte precursor cells are shown in Figure 11.1b.

3. **Microglia**. These are scattered throughout the grey and white matter, as shown in Figure 11.1d. They are phagocytes and rapidly converge on a site of injury or infection within the CNS.

The central fluid-filled spaces of the brain (the cerebral ventricles, see below) and the central canal of the spinal cord are lined with ciliated cells known as **ependymal cells**, which form a cuboidal-columnar epithelium called the **ependyma**.

11.3 The organization of the brain and spinal cord

As Figure 11.2 shows, the brain and spinal cord lie within a bony case formed by the skull and vertebral canal of the spinal column. Both are covered by three membranes called the **meninges**. A tough outer membrane of dense connective tissue called the **dura mater** (or **dura**) covers the whole brain and spinal cord. Attached to the inner face of the dura is the **arachnoid membrane**. Beneath the arachnoid lies the highly vascularized **pia mater** (or **pia**), which is attached to the surface of the brain and spinal cord, following every contour. The narrow space between the pia and arachnoid membranes is filled with a clear fluid called the **cerebrospinal fluid** or CSF, which is actively secreted by the choroid plexuses, vascular structures situated in the fluid-filled spaces of the brain (the **cerebral ventricles**). There is little space between the dura covering the brain and the inside of the skull, but there is a small space between the wall of the vertebral canal and the dura covering the spinal cord. This is the **epidural space** into which local anaesthetics can be administered to produce epidural anaesthesia.

As the space between the skull and the brain is filled with CSF, the brain floats in a fluid-filled container. Moreover, deep infoldings of the dura divide the fluid-filled space between the skull and the brain into smaller compartments. This arrangement restricts the displacement of the brain within the skull during movements of the head and limits the stresses on the blood vessels and the cranial nerves.

The surface of the human brain has many folds called **sulci** (singular **sulcus**) and the smooth regions of the brain surface that lie between the folds are known as **gyri** (singular **gyrus**). Viewing the brain from the dorsal surface reveals a deep cleft known as the longitudinal cerebral fissure that divides the brain into two **cerebral hemispheres**. Each hemisphere can be broadly divided into four lobes: the frontal, parietal, occipital and temporal. The frontal lobe is separated from the parietal lobe by the central sulcus and from the temporal lobe by the lateral cerebral fissure. The temporal lobe lies below the lateral cerebral fissure, but its border with the occipital lobe is poorly defined, as is the border between the occipital lobe and the parietal lobe. Posterior to the cerebral hemispheres below the occipital lobe is a highly convoluted structure known as the **cerebellum** (Figure 11.3).

If the brain is cut in half along the midline between the cerebral hemispheres (a midsagittal section), some details of its internal organization may be seen (Figure 11.4). On the medial surface is a broad, white band known as the **corpus callosum** that interconnects the two hemispheres. Immediately below the corpus callosum is a membranous structure, called the septum pellucidum, that separates two internal spaces

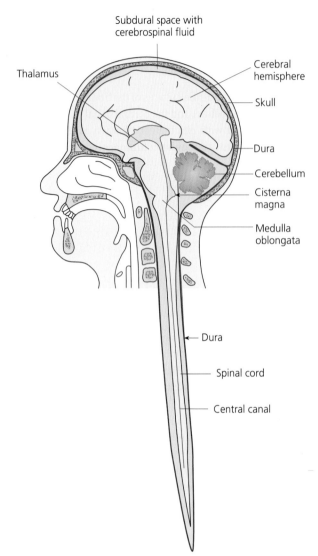

Subdural space with
cerebrospinal fluid

Thalamus

Cerebral
hemisphere

Skull

Dura

Cerebellum

Cisterna
magna

Medulla
oblongata

Dura

Spinal cord

Central canal

Figure 11.2 A diagrammatic representation of a midsagittal view of the CNS.

to the thalamus (Figure 11.4). It merges with a large swelling called the **pons**, which contains fibres that connect the two halves of the cerebellum. Below and behind the pons is the **medulla oblongata** (or medulla), which merges with the spinal cord. As it passes down the spinal column, the spinal cord gives rise to a series of paired nerves that connect the CNS to peripheral organs (see Figure 11.7 below).

If the brain is cut at right angles to the midline, a coronal section is obtained that reveals aspects of its internal structure. The outer part or **cerebral cortex** is greyish in appearance. The cortex and other parts of the brain that have a similar appearance are **grey matter**, which contains large numbers of nerve cell bodies. Beneath the grey matter of the cerebral cortex is the **white matter**, which is composed of bundles of nerve fibres such as those of the corpus callosum and the internal capsule.

An oblique section through the brain reveals a number of other important structures, which are shown diagrammatically in Figure 11.5. The caudate nucleus, the putamen and the globus pallidus together form the **corpus striatum of the basal ganglia** (see Figure 12.14). Between the caudate nucleus and putamen runs the **internal capsule**, which contains nerve fibres connecting the cerebral cortex to the spinal cord. A small region known as the **substantia nigra** lies beneath the thalamus. (The substantia nigra is so named because it contains a characteristic black pigment.) All of these structures play an important role in the control of movement (see Chapter 12 for further details).

known as the lateral cerebral ventricles, which are filled with CSF.

Beneath the septum pellucidum and lateral ventricles is the **thalamus**, a major site for the processing of information from the sense organs. Lying just in front of and below the thalamus is the **hypothalamus**, which plays a vital role in the regulation of the endocrine system via its control of the pituitary gland (see Chapter 16). The midbrain lies posterior and ventral

KEY POINTS:

- Grey matter contains the cell bodies of the neurones.
- White matter mainly consists of myelinated nerve axons and oligodendrocytes.
- The brain is divided into two cerebral hemispheres, which are interconnected by a large band of fibres, the corpus callosum.
- Below the cerebral hemispheres lie the thalamus and hypothalamus.

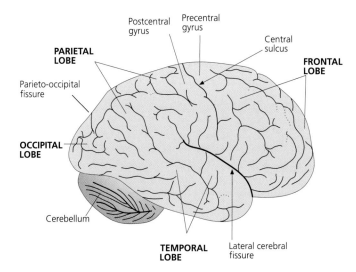

Figure 11.3 label: PARIETAL LOBE, Postcentral gyrus, Precentral gyrus, Central sulcus, FRONTAL LOBE, Parieto-occipital fissure, OCCIPITAL LOBE, Cerebellum, TEMPORAL LOBE, Lateral cerebral fissure

Figure 11.3 A side view of the right side of the human brain. The four lobes of the hemisphere are labelled, together with the central sulcus, the precentral and post-central gyri, and the lateral and parieto-occipital fissures. The cerebellum is shown on the left of the figure below the occipital lobe.

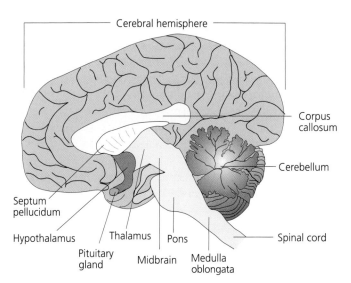

Figure 11.4 labels: Cerebral hemisphere, Corpus callosum, Cerebellum, Spinal cord, Septum pellucidum, Hypothalamus, Thalamus, Pons, Pituitary gland, Midbrain, Medulla oblongata

Figure 11.4 A midsagittal view of the right side of the human brain to show the relationships of the main structures.

The cranial nerves

On the base of the brain there are 12 pairs of nerves that serve the motor and sensory functions of the head (Figure 11.6). These are the **cranial nerves**, which are numbered in roman numerals I–XII. Some contribute to the parasympathetic division of the autonomic nervous system (Chapter 13). These are the oculomotor (III), the facial (VII), the glossopharyngeal (IX) and the vagus (X) nerves. The names and the main functions of all the cranial nerves are given in Table 11.1.

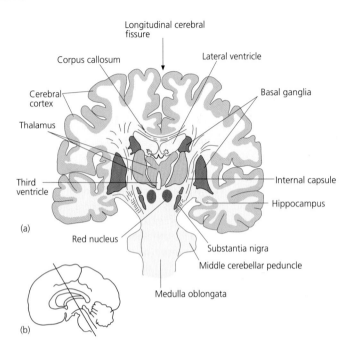

Figure 11.5 (a) An oblique section through the human brain to show the spatial relationship of the cerebral cortex, basal ganglia and thalamus. (b) The plane of section.

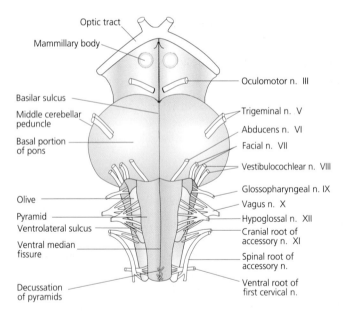

Figure 11.6 A ventral view of the human brain showing the cranial nerves, optic chiasm, the pons and the decussation of the pyramids (crossing over of the nerve fibres of the pyramidal tracts). n. = nerve.

The organization of the spinal cord

While the human brain is a large structure, the spinal cord is delicate – scarcely as thick as a pencil for much of its length. The cross-section of the spinal cord, shown in Figure 11.7, reveals a central region of grey matter surrounded by white matter. The grey matter is roughly shaped in the

Table 11.1 The functions of the cranial nerves

Number	Name	Chief functions
I	Olfactory	Sensory nerve subserving the sense of smell
II	Optic	Sensory nerve subserving vision (output of the retina)
III	Oculomotor	Chiefly motor control of the extrinsic muscles of the eye and the parasympathetic supply for the intrinsic muscles of the iris and ciliary body
IV	Trochlear	Chiefly motor control of the extrinsic muscles of the eye
V	Trigeminal	Sensory and motor; motor control of the jaw and facial sensation
VI	Abducens	Chiefly motor control of the extrinsic muscles of the eye
VII	Facial	Sensory and motor; motor control of the facial muscles and parasympathetic supply to the salivary glands
		Subserves the sense of taste via the chorda tympani
VIII	Vestibulocochlear	Sensory – hearing and balance
IX	Glossopharyngeal	Sensory and motor – control of swallowing and parasympathetic supply to the salivary glands
		Subserves the sense of taste from the back of the tongue (bitter sensations)
X	Vagus	Major parasympathetic outflow to the chest and abdomen
		Afferent inputs from the viscera
XI	Spinal accessory	Motor – control of neck muscles and larynx
XII	Hypoglossal	Motor control of the tongue

form of a butterfly around a central canal, which can be broadly divided into two **dorsal horns** and two **ventral horns**. (The dorsal and ventral horns are also known as the posterior and anterior horns, respectively.) The white matter of the spinal cord contains the sensory and motor nerve fibres that connect the brain and spinal cord.

The spinal cord is divided into segments, each of which gives rise to a pair of spinal nerves that supply a particular region of the body. Each spinal nerve is formed by the fusion of nerve segments known as the **dorsal** and **ventral roots**, as shown in Figure 11.7b. Each dorsal root has an enlargement known as a **dorsal root ganglion**, which contains the cell bodies of the nerve fibres making up the dorsal root. The fibres of the ventral root originate from nerve cells in the ventral horn of the spinal grey matter. To leave the spinal canal, the spinal nerves pierce the dura mater between the vertebrae. Thereafter, they form the peripheral nerve trunks that innervate the muscles and organs of the body. Most people have 31 pairs of spinal nerves, each of which serves a particular region of the body surface known as a **dermatome**. The areas supplied by neighbouring spinal segments overlap so that if a single spinal root is damaged, there is little loss of function. The

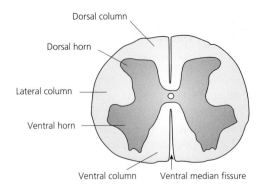

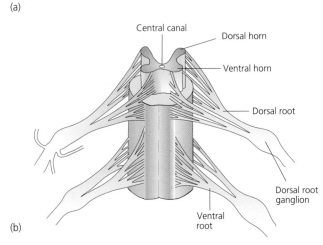

Figure 11.7 These two diagrams illustrate the structure of the spinal cord (a) and the arrangement of the spinal roots (b). In (b), part of the white matter of the spinal cord is shown cut away to reveal the entry of the spinal roots into the central grey matter.

areas supplied by the various spinal nerves are illustrated in Figure 14.6.

Sensory information enters the spinal cord via the dorsal root ganglia. As the sensory fibres carry information from sense organs to the spinal cord, they are known as **afferent nerve fibres**. The ventral root fibres are known as **efferent nerve fibres**. They carry motor information from the spinal cord to the muscles and secretory glands (the **effectors**). The nerves that leave the spinal cord to supply the skeletal muscles are known as somatic nerves, while those that supply the blood vessels and viscera are sympathetic

efferent fibres (see Chapter 13 for further details of the organization and function of the sympathetic fibres).

KEY POINTS:

- The dorsal roots contain sensory nerve fibres while the ventral roots have motor fibres.

- The cell bodies of the dorsal root fibres are found in the dorsal root ganglia. Those of the ventral roots are located in the ventral horn of the spinal cord.

11.4 The structure of peripheral nerve trunks

As mentioned above, both the brain and spinal cord give rise to paired nerves that supply the tissues of the body – the peripheral nerves. The peripheral nerve trunks travel to the structures they innervate following the course of the major blood vessels. Each contains the axons of many CNS neurones surrounded by protective layers of connective tissue (Figure 11.8). The outermost layer of a peripheral nerve is a loose aggregate of connective tissue called the **epineurium**, which serves to anchor the nerve trunk to the adjacent tissue. Within the epineurium, axons run in bundles called **fascicles** and each bundle is surrounded by a tough layer of connective tissue called the **perineurium**.

Within the perineural sheath, individual nerve fibres are protected by a further layer of connective tissue called the **endoneurium**. Individual axons are covered by specialized cells called Schwann cells.

Some nerve trunks transmit information from specific sensory end organs to the CNS (**sensory nerves**) while others transmit signals from the CNS to the muscles and secretory glands (**motor nerves**). Nerve trunks that contain both sensory and motor fibres are called **mixed nerves**. Peripheral nerves also contain sympathetic postganglionic fibres, which innervate the blood vessels and sweat glands. These will be described in Chapter 13.

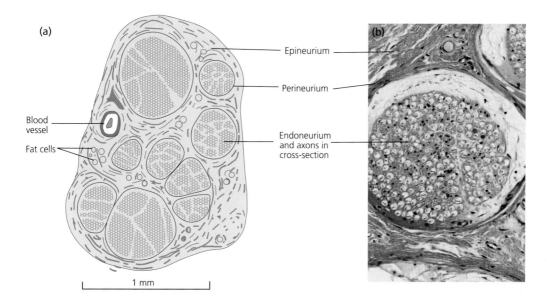

Figure 11.8 (a) A diagrammatic view of a cross section through a small peripheral nerve to show the relationship between the nerve fibres and the surrounding layers of connective tissue (the epineurium, perineurium and endoneurium). Note that at this magnification individual nerve fibres are not clearly resolved. (b) A cross-section through a nerve fascicle of the sciatic nerve. In this case, the larger axons are clearly visible in cross-section.

11.5   The cerebral ventricles and the cerebrospinal fluid

The central canal of the spinal cord and the fluid-filled spaces within the brain (the cerebral ventricles) contain a clear fluid known as cerebrospinal fluid (or CSF), which serves two main functions: it acts as a hydraulic buffer to cushion the brain against damage resulting from movements of the head and it helps to provide a stable ionic environment for neuronal function.

The CSF is not an ultrafiltrate of plasma but is actively secreted by the **choroid plexus**, which consists of a network of capillaries covered by a layer of modified ependymal cells. About 500 ml of CSF are secreted each day. Its composition is compared with that of the plasma in Table 11.2. It has no red cells and very few leukocytes.

The CSF flows from the lateral ventricles into the third ventricle and thence through the cerebral aqueduct to the fourth ventricle, as shown in Figure 11.9. It leaves the cerebral ventricles via apertures in the roof of the fourth ventricle and flows into the subarachnoid space, which lies between the arachnoid membrane and the pia mater. From the subarachnoid space, the CSF passes into the venous blood via the **arachnoid villi**, which project into the venous sinuses as shown in Figure 11.10. The veins connecting the subarachnoid space and the venous sinuses are known as **bridging veins**. They form part of the attachments between the brain and the skull and may become ruptured during head injuries. This results in venous bleeding and the formation of a chronic **subdural haematoma**. Over time, this raises intracranial pressure with consequent alterations to behaviour such as confusion, impaired motor function and headache.

Samples of CSF are normally obtained by inserting a needle between the third and fourth lumbar vertebrae of a subject who is lying on his or her side. Although the tip of the needle lies in the subarachnoid space, there is no risk of damage to the spinal cord, which only extends as far as the first lumbar vertebra. This procedure is known as **lumbar puncture**.

Consequences of raised intracranial pressure

The circulation of the CSF is driven by the difference in pressure between the cerebral ventricles (the site of production) and the venous sinuses. Obstruction of the CSF drainage leads to an increase in intracranial pressure. This can occur following haemorrhage, head injury, inflammation of the aqueduct or the growth of a tumour.

Table 11.2 The composition of the plasma and CSF

	CSF	Plasma	Ratio CSF : plasma
Na^+ (mmol l^{-1})	141	141	1.0
K^+ (mmol l^{-1})	3.0	4.5	0.67
Ca^{2+} (mmol l^{-1})	1.15	1.5	0.77
Mg^{2+} (mmol l^{-1})	1.12	0.8	1.4
Cl^- (mmol l^{-1})	120	105	1.14
HCO_3^- (mmol l^{-1})	23.6	24.9	0.95
Glucose (mmol l^{-1})	3.7	4.5	0.82
Protein (g l^{-1})	0.18	75	0.002
pH	7.35	7.42	–

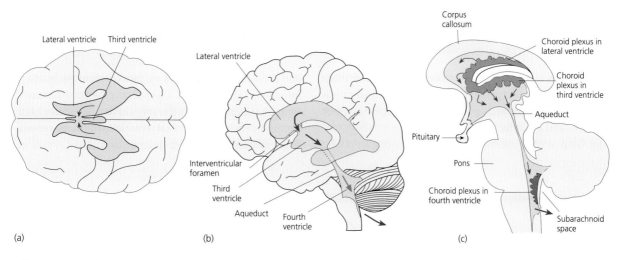

Figure 11.9 This figure shows the sites of formation of the CSF and its circulation within the brain. (a) A view of the brain from above with the position of the lateral and third ventricles indicated in blue. The arrows show the flow of CSF from the lateral ventricles into the third ventricle. (b) A side view of the brain. Once more, the cerebral ventricles are coloured blue and the arrows show the passage of CSF from the cerebral ventricles into the subarachnoid space. (c) The location of the choroid plexus.

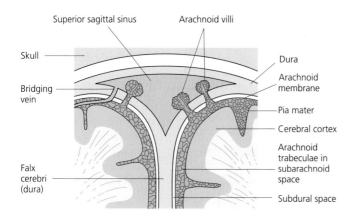

Figure 11.10 The relationship between the meninges, the arachnoid villi and the superior sagittal sinus. The CSF drains into the venous blood via the arachnoid villi (granulations) that penetrate the large venous sinuses formed by folds in the dura.

In an adult, the raised intracranial pressure elicits Cushing's reflex, which is discussed below. In babies and young children (where the skull sutures are not fused), the increased pressure leads to a disproportionate increase in the size of the head known as **hydrocephalus**. If this is severe and remains untreated, brain tissue may be damaged resulting in mental impairment. Hydrocephalus can be treated by placing a catheter in the cerebral ventricles to drain the CSF into the peritoneal cavity. This arrangement is known as a ventriculoperitoneal shunt.

A further cause of raised intracranial pressure is **meningitis** in which the membranes covering the brain (the meninges) are invaded by viruses or bacteria. Under these circumstances, the meninges become inflamed, leading to a potentially life-threatening condition. In meningitis, the intracranial pressure is increased due to restricted drainage of the CSF, giving rise to characteristic clinical signs including headache, vomiting, excessive sensitivity to light (photophobia) and rigidity of the neck. In severe cases, convulsions and coma may occur.

11.6   Cerebral circulation

Blood flows to the brain via the **internal carotid** and **vertebral arteries**. These join together (anastomose) around the optic chiasm to form the **circle of Willis** from which the anterior, middle and posterior cerebral arteries arise. Smaller arteries branch off from the cerebral arteries (Figure 11.11)

and run over the surface of the brain before penetrating into the brain tissue itself to supply the deeper structures. Brain tissue, particularly the grey matter, has a very high capillary density. The cerebral veins drain into large venous sinuses formed by the dura (Figure 11.10).

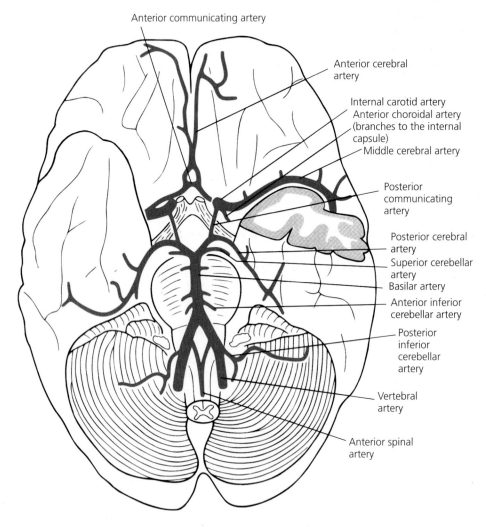

Figure 11.11 A diagrammatic representation of the arterial blood supply to the brain. Each internal carotid artery divides into an anterior and middle cerebral artery. The vertebral arteries join together at the base of the brain to form the basilar artery, which runs forward before giving rise to a pair of posterior cerebral arteries. Communicating arteries run between the cerebral arteries to form a ring of vessels known as the circle of Willis. Part of the right temporal lobe is shown cut away to reveal the course of the middle cerebral artery as it runs along the lateral cerebral fissure.

In contrast to most other organs, the blood flow to the brain is maintained within very closely defined limits and accounts for about 15% of the resting cardiac output or about 0.75 litres per minute. If cardiac output falls, in haemorrhage for example, the cerebral blood flow will be maintained as far as possible by physiological adjustments of the circulation, even though perfusion of many peripheral organs may be compromised. The brain is very sensitive to ischaemia, and loss of consciousness occurs if blood flow is interrupted for only a few seconds. Irreversible brain damage may occur if cerebral blood flow is stopped for more than a few minutes. This is known as a **stroke**. About 80% of strokes are the result of an interruption to the arterial blood supply (mainly as a result of blood clots) and the remainder are caused by rupture of a cerebral blood vessel.

As the brain is encased in the skull, which provides a rigid enclosure, except in very young children, anything that increases the volume of the brain tissue, such as a cerebral tumour or a cerebral haemorrhage, will increase the pressure within the skull (the **intracranial pressure**). If the increase in intracranial pressure is very high, cerebral blood flow will decline and the sufferer will experience mental confusion leading to coma and death unless prompt action is taken. A raised intracranial pressure will tend to force the brainstem into the passage through which the spinal cord passes (the foramen magnum). The resulting compression of the brainstem elicits a marked rise in arterial blood pressure, which acts to offset the fall in cerebral blood flow. This is known as **Cushing's reflex** (sometimes called Cushing's response). In addition to the rise in blood pressure, there is a fall in heart rate mediated by the baroreceptor reflex (see Chapter 19, p. 379). This combination of a marked rise in blood pressure and a slow heart rate is an important clinical indicator of a space-occupying lesion within the skull.

Recommended reading

Anatomy and histology of the nervous system

Brodal, P. (2003). *The Central Nervous System. Structure and Function*, 3rd edn. Oxford University Press: Oxford.

Kiernan, J.A. (2005). *Barr's The Human Nervous System: An Anatomical Viewpoint*, 8th edn. Lippincott, Williams & Wilkins: Baltimore.

Self-assessment questions

Which of the following statements relating to the CNS are true and which are false? Answers are given at the end of the book (p. 755).

1 a) White matter contains large numbers of nerve cell bodies.
   b) The myelin of axons is formed by oligodendrocytes.
   c) The end feet of astrocytes cover the capillaries of the brain.
   d) Nerve cells are protected from changes in the composition of the plasma.

2 a) Axons of peripheral nerves are protected by three layers of connective tissue.
   b) Peripheral axons are always associated with Schwann cells.

   c) All axons are covered by a layer of myelin.
   d) Axons only conduct action potentials from the CNS to a target organ.

3 a) Glial cells generate action potentials.
   b) Astrocytes form a barrier to the diffusion of oxygen between the blood and the brain tissue.
   c) Oligodendrocytes form the myelin sheaths of CNS nerve fibres.
   d) Microglia are macrophages that reside in the brain tissue.

4 a) Each spinal segment gives rise to four spinal nerves.

b) The dorsal root ganglia contain the cell bodies of the primary sensory neurones.

c) The white and grey matter of the spinal cord is arranged in the same way as that of the cerebral hemispheres.

d) The motoneurones of the spinal cord are found in the ventral horn.

5 a) Each cerebral hemisphere is divided into two lobes.

b) Grey matter contains the cell bodies and dendrites of neurones.

c) The cerebral blood flow accounts for about 15% of the resting cardiac output.

d) The CSF is an ultrafiltrate of plasma.

The physiology of motor systems

12

After reading this chapter you should have gained an understanding of:

- The hierarchical nature of control within the motor systems of the body
- The organization of the spinal cord and its role in reflex activity
- The contribution of spinal reflexes to postural control
- The descending pathways that modify the output of the spinal cord
- The role of the motor cortex in the programming and execution of voluntary activity
- The organization of the cerebellum and its role in refined, co-ordinated movements
- The role of the basal ganglia in the planning and execution of defined motor patterns
- The effects of lesions at various levels of the motor hierarchy

12.1 Introduction

Intrauterine movement is detectable by ultrasound from a very early stage of gestation and is felt by the mother for the first time between 16 and 20 weeks ('quickening' of the fetus). By the time of birth, a baby is capable of some co-ordinated movement. During the first 2 years of life, as the brain and spinal cord continue to develop and mature, children learn to defy gravity, by first sitting then standing, and later to walk, run, jump and climb. At the same time, the capacity to perform the precise movements needed for complex manipulations and speech is acquired. In short, co-ordinated, purposeful movement is a fundamental aspect of human existence.

The simplest form of motor act controlled by the nervous system is called a **reflex**. This is a rapidly executed, automatic and stereotyped response to a given stimulus.

As reflexes are not under the direct control of the brain, they are described as involuntary motor acts. Nevertheless, most reflexes involve co-ordination between groups of muscles and this is achieved by interconnections between various groups of neurones. The neurones forming the pathway taken by the nerve impulses responsible for a reflex make up a **reflex arc**. Many voluntary motor acts are guided by the intrinsic properties of reflex arcs but are modified by commands from higher centres in the brain and from sensory inputs.

Two kinds of voluntary motor function can be distinguished: (1) the **maintenance of posture** and (2) **goal-directed movements**. They are inextricably linked in practice. A goal-directed movement will only be performed successfully if the moving limb is first correctly positioned. Similarly, a posture may only be maintained if appropriate compensatory movements are made to counteract any force tending to oppose that posture.

12.2 The hierarchical nature of motor control systems

The term **motor system** refers to the neural pathways that control the sequence and pattern of muscle contractions. As Figure 12.1 shows, the structures responsible for the neural control of posture and movement are distributed throughout the brain and spinal cord.

As skeletal muscles can only contract in response to excitation of the motoneurones that supply them, all motor acts depend on neural circuits that eventually impinge on the motoneurones that form the output of the motor control system. As discussed in Chapter 9, each motoneurone supplies a number of skeletal muscle fibres, and a motoneurone together with the skeletal muscle fibres it innervates constitutes a **motor unit**, which is the basic element of motor control. For this

reason, motoneurones are often referred to as the final common pathway of the motor system. The motoneurones are found in the brainstem and spinal cord, and their excitability is influenced by neural pathways from a variety of brain areas. Thus, there is a kind of hierarchical arrangement of so-called 'motor centres' from the spinal cord up through to the cerebral cortex. Postural control is exerted largely at the level of the brainstem, while goal-directed movements require the participation of the cerebral cortex. The basal ganglia and cerebellum both play an important role in motor control although neither is directly connected with the spinal motoneurones. Instead, they influence the motor cortex by way of the thalamic nuclei (Figure 12.1).

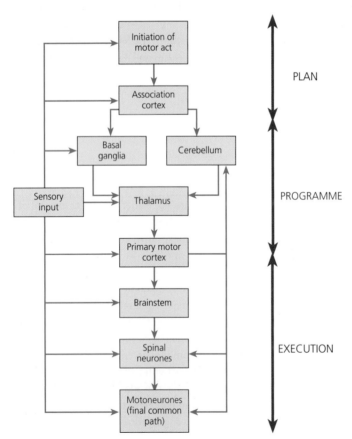

Figure 12.1 A diagrammatic representation of the motor systems of the body showing possible interactions between them.

12.3 Organization of the spinal cord

The motoneurones controlling the skeletal muscles are large neurones whose cell bodies lie in clumps within the brainstem and the ventral (or anterior) horn of the spinal cord. Each motoneurone has a myelinated axon that innervates a motor unit, which may consist of anything between six and 1500 skeletal muscle fibres. These motoneurones are called α-**motoneurones**. They have axons with diameters of 15–20 μm that conduct action potentials very rapidly (70–120 m s⁻¹). They also have long dendrites that receive many synaptic connections (Figure 12.2b), particularly from interneurones and sense organs called **proprioceptors**. The proprioceptors relay information concerning the length and tension in individual muscles (see below). The α-motoneurones also receive synaptic contacts from higher levels of the CNS.

The axons of α-motoneurones collect in bundles that leave the ventral horn and pass through the ventral white matter of the spinal cord before entering the ventral root. Some axons send off branches that turn back into the cord and make excitatory synaptic contact with small interneurones called **Renshaw cells**. These cells

in turn have short axons that synapse with the pool of motoneurones by which they are stimulated. These synapses are inhibitory and bring about recurrent or feedback inhibition.

As Figure 12.2a shows, the motoneurones have a topographical arrangement within the spinal cord: motoneurones supplying the muscles of the trunk are situated in the medial ventral horn while those supplying more distal muscle groups tend to be situated laterally. Muscles that flex the limbs (**flexors**) are under the control of neurones that lie dorsal to those controlling the muscles that extend the limbs (**extensors**).

For movements to be carried out in a functionally appropriate way, it is essential for sensory and motor information to be integrated. All the neural structures involved in the execution of movements are continually informed of the position of the body and of the progress of the movement by the proprioceptors of the muscles and joints, which provide information regarding the position of the limbs and their movements relative to each other and to the surroundings.

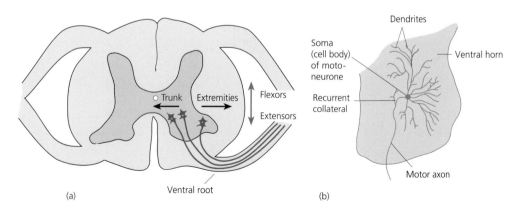

Figure 12.2 (a) A diagrammatic representation of a transverse section of the spinal cord. The localization of motoneurones corresponding to various groups of muscles are indicated with flexors represented more dorsally while extensors are represented ventrally within the cord. The muscles of the trunk are represented medially and the extremities are represented laterally. (b) A drawing of an α-motoneurone in the ventral horn of the spinal cord illustrating the elaborate dendritic tree. Although not shown in this diagram, numerous synaptic connections are made with these dendrites.

The main proprioceptors are the **muscle spindles** and **Golgi tendon organs**. Both provide information with regard to the state of the musculature. Muscle spindles lie within the muscles between and in parallel with the skeletal muscle fibres. They can therefore respond to muscle length. Golgi tendon organs lie within the tendons and are in series with the contractile elements of the muscle. They are sensitive to the force acting on a muscle during contraction.

Muscle spindles

Although muscle spindles are found in most skeletal muscles, they are particularly numerous in those muscles that are responsible for fine motor control, such as those of the eyes, neck and hands. The basic organization of a muscle spindle is illustrated in Figure 12.3. The muscle spindle consists of a small bundle of modified

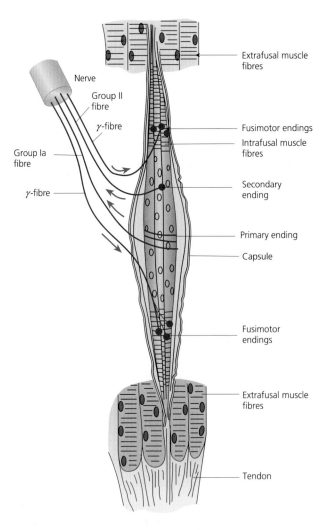

Figure 12.3 The basic organization of a muscle spindle. These sensory organs lie in parallel with the extrafusal muscle fibres and are therefore adapted to monitor muscle length. Note that it is innervated by both motor and sensory nerve fibres.

muscle fibres innervated both by sensory and moto-neurones. The muscle fibres of the muscle spindles are called **intrafusal fibres** while those of the main body of the muscle are the **extrafusal fibres**. The nerve endings and intrafusal fibres of the muscle spindles are enclosed within a capsule.

The motoneurones innervating the intrafusal fibres are known as **gamma-motoneurones** (γ-motoneurones) to distinguish them from the large α-motoneurones, which innervate the extrafusal fibres. The cell bodies of the γ-motoneurones lie in the ventral horn of the spinal cord and their axons leave the spinal cord via the ventral roots. Their axons are smaller than those of the α-motoneurones: their diameters range from 3 to 6 μm and their conduction velocities range between 15 and 30 m s^{-1}.

Within the muscle, the axons of the γ-motoneurones branch to supply several muscle spindles, and within each muscle spindle they branch further to supply several intrafusal fibres. The γ-motoneurones contract the muscle spindles so that their sensitivity to changes in muscle length is maintained. Note, however, that contraction of the intrafusal fibres is too weak to cause movements of the muscle as a whole.

Golgi tendon organs

The Golgi tendon organs are mechanoreceptors that lie within the tendons of muscles immediately beyond their attachments to the muscle fibres (Figure 12.4). Around 10 or 15 muscle fibres are usually connected in series to each Golgi tendon organ, which is then stimulated by the tension produced by this bundle of fibres. Impulses are carried from the tendon organs to the spinal cord.

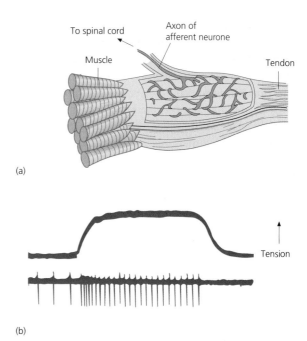

(a)

(b)

Figure 12.4 The Golgi tendon organ. (a) The basic organization of a Golgi tendon organ. (b) The response of a Golgi tendon organ to tension in the muscle; the upper trace shows the tension in the muscle to which the tendon organ is attached and the lower trace shows the firing pattern in the nerve fibre supplying the receptor. Note that the Golgi tendon organ lies in series with the muscle and so is adapted to monitor muscle tension.

KEY POINTS:

- The basic element of motor control is the motor unit, which consists of an α-motoneurone and all the skeletal muscle fibres it controls.

- Proprioceptors are mechanoreceptors situated within muscles and joints. They provide the CNS with information regarding muscle length, position and tension (force).

12.4 Reflex action and reflex arcs

Reflex arcs include at least two neurones, an **afferent** or **sensory** neurone and an **efferent** or **motor** neurone. The fibre of the afferent neurone carries information about the environment from a receptor towards the CNS while the efferent fibre transmits nerve impulses from the CNS to an effector. Reflexes may be (and often are) subject to modulation by activity in the CNS.

In the simplest reflex arc, there are two neurones and just one synapse. Such reflexes are therefore known as **monosynaptic reflexes**. Other reflex arcs have one or more neurones interposed between the afferent and efferent neurones; these are called **interneurones**. If there is one interneurone, the reflex arc will have two synaptic relays and the reflex is called a **disynaptic reflex**. If there are two interneurones, there will be three synaptic relays so the associated reflex would be trisynaptic. If many interneurones are involved, the reflex is called a **polysynaptic reflex**. Examples are the stretch reflex (monosynaptic), the withdrawal reflex (disynaptic) and the scratch reflex (polysynaptic).

The knee jerk is an example of a dynamic stretch reflex

A classic example of a stretch reflex (also known as the myotactic reflex) is the **knee-jerk** or **tendon-tap reflex**, which is used routinely in neurological examinations. A sharp tap applied to the patellar tendon stretches the quadriceps muscle. The stretch stimulates the muscle spindles and impulses are transmitted to the spinal cord. As they enter the spinal cord, the afferent fibres branch; some enter the grey matter of the cord and make monosynaptic contact with the α-motoneurones supplying the quadriceps muscle causing them to discharge in synchrony. The resulting contraction of this muscle abruptly extends the lower leg (hence the name knee jerk). Collaterals of the afferent fibres make synaptic contact with inhibitory interneurones, which in turn inhibit the antagonistic (flexor) muscles of the knee joint.

The stretch reflex arc is illustrated diagrammatically in Figure 12.5a. A similar reflex occurs when the Achilles tendon is struck (the ankle-jerk reflex). In this case, there is a plantar flexion of the foot produced by contraction of the calf muscles.

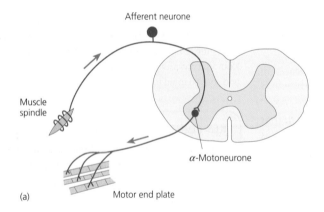

(a)

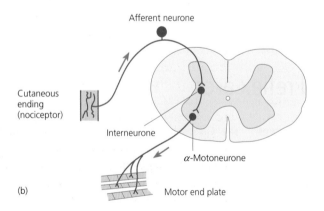

(b)

Figure 12.5 (a) A diagrammatic representation of the stretch reflex arc. Note that this reflex arc comprises only two neurones and one synapse. It is therefore a monosynaptic reflex. (b) A diagrammatic representation of the basic flexor (withdrawal) reflex arc. In this case, there are three neurones and two synapses in the basic arc. The reflex in its simplest form is therefore disynaptic.

The flexion reflex

In this protective reflex, a limb is rapidly withdrawn from a threatening or damaging stimulus. It is more complex than the stretch reflex and usually involves large numbers of interneurones and propriospinal connections arising from many segments of the spinal cord. Withdrawal may be elicited by noxious stimuli applied to a large area of skin or deeper tissues (muscles, joints and viscera) rather than from a single muscle as in the stretch reflex. The receptors responsible are called nociceptors (see Chapter 14, p. 235) and they give rise to the afferent impulses that are responsible for the flexion reflex.

To achieve withdrawal of a limb, the flexor muscles of one or more joints in the limb must contract while the extensor muscles relax. Action potentials in the nociceptor axons synapse on excitatory interneurones, which are connected to the α-motoneurones that supply flexor muscles in the affected limb. At the same time, these action potentials activate interneurones, which inhibit the α-motoneurones supplying the extensor muscles. The excitation of the flexor motoneurones coupled with inhibition of the extensor motoneurones is known as **reciprocal inhibition**.

The synaptic organization of the flexor reflex is illustrated in Figure 12.5b. As the figure shows, the basic reflex is disynaptic, but in a powerful withdrawal reaction it is likely that several spinal segments would be involved and that all the major joints of the limb will show movement. The flexion reflex has a longer latency than the stretch reflex because it arises from a disynaptic arc. It is also non-linear insofar as a weak stimulus will elicit no response while a powerful withdrawal is seen when the stimulus reaches a certain level of intensity.

The crossed extensor reflex

Stimulation of the flexion reflex, as described above, frequently elicits extension of the opposite limb about 250 ms later. This crossed extension reflex assists in the maintenance of posture and balance. The long latency between flexion and crossed extension represents the time taken to recruit interneurones. The reflex arc for a crossed extensor reflex is illustrated in Figure 12.6.

The Golgi tendon reflex

This reflex, which is the result of activation of Golgi tendon organs, complements the tonic stretch reflex and

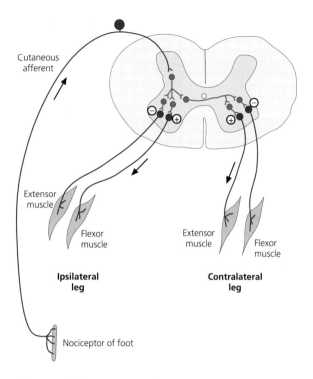

Figure 12.6 The neuronal pathways participating in a crossed extensor reflex arc. Many neurones and synapses are involved in this reflex, which is therefore polysynaptic. − denotes synaptic inhibition and + synaptic excitation.

contributes to the maintenance of posture. The synaptic organization of this reflex is illustrated for the knee joint in Figure 12.7. In this example, the tendon organ is located in the tendon of the rectus femoris muscle and its afferent fibre branches as it enters the spinal cord. One set of branches excites interneurones that inhibit the discharge of the α-motoneurones supplying the rectus femoris muscle while another set activates interneurones that stimulate activity in the α-motoneurones innervating the antagonistic semitendinosus (hamstring) muscles.

During a maintained posture such as standing, the rectus femoris muscle will start to fatigue. As it does so, the force in the patellar tendon, monitored by Golgi tendon organs, will decline. As a result, activity in the afferent fibres will decline and the normal inhibition of the motoneurones supplying the rectus femoris will be removed. Consequently, the muscle will be stimulated to contract more strongly, thereby increasing the force in the patellar tendon once more.

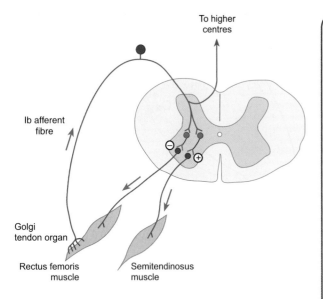

Figure 12.7 A diagrammatic representation of the Golgi tendon organ reflex arc. This shows the synaptic basis of reciprocal inhibition. It is also seen in the crossed extensor reflex illustrated in Figure 12.6.

KEY POINTS:

- The neurones participating in a reflex form a reflex arc, which includes a receptor, an afferent neurone that synapses in the CNS, and an efferent neurone that sends a nerve fibre to a muscle or gland. Interneurones may be present between the afferent and efferent neurones.

- The number of synapses in a reflex arc is used to define the reflex as monosynaptic, disynaptic or polysynaptic.

- The simplest reflexes are the monosynaptic stretch reflexes such as the patellar tendon-tap reflex (the knee-jerk reflex).

- The protective flexion (withdrawal) reflexes are elicited by noxious stimuli. Their reflex arc possesses at least one interneurone, so the most basic flexion reflex is disynaptic.

- Muscle spindles monitor the length of a muscle while tendon organs monitor force.

12.5 The role of the muscle spindle in voluntary motor activity

The activity of the muscle spindles enables the nervous system to compare the lengths of the extrafusal and intrafusal muscle fibres. Whenever the length of the extrafusal fibres exceeds that of the intrafusal fibres, the afferent discharge of a muscle spindle will increase, and whenever the length of the extrafusal fibres is less than that of the intrafusal fibres, the discharge of the spindle afferents will decline. This decline is possible because the spindle afferents normally show a tonic level of discharge. In this way, the muscle spindles can provide feedback control of muscle length.

The role of the muscle spindles as comparators for the maintenance of muscle length is important during goal-directed voluntary movements. Studies of various movements (chewing and finger movements for example) have shown that when voluntary changes in muscle length are initiated by motor areas of the cerebral cortex, the motor command includes changes to the set point

of the muscle spindle system. To achieve this, both α- and γ-motoneurones are activated simultaneously by way of the neuronal pathways descending from higher motor centres. The simultaneous activation of extrafusal fibres (by way of α-motoneurones) and intrafusal fibres (by way of γ-motoneurones) is called **alpha–gamma co-activation**. Its physiological importance appears to lie in the fact that it allows the muscle spindles to be functional at all times during a muscle contraction.

The benefits of incorporating length sensitivity into voluntary activities may be illustrated by the following example. Suppose a heavy weight needs to be lifted. Before lifting, and from previous experience, the brain will estimate roughly how much force will be required to lift the weight and the motor centres will transmit the command to begin the lift. Both extrafusal and intrafusal fibres will be activated simultaneously. If the initial estimate is accurate, the extrafusal fibres will be able to shorten as rapidly as the intrafusal fibres and the activity of the spindle afferents will not change much during the lift. If, however, the weight turns out to be heavier than expected, the estimate of required force will be insufficient and the rate of shortening of the extrafusal fibres will be slower than expected. Nevertheless, the intrafusal fibres will continue to shorten and their central region will become stretched. As a result, the activity in the spindle afferents will increase and will summate with the excitatory drive already arriving at the α-motoneurones via the descending motor pathways. The increased activity in the α-motoneurones will increase the force generated by the muscle until it matches that required to lift the load.

> **KEY POINTS:**
>
> - The role of the muscle spindles as comparators for the maintenance of muscle length is important during goal-directed voluntary movements. Both α- and γ-motoneurones are activated simultaneously by the neuronal pathways descending from higher motor centres.
>
> - The simultaneous activation of extrafusal fibres and intrafusal fibres is called alpha–gamma co-activation and continuously readjusts the sensitivity of muscle spindles as the muscle shortens.

12.6 Effects of damage to the spinal cord

Despite the protection afforded to the spinal cord by the vertebral column, spinal injuries are still relatively common. Motor accidents are the most frequent cause of spinal cord injury, followed by falls and sports injuries (particularly diving). The loss of function that results from such injuries depends on the level of injury and the extent of the damage to the spinal cord. Following damage, there is a loss of sensation due to interruption of ascending spinal pathways and loss of the voluntary control over muscle contraction as a result of damage to descending motor pathways below the level of the lesion. The loss of voluntary motor control is known as **muscle paralysis**. Both spinal reflex activity and functional activities such as breathing, micturition and defecation may be affected. Paralysis may be **spastic** (in which the degree of muscle tone is increased above normal) or **flaccid** (in which the level of tone is reduced and the muscles are 'floppy'). The characteristics of motor dysfunction due to lesions of the corticospinal tract (**upper motoneurone lesions**) and those due to dysfunction of spinal motoneurones (**lower motoneurone lesions**) are discussed further in Box 12.1.

Immediately following injury to the spinal cord, there is a loss of spinal reflexes (areflexia). This is known as **spinal shock** or **neurogenic shock** and involves the descending motor pathways. The clinical manifestations are:

- Flaccid paralysis
- A lack of tendon reflexes
- Loss of autonomic function below the level of the lesion.

Spinal shock may last for several weeks in humans. After this time, there is usually some return of reflex activity as the excitability of the undamaged spinal neurones increases. Occasionally, this excitability becomes excessive and then spasticity of affected muscle groups is seen.

The term **tetraplegia** (sometimes also called quadriplegia) refers to impairment or loss of motor and sensory function in the arms, trunk, legs and pelvic organs. **Paraplegia** refers to impairment of function of the legs and pelvic organs. The arms are spared and the degree of functional impairment will depend on the exact level of the spinal injury. Injuries above the lumbar region will normally result in spastic paralysis of the affected skeletal muscle groups; control of bowel, bladder and sexual functions are also affected. Injuries at or below the lumbar region commonly result in flaccid paralysis of the affected muscle groups and of the muscles controlling bowel, bladder and sexual function.

Box 12.1 Upper and lower motoneurone lesions and sensory deficits following spinal lesions

Voluntary motor acts involve a large number of structures, including the cerebral cortex, corticospinal tract, basal ganglia, cerebellum and spinal cord. When any of these regions is damaged by traumatic injuries or strokes, characteristic changes in the activity of the motor system are apparent.

Damage to the primary motor cortex or corticospinal tract is often referred to as an **upper motoneurone lesion**. Such lesions are characterized by spastic (rigid) paralysis without muscle wasting. There is an increase in muscle tone and the tendon reflexes are exaggerated. The Babinski sign is positive so that in response to stimulation of the sole of the foot, the toes extend upwards instead of downwards as in a normal person. The muscles themselves are weak although the flexors of the arms are stronger than the extensors. In the legs, the reverse is true; the extensors are stronger than the flexors. The muscles do not show fasciculation (muscle twitching in which groups of muscle fibres contract together).

Damage to the spinal cord that involves the motoneurones is called a **lower motoneurone lesion**. A lower motoneurone lesion results in denervation of the affected muscles, which show a loss of muscle tone. Tendon reflexes may be weak or absent. There is a loss of muscle bulk and fasciculation is usually present. Co-ordination is impaired. The Babinski sign is absent (i.e. the toes flex in response to stimulation of the sole of the foot, as in a normal person).

Lesions to the spinal cord affect sensation as well as motor activity. If the spinal cord is completely severed, there is a loss of both motor and sensory function below the level of the lesion. Partial loss of motor activity and sensation occurs when the spinal cord is compressed either by the protrusion of an intervertebral disk or by a spinal tumour. The spinal cord may also be partially cut across as a result of certain traumatic injuries such as stabbing. In such partial transections, the pattern of impaired sensation provides important information regarding the site of the lesion.

KEY POINTS:

- Interruption of the spinal cord causes a loss both of sensation and of voluntary control over muscle contraction (paralysis) below the level of the lesion. Immediately following such an injury, there is a loss of spinal reflexes (spinal shock).

- Spinal shock may last for several weeks and both reflex somatic motor activity and involuntary activities involving bladder and bowel functions may be affected.

12.7   Descending pathways involved in motor control

Although the spinal cord contains the neural networks required for reflex actions, more complex motor behaviours are initiated by pathways that originate at various sites within the brain. Furthermore, the activity of the neural circuitry within the spinal cord is modified and refined by the descending motor control pathways. There are five important brain areas that give rise to descending tracts,

four of which lie within the brainstem and medulla. These are the reticular formation, the vestibular nuclei, the red nucleus and the tectum. Their fibres constitute a set of descending pathways that are sometimes referred to as the **extrapyramidal tracts**. The fifth area lies within the cerebral cortex and gives rise to the **corticospinal tract** (also called the **pyramidal tract**). Figure 12.8 shows the approximate

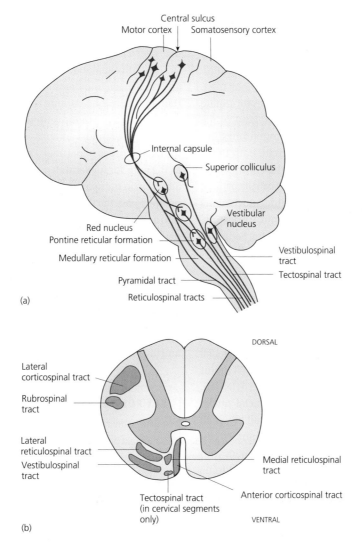

Figure 12.8 The principal motor pathways arising in the brain. (a) The arrangement of the major descending motor tracts and the approximate location of the motor nuclei. (b) The position of the major pyramidal and extrapyramidal descending pathways within the spinal cord.

positions of the major descending tracts and their location within the spinal cord.

Extrapyramidal pathways

The reticular system gives rise to two important descending tracts within the cord, the lateral reticulospinal tract and the medial reticulospinal tract. These tracts are largely uncrossed, and end mainly on interneurones rather than the α-motoneurones themselves. They mainly influence the muscles of the trunk and the proximal parts of the extremities, and are important in the control of certain postural mechanisms and in the startle reaction ('jumping' in response to a sudden and unexpected stimulus).

The vestibular nuclei are located just below the floor of the fourth ventricle close to the cerebellum. The medial and lateral vestibular nuclei give rise to a descending motor pathway called the vestibulospinal tract. The vestibulospinal tract is concerned mostly with the control of posture (acting to make adjustments in response to vestibular signals). Lesions of reticulospinal or vestibulospinal tracts affect the ability to maintain a normal erect posture.

The red nucleus gives rise to the rubrospinal tract, which sends fibres to the spinal cord where they terminate in the lateral part of the grey matter (Figure 12.8). Some make monosynaptic contact with α-motoneurones but most terminate on interneurones that excite both flexor and extensor motoneurones supplying the contralateral limb muscles. Lesions of the red nucleus or rubrospinal tract impair the ability to make voluntary limb movements while having relatively little effect on the control of posture. The fibres of the tectospinal tract end on interneurones that influence the movements of the head and eyes.

Cranial nerve motor nuclei

The neuronal circuitry of the spinal cord is responsible only for the motor activity of parts of the body below the upper part of the neck. Movements of the head and facial muscles are under the control of cranial nerve (CN) motor nuclei. These include the oculomotor nuclei (CN III), the trigeminal motor outflow (CN V), the facial nuclei (CN VII) and the nucleus ambiguus (CN X), all of which supply muscles that are part of a bilateral motor control system. Fibres from these nuclei make contact with interneurones in the brainstem, which are thought to be organized in much the same way as those of the spinal cord supplying the axial and limb muscles.

The corticospinal tract

The corticospinal tract has traditionally been described as the predominant pathway for the control of fine, skilled, manipulative movements of the extremities. While it is undoubtedly of considerable significance, it is now evident that many of its actions appear to be duplicated by the rubrospinal tract (see above), which can, if necessary, take over a large part of its motor function.

The corticospinal tract originates in the cerebral cortex and runs down into the spinal cord. Many of its axons are, therefore, very long. Around 80% of the fibres cross as they run through the extreme ventral surface of the medulla (the medullary pyramids) and pass down to the spinal cord as the lateral corticospinal tract. The uncrossed fibres descend as the anterior (or ventral) corticospinal tract and eventually cross in the spinal cord. As the fibres of the corticospinal tract pass the thalamus and basal ganglia, they fan out into a sheet called the **internal capsule**. Here they seem particularly susceptible to damage following cerebral vascular accidents (strokes), the resulting ischaemic damage giving rise to a characteristic form of paralysis. In humans with corticospinal tract damage, for example following a stroke, there are few overt motor deficits. There is, however, weakness of the hand and finger muscles and a positive **Babinski sign** (a pathological reflex in which the toes point upwards and fan out in response to stroking the sole of the foot).

> **KEY POINTS:**
>
> - The activity of the neural circuitry within the spinal cord is modified and refined by descending motor control pathways. These are the pyramidal (corticospinal) and extrapyramidal (reticulospinal, vestibulospinal, rubrospinal and tectospinal) tracts.
>
> - The fibres of the corticospinal tract form the internal capsule as they pass the thalamus and basal ganglia and cross to the opposite side at the pyramids. Some make monosynaptic contact with the α- and γ-motoneurones of the hand and finger muscles. Loss of precise hand movements is a feature of lesions to the corticospinal tract.

12.8 Goal-directed movements

All through our life, we perform a host of movements that are voluntary and goal-directed in nature. Regions of the brain that are of particular importance in the initiation and refinement of such movements are the motor regions of the cerebral cortex, the cerebellum and the basal ganglia.

The motor cortex – its organization and functions

More than 120 years ago, it was found that electrical stimulation of particular areas of the cerebral cortex of dogs could elicit movements of the limbs on the opposite side of the body (the contralateral side). Subsequent work by W. Penfield and his colleagues in the 1930s showed that stimulation of the precentral gyrus of the right cerebral hemisphere of conscious human subjects elicited movements in the muscles on their left-hand side.

Penfield and his colleagues went on to establish in more detail the exact location of the areas of cortex in which stimulation led to co-ordinated movements.

They found that different groups of muscles were mapped onto the cortex in a strict order, as shown in Figure 12.9. Those areas of the body with especially refined and complex motor abilities, such as the fingers, lips and tongue, have a disproportionately larger representation in the primary motor cortex than those with poorer motor ability, such as the trunk. The motor cortex is made up of an overlapping array of motor points that control the activity of particular muscles or muscle groups. There is a similar somatotopic map on the postcentral gyrus for sensory information (see p. 234).

The orderly representation of parts of the body in the motor cortex is dramatically illustrated in patients suffering from a form of epilepsy known as Jacksonian epilepsy (after the neurologist Hughlings Jackson). Jacksonian convulsions are characterized by twitching movements that begin at an extremity such as the tip of

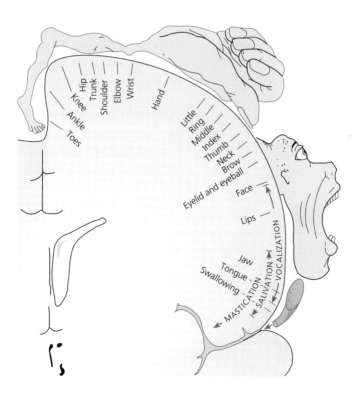

Figure 12.9 Somatotopic organization of the motor cortex (motor homunculus) showing the relative size of the regions representing various parts of the body.

a finger and show a progressive and systematic 'march'. After the initial twitching is seen, there is clonic movement of the affected finger, followed by movement of the hand, then the arm. The process culminates in a generalized convulsion. This progression of abnormal movements reflects the spread of excitation over the cortex from the point at which the over-activity began (the epileptic focus).

Connections of the motor cortex

Outflow from the motor cortex

The corticospinal (or pyramidal) tract is one of the major pathways by which motor signals are transmitted from the motor cortex to the anterior motoneurones of the spinal cord. Nevertheless, the extrapyramidal tracts also carry a significant proportion of the outflow from the motor cortex to the spinal cord. The pyramidal tract itself also contributes to these extrapyramidal pathways via numerous collaterals, which leave the tract within the brain (Figure 12.8). Indeed, each time a signal is transmitted to the spinal cord to elicit a movement, these other brain areas receive strong signals from the pyramidal tract.

Inputs to the motor cortex

The motor areas of the cerebral hemispheres are interconnected and receive inputs from a number of other regions, some of which are illustrated in Figure 12.10.

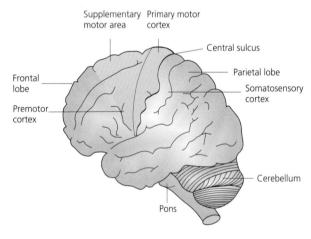

Figure 12.10 The location of the principal motor areas of the cerebral cortex.

The most prominent sensory input is from the somatosensory system. Information reaches the motor areas of the cerebral cortex either directly from the thalamus or indirectly by way of the somatosensory cortex of the postcentral gyrus. The posterior parietal cortex relays both visual and somatosensory information to the motor areas. The cerebellum and basal ganglia also send afferents to the motor cortex by way of the ventral lateral and ventral anterior thalamic nuclei. Figure 12.11 illustrates these pathways diagrammatically.

The pyramidal cells receive information from a wide range of sensory modalities including joint receptors, tendon organs, cutaneous receptors and spindle afferents. It is believed that sensorimotor correlation of this information from the skin and muscles is carried out in the motor cortex and is used to enhance grasping, touching or manipulative movements. The motor cortex is also believed to be involved in the generation of 'force commands' to muscles being used to counter particular loads. These commands adjust the force generated to match that of the load on the basis of information from pressure receptors in the skin and the load sensed by the tendon organs. Certainly, the rates of firing of pyramidal cells during voluntary movement correlate well with the force being produced to generate the movement.

Effects of lesions of the motor cortex

The motor areas of the cortex are often damaged as a result of strokes, and the muscles controlled by the damaged region show a corresponding loss of function (Box 12.1). Rather than a total loss of movement, however, there is a loss of voluntary control of fine movements, with clumsiness and slowness of movement and an unwillingness to use the affected muscles.

Many strokes cause widespread damage not only to the primary motor cortex but also to the neighbouring sensorimotor areas. These areas relay inhibitory signals to the spinal motoneurones via the extrapyramidal pathways, and when they are damaged, there is a release of inhibition that can result in the affected contralateral muscles going into spasm. Spasm is particularly intense if the basal ganglia are also damaged.

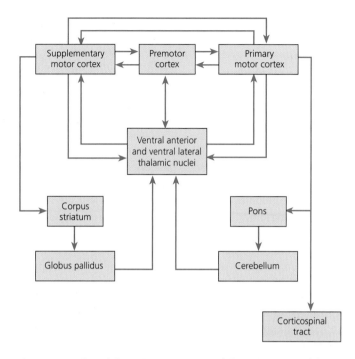

Figure 12.11 A diagrammatic representation of the major connections of the motor areas of the cerebral cortex with other motor areas of the brain.

KEY POINTS:

- The primary motor cortex is arranged as a somatotopic map and those areas of the body that are capable of especially refined and complex movements (i.e. fingers, lips and tongue) have a disproportionately large area of representation.

- The motor cortex receives inputs from many sources but the predominant sensory input is from the somatosensory system. This afferent information is used to refine movements, particularly to match the force generated in specific muscle groups to an imposed load.

- Motor areas of the cerebral cortex are often damaged by strokes, and muscles controlled by the damaged areas show a corresponding loss of function. However, recovery is often good, resulting in little more than clumsiness and a loss of fine muscle control.

12.9 The role of the cerebellum in motor control

Electrical stimulation of the cerebellum causes neither sensation nor significant movement. Damage to this area of the brain, however, is associated with severe abnormalities of motor function. The cerebellum appears to play a particularly vital role in the co-ordination of postural mechanisms and in the control of rapid muscular activities such as running, playing a musical instrument or typing. In contrast to the cerebral cortex, the two halves of the cerebellum each control and receive input from muscles on the same (ipsilateral) side of the body.

The anatomical structures of the cerebellum

The cerebellum (literally means little brain) is located dorsal to the pons and medulla and lies under the occipital lobes of the cerebral hemispheres (Figure 12.10). The cerebellum is usually subdivided into two major lobes, the anterior and posterior lobes, separated by the primary fissure, as shown in Figure 12.12. The lobes are further subdivided into nine transversely orientated lobules each of which is folded extensively, the folds being known as folia.

Like the cerebral cortex, the cerebellum is composed of a thin outer layer of grey matter, the cerebellar cortex, overlying internal white matter. Embedded within the white matter are the paired cerebellar nuclei, which receive afferents from the cerebellar cortex as well as sensory information from the spinal cord. The cerebellar nuclei give rise to all the efferent tracts from the anterior and posterior lobes of the cerebellum. The cerebellum has no direct connections with the cerebral cortex.

The cerebellum is connected to the medulla by the inferior peduncles, which carry information from muscle proprioceptors to the cerebellum. It also receives information from the pons via the middle cerebellar peduncles advising it of voluntary motor activities initiated by the motor cortex. The superior cerebellar peduncles connect the cerebellum and the midbrain. Fibres in these peduncles originate from neurones in the deep cerebellar nuclei and communicate with the motor cortex via the thalamus.

How does the cerebellum contribute to the control of voluntary movement?

Much of our information concerning the role of the cerebellum in the control of motor activity has been gathered from the effects of lesions and other damage and from experimental stimulation of, and recording from, cerebellar neurones. On the basis of such work, it is now believed that the primary role of the cerebellum is to supplement and correlate the activities of the other motor areas. More specifically, it seems to play a role in the control of posture and the correction of rapid movements initiated by the cerebral cortex.

Damage to the posterior cerebellum results in an impairment of postural co-ordination, which is similar to that seen when the vestibular apparatus itself is damaged. For example, a patient with a lesion in the area will feel dizzy, have difficulty standing upright and may develop a staggering gait (**cerebellar ataxia**). Damage to other regions of the cerebellum produces more generalized impairment of muscle control. Ataxia is again a problem, coupled with a loss of muscle tone and a lack of co-ordination. Figure 12.13 shows an example of impaired motor control from a patient with a right-sided cerebellar ataxia. The patient was asked to trace a path between a set of lights with the forefinger of each hand. The left forefinger, which was unaffected, completes the task with ease, but the right forefinger traces the path in a very uncertain manner, with looping and obvious tremor.

Much of the clumsiness of movement seen in patients with cerebellar lesions seems to be the result of a slowness

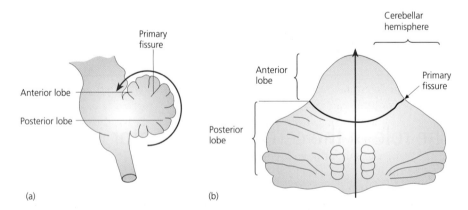

Figure 12.12 The major lobes of the cerebellum: (a) shows a sectional view of the brainstem and cerebellum; (b) shows the surface of the cerebellum 'unfolded' along the axis of the arrow indicated in (a) to show the positions of the anterior and posterior lobes.

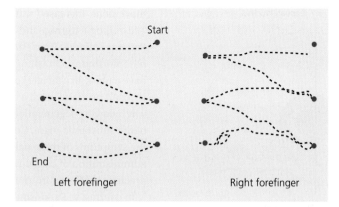

Figure 12.13 Records illustrating cerebellar ataxia. In this task, a flashing light was fixed to each forefinger in turn and the subject asked to trace a path as accurately as possible between two sets of three lights spaced 75 cm apart (shown here as red circles). The movement was captured on photographic film. The patient, who had a right-sided cerebellar lesion, had great difficulty tracing the path with his right hand but no difficulty with his left.

to respond to sensory information about the progress of the movement. There is often an intention tremor preceding goal-directed movements, in which the movement appears to oscillate around the desired position, or the movement may overshoot when the patient reaches for an object, as in the example shown in Figure 12.13. Speech may also become staccato ('scanning speech'), more laboured and less 'automatic'. In general, patients with cerebellar damage seem to require a great degree of conscious control over movements that normally require little thought.

> **KEY POINTS:**
>
> - The cerebellum plays a vital role in the co-ordination of postural mechanisms and in the control of rapid muscular activities.
> - Lesions to the cerebellum may result in dizziness and postural difficulties or a more generalized loss of muscle control as shown by intention tremor, clumsiness and speech problems.

12.10 The basal ganglia

The basal ganglia are the deep nuclei of the cerebral hemispheres. They consist of the caudate nucleus and putamen (known collectively as the **corpus striatum**), the globus pallidus and claustrum. Functionally, the basal ganglia are associated with the substantia nigra of the midbrain (so called because many of its cells are pigmented with melanin) and the red nucleus.

The location of the basal ganglia is illustrated in Figure 12.14. They form an important subcortical link between the frontal lobes and the motor cortex. Their importance in the control of motor function is clear from the severe disturbances of movement seen in patients with lesions of the basal ganglia.

The role of the basal ganglia in the control of movement

Most of the information that has accumulated concerning the actions of the basal ganglia in humans has been derived from studies of the effects of damage to those brain

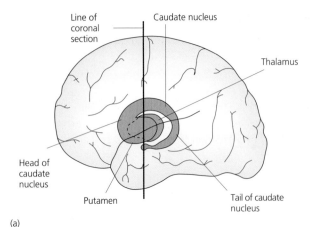

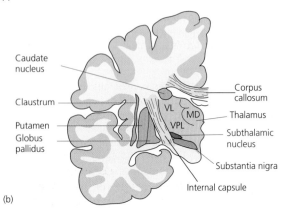

Figure 12.14 The location of the basal ganglia and associated nuclei. (a) The position of the basal ganglia within the cerebral hemispheres. (b) A coronal section taken along the line indicated in (a) to show the detailed location of the various nuclei of the basal ganglia. MD, VL and VPL are the mediodorsal, ventral lateral and posterolateral thalamic nuclei.

areas. A range of movement disorders, called **dyskinesias**, result from damage to the basal ganglia or their connections. The best-known disease of the basal ganglia is **Parkinson's disease** (shaking palsy), a degenerative disorder that results in variable combinations of slowness of movement, increased muscle tone (rigidity), resting (or 'pill-rolling') tremor and impaired postural responses. Patients suffering from Parkinson's disease have difficulty both starting and finishing a movement. They also tend to have a mask-like facial expression. Parkinsonism appears to result primarily from a defect in the nigrostriatal pathway following degeneration of the dopaminergic neurones of the substantia nigra. Considerable success in relieving the symptoms of these patients has been achieved by the administration of carefully controlled doses of L-DOPA, a precursor of the neurotransmitter dopamine.

In contrast to Parkinson's disease, in which movements are restricted, other disorders of the basal ganglia result in the spontaneous production of unwanted movements. These spontaneous movements include **hemiballismus** (a sudden flinging out of limbs on one side of the body as if to prevent a fall or to grab something), **athetosis** (snakelike writhing movements of parts of the body, particularly the hands, which are often seen in cerebral palsy) and **chorea** (a series of rapid, uncontrolled movements of muscles all over the body). In all cases, the disease process results in inappropriate or repetitive execution of normal patterns of movement. The hereditary disorder Huntingdon's chorea is perhaps the best known of this group of diseases.

> **KEY POINTS:**
>
> - The basal ganglia are deep cerebral nuclei; of these, the corpus striatum and globus pallidus are involved in motor control in association with the substantia nigra of the midbrain and the red nucleus.
>
> - A range of movement disorders (dyskinesias) result from damage to the basal ganglia or its connections. Parkinson's disease, in which movement is impoverished, is the most familiar.

 Recommended reading

Neuroanatomy of the motor system

Brodal, P. (2003). *The Central Nervous System. Structure and Function*, 3rd edn. Oxford University Press: Oxford. Chapters 11–14.

Pharmacology of neurodegenerative disorders

Rang, H.P., Dale, M.M., Ritter, J.M. and Flower, R. (2007). *Pharmacology*, 6th edn. Churchill-Livingstone: Edinburgh. Chapter 34.

Physiology

Carpenter, R.H.S. (2002). *Neurophysiology*, 4th edn. Hodder Arnold: London. Chapters 9–12.

Squire, L.R., Bloom, F.E., McConnell, S.K., Roberts, J.L., Spitzer, N.C. and Zigmond, M.J. (2002). *Fundamental Neuroscience*, 2nd edn. Academic Press: San Diego. Chapters 29–35.

Medicine

Donaghy, M. (2005). *Neurology*, 2nd edn. Oxford University Press: Oxford. Chapters 5–9, 26 and 27.

Self-assessment questions

Which of the following statements are true and which are false? Answers are given at the end of the book (p. 755).

1 a) White matter is made up largely of nerve cell bodies.
b) The ventral (anterior) horns of the spinal cord grey matter contain somatic motor neurones.
c) Ascending pathways in the spinal cord convey sensory information.
d) The ventral and dorsal roots of the spinal cord unite to form the spinal nerves.
e) Destruction of the anterior horn cells of the spinal cord results in a loss of sensory input from the area served.

2 a) Muscle spindles play an important role in the regulation of posture and movement.
b) Muscle spindles primarily measure the tension of a muscle while tendon organs primarily measure the length.
c) Muscle spindles receive both afferent and efferent nerve fibres.
d) The sensitivity of Golgi tendon organs is constantly adjusted during a movement.
e) The motor fibres supplying muscle spindles regulate the sensitivity of the spindles during a muscle contraction.

3 a) A monosynaptic reflex arc involves one or more interneurones.
b) Withdrawal reflexes are lost following severing of the spinal cord in the neck.
c) Spinal shock is characterized by flaccid paralysis of all muscles below the level of the spinal injury.
d) The tendon-tap reflex is an example of a crossed extensor reflex.
e) Motoneurones form the final common path for all reflexes.

4 a) The axons of the corticospinal tract mainly synapse directly with motoneurones in the spinal cord.
b) Damage to the pyramidal tract, caused for example by a stroke, results in a loss of fine control of the hand and finger muscles.
c) The muscles attached to the axial skeleton are of principal importance in the maintenance of an upright posture.
d) Motor areas of the cortex receive somatosensory input via the thalamus.
e) Following a cerebral haemorrhage affecting the precentral gyrus of the right hemisphere, the patient feels no sensation on the left side of his body.

5 a) The motor areas of the cortex are situated in the frontal lobe.
b) The corticospinal tract provides the only connection between the motor cortical areas and the spinal cord.
c) The cerebellum has a direct efferent projection to the motor cortex.
d) Intention tremor is a characteristic sign of damage to the cerebellum.
e) Hemiballismus is a sign of cerebellar damage.

13 The autonomic nervous system

After reading this chapter you should have gained an understanding of:

- The anatomical organization of the autonomic nervous system and its separation into the sympathetic and parasympathetic divisions
- How the sympathetic division regulates the activity of the cardiovascular system, visceral organs and secretory glands
- How the parasympathetic nerves regulate the activity of the gut, heart and secretory glands
- The role of acetylcholine and noradrenaline as neurotransmitters in the autonomic nervous system
- Drugs acting on the autonomic nervous system

13.1 Introduction

The autonomic nervous system regulates the operation of the internal organs to support the activity of the body as a whole. It is not a separate nervous system but is the efferent (motor) pathway that links those areas within the brain concerned with the regulation of the internal environment to specific effectors such as blood vessels, the heart and the gut. The autonomic nervous system is not under direct voluntary control, although autonomic responses can be influenced via biofeedback devices (e.g. subjects can learn to lower their heart rate by feedback from a heart rate monitor). The autonomic nerves pass first to **autonomic ganglia**, which are located outside the CNS, before reaching their target organs. The fibres that project from the CNS to the autonomic ganglia are called **preganglionic fibres** and those that connect the ganglia to their target organs are called **postganglionic fibres**.

13.2 Organization of the autonomic nervous system

The autonomic nervous system is divided into two parts:

- The **sympathetic division**, which broadly acts to prepare the body for activity.
- The **parasympathetic division**, which is more discrete in its actions and tends to promote restorative functions such as digestion and a slowing of the heart rate.

The sympathetic nervous system

The sympathetic division originates in the nerve cells of the lateral column of the thoracic and lumbar regions of the spinal cord. These neurones are called **sympathetic preganglionic neurones**. Their axons (sympathetic preganglionic fibres) leave the spinal cord via the ventral root together with the somatic motor fibres. Shortly after the dorsal and ventral roots join together, the sympathetic preganglionic fibres leave the spinal nerve trunk to travel to sympathetic ganglia via the **white rami communicantes**, as shown in Figure 13.1. The preganglionic fibres synapse on sympathetic neurones within the ganglia and the axons of the ganglion neurones give rise to **postganglionic sympathetic fibres** that project to the target organs via the **grey rami communicantes** and the segmental spinal nerves (Figure 13.1). The ganglia themselves act as synaptic relays, distributing information from the higher centres of the nervous system to the various internal organs.

The majority of sympathetic ganglia are found on each side of the vertebral column and are linked together by longitudinal bundles of nerve fibres to form the two sympathetic trunks, as shown on the left-hand side of Figure 13.2. The sympathetic preganglionic fibres to the abdominal organs join to form the **splanchnic nerves**, which pass to the coeliac, superior and inferior mesenteric ganglia where they synapse. Postganglionic sympathetic fibres then pass to the various abdominal organs, as shown in Figure 13.2.

Although the sympathetic preganglionic fibres are myelinated, the sympathetic postganglionic fibres

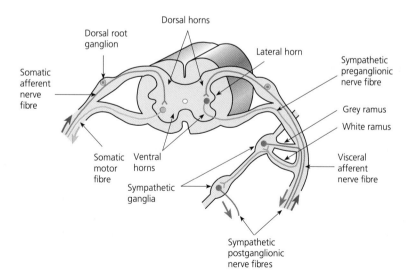

Figure 13.1 A simple diagram comparing the arrangement of the sympathetic preganglionic and postganglionic neurones with the organization of the somatic motor nerves. Note that sympathetic preganglionic fibres may terminate in the sympathetic ganglion of the same segment or pass to another ganglion in the sympathetic chain. Some also terminate in prevertebral ganglia such as the celiac ganglion.

Sympathetic innervation

Parasympathetic innervation

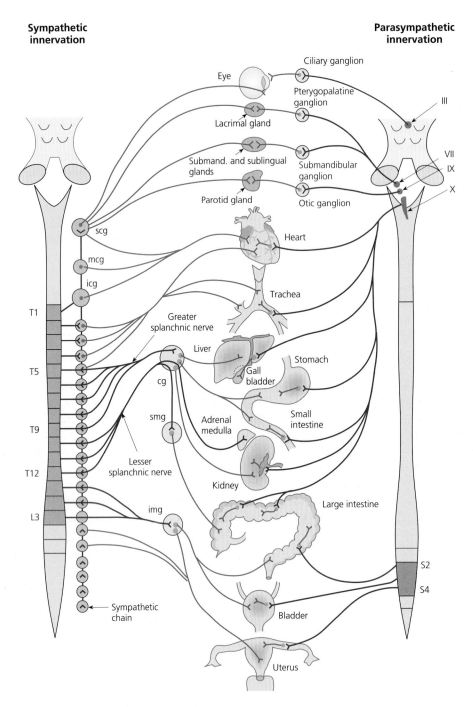

Figure 13.2 A schematic drawing illustrating the organization of the autonomic nervous system. In addition to the innervation of the principal organ systems, segmental sympathetic fibres also innervate blood vessels, piloerector muscles and sweat glands. Parasympathetic preganglionic fibres are found in cranial nerves III (oculomotor), VII (facial), IX (glossopharyngeal) and X (vagus). scg, mcg and icg refer to superior, middle and inferior cervical ganglion; cg, celiac ganglion; smg and img refer to the superior and inferior mesenteric ganglia. Postganglionic fibres are shown in green.

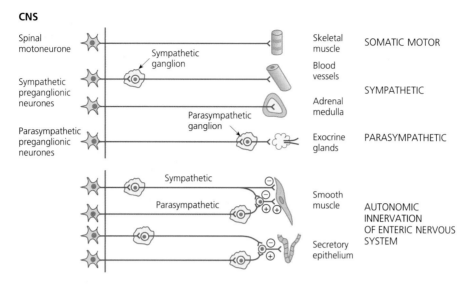

Figure 13.3 A schematic diagram comparing the organization of the sympathetic and parasympathetic divisions of the autonomic nervous system with that of somatic motor innervation. The lower part of the figure shows a simplified plan of the innervation of the gut by sympathetic and parasympathetic nerve fibres. − indicates an inhibitory action, + an excitatory action.

are unmyelinated. This explains the difference in the appearance of grey and white rami. As Figure 13.2 shows, sympathetic postganglionic fibres innervate many organs including the eye, the salivary glands, the gut, the heart and the lungs. They also innervate the smooth muscle of the blood vessels, the sweat glands and the muscles of the skin hairs. As the sympathetic ganglia are located close to the spinal cord, most sympathetic preganglionic fibres are relatively short while the postganglionic fibres are much longer, as illustrated in Figure 13.3.

> **KEY POINTS:**
>
> - The sympathetic division broadly acts to prepare the body for activity ('fight or flight').
> - Increased sympathetic activity is associated with an increased heart rate, vasoconstriction in the visceral organs and vasodilatation in skeletal muscle.

The parasympathetic nervous system

The preganglionic neurones of the parasympathetic division of the autonomic nervous system have their cell bodies in two regions: the brainstem and in sacral segments S3–S4 of the spinal cord. Thus, the parasympathetic preganglionic fibres emerge as part of the **cranial outflow** in cranial nerves III (oculomotor), VII (facial), IX (glossopharyngeal) and X (vagus) and from the **sacral outflow**.

The parasympathetic ganglia are usually located close to the target organ or even embedded within it. Thus, the parasympathetic innervation is characterized by long preganglionic fibres and short postganglionic fibres, in contrast to the organization of the sympathetic nervous system (Figure 13.3). Parasympathetic postganglionic fibres innervate the eye, the salivary glands, the genitalia, the gut, the heart, the lungs and other visceral organs as shown on the right-hand side of Figure 13.2.

> **KEY POINTS:**
>
> - The parasympathetic division tends to promote restorative functions ('rest and digest').
> - Increased parasympathetic activity is associated with increased motility and secretion by the gastrointestinal tract and slowing of the heart rate.

Many organs receive a dual innervation from sympathetic and parasympathetic fibres

Most visceral organs (but not all) are innervated by both divisions of the autonomic nervous system. The specific actions of the autonomic nerves on the various organ systems of the body are discussed at length in the relevant chapters of this book. In many cases, the actions of the sympathetic and parasympathetic divisions are antagonistic so that the actions of the two divisions provide a delicate control over the functions of the viscera. Thus, activation of the sympathetic nerves to the heart increases heart rate and the force of contraction of the heart muscle, while activation of the vagus nerve (parasympathetic) slows the heart. Activation of the parasympathetic supply to the gut enhances its motility and secretory functions, while activation of the sympathetic supply inhibits the digestive functions of the gut and constricts its sphincters.

Some organs only have a sympathetic supply. Examples are the adrenal medulla, the pilomotor muscles of the skin hairs, the sweat glands and the spleen. The parasympathetic supply has exclusive control over the focusing of the eyes by the ciliary muscles. The effects of activation of the sympathetic and parasympathetic divisions of the autonomic nervous system on various organs are summarized in Table 13.1.

Table 13.1 The main actions of sympathetic and parasympathetic stimulation on various organ systems

Organ	Effect of sympathetic activation	Effect of parasympathetic activation
Eye	Pupillary dilatation	Pupillary constriction and accommodation
Lacrimal gland	No effect (vasoconstriction?)	Secretion of tears
Salivary glands	Vasoconstriction, secretion of viscous fluid	Vasodilatation and copious secretion of saliva
Heart	Increased heart rate and force of contraction	Decreased heart rate; no effect on force of contraction
Blood vessels	Mainly vasoconstriction; vasodilatation in skeletal muscle	No effect except for vasodilatation in certain exocrine glands and the external genitalia
Lungs	Bronchial dilatation via circulating adrenaline	Bronchial constriction; secretion of mucus
Liver	Glycogenolysis, gluconeogenesis and release of glucose into blood	No effect on liver but secretion of bile by gall bladder
Adrenal medulla	Secretion of adrenaline and noradrenaline	No innervation
Gastrointestinal tract	Decreased motility and secretion; constriction of sphincters; vasoconstriction	Increased motility and secretion; relaxation of sphincters
Kidneys	Vasoconstriction and decreased urine output	No effect
Urinary bladder	Inhibition of micturition	Initiation of micturition
Genitalia	Ejaculation	Erection
Sweat glands	Secretion of sweat by eccrine glands	No innervation
Hair follicles	Piloerection	No innervation
Metabolism	Increase	No effect

Note: This table summarizes the main effects of activation of the sympathetic or parasympathetic nerves to various organs. The details of specific autonomic reflexes can be found in the relevant chapters of this book.

Autonomic nerves maintain a basal level of tonic activity

The autonomic innervation generally provides a basal level of activity in the tissues it innervates called **tone** (see also Chapter 9, p. 135). The autonomic tone can be either increased or decreased to modulate the activity of specific tissues. For example, the blood vessels are generally in a partially constricted state as a result of **sympathetic tone**. This partial constriction restricts the flow of blood. If sympathetic tone is increased, the affected vessels become more constricted and this results in a decrease in blood flow. Conversely, if sympathetic activity is inhibited, tone decreases and the affected vessels dilate so increasing their blood flow (see Chapter 19). The heart in a resting person is normally under the predominant influence of **vagal tone**. If the vagus nerves are cut, the heart rate rises. During the onset of exercise, the tonic parasympathetic inhibition of the heart declines and sympathetic activation increases. This allows the heart rate to increase.

The enteric nervous system

Sympathetic and parasympathetic nerve fibres act on neurones that are present in the walls of the gastrointestinal tract. These neurones have been considered to form a separate division of the nervous system – **the enteric nervous system**. The enteric neurones are organized as two interconnected plexuses:

* The **submucosal plexus** (also known as Meissner's plexus), which lies in the submucosal layer beneath the muscularis mucosae (see Figure 24.2)
* The **myenteric plexus** (or Auerbach's plexus), which lies between the outer and the inner layers of smooth muscle.

The enteric nervous system can function independently of its autonomic supply, and its neurones play an important part in the regulation of the motility and secretory activity of the digestive system (see Chapter 24).

13.3 Chemical transmission in the autonomic nervous system

Within the autonomic ganglia, the synaptic contacts are highly organized and similar in structure to the other neuronal synapses described in Chapter 6. In the target tissues, however, the synaptic contacts are more diffuse than those of the CNS or those of the neuromuscular junction of skeletal muscle. The postganglionic fibres have varicosities along their length that secrete neurotransmitters into the space adjacent to the target cells rather than onto a clearly defined synaptic region. In multiunit smooth muscle, however, each varicosity is closely associated with an individual smooth muscle cell.

The main neurotransmitters secreted by the neurones of the autonomic nervous system are **acetylcholine** and **noradrenaline** (also known as **norepinephrine**). Within the ganglia of both the sympathetic and parasympathetic divisions, the principal transmitter secreted by the preganglionic fibres is acetylcholine. The postganglionic fibres of the parasympathetic nervous system also secrete

acetylcholine onto their target tissues. The postganglionic sympathetic fibres secrete noradrenaline, except for those fibres that innervate the sweat glands and arrector pili muscles (the muscles that raise the skin hairs and cause 'goose pimples'). These sympathetic postganglionic fibres secrete acetylcholine. The transmitters utilized by the different neurones of the autonomic nervous system are summarized in Figure 13.4.

Acetylcholine activates nicotinic receptors in the autonomic ganglia but muscarinic receptors in the target tissues

The acetylcholine receptors of the postganglionic neurones in autonomic ganglia are called **nicotinic receptors** because they can also be activated by the alkaloid nicotine. They are similar in structure to the nicotinic receptors of the neuromuscular junction but have a different

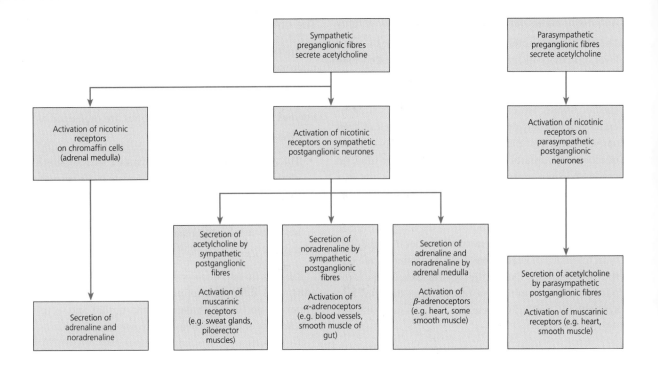

Figure 13.4 A flow chart illustrating the role of the cholinergic and adrenergic innervation in the autonomic nervous system.

response to various drugs and toxins. For example, they can be blocked by a drug called **mecamylamine**, which has no action at the neuromuscular junction, but not by α-bungarotoxin, which is a potent blocker of the nicotinic receptors of the neuromuscular junction. Activation of nicotinic receptors leads to the opening of an ion channel and rapid excitation, as described on p. 113.

The acetylcholine receptors of the target tissues of both parasympathetic and sympathetic postganglionic fibres are **muscarinic receptors** as they can be activated by an alkaloid called muscarine. Muscarinic receptors are also present at sympathetic nerve endings in sweat glands. Muscarinic receptors serve many functions in the body, but they can be classified on the basis of their structure and function into three main classes:

- **M_1 receptors**, which are mainly located on neurones in the CNS and peripheral ganglia. They are also found on gastric parietal cells. Activation of these receptors has an excitatory effect. For example, the secretion of

gastric acid that follows stimulation of the vagus nerve is mediated by M_1 receptors.

- **M_2 receptors**, which occur in the heart and on the nerve terminals of both CNS and peripheral neurones. The slowing of the heart following stimulation of the vagus nerve is mediated by this class of muscarinic receptor.

- **M_3 receptors**, which are located in secretory glands and on smooth muscle. Activation of these receptors generally has an excitatory effect. In visceral smooth muscle, for example, activation of M_3 receptors by acetylcholine leads to contraction.

All muscarinic receptors are linked to G proteins: activation of M_1 and M_3 receptors leads to an increase in the intracellular levels of IP_3 while activation of M_2 receptors results in a decrease in the level of cyclic AMP (see p. 89).

Muscarinic receptors can be inhibited by **atropine** and **hyoscine**. Both drugs are used as in general anaesthesia to block secretions from the salivary glands (a dry mouth is a well-known feature of anaesthetic premedication). The muscarinic antagonists **cyclopentolate** and **tropicamide** are used to dilate the pupils prior to detailed eye examination. The muscarinic agonist **pilocarpine** is used to treat glaucoma.

Adrenergic receptors belong to two main classes

The receptors for noradrenaline are called **adrenoceptors**. Two main classes are known, which are called α-**adrenoceptors** and β-**adrenoceptors**. Each group is further subdivided so that at present five subtypes are recognized: $\alpha_1, \alpha_2, \beta_1, \beta_2$ and β_3. As for muscarinic receptors, the adrenoceptors are coupled to second messenger systems via a G protein: α_1-adrenoceptors activate phospholipase C and increase intracellular levels of IP_3; α_2-adrenoceptors inhibit adenylyl cyclase activity and reduce the intracellular level of cyclic AMP; activation of $\beta1$-, $\beta2$- and $\beta3$-adrenoceptors activates adenylyl cyclase and leads to an increase in intracellular cyclic AMP.

- α_1-**adrenoceptors** are found in the smooth muscle of the blood vessels, bronchi, gastrointestinal tract, uterus and bladder. Activation of these receptors is mainly excitatory and results in the contraction of smooth muscle. However, the smooth muscle of the gut wall (but not that of the sphincters) becomes relaxed after activation of these receptors.

- α_2-**adrenoceptors** are found in the smooth muscle of the blood vessels where their activation causes vasoconstriction.

- β_1-**adrenoceptors** are found in the heart where their activation results in an increased rate and force of contraction. They are also present in the sphincter muscle of the gut where their activation leads to relaxation.

- β_2-**adrenoceptors** are found in the smooth muscle of certain blood vessels where their activation leads to vasodilatation. They are also present in the bronchial smooth muscle where they mediate bronchodilatation.

- β_3-**adrenoceptors** are present in adipose tissue where they initiate lipolysis to release free fatty acids and glycerol into the circulation.

Although the diversity of adrenoceptors is somewhat bewildering, the development of agonists and antagonists that act on specific subtypes of these receptors has been of considerable clinical benefit in the treatment of diseases such as asthma and hypertension. Selective blockers of α_1-adrenoceptors, such as **prazosin**, are used to treat hypertension and heart failure. Blockers of β-adrenoceptors, such as **propranolol** and **atenolol**, are used to treat hypertension, some arrhythmias and heart failure; β-adrenoceptor agonists, such as **salbutamol**, are used to relieve the bronchoconstriction of asthma.

> **KEY POINTS:**
>
> - The main neurotransmitters secreted by the neurones of the autonomic nervous system are acetylcholine and noradrenaline.
>
> - Acetylcholine is the principal transmitter secreted by the preganglionic fibres within the autonomic ganglia of both divisions.
>
> - The parasympathetic postganglionic fibres also secrete acetylcholine onto their target tissues while noradrenaline is the principal neurotransmitter secreted by the postganglionic sympathetic fibres.

The adrenal medulla secretes adrenaline and noradrenaline into the circulation

Although the activation of the autonomic nerves provides a mechanism for the discrete regulation of specific organs, activation of the splanchnic nerve results in the secretion of adrenaline and noradrenaline (also known as epinephrine and norepinephrine) from the adrenal medulla into the circulation. About 80% of the secretion is adrenaline and 20% is noradrenaline. These catecholamines exert a hormonal action on a variety of tissues (see Chapter 16), which forms part of the overall sympathetic response. Their release is always associated with an increase in the secretion of noradrenaline from sympathetic nerve terminals.

The response of a tissue to circulating adrenaline and noradrenaline will depend on the relative proportions of the different types of adrenoceptor it possesses. Adrenaline activates β-adrenoceptors more strongly than α-adrenoceptors, while noradrenaline is more effective at activating α-adrenoceptors than β-adrenoceptors.

During exercise, increased activity in the sympathetic nerves will result in an increased heart rate and vaso-constriction in the splanchnic circulation. The circulating noradrenaline will have a similar effect but the actions of circulating adrenaline will lead to relaxation of the smooth muscle that possesses a high proportion of β-adrenoceptors such as that of the blood vessels of skeletal muscle. The increased levels of circulating adrenaline cause bronchodilatation and a vasodilatation in the skeletal muscle, thus favouring increased gas flow to the alveoli and blood flow to the exercising muscles.

Circulating catecholamines affect virtually every tissue, with the result that the metabolic rate of the body is increased. Indeed, maximal sympathetic stimulation may double the metabolic rate. The major metabolic effect of adrenaline and noradrenaline is to increase the rate of glycogenolysis within cells by activating adenylyl cyclase via β-adrenoceptors, as described in Chapter 7, p. 89. The result is a rapid mobilization of glucose from glycogen and an increased availability of fatty acids for oxidation as a result of lipolysis occurring in adipose tissue. The increased availability of substrates for oxidative metabolism is important both in exercise and during cold stress where an increase in metabolic rate is important for generating the heat required to maintain body temperature.

13.4 Central nervous control of autonomic activity

The activity of the autonomic nervous system varies according to the information it receives from both visceral and somatic afferent fibres. It is also subject to regulation by the higher centres of the brain, notably the hypothalamus.

Autonomic reflexes

The internal organs are innervated by afferent fibres that respond to mechanical and chemical stimuli; these are known as **visceral afferents**. Some visceral afferents reach the spinal cord by way of the dorsal roots and enter the dorsal horn together with the somatic afferents. These fibres synapse at the segmental level and the second-order fibres ascend the spinal cord in the spinothalamic tract. They project to the brainstem, thalamus and hypothalamus. Other visceral afferents, such as those from the arterial baroreceptors, reach the brainstem by way of the vagus nerves.

Information from the visceral afferents elicits specific **visceral reflexes**, which, like the reflexes of the somatic motor system described in Chapter 12, may involve the spinal cord alone or may require the participation of pathways within the brain. Examples of autonomic reflexes are the baroreceptor reflex, the lung inflation reflex and the micturition reflex. These are discussed in detail in Chapters 19, 22 and 23.

In response to a perceived danger, there is a behavioural alerting that may result in aggressive or defensive behaviour. This is known as the **defence reaction**, which has its origin in the hypothalamus. During the defence reaction, there are marked changes in the activity of the autonomic nerves in which normal reflex control is overridden.

The hypothalamus regulates the homeostatic activity of the autonomic nervous system

Both the activity of the autonomic nervous system and the function of the endocrine system are under the control of the hypothalamus, which is the part of the brain mainly concerned with maintaining the homeostasis of the body. If the hypothalamus is destroyed, the homeostatic mechanisms fail. The hypothalamus receives afferents from the retina, the chemical sense organs, somatic senses and from visceral afferents. It also receives many inputs from other parts of the brain including the limbic system and cerebral cortex. Hypothalamic neurones play important roles in thermoregulation, in the regulation of tissue osmolality, in the control of feeding and drinking, and in reproductive activity.

> **KEY POINTS:**
>
> • Many organs receive innervation from both sympathetic and parasympathetic nerve fibres, which act in opposing ways to regulate the activity of the internal organs according to the needs of the body at the time.
>
> • The activity of the autonomic nerves is under the control of neurones in the brainstem and hypothalamus.

Recommended reading

Anatomy

Brodal, P. (2003). *The Central Nervous System. Structure and Function*, 3rd edn. Oxford University Press: Oxford. Chapters 18 and 19.

Pharmacology

Rang, H.P., Dale, M.M., Ritter, J.M. and Flower, R. (2007). *Pharmacology*, 6th edn. Churchill-Livingstone: Edinburgh. Chapters 9–11.

Self-assessment questions

Which of the following statements are true and which are false? Answers are given at the end of the book (p. 755).

1 a) Sympathetic preganglionic neurones are found in the spinal cord.
   b) The sympathetic chain extends from the cervical to the sacral regions of the spinal cord.
   c) Sympathetic preganglionic fibres secrete noradrenaline.
   d) Acetylcholine is secreted by some sympathetic postganglionic fibres.
   e) The blood vessels are regulated entirely by the sympathetic nervous system.

2 a) Parasympathetic preganglionic fibres are found in cranial nerve III (oculomotor).
   b) Stimulation of the vagus nerves increase the heart rate.
   c) Parasympathetic vasoconstrictor fibres are present in the salivary glands.
   d) Parasympathetic postganglionic fibres secrete noradrenaline onto their target organs.
   e) Parasympathetic preganglionic fibres secrete acetylcholine.

3 a) Stimulation of the sympathetic nerves to the eyes causes the pupil to constrict.
   b) Adrenaline secreted by the adrenal medulla enhances glycogen breakdown in the liver.
   c) Stimulation of the vagus nerves increases the motility of the gastro-intestinal tract.

   d) Activation of the sympathetic system causes vasoconstriction in the viscera and skin but vasodilatation in skeletal muscle.
   e) Cutting the vagus nerves leads to a slowing of the heart rate.

4 a) The acetylcholine receptors in both parasympathetic and sympathetic ganglia are muscarinic.
   b) Acetylcholine secreted by parasympathetic postganglionic fibres acts on muscarinic receptors.
   c) Mecamylamine is an agonist at cholinergic receptors.
   d) Atropine is an antagonist at all muscarinic receptors.
   e) Muscarinic antagonists dilate the pupils.

5 a) Noradrenaline secreted by sympathetic postganglionic fibres acts preferentially on β-adrenoceptors.
   b) The adrenal medulla secretes adrenaline in response to stimulation of the splanchnic nerves.
   c) High blood pressure (hypertension) can be treated with β-adrenoceptor blockers.
   d) β-Adrenoceptor antagonists such as atenolol increase heart rate.

14 Sensory systems

After reading this chapter you should have gained an understanding of:

- The principles by which sensory receptors derive information about the environment
- The physiological basis of the senses of touch, pressure, vibration and temperature
- The pathophysiology of pain and itch
- The physiology of the eye and the visual pathways
- The physiology of the ear and the auditory pathways
- The organization of the vestibular system and its role in the sense of balance
- The physiological basis of the senses of smell (olfaction) and taste

14.1 Introduction

We smell the air, taste our food, feel the earth under our feet, and hear and see what is around us. To do all this, and more, we must have some means of converting the physical and chemical properties of the environment into nerve impulses, which are used for signalling between the neurones of the nervous system. The process by which specific properties of the environment become encoded as nerve impulses is called **sensory transduction**. It is carried out by specialized structures called **sensory receptors**, often simply called receptors. Examples include thermoreceptors (which detect temperature), photoreceptors (which detect light) and nociceptors (which detect painful stimuli).

For a particular receptor, there is usually one kind of stimulus to which it will be especially sensitive. Thus, small changes in temperature excite specific temperature receptors in the skin; light stimulates the photoreceptors of the eye and so on. Table 14.1 lists the different sensory receptors that enable our brains to gain information about our external and internal environment. Receptors such as baroreceptors and osmoreceptors that provide information about the body's internal environment are considered in the relevant chapters of this book. This chapter will present a brief account of the mechanisms by which sensory receptors and pathways of the nervous system provide us with information about our external environment.

The organization of sensory pathways in the CNS

Receptors send their information to the CNS (the brain and spinal cord) via **afferent nerve fibres**, often called primary afferents. A given afferent nerve fibre often serves

Table 14.1 Classification of sensory receptors

Receptor type		Example
Mechanoreceptors	Special senses (ear)	Cochlear hair cells
		The hair cells of the vestibular system
	Muscle and joints	Muscle spindles
		Golgi tendon organs
	Skin and viscera	Pacinian corpuscle
		Ruffini ending
		Meissner's corpuscle
		Bare nerve endings
	Cardiovascular	Arterial baroreceptors (sense pressures in the aorta and carotid artery)
		Atrial volume receptors (low-pressure receptors)
Chemoreceptors	Special senses	Olfactory receptors
		Taste receptors
	Skin and viscera	Nociceptors
		Glomus cells (carotid body, sense arterial Po_2)
		Hypothalamic osmoreceptors and glucose receptors
Photoreceptors	Special senses (eye)	Retinal rods and cones
Thermoreceptors	Skin	Warm and cold receptors
	CNS	Temperature-sensing hypothalmic neurones

a number of receptors of the same kind. It will respond to a stimulus over a certain area of space and intensity, which is called its **receptive field**. The receptive fields of neighbouring afferents often overlap, as shown in Figure 14.1. Several primary afferents may then converge to stimulate a second-order sensory neurone, which then conveys the electrical signals (via a number of other neurones) to the brain where the information is processed.

Principles of transduction

Receptors may be bare nerve endings or they may consist of specialized cells in close association with a nerve fibre. Although different receptors respond to environmental stimuli in different ways, in all cases a stimulus ultimately leads to a change in membrane potential, called a **receptor potential**. This conversion of the stimulus into a change in membrane potential is known as sensory transduction. For most sensory receptors, stimulation causes cation-permeable ion channels to open, which leads to the depolarization of the afferent nerve fibre, called a receptor potential. When the threshold is reached, an action potential is generated. The magnitude and duration of a receptor potential governs the number and frequency of action potentials transmitted by the afferent nerve fibres to the CNS. The basic steps in sensory transduction are illustrated in Figure 14.2.

The coding of stimulus intensity and duration

The nervous system needs to establish the location, physical nature and intensity of all kinds of stimuli.

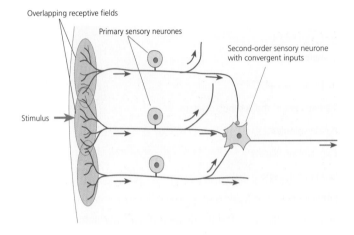

Figure 14.1 A simple diagram illustrating overlap between receptive fields of neighbouring primary afferents and their convergence onto a second-order sensory neurone. The primary afferents also send axon branches to other spinal neurones as indicated by the arrows.

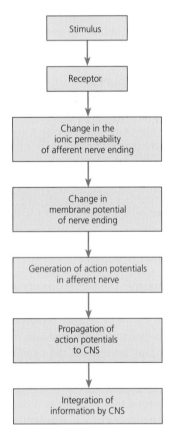

Figure 14.2 The main steps in sensory transduction for cutaneous receptors.

As most receptors respond to very specific stimuli, the primary afferent fibres to which they are connected can be considered as 'labelled lines'. For example, the skin has receptors that respond selectively to touch and others that respond to a small fall or a small rise in temperature. The activation of a specific population of receptors will therefore inform the CNS of the nature and location of the stimulus. The intensity of the stimulus is coded both by the number of active receptors and by the number of action potentials each receptor elicits. The timing and duration of a sequence of action potentials signals its onset and duration.

When a receptor is first stimulated, it often elicits a high-frequency burst of action potentials. The frequency then declines with time, even though the intensity of the stimulus is unchanged. This property is known as adaptation and helps to explain why we stop feeling our clothes against our skin soon after getting dressed.

KEY POINTS:

- The external and internal environment is continuously monitored by sensory receptors. Each kind of receptor is excited most effectively by a specific type of stimulus known as its adequate stimulus.

- The nerve fibres that convey information from the sensory receptors to the CNS are known as afferent or sensory nerve fibres.

- The process by which an environmental stimulus becomes encoded as a sequence of nerve impulses in an afferent nerve fibre is called sensory transduction.

- Different kinds of receptor are activated in different ways, but the first stage in sensory transduction is the generation of a receptor potential.

14.2 The somatosensory system

The skin is the interface between the body and the outside world. It is richly endowed with receptors that sense pressure, touch, temperature, vibration, pain and itch. The muscles and joints also possess sensory receptors that provide information concerning the disposition and movement of the limbs. All of this information is relayed by the afferent nerves of the somatosensory system to the brain and spinal cord.

The various kinds of receptors present in the skin are illustrated in Figure 14.3. Both bare nerve endings and encapsulated receptors are present. Each receptor type is innervated by a particular type of afferent nerve fibre and detects a specific cutaneous sensation (Table 14.2). Pacinian corpuscles and Merkel's discs are innervated by relatively large myelinated nerve fibres called Aβ fibres, while the bare nerve endings that provide temperature and pain sensations are derived either from small myelinated Aδ fibres or from slowly conducting, unmyelinated C-fibres (Table 14.2).

In general, the receptive fields of touch receptors overlap considerably, as illustrated in Figure 14.1. The finer the discrimination required, the higher the density of receptors, the smaller their receptive fields and the greater the degree of overlap. The receptive fields for touch are particularly small at the tips of the fingers and tongue (about $1\,mm^2$) where fine tactile discrimination is required. In other areas, such as the small of the back, the buttocks and the calf, the receptive fields are about a hundred times larger.

The distance between two points on the skin that can just be detected as separate stimuli is closely allied to the density of touch receptors and the size of their receptive fields. This is known as the two-point discrimination threshold. Not surprisingly, the greatest discrimination is at the tips of the fingers, the tip of the tongue and the lips. It is least precise for the skin of the back (Figure 14.4). A loss of precision in two-point discrimination can be used to localize specific neurological lesions.

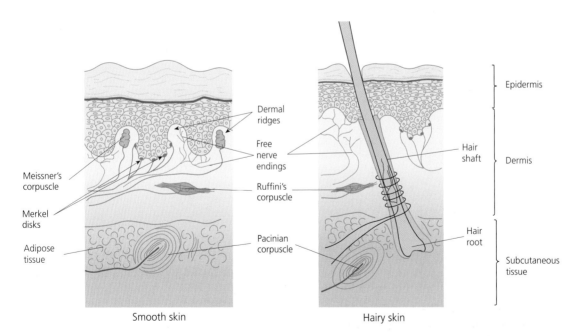

Figure 14.3 The principal types of sensory receptor found in smooth (glabrous) skin and hairy skin. Note the thickness of the epidermis in glabrous skin.

Table 14.2 The receptor types and modalities of the somatosensory system

Modality	Receptor type	Afferent nerve fibre type and conduction velocity
Touch	Rapidly adapting mechanoreceptor, e.g. hair follicle receptors, bare nerve endings, Pacinian corpuscles	$A\beta$ (6–12 μm diameter) 33–75 m s^{-1}
Touch and pressure	Slowly adapting mechanoreceptors, e.g. Merkel's cells, Ruffini end organs	$A\beta$ (6–12 μm diameter) 33–75 m s^{-1}
	Bare nerve endings	$A\delta$ (1–5 μm diameter) 5–30 m s^{-1}
Vibration	Meissner's corpuscles Pacinian corpuscles	$A\beta$ (6–12 μm diameter) 33–75 m s^{-1}
Temperature	Cold receptors	$A\delta$ (1–5 μm diameter) 5–30 m s^{-1}
	Warm receptors	C-fibres (0.2–1.5 μm diameter) 0.5–2.0 m s^{-1}
Nociception (pain)	Bare nerve endings, fast 'pricking' pain	$A\delta$ (1–5 μm diameter) 5–30 m s^{-1}
	Bare nerve endings, slow burning pain, itch	C-fibres (0.2–1.5 μm diameter) 0.5–2.0 m s^{-1}

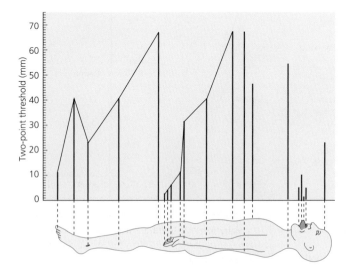

Figure 14.4 The variation of two-point discrimination across the body surface. Each vertical line represents the minimum distance that two points can be distinguished as being separate when they are simultaneously stimulated. Note the fine discrimination achieved by the fingertips, lips and tongue and compare this with the poor discrimination on the thigh, chest and neck.

Thermoreceptors

The skin has two kinds of temperature receptor (thermoreceptor). One type specifically responds to cooling of the skin and another responds to warming. Both are bare nerve endings and the cold receptors are innervated by Aδ myelinated afferents while the warm receptors are innervated by C-fibres.

Cutaneous thermoreceptors are generally insensitive to mechanical and chemical stimuli and maintain a constant rate of discharge for a particular skin temperature. They respond to a change in temperature with an increase or decrease in firing rate, the cold receptors showing a maximal rate of discharge around 25–30°C, while the warm receptors have a maximal rate of discharge around 40°C (Figure 14.5). Thus, a given frequency of discharge from the cold receptors may reflect a temperature that is either above or below the maximum firing rate for a particular receptor. This ambiguity may explain the well known paradoxical sense of cooling when cold hands are being rapidly warmed by immersion in hot water.

Kinaesthesia and haptic touch

People move through their environment; they lift objects, move them around and feel their texture. The constant stimulation of different receptors prevents them adapting. As a result, the brain is provided with more information about an object than would be possible with a single contact. The active exploration of an object to determine its shape and texture is known as **haptic touch**, which is used to great effect by the blind.

We know the position of our limbs even when we are blindfold. This sense is called **kinaesthesia**. Two sources of information provide the brain with information about the disposition of the limbs. Information regarding an intended movement is relayed to the sensory cortex via the motoneurones responsible for the movement while sensory feedback from muscle spindles and Golgi tendon organs (the proprioceptors located in muscles and joints) directly informs the sensory cortex of the actual progress of the movement (see Chapter 12 for further information).

Visceral receptors

The internal organs are much less well innervated than the skin. Nevertheless, all of the internal organs have an afferent innervation, although the activity of these afferents rarely reaches consciousness except as a vague sense of 'fullness' or as pain (see below, p. 235). The afferent fibres reach the spinal cord by way of the visceral

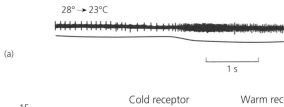

28° → 23°C

(a)

1 s

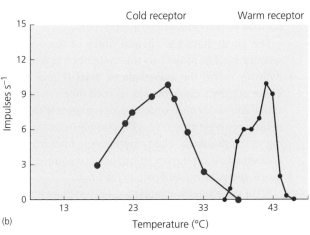

Figure 14.5 Typical response patterns of a warm and a cold skin thermoreceptor. In (a), the action potential discharge of a cold receptor is seen to increase as the skin is cooled from 28° to 23°C; (b) shows the rate of action potential discharge as a function of temperature for a single cold and a single warm thermoreceptor.

nerves that also carry the sympathetic and parasympathetic fibres that provide motor innervation to the viscera (see Chapter 13).

The somatosensory pathways

The cutaneous and visceral afferent fibres enter the spinal cord via the dorsal (posterior) roots. The spinal nerves have a distinct segmental arrangement, and each pair of nerves supplies a particular area of the body surface known as a **dermatome**. The dermatomes are shown in Figure 14.6. The receptive fields of the spinal nerves overlap with that of their immediate neighbours. This overlap is sufficient to prevent the total loss of sensation in a particular area should its main spinal nerve be damaged. The segmental organization of the spinal nerves is revealed when the skin erupts in shingles, following infection by the herpesvirus varicella-zoster virus.

After the large-diameter afferents enter the spinal cord, they travel in the gracile and cuneate fascicles of the **dorsal columns** before they synapse in the medulla oblongata. The second-order fibres leave the dorsal column nuclei as a discrete fibre bundle called the medial lemniscus. The fibres cross the midline before reaching the ventral thalamus. From the thalamus, they project to the somatosensory regions of the cerebral cortex

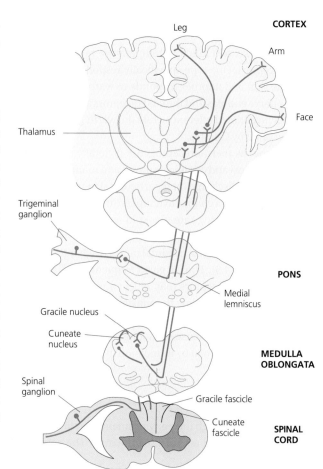

Figure 14.7 The dorsal-column lemniscal pathway for cutaneous sensation.

(Figure 14.7). The dorsal columns relay information concerned with fine discriminatory touch, vibration and position sense (kinaesthetic information).

The small-diameter afferent fibres of the somatosensory system project to the thalamus via a second pathway called the **spinothalamic tract** (Figure 14.8). These afferents enter the spinal cord then synapse on spinal interneurones, which make synapses with other neurones whose axons cross the midline before ascending to the thalamus. The spinothalamic tract is mainly concerned with relaying information regarding crude touch, temperature and pain to the brain.

All of the afferents of a particular type that enter one dorsal root tend to run together in the lower regions of the spinal cord. Initially this segmental organization is

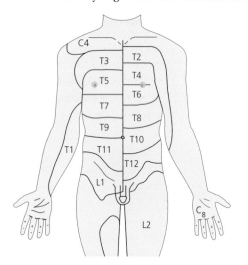

Figure 14.6 The segmental innervation of the trunk, arms and upper leg. The left-hand side of the manikin shows the dermatomes innervated by spinal nerves C4, T1 (inner arm), T3, T5, T7, T9, T11 and L1 while the right-hand side illustrates the innervation for C8 (hand), T2, T4, T6, T8, T10, T12. L2 innervates the thigh. Note the overlaps in the innervation by adjacent spinal nerves.

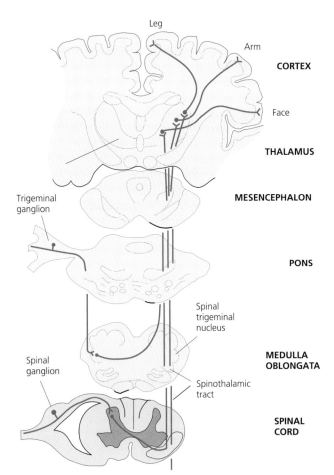

Leg

Arm

CORTEX

Face

THALAMUS

MESENCEPHALON

PONS

Trigeminal
ganglion

Spinal
trigeminal
nucleus

**MEDULLA
OBLONGATA**

Spinal
ganglion

Spinothalamic
tract

**SPINAL
CORD**

Figure 14.8 The spinothalamic tract and its projection to the cerebral cortex.

preserved but, as they ascend, the fibres from the different segments become rearranged so that those from the leg run together, as do those of the trunk, hand and so on. Thus, for example, the afferents from the hand project to cells in a particular part of the dorsal column nuclei and thalamus, while those from the forearm project to an adjacent group of cells. This orderly arrangement provides topographical maps of the body in those parts of the brain that are responsible for integrating information from the different sensory receptors.

The primary sensory cortex

The exploration of the human somatosensory cortex by Wilder Penfield and his colleagues is one of the most remarkable investigations in neurology. While treating patients for epilepsy, Penfield attempted to localize the site of the lesions responsible for the condition by electrically stimulating their cerebral cortex. During this procedure, the patients were conscious but the cut edges of the scalp, skull and dura were infiltrated with local anaesthetic. As the brain has no nociceptive fibres, this procedure did not cause pain but it did elicit specific movements or specific sensations depending on which area of the cortex was stimulated.

Systematic exploration of the postcentral gyrus (see Chapter 11) revealed that it was organized so that the sensations arising in the different regions of the opposite side of the body were represented in an orderly manner, as shown in Figure 14.9. Although the areas of representation are disproportionate with respect to the body surface, they do reflect the degree of importance of the different areas in sensation. Thus, the hands, lips and tongue all have a relatively large area of cortex devoted to them, while the areas devoted to the legs, upper arm and back are relatively small.

KEY POINTS:

- The somatosensory system is concerned with sensations arising from the skin, joints, muscles and viscera. It provides the CNS with information concerning touch, peripheral temperature, limb position (kinaesthesia) and tissue damage (nociception).

- Information from the somatosensory receptors reaches the cerebral cortex by way of the dorsal column–medial lemniscal pathway and by the spinothalamic tract.

- The dorsal column pathway is primarily concerned with fine discriminatory touch and position sense, while the spinothalamic tract is concerned with crude touch, temperature and nociception.

- The postcentral gyrus of the cerebral cortex possesses a topographical map of the contralateral surface of the body.

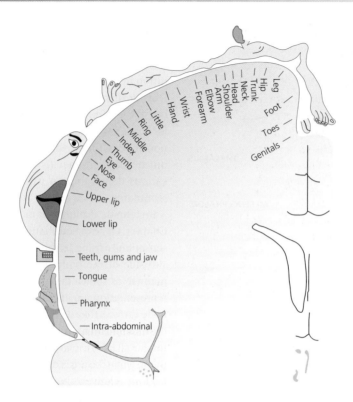

Figure 14.9 The representation of the body surface on the postcentral gyrus revealed by electrical stimulation of the cerebral cortex of conscious subjects.

14.3   Pain

Pain is an unpleasant experience associated with acute tissue damage. It is the main sensation experienced by most people following injury. Pain also accompanies certain organic diseases such as advanced cancer. Nevertheless, pain may arise spontaneously without an obvious organic cause or in response to an earlier injury, long since healed. This kind of pain often has its origin within the CNS itself. Although it is not obviously associated with tissue damage, pain of central origin is no less real to the sufferer.

Unlike most other sensory modalities, pain is almost invariably accompanied by an emotional reaction of some kind such as fear or anxiety. If it is intense, pain elicits autonomic responses such as sweating and an increase in blood pressure and heart rate. Pain may be short lasting and directly related to the injury that elicited

it (**acute pain**) or it may persist for many days or even months (**chronic pain**).

Broadly, pain may be classed under one of three headings:

- **Pricking pain**. This is rapidly appreciated and accurately localized but elicits little by way of autonomic responses. This kind of pain is usually transient and has a sharp, pricking quality. It is sometimes called 'first' or 'fast' pain and is transmitted to the CNS via small myelinated Aδ fibres. It serves an important protective function, as activation of these fibres triggers the reflex withdrawal of the affected region of the body from the source of pain.

- **Burning pain**. This is more intense and less easy to endure than pricking pain. It has a diffuse quality and is more difficult to localize. Burning pain readily evokes autonomic responses including an increased heart rate, elevated blood pressure, dilatation of the pupils and sweating (see Chapter 13). The pattern of breathing may also be altered with rapid, shallow breaths interrupted with periods of apnoea (breath holding) during severe episodes. Burning pain is of slower onset and greater persistence than fast pain. It reaches the CNS via non-myelinated C-fibres and is sometimes called 'second' or 'slow' pain.

- **Deep pain**. This arises when deep structures such as muscles or visceral organs are diseased or injured. Deep pain has an aching quality sometimes with the additional feeling of burning. It is usually difficult to localize and, when it arises from visceral organs, it may be felt at a site other than its origin. This is known as **referred pain** (see below).

Nociceptors are activated by specific substances released from damaged tissue

Pain is detected by bare nerve endings in the skin and certain visceral organs, called nociceptors. The application of pain-provoking stimuli, such as radiant heat, elicits reddening of the skin and other inflammatory changes, which release a number of chemical agents called pain-producing substances or algogens. The pain-producing substances include ATP, bradykinin, histamine, serotonin (5-HT), hydrogen ions and a number of inflammatory mediators such as prostaglandins. They activate the pain endings causing them to send action potentials to the CNS.

CNS pathways in pain perception

Nociceptive fibres are a specific set of small-diameter dorsal root afferents that subserve pain sensation in a particular region of the body. These fibres enter the spinal cord and synapse in the spinal cord (Figure 14.8). The second-order fibres cross the midline and ascend to the thalamus via the spinothalamic tract.

Although details of the central projections of the pain pathways are not fully established, it is known that neurones of the posterior thalamus, which is known to be concerned with pain perception, project to the secondary sensory cortex on the upper wall of the lateral fissure.

Other areas that appear to be involved in the whole pain experience are the reticular formation, the structures of the limbic system such as the amygdala, and the frontal cortex.

It is widely known that the pain from a bruise can be relieved by vigorous rubbing of the skin in the affected area. Pain can also be relieved by the electrical stimulation of the local peripheral nerves. In both cases, the large-diameter afferent fibres responsive to touch and pressure are activated. This activation appears to inhibit the transmission of pain signals by the small, unmyelinated nociceptive afferents. This inhibition occurs at the segmental level in the local networks of the spinal cord and prevents the onward transmission of pain signals to the brain. Transcutaneous electrical nerve stimulation (TENS) is now frequently used to control pain during childbirth and for some other conditions where prolonged use of powerful pain-suppressing drugs is undesirable.

Stressful situations can also produce a profound loss of sensation to pain (analgesia). It has been known for many years that soldiers with very severe battlefield injuries are often surprisingly free of pain. Sportsmen and women may also experience little or no pain from an injury sustained during a highly competitive game only to suffer intense pain once the game has ended. These observations suggest that the brain is able to control the level of pain in some way. It is now believed that descending pathways can inhibit the activity of nociceptive neurones within the spinal cord.

The brain and spinal cord possess a number of peptides that have actions similar to morphine and other opiate drugs, which have powerful analgesic effects. These are the **enkephalins**, the **dynorphins** and the **endorphins**. When these peptides are injected into certain areas of the brain, they have powerful analgesic effects. The anterior hypothalamus and adrenal medulla secrete β-endorphin, along with adrenaline and noradrenaline, as part of the body's overall response to stress. While the exact role of these pain-suppressing peptides is unclear, it is likely that they play a part in the body's ability to modify the perception of pain under certain conditions.

Visceral pain

Although the viscera are not as densely innervated with nociceptive afferents as the skin, they are nevertheless capable of transmitting pain signals. The nociceptive visceral afferents enter the spinal cord via the dorsal roots. They synapse in the dorsal horn and the second-order axons project to the brain, mainly via the spinothalamic tract. The nociceptive nerve endings are activated by a variety of stimuli. When the walls of hollow structures, such as the bile duct, intestine and ureter, are stretched, the nociceptive afferents send impulses to the CNS. For example, when a gallstone is being passed along the bile duct from the gall bladder, the nerve endings of the bile duct are stretched and the nociceptive afferents are activated. Each time the smooth muscle contracts, the sufferer will experience a bout of severe pain known as **colic**, until the stone is passed into the intestine. Similar bouts of spasmodic pain accompany the passage of a kidney stone along a ureter.

Ischaemia also causes intense pain. The classic example is the pain felt following narrowing or occlusion of one of the coronary arteries. Chemical irritants are also a potent source of pain originating in the viscera. This is commonly caused by gastric acid coming into direct contact with the gastric or oesophageal mucosa. This gives rise to a sensation known as heartburn.

Referred pain and projected pain

The pain arising from the viscera is often scarcely felt at the site of origin but may be felt as a diffuse pain at the body surface in the region innervated by the same spinal segments. This separation between the site of origin of the pain and the site at which it is apparently felt is known as referral and the pain is called **referred pain**. The pain arising from cardiac ischaemia is strikingly referred to the base of the neck and radiates down the left arm (Figure 14.10). It may also be felt directly under the sternum.

When a nerve trunk is stimulated somewhere along its length rather than at its termination, there is an unpleasant, confused sensation known as **paraesthesia**. This presumably arises because the CNS attempts to interpret the sequence of action potentials in terms of its normal physiology. Sometimes this sensation is one of explicit pain and, as this is projected to the body surface, this is called **projected pain**. A well-known example is the pain felt when the ulnar nerve in the elbow is knocked giving rise to a tingling sensation in the third and fourth fingers of the affected hand.

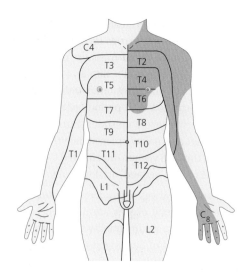

Figure 14.10 Diagram to show the distribution of referred pain from the heart. The dark shading shows the distribution of the referred pain.

Causalgia, neuralgia and phantom limb pain

Gunshot wounds and other traumatic injuries that damage, but do not sever, major peripheral nerves such as the sciatic nerve can sometimes give rise to a severe burning pain in the area served by the nerve. This pain is known as **causalgia** and it may be so great that the patient cannot bear the mildest of stimuli to the affected area. Even a puff of air elicits unbearable pain.

Other kinds of damage to peripheral nerves may also cause severe and unremitting pain that is resistant to treatment. This kind of pain is known as **neuralgia**. The causes include viral infections, neural degeneration and nerve damage from poisons. In **trigeminal neuralgia**, pain of a lacerating quality is felt on one side of the face within the distribution of the trigeminal nerve. Between attacks, the patient is free of pain and no abnormalities of function can be detected. The attacks are not provoked by thermal or nociceptive stimuli, but may be triggered by light mechanical stimulation of the face or lip or by eating food. The pain may be relieved by cutting the appropriate branch of the trigeminal nerve, although this will leave the affected area without normal sensation.

After limb amputations, the nerve fibres in the stump begin to sprout and eventually form a tangle of fibres, fibroblasts and Schwann cells called a **neuroma**. The nerve fibres in a neuroma often become very sensitive to mechanical stimulation that may give rise to severe shooting pains, which may be blocked by injection of local anaesthetic into the mass of the neuroma. Some patients develop a strong illusion that the amputated limb is still present. The phantom limb may appear to be held in a very uncomfortable or painful position.

Drugs used in the management of pain

Pain control is one of the most important aspects of clinical medicine and many strategies are employed to relieve pain according to the exact nature of the pain, the patient's perception of the pain and the extent to which a person's quality of life is affected. These include relaxation and hypnotherapy techniques, the use of acupuncture and TENS devices and even, in the worst cases of intractable pain, surgical section of dorsal root ganglia. Nevertheless, the most commonly used treatments involve the administration of pain-relieving (analgesic) drugs. This section will provide a brief account of some of the more widely used analgesics.

Analgesic drugs may be divided into three major categories:

- Paracetamol
- Non-steroidal anti-inflammatory drugs (NSAIDs)
- Opioid receptor agonists.

In addition to these main types, a number of other drugs are often used either alone or in combination with those above to enhance the analgesic effect or to reduce anxiety. These include:

- Tricyclic antidepressants such as amitriptyline
- Anticonvulsants such as carbamazepine
- NMDA receptor antagonists such as ketamine
- Adrenergic agonists such as clonidine
- 5-HT$_1$ receptor agonists such as sumatriptan.

Paracetamol is frequently used as a first-line analgesic for mild to moderate pain. Its exact mode of action is unclear but it probably inhibits the cyclo-oxygenase (COX) enzymes that are involved in prostaglandin synthesis by the CNS. It may also act directly on peripheral nociceptors. It is analgesic and antipyretic, but has little if any anti-inflammatory effect. The drug itself has few side-effects and is regarded as safe to take during pregnancy and lactation, but one of its metabolites causes serious liver damage and for this reason an overdose of paracetamol can be fatal.

Another widely used and effective group of analgesics are the **NSAIDs** (non-steroidal anti-inflammatory drugs) such as **aspirin** and **ibuprofen**. They are very effective in the relief of musculoskeletal pain, headaches, dental pain, etc., but are less effective in the treatment of visceral pain. As their name suggests, these have anti-inflammatory effects in addition to their analgesic and antipyretic properties. They act by blocking COX enzymes to inhibit the production of prostaglandins in the periphery and hypothalamus. Their adverse effects include gastrointestinal irritation, bronchospasm and skin rashes in susceptible individuals. Aspirin may also cause Reye's syndrome, a degenerative and sometimes fatal condition of the brain that affects children. Apart from aspirin and ibuprofen, other commonly used NSAIDs include **diclofenac**, **fenbufen**, **naproxen**, **celecoxib**, **piroxicam**, **indomethacin** and **phenylbutazone**.

A number of drugs exert their analgesic effect by binding to opioid receptors in the nervous system. They are known collectively as **opioid agonists** and include **codeine**, **pethidine** and **morphine**. Essentially, these drugs mimic the actions of the body's own pain-reducing chemicals, the enkephalins and endorphins (see above). They bind to opioid receptors, which are widely distributed throughout the CNS, and inhibit the transmission of pain by the spinal cord. They also activate the descending analgesic pathways. The opioid analgesics vary in potency from relatively weak agents, such as **codeine**, **dihydrocodeine** and **dextropropoxyphene**, to much stronger drugs such as **morphine**, **diamorphine**, **phenazocine**, **fentanyl**, **methadone** and **pethidine**. The exact choice of drug and the treatment regime employed will depend on the severity of the pain, the progression of the pathological process and the response of the patient.

Opioid drugs have a number of side-effects including constipation (resulting from the actions of the drug on the enteric nervous system) and respiratory depression. The latter may be of particular concern, but can be reversed, if necessary, by a specific opioid antagonist called **naloxone**. Continuous treatment with opioid analgesics may lead to the development of drug dependency, but this is rarely of concern as the strongest of these drugs are most often reserved for patients who are terminally ill.

14.4 Itch

Itch or **pruritus** is a well-known sensation that is associated with the desire to scratch. There are many causes of itch including: skin parasites such as scabies and ringworm, insect bites, chemical irritants derived from plants, for example 'itching powder', and contact with rough cloth or with inorganic materials such as fibre glass. A variety of common skin diseases, such as eczema, also gives rise to the sensation of itch, as do obstructive jaundice and renal failure. The intensity of the sensation is highly variable; it may be so mild as to be scarcely noticeable or so severe as to be as unendurable as chronic pain.

What is the physiological basis of this sensation? Itch, like pain, is transmitted to the CNS from specific bare nerve endings (the itch receptors) in the skin via a subset of peripheral C-fibres. Itch, however, differs from superficial pain in a number of important respects: it can only be elicited from the most superficial layers of the skin, the mucous membranes and the cornea. Increasing the frequency of electrical stimulation of a cutaneous itch spot increases the intensity of the itch without eliciting the sensation of pain. Moreover, skin stripped of the epidermis is very sensitive to pain but is insensitive to itch-provoking stimuli. Finally, the reflexes evoked by pruritic (itch-provoking) stimuli and nociceptive stimuli are different. Pruritic stimuli elicit the well-known scratch reflex, while nociceptive stimuli elicit withdrawal and guarding reflexes. On these grounds, it is reasonable to conclude that itch is a distinct modality of cutaneous sensation.

Exactly how the afferent fibres conveying the sensation of itch are excited remains unknown, but a number of substances, including histamine, can produce itching when applied to the skin. Indeed, antihistamine drugs, such as **loratadine**, **cetirizine** and **fexofenadine**, are able to reduce the sensation of itching associated with allergic reactions. However, as antihistamine drugs are not always effective in controlling itch, other chemical mediators are likely to be involved.

> **KEY POINTS:**
>
> - Pain is the sensation we experience when we injure ourselves or when we have some organic disease. It is an unpleasant experience that is generally associated with actual or impending tissue damage.
>
> - The nociceptors are activated by strong thermal, mechanical or chemical stimuli. Their afferents project to the CNS by way of the spinothalamic tract.
>
> - A variety of drugs are employed in the treatment of pain. The most important are paracetamol, the non-steroidal anti-inflammatory drugs and opioid agonists.

14.5 The physiology of the eye and visual pathways

Man is pre-eminently a visual animal. Those blessed with normal vision largely react to the world as they see it, rather than to its feel or its sounds or smells. Objects are judged at a distance from the light they emit or reflect into the eye. Light itself is a component of the electromagnetic spectrum and the human eye responds to radiation with wavelengths between about 380 nm (deep violet) to 800 nm (deep red).

The anatomy of the eye

The eyes are protected by their location in the bony cavities of the orbits. Only about a third of the eyeball is unprotected by bone. The eyeball itself is roughly spherical and its wall consists of three layers (Figure 14.11): a tough outer coat, the **sclera**, which is white in appearance, a pigmented layer called the **choroid**, which is highly

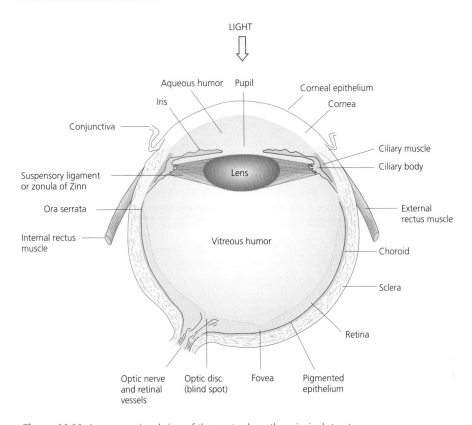

LIGHT

Aqueous humor Pupil
Iris
Conjunctiva
Suspensory ligament
or zonula of Zinn
Ora serrata
Internal rectus
muscle
Lens
Vitreous humor
Corneal epithelium
Cornea
Ciliary muscle
Ciliary body
External
rectus muscle
Choroid
Sclera
Retina
Optic nerve   Optic disc   Fovea   Pigmented
and retinal   (blind spot)   epithelium
vessels

Figure 14.11 A cross-sectional view of the eye to show the principal structures.

vascular, and the **retina**, which contains the photoreceptors (rods and cones) together with an extensive network of nerve cells. The retinal ganglion cells are the output cells of the retina and they send their axons to the brain via the optic nerves.

At the front of the eye, the sclera gives way to the transparent cornea, which consists of a special kind of connective tissue that lacks blood vessels. The pigmented **iris** covers much of the transparent opening of the eye formed by the cornea, leaving a central opening, the **pupil**, to admit light to the photoreceptors of the retina.

The pupil diameter is controlled by two muscles, the circular sphincter pupillae and the radial dilator pupillae of the iris, both of which are innervated by the autonomic nervous system. The sphincter pupillae receives parasympathetic innervation via the ciliary ganglion while the dilator pupillae receives sympathetic innervation via the superior cervical ganglion.

Behind the iris lies the ciliary body, which contains smooth muscle fibres. The lens of the eye is attached to the ciliary body by a circular array of fibres called the **suspensory ligament**. Like the cornea, the lens has no blood vessels and depends on the diffusion of nutrients from the aqueous humour for its nourishment. The lens itself is elastic and can change its shape to bring images into focus on the retina. This process is controlled by the ciliary muscles and is called **accommodation**.

The organization of the retina

The retina is the sensory region of the eye and contains a number of different cell types organized into layers. Just inside the choroid is the first layer of the retina, which is composed of pigmented epithelium. Together with the choroid, this layer is responsible for absorbing excess light and reducing light scatter. The photoreceptors themselves are the rods and cones,

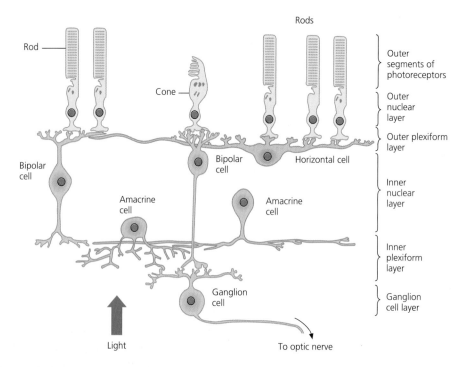

Figure 14.12 A simple diagram to show the cellular organization of the retina. Note that light has to pass through the innermost layers of the retina to reach the photoreceptors.

which lie immediately beneath the pigmented epithe-lium. Bipolar cells, horizontal cells and amacrine cells act as interneurones between the rods and cones and the ganglion cells that are the output cells of the retina. Individual photoreceptors consist of an outer segment, which contains the photosensitive pigment, an inner segment where the cell nucleus is located, and a rod pedicle, which is the site at which the photoreceptors make synaptic contact with the bipolar and amacrine cells of the retina.

A highly schematic diagram of the organization of the retina is shown in Figure 14.12. Note that light passes through the cell layers to reach the photore-ceptors, which are located next to the pigmented epithelium. Rods and cones are distributed through-out the retina, but in the central region, known as the **fovea centralis**, the retina is very thin and consists of a densely packed layer of cones. In the surrounding region, the parafoveal region, both rods and cones are present in abundance, together with the bipolar,

amacrine and horizontal cells connected to the cones of the fovea. In Figure 14.12, one cone is shown con-nected to one ganglion cell. This is only the case for the central region of the retina; elsewhere the signals from a number of photoreceptors converge on a single ganglion cell. In the extreme periphery, as many as 100 rods are connected to a single ganglion cell. The region where the ganglion cell axons pass out of the eye to form the optic nerve (the papilla or **optic disk**) is devoid of photoreceptors.

These variations in photoreceptor density mean that visual acuity (the capacity of the eye to resolve detail) varies across the retina. Objects fixated at the centre of the visual field are seen in great detail while those in the peripheral regions of the field are seen less distinctly. Light falling on the area of the optic disk will not be detected as there are no rods or cones in this region. This is known as the blind spot and it can be demonstrated as described in the legend to Fig-ure 14.13.

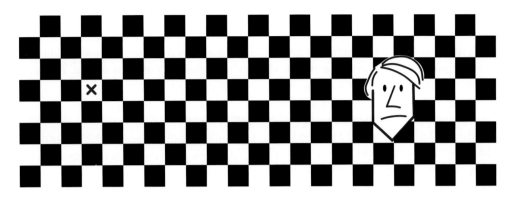

Figure 14.13 Demonstration of the blind spot. Close the left eye and focus on the cross with the right eye. Move the page so that is about 25 cm from you and the face will disappear with no discontinuity in the background.

KEY POINTS:

- The eye consists of an outer coat of connective tissue (the sclera), a highly vascular layer (the choroid) and a photoreceptive layer (the retina).

- Light enters the eye via a clear zone (the cornea) and is focused on the retina by the lens.

- Photoreceptor density is greatest in the region of the fovea, but there are no receptors at the optic disk (the point at which the optic nerve fibres leave the eye). This is known as the blind spot.

The general physiology of the eye

The lacrimal glands and tear fluid

Each orbit is endowed with lacrimal glands, which provide a constant secretion of tear fluid that serves to lubricate the movement of the eyelids and to keep the outer surface of the cornea moist, so providing a good optical surface. Tear fluid has a pH similar to that of plasma (7.4) and is isotonic with blood. It possesses an enzyme, called lysozyme, that has a bactericidal action. Under normal circumstances, about 1 ml of tear fluid is produced each day, most of which is lost by evaporation; the remainder is drained into the nasal cavity via the tear duct. Irritation of the corneal surface (the **conjunctiva**) increases the production of tear fluid and this helps to flush away noxious agents.

Intraocular pressure

The space behind the cornea and surrounding the lens is filled with a clear fluid called the **aqueous humour**, which supplies the lens and cornea with nutrients. The aqueous humour is secreted by the processes of the ciliary body that lie just behind the iris and it flows into the anterior chamber from where it drains into the canal of Schlemm through a fibrous mesh (the trabecular meshwork) situated at the junction between the cornea and the sclera (Figure 14.14). From the canal of Schlemm, the fluid is returned to the venous blood.

The constant production of aqueous humour generates a pressure within the eye known as the **intraocular**

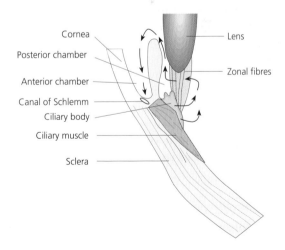

Figure 14.14 Diagrammatic representation of the circulation of the aqueous humour in the anterior chamber of the eye.

pressure, which is about 2 kPa (15 mmHg) in normal individuals. The intraocular pressure maintains the spherical shape of the eye, which is essential for clear image formation. This pressure is normally kept constant as the rate of fluid secretion is matched by the rate of drainage. However, if the drainage is obstructed, the intraocular pressure rises and a condition known as **glaucoma** results with gradual loss of vision, which may be total if the condition remains undiagnosed and untreated for many years. Loss of visual acuity is the result of pressure on the optic nerve. Over time, there is a disruption of the supply of nutrients to the retinal neurons and peripheral optic fibres entering the brain that results in widespread cell death.

Available treatments for glaucoma focus on trying to reduce the rate of production of aqueous humour or to increase its rate of drainage. Drug treatments, surgical procedures or a combination of both are used. These may slow or even halt degeneration of vision but cannot restore vision that has already been lost. Drugs that block adrenergic β-receptors (β-antagonists) such as **timolol** reduce the rate of fluid production by the eye, but these are not suitable for people with heart conditions as they can alter cardiac and lung function. Other drugs with similar actions are prostaglandin analogues and α_2 agonists, such as **clonidine** (see p. 223). Acetylcholine agonists such as **pilocarpine** are also used to treat patients with glaucoma, as these drugs cause pupillary constriction through contraction of the sphincter pupillae muscle of the iris. This increases the ability of aqueous humour to drain from the anterior chamber and thus lowers intraocular pressure.

Surgical treatment options for glaucoma include trabeculoplasty and trabeculectomy. In the former procedure, a highly focused laser beam is used to create a series of tiny holes (usually between 40 and 50) to enhance drainage. Trabeculectomy may be performed in cases of advanced glaucoma. In this procedure, an artificial opening is made in the sclera through which excess fluid is allowed to escape. A tiny piece of the iris may also be removed so that fluid can flow backwards into the eye.

The blink reflex and the dazzle reflex

The eyelids are closed by relaxation of the levator palpabrae muscles coupled with contraction of the orbicularis oculi muscles (see Figure 10.26). The reflex closure of the eyelids can result from corneal irritation due to specks of dust and other debris. This is known as the **blink reflex** or corneal reflex. Very bright light shone directly into the eyes also elicits closure of the eyelids. Lid closure will cut off more than 99% of incident light. This is known as the **dazzle reflex**.

Pupillary reflexes

The pupils are the central dark regions of the eyes through which light passes to reach the retina. If a light is shone directly into one eye, its pupil constricts. This response is known as the **direct pupillary response**. The pupil of the other eye also constricts and this is known as the **consensual response**. Also, the pupils constrict when the eye is focused on a close object. This is known as the **accommodation reflex**, which has the effect of improving the depth of focus. Pupillary constriction is brought about by contraction of the circular muscles of the iris in response to parasympathetic stimulation. Dilatation of the pupils is brought about by increased activity in the sympathetic nerve supply, which causes contraction of the radial dilator muscles of the iris.

While the benefit of the corneal (blink), dazzle and pupillary reflexes in protecting the eye from damage is obvious, they are also of value in the assessment of the physiological state of the brainstem nuclei and in diagnosing brain death following traumatic head injury.

KEY POINTS:

- Tear fluid secreted by the lacrimal glands protects and cleans the surface of the eye.

- Intraocular pressure is maintained by a balance between secretion and drainage of the aqueous humour. Obstruction of fluid drainage gives rise to a condition called glaucoma in which vision is progressively lost.

- The cornea is protected from damage by the blink reflex, which is elicited by irritants on the corneal surface.

- The eyelids also close in response to excessively bright light. This is known as the dazzle reflex.

- The amount of light falling on the retina is controlled by the pupil. The diameter of the pupils is controlled by the autonomic nervous system.

Image formation by the eye

The eye behaves much as a pinhole camera. The image is inverted so that light falling on the retina nearest the nose (the nasal retina) comes from the lateral part of the visual field (the temporal field) while the more lateral part of the retina (the temporal retina) receives light from the central region of the visual field (the nasal field).

As objects may lie at different distances from the eye, the optical elements of the eye must be able to vary its focus in order to form a clear image on the retina. The eyes of a young adult with normal vision are able to focus objects as distant as the stars and as close as 25 cm (the **near point**). This is made possible by the ability of the lens to change its shape and thus its optical power (accommodation). When the lens loses its elasticity, which commonly occurs with advancing age, the power of accommodation is reduced and the nearest point of focus recedes.

The ciliary muscles control the focusing of the eye

The lens is suspended from the ciliary muscle by the suspensory ligament, as shown in Figure 14.15. When the ciliary muscles are relaxed, there is a constant tension on the suspensory ligament exerted by the effect of the intraocular pressure on the sclera. This tension stretches the lens and minimizes its curvature. When the eye switches its focus from a distant to a near object, the ciliary muscle contracts and this opposes the tension in the sclera. As a result, the tension on the suspensory ligament decreases. The reduction in tension allows the lens to become more rounded, so increasing its optical power.

Refractive errors – myopia, hyperopia and astigmatism

When the ciliary muscle of a normal eye is relaxed, the eye itself is focused on infinity so that the parallel rays

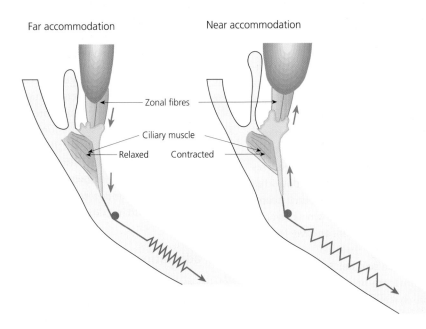

Figure 14.15 A diagram to illustrate the mechanism of accommodation. When the ciliary muscles are relaxed, the tension in the wall of the eye is transmitted to the lens via the zonal fibres of the suspensory ligament. The lens is stretched and adapted for distant accommodation (left). When the ciliary muscles contract, they relieve the tension on the zonal fibres and the lens assumes a more rounded shape suited to focusing objects near the eye (near accommodation).

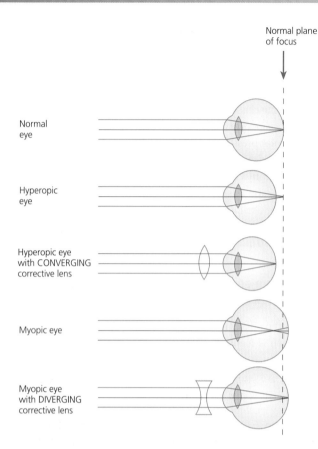

Normal plane of focus

Normal eye

Hyperopic eye

Hyperopic eye with CONVERGING corrective lens

Myopic eye

Myopic eye with DIVERGING corrective lens

Figure 14.16 A simple diagram illustrating the principal refractive errors of the eye and their correction by external lenses.

of light from a distant object will be brought into sharp focus on the retina. If the eyeball is too long, the parallel rays of light from a distant object will be brought into focus in front of the retina and vision will be blurred. This situation can also arise if the lens system is too powerful. In both cases, the eye is able to focus objects much nearer than normal. This optical condition is known as **myopia**. It may be corrected with a diverging (or concave) lens, as shown in Figure 14.16.

If the eye is too short, then parallel light from a distant object is brought into focus behind the retina. This can also arise if the lens system is insufficiently powerful. People with this optical condition have difficulty in bringing near objects into focus and are said to suffer from **hyperopia** or **hypermetropia**. It may be corrected with a converging (or convex) lens, as shown in Figure 14.16.

In some cases, the curvature of the cornea or lens is not uniform in all directions. As a result, the power of the optical system of the eye is different in different planes. This is known as **astigmatism** and it may be corrected by a cylindrical lens placed so that the refraction of light is the same in all planes.

The visual field

The visual field of an eye is that area of space that can be seen at any instant of time. It is measured in a routine clinical procedure called **perimetry**. The eye to be tested is kept focused ('fixated') on a central point. A test light is then gradually moved from the periphery towards the centre and the subject asked to signal when they see it. Tests of this kind show that the field of view is restricted by the nose and by the roof of the orbit so that the visual field is at its maximum laterally and inferiorly. The visual fields of the two eyes overlap extensively in the nasal region. This is the basis of binocular vision, which confers the ability to judge distances precisely. Defects in the visual field can be used to diagnose damage to different parts of the optical pathways.

Photopic and scotopic vision

The cones are the main photoreceptors used during the day when the ambient light levels are high. This is called **photopic vision**. In photopic conditions, visual acuity is high and there is colour vision (afforded by the cones). At night, when light levels are low, the rods are the main photoreceptors. This is known as **scotopic vision** and is characterized by high sensitivity but poor acuity and a lack of colour vision – rods require only low light levels to be stimulated but do not differentiate colours.

Dark adaptation

When one first moves from a well-lit room into a garden lit only by the stars, it is difficult to see anything. After a short time, the surrounding shrubs and trees become increasingly visible so that maximum sensitivity is attained after 20–30 minutes in the dark. The process that occurs during this change in the sensitivity of the eye is called **dark adaptation**.

KEY POINTS:

- The image of the world is brought into focus on the retina by the action of the lens system of the eye.
- The focusing of the lens is controlled by the ciliary muscle, which modulates the tension in the suspensory ligament.
- The major problems in image formation are due to malformation of the eyeball. The most common defects are myopia, hypermetropia (or hyperopia) and astigmatism. They may be corrected with an external lens of appropriate power.

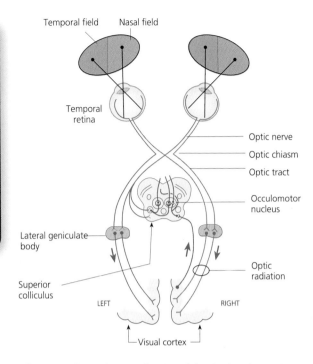

Figure 14.17 A schematic drawing of the visual pathways. Note that the nasal field projects to the temporal retina and that fibres from this region do not cross. Thus, each half of the visual field projects to the opposite hemisphere. The connections between the two hemispheres are not shown. The projections to the brainstem mediate the corneal and pupillary reflexes.

The neurophysiology of vision

The organization of the visual pathways

From the retina, the axons of the ganglion cells pass out of the eye in the optic nerve. At the optic chiasm, the medial bundle of fibres (which carry information from the nasal side of the retina) crosses to the other side of the brain. This partial decussation of the optic nerve allows all the information arriving in one field of vision to project to the opposite side of the brain. Thus, the left visual field projects to the right visual cortex and the right visual field projects to the left visual cortex (Figure 14.17). From the optic chiasm, the fibres pass to the lateral geniculate bodies giving off collateral fibres that pass to the superior colliculi and oculomotor nuclei in the brainstem. From the lateral geniculate bodies, the main optic radiation passes to the primary visual cortex located in the occipital lobe. Nerve fibres connecting the two hemispheres pass through the corpus callosum.

Phototransduction in the retina

The receptors of the retina are the rods and cones. Both contain photosensitive pigments. In the rods, this pigment is known as rhodopsin, which is made up of the aldehyde of vitamin A (11-*cis*-retinal) and a protein called opsin. The cones also have photosensitive pigments containing 11-*cis*-retinal but in their case the aldehyde is conjugated to different photoreceptor proteins that make the different types of cone sensitive to light of different wavelengths.

The sequence of events that immediately follows the absorption of light (**phototransduction**) is shown schematically in Figure 14.18. When a photon of light is captured by one of the visual pigment molecules, the shape of the retinal changes and this triggers a change in the properties of the photoreceptor protein to which it is bound. The activated photoreceptor protein then activates a **G protein** (see Chapter 7) called **transducin**. Transducin activates a phosphodiesterase that breaks down cyclic GMP to 5'GMP. In the dark, the levels of cyclic GMP in the photoreceptors are high. The cyclic GMP binds to the internal surface of ion channels permeable to Na^+ and causes them to open. As a result, the membrane potential of the photoreceptors is low, about $-40\,mV$. When the rhodopsin is activated by light, the

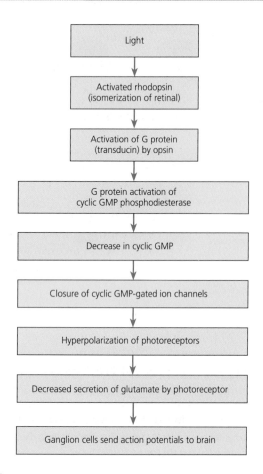

Figure 14.18 The principal steps in phototransduction.

---

levels of cyclic GMP fall and so fewer of the Na$^+$ permeable ion channels are open. The fall in Na$^+$ permeability results in a **hyperpolarization** of the photoreceptor. When the photoreceptors become hyperpolarized following the absorption of light, the secretion of neurotransmitter (glutamate) by the photoreceptors is decreased and the bipolar cells respond with a change in their membrane potential.

The neurones of the primary visual cortex respond to specific features of an image

Many neurones in the visual cortex respond to stimuli presented to both eyes. Moreover, as the visual cortices of the two hemispheres are interconnected via the corpus

callosum, some cortical neurones receive information from both halves of the visual field and can detect small differences between the images in each retina. This is the basis of stereoscopic vision, which provides cues about the distance of objects.

Neurones in the visual cortex often respond to particular features of objects in the visual field. Some respond optimally to a bright or dark bar passing across the retina at a specific angle, for example. Others respond to a light–dark border, and so on. In many cases, the direction of movement is also critical. In this way, the brain is able to identify the specific features of an object, which it then uses to form its internal representation of that object in the visual field. Such representations are crucial to our ability to interpret the visual aspects of the world around us.

Vision is an active process. We learn to interpret the images we see. We associate particular properties with specific visual objects and have a remarkable capacity to identify objects, particularly other people, from quite subtle visual cues. We are able to recognize a person we know at a distance. Yet we do not notice that they appear larger as they approach us, rather we assume that they have a particular size and use that fact as a means of judging how far away they are. This phenomenon is known as **size constancy**.

The importance of the way the brain uses visual cues to form its internal representation of an object is illustrated by many visual illusions. Perhaps the best known is the Müller–Lyer illusion shown in Figure 14.19a. Here, the central line in each half of the figure is the same length but the line with the arrowheads appears shorter. Knowledge of this fact does not diminish the subjective illusion. The fact that the brain judges the size of an object by its context is revealed by the illusion shown in Figure 14.19b. Here, although the central circle is the same size in both parts of the figure, it appears larger when surrounded by small circles.

Colour vision

In full daylight, the cones are the principal photoreceptors and most people experience colour vision – individual objects have their own intrinsic colour. This has obvious advantages in distinguishing the different objects in the environment, but how is this remarkable feat achieved? People with normal colour vision can match the colour of

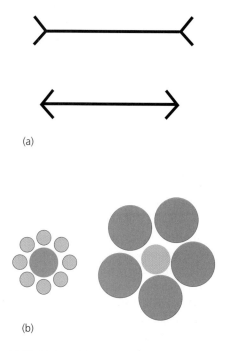

(a)

(b)

Figure 14.19 (a) The Müller–Leyer illusion. Here, the central line appears longer in the figure with outwardly directed fins although it is the same length as that with inwardly directed fins. The length of the lines can be confirmed by measurement with a ruler, yet the illusion does not disappear. (b) Errors of size perception – the Tichener illusion. Here, the central circle is the same diameter in both parts of the figure, although it looks much larger when surrounded by small circles.

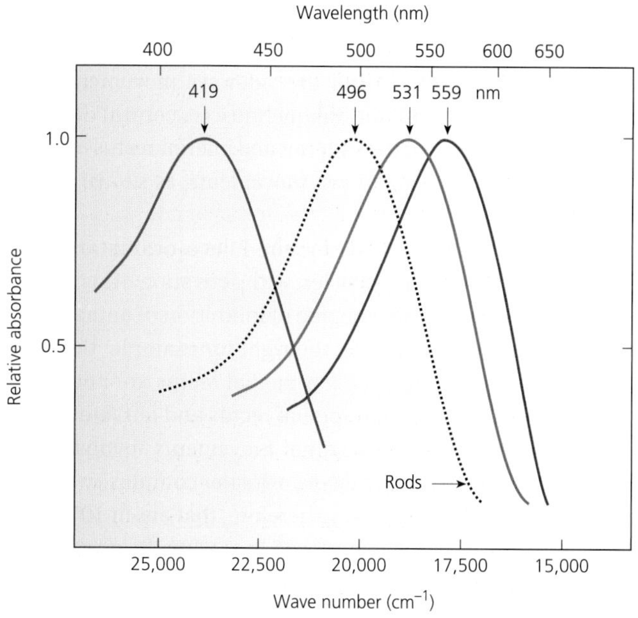

an object by detecting the relative amounts of just three colours – blue, green and red.

There are three different cone pigments, one sensitive to blue light, one to green light and one to red light. The blue-sensitive cones show maximum light absorption at a wavelength of 420 nm, the green-sensitive cones have a maximum absorption at about 530 nm and the red-sensitive cones absorb maximally at 560 nm. The rod pigments have an absorption maximum at 496 nm (Figure 14.20). At least two different pigments are required for any colour vision and the brain must be able to compare the intensity of the signals arriving from different cones. For normal human colour vision, a green light is seen when the green sensitive cones are more strongly stimulated than the red- and blue-sensitive cones, and so on.

Colour blindness

Around 10% of people are unable to distinguish between certain colours and are said to be colour blind. A very small fraction of colour-blind individuals have only rods in their retinas and possess only monochromatic vision. More common forms of colour blindness include a relative insensitivity to either red, green or blue. Colour blindness is usually diagnosed using Ishihara cards, one of which is shown in Figure 14.21.

Figure 14.20 The absorbance spectra of the pigments in the three types of human cone and that of rhodopsin (which is found in the rods).

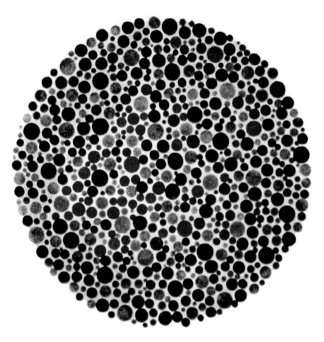

Figure 14.21 An Ishihara card. People with normal colour vision see '42', a red-blind person will only see the '2', while a green-blind person will only see the '4'. A complete test uses many such cards; different cards use different combinations of colours.

KEY POINTS:

- The photoreceptors of the retina contain pigments sensitive to light. When light is absorbed by a pigment molecule, this is signalled to the retinal ganglion cells, which send action potentials to the visual areas of the brain.

- The visual pathways are arranged so that each half of the visual field is represented in the visual cortex of the opposite hemisphere.

- There are three kinds of cone, each of which is sensitive to light of different wavelengths. This confers the ability to discriminate between different colours. People with defective cone pigments are said to be colour blind.

Eye movements

Our eyes constantly scan the world around us. An image appearing in the peripheral field of vision is rapidly centred onto the fovea by a jerky movement of the eyes. These rapid, jerky eye movements are called **saccades**. In contrast, when watching a race or when playing a ball game, the eye follows the object of interest, keeping its position on the retina fairly constant. This smooth tracking of an object is called a **pursuit movement**. These two types of eye movement can be combined, for example when looking out of a moving vehicle. One object is first fixated and followed until the eyes reach the limit of their travel. The eyes then flick to fixate and follow another object and so on. The continuous switching of the point of fixation is seen as a pursuit movement followed by a saccade and then another pursuit movement. This pattern is known as **optokinetic nystagmus**. A nystagmus is a rapid, involuntary movement of the eyes.

If an object such as a pencil is fixated and then moved around the visual field, both eyes track the object. These are known as **conjugate eye movements**. If the pencil is moved first away from the face and then towards it, the two eyes move in mirror image fashion to keep the image in focus on each retina. When the object approaches the eyes, the visual axes converge and as the object moves away they progressively diverge until they are parallel with each other. These eye movements are called **vergence movements**. Double vision (**diplopia**) occurs when the movements of the two eyes are not properly co-ordinated.

The position of each eye within the orbit is controlled by six extraocular muscles. The lateral and medial rectus muscles control the sideways movements, the superior and inferior oblique muscles control diagonal movements, and the superior and inferior rectus muscles control the up and down movements, as shown in Figure 14.22

The eye movements require full co-ordination of all of the extraocular muscles, with activation of synergists and an appropriate degree of inhibition of antagonists. If we look to one side, the right for example, the right lateral rectus and the left medial rectus are both activated while the right medial rectus and left lateral rectus are inhibited. Diagonal movements involve more muscles and require an even greater complexity of control. It is not surprising, therefore, that about 10% of all motoneurones are employed in serving the extraocular eye muscles.

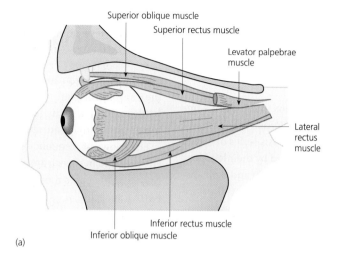

(a)

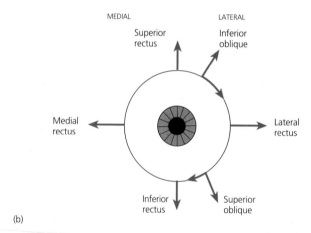

(b)

Figure 14.22 The extraocular muscles of the human eye (a) and the direction of eye movement controlled by each (b).

Defective control of the eye movements is often revealed by a squint in which the eyes fail to look in the same direction and the sufferer experiences double vision. As the majority of the extraocular muscles are innervated by the oculomotor nerve (CN III), damage to this nerve is the most common source of such problems. There are a number of causes, including meningitis, and compression of the nerve by a tumour.

KEY POINTS:

- The eye movements are controlled by six extraocular eye muscles. These muscles are innervated by the third, fourth and sixth cranial nerves.

- The movements of the eyes are driven both by visual information and by information arising from the vestibular system. Their role is to keep objects of importance centred on the central region of the retina.

14.6   The physiology of the ear – hearing and balance

The physical nature of sound

Sound consists of pressure variations in the air or some other medium. These pressure variations originate at a point in space and radiate outwards as a series of waves. Subjectively, sounds are characterized by their **loudness**, their **pitch** (i.e. how low or high they seem) and their specific tonal quality or **timbre**. To be able to extract the maximum information from the sounds in the environment, the auditory system must be able to determine their origin and analyse their specific qualities.

When a sound is propagated, the air pressure alternately increases and decreases above the mean atmospheric pressure. The larger the pressure variation, the higher the energy of the wave and the louder the sound will appear to be. While it is perfectly possible to express loudness in terms of the peak pressure change, it is much more convenient to express the intensity of sounds in relation to an arbitrary standard using the **decibel scale** (abbreviated as dB). This scale has the advantage that equal increments in sound intensity expressed as dB approximately correspond to equal increments in loudness.

$$dB = 10 \cdot \log_{10} \frac{\text{(sound intensity)}}{\text{(reference intensity)}}$$

Because the dB scale is expressed relative to an arbitrary standard, it is sometimes expressed as dB SPL (sound pressure level). If a sound pressure is ten times that of the reference pressure, this will correspond to 20 dB SPL; a sound pressure 1000 times that of the reference pressure will be 60 dB SPL, and so on.

A **pure tone** is a sound consisting of just one frequency. By playing a series of pure tones of varying intensity to a subject, it is possible to determine the threshold of hearing of a given subject for different frequencies. The results can be plotted as intensity (in dB SPL) against the logarithm of the frequency to produce a graphical representation of the auditory threshold, known as an **audiogram**, such as that illustrated in Figure 14.23. As the figure shows, the threshold of hearing varies with frequency. The ear is very sensitive to sounds around 1–3 kHz but its sensitivity declines at lower and higher frequencies. For healthy young people, hearing extends

from about 20 Hz to 20 kHz. In the most sensitive range, the threshold of hearing is close to 20 µPa or 0 dB SPL. The loudest sounds to which the ear can be exposed without irreversible damage are about 140 dB above threshold.

The range of frequencies and intensities of sounds encountered in speech are also indicated in Figure 14.23. Note that the ear is most sensitive to the frequencies of normal speech.

Individual tones gives rise to distinct sensations of pitch – a high-frequency pure tone will subjectively appear as high pitched while a low frequency tone will have a low pitch. Most sounds are not pure tones but mixtures of frequencies, which give the sound its individual quality or **timbre**. If a particular note is sung by different people, or the same note is played by two different instruments, a flute and a violin for example, it is easy to distinguish between them because they sound different despite

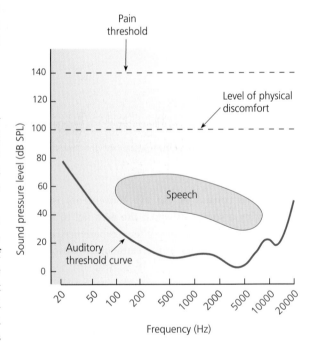

Figure 14.23 A plot of the auditory threshold for human hearing. The shaded area covers the normal range of frequencies and loudness (in dB SPL) of speech. Note that greatest sensitivity of hearing is found in the 2–3 kHz region.

having the same pitch – they have a different mixture of frequencies, and thus different timbres.

> **KEY POINTS:**
>
> - Sounds consist of pressure variations in the air. Subjectively, they are characterized by their pitch, timbre (specific qualities) and loudness.
> - The human ear is most sensitive to frequencies between 1 and 3 kHz, although it can detect sounds ranging from 20 Hz to 20 kHz.

Structure of the auditory system

The auditory system may be arbitrarily divided into the **peripheral auditory system**, which consists of the ear, the auditory nerve and the neurones of the spiral ganglia, and the **central auditory system**, which consists of the neural pathways concerned with the analysis of sound from the cochlear nuclei to the auditory cortex. The ear itself is conventionally divided into the outer ear (the pinna and external auditory meatus), the middle ear (the tympanic membrane, ossicles and associated muscles) and the inner ear (the cochlea and auditory nerve). These structures are illustrated in Figure 14.24.

The **outer ear** consists of the **pinna** (or auricle), which is highly convoluted. These convolutions play an important role in helping to determine the direction of a sound. The pinna leads to the **external auditory meatus** or **auditory canal**, which is about 25 mm in length and about 7 mm in diameter. The **middle ear** is an air-filled space between the **tympanic membrane** or **ear drum** and the **oval window** of the cochlea. These two membranes are coupled via three small bones, the **ossicles**. The middle ear is connected to the pharynx via the **Eustachian tube** (or auditory tube). As long as the Eustachian tube is not blocked, the middle ear will be maintained at atmospheric pressure and the

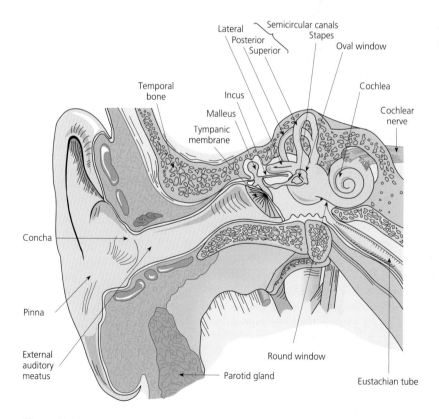

Figure 14.24 Diagrammatic representation of the structure of the human ear.

pressure on either side of the tympanic membrane will be equalized.

The tympanic membrane is roughly cone-shaped and is connected to the first of the ossicles, the **malleus** (or hammer). The malleus is connected to the second of the ossicles, the **incus** (or anvil), which is itself connected to the **stapes** (or stirrup). The footplate of the stapes connects directly with the oval window of the cochlea. The ossicles of the middle ear help to ensure the efficient transmission of airborne sounds to the fluid-filled cochlea. Movements of the tympanic membrane occurring in response to sounds are transmitted to the cochlea by the vibration of the ossicles. If the function of the middle ear is impaired, for example by fluid accumulation or as a result of fixation of the footplate of the stapes to the bone surrounding the oval window, the threshold of hearing will be elevated. This type of deafness is known as **conductive hearing loss** and will be discussed further on p. 255.

The **inner ear** consists of the organ of hearing, the **cochlea**, and the organ of balance – the **semicircular canals**, **saccule** and **utricle** (see p. 256). It is located in the temporal bone and consists of a series of passages called the **bony labyrinth**, containing a further series of sacs and tubes, the **membranous labyrinth**. The fluid of the bony labyrinth is called **perilymph** and is similar in composition to cerebrospinal fluid in having a high sodium concentration and a low potassium concentration. The fluid within the membranous labyrinth is called **endolymph** and has a high potassium concentration and a low sodium concentration.

As Figure 14.24 shows, the cochlea is a coiled structure, which, in humans, is about 1 cm in diameter at the base and 5 mm in height. In cross-section (Figure 14.25), it is seen to consist of three tubes, which spiral together for about two and a half turns. These are the **scala vestibuli** (filled with perilymph), the **scala media** (filled with endolymph) and the **scala tympani** (also filled with perilymph). The scala vestibuli and scala tympani communicate at the apex of the cochlear spiral at the **helicotrema**. The scala media is separated from the scala vestibuli by **Reissner's membrane** and from the scala tympani by the **basilar membrane**, on which sits the organ of Corti, the true organ of hearing. The footplate of the stapes is connected with the fluid of the scala vestibuli via the **oval window**. The flexible membrane of the **round window** separates the fluid of the scala tympani from the cavity of the middle ear.

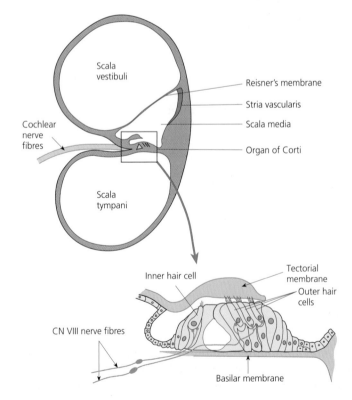

Scala vestibuli

Reisner's membrane

Stria vascularis

Scala media

Organ of Corti

Cochlear nerve fibres

Scala tympani

Inner hair cell

Tectorial membrane

Outer hair cells

CN VIII nerve fibres

Basilar membrane

Figure 14.25 A schematic cross-sectional drawing of the structure of the cochlea illustrating the position and structure of the organ of Corti.

KEY POINTS:

- The auditory system consists of the ear and the auditory pathways.
- The ear is divided into the outer ear, the middle ear and the inner ear.
- The inner ear consists of the cochlea, which is the organ of hearing, and the vestibular system, which is concerned with balance.

The mechanism of sound transduction

The movement of the stapes in response to sound results in pressure changes within the cochlea. As a result, a series of travelling waves is set up in the basilar membrane itself. These waves begin at the basal end by the oval window and progressively grow in amplitude as they proceed along the basilar membrane until they reach a peak, whereupon they rapidly decline, as shown in Figure 14.26b. The position of this peak depends on the frequency of sound waves impinging on the ear. The peak amplitude for high frequencies is near to the base of the cochlea while low frequency sounds elicit the largest motion nearer to the helicotrema (Figure 14.26c). The frequency of a sound wave is thus mapped onto the basilar membrane by the point of maximum displacement and the constituent frequencies of complex sounds are mapped onto different regions of the basilar membrane.

How is the displacement of the basilar membrane by the travelling wave converted into nerve impulses? The cells responsible for this process are the **hair cells** of the organ of Corti. If these are absent (as in some genetic defects) or destroyed, the affected person will be deaf. The hair cells are arranged in two groups: the outer hair cells, which are arranged in three rows, and a single row of inner hair cells. The **stereocilia** project from the upper surface of the hair cells (Figure 14.27) with their tips near to the gelatinous tectorial membrane.

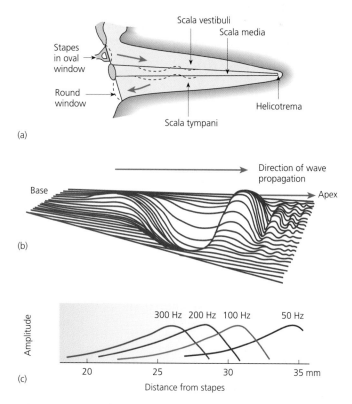

Figure 14.26 (a) A diagrammatic representation of the way in which sound energy elicits the travelling wave in the basilar membrane. (b) Diagram showing how the travelling wave progressively reaches a maximum as it passes along the basilar membrane before abruptly dying out. (c) The peak amplitude of the travelling wave is reached at different places along the basilar membrane for different frequencies of sound.

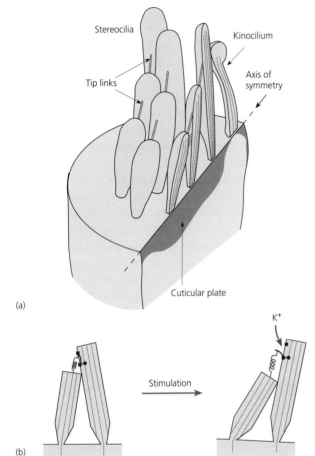

(a)

(b)

Figure 14.27 (a) The arrangement of the stereocilia on the inner hair cells. Note the thin protein filaments linking adjacent cilia (tip links). (b) Diagram showing how the tip links open the transduction channels when the cilia bend in response to movement of the basilar membrane.

In the absence of stimulation, the hair cells have a membrane potential of about –60 mV. Displacement of the sterocilia by a travelling wave opens specific ion channels in the stereocilia, known as transduction channels. These are very permeable to potassium ions. Potassium therefore moves into the hair cell from the surrounding endolymph (which is rich in potassium) to give rise to a small depolarization or receptor potential in the hair cells. This depolarization will result in the opening of voltage-gated calcium channels in the basolateral surface of the hair cell and the secretion of neurotransmitter onto the afferent nerve endings of the cochlear nerve, so exciting

them. For any particular hair cell, the specific frequency that most easily elicits a receptor potential is known as the characteristic frequency for that cell (Figure 14.28).

The afferent fibres have their cell bodies in the spiral ganglion and a specific set of afferent fibres innervate a particular region of the basilar membrane. As the basilar

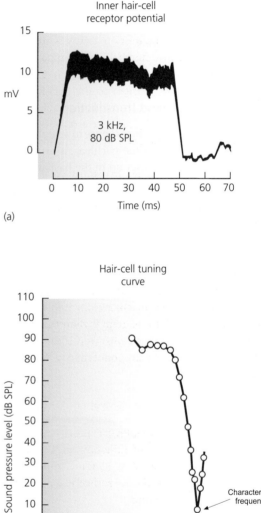

(a)

(b)

Figure 14.28 (a) A receptor potential observed in a mammalian inner hair cell for a 3 kHz tone of 80 dB SPL. (b) The variation in sound intensity needed to elicit a receptor potential of a fixed amplitude for different frequencies. In this case, the cell was most sensitive to a pure tone at 18 kHz.

membrane is tuned so that a sound of a given frequency will elicit maximum motion in a particular region, specific sounds will excite a specific set of afferent fibres. This is known as tonotopic mapping. Sound intensity is coded by the frequency of action potentials generated in the afferent fibres and by the number of fibres excited.

Central auditory processing

The organization of the auditory pathway is shown in Figure 14.29. The cochlear nerve divides as it enters the cochlear nucleus. As the arrangement of the cochlear nerve fibres reflects their origin in the cochlea, this organization is preserved in the cochlear nucleus and the neurones of each subdivision are arranged so that they respond to different frequencies in a tonotopic order.

From the cochlear nuclei, neurones pass to the olivary nuclei, which is the first part of the auditory system to receive inputs from both ears. From each olivary nucleus, fibres project to the inferior colliculus via the lateral lemniscus and thence to the medial geniculate body of the thalamus. From the medial geniculate body, auditory fibres project to the primary auditory cortex, which is located on the upper aspect of the temporal lobe. The area of cortex devoted to hearing in the left hemisphere is greater than that in the right and is closely associated with Wernicke's area – a region of the brain specifically concerned with speech (see Chapter 15).

Hearing deficits and their clinical evaluation

Hearing is a complex physiological process that can be disrupted in a variety of different ways. Broadly, hearing deficits are classified as conductive deafness, sensorineural deafness and central deafness according to the site of the primary lesion.

Conductive deafness

This hearing loss results from a defect in the middle or outer ear that reduces or prevents the transmission of sound to the inner ear. Conductive hearing loss can result from various causes, including middle-ear infections (which can lead to a condition known as 'glue ear'), serous otitis media, in which fluid accumulates in the middle ear, and otosclerosis, in which the movement of

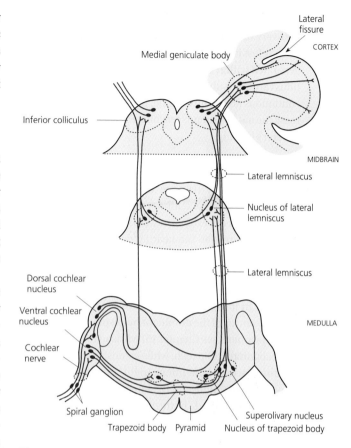

Figure 14.29 The organization of the ascending pathways of the auditory system.

the footplate of the stapes is impeded by the growth of bone around the oval window.

Sensorineural hearing loss

This is the result of damage to the cochlea or auditory nerve. It is quite commonly the result of traumatic damage to the cochlea by high-intensity sounds caused by industrial processes such as the riveting of steel plates ('boilermakers disease') or by sounds resulting from overamplified music. Age-related hearing loss (known as **presbyacusis**) is a type of sensorineural deafness that specifically affects the high frequencies. Loss of hair cells can be caused by ototoxic drugs such as the aminoglycoside antibiotics (e.g. streptomycin and neomycin) and certain diuretics (e.g. furosemide (frusemide)). The excessive growth of Schwann cells (a Schwannoma) in the cochlear nerve can result in compression of the nerve and can lead to deafness. A further very distressing but common cause of sensorineural hearing loss is **tinnitus** – the unremitting sensation of sound generated within the ear itself. The characteristics of tinnitus vary from subject to subject and include high-pitched, continuous notes and buzzing, pulsing sounds. All tend to mask the natural sounds reaching the ear, so impairing the sufferer's hearing.

Central deafness

This is caused by damage to the auditory pathways above the level of the auditory nerve. Because the auditory pathways are extensively crossed at all levels above the cochlear nuclei, the hearing deficits resulting from such damage are usually very subtle. Generally speaking, unilateral damage to the auditory cortex does not result in deafness although the affected person may have difficulty in localizing sounds. Extensive damage to the auditory cortex of the dominant hemisphere (usually the left hemisphere, see Chapter 15) will lead to difficulties in the recognition of speech, while extensive damage to the auditory cortex of the minor (right) hemisphere affects recognition of timbre and the interpretation of temporal sequences of sound, both of which are important in music and speech.

The vestibular system and the sense of balance

The sense of balance plays an important role in the maintenance of normal posture and in stabilizing the retinal image, particularly during locomotion. The vestibular system is responsible for detecting linear and rotational acceleration movements of the head, and plays a part in controlling eye movements. People who have lost the function of their vestibular system have difficulty in walking over an irregular surface when they are deprived of visual cues by a blindfold. Furthermore, their visual world appears to move up and down rather than remaining stable during walking.

Structure of the vestibular system

The vestibular portion of the membranous labyrinth is the organ of balance. The whole structure is about 1 cm in diameter and, as illustrated in Figure 14.30, it consists of two chambers (the utricle and saccule) and three semicircular canals. The nerve supply is via the vestibular branch of the eighth cranial nerve.

The utricle and saccule are arranged horizontally and vertically, respectively, while the three semicircular canals are arranged at right angles to each other, with the lateral canal inclined at an angle of about 30° from the horizontal.

KEY POINTS:

- The outer and middle ear focus sound energy onto the oval window. This allows the efficient transfer of sound energy to the cochlea.

- Incident sound waves cause pressure waves to be set up in the fluids of the cochlea. These evoke a travelling wave in the basilar membrane that activates the hair cells responsible for sound transduction.

- The basilar membrane is tuned so that the constituent frequencies of a sound are represented as a map on the basilar membrane. High frequencies are represented near the oval window, while low frequencies are represented near to the apex of the cochlea.

- Hearing deficits are classified as conductive deafness, sensorineural deafness or central deafness.

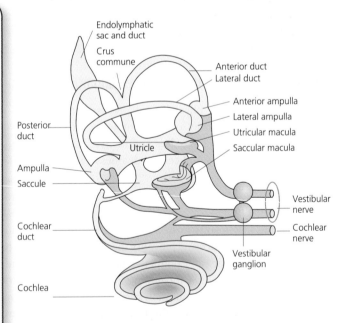

Figure 14.30 A schematic drawing showing the principal structures of the membranous labyrinth and their neural connections (shown in blue).

The other two canals lie in vertical planes. Near the utricle, each canal has an enlargement known as the **ampulla**. The fluid within the semicircular canals, utricle and saccule is endolymph and the whole structure floats in the perilymph, which is contained within the bony labyrinth.

The receptors of the vestibular system are hair cells, which operate in a similar fashion to those of the cochlea described earlier. Bending of the hair cells signals positional changes of the head. The hair cells of the semicircular canals respond to angular or rotational movements of the head body, while those of the utricle and saccule are primarily responsible for detecting the orientation of the head with respect to gravity.

The semicircular canals

The sensory portion of the semicircular canals is located within the ampulla. A cross-section of the ampulla reveals that the wall of the canal projects inward to form the **crista ampullaris** (Figure 14.31a). The hair cells are located on the epithelial layer that covers the crista and the hairs that project from their upper surface are embedded into a large gelatinous mass, the **cupula**, which is in loose contact with the wall of the ampulla at its free end. As a result, it forms a compliant seal that closes the lumen of the canal, preventing free circulation of the endolymph.

When the head is turned, the walls of the labyrinth must move as they are attached to the skull, but the endolymph in the semicircular canals tends to lag behind by virtue of its inertia. The effect is to deflect the cupula away from the direction of movement, as shown in Figure 14.32. This deflection causes the hairs of the hair cells to bend and

changes the rate of discharge of the afferent fibres of the vestibular branch of the eighth cranial nerve. For a simple turning movement to the left in a horizontal plane, the hair cells in the left horizontal canal excite their afferent fibres while those in the right canal inhibit theirs. The outputs of the two cristae act in a push–pull manner to signal the angular movement of the head. Similar coupled reactions will occur in the other pairs of canals in response to movements in other planes, and the arrangement of the semicircular canals in space is such that, whatever the angular movement, at least one pair of the semicircular canals will be stimulated.

Signals from the semicircular canals control eye movements

The information derived from the semicircular canals is used to control eye movements (see also p. 248). Direct stimulation of the ampullary nerves elicits specific movements of the eyes. For example, stimulation of the afferent fibres that serve the left horizontal canal causes the eyes to turn to the right. This is one of a group of **vestibulo-ocular reflexes**. The eye movements elicited by activation of the ampullary receptors are specifically adapted to permit the gaze to remain steady during movement of the head – for example during walking.

The utricle and saccule

The sensory epithelium of the utricle and saccule consists of a layer of hair cells covered by a gelatinous membrane containing small crystals of calcium carbonate called **otoliths** (Figure 14.31b). This membrane (the otolith

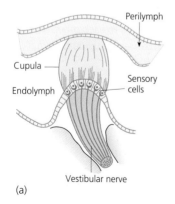
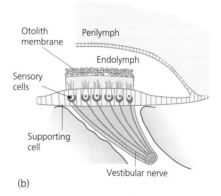

(a) (b)

Figure 14.31 The structure of the ampulla of the semicircular canals (a) and of the saccule (b). Note that the sensory hairs of the crista ampullaris project into a gelatinous mass called the cupula that forms a barrier across the ampulla, preventing the endolymph from circulating freely. The hair cells of the saccule project into a gelatinous sheet covered with crystals of calcium carbonate (otoliths).

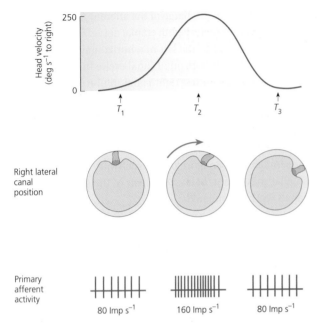

Figure 14.32 The response of vestibular afferent fibres following displacement of the head to the left. Immediately following the movement, the cupula is displaced and the hair cells excite the afferent nerve endings, which increases their rate of action potential discharge. Once the movement ceases, the cupula resumes its normal position and the action potential discharge returns to its resting level. Imp, impulses.

membrane) is able to move relative to the hair cells as the orientation of the head is changed. The sensory epithelium of the utricle is in the horizontal plane while that of the saccule lies in the vertical plane. Movement of the otolith membrane caused by tilting of the head leads to deflection of the stereocilia and excitation of the vestibular afferents, which, in turn, provide the brain with sufficient information to define the position of the head with respect to gravity. Information from the utricle and saccule is important in the control of posture.

Disorders of the vestibular system

From the previous discussion, it should be clear that the vestibular system is concerned with maintaining posture and stabilizing the visual field on the retina. People with unilateral damage to the vestibular system have a sense of turning and of vertigo, and may have abnormal eye movements. (Vertigo is an illusion in which the subject feels that either they or their surroundings are moving even though they are, in fact, stationary). Remarkably, if the vestibular system is damaged on both sides, affected subjects are essentially unaware that they have a sensory deficit. They are able to stand, walk and run in an apparently normal fashion. If, however, they are blindfolded, they become unsure of their posture and will fall over if they are asked to walk over a compliant surface such as a mattress. A normal subject would have no such difficulty.

Overproduction of endolymph may result in a condition known as **Menière's disease**, which affects both hearing and balance. There is a loss of sensitivity to low-frequency sounds accompanied by attacks of dizziness or vertigo that may be so severe that the sufferer is unable to stand. These attacks are frequently accompanied by nausea and vomiting.

Motion sickness does not reflect damage to the vestibular system, rather it is caused by a conflict between the information arising from the vestibular system and that provided from other sensory systems such as vision and proprioception. Indeed, subjects with bilateral damage to the vestibular system do not appear to suffer from motion sickness.

KEY POINTS:

- The vestibular system on each side consists of three semicircular canals and two chambers, the utricle and the saccule.

- The three semicircular canals are arranged at right angles to each other, while the utricle and saccule are arranged horizontally and vertically, respectively.

- The semicircular canals are arranged to signal angular accelerations of the head, while the utricle and saccule signal linear accelerations such as gravity.

- The information derived from the semicircular canals is used to control eye movements with the aim of stabilizing the visual field on the retina. Information from the utricle and saccule plays an important role in the maintenance of posture.

14.7   The chemical senses – smell and taste

The chemical senses, smell (or olfaction) and taste (or gustation), are amongst the most basic responses of living organisms to their environment. In humans, the sense of taste plays a vital role in the selection of foods and avoidance of poisons, and both smell and taste play a significant role in the enjoyment of food. Indeed, the taste of food is profoundly disturbed when the sense of smell is temporarily impaired during a severe head cold.

The sense of smell (olfaction)

Although our sense of smell is not highly developed compared with many other mammals, humans are able to discriminate between the odours of many thousands of different substances. The sensory organ for the sense of smell is the olfactory epithelium, which lies high in the nasal cavity above the turbinate bones (Figure 14.33). It is about 2–3 cm² in area on each side and consists of ciliated receptor cells and supporting cells. It is covered by a layer of mucus secreted by Bowman's glands, which lie beneath the epithelial layer. The axons from the olfactory receptor cells pass through the cribriform plate to make their synaptic contacts in the glomeruli of the olfactory bulbs.

Olfactory transduction depends on activation of specific G protein-linked receptors

To excite an olfactory receptor, a substance must be both volatile and able to dissolve in the layer of mucus that covers the olfactory epithelium. The olfactory receptor molecules are binding proteins located on the cilia of the olfactory cells. Each olfactory receptor protein is coupled to a G protein that activates adenylyl cyclase. When an odorant molecule is bound by an appropriate receptor molecule, there is an increase in the level of cyclic AMP in the receptor cell and this opens a cation-selective channel that leads to the depolarization of the olfactory receptor. If the depolarization reaches the threshold for action potential generation, an action potential will be propagated to the olfactory bulb. From

here, olfactory information travels to the olfactory cortex via the lateral olfactory tract. Individual olfactory receptors respond to more than one odiferous substance, although a receptor is usually excited best by one odour. It is therefore likely that olfactory information is coded in the pattern of incoming information, which the brain learns to interpret.

The sense of taste (gustation)

The tongue is the principal organ of taste. All over its surface, there are small projections called **papillae**, which give

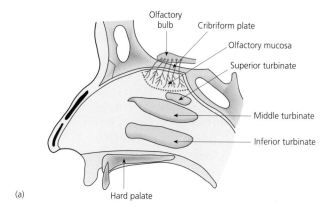

(a)

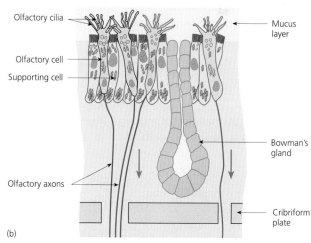

(b)

Figure 14.33 A sectional view of the nose to show the location (a) and detailed arrangement (b) of the olfactory epithelium.

the tongue its roughness. Four different types of papillae can be identified: filiform, folate, fungiform and vallate. The organs of taste are the **taste buds**, which are found only on the folate, fungiform and vallate papillae. The taste buds are located just below the surface epithelium and communicate with the surface via a small opening called a taste pore (Figure 14.34). Each taste bud consists of a group of taste cells and supporting cells. The taste cells are innervated by fibres from the facial and glossopharyngeal nerves. A few taste buds are scattered around the oral cavity and on the soft palate.

There are four distinct modalities of taste: salty, sour, sweet and bitter. The complexity of the taste of foodstuffs arises partly from the mixed sensations arising from stimulation of the different modalities of taste but chiefly from the additional stimulation of the olfactory receptors. Different parts of the tongue show different sensitivities to these taste modalities (Figure 14.34).

The mechanisms of sensory transduction vary between the different taste modalities, but in all cases

the final steps in the transduction process are the secretion of neurotransmitter by the taste cell and excitation of gustatory afferent fibres. Taste sensations reach the cerebral cortex primarily via the facial and glossopharyngeal nerves. The brain must then interpret the pattern of activity from the active population of taste cells to discriminate the taste of a substance that is present in the mouth.

KEY POINTS:

- The chemical senses are olfaction (sense of smell) and gustation (sense of taste).

- The olfactory receptors are located in the epithelium above the third turbinate in the nose.

- The taste receptors are primarily found on the tongue. There are four modalities of taste: salty, sour, sweet and bitter. Individual taste cells respond preferentially to one modality of taste.

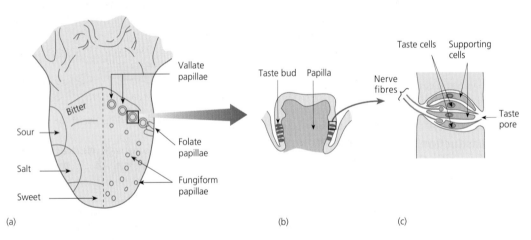

(a)   (b)   (c)

Figure 14.34 The regions of the tongue that show greatest sensitivity to the four modalities of taste (a). A cross-sectional view of a vallate papilla showing the location of the taste buds (b) and the detailed structure of a single taste bud (c).

Recommended reading

Anatomy of sensory systems

Brodal, P. (2003). *The Central Nervous System. Structure and Function*, 3rd edn. Oxford University Press: Oxford. Chapters 5–10.

Physiology

Carpenter, R.H.S. (2002). *Neurophysiology*, 4th edn. Hodder Arnold: London. Chapters 4–8.

Gregory, R.L. (1998). *Eye and Brain: the Psychology of Seeing*, 5th edn. Oxford University Press: Oxford.

Melzack, R. and Wall, P.D. (1996). *The Challenge of Pain*, 2nd edn. Penguin: London.

Pocock, G. and Richards, C.D. (2006). *Human Physiology: the Basis of Medicine,* 6th edn. Oxford University Press: Oxford. Chapter 8.

Medicine

Donaghy, M. (2005). *Neurology*, 2nd edn. Oxford University Press: Oxford. Chapters 6 and 14–17.

Self-assessment questions

Which of the following statements are true and which are false? Answers are given at the end of the book (p. 755).

1. a) Afferent nerve fibres carry information away from the CNS.
    b) Particular types of receptor respond most easily to a specific quality of a stimulus.
    c) The frequency of a train of action potentials in an afferent fibre reflects the intensity of the stimulus given to its receptor.
    d) The first step in sensory transduction is the generation of a receptor potential.
    e) Myelinated axons conduct impulses more rapidly than unmyelinated axons.

2. a) Thermoreceptors respond either to warm or to cold stimuli.
    b) Nociceptors are encapsulated receptors.
    c) Skin thermoreceptors are bare nerve endings.
    d) Sensory information from the touch receptors of the skin reaches the brain via the dorsal column nuclei.
    e) The spinothalamic tract relays information from the skin thermoreceptors.

3. a) Pain arising from the visceral organs may be referred.
    b) Pain receptors may be activated following the release of bradykinin.

    c) Somatosensory information reaches consciousness in the precentral gyrus.
    d) Pain always elicits autonomic responses.
    e) C-fibres conduct impulses more slowly than Aβ fibres.

4. a) The outer layer of the eye is called the iris.
    b) Extraocular muscles control the diameter of the pupil.
    c) The cornea contains no blood vessels.
    d) The pupils dilate in response to sympathetic stimulation of the radial muscles of the iris.
    e) Myopia can be corrected by a diverging lens.

5. a) In the retina, receptor density is greatest at the optic disk.
    b) Full colour vision is possible in dim light.
    c) Cones provide colour vision.
    d) Intraocular pressure is normally about 6 kPa (45 mmHg).

6 a) The range of human hearing is from 20 Hz to 20 kHz.
   b) The ear is most sensitive to frequencies between about 1 and 3 kHz.
   c) The ossicles of the middle ear are essential for the efficient transmission of airborne sounds to the cochlea.
   d) The loudness of a sound is determined by its wavelength.
   e) Reissner's membrane houses the organ of Corti.

7 a) A hearing defect caused by a problem in the middle ear is called sensorineural hearing loss.
   b) The Eustachian tube connects the middle ear with the pharynx.

   c) Hair cells sensitive to low-pitched tones are located close to the oval window.
   d) The endolymph of the scala media is similar in composition to plasma.
   e) The semicircular canals contain endolymph.

8 a) The utricle and saccule are important in the control of posture.
   b) The semicircular canals are important in the control of eye movements.
   c) The utricle and saccule detect angular accelerations of the head.
   d) Taste receptors are found only on the tongue.
   e) The olfactory epithelium lines the nostrils.

Some aspects of higher nervous function 15

After reading this chapter you should have gained an understanding of:

- The functions of the association cortex in the cerebral hemispheres
- The separate roles of the two hemispheres
- The organization of the pathways that control speech
- How the EEG is used to monitor cerebral function in humans
- The physiology of sleep and other circadian rhythms
- The physiological basis of learning and memory

15.1 Introduction

The brain is not simply concerned with controlling movements and regulating the internal environment. It is also concerned with assessing aspects of the world around us so that we may interact appropriately with our surroundings and with each other. For example, complex behaviours are learned, experiences are remembered and our memories can be retrieved in order to determine a course of action. Furthermore, we communicate with each other using spoken and written language. To achieve all this, we require a certain level of awareness of ourselves and our environment that we call consciousness. This chapter will explore some aspects of these very complex phenomena.

15.2 The specific functions of the left and right hemispheres

The cerebral hemispheres are the largest and most obvious structures of the human brain. As discussed in earlier chapters, they receive information from the primary senses (somatic sensations, taste, smell, hearing and vision) and control co-ordinated motor activity. The main somatosensory and motor pathways are crossed so that the left hemisphere receives information from, and controls the motor activity of, the right side of the body while the right hemisphere is concerned with the left side.

However, the two cerebral hemispheres are not symmetrical, either morphologically or in their functions. It is commonly recognized that people prefer to use one hand rather than another. Indeed, most people (about 90% of the population) prefer to use their right hand for writing, holding a knife or tennis racquet and so on. What is not always appreciated is that most people also prefer to use one foot rather than the other (for example when kicking a ball) and they generally pay more attention to information arriving from one eye and one ear. Therefore, the two halves of the brain are not equal with respect to the information to which they pay attention or with respect to the activities they control.

Much of our present understanding of the ways in which the two halves of the brain operate has come from studies of people who have suffered brain injuries of some kind. For example, as long ago as the nineteenth century, it was reported that people who have suffered a stroke affecting the left side of the brain often show loss of speech as well as right-sided paralysis. Later studies showed that specific deficits in the understanding of language, reading and writing were also associated with damage to the left cerebral hemisphere. Other patients with damaged left hemispheres were unable to carry out specific, directed actions such as brushing their hair although the muscles involved were not paralysed. Thus, as the clinical evidence accumulated, it became apparent that the two hemispheres must have different capacities. The idea developed that one hemisphere, the left in right-handed people, was the leading or **dominant hemisphere** while the other was subordinate and lacked a specific role in higher nervous functions. This idea has had to be modified in the light of current evidence and it is now clear that both hemispheres have very specific functions in addition to the role they play in sensation and motor activity.

The association areas of the frontal cortex

Early exploration of the cerebral cortex of man showed that, while stimulation of some areas elicited specific sensations or motor responses, stimulation of others had no detectable effect. These areas were called 'silent'

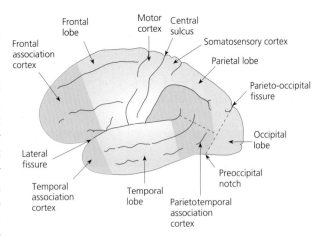

Figure 15.1 A side view of the human brain showing the association areas of the frontal, parietal and temporal lobes.

areas and were loosely considered as **association cortex** (Figure 15.1). Over the last century, the function of these areas has gradually become clearer.

Compared with the brains of other animals, including those of the higher apes, the frontal lobes of the human brain are very large relative to the size of the brain as a whole. What is their role? Certain aspects of their function became clear during the nineteenth century. In 1848, an American mining engineer called Phineas Gage received a devastating injury to his frontal lobes (Figure 15.2). He was packing down an explosive charge when it detonated,

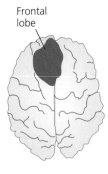

Figure 15.2 The skull of Phineas Gage showing the course of the iron bar (left) and the likely extent of the damage to his frontal cortex.

driving the tamping iron through his upper jaw, into his skull and through his frontal lobes. Remarkably, he survived the accident. Although he was able to live an essentially normal life, his personality had irrevocably changed. He became irascible, unpredictable and much less inhibited in his social behaviour – indeed, his friends remarked that he was 'no longer Gage'. He also later suffered from increasingly violent epileptic seizures.

Later work on monkeys showed that lesions to the frontal lobes appeared to reduce anxiety. This discovery was subsequently exploited clinically in an attempt to help patients suffering severe and debilitating depression. By cutting the fibres connecting the frontal lobes to the thalamus and other areas of the cortex (a **frontal leucotomy**), it was hoped to reduce the feeling of desperate anxiety and permit patients to resume a normal life. Early results were encouraging in that the affected patients seemed less anxious than before, but it later became clear that the procedure could result in severe personality changes and epilepsy, which perhaps might have been predicted from the case of Phineas Gage. In recent years, this procedure has become replaced by less drastic and more reversible drug therapies. Nevertheless, the clinical and experimental observations have demonstrated that the association areas of the frontal lobes are involved in the determination of all aspects of an individual's personality.

Damage to the parietal lobes leads to loss of higher-level motor and sensory performance

The parietal lobes (Figure 15.1) also contain association areas. Damage to these parts of the brain is associated with specific deficits known as agnosias and apraxias.

Agnosia is a failure to recognize an object even though there is no specific sensory deficit. It reflects an inability of the brain to integrate sensory information in a normal way and is regarded as a failure of 'higher-level' sensory performance. An example is visual object agnosia in which the visual pathways appear to be essentially normal but recognition of objects does not occur. If an affected person is allowed to explore the object with another sense, by touch for example, they can often name it. Nevertheless, they may not be able to appreciate its qualities as a physical object. Thus, they may see a chair but not avoid it as they cross a room.

Apraxia is the loss of ability to perform specific purposeful movements even though there is no paralysis or loss of sensation. An affected person may be unable to perform a complex motor task on command, for example waving someone goodbye, but may be perfectly capable of carrying out the same act spontaneously.

The most bizarre effect of lesions to the parietal lobes is seen when the lesion affects the posterior part of the right hemisphere around the border of the parietal and occipital lobes. These lesions lead to neglect of the left side of the body. Affected individuals ignore the left side of their own bodies, leaving them unwashed and uncared for. They ignore the food on the left side of their plates and will only copy the right side of a simple drawing (Figure 15.3).

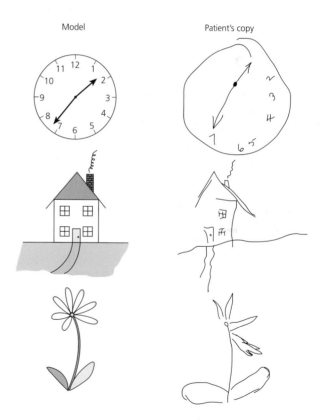

Figure 15.3 Drawing by a patient suffering from a stroke in the posterior region of the right hemisphere leading to the neglect syndrome. The model is on the left of the figure and the patient's drawing is shown on the right. Note the failure to complete the left-hand parts of the originals.

The corpus callosum integrates the activity of the two cerebral hemispheres

Sensory information from the right half of the body is represented in the somatosensory cortex of the left hemisphere and vice versa. Equally, the left motor cortex controls the motor activity of the right side of the body. Despite this apparent segregation, the brain acts as a whole, integrating all aspects of neural function. This is possible because, although the primary motor and sensory pathways are crossed, there are many interconnections between the two halves of the brain, known as **commissures**. As a result, each side of the brain is constantly informed of the activities of the other.

The largest of the commissures is the vast band of fibres that connects the two cerebral hemispheres known as the **corpus callosum** (see p. 185). Most of the nerve fibres of the corpus callosum project to comparable functional areas on the contralateral side. It is known that epileptic discharges can spread from one hemisphere to the other via the corpus callosum and that major epileptic attacks can involve both sides of the brain. In the 1960s, an experimental surgical procedure was performed in an attempt to cure severe bilateral epilepsy of this kind. The corpus callosum was cut – the **split-brain operation**. This had the desired end result – a reduction in the frequency and severity of the epileptic attacks. It also offered the opportunity of careful and detailed study of the functions of the two hemispheres of the human brain.

Each hemisphere is specialized to perform a specific set of higher nervous functions

The key to the first experiments on split-brain patients was to exploit the fact that the visual pathway is only partially crossed at the optic chiasm, while speech is located in the left hemisphere. Fibres from the temporal retina remain uncrossed, while those from the nasal region of the retina project to the contralateral visual cortex, as shown in Figure 15.4. The effect of this arrangement is that the right visual field is represented in the left visual cortex, while the left visual field is represented in the right visual cortex. By projecting words and images onto a screen in such a way that they would appear in either the

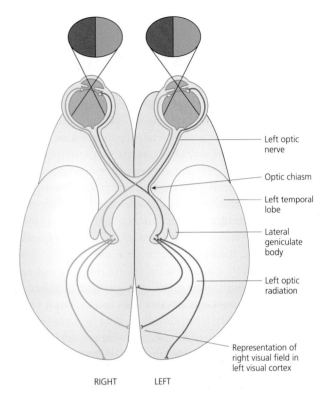

Left optic nerve

Optic chiasm

Left temporal lobe

Lateral geniculate body

Left optic radiation

Representation of right visual field in left visual cortex

RIGHT   LEFT

Figure 15.4 A ventral view of the human brain showing the organization of the visual pathways. Note that the left cortex 'sees' the right visual field and vice versa.

right or the left visual field, R.W. Sperry and his colleagues were able to investigate the specific capabilities of each hemisphere. The subject could then be asked what they had seen.

If, say, the word 'BAND' was briefly projected onto the right visual field, the patient was able to report this to the investigator (Figure 15.5) as this word was represented in the visual cortex of the left hemisphere, which also controls speech. If the left visual field had a qualifying word such as 'HAT', the subject was unaware of the fact. If asked what type of band had been mentioned, they were reduced to guessing. If a word such as 'BOOK' was projected onto the left field, they could only guess at the answer. If, however, the patient was then allowed to choose something that matched the word that had been projected onto the left visual field, they then made a suitable choice with their left hand (i.e. the hand that is

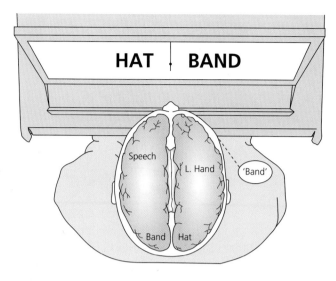

Figure 15.5 The response of a 'split-brain' patient to words projected onto the left and right visual fields.

controlled from the right hemisphere). This series of experiments showed that, while speech was totally lateralized to the left hemisphere, the right hemisphere was aware of the environment, was capable of logical choice and possessed simple language comprehension.

Further studies showed that the right hemisphere could identify objects held in the left hand by their shape and texture (this is called **stereognosis**) as efficiently as the right hand. Indeed, the left hand (and therefore the right hemisphere) proved to be much better at solving complex spatial problems. Overall, Sperry concluded that, far from being subordinate to the left hemisphere, the right hemisphere was conscious and was better at solving spatial problems and non-verbal reasoning. His conclusions are summarized in Figure 15.6.

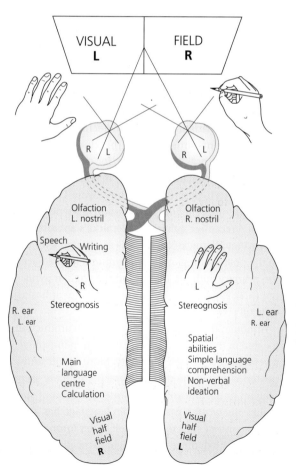

KEY POINTS:

- In addition to their role in motor control, the frontal lobes play an important part in shaping the personality of an individual.

- Lesions in the parietal lobe result in defects of sensory integration known as agnosias and in an inability to perform certain purposeful acts (apraxia).

- Severance of the corpus callosum, which interconnects the two hemispheres, has shown that the two hemispheres have very specific capabilities in addition to their role in sensation and motor activity.

- Speech and language abilities are mainly located in the left hemisphere, together with logical reasoning. The right hemisphere is better at solving spatial problems and non-verbal tasks.

Figure 15.6 The functional specialization of the two cerebral hemispheres. (Note that the areas of the cortex associated with the primary senses and those involved in motor control are not represented in this diagram.)

15.3 Speech

Speech is a very complex skill that is unique to humans. It involves knowledge of the vocabulary and grammatical rules of at least one language and requires very precise motor acts to permit the production of specific sounds in their correct order. It also requires precise regulation of the flow of air through the larynx and mouth. It is, therefore, not surprising to find that large areas of the brain are devoted to speech and its comprehension. Because it is specifically a human characteristic, our knowledge of the systems that govern the production and comprehension of speech has largely come from careful neurological observations.

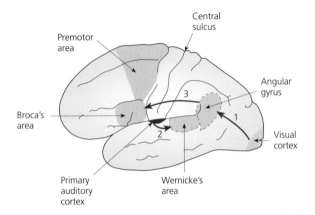

Figure 15.7 A diagram of the left hemisphere to show the location of the principal regions involved in the control of speech. The blue arrows indicate the flow of information from the sensory regions of the brain to Broca'a area, which controls the production of speech.

The principal speech areas are located in the frontal and temporal lobes

As mentioned earlier, speech is known to be lateralized to the left hemisphere in the majority of people. Damage to this hemisphere, such as that caused by a stroke for example, can result in difficulties with speech production known as **aphasias**. Characteristically, the speech of such people is slow, halting and telegraphic in quality. Such a patient is able to name objects and describe their attributes but has difficulty with the small parts of speech that play such an important role in grammar (e.g. 'if', 'is', 'the' and so on). They also have difficulty in writing. However, these patients retain a good understanding of both spoken and written language, so this type of aphasia is sometimes called **expressive aphasia**. It is also known as **Broca's aphasia** after Paul Broca who carried out much of the early investigation into the condition.

Broca was able to examine the brains of several aphasic patients at post mortem. He discovered that there was extensive damage to the frontal lobe of the left hemisphere, particularly in the region that lies just anterior to the motor area responsible for the control of the lips and tongue (Figure 15.7). This is now known as Broca's area. Patients with Broca's aphasia do not have a paralysis of the lips and tongue and can sing wordlessly. What they have lost is the ability to use the apparatus of speech to form words, phrases and sentences.

Another type of aphasia, discovered by Carl Wernicke, was characterized by free-flowing speech that had little or no informational content (technically known as **jargon**). Patients suffering from this kind of aphasia tend to make up words (**neologisms**), for example 'lork' and 'flieber', and often have difficulty in choosing the appropriate words to describe what they mean. They have difficulty in comprehending speech and this type of aphasia is called **receptive aphasia** or **Wernicke's aphasia** after its discoverer. Wernicke's aphasia is associated with lesions to the posterior region of the temporal lobe adjacent to the primary auditory cortex and the angular gyrus (Figure 15.7). Patients suffering from Wernicke's aphasia have difficulty with reading and writing.

Although the areas that control speech production are located in the left hemisphere, the right hemisphere has a very basic language capability. More significantly, the posterior part of the right cerebral hemisphere seems to play an important role in the interpretation of speech. Unlike the written word, spoken language has an emotional content that reveals itself in its intonation, and patients with damage to their right hemisphere will often speak in a flat monotone.

Table 15.1 The distribution of handedness and the location of speech in the cerebral hemispheres

Location of speech	Right-handed individuals	Left-handed individuals	Total
Left hemisphere	85.5%	7%	92.5%
Right hemisphere	4.5%	1.5%	6.0%
Both hemispheres	0%	1.5%	1.5%
Total	90%	10%	100%

The table shows the location of speech as a percentage of the total adult population. Thus, 85.5% of the population are right handed and have their speech localized to their left hemispheres and so on.

In the majority of people, speech is controlled from the left hemisphere

Although in most people speech is controlled by the left hemisphere, it is not universally so. It is of some importance that a neurosurgeon is aware of which hemisphere controls speech before embarking on an operation. This is established by the **Wada test**, which temporarily anaesthetizes one hemisphere. The patient lies on their back while a cannula is inserted into the carotid artery on one side. The patient is then asked to count backwards from 100 and to keep both arms raised. A small quantity of a short-lasting anaesthetic is then injected. As the anaesthetic reaches the brain, the arm on the side opposite that of the injection falls and the patient may stop counting for a few seconds or for several minutes depending on whether or not speech is localized to the side receiving the injection.

These tests reveal which hemisphere controls speech in a person who has no aphasia. In 95% of right-handers, speech is controlled from the left hemisphere. This is also true of about 70% of left-handers. Speech is localized to the right hemisphere in about 15% of left-handers and the remainder show evidence of speech being controlled from both hemispheres (Table 15.1). This distribution of the speech areas has recently been confirmed using a non-invasive imaging technique called Doppler ultrasonography.

KEY POINTS:

- Speech is localized to the left hemisphere in the majority of the population, irrespective of whether they are right or left handed.

- The neural patterns of speech originate in the temporal lobe in Wernicke's area, which is adjacent to the auditory cortex. The neural codes for speech pass to Broca's area for execution of the appropriate sequence of motor acts.

- Damage to the speech areas results in aphasia.

15.4 The EEG can be used to monitor the activity of the brain

The brain is constantly involved in the control of a huge range of activities both when awake and during sleep. Its activity can be monitored indirectly by placing electrodes on the scalp. If this is done in such a way as to minimize electrical interference from the muscles of the head and neck, small oscillations are seen that can reflect the spontaneous activity of the brain (Figure 15.8). These electrical oscillations are known as an **electroencephalogram** or **EEG**. For normal subjects, the amplitude of the EEG waves ranges from 10 to 150 μV.

The appearance of the EEG varies according to the position of the electrodes, the behavioural state of the

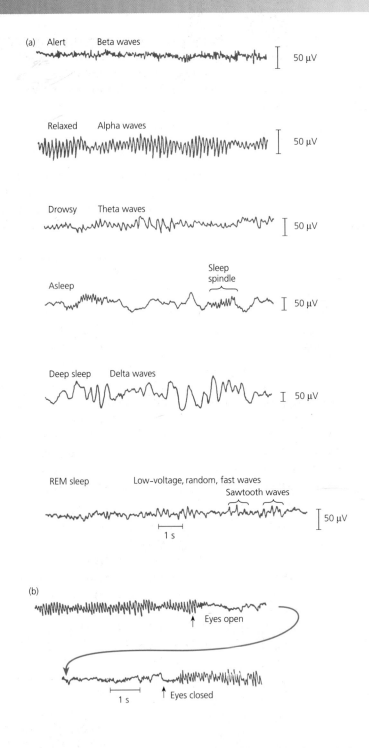

Figure 15.8 Typical stretches of the EEG for various stages of awareness. Panel (a) shows the changes in the EEG as a subject falls asleep. Note that during deep sleep, the EEG changes to a pattern similar to that seen in the normal awake state but which has characteristic 'sawtooth' waves. This is the REM phase of sleep. Panel (b) shows alpha block following opening of the eyes. Note the sudden loss of the alpha waves when the eyes open and their resumption when the eyes close again. The two lines in (b) are of a continuous stretch of record from the occipital region of the left hemisphere.

subject (i.e. whether awake or asleep), the subject's age and whether there is any organic disease.

When a subject is awake and alert, the EEG consists of high-frequency waves (20–50 Hz), which have a low amplitude (about 10–20 μV). These are known as **beta waves** and they appear to originate in the cerebral cortex. As a subject closes his or her eyes, this low amplitude–high frequency pattern is replaced by a higher amplitude, lower frequency pattern known as the alpha rhythm. The **alpha waves** have an amplitude of 20–40 μV and contain one predominant frequency in the range of 8–12 Hz. The alpha rhythm is disrupted if the subject concentrates their attention on a problem or opens their eyes. This is called alpha block and is illustrated in Figure 15.8. As the subject becomes more drowsy and falls asleep, the alpha rhythm disappears and is replaced by slower waves of greater amplitude known as **theta waves** (40–80 μV and 4–7 Hz). These are interspersed with brief periods of high frequency activity known as 'sleep spindles'. In very deep sleep, the EEG waves are slower still (**delta waves**). They have a frequency of less than 3 Hz and a relatively high amplitude (100–120 μV). The characteristics of the principal EEG waves are summarized in Table 15.2.

Epilepsy

EEG recording is of value in the diagnosis of **epilepsy**, a condition characterized by disordered electrical activity in a part of the brain. Epilepsy is caused by abnormal functioning of brain tissue that may arise from traumatic injury, from infections or as a result of ischaemic damage to the brain during birth. There may be no obvious precipitating cause (idiopathic epilepsy) and some cases of epilepsy appear to be of familial origin. Epileptic attacks may be accompanied by gross seizures and loss of consciousness (**grand mal**) or by other mental changes such as the lapses of concentration typical of **petit mal**.

The EEG changes are characteristic of particular disease processes. The EEG is not only useful in diagnosing the presence of epilepsy; it can also be used to locate the site that triggers an epileptic attack (the **epileptic focus**). Some examples of the appearance of the EEG during an epileptic attack are shown in Figure 15.9.

A number of drugs may be used to control epileptic seizures. They are collectively known as **anticonvulsants**. Patients vary considerably in their response to these drugs and it often takes time to find the best drug or drug combination and the dose required to achieve optimal, long-term control of seizures. The more widely used anticonvulsant drugs include **phenytoin**, **carbamazepine**, **sodium valproate**, **topiramate** and **clonazepam**. In most cases, the mode of action of the drug is not entirely clear. However, phenytoin and carbamazepine are believed to block the repetitive firing of neurones by inhibiting voltage-sensitive sodium channels and sodium valproate

Table 15.2 The characteristics of the principal waves of an EEG

Wave type	Frequency (Hz)	Amplitude (μV)	Notes
Alpha (α)	8–12	20–40	Best seen over the occipital pole when the eyes are closed
Beta (β)	20–50	10–20	Normal awake pattern
Theta (θ)	4–7	40–80	Normal in children and in early sleep
			Evidence of organic disease when seen in awake adults
Delta (δ)	<3	100–150	Seen during deep sleep
			Evidence of organic disease when seen in awake adults

Note that the higher the frequency, the lower the amplitude, i.e. the higher frequencies are less synchronized than the low frequencies.

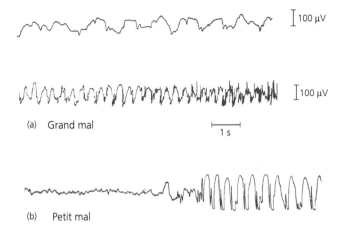

(a) Grand mal

1 s

(b) Petit mal

Figure 15.9 The change in the EEG that can be seen during an epileptic attack. In (a), the EEG has a preponderance of delta wave activity, which is replaced by large spikes during a convulsive seizure. In (b) is shown the change in the EEG of another patient during a petit mal seizure. Note the characteristic spike and wave complex.

may enhance GABA-mediated neurotransmission. Many of the anticonvulsant drugs have been shown to be teratogenic, that is they increase the risk of birth defects if taken during pregnancy.

Recently, a new treatment for epilepsy has been developed that involves intermittent stimulation of the left vagus nerve with an implanted device rather like a pacemaker. It can reduce the frequency of epileptic seizures in patients for whom drug treatments have proved unsatisfactory.

> **KEY POINTS:**
>
> - The EEG can be used to monitor the spontaneous electrical activity of the brain.
> - The waves of the EEG are of low amplitude (10–150 μV) and their frequency and amplitude depend on the state of arousal of a person.
> - The EEG is helpful in the diagnosis of epilepsy and can be used to determine the site of an epileptic focus.

15.5 Sleep

Everyone knows intuitively what sleep is and most people are familiar with the physical and psychological consequences of having too little sleep. Nevertheless, although sleep is the subject of intensive research, knowledge of exactly how and why we sleep is still incomplete. This section will explore some of the important information obtained from sleep studies regarding the processes that occur during sleep.

Key questions regarding the physiology of sleep include:

- What factors control the sleep–wakefulness cycle?
- What is happening in the brain during sleep?
- What is the purpose of sleep?

- How are drugs used to influence the sleep–wakefulness cycle?

Sleep is often defined as a state of partial unconsciousness from which a person may be aroused by stimulation. Thus, sleep differs from coma from which arousal is not possible. Sleep occurs in a sequence of stages during which the activity of the brain alters. These stages are accompanied by characteristic physiological changes. As outlined above, the appearance of the EEG can be used to follow the main stages of sleep: the EEG waves become slower and of greater amplitude as a person passes from the wakeful state to deep sleep (Figure 15.8). Stage I sleep is the state of drowsiness that usually precedes falling asleep. As sleep sets in, the EEG is dominated by theta waves (stage II sleep) and there may be periods of fast EEG activity with short periods of fast rhythmic waves called **sleep spindles** (Figure 15.8). As sleep becomes deeper (stage III), the sleep spindles are associated with occasional large waves to form K complexes. In deep sleep (stage IV), the EEG is dominated by large-amplitude, low-frequency delta waves.

The depth of sleep, as judged by how easy it is to arouse someone with a standard stimulus (e.g. a sound), correlates well with the appearance of the EEG. The deep phase of sleep (stage IV) associated with delta wave activity is also called **slow-wave sleep** or **ortho-sleep**. It is associated with a variety of characteristic physiological changes including: slowing of the heart rate, a fall in blood pressure, slowing of respiration, reduced muscle tone, a fall in body temperature and an absence of rapid eye movements. Deep sleep is the sleep stage in which sleep-walking, night terrors and bed-wetting are most likely to occur.

After a person has been in deep (slow-wave) sleep for 1 or 2 hours, the EEG assumes a pattern similar to the wakeful state, with fast beta wave-like activity (Figure 15.8). This change in EEG pattern (also called stage V sleep) is associated with jerky movements of the eyes, an increase in heart rate and respiratory rate and sometimes (in men) penile erection. During these changes, the subject is as difficult to arouse as they are during slow-wave sleep. As the EEG looks like that of a person who is awake, this phase has been called **paradoxical sleep**, although it is more often called rapid eye movement or **REM sleep**. During a single night, slow-wave sleep is interrupted by four to six episodes of REM sleep, each of which lasts about 20 minutes. REM sleep is associated with dreaming – if a subject is woken during a REM episode, they are much more likely to report having dreams than they are if they are awoken during periods of slow-wave sleep.

The amount of time spent in sleep varies with age as does the proportion of sleep spent as REM sleep. As Figure 15.10 shows, very young children spend a large part of the day asleep and REM sleep accounts for nearly half of all their sleep. By the age of 20 years, only about 8 hours are spent in sleep of which less than a quarter is spent as REM sleep. The requirement for both kinds of sleep declines further with age and, by the age of 60, many people require as little as 6 hours.

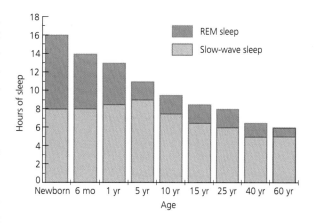

Figure 15.10 A histogram to show the pattern of sleep changes with age. Note the decline in the total hours of sleep and in the proportion of REM sleep with increasing age.

Why do we sleep?

The answer to this question still eludes us. The notion that sleep is required to restore the body is an old one but exactly what needs to be restored is not clear. What is clear is that sleep deprivation leads to irritability and a decline in intellectual performance. If deprivation is prolonged, then severe mental changes can occur bordering on psychosis. If people are allowed deep sleep but are deprived of REM sleep by the simple expedient of waking them every time a REM episode begins, they also become anxious and irritable. Following such a period, their sleep tends to have a higher frequency of REM episodes than normal, a finding which suggests that REM sleep serves an important function.

The sleep–wakefulness cycle is actively controlled from the thalamus, hypothalamus and brainstem

There is still much to discover regarding the mechanisms that control the sleep–wakefulness cycle. The following is a brief summary of the information that has been obtained from human and animal experiments. In the 1930s, R. Hess found that electrical stimulation of the thalamus initiates sleep behaviour in cats. A few years later, it was shown that sleeping animals can be aroused by electrical stimulation of a region of the brainstem known as the ascending reticular formation. This observation suggested that sleep is controlled, at least in part, by the activity of the brainstem. Later work has suggested

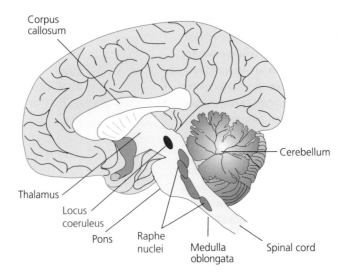

Figure 15.11 A diagram to illustrate the location of some of the brain stem nuclei believed to be concerned with control of the sleep–wakefulness cycle. The locus coeruleus are paired nuclei that lie beneath the fourth ventricle. Their neurones contain noradrenaline. The raphe nuclei form a narrow band along the midline of the medulla, pons and midbrain. Their neurones contain serotonin.

that some specific groups of nerve cells containing serotonin and others that contain noradrenaline play a role in the regulation of sleep.

Serotonergic neurones are found in a series of nuclei called the raphe nuclei, which are scattered along the ventral region of the brainstem (Figure 15.11). If these neurones are depleted of their serotonin, the affected animals are unable to sleep at all. Sleep can, however, be restored if 5-hydroxytryptophan (a precursor of serotonin) is administered. Noradrenergic neurones are located in the locus coeruleus, which lies beneath the cerebellum (Figure 15.11). If these neurones are depleted of their noradrenaline by administration of a metabolic inhibitor, the affected animals are able to enter slow-wave sleep but do not have normal REM sleep episodes.

An attractive, if simplified, explanation of these results is that activity in the serotonergic neurones initiates sleep by lowering the activity of the brainstem reticular formation. This leads to a lessening of cortical activity and results in sleep. Subsequent activity in the noradrenergic neurones gives rise to the REM episodes. However, as continued administration of inhibitors of serotonin synthesis does not cause permanent insomnia, even though serotonin levels remain low, this suggests that the

system controlling the sleep–wakefulness cycle is much more complex than this model suggests. Recent work has suggested that other neurotransmitters are involved.

Sleep disorders and drug treatments

A number of disorders are associated with sleep in addition to disorders such as sleep-walking, night terrors and bed-wetting, which are common in children but normally resolve as the child gets older. These include:

- Insomnia, an inability to fall asleep or to remain asleep. Chronic insomnia leads to daytime fatigue and impaired mental and physical performance.

- Narcolepsy, a condition in which the person may fall asleep spontaneously at any time, lapsing directly into REM sleep.

- Sleep apnoea, in which the person stops breathing, usually due to closure of the upper airway, and wakes many times during a night. Use of a positive-pressure breathing mask during sleep is often of benefit in treating this condition.

Insomnia is a very commonly reported sleep disorder, which may be triggered by a number of physical or psychological factors. A variety of drugs are used to treat insomnia. They are collectively known as **hypnotics** and amongst the most commonly used are a group of drugs called **benzodiazepines**. These act to augment the inhibitory effects of GABA at neural synapses and include **flurazepam**, **temazepam**, **quazepam**, **triazolam** and **zolpidem**. Other sedative drugs include melatonin receptor agonists such as **ramelteon**, antihistamines (e.g. **doxylamine** and **diphenhydramine**) and some antidepressants (e.g. **trimipramine** and **trazodone**).

Narcolepsy is a very distressing condition, which is often very difficult to treat. CNS stimulants such as amphetamines (e.g. **dexedrine**, **desoxyn**, **adderall** and **ritalin**) help to improve alertness in some sufferers. Antidepressants, known as monoamine oxidase inhibitors, may also be of value in the treatment of this condition. These include **phenelzine** and **selegiline**.

> **KEY POINTS:**
>
> - The sleep–wakefulness cycle occurs on a daily basis in most people. It appears to be controlled by activity in many different brain areas, including the ascending reticular formation.
>
> - The stages of sleep can be followed by monitoring an EEG. During deep sleep, the EEG is dominated by slow-wave activity (slow-wave sleep), but this pattern is interrupted several times a night by bouts of rapid eye movement (REM) sleep in which the EEG is desynchronized.
>
> - During slow-wave sleep, there is a slowing of the heart rate and respiratory rate, blood pressure falls and there is extensive relaxation of the somatic muscles.
>
> - During REM sleep, the heart rate and respiratory rate increase, penile erection may occur and rapid movements of the eyes occur. REM episodes are associated with dreaming.

15.6   Circadian rhythms

Although the sleep–wakefulness cycle is the most familiar example of a daily cyclical variation in activity, many aspects of normal physiology are related to time. Indeed, more than 200 measurable physiological parameters have been shown to exhibit rhythmicity. Cycles of periodicity shorter than 24 hours are called **ultradian** rhythms (e.g. the heart beat, the respiratory rhythm) while those longer than 24 hours are called **infradian** rhythms (e.g. the menstrual cycle, gestation). Most biological variables, however, show a periodicity that roughly approximates to the 24-hour day. These are called **circadian rhythms** (from the Latin *circa*, meaning 'around' and *dies* meaning 'a day'). Familiar examples include the core temperature, pulse rate, systemic arterial blood pressure, peak expiratory flow rate, renal activity (as measured by the excretion of potassium) and the secretion of a number of hormones (e.g. the secretion of cortisol by the adrenal cortex).

It is thought that internal biological clocks exist in the brain and other tissues, regulating many physiological processes to a roughly 24-hour periodicity, and that external environmental clues, both physical and social, are able to entrain these biological clocks to a strict 24-hour cycle. The external clues are sometimes called '**zeitgebers**' (the German for 'time-givers').

The exact nature of the intrinsic biological clocks is not clear, but recent experiments using laboratory rats suggest that the body possesses cell groups that may act as rhythm generators in many tissues including the liver, white blood cells, salivary glands and certain endocrine organs. These peripheral 'clocks' appear to be controlled, in turn, by cells in a region of the hypothalamus called the **suprachiasmatic nucleus** (**SCN**), sometimes referred to as the 'master clock' (Figure 15.12). Evidence for this comes from a variety of experiments. Brain slices or cultured neurones from the SCN retain synchronized, rhythmical firing patterns even though they are isolated from the rest of the brain. Furthermore, destruction of the SCN in laboratory rats causes a loss of circadian function, which may be restored by the implantation in the SCN of cells from fetal rat brain. Circadian periodicity of cell division rates has also been demonstrated in cells isolated from

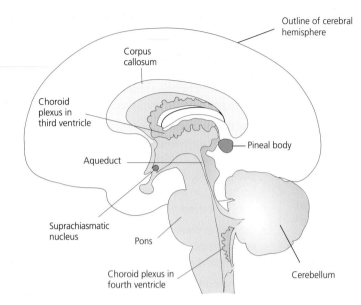

Figure 15.12 A diagram of the brain to illustrate the locations of the suprachiasmatic nucleus (SCN) and the pineal gland.

other tissues including skin and salivary glands, and, most recently, genes have been identified that appear to be involved in the timing of cellular activity. Mutation or loss of these genes can lead to disruption of circadian rhythms.

The **pineal gland** has also been implicated in the regulation of circadian rhythmicity. This gland, which is attached to the third ventricle close to the thalamus (Figure 15.12), secretes a hormone called **melatonin**, which is synthesized from serotonin. The synthesis and secretion of melatonin is increased during darkness and inhibited by daylight. Melatonin appears to induce drowsiness and loss of alertness, perhaps by modifying activity of the SCN neurones. Alterations in the secretion of melatonin, or other related neurochemicals, may be responsible for the condition known as **SAD** (**seasonal affective disorder**) in which the sufferer feels depressed and lethargic during the dark winter months and for which light therapy is often a successful treatment.

Although the physiological mechanisms that underlie biological rhythmicity are not established, there is no doubt that biological rhythms are of considerable significance both for normal daily activity and for clinical medicine. Both shift work (particularly that which involves rotating day and night work) and travel through time zones with its consequent jet lag disrupt the normal circadian rhythms. This makes it difficult for an individual to maintain an active state at times when their natural biological rhythms are preparing the body for inactivity, rest or sleep. Unsurprisingly, more errors of judgement are made at work in the early hours of the morning than at any other time. Conversely, when a worker is going home to rest or sleep following a night shift, the natural rhythms are stimulating a number of physiological processes in preparation for the activities of the day. Cortisol levels rise, along with core temperature, heart rate and blood pressure. These changes interfere with normal sleep patterns.

Many of the physiological parameters that are routinely monitored by clinicians are influenced by the time of day. Heart rate, blood pressure and temperature are obvious examples, but there are many others including blood cell numbers, respiratory values, enzyme activities and blood gas levels. In addition, it is now clearly established that the effectiveness of many drugs differs according to the time of administration as a result of differences in gastric emptying, gut motility, renal clearance rates and the metabolic activity of cells. Indeed, the subject of **chronopharmacology** is becoming an increasingly important aspect of many drug therapies.

KEY POINTS:

- Internal biological clocks exist in the brain and other tissues and regulate the activity of many physiological processes including heart rate, blood pressure, body core temperature and the blood levels of many hormones. Environmental influences appear to entrain these biological clocks to a strict 24-hour (circadian) cycle.

- The pineal gland has been implicated in the regulation of circadian rhythms. This gland secretes a hormone called melatonin. The synthesis and secretion of melatonin is increased during darkness and inhibited by daylight.

15.7 Learning and memory

To survive in the world, all complex animals, including humans, need to learn about their environment. The process of learning is concerned with establishing a store of information that can be used to guide future behaviour. The store of information gained through learning is known as memory. The importance of memory to normal human activity is evident in those suffering from various neurodegenerative diseases, notably Alzheimer's disease (see Box 15.1).

There are essentially three types of learning: simple learning, associative learning and complex learning. **Simple learning** is concerned with the modification of a behavioural response to a repeated stimulus. The response may become weaker as the stimulus is perceived to be of no importance (habituation) or stronger if an unpleasant stimulus is given (sensitization). In **associative learning**, a neutral stimulus is associated with some more important matter. For example, the chiming of a clock can be a reminder that lunch is due, or that it is time to go home. **Complex learning** is diverse in its nature. It includes **imprinting** (the process by which young birds learn to recognize their parents by some specific characteristic), **latent learning** in which experience of a particular environment can hasten the learning of a specific task, and **observational learning** (**copying**).

Cellular mechanisms of learning

All behaviour is determined by the synaptic activity of the CNS and particularly that of the brain. In view of this, it would seem logical to look for changes in synaptic connectivity following some learnt task. This approach has proved feasible in certain lower animals that have simple nervous systems and stereotyped behavioural patterns. Changes in the strength of certain synaptic connections have been shown, for example during the habituation and sensitization of the gill withdrawal reflex of the sea-slug *Aplysia*.

In mammals, the CNS is so complex that this direct association is less easy to demonstrate. Nevertheless, in some brain regions, long-lasting changes in the strength of certain synaptic connections have been reliably found following specific patterns of stimulation. One region that has been the subject of intensive study is the **hippocampus**, which has been implicated in human memory (see below). The efficacy of synaptic connections between neurones of the hippocampus is readily modified by previous synaptic activity. This is precisely what is required of synaptic connections that are involved in learning. The basic neural circuit of the hippocampus consists of a trisynaptic pathway, as shown in Figure 15.13. Brief, patterned stimulation leads to a very long-lasting increase in the efficacy of synaptic transmission in several of these pathways, which lasts for many minutes or even hours (Figure 15.13). This phenomenon has become known as long-term potentiation or LTP.

Since its discovery by T. Bliss and T. Lømo in 1971, hippocampal LTP has become the focus of a great deal of experimental work to determine exactly how the changes in synaptic efficacy are brought about. However, further discussion of the mechanisms involved is outside the scope of this book.

Box 15.1 Alzheimer's disease

Spontaneous and progressive degeneration of neurones in specific areas of the brain or spinal cord is responsible for a number of disorders of the CNS. These include Parkinson's disease, Huntington's chorea, Creutzfeldt–Jakob disease (CJD) and Alzheimer's disease. Alzheimer's disease is the most common cause of dementia in the elderly, with more than 30% of those over 85 years of age showing signs of the disorder, but it may also affect people in their thirties. The disease is characterized by progressive and inexorable disorientation and impairment of memory (generally over a period of 5–15 years) as well as by other defects such as disturbances of language, visuospatial and locomotor function.

The precise causes of Alzheimer's disease are not yet established, but post-mortem examination of the brains of patients has revealed some characteristic anatomical and microscopic histological changes. Gross inspection of an affected brain shows atrophy with generalized loss of neural tissue, narrowing of the cerebral cortical gyri and widening of the sulci. There is also dilatation of the cerebral ventricles. These changes are most obvious in the frontal, temporal and parietal lobes of the brain. The neocortex, basal forebrain and hippocampal areas are particularly affected. The latter area is known to be associated with the formation of memories (see main text).

Microscopic examination of affected tissue reveals the presence of senile plaques, which consist of a core of abnormal protein surrounded by neurofibrillary tangles within the neuronal cytoplasm. The abnormal protein of the senile plaque is known as β-amyloid protein (Aβ), which is formed from the breakdown of an amyloid precursor protein. The neurofibrillary tangles are coarse filamentous aggregates of insoluble filaments made of a protein known as Tau protein. In Alzheimer's disease, an abnormal and insoluble form of Tau protein accumulates as aggregates within the neurones and disrupts normal

microtubule assembly and maintenance. Aβ is believed to increase the formation of neurofibrillary tangles derived from Tau. Plaques and tangles are seen in the brains of normal elderly people but their increased prevalence and specific location within areas such as the hippocampus is diagnostic of Alzheimer's disease.

Although the histological changes characteristic of Alzheimer's disease are well documented, the underlying causes are not established. A number of factors have been identified that may have a role in the development of the disease: around 10% of Alzheimer's disease cases are familial in origin. Furthermore, there is a clear association between Down syndrome (trisomy 21) and neuronal degeneration similar to that seen in Alzheimer's disease. Thus, it is likely that there is a genetic component to this disease. Indeed, it has been shown that mutations at four chromosomal loci are associated with Alzheimer's disease. Unsurprisingly, one of these is located on chromosome 21. The affected gene encodes the amyloid precursor protein whose abnormal breakdown generates the β-amyloid plaques. Further gene mutations also occur at loci on chromosomes 14 and 1, the so-called presenilin genes that are apparently associated with increased production of amyloid in the CNS and with an enhanced rate of apoptosis (programmed cell death).

There is no effective treatment for Alzheimer's disease at present although some drugs have been shown to slow the progress of the disease or to bring about small or temporary improvements in mental function. These include anticholinesterase drugs such as **Aricept**, **Exelon** and **Reminyl** and a glutamate antagonist called **Ebixa**. Considerable research effort is being directed towards the development of drugs that may be able to inhibit the activity of enzymes involved in the formation of β-amyloid.

Memory

Human memory does not act like a tape recorder or a computer hard disk in which our experiences are recorded in an orderly sequence, which is then available

for total recall. Only certain aspects of our experience are remembered; other items (such as phone messages) may be remembered for a short time and then forgotten. Still other trivial incidents are soon beyond recall – the

(a)

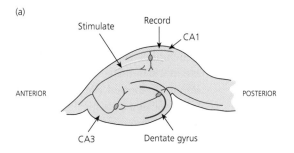

(b)

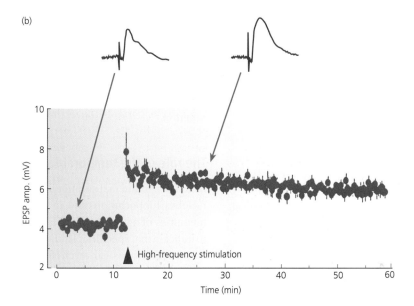

Figure 15.13 Long-term potentiation of synaptic transmission in the hippocampus. (a) The organization of the hippocampus and the position of the electrodes for the records shown in (b). Before the high-frequency stimulation, the excitatory postsynaptic poteatials (EPSPs) evoked by stimulation were about 4 mV in amplitude. After the period of high-frequency stimulation, they were about 6 mV (see inset examples). The increase in amplitude lasted for the duration of the experiment. (Courtesy of T.V.P. Bliss and D.A. Richards.) CA1 and CA3 refer to specific regions of the hippocampus.

precise location of your pen or your spectacle case for example. Even our long-term memories are not exact – no two people will give identical versions of an event they both witnessed. More usually, the salient points are recalled. This shows that memory is a representation of our past experience, not a record of it.

Memory must first begin with our sensory impressions – and at this early stage it will consist of the information passing through the sensory pathways. This will include that part of the sensory experience that reaches the association cortex where information from the various senses is integrated to form our image of the world. This type of memory, sometimes called **immediate memory**, is very short-lived as it is constantly being updated. For this reason, it is assumed that it is encoded in the electrical activity of networks of neurones.

More enduring information storage is classified as short-term memory and long-term memory. Information stored in **short-term memory** may either become incorporated into our permanent, long-term memory store or discarded.

Examples of short-term memory include the remembering of an appointment, the rehearsal of a telephone number before dialling or the need to buy groceries – once the task has been executed then the incident is rapidly forgotten unless it is given some special significance.

Our **long-term memories** include our name, the names and appearance of our family and friends, important events in our lives and so on. In the absence of brain damage or disease, it lasts for our lifetime. This stage of memory requires the remodelling of specific neural connections. The various stages of establishing a memory are summarized in Figure 15.14.

Long-term memory is conveniently divided into procedural memory (also called reflexive memory) and declarative memory. **Procedural memory** has an automatic quality. It is acquired over time by the repetition of certain tasks and is evidenced by an improvement in the performance of those tasks. It is 'knowing how' to do something. Examples are speech in which the vocabulary and grammar of a language are acquired through experience, playing a game, riding a bike or mastering a musical instrument. Once learnt, the performance of these tasks does not require a conscious effort of recall. **Declarative memory** is further subdivided into **semantic memory** (factual knowledge) and memory for events or **episodic memory** (our personal autobiography). Procedural memories are very resistant to disruption, while episodic memories (which relate to specific events) are relatively easily lost.

The neural basis of memory

One of the key questions regarding memory is: where is memory located within the brain? Early animal studies, in which the cerebral cortex was partially or totally removed, showed no evidence of a specific memory being located at a specific site. In these experiments, animals were taught to find their way around a maze and their performance after part of their cortex was removed was compared with their performance before the operation. The results indicated that the loss of performance was related to the total area of cortex removed rather than to removal of a specific area. This suggested that memory is stored in many parallel pathways rather than at a specific site. This is, perhaps, not surprising, as many parts of the brain will be involved in any task that requires extensive co-ordination of sensory and motor function.

The role of the temporal lobe

It has been found that stimulation of points in the temporal lobes, hippocampus and amygdala can evoke specific memories of a vivid nature. This was originally thought to indicate that specific episodic memories are located in the temporal lobes. However, careful re-evaluation of this evidence suggests that this is not the case. The reported experiences tended to have a dream-like quality. Moreover, stimulation of the same site did not always elicit the

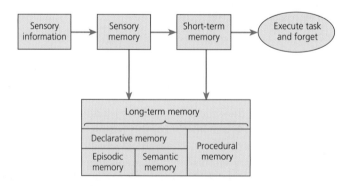

Figure 15.14 A simple schematic drawing showing one view of the organization of memory.

same memory and excision of the area under the electrode did not result in the loss of the memory item that had been elicited by stimulation.

How is memory retrieved?

At present, this question cannot be answered as we do not know where specific memories are laid down in the brain. For procedural memory, it seems plausible that recall begins with the triggering of specific sequences of nerve activity by certain cues. The more a specific act has been rehearsed and refined, the more accurate the performance. This is obviously true of motor acts such as walking, running, etc., but is also true of the use of language, reading music and other similar tasks. The execution of the learnt pattern is therefore the act of recall in this case.

The recall of declarative memory has been studied by observing how people recount stories they have been told. Rather than simply retelling the original with all its detail, they tend to shorten the story to its essentials – in essence, they reconstruct the story. This suggests that the recall of such memories is a specific decoding process.

A deficit in recall is called **amnesia** and is often observed in patients who have received a head injury. If the amnesia is of events before the injury, it is known as **retrograde amnesia**; if it is of events after the injury, it is **anterograde amnesia**. Total amnesia is rare and generally very short-lived so that the amnesia following a head injury gradually passes and the period of amnesia is localized to the period immediately surrounding the incident.

> **KEY POINTS:**
>
> - Learning is the laying down of a store of knowledge that can be used as a guide to future activity. Memory is the name given to that store.
>
> - All learning must involve changes to the connections between the neurones involved in specific tasks.
>
> - Memory is not, in general, localized to very specific sites but is usually distributed through a number of neural networks, each of which is involved in the task. Memory can be divided into short-term memory and long-term memory, which involves the remodelling of specific neural connections.
>
> - Loss of memory is called amnesia.

 # Recommended reading

Carpenter, R.H.S. (2002). *Neurophysiology*, 4th edn. Hodder Arnold: London. Chapters 13 and 14.

Sacks, O. (1986). *The Man Who Mistook His Wife for a Hat*. Pan Books: London.

Springer, S.P. and Deutsch, G. (1989). *Left Brain, Right Brain*, 3rd edn. W.H. Freeman & Co: New York.

Squire, L.P. and Kandel, E.R. (2002). *Memory: From Mind to Molecules*. Freeman: New York.

Medicine

Donaghy, M. (2005). *Neurology*, 2nd edn. Oxford University Press: Oxford. Chapters 21–24.

Self-assessment questions

Which of the following statements are true and which are false? Answers are given at the end of the book (p. 755).

1   a) The frontal lobes are involved in motor control.
    b) Damage to the parietal lobes can lead to a failure of object recognition.
    c) In left-handed people, the faculty of speech is located in the right hemisphere.
    d) Wernicke's area is concerned with the comprehension of speech.
    e) A patient with Broca's aphasia will have paralysis of the lips and tongue.

2   a) The normal EEG of an awake person is dominated by alpha waves.
    b) During deep sleep, an EEG is always dominated by delta waves.

    c) An EEG can be used to monitor the health of the brain.
    d) Heart rate increases during REM sleep.
    e) Delta waves are of smaller amplitude than beta waves.

3   a) Melatonin is secreted by the pituitary gland.
    b) Melatonin secretion is inhibited by daylight.
    c) Cortisol secretion is at its lowest at around 3 a.m.
    d) The menstrual cycle is an example of an ultradian rhythm.
    e) The suprachiasmatic nucleus is located in the hypothalamus.

The endocrine system

16

After reading this chapter you should have gained an understanding of:

- The concept of glands, hormones and target tissues
- The role of the hypothalamus in the regulation of pituitary function
- The main actions of the hormones of the anterior pituitary gland
- The actions of the posterior pituitary gland hormones: oxytocin and vasopressin
- The actions of growth hormone
- The storage and secretion of the thyroid hormones and their role in metabolism
- The role of the adrenal cortical steroid hormones in metabolism and fluid balance
- The adrenal medullary hormones, their role in metabolism and their effects on the cardiovascular system
- The hormonal regulation of mineral metabolism with specific reference to whole-body handling of calcium and phosphate
- The mechanisms involved in the maintenance of plasma glucose levels including glycogen storage, glycogenolysis and gluconeogenesis
- The actions of hormones on plasma glucose regulation
- Diabetes mellitus and its treatment

16.1 Introduction

The endocrine system provides a means of communication within the body that allows cells and tissues distant from one another to send and receive specific signals. Unlike the nervous system, which uses electrical signals to relay information in a very discrete manner to individual cells, the endocrine system employs chemical messengers, which are secreted into the bloodstream from where they can influence many cells simultaneously. These chemical messengers are known as **hormones** (see Chapter 7, p. 82). They provide a slower means of communication than nerve cells, but they are able to influence the activity of many different cells at the same time. It is not surprising, therefore, that hormones are of particular importance in the regulation of co-ordinated processes such as growth, development, metabolism, the maintenance of a stable internal environment (homeostasis) and reproduction.

Unlike the glands associated with the gastrointestinal system or the sweat glands, hormones are synthesized and secreted by glands that have no ducts; the hormones being directly secreted into the blood. Such glands are known as **endocrine glands**. In addition to the classical endocrine glands discussed in this chapter, it is now known that a number of other organs also contain cells that secrete hormones. Examples are the heart, which

secretes atrial natriuretic peptide (ANP), the brain, which secretes specific hormones from the hypothalamus, and the gastrointestinal tract, which secretes a variety of hormones (see Chapter 24). The anatomical location of the principal endocrine glands is shown in Figure 16.1, which also indicates the principal hormones secreted by each gland.

The cells that respond to a specific hormone are called its **target cells**. They possess specific receptors for the hormones to which they respond (see Chapter 7). For example, the cells of breast, vaginal and endometrial tissue of the uterus possess receptors specific to hormones known as oestrogens. Endocrinology, therefore, is the study of the endocrine glands and other hormone-secreting tissues, their hormones and their target cells.

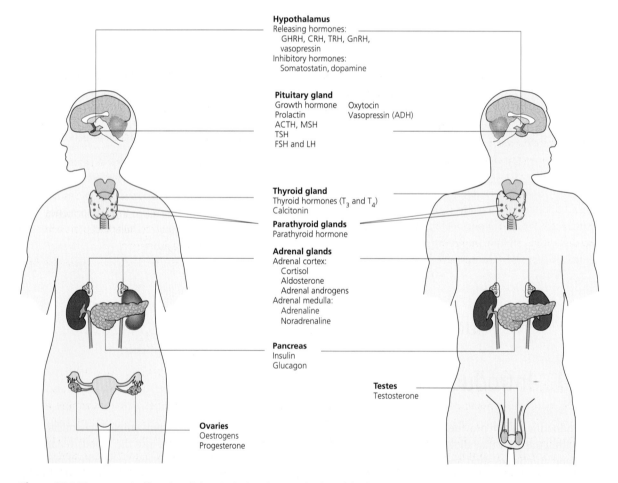

Figure 16.1 The anatomical location of the principal endocrine glands and the hormones they secrete.

16.2 The chemical nature of hormones

Hormones are divided into three categories according to their chemical properties. These are:

- The **steroids**, which are derived from cholesterol. These include hormones secreted by the ovaries, testes and adrenal cortex.
- The **peptides**, which form the largest group and include hormones of the hypothalamus, pituitary gland, pancreas and many others.
- **Derivatives of specific amino acids**, which include the catecholamines secreted by the adrenal medulla and the thyroid hormones, both of which are synthesized from tyrosine.

The chemical nature of hormones influences both their mechanism of action and the way in which they travel in the blood. Peptide hormones and the catecholamines are water soluble and travel in free solution, but the steroids and thyroid hormones are hydrophobic and are transported to their target tissues bound to specific binding proteins, such as cortisol-binding globulin, or to plasma proteins, such as albumin. The steroid and thyroid hormones are cleared much more slowly from the circulation than those carried in free solution, and their effects are more long lasting.

In addition to the circulating hormones, other chemical agents act as local hormones; examples are histamine, the prostaglandins and the inorganic gas nitric oxide. These agents are secreted in response to local factors (e.g. tissue injury) and act on neighbouring cells, reaching their targets by diffusion over relatively short distances (up to a few mm). This is known as **paracrine signalling** and such chemicals are sometimes called paracrine hormones. In some situations, a paracrine agent acts on the cells that secreted it. This is known as **autocrine signalling**. Further details of the principal mechanisms by which hormones and paracrine agents act on their target cells are given in Chapter 7.

KEY POINTS:

- Hormones are released by endocrine glands and travel in the blood to act on target cells elsewhere in the body.

- Hormones fall into three broad classes of chemical compound: steroids (derivatives of cholesterol), peptides (the largest group) and those derived from single amino acids (the thyroid hormones and catecholamines).

- Hormones act by binding to specific receptor proteins in their target cells. Peptides and catecholamines bind to receptors on the plasma membrane, while steroids and thyroid hormones enter cells and bind to intracellular receptors and act to modulate gene expression.

16.3 Measurement of hormone levels in body fluids

In clinical practice, it is often helpful to measure the concentration of hormones in the blood or other body fluids in order to assess endocrine function or make a diagnosis. Modern assay techniques have made it possible to measure the very small quantities of hormones that are normally present in the body fluids. Radioimmunoassays are now available for all the polypeptide, thyroid and steroid hormones while, more recently, sensitive chemiluminescence assays have been developed for measuring peptides. Mass spectrometry is also used to measure steroid hormones. Some hormone assays have been developed for home use, examples of which include the home pregnancy test, which detects human chorionic gonadotrophin (hCG) in the urine, and fertility predictors, which measure fluctuating levels of pituitary and ovarian hormones.

16.4 Patterns of hormone secretion – circadian rhythms and feedback control

A great many physiological parameters, including core temperature, heart rate and blood pressure, vary according to the time of day (a circadian rhythm, see Chapter 15, p. 275). The secretion of many hormones shows a similar pattern. The pituitary hormones adrenocorticotrophic hormone (ACTH), prolactin and growth hormone all exhibit circadian rhythms. Furthermore, many hormones show significant alterations in their rate of secretion in response to stress of all kinds. For these reasons, it is important to take these sources of variation into consideration when interpreting hormone measurements in clinical practice.

The secretion of many hormones is closely regulated by feedback mechanisms. In most cases, the secretion of one hormone is inhibited by a second hormone secreted by the target tissue itself. For example, the thyroid gland secretes thyroid hormones (T_3 and T_4) in response to thyroid-stimulating hormone (TSH) secreted by the anterior pituitary gland. T_3 and T_4 in turn inhibit the secretion of TSH so that the secretion of TSH, and thus the plasma concentration of the thyroid hormones, is kept within narrow limits. Many of the anterior pituitary hormones are controlled by similar feedback loops. This is known as **negative feedback**. Occasionally, the secretion of one hormone is *increased* by a hormone secreted by the target tissue. This is called **positive feedback**. An important example is the preovulatory surge in gonadotrophins, which is brought about by the positive feedback effects of oestrogens on the anterior pituitary (see Chapter 28).

> **KEY POINTS:**
>
> - The secretion of most hormones is under negative feedback control.
> - During the preovulatory phase of the female reproductive cycle, the secretion of the gonadotrophins is regulated by positive feedback.

16.5 The pituitary gland and the hypothalamus

Figure 16.2 illustrates the anatomical relationship between the pituitary gland and the part of the brain to which it is attached by the **pituitary stalk** – the hypothalamus. (The pituitary stalk is also known as the infundibulum.) The pituitary gland is situated in a depression of the sphenoid bone at the base of the skull and consists of two anatomically distinct portions, the **anterior lobe** and the **posterior lobe**. The posterior lobe consists of a thin sliver of tissue known as the **intermediate lobe** and the **neural lobe**, which is connected to the brain via nerve fibres (see below). The pituitary gland is also known as the hypophysis, so the anterior lobe is sometimes called the **adenohypophysis** and the neural lobe the **neurohypophysis**.

The anterior and intermediate lobes are derived from non-neural embryonic tissue that grows upwards from the buccal cavity. It contains at least five different types of cell that secrete a variety of peptide hormones. The most important of these, together with their alternative names, the cell types that secrete them and their principal targets, are listed below:

- **Growth hormone** (GH or somatotrophin) is secreted by cells called somatotrophs. GH has powerful effects on growth and metabolism, which are discussed below (p. 291).
- **Prolactin** is secreted by lactotrophs. It induces mammary gland development and milk synthesis. Its actions will be discussed further in Chapter 29.
- **Adrenocorticotrophic hormone** (ACTH or corticotrophin) is secreted by corticotrophs. ACTH stimulates

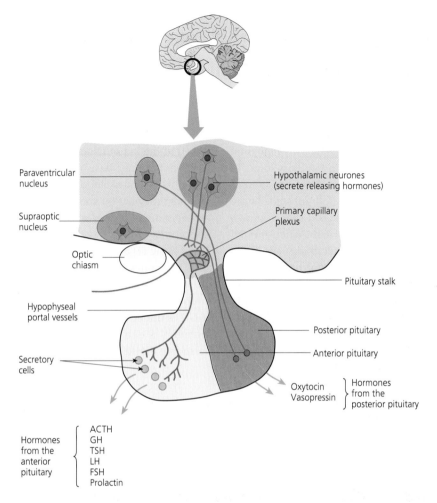

Figure 16.2 The relationship between the hypothalamus and the pituitary gland. Note the prominent portal system that links the hypothalamus to the anterior pituitary gland (the hypophyseal portal vessels). The anterior pituitary has no direct neural connection with the hypothalamus. In contrast, nerve fibres from the paraventricular and supraoptic nuclei pass directly to the posterior pituitary where they secrete oxytocin and vasopressin (ADH) into the bloodstream.

the secretion of the adrenal cortical hormones and also maintains the structure and function of the gland (i.e. it has a trophic effect). Its actions are discussed below on p. 306. The corticotrophs also secrete melanocyte-stimulating hormone (MSH), which may have a role in regulating the activity of melanocytes in the skin.

- **Thyroid-stimulating hormone** (TSH or thyrotrophin) is secreted by thyrotrophs. TSH stimulates the secretion of thyroid hormone and exerts a trophic effect on the thyroid gland (see p. 296).

- The gonadotrophins **follicle-stimulating hormone** (FSH) and **luteinizing hormone** (LH) are secreted by gonadotrophs. Both are involved in the regulation of reproductive function in males and females. Further details of their actions may be found in Chapter 28.

> **KEY POINT:**
>
> - The anterior pituitary gland secretes five peptide hormones and a number of related peptides.

16.6 The role of the hypothalamic hormones in the control of anterior pituitary secretion

The anterior pituitary is functionally connected to the hypothalamus by blood vessels that run in the pituitary stalk and form the **hypothalamic–hypophyseal portal system**. The vessels begin as capillary loops in the hypothalamus, which merge to form a system of parallel veins that pass down the pituitary stalk. In the anterior lobe, these veins form vascular spaces known as sinusoids, which are lined with cells. When stained with suitable dyes, the hormone-secreting cells of the pituitary appear as acidophils (somatotrophs and lactotrophs), basophils (corticotrophs, thyrotrophs and gonadotrophs) and chromophobes (non-secreting cells), as shown in Figure 16.3. Due to their anatomical location, all receive blood flowing from the hypothalamus.

Specific hormones are synthesized in the cell bodies of neurones that lie within the median eminence region of the hypothalamus. In response to appropriate neural stimuli, these hormones are secreted into the hypophyseal portal blood. When they reach the anterior pituitary, they act on specific pituitary cells to modify the secretion of one, or sometimes several, of the anterior pituitary hormones. The rates of secretion of TSH, FSH, LH, ACTH and MSH are all increased by hypothalamic hormones (known as releasing hormones), while the secretion of prolactin is inhibited by a hypothalamic inhibitory hormone called dopamine. The secretion of GH is under dual control by the hypothalamus. Its secretion is enhanced by growth hormone-releasing hormone (GHRH) and inhibited by somatostatin, a

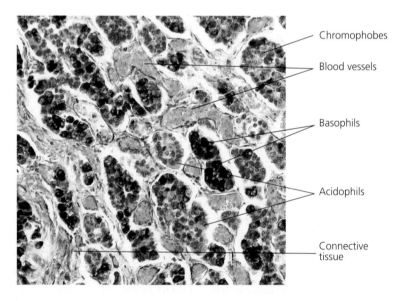

Figure 16.3 A section of the anterior pituitary gland stained with Malory's triple stain. The acidophils have an orange appearance; they include the somatotrophs (which secrete growth hormone) and lactotrophs (which secrete prolactin). The basophils appear dark blue and include the corticotrophs (which secrete ACTH), thyrotrophs and gonadotrophs. The blood vessels appear yellow.

Table 16.1 The hypothalamic releasing and inhibitory hormones

Hypothalamic hormone	Alternative name(s)	Stimulates	Inhibits
Vasopressin	Antidiuretic hormone (ADH)	ACTH	–
Corticotrophin-releasing hormone	CRH	ACTH	–
Luteinizing hormone-releasing hormone	LHRH	LH, FSH	–
	FSH-releasing factor		
	Gonadotropin-releasing hormone (GnRH)		
Thyrotrophin-releasing hormone	TRH	TSH, prolactin	–
Growth hormone-releasing hormone	GHRH	GH	–
Growth hormone-inhibiting factor	Somatostatin	–	GH
	GHIH		Prolactin
			TSH
Dopamine	Prolactin-inhibiting factor (PIF)	–	Prolactin

growth hormone-inhibiting hormone (GHIH). Table 16.1 lists all the hypothalamic-regulating hormones and their target hormones. Many of the anterior pituitary hormones themselves stimulate the secretory activity of other endocrine glands. The hypothalamus, pituitary and the products of the target tissues therefore form a complex functional unit.

> **KEY POINTS:**
>
> - A portal system of blood vessels (the hypophyseal portal vessels) carries hormones from the median eminence region of the hypothalamus to the anterior pituitary.
> - The secretion of the anterior pituitary hormones is regulated by feedback mechanisms involving many of its target organs.

16.7 The posterior pituitary hormones

The posterior pituitary gland secretes two hormones, **oxytocin** and **vasopressin** (also known as **antidiuretic hormone or ADH**). Both are peptide hormones synthesized by neurones whose cell bodies lie within the paraventricular and supraoptic regions of the hypothalamus (Figure 16.4).

Their secretion forms the effector arm of specific neuro-endocrine reflexes, which are outlined below. Like the hormones of the anterior pituitary, oxytocin and vasopressin exert their effects by binding to receptors located on the plasma membranes of their target cells.

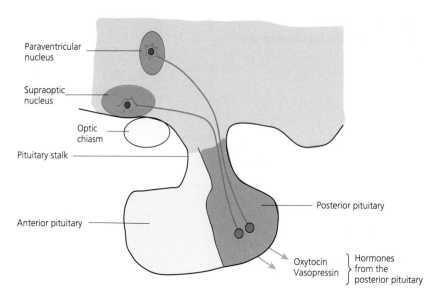

Figure 16.4 A diagrammatic representation of the relationship between the supraoptic and paraventricular nuclei and the posterior pituitary gland (the neurohypophysis). The neurosecretory fibres originate in these nuclei and terminate in the posterior pituitary gland itself.

Actions of vasopressin (ADH)

This hormone has two distinct actions. Firstly, as its name suggests, it is a potent vasoconstrictor, acting in particular on the smooth muscle of arteries supplying the skin, kidneys and the gastrointestinal tract. This effect is important during severe haemorrhage or dehydration when it is essential to maintain blood flow to the brain and other vital organs (see also Chapter 19). Secondly, vasopressin plays an important role in maintaining the normal hydration status of the body through its water-conserving action on the collecting ducts of the kidneys. This antidiuretic action gives vasopressin its alternative name of **antidiuretic hormone** or **ADH**. The action of ADH on the collecting ducts is described in Chapter 23. Put briefly, ADH facilitates the reabsorption of water from the final third of the distal convoluted tubules and the collecting ducts of the kidneys by increasing their permeability to water. As a result, urinary volume is reduced and water is conserved by the body.

When the osmolality of the plasma rises above its normal value, osmoreceptors in the hypothalamus trigger the secretion of vasopressin by the cells of the paraventricular and supraoptic nuclei. Other stimuli that trigger the secretion of vasopressin are a fall in the effective circulating volume (ECV; see p. 712) and central venous pressure (which are sensed by receptors in the atria and great veins), pain, trauma or stress of any kind. In these situations, patients may retain water despite being well hydrated and, as a result, they may become hyponatraemic (sodium depleted). This has important implications for the maintenance of normal fluid balance during the postoperative period.

Disorders of vasopressin secretion

Abnormally high plasma concentrations of vasopressin may result from brain traumas, vasopressin-secreting tumours or certain drug treatments (e.g. the antiepileptic drug phenytoin). The urine of such patients is highly concentrated and as a consequence there is water retention, a reduced plasma osmolality and low plasma sodium (hyponatraemia).

A lack of vasopressin results in a condition called **diabetes insipidus** in which the sufferer is unable to concentrate his or her urine even in dehydration. Thus, copious amounts of very dilute urine are produced continually and there is intense thirst. Pituitary tumours, head injuries or congenital hypopituitarism may cause this condition. Vasopressin deficiency may be treated successfully by the administration of a synthetic form of the hormone, usually by nasal spray.

Actions of oxytocin

This hormone has two main functions in women. During lactation, it stimulates the milk ejection reflex in response to suckling (milk 'let-down'). Oxytocin also promotes contractions of the uterus during labour. These actions are discussed further in Chapter 28. In males, oxytocin appears to have a role in penile erection, ejaculation and sperm progression. A lack of oxytocin will result in a failure to breastfeed an infant. Oxytocin excess has not been demonstrated.

> **KEY POINTS:**
>
> - The posterior pituitary gland is a downgrowth of the hypothalamus. It secretes two peptide hormones, vasopressin and oxytocin, which are structurally similar but have very different actions.
>
> - Vasopressin stimulates the reabsorption of water from the collecting ducts of the kidneys, thereby conserving water. It also constricts vascular smooth muscle.
>
> - Oxytocin stimulates the ejection of milk from the lactating breast and increases the contractile activity of the uterine myometrium during parturition. In males, oxytocin appears to play a role in erection, ejaculation and sperm progression.

16.8 Growth hormone

Growth hormone (GH) is a large peptide secreted by the somatotrophs of the anterior pituitary. As described earlier, the secretion of GH is stimulated by hypothalamic GHRH and inhibited by somatostatin. However, the positive effects of GHRH are normally dominant. GH is released in discrete pulses, the peaks and troughs of GH secretion reflecting increased rates of GHRH and somatostatin secretion, respectively. Furthermore, over a 24-hour period, the rate of secretion fluctuates significantly, with prominent peaks occurring during slow-wave sleep, as illustrated in Figure 16.5. (See Chapter 15 for a discussion of the stages of sleep.)

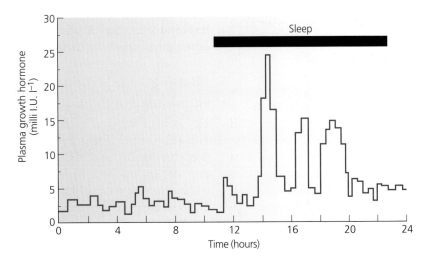

Figure 16.5 Diurnal fluctuations in growth hormone secretion in a normal 7-year-old child. Note the marked pulsatile release of the hormone during sleep.

This circadian pattern is particularly noticeable in childhood.

A number of other physical and psychological factors also influence the secretion of GH and should be considered when carrying out measurements of plasma GH levels in clinical practice. They include anxiety, pain, surgery, cold, haemorrhage, fever, fasting and strenuous exercise. The significance of raised plasma GH in these conditions is not clear, but the glucose-sparing actions of the hormone (see later) may be of value in stressful situations such as these.

Actions of growth hormone

GH is best known for its effect on the growth in stature of children. However, GH also exerts a wide range of additional metabolic effects in both children and adults. Indeed, except for neurones, all cells respond to GH.

Growth-promoting actions of growth hormone

Overall growth of the body involves an increase in size and weight of the body tissues with the deposition of additional protein. A number of hormones are involved

in the regulation of growth and development. Thyroid hormone, for example, is needed for the normal development and maturation of the skeleton and the nervous system during fetal life and early childhood, while the sex steroids underpin the growth spurt that occurs at the time of puberty in adolescents. GH, however, appears to be the predominant regulator of growth between the ages of about 3 years and puberty. Figure 16.6 shows growth patterns seen during childhood and adolescence, and illustrates the differences between girls and boys in relation to the timing of the growth spurt and the overall height achieved. Although there is a very small additional increase in height after puberty, for practical purposes it may be considered that the average boy stops growing at around 17.5 years of age and the average girl at around 15.5 years of age, with a 2-year variability range on either side.

Most body measurements follow approximately the growth curves described for height. The skeleton and muscles grow in this manner, as do many internal organs such as the liver, spleen and kidneys. Certain tissues, however, show different growth characteristics. The reproductive organs, for example, show a significant growth spurt during puberty, while the brain, skull, eyes and

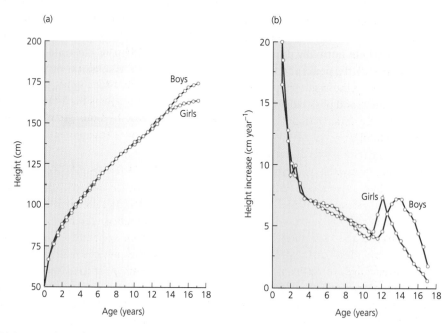

Figure 16.6 (a) A comparison of average growth curves for boys and girls from early childhood to adulthood. (b) The rate at which height increases as a function of age. Note the progressive decline in growth rate until the adolescent growth spurt, which is earlier in girls compared with boys. This is the main reason for the average difference in final height between men and women.

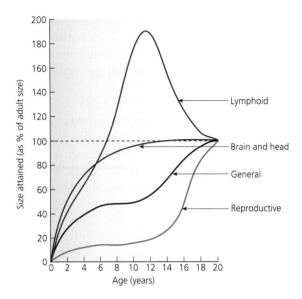

Figure 16.7 Typical growth curves for individual body tissues.

ears develop earlier than any other part of the body. The lymphoid tissue reaches its maximal mass before adolescence and then declines to its adult value. The thymus gland is prominent during childhood (when it plays an important part in the maturation of the immune system), but then atrophies after puberty. Figure 16.7 illustrates the growth curves for particular body tissues.

In order to appreciate how GH influences the growth of the skeleton to achieve an increase in stature, it is necessary to understand the nature and physiology of bone tissue. This is discussed in more detail in Chapter 10 but a brief summary may be helpful here.

Bone consists of an organic matrix (osteoid) strengthened by calcium and phosphate crystals, and three types of cells: the osteoblasts (which secrete osteoid), osteocytes (mature bone cells) and osteoclasts (which secrete enzymes that can dissolve bone to allow remodelling). The skeleton is first laid down as a cartilage model, which is subsequently replaced by bone (ossification), and growth in stature is achieved largely through an increase in length of the long bones of the limbs. The addition of new material during growth takes place at the epiphyseal growth plates, which lie between the main shaft of the bone (the diaphysis) and the two ends (the epiphyses). Lying beneath these plates are stem cells (chondroblasts) that give rise to proliferating clones of cartilage

cells (chondrocytes), which extend into the diaphysis of the bone. As they progress, they mature, enlarge and later degenerate as shown in Figure 10.18. Calcification begins in the innermost layer of cells. Here, osteogenic cells differentiate into osteoblasts that secrete osteoid and lay down bone. At the same time, osteoclasts erode bone within the diaphysis so that the marrow cavity can increase in size as the bone grows. The features of a typical long bone are shown in Figure 10.14.

The growth-promoting actions of GH are both direct and indirect. GH seems to exert a direct stimulatory effect on the chondrocytes, increasing their rate of differentiation and therefore the rate of cartilage formation. Other actions of GH are mediated by two peptide hormone intermediaries called **insulin-like growth factors (IGFs)** because of their structural and functional similarities to insulin. These are IGF-1 and IGF-2, which are synthesized in response to GH, chiefly by the liver but also by other tissues including cartilage and fat. The IGFs, particularly IGF-1, stimulate the clonal expansion of chondrocytes and the formation and maturation of osteoblasts in the growth plates of long bones. They also increase the rate of secretion of collagen matrix by the cartilage cells. IGF-1 also stimulates a variety of cellular processes that are necessary for tissue growth. For example, in many cells (including fibroblasts, muscle and liver), it stimulates the production of DNA and increases the rate of cell division. The growth-promoting effects of GH seem to be crucial for normal growth during childhood, particularly from the age of around 3 years until the end of adolescence. IGF-1 levels reflect the rate of growth during this time and show a marked increase at puberty.

Metabolic actions of growth hormone

In addition to its direct and indirect effects on growth, GH exerts important direct effects on the metabolism of fats, proteins and carbohydrates throughout life. It stimulates the uptake of amino acids and promotes protein synthesis (especially by the cells of the liver, muscle and adipose tissue). The net effects of GH on protein metabolism are an increase in the rate of protein synthesis, a decrease in plasma amino acid concentration and a positive nitrogen balance (the difference between the amount of nitrogen ingested as food and the amount excreted as nitrogenous waste). These effects contribute to the overall promotion of growth by GH.

The effects of GH on carbohydrate and fat metabolism are essentially diabetogenic (i.e. they oppose the actions of insulin). Plasma glucose is elevated in response to GH through a reduction in cellular uptake of glucose and an increase in the rate of conversion of glycogen to glucose by the liver. At the same time, GH promotes the breakdown of stored fat and the release of free fatty acids into the plasma. In this way, a non-carbohydrate source of energy is provided and glucose is spared for use by the brain, which relies almost entirely on glucose for its energy metabolism. This is of particular importance during fasting. Indeed, hypoglycaemia is a very potent metabolic stimulus for GH secretion.

Other metabolic factors known to increase the secretion of GH are a rise in plasma amino acid levels and a fall in the plasma free fatty acid concentration. The effects of all these metabolic stimuli are mediated by changes in the output of hypothalamic GHRH and somatostatin. The secretion of GH appears to be influenced by the negative feedback actions of IGF-1 and GH itself. As GH secretion rises, GHRH output by the hypothalamus is depressed while IGF-1 appears to inhibit GH secretion via a feedback action on GH synthesis by the pituitary gland. The direct and indirect actions of GH and the factors that influence its secretion are summarized in Figure 16.8.

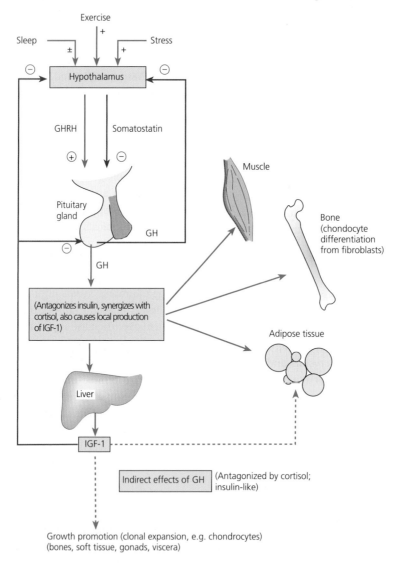

Figure 16.8 A schematic diagram showing the principal actions of growth hormone and the factors that regulate its secretion.

Disorders of growth hormone secretion

The effects of inadequate or excess GH production may be predicted from the preceding discussion. Insufficient secretion of GH during adulthood is not life-threatening and is not usually treated. A lack of GH during childhood, however, results in a condition called pituitary dwarfism in which the child is very short in stature (rarely growing beyond 120–130 cm) although the proportions of the body are normal. GH-deficient children can often be treated successfully by injections of human GH manufactured using recombinant DNA technology. As illustrated in Figure 16.9, considerable catch-up growth may be achieved following this treatment. Occasionally, pituitary dwarfism is caused by a lack of functional GH receptors rather than a lack of the hormone itself. This condition is known as Laron dwarfism.

Pituitary gigantism is a very rare condition caused by oversecretion of GH during childhood (usually as the result of a GH-secreting pituitary tumour). A further condition characterized by excessive growth in stature is Sotos' syndrome, a form of cerebral gigantism, which is caused by an overreaction to GH of its target tissues. Figure 16.10 illustrates the extremes of growth that can occur in pituitary dwarfism and gigantism.

GH-secreting pituitary tumours are much more common in adulthood than in childhood and result in a condition called **acromegaly**. As long bones can no longer increase in length due to the closure of the epiphyseal growth plates, the bones start to thicken in response to elevated plasma GH. This gives the patient a characteristic appearance with large hands and feet, thickened jaw and frontal bones, and coarsened facial features, as illustrated in Figure 16.11. The elevated GH levels of acromegaly also result in a variety of soft tissue and metabolic changes, including an increased risk of diabetes as a result of the anti-insulin effects of GH.

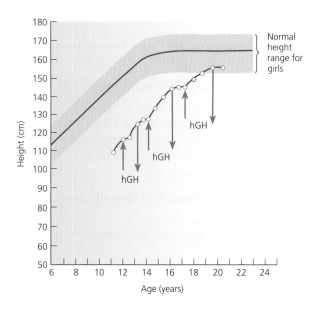

Figure 16.9 The pattern of growth in a girl with isolated pituitary GH deficiency, who was treated with three periods of exogenous human GH (hGH) administration. Note the 'catch-up' growth seen during the treatment periods. The upward- and downward-pointing arrows represent the start and end of each period of treatment. The range of heights for normal girls is indicated by the area shown in red.

Figure 16.10 A pituitary giant and a pituitary dwarf standing next to the well-known British television personality David Frost.

(a)

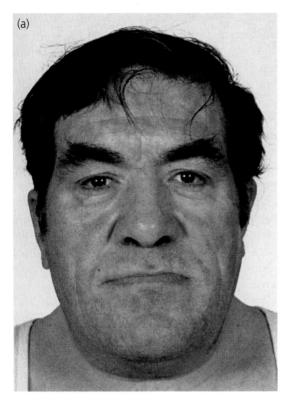

(b)

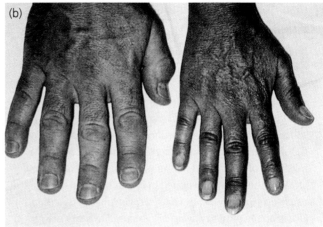

Figure 16.11 A patient with acromegaly caused by the over-production of GH during adulthood. Note the characteristic coarsened features of the face (a) and the enlarged hand (shown on the left in panel (b)).

KEY POINTS:

- Growth hormone (GH) is secreted by the somatotrophs of the anterior pituitary gland in response to GHRH from the hypothalamus. Its secretion is inhibited by somatostatin.

- GH is crucial to normal skeletal growth between the ages of about 3 years and puberty. Skeletal growth occurs in response to IGFs secreted by the liver in response to GH. These factors encourage cartilage cells to divide and enhance the deposition of cartilage at the growth plates.

- The most important metabolic stimulus for GH secretion is low blood glucose.

- GH deficiency in childhood results in pituitary dwarfism while excess GH output results in gigantism. Excessive secretion of GH during adulthood causes acromegaly.

16.9 The thyroid gland

The thyroid gland is situated just below the larynx and adheres to the trachea. It weighs between 10 and 20 g in adults and consists of two lobes connected by a thinner region of tissue, the isthmus (Figure 16.12). It receives its main blood supply from the superior and inferior thyroid arteries, which arise from the external carotid

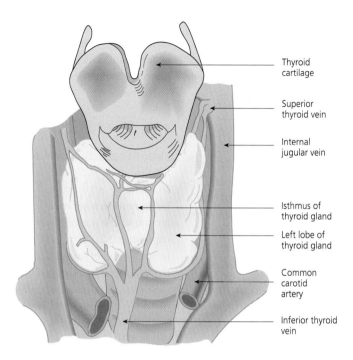

Thyroid
cartilage

Superior
thyroid vein

Internal
jugular vein

Isthmus of
thyroid gland

Left lobe of
thyroid gland

Common
carotid
artery

Inferior thyroid
vein

Figure 16.12 A frontal view showing the location of the thyroid gland and its venous drainage.

and subclavian arteries, respectively. The superior and middle thyroid veins drain into the internal jugular vein while the inferior thyroid vein drains into the left brachiocephalic vein.

The principal histological features of the thyroid gland are illustrated in Figure 16.13. The functional unit of the gland is the follicle of which many thousands are present, ranging in diameter from 20 to 900 μm. Each follicle is surrounded by a basement membrane and consists of a central cavity lined by an epithelial layer of follicular cells, whose character and appearance vary according to the activity of the gland. The central cavity contains colloid, consisting largely of an iodinated glycoprotein called thyroglobulin, which is synthesized by the follicular cells and from which two iodine-containing hormones are formed. These are thyroxine (T_4) and tri-iodothyronine (T_3) and they are concerned mainly with the regulation of metabolism and oxygen utilization and with the control of normal growth and development in the fetus and during childhood.

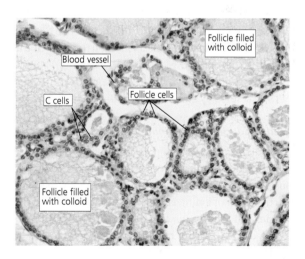

Follicle filled
with colloid

Blood vessel

Follicle cells

C cells

Follicle filled
with colloid

Figure 16.13 The histological appearance of the thyroid gland. Note the follicular cells surrounding the large areas of colloid. The C cells secrete calcitonin and are distinguished from the follicular cells by their pale staining and relatively large, pale nuclei.

Scattered between the follicles are parafollicular cells (also known as C cells), which secrete a peptide hormone called **calcitonin**. This hormone has a role in the regulation of plasma calcium levels and its actions are discussed below (p. 314).

Synthesis of the thyroid hormones

Thyroglobulin is synthesized by the follicular cells from peptide units combined with carbohydrates. The completed glycoproteins are then packaged into small vesicles, which are released into the follicular lumen by exocytosis and added to the colloid store.

Three further steps are then involved in the synthesis of the thyroid hormones from thyroglobulin. These are:

- Iodide trapping (the uptake and concentration of iodide by the gland)
- Iodide oxidation and incorporation into tyrosine within the thyroglobulin of the colloid
- Coupling of iodinated tyrosine molecules to form T_3 and T_4.

Iodide trapping

Thyroid hormones are the only substances in the body that contain iodine and, for normal thyroid hormone synthesis, a minimum of 75 mg of dietary iodine is required each day. Iodide from food and water is absorbed into the blood from the small intestine and is then taken up by the follicular cells of the thyroid gland by a sodium–iodide cotransport mechanism. Iodide transport is stimulated by TSH secreted by the anterior pituitary.

Iodide oxidation and iodination of thyroglobulin

The iodide trapped by the follicular cells is oxidized to reactive iodine through the action of an enzyme located close to the apical membrane of the follicular cell. This reactive form of iodine is released into the lumen of the follicle, which contains the colloid, where it becomes attached to a tyrosine residue within the thyroglobulin. This forms mono-iodotyrosine (MIT). The addition of a second iodine atom forms di-iodotyrosine (DIT). Coupling reactions then occur between the mono- and di-iodotyrosines to form the active hormones tri-iodothyronine (T_3) and thyroxine (T_4), which remain attached to thyroglobulin stored within the lumen of the thyroid follicle. In contrast to most endocrine glands, which do not store appreciable amounts of hormone, the thyroid gland contains several weeks' supply of thyroid hormones.

Secretion of thyroid hormones

Before T_3 and T_4 can be released into the circulation, they must be separated from the thyroglobulin. To do so, colloid is first taken up into the follicular cells where lysosomal enzymes hydrolyze the thyroglobulin to liberate the active hormones. T_3 and T_4 then diffuse into the capillaries that surround the follicle. Thyroglobulin itself is not normally released into the circulation unless the thyroid gland becomes inflamed or damaged. The thyroid hormones travel in the bloodstream largely bound to plasma proteins, especially thyronine-binding globulin, albumin and prealbumin.

Regulation of thyroid hormone secretion

As explained earlier, the activity of the thyroid gland is controlled by TSH from the anterior pituitary gland. The secretion of TSH is, in turn, regulated by hypothalamic thyrotrophin-releasing hormone (TRH). TSH regulates most aspects of the synthesis and secretion of thyroid hormones, including iodide trapping, oxidation, thyroglobulin synthesis and the iodination of thyroglobulin. It also stimulates coupling of MIT and DIT and all the events leading to the secretion of T_3 and thyroxine. The net result of the actions of TSH is to increase the synthesis of fresh thyroid hormone for storage within the follicles as well as to increase the secretion of thyroid hormones into the circulation. TSH is a trophic hormone, which means that it maintains the thyroid gland and its blood supply. In the absence of TSH, the thyroid gland rapidly atrophies (degenerates).

Chemical and nervous mechanisms play a role in regulating the output of TRH and thus TSH and the thyroid hormones. Cold stress is known to stimulate the secretion of thyroid hormones. Within 24 hours of entering a cold environment, there is a rise in circulating levels of thyroxine, which peak a few days later. The circulating levels of thyroid hormone influence the rate of TSH secretion by means of their negative feedback effects on the hypothalamus and the anterior pituitary. Other hormones also appear to alter the output of TSH.

Oestrogens, for example, increase the responsiveness of the TSH-secreting cells of the anterior pituitary to TRH.

KEY POINTS:

- The follicular cells of the thyroid gland secrete two iodine-containing hormones, tri-iodothyronine (T_3) and thyroxine (T_4). A further hormone, calcitonin, is secreted by parafollicular cells.

- T_3 and T_4 play an important role in the control of metabolic rate, maturation of the skeleton and development of the CNS.

- Iodide is concentrated in the follicular cells, where it is oxidized to iodine for incorporation into thyroglobulin.

- T_3 and T_4 travel in the plasma bound to carrier proteins. In the tissues, most of the T_4 is converted to T_3.

- TSH secreted by the anterior pituitary controls all aspects of the activity of the thyroid gland. Its rate of secretion is inhibited by the negative feedback effects of thyroid hormone. TSH secretion is itself regulated by hypothalamic TRH.

Actions of the thyroid hormones

Thyroid hormones have effects on virtually every tissue and also promote an increase in the rate of oxygen consumption and heat production by the body as a whole (see also Chapters 32 and 33). As the thyroid hormones act by modifying gene expression, the biological effects of the thyroid hormones normally take between a few hours and several days to appear. They are summarized in Figure 16.14.

Heat production (calorigenesis)

Most tissues increase their oxygen consumption and heat production in response to thyroid hormone. The principal exceptions are brain, spleen, testes, uterus and anterior pituitary gland, all of which have few receptors for thyroid hormones. This calorigenic action is important in thermoregulation, particularly when adapting to life in a cold environment.

Basal metabolic rate (BMR) is a useful indicator of the thyroid activity of an individual. A high BMR may indicate an overactive thyroid while a reduced BMR may signify hypothyroidism (an underactive thyroid gland). Removal of part or all of the thyroid gland (thyroidectomy) will cause BMR to fall. The precise cellular mechanisms that account for the calorigenic action of thyroid hormones are not known. However, there is an increase in both size and number of mitochondria in responsive cells and an increase in the concentrations of many of the enzymes that participate in cellular respiration.

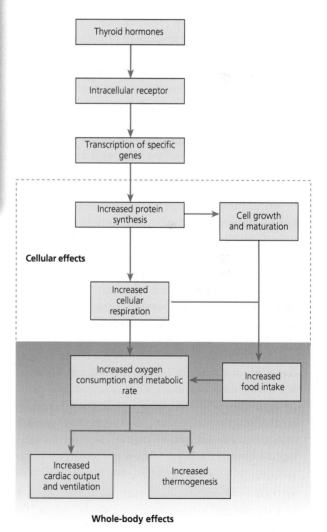

Figure 16.14 A schematic drawing summarizing the principal physiological actions of the thyroid hormones.

The increase in oxygen consumption is accompanied by an increase in body temperature, but this is offset to some extent by a compensatory increase in heat loss through increased blood flow to the cutaneous circulation, by sweating and by an increase in respiratory rate. Cardiac output is also increased in response to thyroid hormones, through an elevation in both heart rate and stroke volume (see Chapter 17).

Actions of thyroid hormones on fats, carbohydrates and proteins

Thyroid hormones modify the metabolism of fats and carbohydrates to provide substrates for energy metabolism. They stimulate the breakdown of stored fat, thereby increasing plasma levels of free fatty acids, and increase the rate of oxidation of free fatty acids. The net result of these actions is a depletion of the body's fat stores, a fall in weight and a fall in plasma levels of cholesterol and other lipids.

Thyroid hormones also affect carbohydrate metabolism in a number of ways. They enhance the rate of intestinal absorption of glucose and the rate of glucose uptake by muscle and adipose cells. Low plasma concentrations of thyroid hormones appear to potentiate the effects of insulin, stimulating both glycogen synthesis and glucose utilization. These actions are hypoglycaemic (i.e. they cause a reduction in plasma glucose). Higher concentrations, however, increase the rates of both glycogen breakdown and gluconeogenesis. High concentrations of thyroid hormone are, therefore, hyperglycaemic (i.e. they bring about an elevation in plasma glucose).

The actions of thyroid hormones on protein metabolism are rather complex and depend upon their plasma concentrations. Low concentrations of T_3 and T_4 stimulate the uptake of amino acids into cells and their incorporation into specific structural and functional proteins (anabolism). Protein synthesis is depressed in hypothyroid individuals. Conversely, high levels of thyroid hormone stimulate the breakdown of protein (catabolism) and can lead to severe weight loss and muscle weakness (thyrotoxic myopathy).

Whole-body actions of thyroid hormones

Thyroid hormones have important effects on growth and maturation. The thyroid gland is capable of synthesizing and secreting thyroid hormones, in response to fetal TSH, by week 11 or 12 of gestation. These hormones are essential for the normal differentiation and maturation of fetal tissues, particularly those of the skeleton and the nervous system. After birth and until puberty, thyroid hormones also stimulate the linear growth of bone. They promote ossification of the bones and maturation of the epiphyseal growth regions (see p. 157).

Adequate thyroid hormone levels during the late fetal and early postnatal periods are vital for the normal development of the CNS. Insufficient thyroid hormone at this time results in severe mental retardation (cretinism), which, if not recognized and treated quickly, is irreversible (see below). In the developed world, newborn babies are now tested routinely for congenital hypothyroidism so that hormone replacement therapy can be administered if necessary. The exact role of thyroid hormones in the maturation of the nervous system is unclear, but it is known that in their absence there is a reduction in both the size and number of cortical neurones, deficient myelination of nerve fibres and a reduction in the blood supply to the brain.

Many of the actions of thyroid hormones are shared by the sympathetic nervous system. This physiological link is emphasized by the fact that drugs that block adrenergic β-receptors (see p. 223) also lessen many of the cardiovascular and central nervous manifestations of hyperthyroidism.

KEY POINTS:

- Most tissues increase their oxygen consumption in response to thyroid hormones. This increase in metabolic rate is important in the maintenance of body temperature.

- The metabolic actions of thyroid hormones are somewhat complex and dose dependent: low concentrations tend to be hypoglycaemic while higher concentrations stimulate glycogenolysis and gluconeogenesis. Low levels of thyroid hormone stimulate protein synthesis while higher concentrations are catabolic.

Principal disorders of thyroid hormone secretion

Defects of thyroid function are among the most common endocrine disorders, affecting 1–2% of the adult population. An overactive thyroid leads to **hyperthyroidism** while an underactive thyroid causes **hypothyroidism**. The principal consequences of both of these extremes may largely be predicted from the preceding account of normal thyroid function.

Clinical features of hyperthyroidism

Most of the manifestations of an overactive thyroid gland are related to the increase in oxygen consumption and increased utilization of metabolic fuels associated with a raised metabolic rate, and to a parallel increase in sympathetic nervous activity. The diagnosis is confirmed by measurement of plasma levels of T_3 and T_4, which become elevated. The most common cause of hyperthyroidism is Graves' disease. This is an autoimmune disorder in which an abnormal antibody stimulates the TSH receptors. The disease is characterized by a swelling of the thyroid gland (goitre) and exophthalmos, which is well illustrated in Figure 16.15.

Treatments for hyperthyroidism include surgical removal of part, or all, of the gland and the ingestion of radioactive iodine to selectively destroy the most active thyroid cells. The output of thyroid hormones can also be reduced by a number of antithyroid drugs. The most widely used are the **thionamides**. These prevent the synthesis of thyroid hormones by inhibiting the reactions needed for the oxidation of iodide and the coupling reactions that produce di-iodothyronine. Examples include **carbimazole** and **propyluracil**. These drugs are given orally and accumulate in the thyroid gland. Their effects are usually delayed by 3–4 weeks, the time taken for the hormone already stored within the gland to be used up.

Clinical features of hypothyroidism

Hypothyroidism may result from iodine deficiency, from a primary defect of the thyroid gland or as a consequence of a reduction in the secretion of either TRH or TSH. The most common cause is an autoimmune disease called **Hashimoto's thyroiditis**, which causes destruction of the thyroid gland. In adults, the full-

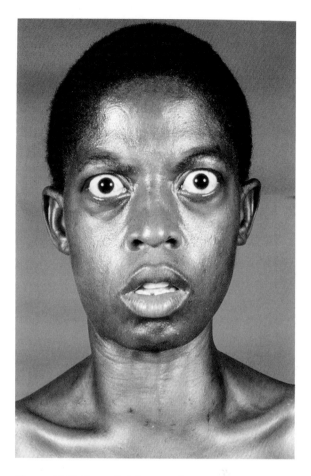

Figure 16.15 The typical features of a patient with hyperthyroidism. Note the bulging eyes (exophthalmos) and retraction of the eyelids.

blown hypothyroid syndrome is called **myxoedema** (literally 'mucous swelling'). The diagnosis is confirmed by low plasma levels of T_3 and T_4. Depending upon the cause, hypothyroidism may be reversed by iodine supplements or by the administration of synthetic thyroid hormones. Synthetic T_4, administered orally, is the treatment of choice and its efficacy is assessed by monitoring plasma TSH levels, which fall to normal (as a result of negative feedback inhibition) once the optimum dose is achieved. Synthetic T_3 is more rapidly acting and may be used in cases of hypothyroid coma. Figure 16.16 shows the typical physical appearance of a patient with

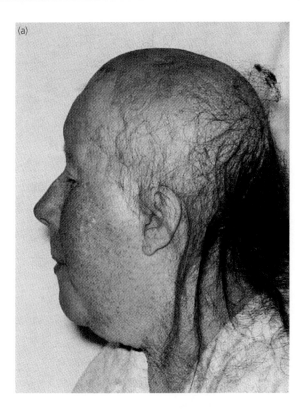
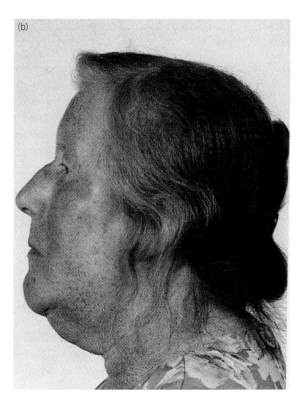

Figure 16.16 (a) A patient suffering from hypothyroidism (myxoedema) manifesting typical facial puffiness, thinning hair and dull appearance. (b) The same patient after treatment with thyroid hormone, which has reversed the physical symptoms evident in (a).

hypothyroidism before and after hormone replacement therapy.

Thyroid hormones are essential for normal development of the skeleton and the nervous system. Congenital hypothyroidism (cretinism) affects approximately 1 in 4000 infants but may be treated successfully with hormone replacement if diagnosed in time (see above). Figure 16.17 shows examples of a hypothyroid infant, a 17-year-old cretin (untreated hypothyroidism) and a normal 6-year-old child.

KEY POINTS:

- Excessive thyroid hormone secretion (hyperthyroidism) results in an increased BMR, sweating, increased heart rate and weight loss.

- Insufficient secretion of thyroid hormone (hypothyroidism) in early childhood results in cretinism; in adults it is manifest as a reduced BMR with bradycardia, lethargy, cold sensitivity and a range of other metabolic abnormalities.

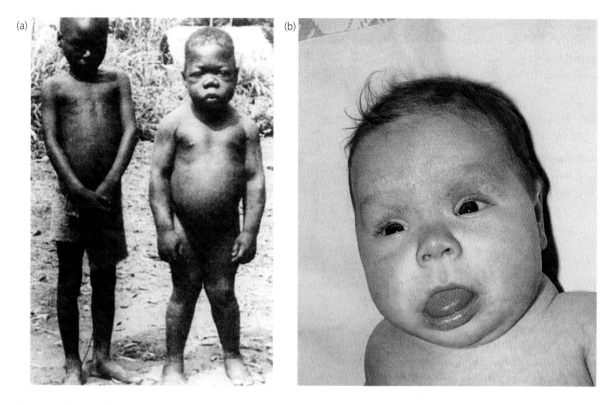

(a)

(b)

Figure 16.17 Typical examples of children suffering from hypothyroidism. (a) A 17-year-old girl cretin (on the right) standing next to a normal 6-year-old child. (b) The appearance of a baby suffering from congenital hypothyroidism, which was successfully treated by hormone replacement.

16.10 The adrenal glands

The adrenal glands are located above the kidneys. They are enclosed in a fibrous capsule surrounded by fat (Figure 16.18). Each adult adrenal gland weighs between 6 and 10 g and is composed of two distinct types of tissue: an outer region, the adrenal cortex, and an inner medulla, which is paler in colour. The adrenal glands receive a rich arterial blood supply from branches of the aorta, the renal arteries and the phrenic arteries. Blood flows from the outer cortex to the medulla, where it drains inwardly into venules. The right adrenal vein drains directly into the inferior vena cava while the left drains into the left renal vein.

The adrenal cortex secretes steroid hormones while the medulla secretes the catecholamines **adrenaline** (epinephrine) and **noradrenaline** (norepinephrine). The secretion of the adrenal cortical hormones is regulated by

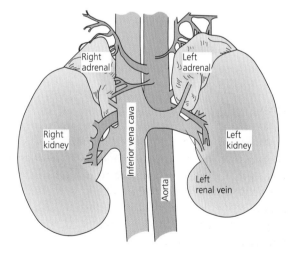

Figure 16.18 The anatomical location of the adrenal glands and the organization of their blood supply. Note that the arterial supply is via many small arteries that originate from the aorta. The venous drainage is via a large central vein.

ACTH secreted by the anterior pituitary while the adrenal medullary hormones are secreted in response to activation of the sympathetic nerve supply to the gland. Thus, each adrenal gland is, in effect, structurally and functionally two endocrine glands in one.

> **KEY POINTS:**
> - The adrenal glands are composite glands that consist of an outer cortex and an inner medulla.
> - The adrenal cortex secretes steroid hormones.
> - The adrenal medulla secretes adrenaline and noradrenaline.

The adrenal cortex

There are three morphologically distinct zones of cells within the adrenal cortex (Figure 16.19). Each zone secretes different steroid hormones (adrenocorticosteroids) synthesized from cholesterol. The cells of the outer **zona glomerulosa** secrete **mineralocorticoids**, which control electrolyte balance. The cells of the middle **zona fasciculata** secrete **glucocorticoids**, which play an important role in regulating glucose metabolism, and the cells of the inner **zona reticularis** secrete **sex steroids**. The zona

fasciculata is the biggest zone, occupying around 75% of the adrenal cortex.

Hormones of the adrenal cortex

Aldosterone is the principal **mineralocorticoid** secreted by the adrenal cortex. Like other steroid hormones, aldosterone is not stored within the gland but is synthesized as required and diffuses out of the cells of the zona glomerulosa immediately after synthesis.

In humans, the principal **glucocorticoid** is **cortisol**, but **cortisone** and **corticosterone** are also produced in small amounts. Like aldosterone, cortisol is not stored in the gland but is synthesized and secreted on demand. Cortisol binds to several different plasma proteins so that only 5–10% of the circulating hormone is in a free (active) form.

The principal sex steroids produced by the cells of the zona reticularis are **dehydroepiandrosterone** and **androstenedione**. These are weak androgens but are converted to the more powerful testosterone when they reach peripheral tissues. In males, the quantities of androgens secreted by the adrenal cortex are insignificant in comparison with those produced by the testes. In females, however, the adrenal glands secrete about half of the total androgenic hormone requirement. Similarly, the oestrogenic hormones secreted by the adrenal cortex are quantitatively insignificant in women (at least until after the menopause).

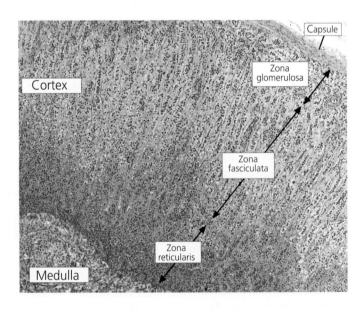

Figure 16.19 A section through the cortex and medulla of the human adrenal gland. Note the three zones of the adrenal cortex, the cells of which secrete steroid hormones. The chromaffin cells of the medulla secrete the catecholamines adrenaline and noradrenaline.

Principal actions of the corticosteroids

Effects of aldosterone

The actions of aldosterone are discussed in Chapter 23. Normally between 0.1 and 0.4 μmol of aldosterone are secreted each day. In the absence of this hormone, death occurs within a few days unless prevented by the therapeutic administration of salt or by hormone replacement. Aldosterone acts to conserve body sodium by stimulating the reabsorption of filtered sodium in exchange for potassium in the distal nephron. Failure of aldosterone secretion results in a fall in plasma sodium and chloride levels and a rise in potassium. The extracellular fluid volume and the blood volume fall, with a subsequent drop in cardiac output (see Chapter 17), which, if uncorrected, may prove fatal. Aldosterone secretion is regulated by the renin–angiotensin system (see pp. 480–481).

Effects of cortisol

Around 30–80 μmol of cortisol are secreted into the bloodstream each day. Cortisol is essential to life. It has a wide range of physiological effects throughout the body, the most important of which concern the metabolism of carbohydrates, proteins and, to a lesser extent, fats. Cortisol also plays a crucial role in the body's response to a variety of stressful stimuli: it has immunosuppressive, anti-inflammatory, antiallergic and aldosterone-like actions.

The metabolic actions of cortisol are summarized in Figure 16.20. In general terms, the actions of cortisol oppose those of insulin, that is they promote an increase in plasma glucose concentration. This increase is achieved in two ways. Firstly, cortisol stimulates the mobilization of amino acids from protein (in particular from muscle tissue), thus providing the liver with the raw materials for gluconeogenesis (the synthesis of glucose from non-carbohydrate sources). As a result, glycogen stores are initially built up while any excess glucose is released into the plasma. Secondly, cortisol inhibits the uptake and utilization of glucose by those tissues in which glucose uptake is insulin-dependent. Cortisol also increases the appetite and stimulates lipolysis (fat breakdown) in adipose tissue and the release of fatty acids into the blood for those tissues that can utilize them.

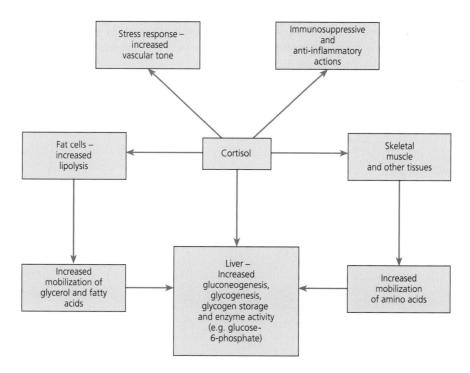

Figure 16.20 The principal physiological actions of cortisol.

Stress may broadly be defined as anything that challenges the normal homeostatic balance of mind or body and includes physical trauma, extreme environmental conditions, infection, and mental and emotional trauma. Cortisol counteracts many of the effects of stress on the body. For example, cortisol increases vascular tone and blocks the processes that lead to inflammation in damaged tissue. For this reason, the hormone has proved to be of value in the treatment of inflammatory conditions such as rheumatoid arthritis and in the treatment of severe asthma. High levels of cortisol also suppress the normal immune response to infection and promote a feeling of euphoria, an effect that may have a role in helping to mitigate the effects of stress. This also explains why the sudden withdrawal of prescribed steroidal treatments can lead to severe depression. Finally, fetal cortisol plays a vital role in the maturation of many organs and is particularly important in the stimulation of pulmonary surfactant production (see Chapter 29).

Adrenocortical sex steroids

Unlike the glucocorticoids and mineralocorticoids, the adrenal sex steroids are not essential for life. The major androgen in males is testosterone secreted by the testes, the adrenal cortex being a minor source. In females, testosterone is formed in peripheral tissues from androstenedione secreted by the adrenal medulla and ovaries. The androgens play a role in the develop-

ment of secondary sexual characteristics such as the growth of axillary and pubic hair and may have a role in the growth spurt seen in middle childhood (around the age of 7).

The regulation of steroid hormone secretion by the adrenal cortex

The regulation of aldosterone secretion is discussed fully in Chapter 23. The synthesis and secretion of the glucocorticoids and, to a lesser extent, the sex steroids are controlled by ACTH secreted by the anterior pituitary gland in response to hypothalamic corticotrophin-releasing hormone (CRH, see pp. 288–289).

A typical negative feedback system regulates glucocorticoid secretion. This means that an increase in plasma cortisol concentration will inhibit the output of CRH and ACTH and thereby reduce the secretion of cortisol. ACTH, and thus cortisol, shows a distinct circadian (24-hour) pattern of secretion, related to the sleep–wakefulness cycle. The concentration of cortisol in the plasma is at its lowest at around 3 a.m. but rises steeply to reach a maximum between 6 and 8 a.m. before slowly falling during the rest of the day (Figure 16.21). This normal rhythm of cortisol release is interrupted by acute stress of any kind as a result of the direct stimulation of CRH secretion. Figure 16.22 summarizes the mechanisms involved in the regulation of cortisol secretion.

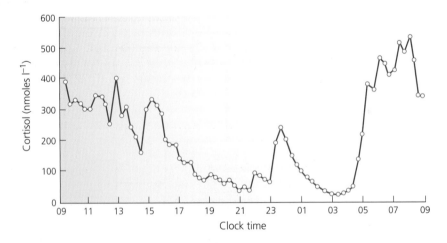

Figure 16.21 The diurnal variation in plasma cortisol. The plasma levels rise during the early part of the day and are lowest around 3 a.m. This reflects the pattern of ACTH secretion by the anterior pituitary.

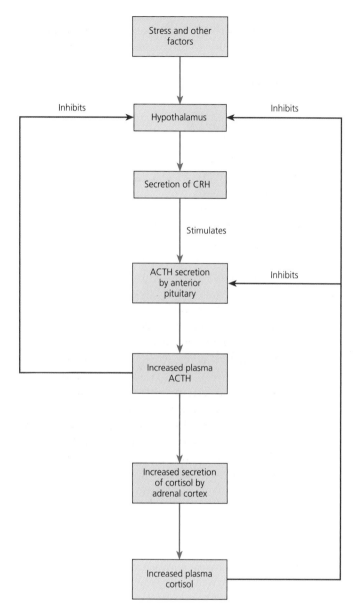

Figure 16.22 A flow chart showing the factors that regulate the secretion of glucocorticoids. Note that the red arrows represent negative feedback inhibition. Circulating glucocorticoids inhibit the secretion of corticotrophin-releasing hormone (CRH) by the hypothalamus and the secretion of ACTH by the anterior pituitary. Circulating ACTH also probably inhibits the secretion of CRH.

KEY POINTS:

- Aldosterone is the chief mineralocorticoid and acts to conserve body sodium by stimulating its reabsorption in exchange for potassium in the distal nephron.

- Cortisol is the dominant glucocorticoid and is essential to life. It stimulates gluconeogenesis and glycogen production as well as lipolysis.

- Glucocorticoid secretion is controlled by pituitary ACTH, which shows a distinct circadian rhythm.

Principal disorders of adrenal cortical hormone secretion

Overproduction of cortisol (Cushing's syndrome)

Excessive cortisol secretion may occur as the result of an adrenal tumour, an ACTH-secreting tumour of the anterior pituitary or hypersecretion of CRH from the hypothalamus. It may also be caused by the ectopic secretion of ACTH (e.g. by certain types of lung tumour). Prolonged treatment with steroid medications may have a similar effect.

Patients with Cushing's syndrome have a characteristic appearance, which can be seen in Figure 16.23. There is a redistribution of fat with increased deposition around the trunk, neck ('buffalo hump') and face, but wasting of muscle tissue, particularly in the limbs, as a result of increased protein mobilization. These changes contribute to the 'melon on toothpicks' appearance that is typical of this condition. Other characteristics of excess cortisol include abnormal pigmentation, changes in carbohydrate and protein metabolism, thin skin with easy bruising, purple abdominal striae (stripes), poor wound healing and predisposition to infection. Treatment is usually by surgical removal of a pituitary or adrenal tumour or, if the tumour is inoperable, by the administration of drugs such as **ketoconazole** and **aminoglutethimide**, which inhibit the synthesis of cortisol by the adrenal cortex.

Overproduction of aldosterone

Excessive secretion of aldosterone may result either from hyperactivity of the cells of the zona glomerulosa (**Conn's syndrome**) or from excessive renin secretion (e.g. as a result of renal artery stenosis). The consequences of aldosterone excess include sodium retention, loss of potassium (hypokalaemia) and raised plasma pH (alkalosis). As a result of sodium retention, the plasma volume is increased and this may cause hypertension. Treatments include surgical removal of an aldosterone-secreting tumour, potassium replacement and the administration of diuretics such as **spironolactone** or **amiloride**.

Overproduction of adrenal androgens

Excess androgen secretion by the adrenal cortex is most often the result of overproduction of ACTH, either by the pituitary itself or by an ACTH-secreting tumour. Occasionally it is caused by an adrenal tumour or an abnormality of steroid synthesis by the gland. Androgen excess

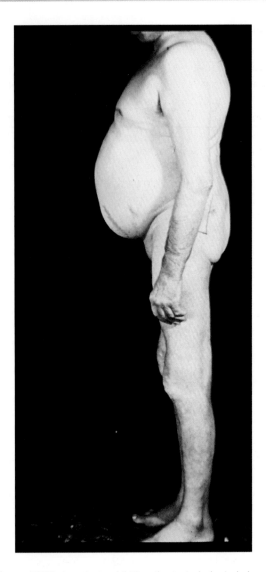

Figure 16.23 A patient exhibiting the typical physical characteristics associated with Cushing's syndrome. Note the thin legs, large torso and purple marks (striae) on the lower abdomen.

brings about a number of distressing physical changes in both men and women. The principal signs are acne, frontal baldness and hirsutism (excessive growth of facial and body hair). Males may experience a reduction in testicular volume due to negative feedback inhibition of gonadotrophin output by the adrenal androgens, while women may show enlargement of the clitoris. Excessive adrenal sex steroid secretion in children may bring about precocious puberty (sexual maturation before the age of 8).

Underproduction of adrenal cortical hormones

The major cause of adrenal cortical insufficiency is **Addison's disease**, a comparatively rare condition caused by damage to the adrenal glands, autoimmune disease or pituitary damage. The secretion of both cortisol and aldosterone is depressed, giving rise to a wide range of symptoms, many of which may be predicted from the preceding account of the actions of these hormones. People with Addison's disease usually show progressive weakness, lassitude and weight loss. Their plasma glucose and sodium levels fall and hypotension and dehydration are common. A characteristic sign is increased pigmentation of the skin and mucosal membranes of the mouth. The usual treatment for this condition is corticosteroid replacement therapy.

KEY POINTS:

- Excess cortisol secretion gives rise to Cushing's syndrome in which there is wasting of muscle tissue and a redistribution of body fat.

- Excess aldosterone leads to sodium retention, loss of potassium (hypokalaemia) and alkalosis (raised plasma pH). As a result of sodium retention, the plasma volume is increased and this may cause hypertension.

- Androgen excess results in distressing physical changes in both males and females. In children, excessive adrenal sex steroid secretion may bring about precocious puberty.

- Addison's disease is the result of insufficient secretion of aldosterone and cortisol by the adrenal cortex. There is progressive weakness, lassitude and weight loss.

The adrenal medulla

The inner (medullary) region of the adrenal gland secretes the catecholamine hormones adrenaline and noradrenaline into the circulation in response to stimulation of the splanchnic nerve fibres that innervate the gland. About 80% of the secreted catecholamine is adrenaline and 20% is noradrenaline. Secretion occurs as part of a generalized sympathetic stimulation and is of particular importance in preparing the body for coping with acute stress (the fight–flight–fright response). Circulating catecholamines are inactivated within minutes by the enzymes catechol-O-methyl transferase and monoamine oxidase in tissues such as the liver, kidneys and brain.

Actions of adrenaline and noradrenaline

Adrenaline (epinephrine) and noradrenaline (norepinephrine) have somewhat different effects, which are summarized in Table 16.2. Both hormones raise the systolic blood pressure by stimulating both heart rate and stroke volume (and therefore cardiac output). Adrenaline, however, reduces diastolic pressure as a result of causing vasodilatation in certain blood vessels, particularly those of skeletal muscle, while noradrenaline opposes this effect by causing a more generalized vasoconstriction (see below). Adrenaline also acts as a bronchodilator and reduces gut motility. Adrenaline exerts important metabolic effects, which are similar to those of the thyroid hormones (see above). It promotes the breakdown of glycogen in the liver, lipolysis, oxygen consumption and calorigenesis. Noradrenaline is a potent stimulator of lipolysis but has little effect on glycogen breakdown.

Adrenaline interacts primarily with β-adrenoceptors while noradrenaline binds preferentially to α- and β_1-adrenoceptors. In clinical medicine, both agonists and antagonists of the α and β receptors are widely used to treat specific disorders (see p. 223). For example, β-blockers (antagonists), such as atenolol, have been prescribed to reduce cardiac output in the treatment of high blood pressure and β_2-agonists, such as salbutamol, are administered to asthmatic patients to bring about bronchodilatation. For more information on α and β receptors see Chapter 13.

Control of adrenal medullary secretion

A number of stimuli increase the rate of catecholamine secretion from the adrenal medulla. These include hypoglycaemia, cold, haemorrhage and hypotension. Secretion may also accompany emotional reactions such as fear, anger, pain and sexual arousal. All of these responses are mediated by the splanchnic nerve supply to the adrenal medulla.

Disorders of adrenal medullary secretion

While underproduction of adrenal medullary hormones is not clinically significant, excessive secretion of catecholamines can have serious consequences. It may

Table 16.2 The efficacy of the catecholamine hormones in modulating various physiological processes

Adrenaline > Noradrenaline	Noradrenaline > Adrenaline
↑ Glycogenolysis (β_2)	↑ Gluconeogenesis (α_1)
↑ Lipolysis (β_3)	
↑ Calorigenesis (β_1)	
↑ Insulin secretion (β_2)	↓ Insulin secretion (α_2)
↑ Glucagon secretion (β_2)	
↑ K$^+$ uptake by muscle (β_2)	
↑ Heart rate (β_1)	
↑ Contractility of cardiac muscle (β_1)	
↓ Arteriolar tone in skeletal muscle (β_2)	↑ Arteriolar tone in non-muscle vascular beds (α_1) leading to vasoconstriction and ↑ blood pressure
	↑ Tone in gastrointestinal sphincters (α_1)
↓ Tone in non-sphincter gastrointestinal smooth muscle (β_1)	↓ Tone in non-sphincter smooth muscle (α_1)
↓ Tone in bronchial smooth muscle (bronchodilatation) (β_2)	↑ Tone in bronchial smooth muscle (bronchoconstriction) (α_1)

The adrenoceptors mediating the various effects are shown in brackets. An *up arrow* indicates an increase, and a *down arrow* a decrease, in the specified physiological process. For a discussion of adrenoceptor subtypes, see Chapter 13.

arise as the result of a tumour of the medullary tissue called a **phaeochromocytoma**. The principal symptoms of this condition are severe hypertension, which is often episodic, hyperglycaemia, a raised metabolic rate, chest tightness, tremor, sweating, weakness and anxiety. Surgical removal of the tumour, although a difficult procedure, is the usual treatment, although many of the symptoms can be alleviated by the use of drugs such as **phenoxybenzamine** and **propranolol**, which block the catecholamine receptors.

> **KEY POINTS:**
> - The adrenal medulla secretes adrenaline and noradrenaline as part of a general sympathetic response to stress.
> - Both hormones cause increases in heart rate, contractility and cardiac output.

16.11 Hormonal regulation of plasma calcium and phosphate

Calcium and phosphate are required for a wide variety of cellular functions, so it is essential that the plasma levels of these minerals are maintained within narrow limits.

The endocrine system plays a key role in this regulation via parathyroid hormone, vitamin D metabolites and calcitonin.

Calcium balance

The adult human body contains around 1 kg of calcium, most of which is found in the bones in the form of calcium phosphate crystals. The skeleton therefore acts as a large reservoir of calcium that can be mobilized by various hormones. Calcium is also a major component of the teeth and connective tissue. It is involved in a number of important regulatory functions, including blood clotting, muscle contraction, cell–cell adhesion, the control of neural excitability and the secretion of both neurotransmitters and hormones. Total plasma calcium is 2.3–2.4 mmol l^{-1}, of which around half is free (ionized) and half is bound to plasma proteins or to anions such as phosphate. By contrast, the free intracellular concentration of calcium is maintained at about 0.1 μmol l^{-1} by a variety of mechanisms (see Chapters 3 and 4).

To maintain the calcium balance of the body, the calcium intake from the diet must match the losses in the faeces and urine, as shown in Figure 16.24. Daily oral intake of calcium usually lies between 200 and 1500 mg per day. For a typical intake of 1000 mg, around 350 mg will be absorbed into the extracellular fluid from the small intestine. However, as the intestinal secretions contain calcium, net absorption is only around 150 mg per day, so that 850 mg of calcium is lost from the body each day in the faeces. Furthermore, about 150 mg of calcium is excreted in the urine each day and small amounts are also lost in the sweat and saliva. Thus, the kidneys and the intestine are important organs in the regulation of the entry and exit of calcium to and from the plasma.

In addition, the skeleton provides a large reservoir of calcium. Although about 99% of the skeletal calcium is in the form of stable bone, which is not readily exchangeable with the plasma, the remaining 1% is in the form of simple calcium phosphate salts, which can readily be released into the plasma if necessary. Similarly, calcium may be added to this pool from the plasma if calcium intake is in excess of that which can be excreted by the kidneys. The skeleton therefore acts as a kind of 'buffer' for plasma calcium.

Phosphate balance

Like calcium, phosphate is required for a number of cellular functions. It is an important component of cell membranes and is needed for the synthesis of DNA and RNA. Phosphate metabolites also play a key role in energy metabolism. The plasma level of inorganic phosphate is around 2.3 mmol l^{-1}, although its concentration is not as closely regulated as that of calcium. The total phosphate content of a 70 kg ('standard') man is about 770 g of which 75–90% is contained within the skeleton in combination with calcium. Daily intake of phosphate is around 1200 mg, of which about a third is excreted in the faeces. The balance is excreted in the urine where it helps to buffer hydrogen ions secreted by the distal tubule.

Hormones involved in the regulation of plasma calcium and phosphate

Three hormones are responsible for the minute-to-minute regulation of plasma mineral levels. These are:

- 1,25-dihydroxycholecalciferol (a metabolite of vitamin D)
- Parathyroid hormone (PTH)
- Calcitonin.

They are released in response to a variety of physiological stimuli related to changes in plasma calcium, and exert their actions on bone, kidney and the intestine.

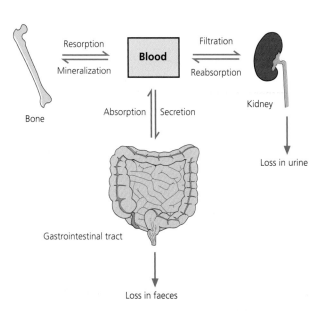

Figure 16.24 The main factors involved in calcium homeostasis.

The role of vitamin D in calcium metabolism

Vitamin D (or cholecalciferol) is a fat-soluble vitamin (see Chapter 31) obtained both from the diet and by the action of sunlight on 7-dehydrocholesterol in the skin. It is metabolized to a hormone called 1,25-dihydroxy-cholecalciferol (also known as **calcitriol** or 1,25-DHCC) in the liver and kidneys. The main action of 1,25-DHCC is to stimulate the absorption of ingested calcium from the gut through a direct effect on calcium transport across the intestinal mucosa. Phosphate absorption is also enhanced. The raised plasma calcium levels resulting from the stimulation of intestinal absorption create favourable conditions for the mineralization of bone. However, 1,25-DHCC also stimulates the activity of both osteoblasts and osteoclasts (see Chapter 10), the effect of which is to facilitate the remodelling of bone.

Effects of vitamin D deficiency and excess

The abnormalities caused by a deficiency or an excess of vitamin D are discussed in Chapter 31. Briefly, vitamin D deficiency gives rise to a failure of the normal calcification of bone. In children, the condition is called **rickets** and in adults, osteomalacia. Figure 31.2 shows a child with rickets and illustrates the characteristic deformities of the legs. In osteomalacia, there is bone pain and an increased susceptibility to fracture.

Large doses of vitamin D may give rise to a condition called **vitamin D intoxication**. Its principal symptoms, which are nausea, vomiting and dehydration, are caused by an elevated plasma calcium level. In the long-term, there may be impairment of renal function as soft tissue within the kidneys becomes calcified.

Parathyroid hormone

The **parathyroid glands** contain two distinct cell types: **chief cells**, which secrete parathyroid hormone (PTH), and **oxyphil** cells. Both cell types can be seen in Figure 16.25. As Figure 16.26 shows, the four parathyroid glands are closely associated with the thyroid gland. Each gland is around 3–8 mm in length, 2–5 mm in width and about 1.5 mm in thickness.

The actions of parathyroid hormone

The actions of PTH are summarized in Figure 16.27. PTH is secreted continuously at a low rate and normal levels of the hormone are necessary for normal osteoblast

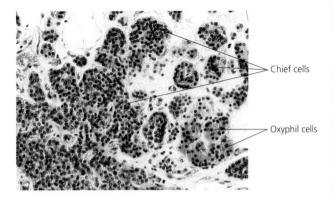

Figure 16.25 A section of a parathyroid gland showing the dark-staining chief cells (which secrete parathyroid hormone) and the lighter-staining oxyphil cells.

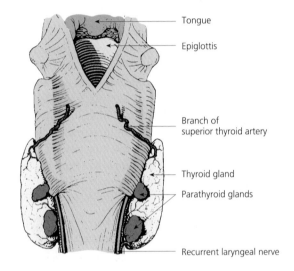

Figure 16.26 The anatomical location of the parathyroid glands and their arterial blood supply. This view is from the dorsal (posterior) aspect of the pharynx.

function and the calcification of osteoid. The main stimulus for an increase in the rate of PTH secretion is a fall in plasma calcium concentration (hypocalcaemia). As might be expected, hypercalcaemia (an elevated plasma calcium concentration) inhibits the secretion of PTH (another example of negative feedback regulation).

Although PTH does not seem to have a direct action on the intestine, it does stimulate the kidneys to synthesize 1,25-DHCC from vitamin D, which then stimulates the intestinal absorption of dietary calcium as described above. Overall, these direct and indirect actions of PTH bring about an increase in plasma calcium concentration.

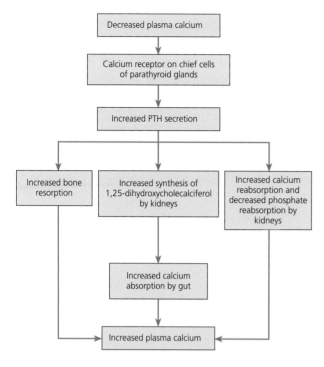

Figure 16.27 A flow diagram summarizing the principal actions of parathyroid hormone.

High levels of PTH have a rather complex effect on bone metabolism. Initially, there is a rapid loss of calcium from the readily releasable pool (see above), but in the longer-term, high circulating levels of PTH stimulate the resorption of stable bone by osteoclasts. In this way, large amounts of calcium and phosphate are added to the extracellular fluid. PTH also stimulates the reabsorption of calcium by the distal tubules of the renal nephrons.

Disorders of parathyroid hormone secretion

Effects of excess PTH secretion (primary hyperparathyroidism). Over-secretion of PTH occurs when a tumour of the parathyroid glands develops. PTH may also be secreted by other types of malignant tumours such as certain lung tumours. The principal result of PTH excess is an increase in plasma calcium (**hypercalcaemia**). Kidney stones are a common finding and can lead to severely impaired renal function, which may cause death if left untreated. Hypercalcaemia also causes fatigue, weakness, mental changes,

intense thirst, constipation and anorexia. Surgical removal of the parathyroid tumour is the usual treatment for this condition.

The skeletal effects of hyperparathyroidism are variable but there is often significant demineralization of bone, which causes pain, fractures of the long bones and compression fractures of the spine. Cysts composed of osteoclasts ('brown tumours') may also be present.

Effects of PTH insufficiency (hypoparathyroidism). Hypoparathyroidism, with a reduction in PTH secretion, may result from loss of function or damage to the glands. It may also arise following accidental removal of the glands during thyroid surgery. As PTH secretion falls, there is a gradual decline in the plasma calcium concentration. The resulting hypocalcaemia causes a number of characteristic neurological symptoms. There is abnormal excitability of nerves and skeletal muscles (**tetany**), which results in muscular spasms, particularly in the hands and feet, feelings of pins and needles (**paraesthesia**), especially of the skin around the mouth, and, occasionally, even convulsions.

Treatment of the disorder includes an initial transfusion of calcium to restore normal levels, followed by the administration of vitamin D metabolites, which will help to increase intestinal absorption of calcium.

KEY POINTS:

- Plasma levels of calcium and phosphate are regulated by the actions of parathyroid hormone (PTH), the active metabolites of vitamin D and calcitonin.

- Dihydroxycholecalciferol is synthesized from vitamin D. Its main effect is to enhance the absorption of dietary calcium by the intestine.

- Vitamin D deficiency causes rickets in children and osteomalacia in adults.

- PTH is secreted by the chief cells of the parathyroid glands. It is released in response to a fall in plasma calcium and acts to maintain normal plasma calcium.

Calcitonin

Calcitonin is a peptide hormone secreted from the parafollicular or C cells that lie between the follicles of the thyroid gland (see Figure 16.13). It is known that calcitonin is able to reduce the concentration of both calcium and phosphate in the plasma, but its precise role in the physiological regulation of these minerals is not clear. Indeed, the plasma levels of these minerals generally remain normal in patients suffering from calcitonin-secreting tumours of the C cells.

The actions of calcitonin

Figure 16.28 summarizes the actions of calcitonin and the regulation of its secretion. Overall, these bring about a fall in plasma calcium. The principal action of the hormone is to inhibit the activity of osteoclasts in the bone tissue. This has the effect of reducing the rate of bone resorption and the release of calcium and phosphate into the plasma. Calcitonin is sometimes given to patients suffering from hypercalcaemia associated with malignancy and in Paget's disease, in

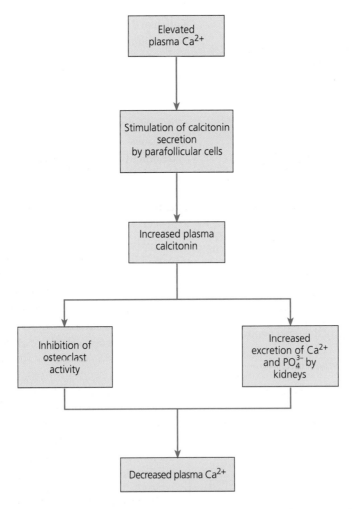

Figure 16.28 A schematic drawing showing the principal actions of calcitonin and the factors thought to regulate its secretion.

which there is excessive turnover of bone. Calcitonin may also increase the rate of renal excretion of calcium and phosphate.

KEY POINTS:

- Calcitonin is secreted by the parafollicular cells of the thyroid gland. It is a hypocalcaemic agent, secreted in response to elevated plasma calcium.

- It is sometimes given to individuals suffering from Paget's disease to help lower plasma calcium.

Other hormones involved in the regulation of plasma calcium

Although the hormones discussed above are the chief regulators of plasma calcium and phosphate, both are also influenced by a number of other hormones including:

- Growth hormone
- Adrenal glucocorticoids
- Thyroid hormones
- Sex steroids, particularly the oestrogens.

Oestrogens appear to play a very important role in the regulation of plasma mineral levels and in the maintenance of the skeleton. They inhibit the PTH-mediated resorption of bone and stimulate the activity of osteoblasts. Following the menopause, or after removal of the ovaries, the dramatic fall in the secretion of oestrogens results in an increased rate of bone resorption that can lead to a condition known as **osteoporosis**. This is characterized by a significant reduction in bone density, increased bone fragility and an increased susceptibility to vertebral compression fractures and fractures of the wrist and hip. Hormone replacement therapy (HRT) has been shown to reduce the rate of progress of postmenopausal osteoporosis.

16.12 The hormonal regulation of plasma glucose

To carry out their normal metabolism, the cells of the body must receive a continuous supply of glucose, the major fuel used to produce ATP for energy-requiring processes. Certain tissues, including the CNS, the retina and the germinal epithelium, rely almost entirely on glucose metabolism for the generation of ATP. The nervous system alone requires around 110 g of glucose each day to meet its metabolic needs. It is therefore vital that plasma glucose levels are maintained within the normal range (4–8 mmol l^{-1}) and a variety of hormones ensure that plasma glucose remains within this range despite wide variations in dietary intake. The pancreatic hormones **insulin** and **glucagon** provide short-term (minute to minute) regulation while a number of other hormones are important in longer-term control.

How does the body handle glucose?

To understand fully the mechanisms involved in the hormonal control of plasma glucose, it is necessary to con-sider briefly the ways in which the body handles glucose. Following its absorption from the small intestine, glucose travels directly to the liver via the portal vein. Here it may be broken down to intermediates (**glycolysis**), which may be used as energy sources, or converted to **glycogen** (**glycogenesis**), which is the stored form of glucose. When plasma glucose levels fall, glycogen is broken down to glucose (**glycogenolysis**). The relative amounts utilized or stored in the liver will depend upon the plasma glucose concentration and the needs of the tissues at the time. In addition to the liver, small amounts of glycogen are also stored in the kidneys, skeletal muscle, skin and certain glands. When the body's glycogen stores are full, any excess glucose in the plasma is converted to fatty acids and stored in fat cells (adipocytes).

When plasma glucose is elevated, two additional pathways of glucose metabolism become significant. These are:

- The protein glycosylation pathway, in which certain proteins, especially haemoglobin, become glycosylated (i.e. the proteins have sugar residues attached to them).

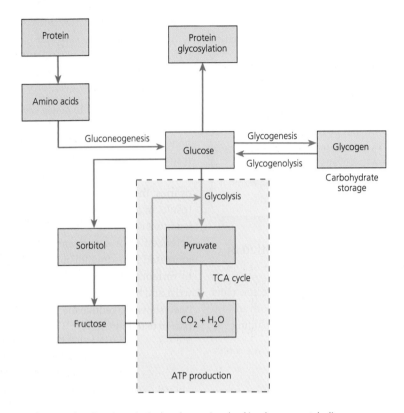

Figure 16.29 A schematic diagram showing the principal pathways involved in glucose metabolism.

- The polyol pathway, in which glucose is converted to sorbitol, which is then oxidized slowly to fructose. This happens chiefly in the retina, lens, kidney, aorta and Schwann cells.

In addition to storing glucose in the form of glycogen, cells of the liver and kidneys are able to synthesize glucose from non-carbohydrate sources such as glycerol, lactate and certain amino acids. This process is known as **gluconeogenesis**. Under most conditions, through its roles in glycogenolysis and gluconeogenesis, the liver is the chief source of glucose for the general circulating pool. The principal pathways involved in glucose metabolism are summarized in Figure 16.29.

The hormonal regulation of plasma glucose

In view of its importance in brain metabolism, it is essential that plasma glucose is not allowed to fall too low. Of the many hormones that contribute to plasma glucose regulation, only one, insulin, acts to reduce plasma glucose. All the others act to raise plasma glucose, that is they are glucogenic in their effects. Glucogenic hormones include glucagon, adrenaline and noradrenaline, cortisol, GH and thyroid hormone.

KEY POINTS:

- Glucose is an essential metabolic substrate, particularly for the CNS. Plasma glucose is closely regulated between 4 and 8 mmol l^{-1} by a number of hormones, including insulin and glucagon.

- Glucose is stored in the form of glycogen and can be synthesized from amino acids (gluconeogenesis).

- Plasma glucose is determined by the balance between intestinal absorption, uptake by the tissues and gluconeogenesis synthesis by the liver.

Insulin and glucagon provide short-term regulation of plasma glucose

Although the bulk of the pancreatic tissue is concerned with the exocrine secretion of digestive enzymes (see Chapter 26), around 2% of its total mass is occupied by small patches of endocrine tissue called the **islets of Langerhans** (Figure 16.30). The principal cell types within the islets are α-cells and β-cells, which synthesize, store and secrete glucagon and insulin, respectively. Glucagon and insulin have opposing actions that, together, maintain minute-to-minute control of plasma glucose.

The regulation of insulin secretion

Insulin was the first protein to have its amino acid sequence determined, a feat achieved by Frederick Sanger in 1953, for which he later received the Nobel Prize for Chemistry. Once secreted into the blood, insulin travels in free solution. As it is taken up rapidly by the tissues, insulin has a short half-life in the plasma, around 5 minutes. It is avidly

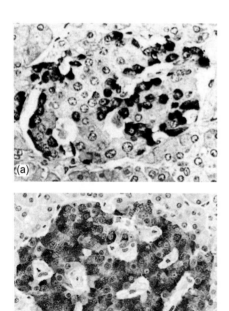

Figure 16.30 Pancreatic islets stained with an antibody to show cells that contain glucagon (a) and insulin (b). The hormone-containing cells stain black or brown. Note the different appearance of the surrounding acinar tissue.

taken up by its target tissues, particularly the liver, kidneys, muscle and fat, so that very little appears in the urine.

The most powerful stimulus for insulin secretion is glucose. At relatively low plasma glucose concentrations (around $3-4\,\mathrm{mmol\,l^{-1}}$), insulin is secreted at a very low rate. Following a carbohydrate-containing meal, insulin secretion increases as plasma glucose rises (Figure 16.31). Typically, there is a three- to tenfold increase in plasma insulin that peaks 30–60 minutes after eating begins. As the figure shows, insulin release normally occurs in two stages (i.e. it is biphasic). There is an early rise in secretion, reflecting the release of available insulin, and a later rise that depends on the synthesis of new insulin in response to the glucose load. The close link between plasma glucose and insulin secretion prevents the plasma glucose level from rising to excessively high values during and after a meal.

Insulin secretion is reduced by sympathetic stimulation and circulating catecholamines, while parasympathetic stimulation increases insulin secretion. Some amino acids (particularly leucine) stimulate the secretion of insulin, which in turn increases the rate of amino acid uptake into cells and the overall rate of protein synthesis.

The hypoglycaemic actions of insulin

Insulin binds to plasma membrane receptors of cells in the liver, fat and muscle (its principal target tissues). This hormone–receptor interaction results in the insertion into the membrane of glucose carriers. As a result, there is an increase in the rate of glucose uptake into these insulin-responsive cells. Once inside the cells, glucose may be utilized for metabolism or converted to glycogen and fat, as described earlier. As plasma glucose falls, insulin secretion is inhibited.

KEY POINTS:

- Insulin is synthesized by the β-cells of the pancreatic islets and is secreted chiefly in response to a rise in plasma glucose and amino acids.

- It stimulates the uptake and utilization of glucose by cells (particularly liver, adipose and skeletal muscle), thereby causing a fall in plasma glucose.

- Insulin stimulates both glycogen and protein synthesis.

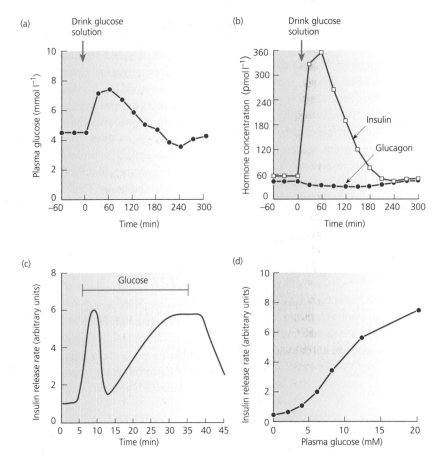

Figure 16.31 The relationship between plasma glucose and insulin secretion: (a) and (b) show the changes in plasma glucose, insulin and glucagon following intake of a solution of glucose; (c) shows how insulin secretion (expressed in arbitrary units) varies with time following a glucose load; (d) plots the rate of insulin secretion as a function of the plasma glucose concentration.

The regulation of glucagon secretion

Glucagon is a small peptide hormone, which, like insulin, has a short half-life in the plasma (around 6 minutes). It is secreted in response to a fall in plasma glucose and its actions are hyperglycaemic (i.e. they raise plasma glucose levels). A number of amino acids (especially arginine and alanine) also stimulate the secretion of glucagon, as does both sympathetic and parasympathetic stimulation and the gut hormone cholecystokinin (see Chapter 24). Insulin appears to inhibit the secretion of glucagon.

The hyperglycaemic actions of glucagon

Glucagon acts to raise plasma glucose levels and to provide adequate levels of other energy substrates during periods of fasting. It has effects on carbohydrate, protein and fat metabolism. The principal target organ of glucagon is the liver, where it inhibits glycogen synthesis and stimulates glycogenolysis with the result that glucose is released into the plasma. At the same time, gluconeogenesis is also stimulated, further contributing to the elevation in plasma glucose. Glucagon also has a significant lipolytic effect, mobilizing fatty acids and glycerol from adipose tissue. In this way, a ready supply of non-carbohydrate metabolic substrates is ensured, helping to spare glucose for use by the brain.

> **KEY POINTS:**
>
> - Glucagon is secreted by the α-cells of the islets of Langerhans and is the most potent hyperglycaemic hormone.
> - It promotes the release of glucose into the blood.
> - Hypoglycaemia is the principal stimulus for the secretion of glucagon, which then stimulates the breakdown of glycogen and fats. It also stimulates gluconeogenesis.

Other hormones involved in the regulation of plasma glucose

Although insulin and glucagon are the principal short-term regulators of plasma glucose, a variety of other hormones also play a role in the maintenance of plasma glucose. These are cortisol, GH, the adrenal medullary hormones and the thyroid hormones. The actions of all of these are described in more detail earlier in this chapter so they will be summarized only briefly here. Their role in the regulation of plasma glucose is summarized in Figure 16.32.

The role of cortisol in the regulation of plasma glucose

Cortisol is a steroid hormone and consequently its actions have a prolonged time course, with a delay between stimulus and effect. For this reason, cortisol exerts a longer-term influence on plasma glucose rather than being involved in its short-term regulation. It is secreted in response to a fall in plasma glucose and, overall, its effects serve to elevate plasma glucose and to promote the release of fatty acids by lipolysis. The hyperglycaemic effect is achieved through an anti-insulin action in peripheral tissues (reducing the cellular uptake of glucose) and the stimulation of gluconeogenesis.

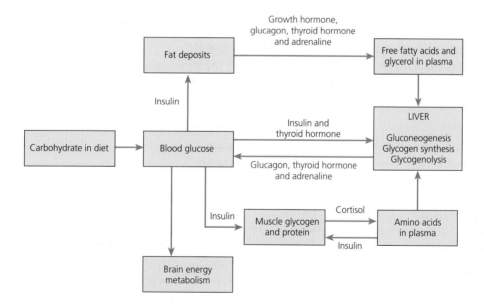

Figure 16.32 An overview of the hormonal regulation of plasma glucose concentration. During the absorptive state, insulin promotes the uptake of glucose by the liver, muscle and adipose tissue. In the postabsorptive state, glucose levels are maintained by glycogenolysis in the liver (which is stimulated by glucagon, thyroid hormone and adrenaline) and by gluconeogenesis, which is regulated by cortisol and glucagon. Lipolysis makes fatty acids available for oxidation and this process is promoted by growth hormone, glucagon, thyroid hormone and adrenaline.

The role of growth hormone in the regulation of plasma glucose

The actions of GH on glucose metabolism are most significant in times of fasting or starvation. GH is a 'glucose-sparing' hormone, which acts to reduce glucose uptake by tissues such as muscle. It also stimulates the breakdown of hepatic glycogen and increases the breakdown of stored fats (lipolysis). In this way, fatty acids are provided for metabolism while glucose is spared for use by the brain.

The role of the adrenal medullary hormones in the regulation of plasma glucose

Adrenaline and noradrenaline (the catecholamines) work alongside the other hyperglycaemic hormones to maintain plasma glucose levels during a period of hypoglycaemia and to spare the available glucose for use by the brain. The catecholamines become particularly important if plasma glucose falls to very low levels or if the secretion of glucagon is impaired. As well as stimulating the conversion of glycogen to glucose, adrenaline and noradrenaline act to stimulate the output of hyperglycaemic hormones and to suppress the secretion of insulin. They also stimulate lipolysis so that free fatty acids and glycerol are available for metabolism by non-glucose-dependent tissues.

The role of thyroid hormones in the regulation of plasma glucose

As explained earlier (see p. 300) the effects of thyroid hormones on carbohydrate metabolism are dependent on the concentration of thyroid hormone in the circulation. In general, it appears that low concentrations of thyroid hormone enhance glycogen synthesis and thereby tend to reduce plasma glucose, while higher concentrations enhance glycogenolysis and gluconeogenesis, so increasing plasma glucose. Thyroid hormones are secreted in response to stresses of all kinds, including hypoglycaemia and starvation. Under such conditions, they are important (together with other hyperglycaemic hormones, particularly the catecholamines), in ensuring that glucose reserves are mobilized and that glucose-dependent tissues receive an adequate supply.

> **KEY POINTS:**
> - During prolonged periods of fasting, hyperglycaemic hormones, such as cortisol, catecholamines, GH and thyroid hormone, play a significant role in the regulation of plasma glucose.
> - They help to maintain glycogen stores, and stimulate gluconeogenesis. They also mobilize fatty acids to provide non-carbohydrate metabolic substrates for those tissues able to use them.

Normal variation of plasma glucose

Despite wide variations in food intake during the day or over a period of days, a variety of hormones act to maintain plasma glucose within a narrow range. Immediately prior to a meal, for example on getting up in the morning, plasma glucose is likely to be relatively low (around $4\,\text{mmol}\,\text{l}^{-1}$). Consequently, the secretion of glucagon and other hyperglycaemic hormones will be elevated as the body conserves glucose to maintain an adequate supply to the brain. If a glucose-rich meal (for example cereals, orange juice and toast) is then eaten, plasma glucose will start to rise as glucose is absorbed from the small intestine.

During the 90 minutes or so following the start of the meal, the body is said to be in the **absorptive state** of metabolism and during this time plasma glucose may rise to 7 or $8\,\text{mmol}\,\text{l}^{-1}$. Two hours or so after the start of the meal, plasma glucose starts to fall once more as glucose is utilized by the cells or converted to glycogen and fat. The fall in plasma glucose is due predominantly to the actions of insulin, which is secreted from the pancreatic β-cells in response to the rise in plasma glucose through the mechanisms described earlier.

Within 3–4 hours of finishing the meal, the body enters the **postabsorptive state** of metabolism and begins to defend itself against hypoglycaemia once more. Plasma glucose will have fallen back to around $4.5\,\text{mmol}\,\text{l}^{-1}$ and insulin secretion will be low. Glucagon secretion rises in response to the fall in plasma glucose and stimulates the breakdown of stored hepatic glycogen to release glucose into the bloodstream. Glycogen stored in other tissues, such as muscle, will also be broken down but the glucose released will be metabolized directly, thus helping to conserve plasma glucose. At the same time, in response to cortisol secreted by the adrenal cortex, amino acids will be mobilized from muscle and converted to glucose (gluconeogenesis).

GH, the catecholamines and glucagon all stimulate lipolysis in fat cells. The free fatty acids released can be used by the liver and muscle as a metabolic fuel, in preference to glucose. Ketone bodies, produced in the liver by the metabolism of fatty acids, provide an additional energy source for muscle and, in longer periods of fasting, for the brain. As a result of the mechanisms outlined above, plasma glucose rarely falls below about 3 mmol l^{-1} in a healthy person, even during prolonged periods of fasting.

> **KEY POINTS:**
>
> - In the absorptive phase of metabolism (just after a meal), plasma glucose is high, insulin levels are high and the secretion of hyperglycaemic hormones is inhibited. During this time, glycogen stores are replenished and glucose uptake is high.
>
> - In the postabsorptive state (from 3 hours or so after a meal until the next intake of food), counter-regulatory mechanisms are initiated. Levels of the hyperglycaemic hormones are enhanced, and there is an increase in the mobilization of fats (and, in prolonged fasting, amino acids) so that glucose is spared for use by the CNS.

Diabetes mellitus

Failure of the pancreatic β-cells to secrete sufficient insulin gives rise to a condition called **diabetes mellitus**, which is characterized by high levels of glucose in the bloodstream (hyperglycaemia). It is a relatively common disorder, affecting almost 2% of the population. In Western populations, the disease appears to be increasing in incidence, particularly among young people.

A diagnosis of diabetes mellitus is usually made if the fasting plasma glucose remains consistently above 8 mmol l^{-1}. Confirmation of the diagnosis usually requires an **oral glucose tolerance test**. In this test, the patient's plasma glucose is measured and he or she is then given a glucose solution to drink (corresponding to an intake of 75 g of glucose). Every 30 minutes after the drink, further blood samples are then taken to determine the changes in plasma glucose. In a normal healthy person, plasma

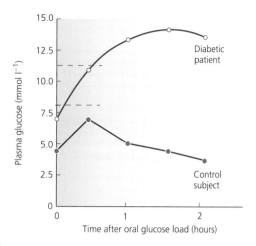

Figure 16.33 Changes in plasma glucose following a glucose load in a diabetic patient compared with those seen in a normal subject. The glucose load was administered at time 0. The dotted lines show the range of plasma glucose levels that prompt a diagnosis of diabetes.

glucose will return to normal within 2 hours, but in a person suffering from diabetes mellitus, plasma glucose will still exceed 11 mmol l^{-1} 2 hours after the glucose load. These changes are illustrated in Figure 16.33.

There are two principal forms of diabetes mellitus, known as type 1 and type 2 diabetes. In **type 1 diabetes mellitus**, the pancreas fails to secrete normal amounts of insulin. Type 1 diabetes usually develops before the age of 40 and is believed to be due to the autoimmune destruction of the β-cells by islet-cell antibodies. This form of the disease is treated by administering insulin and, for this reason, it is called **insulin-dependent diabetes**. In **type 2 diabetes mellitus**, older patients develop a non-insulin-dependent form of diabetes. Insulin is secreted (though often in reduced amounts), but there is a loss of sensitivity to the hormone. The exact causes of this type of diabetes are unknown but there is a strong genetic component and obesity is a risk factor. Type 2 diabetes may be treated by careful management of the diet and the use of antidiabetic drugs (see below).

The classical symptoms of the disease include:

- Elevated blood glucose leading to glycosuria (glucose in the urine)

- Polyuria (the passage of copious volumes of urine) as a result of the osmotic diuresis

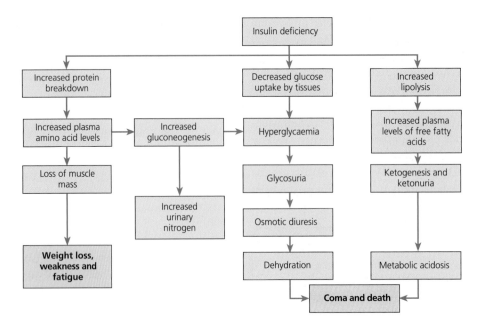

Figure 16.34 The acute and chronic consequences of insulin deficiency.

- Polydipsia (intense thirst) as a result of increased water loss via the urine
- Increased production of ketones as a result of fat breakdown, which may lead to their appearance in the urine and ketoacidosis, a serious medical problem
- Weight loss (in most patients).

These effects of insulin deficiency are summarized in Figure 16.34.

If diabetes remains undiagnosed or is poorly controlled so that tissues are exposed to prolonged periods of hyperglycaemia, there are a number of potentially serious, long-term complications. These include:

- Changes to the lens of the eye (cataract)
- Degenerative changes to the retina (retinopathy)
- Degeneration of peripheral nerves (peripheral neuropathy)
- Thickening of the filtration membrane of the renal nephron (nephropathy)
- Peripheral vascular lesions
- Chronic skin infections.

Diagnosed diabetics are routinely monitored for any changes in vision and renal function, and receive regular foot care to prevent damage of the kind illustrated in Figure 16.35.

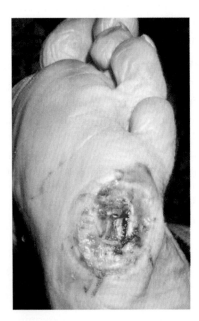

Figure 16.35 The foot of a diabetic showing severe ulceration.

Management and treatment of diabetes

Regular monitoring of blood glucose is of paramount importance in individuals diagnosed as diabetic. Diabetics normally monitor their own blood glucose levels several times a day (more often if there has been exposure to situations that may alter glucose utilization such as heavy exercise or stress of any kind) and regulate their food intake and insulin therapy accordingly.

The effectiveness of long-term glycaemic control is monitored using measurements of the blood concentration of glycosylated haemoglobin (HbA_{1c}). Haemoglobin (Hb) is glycosylated by a non-enzymic process that is dependent upon the prevailing blood concentration of glucose. Provided that the lifespan of a person's red blood cells is normal, the level of glycosylated Hb in the blood gives an indication of the average blood glucose concentration over the preceding 60 days or so (half the typical lifespan of a red cell, see Chapter 20). Non-diabetic individuals with normal plasma glucose concentrations (normoglycaemia) will have an HbA_{1c} level of between 3.5 and 5.5%. Diabetics usually aim for values below about 7%. The figures in Table 16.3 show how the mean (average) blood glucose concentration influences the measured level of HbA_{1c}.

Oral antidiabetic drugs

Three principal groups of drugs are regularly used in the treatment of type 2 diabetes for those patients in whom diet fails to control hyperglycaemia. These are:

- **Sulphonylureas**, e.g. glipizide, glicazide, tolbutamide and glibenclamide. These drugs stimulate insulin release from the pancreatic islets so, if they are to be of any value, the patient must have at least some functional β-cells.

- **Biguanides**, e.g. metformin. These agents act on peripheral tissues to stimulate the uptake of glucose. Their mechanism of action is unknown.

- **Glitazones**, e.g. rosiglitazone and pioglitazone. This is a comparatively new group of drugs whose action is to increase the sensitivity of target cells to insulin by increasing the transcription of certain insulin-sensitive genes. They are often used in conjunction with metformin or one of the sulphonylureas.

Insulin preparations

Many type 1 diabetics are treated with human insulin administered by multiple, daily, subcutaneous injections or by continuous, subcutaneous infusion. A variety of different preparations is available, each having a different duration of action. Regimens of insulin administration are designed to achieve the best possible glycaemic control in individual patients. Most use a regimen involving a short-acting insulin preparation mixed with twice-daily injections of an intermediate-acting insulin preparation administered before breakfast and before the evening meal. Occasionally, an intermediate-acting insulin preparation may be used to provide a background level of insulin, with a short-acting form of insulin taken three times a day before meals.

Illness, injury, surgery or stress of any kind presents a challenge to glycaemic control, presumably because of changes in the rates of secretion of hyperglycaemic hormones such as cortisol and the catecholamines. In these circumstances, diabetics will be monitored very closely and insulin is administered on a 'sliding scale' that allows for continual variation in the insulin dose in response to the prevailing level of plasma glucose.

Consequences of hypoglycaemia

Hypoglycaemia is defined as a plasma glucose concentration below 2.5 mmol l^{-1}. It may arise from a number of causes, including insulin overdose in diabetics, an insulin-secreting tumour of the islet cells, lack of GH or cortisol, severe liver disease, very severe exercise or inherited defects of enzymes required for gluconeogenesis. The consequences of hypoglycaemia fall into two

Table 16.3 The relationship between plasma glucose concentration and glycosylation of haemoglobin

Blood glucose (mmol l^{-1})	HbA$_{1c}$ (% of total Hb)
18	13
13	10
10	8
8	7
5	5

categories: those associated with activation of the autonomic nervous system and those caused by altered brain function.

A hypoglycaemic episode is often preceded by a feeling of hunger mediated by the parasympathetic nervous system. Later, symptoms of sympathetic activation begin to appear such as tachycardia, sweating, pallor and anxiety. Because the brain relies on glucose as its principal source of energy, hypoglycaemia often elicits symptoms that are related to altered cerebral function: headache, difficulties with problem solving, confusion, irritability, convulsions and, eventually, coma. If the plasma glucose falls below $1\,\mathrm{mmol\,l^{-1}}$, there may be irreversible neuronal damage and the patient may die. Treatment is the ingestion of glucose as soon as the symptoms appear.

KEY POINTS:

- A lack of insulin or of insulin receptors gives rise to diabetes mellitus in which there is hyperglycaemia leading to polyuria, excessive thirst and weight loss.

- Diabetes mellitus, is classified as type 1 or type 2. Type 1 diabetes mellitus is insulin dependent and can be controlled by injections of insulin. Type 2 is a non-insulin-dependent form of diabetes that occurs in older people. It can be controlled by regulating the diet, often in combination with oral antidiabetic drugs such as tolbutamide.

Other hormone-secreting tissues

In addition to insulin and glucagon, a large number of peptide hormones are secreted by the gastrointestinal tract to regulate the activity of the stomach, intestine and accessory organs. Their actions are discussed in Chapter 24, so only a brief summary will be given here. **Gastrin** is secreted by the G cells of the gastric glands in response to the intake of food. It stimulates the secretion of gastric acid and gastric motility. A number of hormones are secreted by enteroendocrine cells within the small intestine. These include **cholecystokinin** (secreted by I cells in response to a meal containing fat), **secretin** (secreted by S cells in response to low pH in the duodenum) and **vasoactive intestinal polypeptide**. Cholecystokinin (CCK) stimulates contraction of the gall bladder and the secretion of enzyme-rich fluid from the pancreas. Secretin and vasoactive intestinal polypeptide (VIP) stimulate the secretion of pancreatic alkaline fluid.

The kidneys secrete **erythropoietin** in response to a low partial pressure of oxygen in the arterial blood. This hormone stimulates the proliferation of red blood cells. The heart secretes **atrial natriuretic peptide** (ANP) in response to an increase in the circulating volume of blood. ANP increases the excretion of sodium, so reducing the extracellular volume. Adipose tissue secretes the hormone **leptin**, which plays a role in the regulation of food intake. The liver secretes **IGF-1** and **IGF-2** and a hormone called **hepcidin**, which regulates iron uptake by the body. It also secretes **thrombopoietin**, a hormone that acts on the bone marrow to regulate the production of platelets.

Recommended reading

Histology

Young, B., Lowe, J.S., Stevens, A., Heath, J.W. and Deakin, P.J. (2006). *Wheater's Functional Histology: A Text and Colour Atlas*, 5th edn. Churchill-Livingstone: Edinburgh. Chapter 17.

Pharmacology

Rang H.P., Dale, M.M., Ritter, J.M. and Flower, R. (2007). *Pharmacology*, 6th edn. Churchill-Livingstone: Edinburgh. Chapters 25–29.

Endocrine physiology

Brook, C. and Marshall, N. (2001). *Essential Endocrinology*, 4th edn. Blackwell Science: Oxford.

Griffin, J.E. and Ojeda, S.R. (2004). *Textbook of Endocrine Physiology*, 5th edn. Oxford University Press: Oxford.

Self-assessment questions

Which of the following statements are true and which are false? Answers are given at the end of the book (p. 755).

1 a) Hypothalamic releasing hormones are synthesized and secreted by neurones.
  b) Blood flows from the anterior pituitary to the hypothalamus in portal vessels.
  c) The hypothalamic releasing hormones reach the general circulation in significant amounts.
  d) Loss of dopaminergic neurones in the hypothalamus is likely to lead to a rise in the secretion of prolactin.
  e) Growth hormone secretion is regulated by a single hypothalamic hormone.

2 A 10-year-old child in whom anterior pituitary function is deficient is likely to:
  a) develop acromegaly.
  b) be of short stature but have relatively normal body proportions.
  c) to be in constant danger of becoming dehydrated.
  d) become sexually mature at a later age than normal.
  e) have a low basal metabolic rate.

3 a) Chromaffin cells are found in the adrenal medulla.
  b) Excess secretion of catecholamines will lead to hypertension.
  c) The heart rate is reduced by circulating adrenaline and noradrenaline.
  d) Increased catecholamine secretion stimulates lipolysis.
  e) Adrenaline and noradrenaline are secreted by the adrenal cortex.

4 a) Normal plasma levels of parathyroid hormone (PTH) stimulate osteoblast activity.
  b) Normal plasma levels of PTH decrease calcium excretion from the body.
  c) PTH directly increases calcium absorption by the gut.
  d) PTH is secreted in response to elevated plasma calcium levels.
  e) High levels of circulating PTH demineralize bone and elevate plasma calcium.

5 a) Oxytocin stimulates the synthesis of milk by the mammary glands.
  b) Lack of ADH (vasopressin) will result in excessive production of urine.
  c) Both oxytocin and ADH are secreted in response to neuroendocrine reflexes.
  d) The secretion of oxytocin and ADH are regulated by releasing hormones secreted by the hypothalamus.
  e) ADH increases water reabsorption by cells of the collecting ducts of the kidneys.

6 a) The adrenal cortex secretes both peptide and steroid hormones.
  b) The adrenal cortex will atrophy following removal of the anterior pituitary gland.
  c) Aldosterone plays a role in the regulation of plasma calcium.
  d) The secretion of cortisol peaks at around 6 a.m. each day.
  e) Cortisol is a hyperglycaemic hormone.

7 a) Thyroid hormones are essential for the early development and maturation of the CNS.
  b) T_3 and T_4 stimulate the secretion of TSH by the anterior pituitary.
  c) People who have an underactive thyroid gland have a low BMR.
  d) A resting pulse rate of 65 per minute would suggest a diagnosis of thyrotoxicosis.
  e) Most of the iodide in the body is present in the thyroid gland.

8 Concerning some general principles of glucose metabolism:
  a) Gluconeogenesis is the formation of glucose from glycogen.
  b) Glycogenolysis and gluconeogenesis are functions of the liver.
  c) Glucose may be synthesized from fat reserves.
  d) Gluconeogenesis is stimulated when plasma glucose is low.

e) Glycogenesis is stimulated when cellular ATP reserves are low.

9 Concerning the absorptive and postabsorptive states:
a) During the absorptive state, glucose is the major energy source.
b) Events of the absorptive state are controlled by insulin.
c) In the postabsorptive state, glucagon secretion is inhibited.
d) In the postabsorptive state, glycogen and fat reserves are mobilized.
e) Hypoglycaemia inhibits the secretion of growth hormone.

10 A person has been on hunger strike for a week. Compared with normal he has:
a) increased release of fatty acids from adipose tissue.
b) ketosis and ketonuria.
c) elevated plasma glucose.
d) increased activity of hepatic glycogen synthetase.

e) increased plasma catecholamine levels.

11 Shortly after a carbohydrate-rich meal, the following metabolic processes are likely to be enhanced:
a) glycogenesis
b) glycogenolysis
c) cortisol secretion

Shortly after waking up, the following are likely to be enhanced:
d) growth hormone secretion
e) gluconeogenesis
f) lipolysis

12 Diabetes mellitus:
a) may be caused by a reduction in functional insulin receptors.
b) is characterized by a fall in urine output.
c) is associated with an increase in lipolysis.
d) leads to hypoglycaemia.
e) is a common symptom of acromegaly.

The heart

17

After reading this chapter you should have gained an understanding of:

- The basic anatomy of the heart
- The physiological properties of cardiac (heart) muscle
- The arrangement and properties of the coronary circulation
- How the heartbeat is initiated and regulated
- The cardiac cycle and the heart sounds
- The concept of cardiac output and the main factors that regulate it
- The causes and treatment of heart failure
- The mode of action of the principal drugs that act on the heart

17.1 Introduction

In single-celled or very simple organisms, the exchange of nutrients and waste products between cells and their environment can be accomplished by passive diffusion across the plasma membrane. In more complex animals, however, the distance separating cells from the environment is considerable and diffusion alone would be unable to meet the metabolic requirements. To overcome this problem, animals have evolved a circulatory system to deliver nutrients to cells and to remove metabolic waste products. The circulatory system consists of a pump (the heart), a set of tubes (the blood vessels) and the blood itself. The system is also known as the **cardiovascular system**. A detailed discussion of the blood vessels and circulation may be found in Chapter 19 and the properties of the blood are discussed in Chapter 20. This chapter is concerned with the anatomy and physiology of the heart and will conclude with a brief discussion of heart failure and some important drugs that act on the heart.

17.2 The organization of the circulation

In essence, the cardiovascular system is arranged as two circuits in series. Blood is pumped from the right side of the heart through the lungs. This is the **pulmonary circulation**. The blood returning from the lungs is then pumped around the rest of the body by the left side of the heart. This is the **systemic circulation**. The overall arrangement of the circulation is illustrated diagrammatically in Figure 17.1.

The pumping action of the left ventricle raises the pressure in the aorta above that of the large veins; it is this pressure difference that provides the force to drive the blood around the systemic circulation. Similarly, blood flows through the pulmonary circulation because the pumping action of the right ventricle raises the pressure in the pulmonary arteries above that in the pulmonary veins.

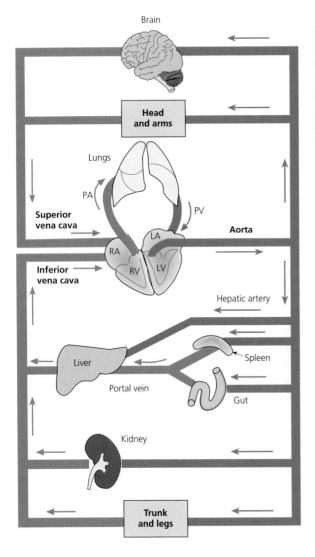

> **KEY POINTS:**
>
> - The pulmonary and systemic circulations are arranged in series.
> - The right side of the heart pumps blood through the lungs (the pulmonary circulation) and the left side of the heart pumps blood round the rest of the body (the systemic circulation).

Figure 17.1 A schematic drawing of the circulation. The arrows indicate the direction of blood flow. Note that the blood returning to the heart enters the right atrium. It then enters the right ventricle, which pumps the blood through the lungs. After leaving the lungs, the blood enters the left atrium and then passes to the left ventricle, which pumps it through the rest of the body via the systemic circulation. Thus, the pulmonary circulation is in series with the systemic circulation. PA, PV – pulmonary artery and pulmonary vein; RA, LA – right and left atria; RV, LV – right and left ventricles.

17.3 The gross anatomy of the heart

The heart lies in the lower part of the left side of the chest (the thoracic cavity). It is roughly the size of its owner's clenched fist and, in adults, weighs between 250 and 350 g. It consists of four muscular chambers. These are the right and left atria (the upper chambers) and the right and left ventricles (the lower chambers), as can be seen in Figure 17.2. The heart lies within a tough fibrous sac called the **pericardium**, which is attached to the diaphragm. This means that the apex of the heart is relatively fixed and when the ventricles contract, the atria move towards the apex and expand. The muscle of the heart wall is known as the **myocardium**. The relative thickness of the myocardium reflects the force generated by the contraction of the different chambers: it is thickest in the left ventricle, thinner in the right ventricle and thinnest in the atria. The individual cells that form the myocardium are called **cardiac myocytes**. Their detailed structure and physiology are discussed below. The right and left ventricles are separated by a muscular sheet called the **interventricular septum**. This physical separation of the left and right sides of the heart ensures that there is no mixing of deoxygenated and oxygenated blood.

The atria are separated from the ventricles by a fibrous skeleton on which the four heart valves are located. These valves ensure the unidirectional flow of blood from the atria to the ventricles and from the ventricles to the aorta and pulmonary arteries. They are illustrated in Figures 17.2 and 17.3. Backflow of blood from the ventricles to the atria is prevented by the atrioventricular valves. On the right side of the heart is the **tricuspid valve**, which consists of three flexible, roughly triangular flaps of connective tissue. On the left side is the **bicuspid valve** (also known as the **mitral valve**), which consists of two connective tissue flaps. The free edges of the tricuspid and bicuspid valves are connected to a set of conical muscles (the **papillary muscles**) that project from the ventricular wall by a set of thin tendons (the **chordae tendinae**). These are shown in the section of the heart illustrated in Figure 17.3. The chordae tendinae prevent the valves from being pushed up into the atria when the ventricles contract.

Blood is prevented from flowing back into the right ventricle from the pulmonary artery by the **pulmonary**

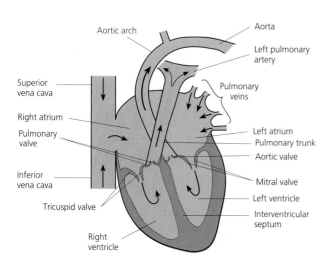

Figure 17.2 A simple diagram to illustrate the arrangement of the chambers and the direction of blood flow through the heart. The blood passing through the right side of the heart contains deoxygenated blood while that passing through the left side of the heart contains oxygenated blood. The arrows indicate the direction of blood flow.

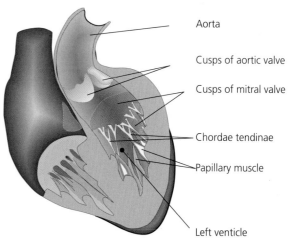

Figure 17.3 Drawing of the heart with the anterior wall of the left heart cut away to show the aortic valve, mitral valve, papillary muscles and chordae tendinae.

valve and into the left ventricle from the aorta by the **aortic valve**. The pulmonary and aortic valves are also known as the **semilunar valves**. The valves and the four chambers of the heart are lined by a thin layer of epithelial tissue called the **endocardium.** It is continuous with the epithelial lining of the blood vessels (vascular endothelium) that leave and enter the heart chambers. The actions of the valves during the cardiac cycle are discussed further in section 17.7 (pp. 337–338).

The major blood vessels

A simple diagram of the heart and the blood vessels that enter and leave the four chambers is shown schematically in Figure 17.2. The right ventricle gives rise to the pulmonary trunk, which divides after a short distance to form the right and left pulmonary arteries. These vessels carry deoxygenated blood to the lungs where gas exchange occurs. Oxygenated blood is then returned to the left atrium by the four pulmonary veins.

The aorta leaves the left ventricle and emerges from beneath the right side of the pulmonary trunk, as shown in Figure 17.4. The ascending aorta gives rise to the brachiocephalic artery before curving backwards and to the left over the right pulmonary artery to form the aortic arch. The brachiocephalic artery subsequently branches to form the common carotid and subclavian arteries. When it reaches the left side of the fourth thoracic vertebra, the aorta descends towards the lower body. Deoxygenated blood returns from the body tissues to the right atrium via the **venae cavae** (or **great veins**). The **superior vena cava** returns blood from the upper parts of the body while the **inferior vena cava** returns blood from lower regions.

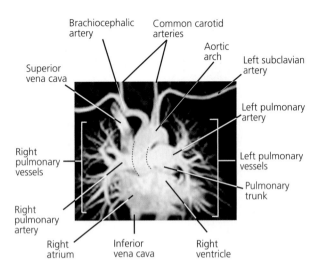

Figure 17.4 An enlarged image of the chest region from Figure 19.2 to show the position of the heart and the origin of the aorta and pulmonary arteries more clearly. This image was formed by magnetic resonance imaging.

KEY POINTS:

- Blood returns to the heart from the systemic circulation via the superior and inferior venae cavae.

- The heart itself consists of four chambers, two atria and two ventricles.

- The atria are separated from the ventricles by two valves that prevent the reflux of blood into the atria when the ventricles contract: the mitral valve (on the left side) and tricuspid valve (on the right).

- Reflux of blood from the pulmonary artery and aorta into the ventricles is prevented by the pulmonary and aortic valves.

17.4   The structure of cardiac muscle

The muscle of the heart (the myocardium) consists of individual cells physically linked together in a three-dimensional matrix by junctions called **intercalated disks** (Figure 17.5). The individual cardiac muscle cells are rod-shaped cells known as **cardiac myocytes**, which are about 15 μm in diameter and 80–100 μm in length. They usually have only one centrally located nucleus and their mitochondria are distributed throughout the cytoplasm. Adjacent cells are electrically coupled by gap junctions that permit electrical activity to spread from one cell to another

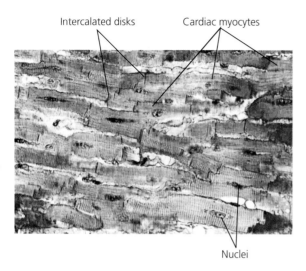

Intercalated disks   Cardiac myocytes

Nuclei

Figure 17.5 A section of heart muscle stained to show the main structures of the cardiac myocytes. The striated nature of the myocytes is clearly visible, as are the intercalated disks. The elongated nuclei are approximately 10 μm in length.

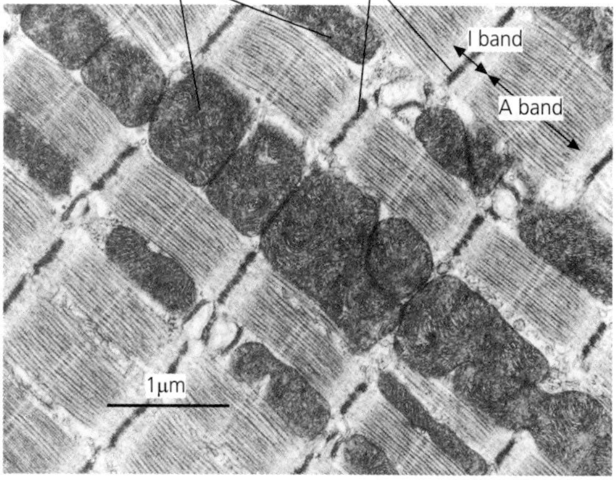

Mitochondria   Z lines

I band

A band

1 μm

Figure 17.6 An electron micrograph that shows the detailed structure of the sarcomeres. Note the abundant mitochondria.

so that the myocardium forms a network or functional syncytium. This adaptation allows each wave of excitation to spread across the whole of the heart to elicit a co-ordinated contraction of the heart itself (see below).

As for skeletal muscle (see Chapter 9), when cardiac muscle is viewed by polarized light, the individual cells are seen to possess alternating dark and light zones within their cytoplasm. The regions that appear dark do so because they refract the polarized light. This property is called **anisotropy** and the corresponding band is known as an **A band**. The light regions do not refract polarized light and are said to be **isotropic**. These regions are called **I bands**. Each I band is divided by a characteristic line known as a **Z line** and the unit between successive Z lines is a **sarcomere**. This repeating pattern can also be seen after staining cardiac muscle with certain dyes and in electron micrographs, such as that

shown in Figure 17.6. The sarcomere is the fundamental contractile unit of cardiac muscle. As each sarcomere is only 1.5–2 μm in length, each myocyte contains many sarcomeres placed end to end.

> **KEY POINTS:**
>
> - Cardiac muscle is made up of many individual cells (cardiac myocytes) linked together via intercalated disks.
>
> - Adjacent cardiac myocytes are electrically coupled via gap junctions. This allows cardiac muscle to act as a functional syncytium so that each wave of excitation is able to spread across the whole of the heart.

17.5 The coronary circulation

The myocardium is a highly metabolically active tissue that requires a rich supply of blood to provide it with oxygen and nutrients, and to remove the waste products

of cellular metabolism. Its blood supply is provided by the **right** and **left coronary arteries**, which arise from the base of the aorta as it leaves the left ventricle. They circle around the heart in the **coronary sulcus**, lying between the atria and the ventricles. The right coronary artery branches to form the posterior interventricular and marginal arteries that mainly supply the right side of the heart, while the left coronary artery carries blood to the left atrium and left ventricle via its branches, the anterior **interventricular** and **circumflex arteries**. The cardiac veins drain the myocardium and empty into the right atrium via the coronary sinus. A diagrammatic representation of the coronary circulation is illustrated in Figure 17.7. Figure 17.8 shows an angiogram

of the coronary vessels with their extensive ventricular branches.

Blood flow through the coronary circulation varies considerably as the heart contracts and relaxes during the cardiac cycle. During systole, the heart muscle contracts and compresses the coronary vessels. During diastole, the heart muscle relaxes, so that the coronary vessels are released from the compression they underwent during systole. As a result, coronary blood flow is restricted during systole and is at its highest during early diastole, as shown in Figure 17.9.

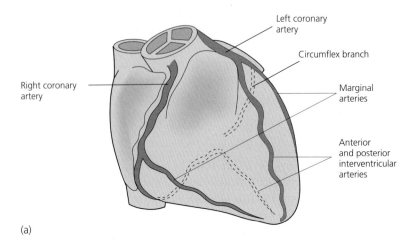

(a)

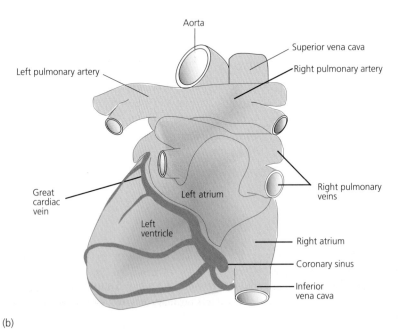

(b)

Figure 17.7 The organization of the coronary blood supply. (a) The arterial supply; note the origin of the coronary arteries from the aortic sinuses at the base of the aorta. (b) The venous drainage of the coronary circulation viewed from the posterior aspect.

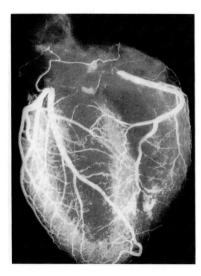

Figure 17.8 Angiogram of the coronary arteries from the back. Note the rich blood supply to the ventricles, especially the left, compared with the relatively sparse supply to the atria.

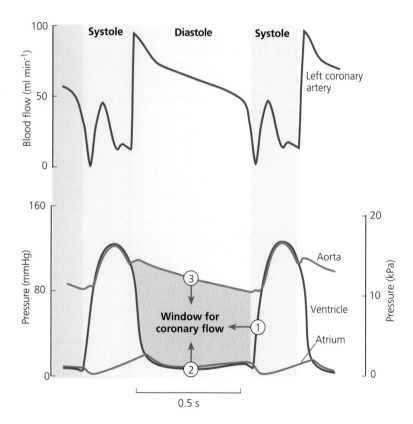

Figure 17.9 The blood flow through the left coronary artery during the cardiac cycle and the pressure changes that influence it. Blood flow is at its lowest during early systole but rises as the aortic pressure begins to rise. It falls during late systole as the coronary vessels become compressed before rising sharply in early diastole. Most coronary blood flow occurs during diastole (shown in purple in the lower panel of the figure). A shorter diastolic interval (1), a rise in ventricular end-diastolic pressure (2) or a fall in arterial pressure (3) will all result in a reduction of coronary blood flow.

The work done by the cardiac muscle in response to the demands of the circulation has a profound influence on the metabolic requirements of the myocardium: the harder the heart has to work, the more nutrients and oxygen it will need. Indeed, blood flow through the coronary circulation can increase from its resting level of about $75\,ml\,min^{-1}\,(100\,g\,tissue)^{-1}$ to as much as $400\,ml\,min^{-1}\,(100\,g\,tissue)^{-1}$ when the heart is working at its maximum capacity. The exact mechanisms that account for this matching of coronary blood flow to metabolic requirements are not entirely clear, but the most important determinant of blood flow appears to be **metabolic hyperaemia** (see p. 378), in which the breakdown products of cellular metabolism, such as adenosine, carbon dioxide, hydrogen ions and potassium, cause vasodilatation of the coronary arterioles.

> **KEY POINTS:**
> - Blood flow through the coronary arteries varies greatly during the cardiac cycle.
> - Flow to the myocardium is at its peak during early diastole when the mechanical compression of coronary vessels is minimal and aortic pressure is still high.
> - During the isovolumetric contraction phase of the cycle, the coronary blood vessels become compressed as the ventricular pressure rises. As a result, coronary blood flow declines to its minimum value.
> - Coronary blood flow is regulated by metabolic hyperaemia.

17.6 The origin of the heartbeat

Muscle contraction is initiated by electrical activity. The heart is no exception. However, unlike skeletal muscle, which can only contract when stimulated by its specific motor innervation, the rhythmic contractions of the heart are maintained by electrical signals generated within the heart itself. The heart displays **autorhythmicity**. Indeed, under appropriate conditions, the heart will continue to beat following removal from the body. It may be transplanted from one human to another and can continue to function normally in the absence of any extrinsic nerves. Even isolated cardiac muscle cells (myocytes) continue to show spontaneous contractile activity when kept in a warm nutrient medium. This behaviour may be viewed directly under a microscope. However, under such conditions it is also evident that heart cells all beat at their own rate when isolated from one another. Clearly this is not the case in the intact heart where there is highly synchronized contraction of the heart muscle. The atria contract first, followed by the ventricles, to bring about each beat of the heart. Each contraction is followed by a period of relaxation. Only in this way can the heart behave as an effective pump. Co-ordinated contraction of the myocardium is achieved by means of specialized conducting tissue.

The role of the sinoatrial node

In the normal heart, the electrical activity that will give rise to the heartbeat is initiated by the **sinoatrial node** (SA node or SAN), which is a small crescent-shaped sliver of tissue located in the wall of the right atrium, near the junction with the superior vena cava. The cells of the SA node act as the natural 'pacemaker' of the heart. From here, a wave of electrical excitation spreads across the whole of the heart to bring about a co-ordinated wave of contraction. In certain disease states, myocytes in other parts of the heart can show independent pacemaker activity, which causes irregular beating of the heart known as **arrhythmia**.

As for skeletal muscle, the contraction of a cardiac muscle fibre is associated with an action potential. The ionic mechanisms that govern the generation of resting and action potentials in excitable cells are described in Chapters 5 and 8. The pacemaker cells of the SA node spontaneously generate action potentials roughly every 600 ms (or about 100 times a minute). The resting membrane potential of these cells is unstable and fluctuates spontaneously, as shown in Figure 17.10. Immediately after an action potential, the membrane potential is at its most negative (around –60 mV). It then

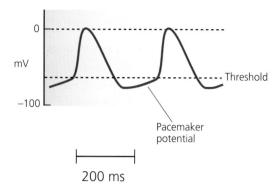

Figure 17.10 The action potentials of the pacemaker cells of the sinoatrial node of the frog heart. Note the absence of a stable resting membrane potential and the slow depolarization (the pacemaker potential) that occurs between successive action potentials.

falls slowly (i.e. it becomes less negative) as a result of the slow movement of sodium ions (which are positively charged) into the cell, until it reaches a value of about –50 mV, which is the threshold for the generation of an action potential. The slow depolarization that precedes the action potential is called the **pacemaker potential**. The slope of the pacemaker potential determines how quickly the membrane potential falls towards the threshold and is an important factor in determining the heart rate. Autonomic nerves, certain hormones (e.g. adrenaline) and those drugs that alter the heart rate all act by influencing the rate at which the pacemaker potential reaches the threshold for excitation.

Conduction of the impulse throughout the myocardium

The action potential initiated in the SA node spreads throughout the entire myocardium, via gap junctions between adjacent cells. Excitation spreads first across the whole of both atria. It then passes to the ventricles via the atrioventricular (AV) node, which forms a bridge of conducting tissue between the atria and the ventricles. Conduction through the AV node is relatively slow and the spread of excitation to the ventricles is delayed by about 0.1 s at this point. This ensures that the atria have time to contract fully before the ventricular muscle is excited.

Conduction through the remainder of the myocardium is quite rapid. It occurs via the **bundle of His**, which divides into the right and left bundle branches that supply the right and left ventricles, as shown in Figure 17.11. The bundle branches consist of specialized, large-diameter myocytes arranged end to end to permit the rapid conduction of the wave of excitation from the AV node to an extensive network of large fibres known as **Purkinje fibres** that lies just beneath the endocardium (Figure 17.12). The Purkinje fibres then spread the excitation rapidly to the ventricular myocytes.

KEY POINTS:

- The heart beats spontaneously and shows an inherent rhythmicity that is independent of any extrinsic nerve supply.

- Excitation is initiated by a group of specialized cells in the sinoatrial node that lies close to the point of entry of the great veins into the right atrium.

- A wave of depolarization is then conducted across the atria via the atrial myocytes.

- When the excitation reaches the atrioventricular node, there is a slight delay before the conducting tissue and Purkinje fibres spread the excitation throughout the ventricular myocardium.

Action potentials in myocytes from atria, ventricles and conducting fibres

Figure 17.11 shows the action potentials of myocytes in various parts of the myocardium and illustrates some important differences between them. As explained earlier, the cells of the SA node have a slowly developing action potential of relatively short duration (150–200 ms). Myocytes from the atria, ventricles and conducting system have action potentials with different characteristics. Although these action potentials vary in duration, they all show a fast initial upstroke followed by a plateau phase of depolarization prior to repolarization. The plateau phase is due to the movement of positively charged calcium ions into the cell and is longest in the Purkinje fibres (Figure 17.11).

(a)

(b)

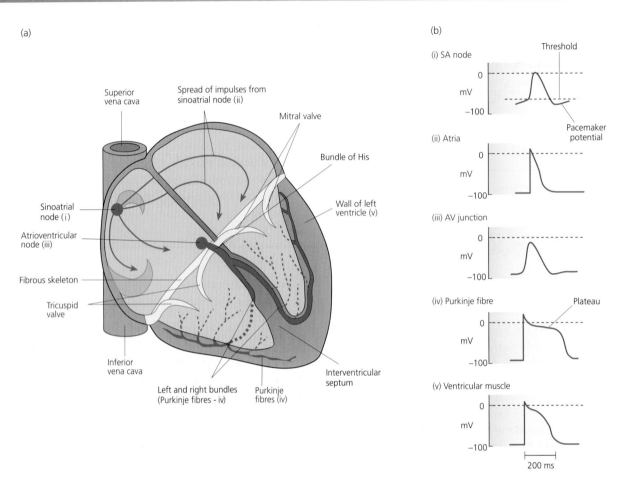

Figure 17.11 A diagram showing the conducting tissue of the heart (a) and the characteristic appearance of action potentials recorded from various types of cardiac myocyte (b). Note the presence of the pacemaker potential in the cells of the SA node, lack of a specific conducting pathway in the atria and elaborate branching of the conducting tissue that supplies the ventricles (shown in blue). The atria are separated from the ventricles by fibrous tissue (shown in yellow) that does not propagate action potentials.

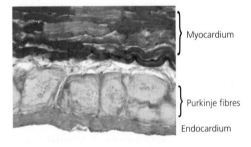

Figure 17.12 A section through the wall of a ventricle that shows the innermost layer, the endocardium, Purkinje fibres and a mass of densely staining muscle tissue. Note the large diameter of the Purkinje fibres and their pale staining.

Like the membrane of a nerve axon, the myocyte membrane is refractory (non-excitable) during, and immediately following, the passage of an action potential. The long action potential of the Purkinje fibres thus ensures that the ventricular myocytes cannot be re-excited until they have fully relaxed in early diastole. The electrical activity of the heart can be monitored using an electrocardiogram (ECG). The principles of this recording technique are explained in Chapter 18.

KEY POINTS:

- The myocytes of the atria, ventricles and conducting system have action potentials with different characteristics.

- Although these action potentials vary in duration, they all show a fast initial upstroke followed by a plateau phase of depolarization prior to repolarization.

- The plateau phase ensures that the action potential lasts almost as long as the contraction of the cell and ensures the unidirectional excitation of the heart muscle.

17.7 The heart as a pump – the cardiac cycle and the heart sounds

The electrical events described in the previous section govern the mechanical activity of the heart. Excitation causes the myocardial cells to contract while repolarization causes them to relax. This alternating contraction and relaxation of the myocardium allows the heart to act as a pump, propelling blood from the venous to the arterial systems of the systemic and pulmonary circulations. This repeating pattern of contraction and relaxation is called the **cardiac cycle** and consists of two phases:

- **Diastole** during which the chambers of the heart relax and fill with blood

- **Systole** during which the chambers contract and eject blood into the pulmonary and systemic circulations.

In a person at rest, the heart rate is around 70 beats per minute and each cardiac cycle lasts for 0.8–0.9 s; systole lasts for approximately 0.3 s and diastole for 0.5 s.

During diastole, the entire myocardium is relaxed. The right atrium and right ventricle fill with deoxygenated blood returning from the systemic circulation via the great veins (the venous return) while the left atrium and left ventricle fill with oxygenated blood arriving from the lungs via the pulmonary veins. While the atria are relaxed (**atrial diastole**), blood flows from the atria into the ventricles through the open AV valves. The atria contract (**atrial systole**) before the ventricles and force some additional blood into the ventricles. The ventricles contract 0.1–0.2 s later (**ventricular systole**). The

volume of blood in a ventricle just before it begins to contract is called the **end-diastolic volume**. The volume of blood ejected by a ventricle during systole is called the **stroke volume**, which is normally about two-thirds of the end-diastolic volume. The volume of blood remaining in a ventricle after systole is called the **end-systolic volume**. In an adult at rest, the stroke volume is normally around 70–75 ml and the end-systolic volume is around 50 ml.

KEY POINTS:

- The alternating contraction and relaxation of the heart muscle is known as the cardiac cycle. During diastole, the chambers of the heart are relaxed and fill with blood. During systole, they contract to eject blood.

- Towards the end of diastole, the atria contract, increasing the volume of blood contained in the ventricles by about 20%. The total amount of blood filling the ventricles at the end of diastole is called the end-diastolic volume.

- Nearly two-thirds of the blood in the ventricles is expelled during systole. The volume ejected by each ventricle is known as the stroke volume and is roughly 70 ml in an adult at rest.

The action of the heart valves during the cardiac cycle

The ability of the ventricles to fill under low pressure and to eject blood against high arterial pressures is critically dependent on the precise operation of the AV and semilunar valves. Opening and closing of the AV valves occurs as a result of the pressure differences between the atria and ventricles that occur during the cardiac cycle. These pressure changes are illustrated in Figure 17.13. When the ventricles are relaxed, atrial pressures exceed ventricular pressures because blood is entering the atria from the venous circulation. Consequently, the AV valves remain open during diastole. When the atria contract at the start of systole, an additional volume of blood is forced into the ventricles. (This increase in the volume of blood contained in the ventricle occurs just before the R wave of the ECG; Figure 17.13.) As the ventricles start to contract, the ventricular pressures begin to rise. As soon as the pressure within the ventricles exceeds the atrial pressure, the AV valves close and blood is prevented from flowing back into the atria. As the total surface area of the cusps of the AV valves is much greater than that of the opening they cover, the upper (atrial) surfaces of the valves are pressed firmly together. This ensures that the valves remain firmly closed while the ventricles change size in systole.

During ventricular systole, the aortic and pulmonary valves open only when the pressures in the ventricles exceed those in the aorta and pulmonary artery. During this phase of the cardiac cycle, blood flows from the left ventricle into the aorta and from the right ventricle into the pulmonary artery. As the ventricles begin to relax at the start of diastole, ventricular pressures fall below those of the aorta and pulmonary artery and the aortic and pulmonary valves close. This prevents reflux of blood from the arterial system into the ventricles during diastole.

Note that, during diastole, the AV valves are open and the aortic and pulmonary valves are closed. In the ejection phase of systole, the situation is reversed, with the aortic and pulmonary valves open and the AV valves closed.

Closure of the heart valves gives rise to the heart sounds

Closure of the heart valves sets up oscillations in the blood within the chambers of the heart and in the wall of the heart itself. These oscillations may be heard with a stethoscope during clinical examination of the chest. In healthy adults, two heart sounds are normally audible. The **first heart sound** begins to be heard at the onset of ventricular systole and is associated with closure of the AV valves. This sound begins immediately after the

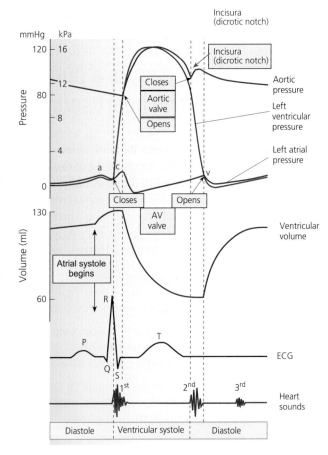

Figure 17.13 A simplified diagram to show the major mechanical and electrical events of the cardiac cycle. The pressure changes shown are for the left side of the heart and reflect the underlying mechanical events. The ECG and heart sounds are shown at the bottom of the figure. It is particularly important to note the relative timing of the various events. Thus, for example, the QRS complex (which reflects ventricular depolarization) largely precedes ventricular contraction, while the first heart sound is heard as the atrioventricular (AV) valves close following the start of the rise in intraventricular pressure. See text for further explanation.

R wave of the ECG (see Chapter 18). The **second heart sound** is heard when the aortic and pulmonary valves close at the start of ventricular relaxation. It is quieter than the first sound and has a more 'snapping' quality (hence the commonly used terms 'lub' and 'dup' for the first and second heart sounds). The second heart sound coincides with the end of the T wave of the ECG (see the lower part of Figure 17.13).

Two further heart sounds may be detected in some normal individuals, particularly with the aid of electronic amplification to provide a graphical record known as a **phonocardiogram**. The third heart sound occurs roughly 140–160 ms after the second sound, as the AV valves open and blood starts to move rapidly from the atria into the ventricles, creating turbulent flow. It is most often audible in children in whom the chest wall is relatively thin. A fourth heart sound may occasionally be heard. It occurs just before the first heart sound and is thought to be caused by oscillations in the blood flow set up by atrial contraction. Abnormal heart sounds are usually (though not invariably) the result of disease and are called **murmurs**.

KEY POINTS:

- During the cardiac cycle, various sounds can be heard using a stethoscope applied to the chest. These are called the heart sounds.

- The first heart sound arises as the atrioventricular valves close and the second heart sound corresponds to closure of the pulmonary and aortic valves.

- Two other heart sounds can occasionally be heard in normal subjects, the third and fourth heart sounds.

- Abnormal heart sounds are often the result of disease and are called murmurs.

Pressure changes during the cardiac cycle

Figure 17.13 illustrates the major events occurring during a single cardiac cycle for the left side of the heart (the period between the end of one contraction of the heart and the end of the next). Pressure changes in the aorta, left ventricle and left atrium are shown, in addition to the ventricular volume, the ECG and the phonocardiogram. The pattern in the right side of the heart is similar but the pressures are lower.

Pressure changes in the atria

Three major increases in pressure, known as the a, c and v waves, can be seen in the left atrium in Figure 17.13. Atrial contraction is preceded by depolarization of the atrial myocardium (the P wave of the ECG) and the a wave reflects the increase in atrial pressure that occurs during contraction. The c wave is due to the bulging of the mitral valve into the left atrium as the AV valves close at the start of ventricular contraction. The v wave reflects the slow rise in atrial pressure seen as blood continues to flow from the pulmonary veins into the left atrium during ventricular contraction. The pressure in the left atrium falls as soon as the AV valves open at the end of ventricular systole. As there are no valves between the right atrium and the internal jugular vein, pressure changes in the right atrium are reflected in the **jugular pulse**, which is illustrated in Figure 17.14. Unlike the arterial pulse, the jugular pulse cannot be felt by the finger tips (the pressures are too low) but the venous pulsations can often be observed in the neck when a subject is semirecumbent. It is more prominent when the venous pressure is higher than normal – when there is fluid overload or right-side heart failure for example.

Pressure and volume changes in the ventricles

Ventricular contraction starts at the peak of the R wave of the ECG (the QRS complex represents ventricular depolarization and precedes ventricular contraction; see Chapter 18). The AV valves close at the start of contraction to produce the first heart sound, as shown on the phonocardiogram record in Figure 17.13. For a short time (0.02–0.03 s), the pressures in the pulmonary artery and aorta keep the pulmonary and aortic valves closed. Consequently, there is no change in volume even though the ventricles are contracting. This is called the period of isovolumetric contraction. During this phase, the pressures within the ventricles rise until they exceed those in the pulmonary artery and aorta when the pulmonary and aortic valves open.

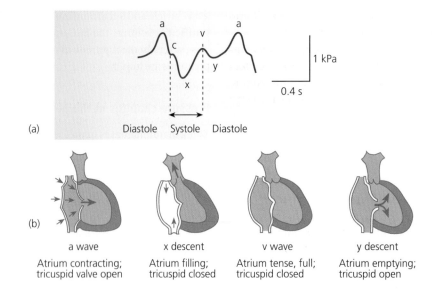

(a)

Diastole Systole Diastole

(b)

| a wave | x descent | v wave | y descent |
| Atrium contracting; tricuspid valve open | Atrium filling; tricuspid closed | Atrium tense, full; tricuspid closed | Atrium emptying; tricuspid open |

Figure 17.14 The origin of pressure waves recorded from the internal jugular vein. (a) A normal jugular pulse. (b) The events that give rise to the component waves of the jugular pulse.

The ejection of blood during ventricular systole is rapid at first but progressively slows towards the end of systole – this can clearly be seen in the change in ventricular volume shown in Figure 17.13. During the phase of rapid ejection, ventricular pressure continues to rise steeply until it reaches a maximum. For about the last quarter of ventricular systole, very little blood flows into the aorta and pulmonary artery even though the ventricles remain contracted.

At the end of ventricular systole, the ventricles repolarize (the T wave of the ECG trace) and begin to relax (diastole) so that the pressures within the ventricles fall rapidly. As the pressures in the ventricles fall below those in the aorta and pulmonary artery, the aortic and pulmonary valves close. This prevents the backflow of blood into the ventricles. The closure of these valves produces the second heart sound, which is shown in the bottom trace of Figure 17.13.

During the first part of ventricular diastole, the intraventricular pressures continue to fall but the volume of blood in the ventricles does not change as the AV valves are also closed. This period is known as the phase of isovolumetric relaxation and lasts for 0.03–0.06 s before the ventricular pressures fall below those in the atria, permitting the mitral and tricuspid valves to open. As the

ventricles continue to relax, they start to fill with blood (Figure 17.13). About two-thirds of the way through ventricular diastole, the atria depolarize (this is seen as the P wave of the ECG) and then contract, forcing additional blood into the ventricles. This may be seen clearly in Figure 17.13, which shows a second component to the rising phase of the ventricular volume curve corresponding to atrial systole. This amounts to about 20% of the end diastolic volume.

Pressure changes in the aorta

Blood begins to flow rapidly into the aorta from the left ventricle as soon as the left ventricular pressure exceeds the aortic pressure, so allowing the aortic valve to open. The blood forced from the ventricles stretches the walls of the aorta, and the aortic pressure increases to around 120 mmHg (16 kPa), which is the **systolic blood pressure** (see Chapter 19). As the flow of blood from the ventricles falls, the pressure in the aorta begins to decline (Figure 17.13). When the aortic valve closes, there is a brief surge in pressure that gives rise to the **incisura** or **dicrotic notch**. After this, the aortic pressure falls slowly throughout diastole, as the stretched elastic tissue of the aorta forces blood into the systemic circulation. By the time the ventricles contract again, the aortic pressure has fallen to around 80 mmHg (10.6 kPa), the **diastolic pressure**.

The pressure curve for the pulmonary artery has similar characteristics to that of the aorta except that the pressures are much lower: the systolic pressure in the pulmonary artery is about 25 mmHg (3.3 kPa) and the diastolic pressure is about 8 mmHg (1 kPa).

> **KEY POINTS:**
>
> - Blood flow in the arteries is pulsatile and reflects the events of the cardiac cycle.
> - Pressure at the peak of ejection is called the systolic pressure while the lowest pressure occurs at the end of diastole (the diastolic pressure).

17.8 Cardiac output

The volume of blood pumped from one ventricle each minute is known as the **cardiac output**. It is the product of the heart rate (in beats per minute) and the stroke volume (ml). Thus:

Cardiac output = Heart rate × Stroke volume

So, for a person with a heart rate of 70 beats per minute and a stroke volume of 70 ml, cardiac output would be 4900 ml min^{-1} (or 4.9 l min^{-1}). The cardiac output can be directly measured using the Fick principle (see Box 17.1).

The **venous return** is the volume of blood returning to the heart from the systemic circulation every minute and it is essential that, except for very small transient alterations, the cardiac output is equal to the venous return and vice versa. In other words, the heart must be able to pump a volume equivalent to that which it receives. Typically, resting cardiac output in an adult lies between 4 and 7 l min^{-1}, but throughout life the cardiac output varies according to the oxygen requirements of the body tissues. Cardiac output is reduced during sleep, for example, and raised in fear, during periods of excitement or following a heavy meal. A much larger rise in cardiac output occurs during exercise. Indeed, during very strenuous activity, cardiac output may increase to five or six times its resting value. This increase is mainly due to a faster heart rate, but stroke volume is also increased (see Chapter 32). Regulation of the cardiac output is achieved by autonomic nerves, by certain hormones and by mechanisms that are intrinsic to the cardiovascular system. These are considered below.

Nervous and hormonal control of heart rate

Although cardiac muscle is able to contract without stimulation by extrinsic nerves (it has inherent rhythmicity), it is supplied with parasympathetic and sympathetic autonomic nerves, which are both able to influence the heart rate. Physiological changes in the heart rate are known as **chronotropic effects**. The parasympathetic supply to the heart is via the vagus nerves, which, when activated, slow the heart (negative chronotropy). Stimulation of the sympathetic nerves increases the heart rate (positive chronotropy).

Clinically, a resting heart rate that is below 60 beats min^{-1} is called a **bradycardia** while a resting heart rate faster than 100 beats min^{-1} is called a **tachycardia**. These terms are also often used more loosely to refer to any slowing (bradycardia) or speeding up (tachycardia) of the heart rate.

The resting heart is dominated by the parasympathetic innervation

Vagal nerve fibres synapse with postganglionic parasympathetic neurones in the heart itself (see Chapter 13) and the short postganglionic fibres synapse mainly on the cells of the SA and AV nodes. They release the neurotransmitter acetylcholine from their nerve terminals, which act to slow down the heart rate and slow the rate of conduction of the cardiac impulse from the atria to the ventricles by reducing the excitability of the atrioventricular node.

Box 17.1 Application of the Fick principle to the determination of cardiac output

A variety of methods have been devised to measure the cardiac output. The easiest to understand is known as the Fick principle, which states that the total uptake or release of a substance by an organ is equal to the blood flow through that organ multiplied by the difference between the arterial and venous concentrations of the substance. To measure cardiac output (CO), this principle is applied to the uptake of oxygen by blood flowing through the lungs (remember that the output of the right and left sides of the heart are the same and that the entire cardiac output flows through the lungs). The Fick equation for the uptake of oxygen by the blood as it passes through the lungs is:

$$CO = \frac{\text{Oxygen uptake (ml min}^{-1})}{Cv_{O_2} - Ca_{O_2}}$$

where Cv_{O_2} is the oxygen content (NOT the partial pressure) of the blood in the pulmonary veins (which is the same as that of normal arterial blood) and Ca_{O_2} is the oxygen content of the blood in the pulmonary arteries (which is the same as fully mixed venous blood).

In practice, oxygen consumption is determined using spirometry (see Chapter 22) and the oxygen content of pulmonary venous blood can be obtained from a sample of blood obtained from the radial, brachial or femoral artery. The oxygen content of pulmonary arterial blood is more difficult to determine as it requires a sample of fully mixed venous blood, which can only be obtained from the right ventricle or pulmonary artery itself. Normally, a catheter is introduced through the antecubital vein and passed into the outflow of the right ventricle or into the pulmonary artery itself. The blood samples are then analysed and the cardiac output calculated using the equation given above.

A worked example

Assume that a person consumes 250 ml of oxygen each minute from the inspired air, that the oxygen content of the pulmonary arterial blood is 15 ml per 100 ml of blood and that the oxygen content of the pulmonary venous blood is 20 ml per 100 ml blood. Each 100 ml of blood passing through the lungs must therefore have taken up 5 ml of oxygen, which is the equivalent of each ml of blood taking up 0.05 ml of oxygen. Entering these values into the Fick equation:

$$CO = 250/0.05$$
$$= 5000 \text{ ml min}^{-1} \text{ or } 5.0 \text{ l min}^{-1}$$

A healthy adult heart normally beats at around 60–80 beats min^{-1} at rest. However, a heart that has been denervated (i.e. has been separated from its nerve supply) will beat much faster, at about 100 beats min^{-1}. This is the intrinsic rate of discharge of the cells of the SA node. It is evident from this observation that the parasympathetic nerve supply to the heart normally exerts a tonic inhibitory effect on the SA node to slow the intrinsic heart rate. Furthermore, drugs such as atropine, which antagonizes the effects of acetylcholine on the heart, cause the heart rate to increase.

The acetylcholine secreted by the nerve terminals of the vagal nerve fibres increases the permeability of the SA node cells to potassium ions. This has two effects: it decreases the slope of the pacemaker potential and hyperpolarizes the membrane potential. These changes increase the time it takes for the pacemaker potential to reach the threshold potential at which an action potential is triggered. As the interval between successive action potentials lengthens, the heart rate falls. These effects are illustrated in Figure 17.15a and c.

Stimulation of the sympathetic nerves increases the heart rate

The sympathetic preganglionic nerves that supply the heart synapse in the ganglia of the thoracic sympathetic chain (see Chapter 13). From here, they project to the heart via long postganglionic fibres. When stimulated, the nerve endings of these fibres secrete noradrenaline whose effect is to increase the slope of the pacemaker potential so that the threshold for action potential generation is reached more quickly. The time between successive action potentials is reduced and the heart

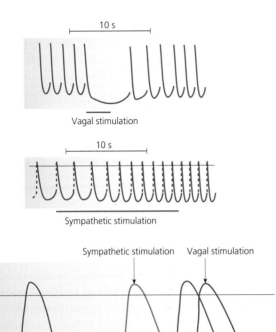

Figure 17.15 The effect of stimulation of the sympathetic and the parasympathetic (vagal) nerves on the pacemaker activity of the frog heart. (a) The effect of vagal stimulation. Note that the hyperpolarization of the pacemaker cell stops the heart. After the period of stimulation, the slope of the pacemaker potential is reduced and the heart rate is slowed compared with the period before stimulation. (b) The effect of stimulating the sympathetic nerves. Note the increased slope of the pacemaker potential and increased heart rate following stimulation. (c) A diagrammatic representation of the effects of sympathetic and parasympathetic stimulation on the pacemaker potential. Note that strong vagal (parasympathetic) stimulation both hyperpolarizes the membrane potential and reduces the slope of the pacemaker potential.

speeds up. These effects are illustrated in Figure 17.15b. The conduction time through the AV node is also reduced by noradrenaline.

To summarize: at a given time, the heart rate is largely determined by the balance between the activity in parasympathetic and sympathetic nerves supplying the heart. When physiological circumstances require the heart to beat more rapidly, as in exercise, the activity of the parasympathetic nerves is inhibited, while that of the sympathetic nerves is enhanced. In a healthy person, maximal sympathetic stimulation can increase the resting heart rate to around 180–200 beats min^{-1}. During periods of rest, the parasympathetic activity of the vagus dominates and the heart rate falls to around 70 beats min^{-1}.

Hormonal control of the heart rate

Although the autonomic nervous system is primarily responsible for regulating the heart rate, a number of hormones also exert chronotropic effects. In particular, adrenaline and noradrenaline from the adrenal medulla increase the heart rate and contribute to the tachycardia that occurs during physical or psychological stress. Thyroid hormone also exerts a positive chronotropic effect (see Chapter 16).

> **KEY POINTS:**
>
> - Heart rate is governed by the influence of the autonomic nerves on the rate of discharge of the pacemaker cells of the sinoatrial node.
>
> - Parasympathetic stimulation slows the heart rate (negative chronotropy).
>
> - Sympathetic stimulation and circulating catecholamines from the adrenal medulla both increase the heart rate (positive chronotropy).

The regulation of stroke volume

Stroke volume is regulated by two different mechanisms:

- **Intrinsic regulation** of the force of contraction of the ventricular myocardium
- **Extrinsic regulation** via autonomic nerves and circulating hormones.

Intrinsic regulation of stroke volume: the Frank–Starling relationship

The force with which the heart contracts during systole is largely determined by the degree of stretch of the myocardial fibres at the end of diastole. For the normal working range of the heart, cardiac muscle responds to increased stretch with a more forceful contraction. This property of the heart muscle forms the basis of **Starling's Law of the Heart**, which states that 'the energy of contraction of the ventricle is a function of the initial length of the muscle fibres comprising its walls'. Starling's Law is now more often referred to as the **Frank–Starling relationship**, honouring both physiologists who first discovered it.

The underlying mechanism is as follows:

- As blood fills the heart during diastole, the pressure within the ventricle gradually rises until the end-diastolic volume is reached.
- The greater the end-diastolic volume, the greater the end-diastolic pressure. This is known as the **preload**.
- The end-diastolic pressure determines how much the cardiac muscle fibres are stretched.
- The greater the degree of stretch, the more forcefully the heart contracts during the subsequent systole.

Figure 17.16a illustrates the increase in stroke volume that occurs as the end-diastolic volume increases. The end-diastolic pressure in the right side of the heart is largely determined by the central venous pressure (see below). For a normal heart, the cardiac output (stroke volume × heart rate, see p. 341) increases as central venous pressure rises (Figure 17.16b).

The importance of this relationship lies in the fact that, during systole, the ventricle will always eject the volume

of blood that entered it during diastole. In other words, across its normal working range, *the heart automatically adjusts its cardiac output to match its venous return.* The Frank–Starling mechanism normally ensures that the outputs from the right and left ventricles are closely matched. Such matching is essential, as even a small difference between outputs will quickly result in potentially serious alterations in the distribution of blood between the pulmonary and systemic circulations.

If output from the right ventricle exceeds that from the left, after a few heart beats the volume of blood in the pulmonary circulation will be increased slightly and pressure in the pulmonary veins will rise. As a result, venous return to the left side of the heart will increase, resulting in a greater end-diastolic volume in the left ventricle. This in turn will lead to an increase in left ventricular stroke volume, through the Frank–Starling mechanism, which will restore the balance between the outputs of the two ventricles.

The pressure in the aorta opposes the ejection of blood from the ventricles and represents the load against which the heart must pump the blood. For this reason, it is known as the **afterload**. An increase in afterload will cause a fall in stroke volume. However, the fall will normally be only transient because of the Frank–Starling mechanism: the venous return from the lungs will remain the same so, as stroke volume falls, the end-diastolic volume of the left ventricle will increase and the ventricle becomes more distended (i.e. there will be an increase in degree of stretch of the myocardial fibres resulting in an increased preload). As a result, the myocardium contracts more forcefully and

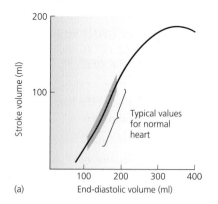
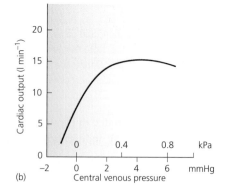

Figure 17.16 The Frank–Starling relationship. (a) The relationship between end-diastolic volume and stroke volume determined using an isolated heart–lung preparation. (b) The relationship between cardiac output and central venous pressure in the intact heart (remember that the higher the end-diastolic volume, the greater the central venous pressure).

more blood is ejected. This restores the end-diastolic volume and stroke volume to their normal values.

What factors determine the end-diastolic volume?

The end-diastolic volume is an important determinant of stroke volume and thus of cardiac output and cardiac work. It is therefore important to understand the factors that affect this volume. Essentially, these fall into two categories:

- Factors affecting pressure outside the heart (the intrathoracic pressure)
- Factors affecting pressure inside the heart (the central venous pressure and the force with which the atria contract).

The **intrathoracic pressure** is altered by breathing (see Chapter 22). During inspiration, the diaphragm contracts so that the volume of the chest is increased and the volume of the abdominal cavity is reduced. These volume changes result in an increase in intra-abdominal pressure and a fall in intrathoracic pressure. The pressure difference favours the flow of blood from the abdominal veins to the thoracic veins and enhances filling of the right ventricle. End-diastolic volume thus tends to be greater during inspiration than during expiration.

Central venous pressure (**CVP**) is the term used to describe the pressure inside the right atrium at the end of diastole. It is equal to the pressure in the superior and inferior venae cavae as they enter the right atrium and it is the main determinant of ventricular filling (i.e. the end-diastolic volume). It therefore follows from the Frank–Starling relationship that CVP is an important determinant of cardiac output. The relationship between CVP and cardiac output is shown in Figure 17.16b.

The CVP may be influenced by a number of factors, for example:

- Gravity
- Respiration (see above)
- Compression of the deep veins during movement (the muscle pump), which displaces blood from the limbs into the central veins
- Peripheral venous tone (i.e. the degree of constriction of the veins); strong sympathetic stimulation, for example, will cause veins to constrict and increase venous return and cardiac output

- Blood volume; after a major haemorrhage blood volume is lower and venous return falls, which results in a lower CVP.

Extrinsic regulation of the stroke volume by sympathetic nerves and circulating catecholamines

Although the initial length of the heart muscle fibres is a major determinant of their **contractility** (force of contraction), extrinsic factors also contribute. Such factors are said to exert an **inotropic** effect on the heart muscle. Agents that enhance the force of contraction and thus the stroke volume are said to exert a positive inotropic effect while those that reduce the stroke volume exert a negative inotropic effect.

Positive inotropic agents

The sympathetic nerve supply. The myocardium receives a rich innervation of sympathetic nerve fibres. Increased activity in these sympathetic fibres exerts a positive inotropic effect on the myocardium of both the atria and the ventricles so that, for a given end-diastolic volume, the force of contraction of the heart, and thus the stroke volume, will be enhanced as the ventricles are emptied more completely. This positive inotropic effect is illustrated in Figure 17.17.

Circulating catecholamines. An increase in the secretion of adrenaline and noradrenaline (epinephrine and norepinephrine) from the adrenal medulla will exert a positive inotropic effect on the myocardium. These hormones reach all parts of the heart by way of the coronary vessels.

At the same time as increasing the contractility of the myocardium, the circulating catecholamines and increased sympathetic activity also shorten the duration of systole by enhancing the rate at which the heart contracts. This adaptation helps to maintain the stroke volume as the heart rate increases.

Other circulating positive inotropic agents. A number of other substances circulating in the blood can exert effects on the contractility of cardiac muscle. These include calcium ions, thyroxine and certain drugs such as caffeine, theophylline and the cardiac glycoside **digoxin**. All are thought to act by raising the level of free intracellular calcium, either directly (in the case of calcium itself)

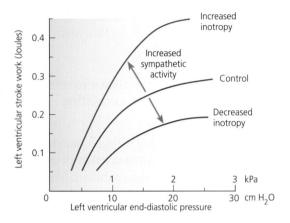

Figure 17.17 The relationship between the force of contraction (expressed here as stroke work) and end-diastolic pressure under control conditions is shown in red. When the inotropic state of the heart is enhanced by sympathetic activity, the relationship is steeper so that the heart contracts more forcefully for a given left end-diastolic pressure (blue line). Negative inotropic agents have the reverse effect, weakening the force of contraction for a given end-diastolic pressure (purple line). (Remember that the higher the end-diastolic volume, the greater the end-diastolic pressure.)

or indirectly, by stimulating the release of calcium from intracellular stores (see Chapter 7).

Negative inotropic agents

Negative inotropic agents are substances that reduce the force with which the heart contracts. As there is virtually no parasympathetic innervation to the ventricular myocardium in humans, parasympathetic activation has no direct effect on the stroke volume for a given end-diastolic volume. However, a number of drugs do exert a negative inotropic effect on the heart. These include **propranolol** (a β-blocker), calcium-channel blockers and many general anaesthetics, including the barbiturates (e.g. thiopentone) and halothane. A low blood pH (acidaemia) also has a negative inotropic effect.

Variation of cardiac output in the normal and denervated heart

The cardiac output is the product of the heart rate and the stroke volume (see p. 341). In a healthy young adult, the heart rate can increase from around 70 beats min^{-1} at rest to as much as 200 beats min^{-1} during heavy

exercise. The stroke volume is normally about 70 ml but it too can increase during exercise by about 30–40%. As a consequence of these changes, cardiac output can vary between 5 l min^{-1} at rest up to about 25 l min^{-1} in severe exercise. In trained athletes, the range is much wider. As a result of the physical training, the heart muscle is able to generate a greater force of contraction. Consequently the resting heart rate may be as low as 40–50 beats min^{-1} while stroke volume is about 120 ml. During maximal effort, the cardiac output of such individuals can exceed 35 l min^{-1}. The effects of the sympathetic and parasympathetic divisions of the autonomic nervous system on cardiac output are summarized in Figure 17.18.

Heart transplant patients are able to increase their cardiac output to meet the demands of heavy exercise almost as efficiently as healthy people, despite the lack of sympathetic innervation to the heart itself. Part of this increase is due to the sympathetic effects on the heart rate of catecholamines released from the adrenal medulla, but to a large extent the increased cardiac output reflects the intrinsic ability of the heart muscle to contract more forcefully as it is stretched by the increase in venous return that accompanies the exercise.

KEY POINTS:

- Cardiac output is the product of heart rate and stroke volume and varies from 5 l min^{-1} at rest to over 25 l min^{-1} in severe exercise. The stroke volume is regulated by intrinsic and extrinsic mechanisms.

- The intrinsic regulation is expressed by Starling's Law of the Heart, which states that 'the energy of contraction of the ventricle is a function of the initial length of the muscle fibres comprising its walls'. The degree of stretch is determined by the end-diastolic pressure. Over any significant period, the output of the heart matches the venous return.

- Extrinsic regulation of the stroke volume is largely accomplished by the activity of the sympathetic nerves supplying the heart and circulating catecholamines secreted from the adrenal medulla. Both cause an increase in the force of contraction during systole for any given end-diastolic volume. This is known as a positive inotropic effect.

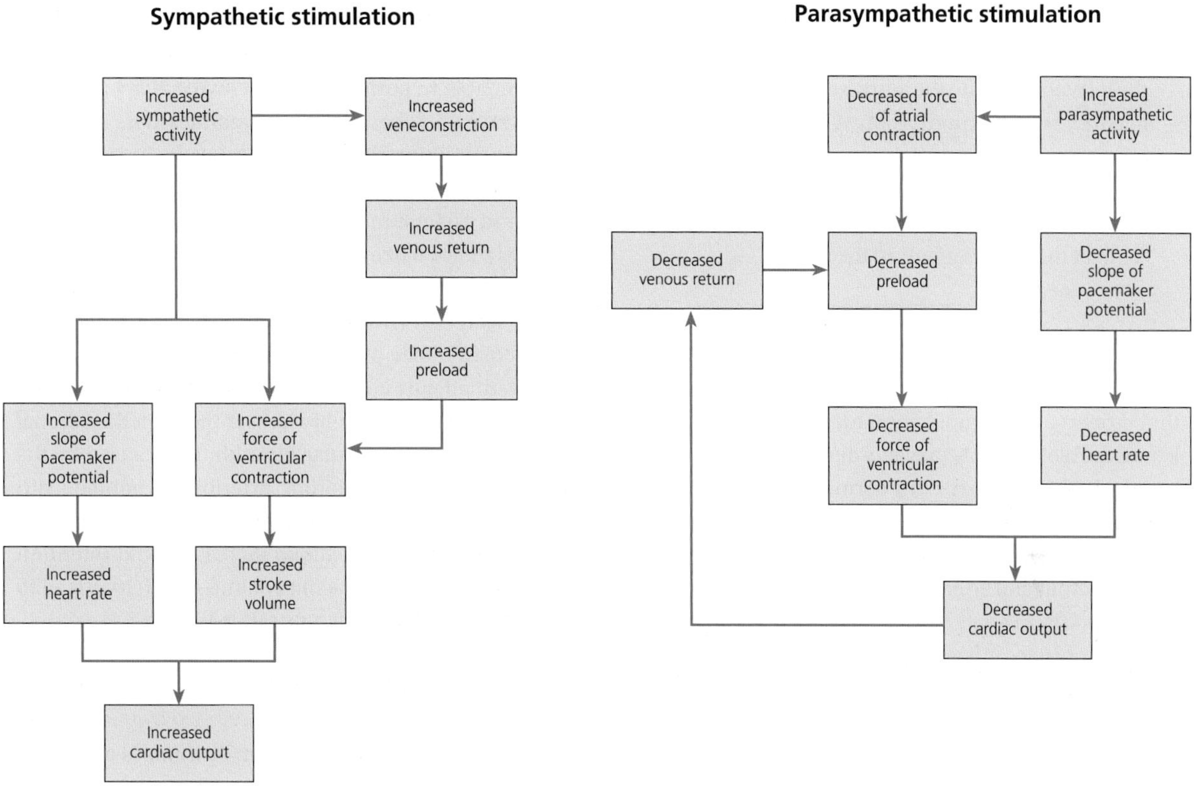

Sympathetic stimulation

Parasympathetic stimulation

Figure 17.18 A flow diagram summarizing the effects of sympathetic and parasympathetic activation on the cardiac output.

17.9 Heart failure

Heart failure occurs when the heart is unable to pump sufficient blood at normal filling pressures to meet the metabolic demands of the body. Failure may involve one of the two ventricles or both the left and right ventricles simultaneously. People with heart failure are unable to exercise normally and suffer from excessive fatigue because their cardiac output does not increase in proportion to the work performed, as it would in a healthy person.

The causes of heart failure fall into two main categories: cardiac and extracardiac. Heart failure of cardiac origin may result from the loss of healthy muscle mass caused by damage to the heart (myocardial infarction) or myocarditis (inflammation of the heart muscle). It may also result from conditions in which the filling or emptying of the heart is impaired, for example in aortic stenosis (narrowing of the aortic valve).

Extracardiac causes of heart failure refer to conditions that increase either the preload or afterload on the heart:

1. **Increased preload**. The heart performs more work when it pumps a given volume of blood from distended ventricles. As a result, a situation can develop where the coronary arteries are unable to supply sufficient oxygen to meet the requirements of the heart (myocardial ischaemia) and the initial heart failure is exacerbated. The preload is the tension that exists in the walls of the heart as a result of diastolic filling (see p. 344). It is therefore determined by the end-diastolic pressure. The preload may become excessively elevated in patients with renal failure who are likely to have an increased blood volume as a result of sodium and water retention.

2. **Increased afterload**. The afterload is the pressure that the heart must overcome in order to pump blood from the left ventricle into the aorta (see p. 344). It is elevated in patients with hypertension. Whenever the afterload is increased, the work of the heart is also increased for any given stroke volume. In this situation, the coronary arteries may be unable to supply sufficient oxygen to meet the requirements of the heart.

Acute heart failure

Consider the situation when a blood clot occludes one of the coronary arteries supplying the left side of the heart. The loss of the blood supply prevents the affected region of the myocardium from contracting normally so that the left ventricle does not pump the blood it contains as efficiently as it should (left-sided heart failure). The first effects are a fall in cardiac output and arterial blood pressure. These changes are followed by a variety of compensatory mechanisms that act to restore the cardiac output as far as possible.

The fall in arterial blood pressure will be detected by the arterial baroreceptors causing a reflex increase in sympathetic stimulation of the heart and constriction of the blood vessels. The secretion of adrenaline and noradrenaline by the adrenal medulla is also increased. Consequently, the heart rate will increase. This effect is supported by an increase in the contractility of the part of the myocardium that is still being normally perfused with blood. As a result of these two mechanisms, cardiac output increases although it may still be well below normal. Generalized vasoconstriction in response to sympathetic stimulation will, at the same time, serve to increase blood pressure and ensure that blood flow to the brain is maintained.

These changes occur within about 30 s of the occurrence of the coronary thrombosis and give rise to the characteristic symptoms of a heart attack: tachycardia, severe pain in the chest, which may radiate to the left arm (**angina pectoris**), pallor (resulting from vasoconstriction of the skin vessels) and sweating ('cold sweat').

Chronic heart failure

Chronic heart failure is characterized by the poor cardiac output that results from ischaemic damage to the myocardium, incompetence of the heart valves or chronic hypertension. This can be a relatively stable condition in which the patient may scarcely be aware of the situation until he or she undertakes exercise, when dyspnoea and fatigue set in rapidly, or it may be so severe that the person is unable to perform any physical activity at all and may even suffer chest pain and breathlessness at rest.

In heart failure, the cardiac output is diverted away from the skin and visceral organs in favour of the heart, brain and skeletal muscle. The reduction in renal blood flow has profound consequences: following the vasoconstriction of the afferent arterioles, there is an increase in the secretion of renin leading to elevated levels of angiotensin II (see pp. 480–481). This hormone has two important effects: it is a powerful vasoconstrictor and it stimulates the secretion of aldosterone from the adrenal cortex. The low glomerular filtration rate caused by the constriction of the afferent arterioles, coupled with the increased sodium retention caused by the elevated levels of aldosterone, leads to increased fluid retention and expansion of the plasma volume, which may lead to oedema by the mechanisms described below.

Oedema in cardiac failure

In chronic heart failure, the plasma volume is expanded through the mechanisms described above. This lowers the oncotic pressure (by diluting the plasma proteins) and increases the capillary pressure so that the Starling forces increasingly favour the movement of fluid out of the blood into the tissues (see p. 383). As a result, the tissues accumulate fluid and oedema results.

The oedema may become evident in the limbs or in the lungs depending upon which side of the heart is affected. In right-sided heart failure, the increased systemic capillary pressure leads to oedema in the periphery. This is most noticeable in the ankles. In acute left-sided heart failure (the most common form of ischaemic heart disease), there is a raised pressure in the left atrium and pulmonary vessels. The raised pressure in the pulmonary veins causes them to become engorged and there is increased back pressure on the pulmonary capillaries. The increased pressure in the pulmonary capillaries leads to greater transfer of fluid into the pulmonary interstitium and ultimately to pulmonary oedema. In this situation, gas exchange in the lungs is impaired and the oxygen content in the arterial blood is reduced, further compounding the effects of low cardiac output. A vicious cycle develops leading eventually to death. The factors that lead to oedema during chronic heart failure are summarized in Figure 17.19.

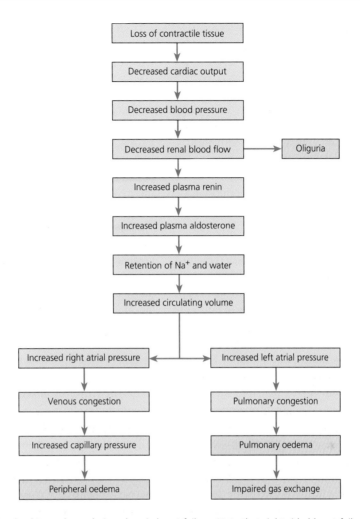

Figure 17.19 The factors that lead to oedema during chronic heart failure. Note that right-sided heart failure results in peripheral oedema, which is most obvious in the ankles. Pulmonary oedema reflects left-sided heart failure.

KEY POINTS:

- Heart failure occurs when the heart is unable to pump sufficient blood at normal filling pressures to meet the metabolic demands of the body. Failure may involve either of the two ventricles or both ventricles simultaneously.

- It may result from ischaemic damage to the myocardium, resulting in loss of contractility, or from incompetence or narrowing of the heart valves, resulting in increased preload. It may also be caused by excessive peripheral resistance resulting in increased afterload.

- When heart failure occurs, there is increased sympathetic activity – increased heart rate, increased contractility of the unaffected myocardium and vasoconstriction. These changes help to compensate for the reduced cardiac output.

17.10   Drugs that act on the heart

A large number of drugs are used in the treatment of conditions affecting the function of the heart (Table 17.1). These include drugs used in the treatment of angina pectoris, acute myocardial infarction, cardiac arrhythmias, hypertension and heart failure. Antiarrhythmic agents are discussed in Chapter 18, following discussion of the ECG, and those used in the treatment of hypertension are considered in Chapter 19. The other classes of cardioactive drugs are discussed briefly below.

Angina pectoris is a condition of myocardial ischaemia that usually results from the deposition of atheromatous plaques within the coronary arteries. As the arteries progressively narrow, there is a reduction in the blood supply to the cardiac muscle so that the supply of nutrients becomes insufficient to meet the needs of the tissue. This is especially evident during exercise when the metabolic demands of the myocardium are increased. Furthermore, waste products of muscle metabolism accumulate in the tissue and cause the intense pain and chest tightness that are characteristic of angina. Tobacco smoking is an important factor contributing to the development of atheroma and for some, giving up smoking can lead to a marked improvement in the symptoms of angina. For many others, however, drug treatment is necessary.

The principal aim of medications used in the treatment of angina is to reduce the work of the heart so that its limited blood supply is able to meet its oxygen demands. In most cases, the drugs of first choice are **nitrates** whose main effect is to bring about vasodilatation of peripheral blood vessels (especially veins), which causes pooling of blood in the venous part of the circulation and a reduction in venous return. As explained earlier, this will lead to a reduction in end-diastolic volume and thus cardiac output. Consequently, the work done by the heart will fall. **Glyceryl trinitrate** (**GTN**) given sublingually (under the tongue) as a tablet or spray is very effective in reducing the pain of acute angina. However, the effects are short lived (around 30 minutes) and it may be necessary to administer it as a transdermal patch, which remains effective for up to 24 hours. Some long-acting nitrate drugs are now also available and include **isosorbide dinitrate** and **isosorbide mononitrate**.

Side-effects of vasodilators such as the nitrates are predictable and include headaches (caused by vasodilatation of the cerebral blood vessels), hypotension and fainting. There may also be an increase in heart rate (reflex tachycardia) in response to the fall in cardiac output and if this occurs it may be necessary to combine the nitrate therapy with drugs such as **β-blockers** that slow the heart rate.

Calcium antagonists are used extensively in the treatment of angina. This group of drugs includes **diltiazem**, **verapamil**, **amilodipine** and **nifedipine** (although this is more rarely used in the treatment of angina due to uncertainty about its safety in these patients). They work by blocking calcium channels in the smooth muscle of artery walls to bring about relaxation and vasodilatation.

In patients who have developed a tolerance to the drug treatments described above or for whom they are ineffective, **nicorandil** may be beneficial. It brings about vasodilatation by activating potassium channels in the vascular smooth muscle (in particular of the veins), thereby reducing venous return and the preload on the heart.

Some patients have a form of angina called 'unstable angina' in which platelet aggregation occurs around eroded or ruptured atheromatous plaques. In such cases, there is a high risk of coronary thrombosis with subsequent myocardial infarction. To reduce this risk, **antiplatelet drugs**, such as **aspirin**, **clopidogrel** and **tirofiban**, are often used in combination with β-blockers and heparin to treat these patients.

Despite the variety of drugs available to treat angina, some patients will fail to respond to medication and the narrowing of their coronary arteries will progress to the point at which coronary artery bypass grafting (CABG) or percutaneous transarterial coronary angioplasty (PTCA) becomes necessary. In CABG, a section of blood vessel from elsewhere in the body (usually either the saphenous vein or the internal mammary artery) is grafted between the aorta and a point beyond the narrowing of the affected coronary artery. This should improve blood flow to the region of myocardium supplied by that artery. In PTCA, a balloon catheter is inserted into the affected artery and expands a mesh tube against the artery wall

Table 17.1 Drugs that act on the heart

Drug	Comments
Nitrates	Given orally for angina pectoris
Glyceryl trinitrate	
Isosorbide dinitrate	
Isosorbide mononitrate	
Calcium channel antagonists	Given orally for angina pectoris
Diltiazem	
Verapamil	
Amilodipine	
Nifedipine	
Potassium channel agonist	
Nicorandil	Given orally for angina pectoris, provides alternative therapy for patients for whom the drugs listed above are ineffective or contraindicated
Antiplatelet drugs	Often given with β-blockers and heparin in the treatment of unstable angina
Aspirin	
Clopidogrel	
Tirofiban	
Cardiac glycosides	Slow the heart and exert a positive inotropic effect
Digoxin	Used to treat heart failure
Ouabain	
Duretics	
Furosemide (frusemide)	Does not act directly on the heart but reduces the circulating volume. Used in combination with other drugs to treat heart failure
Antiarrhythmics	Also known as antidysrythmics. Used to treat irregular heart beat
Class 1	Subdivided into three subclasses; reduce rate of depolarization
IA quinidine, procainamide	(see Chapter 18 p. 363)
IB lignocaine	
IC flecainide	
Class II	All are β-adrenoceptor antagonists
Propranolol	
Atenolol	
Timolol	
Class III	Slow the heart by prolonging the refractory period
Amiodarone	(sotalol is also a β-adrenoceptor antagonist)
Sotalol	
Class IV	All are calcium channel antagonists and are listed above
Adenosine	Given to stop supraventricular tachycardia – now largely replaces verapamil

to compress the atheromatous plaque and widen the vessel.

Acute myocardial infarction (MI) occurs when a coronary artery becomes completely occluded, most commonly by a blood clot (thrombosis). The area of myocardium supplied by the occluded artery will become ischaemic and the tissues irreversibly damaged. Death may occur immediately or within hours of a severe MI, usually as a result of arrhythmias such as ventricular fibrillation (see Chapter 18). The rapid implementation of appropriate drug treatments can substantially increase a patient's chances of survival.

Drugs used in the treatment of acute MI are designed to limit the size of the infarct and to address the major complications of the MI, including hypoxia, pain, nausea and vomiting. In most cases, the following treatments will be appropriate:

- Aspirin in a low dose to inhibit platelet aggregation and prevent further clot formation

- Oxygen to reduce the risk of hypoxia

- Morphine or diamorphine (opiate analgesics) for pain relief

- Antiemetics such as cyclizine or metoclopramide

- Sedatives such as diazepam may be needed if the patient is very distressed, although opiate analgesics also have a sedative effect

- Thrombolytic ('clot busting') drugs such as streptokinase significantly reduce the mortality from MI and, unless contraindicated for the patient, are given as soon as possible after the onset of the infarct

- Anticoagulant drugs such as heparin may be helpful in reducing the risk of the patient developing a deep-vein thrombosis after acute MI

- β-Antagonists and angiotensin-converting enzyme (ACE) inhibitors have been shown to limit the size of the infarct and to reduce mortality from MI.

Drug treatment of heart failure

As the main problem in cardiac failure is the inability of the heart to pump sufficient blood to meet the needs of the circulation, the first priority is to minimize the demands on the circulation. This can be achieved by rest. Following a mild coronary thrombosis, the physiological compensatory mechanisms discussed earlier help to maintain an adequate cardiac output and blood pressure. Over the ensuing months, the remaining heart muscle hypertrophies and there is an increase in the number of blood vessels in those areas of the myocardium that became hypoxic as a result of the reduction in blood supply. These adaptations improve the contractility of the heart and eventually a normal pattern of life can be resumed.

In severe heart failure, the situation is more complicated and treatment has three aims:

- To reduce the work of the heart; this can be achieved by rest, as mentioned above

- To reduce the circulating volume and the resultant cardiac dilatation; this can be achieved by administration of diuretics such as **furosemide** (frusemide)

- To improve myocardial contractility; this is often attempted by the administration of drugs that have a positive inotropic effect on the myocardium. Examples are the cardiac glycosides (e.g. **digoxin** and **ouabain**) and β_1-adrenoceptor agonists such as **dobutamine**.

 Recommended reading

Anatomy

MacKinnon, P.C.B. and Morris J.F. (2005). *Oxford Textbook of Functional Anatomy*, Vol. 2. *Thorax and Abdomen*. Oxford University Press: Oxford. pp. 67–79.

Histology of the heart and blood vessels

Junquieira, L.C. and Carneiro, J. (2005). *Basic Histology*, 11th edn. McGraw-Hill: New York. Chapter 11.

Physiology of the circulatory system

Levick, J.R. (2003). *An Introduction to Cardiovascular Physiology*, 4th edn. Hodder Arnold: London.

Pocock, G. and Richards, C.D. (2006). *Human Physiology: The Basis of Medicine*, 3rd edn. Oxford University Press: Oxford. pp. 266–267; 273–283; 299–304; 591–593; 600–603.

Pharmacology of the heart and circulation

Grahame-Smith, D.G. and Aronson, J.K. (2002). *Oxford Textbook of Clinical Pharmacology*, 3rd edn. Oxford University Press: Oxford. Chapter 23.

Rang H.P., Dale, M.M., Ritter, J.M. and Flower, R. (2007). *Pharmacology*, 6th edn, Churchill-Livingstone: Edinburgh. Chapters 17 and 18.

Self-assessment questions

Which of the following statements are true and which are false? Answers are given at the end of the book (p.755).

1
a) The action potentials in the heart are about 100 times longer than those of skeletal muscle.
b) The cells of the sinoatrial node have a steady resting potential of −90 mV.
c) The cardiac action potential is conducted through the myocardium entirely via specialized conducting fibres.
d) The spread of cardiac excitation is delayed by about 0.1 s at the atrio-ventricular node.
e) The conducting tissue of the heart is made up of specialized myocytes known as Purkinje fibres.

2
a) During ventricular diastole, the pressure in the left ventricle is close to zero.
b) During ventricular systole, the pressure in the left ventricle reaches a maximum of about 16 kPa (120 mmHg).
c) During ventricular systole, all the blood in the ventricles is ejected.
d) During the initial stage of ventricular contraction, the volume of the ventricle does not change.
e) The mitral valve closes because the pressure in the left ventricle exceeds that in the left atrium.

3
a) The first heart sound corresponds to the closure of the mitral and tricuspid valves.
b) The first heart sound occurs just before the R wave of the ECG.
c) The second heart sound is due to closure of the aortic and pulmonary valves.
d) The second heart sound occurs during the T wave.

4
a) The cardiac output is, on average, the same for both left and right sides of the heart.
b) Cardiac output can be increased by circulating catecholamines.
c) The cardiac output is determined by the arterial blood pressure.
d) The end-diastolic volume is an important factor in determining the stroke volume.
e) Strong sympathetic stimulation has a negative inotropic effect.
f) Stimulation of the vagus nerve increases heart rate.

18 The electrocardiogram

After reading this chapter you should have gained an understanding of:

- The basic features of the electrocardiogram (ECG)
- The use of the ECG to assess the electrical activity of the heart
- Common abnormalities in the ECG
- The treatment of arrhythmias

18.1 Introduction to the electrocardiogram

The cardiac muscle cells (myocytes) are joined end to end to form a functional network or syncytium. The previous chapter describes how the heart beat is initiated by the pacemaker activity of the sinoatrial (SA) node. From the SA node, the excitation spreads across the atria from right to left before passing to the ventricles via the atrioventricular (AV) node. From the AV node, the excitation spreads via the bundle of His and its left and right branches to excite the ventricles. As the wave of excitation proceeds, some parts of the heart are depolarized while other regions remain at rest. Local electrical circuits are established between these regions, and small currents then flow through the cells and extracellular fluid. These electrical currents give rise to small differences in potential that can be detected by appropriately positioned electrodes on the body surface. This is the basis of the **electrocardiogram** or **ECG**, which provides a non-invasive technique for monitoring the electrical activity of the heart.

The ECG can be used to gain information about the following aspects of cardiac function:

- The heart rate
- The size of the muscle mass of the individual chambers of the heart
- Disorders in the conduction of the wave of excitation throughout the heart
- The site of any abnormal pacemaker activity
- The health of the myocardium.

KEY POINTS:

- The ECG records the electrical activity of the heart throughout the cardiac cycle.
- It provides information about the spread of excitation and can detect a number of cardiac abnormalities including arrhythmias, heart block and myocardial ischaemia.

18.2 Recording the ECG

The ECG is recorded by placing electrodes at different points on the body surface and measuring voltage differences between these points with the aid of an electronic amplifier. The resulting signal can be viewed on a visual display or recorded on a paper chart. The position of any pair of electrodes on the body surface will detect a particular portion of the current flow that occurs during the depolarization and repolarization of the heart muscle. A complete picture of the spread of excitation requires information from a number of electrode placements, known as **leads**. There are two types of ECG leads: bipolar and unipolar. **Bipolar leads** record the voltage between electrodes placed on the wrists and left ankle (with the right ankle acting as the point of reference – the earth or ground electrode). **Unipolar leads** record the voltage between a single electrode placed on the body surface and a reference electrode.

The three standard bipolar limb leads are, by convention, known as limb leads I, II and III. These are illustrated in Figure 18.1.

- In **lead I**, the positive terminal of the amplifier is connected to the left arm and the negative terminal to the right arm. This pair of electrodes records the electrical changes that occur between the right and left sides of the heart.

- In **lead II**, the right arm is connected to the negative terminal and the left leg to the positive terminal. This pair of electrodes records the electrical changes between the right upper portion of the heart and the tip of the ventricles.

- In **lead III**, the left leg is the positive terminal and the left arm the negative terminal. This lead records the electrical changes that occur between the left atrium and the tip of the ventricles.

There are two types of unipolar leads used in electrocardiography, the **augmented limb leads** and the **chest (or precordial) leads**. When the ECG is recorded from the unipolar chest leads, six electrodes record the electrical activity from specific points on the chest at the level of the heart, as shown in Figure 18.2. These positions are known as leads V1 to V6. The exact locations of the electrode positions (or leads) are described in the legend to Figure 18.2. A reference electrode can be produced by joining the three limb leads.

The voltage recorded from the limb leads can be increased using the **augmented limb leads**. In this configuration, two of the leads are used to produce a reference electrode. The signal is then recorded as the difference between these two electrodes and the remaining limb electrode. Lead aV_R is recorded from the right arm with the reference electrode formed by adding the signals from left arm and left ankle. Lead aV_L is formed by recording from the left arm with the reference electrode formed by adding the signals from right arm and left ankle. Lead aV_F is formed by recording the signal from the left ankle with the reference electrode formed by adding the signals from the two arms.

KEY POINTS:

- A 12-lead ECG record uses three limb leads (leads I, II and III), three augmented limb leads (aV_L, aV_R and aV_F) and six chest leads (V1–V6).

- Each lead provides information about the activity of the heart viewed from a specific point on the body surface.

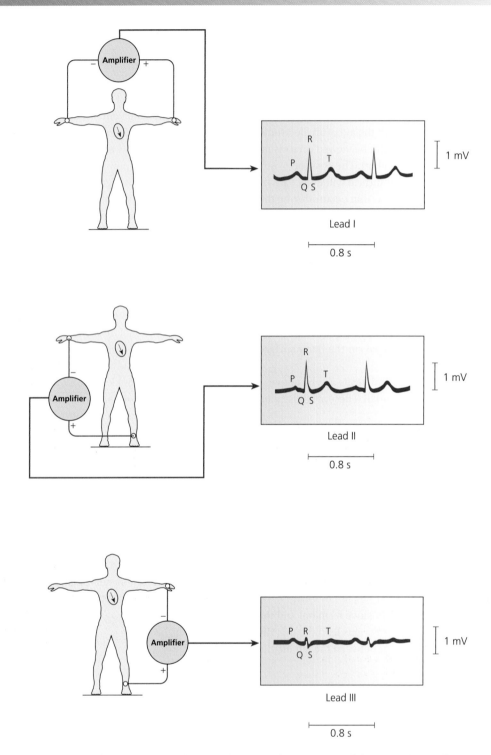

Figure 18.1 The arrangement of the limb leads used to record the ECG. The appearance of the principal waves for the various leads is also shown. As lead II is oriented along the main atrioventricular axis of the heart, it usually gives rise to the most prominent R wave.

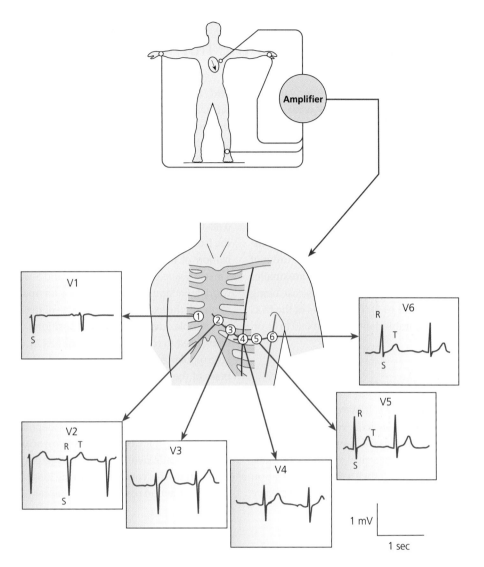

Figure 18.2 Unipolar recording of the ECG with standard chest leads. The limb leads are connected together to provide a virtual earth, as shown in the inset figure at the top. The exploring electrode is then placed in one of six positions on the chest (leads V1 to V6) as shown. Recordings from leads V1 and V2 usually show a pronounced S wave while leads V5 and V6 show a large R wave. Lead V1 is placed at the right margin of the sternum in the 4th intercostal space. Lead V2 is placed at the left margin of the sternum in the 4th intercostal space. Lead V3 is placed midway been leads V2 and V4. Lead V4 is placed on the midclavicular line (shown here in red) in the 5th intercostal space. Lead V5 is placed at the same level as lead V4 in the anterior axillary line. Lead V6 is also placed at the same level as lead V4 but in the midaxillary line (shown here in green).

18.3 The characteristics of the normal ECG

The normal ECG shows three main deflections in each cardiac cycle (Figure 18.3). These are: the **P wave**, which corresponds to electrical currents generated as the atria depolarize prior to contraction; the **QRS complex**, which corresponds to ventricular depolarization; and the **T wave**, which corresponds to ventricular repolarization. The exact appearance of the different components depends on the lead from which they are recorded. The electrical signals are very small and specialist equipment is required to record them. The largest QRS complexes recorded by the chest leads are only of the order of 2–3 mV. The signals recorded by the limb leads are smaller still, the amplitude of a typical R wave in this case being 0.5–1 mV (Figures 18.1 and 18.2). Those parts of the ECG trace that have no measurable deflection are said to lie on the **isoelectric line**.

The origin of the ECG waves

Although each cardiac cycle is initiated by depolarization of the SA node, this electrical event is not seen in the ECG trace because the mass of tissue involved is very small. The first discernible electrical event, the **P wave**, lasts about 0.08 s and coincides with the depolarization of the atria. The interval between the start of the P wave

and the start of the QRS complex is known as the **PR interval**. The PR interval lasts for 0.12–0.2 s and includes the time taken to depolarize the AV node, the bundle branches and the Purkinje system. The atria contract during the flat part of the record between the P wave and the QRS complex.

The next electrical event of the cardiac cycle is reflected in the **QRS complex** of the ECG trace, which lasts about 0.08–0.1 s. This is normally seen as a large deflection from the isoelectric line because it is produced by the depolarization of the ventricles, which are the largest mass of muscle tissue in the heart. The Q and S waves are downward deflections and the R wave is an upward deflection, although the exact pattern and size of the components of the complex depend upon the position of the electrodes being used to record the ECG (Figures 18.1 and 18.2). The atria repolarize during the QRS complex but, as the muscle mass of the ventricles is so much larger than that of the atria, this event is not seen as a separate wave in the ECG.

During the interval between the S and the T waves, the entire ventricular myocardium is depolarized and the ventricles contract. Because all the myocardial cells are at about the same potential, the **ST segment** lies on the isoelectric line. This corresponds to the long plateau phase of the cardiac action potential (see Chapter 17, pp. 335–336).

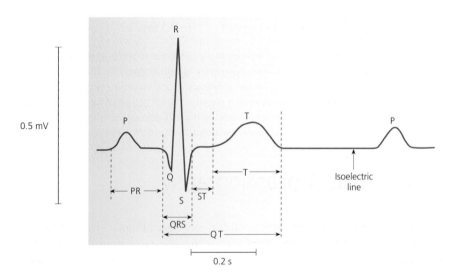

Figure 18.3 A generic ECG trace drawn from the ECG recorded with the limb leads. The labels show the principal waves and the intervals that are normally reported. The P–P interval gives the heart rate.

Table 18.1 Standard ECG intervals

Measurement	Duration	Comments
P wave	<0.12 s	Normally smooth and rounded
PR interval	0.12–0.21s	Interval varies with heart rate
QRS width	0.07–0.11s	Amplitude and polarity of R wave in different leads can be used to determine the main cardiac axis
QT interval	0.3–0.4s	Varies with heart rate
R–R interval	c. 0.75–0.85s	Can be used to calculate the heart rate as follows: beats min^{-1} = 60/R–R interval

The final major event of the ECG trace is the **T wave**, which normally appears as a broad upward deflection and represents the repolarization of the ventricular myocardium, which precedes ventricular relaxation. The action potential of the ventricular muscle usually lasts for 0.2–0.3 s so that the interval between the start of the QRS complex and the end of the T wave is around 0.3 s. The T wave is relatively broad because some ventricular fibres begin to repolarize earlier than others, and thus the whole process of repolarization is rather prolonged. Very occasionally the T wave is followed by a low amplitude wave of uncertain origin known as the U wave. The standard values for the ECG intervals are given in Table 18.1.

Why does the T wave have the same polarity as the R wave?

The polarity of an ECG wave reflects the predominant direction of current flow sensed by a particular ECG electrode. During ventricular depolarization, the cardiac cells closest to the Purkinje cells depolarize first and the wave of depolarization passes from the inside of the ventricles to the outside. However, the cells on the outer surface of the ventricles have shorter action potentials than those on the inside and begin to repolarize first. The wave of repolarization thus passes from the outside to the inside of the ventricles. Consequently, the T wave has the same polarity as the R wave.

> **KEY POINTS:**
>
> - The ECG is recorded by placing electrodes at different points on the body surface and measuring voltage differences between them. Standard leads are used for this purpose and the appearance of the ECG trace depends on the position of the particular leads.
>
> - The P wave of the ECG is due to atrial depolarization, the QRS complex to ventricular depolarization, and the T wave to ventricular repolarization. Atrial repolarization is hidden within the QRS complex.

18.4 Clinical aspects of electrocardiography

As the ECG provides a means of recording the electrical activity of the heart *in situ*, it is of clinical value in the investigation of cardiac arrhythmias and myocardial ischaemia. Among the most common arrhythmias are atrial fibrillation, extra contractions of the ventricles (called **ventricular extrasystoles** or ectopic beats) and progressive stages of heart block, in which excitation of the ventricles by the atria is impaired. The ECG may also be used to determine the electrical axis of the heart as a whole, which can itself give information about certain pathological cardiac conditions such as right ventricular hypertrophy.

Each action potential originating in the SA node initiates one beat of the heart. This is known as **sinus rhythm** and is the normal state of affairs. Any deviation from the normal sinus rhythm is known as an **arrhythmia**. While some arrhythmias are of no clinical significance, others may reflect serious and life-threatening disorders of the myocardium. Indeed, many arrhythmias arise as a consequence of ischaemic heart disease.

The heart rate commonly varies with the respiratory cycle in healthy young people (sinus arrhythmia). In highly trained athletes, the heart rate is usually slower than normal (although stroke volume is higher). This is known as **sinus bradycardia** (less than 60 beats min^{-1}). Sinus bradycardia is also seen in fainting attacks, hypothermia and hypothyroidism. A rapid heart beat (>100 beats min^{-1}) is known as **sinus tachycardia** and is associated with exercise, stress, haemorrhage and hyperthyroidism. (Note: a bradycardia is a heart rate below 60 beats min^{-1} and a tachycardia is a heart rate greater than 100 beats min^{-1})

If the rhythmical activity of the SA node activity is much slower than normal, the fundamental rhythm may be taken over by another part of the heart. These rhythms are known as **escape rhythms** and are named after their site of origin. Those that originate in the atria are called atrial rhythms, those that originate in or close to the AV node are junctional rhythms, and those that originate in the ventricles are ventricular rhythms.

Sick sinus syndrome

In some people, particularly the elderly, the SA node fails to excite the atria in a regular manner, resulting in a slow resting heart rate that does not increase appropriately with exercise. This is called **sick sinus syndrome** and has many causes. Arrhythmias caused by failure of the SA node to excite the heart in the normal way can often be treated by the implantation of an artificial pacemaker.

> **KEY POINTS:**
> - The ECG provides information about the heart rate, the size of the muscle mass of the individual chambers of the heart and the spread of excitation.
> - It also provides information regarding the site of any abnormal pacemaker activity.

Defects of conduction

Defects in the conduction of the electrical activity can, in principle, occur at any point in the pathway between the SA node and the ventricular muscle. These problems are known as **heart block**. Problems with conduction from the atria to the ventricles are known as AV block and are classified according to their severity as first-, second- or third-degree heart block. Impairment of conduction through the bundle of His is known as **bundle block** and may affect the main bundle or any of its branches.

In **first-degree heart block**, each wave of depolarization originating in the SA node is conducted to the ventricles but there is an abnormally long PR interval (>0.2 s), indicating that somewhere in the pathway a delay has occurred (see Figure 18.5a). This is usually at the AV node. First-degree heart block is not usually a problem but may be a sign of some other disease process (e.g. coronary artery disease or electrolyte disturbance). In **second-degree heart block**, there is an intermittent failure of excitation to pass to the ventricles. In **third-degree heart block**, the wave of excitation originating in the SA node fails to excite the ventricles (AV block). The ventricles then show a slow intrinsic rhythm of their own (an escape rhythm) with abnormally shaped, broad QRS complexes that are not associated with P waves (Figure 18.4). The QRS complex becomes broader than normal because the conduction of the wave of excitation through the cardiac muscle is significantly slower than that through the Purkinje fibres.

In some situations, the excitation reaches the AV node and passes through the bundle of His only to be delayed or fail to pass from one or other branch to the ventricular muscle. This is known as **bundle branch block**. The QRS complex is broadened (>0.12 s) as the wave of excitation will spread more slowly across the ventricles than it normally would via the Purkinje fibre system.

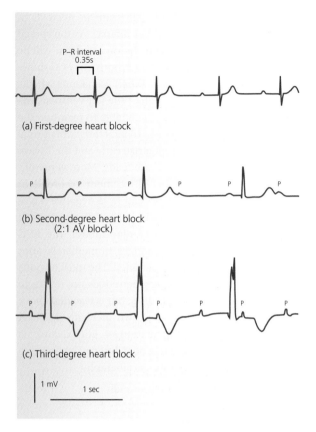

Figure 18.4 The ECG seen in various types of heart block. (a) First-degree heart block; note the long PR interval. (b) Second-degree heart block in which every second P wave elicits an R wave – this is 2:1 block. Other regular patterns may also be seen, such as 3:1 or 4:1 block. (c) Third-degree heart block – here there is complete dissociation of the QRS complex from the P wave. Moreover, the QRS complex is broader than normal and has an abnormal appearance.

KEY POINTS:

- In first-degree heart block, the PR interval is prolonged.

- In second-degree heart block, there is an intermittent failure of atrial excitation to pass to the ventricles.

- In third-degree heart block, there is complete dissociation of the QRS complex from the P wave.

- In bundle branch block-the QRS complex is broadened.

Arrhythmias of atrial or ventricular origin

As mentioned above, abnormal rhythms may begin in the atria – away from the SA node, they may begin at the AV node (nodal or junctional rhythms) or they may begin in the ventricles. Arrhythmias originating in the atria or nodal region are often called **supraventricular rhythms**. When the cycle of excitation begins at a site remote from the SA node, the resulting rhythm is slower than normal (bradycardia). Sometimes the atria or ventricles contract earlier than expected. This is known as an **extrasystole** (an ectopic beat). For both escape rhythms and extrasystoles, the appearance of the ECG trace will depend on the origin of the abnormal excitation (Figure 18.5).

When the atria depolarize more frequently than about 220 times a minute, **atrial flutter** is present. The ECG baseline is not flat but often shows a continuous, regular sawtooth pattern. As the AV node cannot be activated more than about 200 times a minute, the ventricles are activated at a lower rate. In some people, the ventricles are activated in a completely regular manner with, for example, a 2:1 or 4:1 ratio of P waves to QRS complexes. In others, the ventricles may be activated in a very irregular manner.

If the atria are activated more than about 350 times a minute, they do not contract in a co-ordinated way. The individual muscle fibre bundles contract asynchronously. The ECG trace shows no P waves, only an irregular baseline (Figure 18.5c). This is known as **atrial fibrillation** and the ventricles show an irregular rhythm with an abnormal QRS complex. If the site of abnormal activity is in the wall of one of the ventricles, the rhythm is known as **ventricular tachycardia** and the QRS complex will be wide and abnormal in appearance, as seen in Figure 18.5d. When the ventricular muscle fibres fail to contract in a concerted way, the heart is in a state of **ventricular fibrillation**, and is unable to pump the blood around the body.

As the quantity of blood forced from the atria into the ventricles contributes only about 20% of the end-diastolic volume, neither atrial flutter nor atrial fibrillation is immediately life-threatening. In contrast, ventricular fibrillation requires urgent action to restore normal function by means of a **defibrillator**, which uses a strong electrical shock to terminate the fibrillation and reset a normal heart beat.

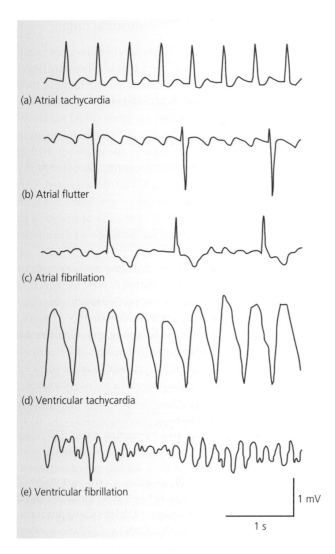

(a) Atrial tachycardia

(b) Atrial flutter

(c) Atrial fibrillation

(d) Ventricular tachycardia

(e) Ventricular fibrillation

1 mV

1 s

Figure 18.5 Abnormal cardiac rhythms. (a) Atrial tachycardia. (b) Atrial flutter – note the sawtooth pattern of the P waves and the broad QRS complexes. The ventricular rate is 45 b.p.m. (c) Atrial fibrillation – note the irregular baseline and the highly abnormal QRST pattern. (d) Ventricular tachycardia – a regular but highly abnormal pattern of excitation is seen as the point of excitation lies in the ventricles themselves. (e) Ventricular fibrillation – note the highly irregular pattern with no identifiable waves present. This is a life-threatening condition.

ECG changes in ischaemia and infarction

Cardiac ischaemia refers to a situation in which the blood supply is inadequate to meet the metabolic requirements of the heart muscle. If this situation is prolonged, the affected tissue becomes damaged and may die (a 'heart attack'). This will result in necrosis of the affected area (a **cardiac infarct**). As the spread of excitation throughout the myocardium will be affected in both of these conditions, the ECG shows characteristic changes. When ischaemia affects the inner part of the wall of the left ventricle, the affected myocytes may be unable to maintain the prolonged action potential characteristic of normal cells, they begin to repolarize early and the T wave is inverted in many of the leads. More generally, during ischaemia the ST segment recorded in long axis leads, such as lead II, is negative (i.e. it lies below the isoelectric line). If the ischaemia progresses to cause a cardiac infarct, the ECG shows characteristic changes immediately after the damage has been sustained. The most obvious change is that the ST segment does not return to baseline as it normally does, but is elevated above the isoelectric line, as shown in Figure 18.6c. A more detailed interpretation of ECG records is beyond the scope of this book but can be found in specialist texts of electrocardiography.

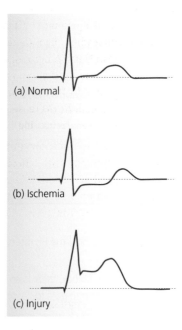

(a) Normal

(b) Ischemia

(c) Injury

Figure 18.6 Characteristic changes compared with normal (a) in the QRST waves seen in (b) ischaemia (note that the ST segment lies below the isoelectric line) and (c) infarct (injury – note that the ST segment lies above the isoelectric line).

KEY POINTS:

- If the normal pattern of excitation is impaired, the ECG shows characteristic changes.
- In atrial flutter, the P waves have a very high frequency with a sawtooth pattern.
- In atrial fibrillation, the baseline is irregular and the appearance of the QRS complex is highly abnormal.
- In ventricular tachycardia, there is a regular but highly abnormal QRS complex.
- In ventricular fibrillation, the ECG shows a highly irregular pattern with no identifiable waves present.

18.5 Antiarrhythmic drugs

A large number of drugs have been used to treat arrhythmias (sometimes called antidysrhythmic drugs), although it is now clear that care needs to be exercised in selecting drugs for particular types of arrhythmias. The most widely used drugs can be divided into four main classes:

- **Class I** antiarrhythmic drugs are sodium channel blockers and they reduce the rate at which the cardiac muscle cells depolarize. The class is divided into three subclasses: Class IA, which includes **quinidine**, **procainamide** and **disopyramide**; Class IB, which includes the local anaesthetic **lignocaine** (lignocaine is known as lidocaine in North America), which is given intravenously when it is used as an antiarrhythmic as it is metabolized rapidly by the liver; and Class IC drugs, which cause a general decrease in excitability; examples are **flecainide** and **encainide**.

- **Class II** drugs are β-adrenoceptor antagonists. This group includes **propranolol** (the best-known drug in this class), **atenolol** and **timolol**. They counter the effects of excessive adrenaline secretion and are helpful in patients recovering from heart attacks.

- **Class III** drugs are agents that slow the heart rate by prolonging the refractory period of the cardiac action potential. These drugs are used to slow tachycardias. The classic example is **amiodarone** (although it has a number of serious side-effects). **Sotalol** is a β-adrenoceptor antagonist that has a class III action.

- **Class IV** drugs are all calcium channel antagonists. These drugs slow conduction in the SA and AV nodes and this helps to slow the heart rate and prevent supraventricular tachycardias. **Verapamil** is the most widely used drug of this type but **diltiazem** is also used.

In addition to the drugs listed above, **adenosine** is sometimes given to stop a supraventricular tachycardia, as a safer alternative to verapamil. Adenosine is produced naturally by the body as a by-product of ATP metabolism and has a number of powerful pharmacological actions including vasodilatation, inhibition of platelet aggregation and bronchoconstriction.

 ## Recommended reading

Physiology

Levick, J.R. (2003). *An Introduction to Cardiovascular Physiology,* 4th edn. Hodder Arnold: London. Chapter 5.

Pocock, G. and Richards, C.D. (2006). *Human Physiology: The Basis of Medicine,* 3rd edn. Oxford University Press: Oxford. Chapter 15, pp. 266–273 and Chapter 31, pp. 593–600.

Pharmacology of the heart and circulation

Grahame-Smith, D.G. and Aronson, J.K. (2002). *Oxford Textbook of Clinical Pharmacology,* 3rd edn. Oxford University Press: Oxford. Chapter 23.

Rang, H.P., Dale, M.M., Ritter, J.M. and Flower, R. (2007). *Pharmacology,* 6th edn. Churchill-Livingstone: Edinburgh. Chapter 17.

Clinical medicine

Hampton, J.R. (1998). *The ECG Made Easy,* 5th edn. Churchill-Livingstone: Edinburgh.

Self-assessment questions

Which of the following statements are true and which are false? Answers are given at the end of the book (p. 755).

1 a) The P wave of the ECG reflects atrial contraction.

b) The QRST complex of the ECG reflects the time during which ventricular fibres are depolarized.

c) The peak amplitude of the R wave is about 1 mV.

d) The T wave reflects the depolarization of the ventricular fibres.

e) The PR interval is normally about 0.1 s.

2 a) In atrial fibrillation, the ECG shows a sawtooth pattern.

b) The QRS complex is abnormal in atrial fibrillation.

c) In second-degree heart block, each P wave is followed by a QRS complex.

d) The QRS complex in third-degree heart block is abnormally wide.

e) The ST segment of the ECG lies below the isopotential line following a heart attack.

The circulation

After reading this chapter you should have gained an understanding of:

- The anatomy of the circulation
- The relationship between pressure and flow in the circulation
- Arterial blood pressure and its measurement
- The control of the blood vessels:
  - Autoregulation
  - The role of autonomic nerves
  - Endocrine control
- The microcirculation and tissue fluid exchange
- Hypertension and its treatment

19.1 Introduction

In unicellular organisms and simple animals such as sponges, the exchange of nutrients and waste products between the cells and the environment can be accomplished by diffusion. However, in more complex animals, such a simple process is inadequate as most cells are separated from the external environment by a considerable distance. More complex animals have evolved a circulatory system that overcomes this problem. Blood is pumped around the body and cells exchange their waste products for nutrients with the circulating blood. Oxygen from the lungs and nutrients from the gastrointestinal tract are transported to the cells via the blood while carbon dioxide and other waste products are transported to the lungs and kidneys for excretion.

The overall organization of the circulation is shown in Figure 17.1, which shows that the circulation consists of a pump (the heart) and a series of interconnected pipes (the blood vessels). The right side of the heart pumps blood through the lungs (**the pulmonary circulation**) and the left side of the heart pumps blood around the body (**the systemic circulation**). The anatomy and physiology of the heart are discussed in Chapter 17. This chapter is concerned with the systemic circulation and deals with the anatomical arrangement, structure and functions of the systemic blood vessels. The detailed anatomy and physiology of the pulmonary circulation are discussed in Chapter 22.

> **KEY POINTS:**
>
> • The circulation is organized so that the right side of the heart pumps blood through the lungs (the pulmonary circulation) and the left side of the heart pumps blood around the rest of the body (the systemic circulation).
>
> • Thus, the two circulations are arranged in series.

19.2 The anatomy of the systemic circulation

The blood vessels are divided into four broad categories:

- Arteries, which distribute the blood to the tissues
- Arterioles, which regulate the flow of blood to specific organs according to the requirements of the body
- Capillaries, which are the exchange vessels
- Venules and veins, which return the blood to the heart.

The major arteries

From the left side of the heart, the blood first passes into the **ascending aorta**, which initially passes to the right before curving backwards and downwards to the left, to pass behind the lungs and in front of the spine. The heart itself is supplied by the **coronary arteries**, which branch from the aorta as it leaves the left ventricle.

As the aorta curves to form the **aortic arch**, it first gives rise to the **brachiocephalic** artery, then the **left common carotid** and **left subclavian arteries**. The brachiocephalic artery subsequently branches to form the **right common carotid** and **right subclavian arteries**. The common carotid arteries supply the head and neck, while the subclavian arteries supply the arms and part of the body wall. At the axilla, the subclavian artery becomes the **axillary artery**, which branches to supply the chest wall and shoulder. As the axillary artery passes into the upper arm, it becomes the **brachial artery**, which passes between the biceps and brachialis muscles. On the inside of the elbow (the antecubital fossa), the brachial artery lies close to the surface where its pulse can be clearly felt. This is the point at which a stethoscope is applied during measurements of blood pressure

(see Box 19.1). Below the elbow, the brachial artery divides to give rise to the **radial artery** and the **ulnar artery**.

As the aorta descends, it gives rise to the **bronchial** and **intercostal arteries** of the thorax. In all, there are nine pairs of intercostal arteries, each of which passes around the chest between the ribs. As the aorta traverses the lower part of the thorax, it gives rise to the **superior phrenic arteries** that supply the diaphragm. On entering the abdominal cavity, the aorta gives rise to the **coeliac trunk**, which supplies the gastrointestinal tract and spleen. Paired arteries supply the adrenal glands and kidneys. The **renal arteries** are large-diameter vessels that enter the kidneys via the renal hilus (see Figure 23.2). The main branches of the aorta are shown in Figure 19.1 while a functional magnetic resonance image of the arterial tree is shown in Figure 19.2.

Below the renal arteries, the aorta divides to form the **common iliac arteries**, which are the terminal branches of the aorta. At a level corresponding to the junction between the final lumbar vertebra and the sacrum, the common iliac arteries divide into the **external iliac** and **internal iliac** arteries, which supply the legs. The external iliac artery branches to give rise to the deep and superficial **femoral arteries**, which supply the muscles of the thigh. The superficial femoral artery merges with the **popliteal artery**, which subsequently divides into the **anterior** and **posterior tibial arteries**.

The major veins

The systemic arteries distribute oxygenated blood to the main organ systems. Deoxygenated blood is returned to

Box 19.1 Measurement of blood pressure by auscultation

Auscultation means 'listening to'. So measuring blood pressure by auscultation means making use of the sounds that are heard when the blood flow through an artery is gradually restored after it has been occluded by an inflatable rubber cuff. The device used to record the pressures is known as a **sphygmomanometer**.

Normally, the blood pressure of the brachial artery is measured. Initially, an inflatable rubber cuff within a cotton sleeve is placed around the upper arm of the person whose blood pressure is to be measured. The cuff is inflated until the radial pulse can no longer be felt so that the pressure within the cuff is in excess of the systolic pressure. A stethoscope is then positioned over the inside of the elbow. The pressure in the cuff is then gradually lowered. Initially no sounds will be heard through the stethoscope as the pressure of the cuff occludes blood flow through the artery, but as the pressure within the cuff is lowered, there comes a point where the pressure within the artery is just sufficient to overcome the pressure exerted by the cuff. At the peak of systole, there is a brief spurt of blood into the artery below the point at which it is occluded and this causes vibration of the vessel wall,

which can be detected as a tapping sound through the stethoscope. The pressure at which this tapping sound is first heard is conventionally accepted to represent peak systolic pressure. Cuff pressure is then lowered further. As more and more blood passes through the artery, the sounds heard through the stethoscope become louder at first, but as the diastolic pressure is approached, the artery remains open for almost all of the cardiac cycle so the blood flow starts to become less turbulent and more streamlined. Streamlined flow creates less vibration and therefore less noise in the artery, so the sounds diminish in volume fairly abruptly as diastolic pressure is reached. The pressure is allowed to fall still further until the sounds disappear. By convention, the point at which complete silence occurs is taken as diastolic pressure.

Normal systolic pressure measured in this way is usually less than 150 mmHg in a healthy adult and diastolic pressure should be less than 90 mmHg. In young adults and children, the pressures tend to be lower. In elderly people, there tends to be an increase in systolic pressure without a proportionate increase in diastolic pressure.

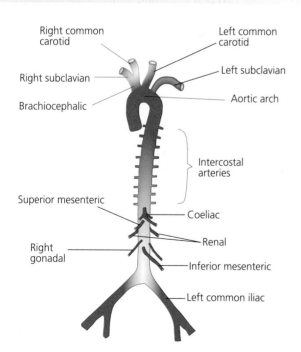

Figure 19.1 A simplified drawing showing the principal branches of the aorta, with the major arteries identified.

Labels: Right common carotid; Left common carotid; Right subclavian; Left subclavian; Brachiocephalic; Aortic arch; Intercostal arteries; Superior mesenteric; Coeliac; Right gonadal; Renal; Inferior mesenteric; Left common iliac

the heart via the systemic veins. More blood is contained within the veins than in the arteries, partly because the walls of the veins are thinner and more easily distended than those of the arteries (see below) and partly because there are more veins than there are arteries. (There are approximately two veins for every artery of moderate size.) The veins thus act as a store of blood that can be mobilized when required (see below).

The arrangement of the principal veins is shown in Figure 19.3. Blood returning to the heart from the head, neck and arms reaches the right side of the heart via the **superior vena cava**. The blood returning from the lower part of the body (the lower limbs and abdomen) reaches the right side of the heart via the **inferior vena cava**.

The superior vena cava arises from the fusion of two large veins called the **brachiocephalic veins**. Each brachiocephalic vein arises from the junction of the **internal jugular vein** (which receives blood from the brain, the face and the neck) and the **subclavian vein**, which returns blood from the arms (see below). The **external jugular vein** is one of the major tributaries of the internal jugular

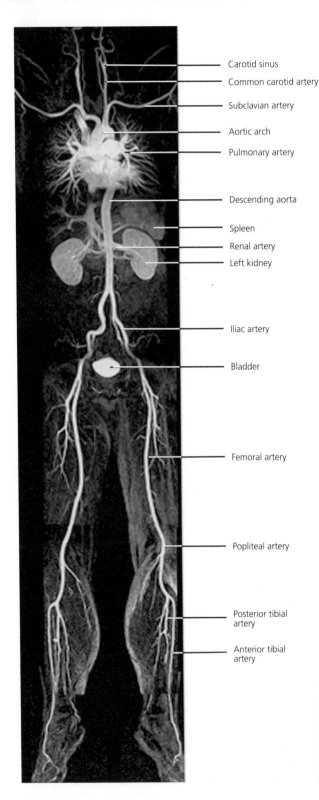

Carotid sinus

Common carotid artery

Subclavian artery

Aortic arch

Pulmonary artery

Descending aorta

Spleen

Renal artery

Left kidney

Iliac artery

Bladder

Femoral artery

Popliteal artery

Posterior tibial artery

Anterior tibial artery

vein. It is a large, superficial vein that passes from the angle of the jaw down towards the middle of the collar bone (clavicle).

The fingers are drained by veins that run on each side. These are the **digital veins**, which drain into the **dorsal venous network** on the back of the hand, which also receives blood from the palm. Blood passing into the dorsal venous arch is returned to the heart via the **basilic vein** and the **cephalic vein**. The basilic vein runs along the inner (medial) aspect of the arm while the cephalic vein courses along the outer aspect. The two veins are linked by the **median cubital vein**, which is clearly visible in most people as it crosses the cubital fossa. This vein is frequently used for making intravenous injections and taking blood by venepuncture. The basilic vein is joined by other veins draining the deep structures of the upper arm to form the **axillary vein**, which merges into the subclavian vein.

The inferior vena cava lies in front of the spinal column on the right-hand side. It arises from the fusion of the **common iliac veins**, which receive blood from the lower limbs and abdomen. The vena cava also receives blood from the **hepatic vein**, which drains the liver, the **renal veins** and a number of other veins draining structures supplied by the aorta. Deep within the abdomen and the legs, the major veins are named according to the position of their corresponding arteries (e.g. the femoral vein, the internal and external iliac veins). On the lateral aspect of the lower leg, the more superficial veins drain into the short or **small saphenous vein**, which runs on the back of the calf before draining into the **popliteal vein**. The small, superficial veins on the medial aspect of the leg drain into the large or **great saphenous vein**, which passes from the instep up the inner side of the leg before it empties into the femoral vein.

Figure 19.2 A functional magnetic resonance image (fMRI) of the circulation of a healthy young adult male. A magnetic contrast medium was injected prior to acquiring the MRI in which the heart and the main arterial vessels are clearly seen. The contrast medium has accumulated in the bladder during the imaging process.

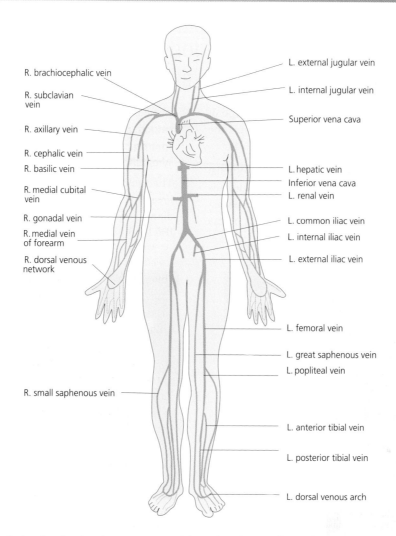

R. brachiocephalic vein

R. subclavian vein

R. axillary vein

R. cephalic vein

R. basilic vein

R. medial cubital vein

R. gonadal vein

R. medial vein of forearm

R. dorsal venous network

R. small saphenous vein

L. external jugular vein

L. internal jugular vein

Superior vena cava

L. hepatic vein

Inferior vena cava

L. renal vein

L. common iliac vein

L. internal iliac vein

L. external iliac vein

L. femoral vein

L. great saphenous vein

L. popliteal vein

L. anterior tibial vein

L. posterior tibial vein

L. dorsal venous arch

Figure 19.3 A schematic drawing showing the arrangement of the major veins. L, Left; R, right.

19.3 The pulmonary circulation

The lungs receive blood from two sources: the bronchial circulation and the pulmonary circulation. These are discussed fully in Chapter 22 (pp. 448–449) but a brief outline follows. The bronchial arteries arise chiefly from the aortic arch, the thoracic aorta and the intercostal arteries. These arteries supply oxygenated blood to the smooth muscle of the principal airways and the interstitial lung tissue. The blood from the bronchial circulation drains into the right atrium or the pulmonary veins.

The output of the right ventricle passes into the **pulmonary artery**, which subsequently branches to supply the individual lobes of the lung. The pulmonary arteries branch along with the bronchial tree until they reach the respiratory bronchioles. Here, they form a dense capillary network, which drains into pulmonary venules. Small veins merge with larger veins until two large **pulmonary veins** emerge from each lung to empty into the left atrium.

19.4 The structure of the blood vessels

The walls of the main arteries and veins consist of three layers whose thickness varies according to the type of vessel: from the inside these are the tunica intima, the tunica media and the tunica adventitia (Figure 19.4).

- The **tunica intima** consists of a layer of flat endothelial cells overlying a thin layer of connective tissue. It is separated from the tunica media by the internal elastic lamina. The endothelial cells of the tunica intima are in direct contact with the blood.

- The **tunica media** provides the mechanical strength of the blood vessel and consists of a circular layer of smooth muscle containing elastin and collagen. The smooth muscle of the tunica media is innervated by sympathetic nerve fibres.

- The **tunica adventitia** consists of a loosely formed layer of elastic and collagenous fibres oriented along the length of the vessel that serve to anchor the blood vessel in place. The tunica adventitia is

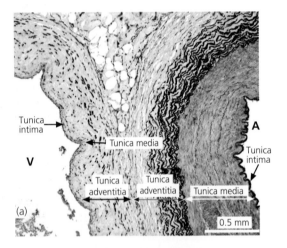

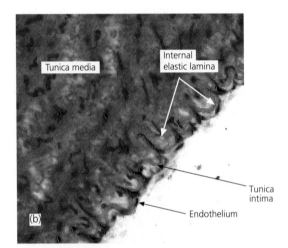

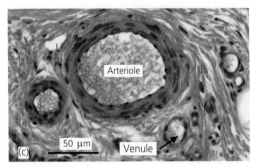

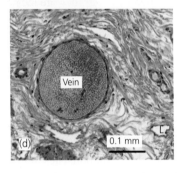

Figure 19.4 Blood vessels in cross-section. (a) Part of the walls of the femoral artery (A) on the right and femoral vein (V) on the left stained to show the elastic fibres in black. The three principal layers are labelled. Note the difference in wall structure between the artery and the vein. The tunica media is stained purple and is very thick in the artery but thin in the vein. (b) The wall of a muscular artery enlarged to show the tunica intima and the endothelium with its flattened nuclei. The wavy red line is the internal elastic lamina. (c) A cross-section showing an arteriole and a venule. A smaller arteriole can be seen on the left. Note the difference in wall thickness between the arterioles and the venule. (d) A cross-section of a small vein. Note the thin wall. A small lymphatic (L) can be seen on the lower right of the field.

separated from the tunica media by the external elastic lamina.

The arteries are the primary **distribution vessels** and may be subdivided into two groups:

- The **elastic arteries** are large vessels of 1–2 cm diameter. They include the aorta and pulmonary arteries together with their major branches. Although the walls of the elastic arteries are relatively thick, they contain a high proportion of elastic tissue that becomes stretched during systole. The stretched elastic tissue provides the energy needed to drive the blood around the circulation in diastole.

- **Muscular arteries** arise from the elastic arteries. In man, the larger muscular arteries are about 10 mm in diameter while the smallest are about 1 mm. The tunica media of the muscular arteries contains a higher proportion of smooth muscle than that of the elastic arteries. This makes them very resistant to collapse at the sharp bends that occur at the joints. Examples of muscular arteries are the cerebral, popliteal (in the legs) and brachial (in the upper arm) arteries.

The muscular arteries give rise to the **arterioles**, which possess a thick layer of smooth muscle relative to their size (Figure 19.4c). This layer of smooth muscle allows the arterioles to change their diameter to a considerable degree and enables them to regulate the flow of blood to the vascular bed in which they are located. For this reason, the arterioles are known as **resistance vessels**. The arterioles branch repeatedly and the final branches (the terminal arterioles) give rise to **capillaries**, which have thin walls and are 5–8 µm in diameter.

The capillaries are the principal **exchange vessels**. They have no smooth muscle in their walls and consist of a single layer of endothelial cells. The capillaries coalesce to form **postcapillary venules** of about 20 µm diameter, which also lack smooth muscle. In turn, these give rise to the **true venules** and the **veins**, which merge to form the great veins that return blood to the heart.

The walls of the veins and venules are similar in structure to those of the arteries but they are much thinner in relation to the overall diameter of the vessel (Figure 19.4a and d). Consequently, the veins are much more distensible than the arteries. Unlike other blood vessels, the larger veins of the limbs possess valves at intervals along their length. These are arranged so that blood can pass freely towards the heart while back flow is prevented.

In a few tissues, notably the skin, there are some direct connections between the arterioles and venules. These specialized vessels are known as arteriovenous shunt vessels (or **anastomoses**) and they have relatively thick muscular walls, richly supplied by sympathetic nerve fibres. When these vessels are open, some blood can pass directly from the arterioles to the venules without passing through the capillaries.

> **KEY POINTS:**
>
> - The principal types of blood vessel are the arteries, the arterioles, the capillaries, the venules and the veins.
> - Except for the capillaries and the smallest venules, the walls of the blood vessels have three layers, the tunica intima, the tunica media and the tunica adventitia.

19.5 Arterial blood pressure

For the systemic circulation as a whole, the driving force for blood flow is the difference between the arterial blood pressure and the pressure in the right atrium – the **central venous pressure**. As blood is pumped into the aorta only during the ejection phase of ventricular contraction, the pressure in the arterial system varies with the cardiac cycle. The pressure at the peak of ejection is called the **systolic pressure** while at its lowest point, during ventricular relaxation, it is known as the **diastolic pressure**. During systole, the arterial pressure rises rapidly as the rate at which blood is being pumped into the arterial tree is greater than the rate at which it can be distributed to the tissues. As a result, the pressure rises and the walls of the elastic arteries become distended.

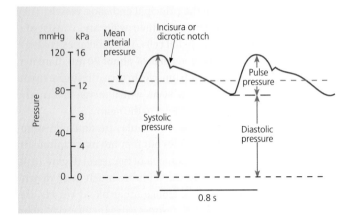

Figure 19.5 A diagram of the arterial pressure wave showing the systolic, diastolic and pulse pressures. The mean arterial pressure is shown by the light blue dotted horizontal line and is equal to the diastolic pressure plus one-third of the pulse pressure.

The flow of blood into the aorta declines as the heart begins to relax and the pressure within the left ventricle falls. When the pressure in the aorta exceeds that in the ventricles, the aortic valve closes and this generates a small pressure wave known as the dicrotic wave, which is preceded by a slight dip in pressure called the **dicrotic notch** or **incisura** (Figure 19.5). Following this, the pressure declines to its diastolic value before the next systole causes another pulse wave. The pressure wave in the arteries travels as a pulse and can be felt at various parts of the body as the **arterial pulse**.

How is blood pressure measured?

Blood pressure is measured indirectly by **auscultation**. This method relies on the fact that streamlined flow is silent whereas turbulent blood flow creates sounds within the blood vessels that may be heard by means of a stethoscope. Further details of the auscultatory method for measuring blood pressure are given in Box 19.1 (p. 367). Various automatic devices are now available for monitoring arterial blood pressure but careful auscultation still provides the most accurate measurements.

What is normal systemic arterial blood pressure?

In a healthy young adult at rest, systolic pressure is around 120 mmHg (16 kPa) while diastolic pressure is around 80 mmHg (10.7 kPa). This is normally written as 120/80 mmHg (16/10.7 kPa). The difference between the systolic and diastolic pressures is normally about 40 mmHg (5.3 kPa). This is called the **pulse pressure** (Figure 19.5).

Although the figure of 120/80 mmHg is a useful one to remember, it is also important to realize that a number of factors will influence the blood pressure, even at rest. Probably the most obvious effect is that of age. Mean blood pressure tends to increase with age so that by the age of 70, blood pressure averages 180/90 mmHg. This increase in arterial pressure is due to a reduction in the elasticity of the arteries (**arteriosclerosis** or hardening of the arteries). Consistently high blood pressure (diastolic pressure greater than 100 mmHg) is known as **hypertension** and is very common. The vascular complications associated with hypertension include stroke, heart disease and chronic renal failure. For this reason, regular screening is essential to avoid serious organ damage. For a more detailed discussion of hypertension, its causes and treatment, see pp. 384–386.

The **mean arterial pressure** (MAP) is a time-weighted average of the arterial pressure over the whole cardiac cycle. It is not a simple arithmetic average of the diastolic and systolic pressures because the arterial blood spends relatively longer near the diastolic pressure than the systolic (Figure 19.5). For most working purposes, however, an approximation to MAP can be obtained by applying the following simple equation:

$$MAP = \text{Diastolic pressure} + (1/3 \times \text{pulse pressure})$$

For example, if the systolic pressure is 110 mmHg and the diastolic pressure is 80 mmHg:

$$MAP = 80 + 1/3 (110 - 80) \text{ mmHg} = 90 \text{ mmHg}$$

The mean arterial pressure gives a measure of the average perfusion pressure of the systemic circulation.

Short-term rises in arterial blood pressure can be brought about by so-called pressor stimuli, such as pain, fear, anger and sexual arousal. Conversely, pressure falls significantly during sleep, sometimes to as little as 70/40 mmHg, and, to a much lesser and more gradual extent, during normal pregnancy. Gravity also affects blood pressure. On rising from a lying to a standing position, there is a transient fall in blood pressure followed by a small reflex rise (see Figure 19.12).

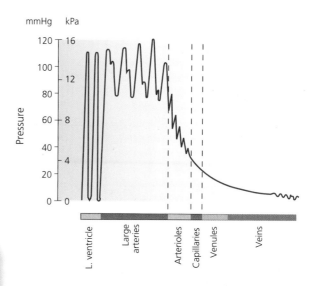

Figure 19.6 The change in pressure as the blood flows through the vessels of the systemic circulation. Note that the greatest fall in pressure occurs as the blood traverses the arterioles, which are the main site of vascular resistance. The greatest oscillations in pressure are seen in the left ventricle. Small pressure waves can be recorded from the large veins. These constitute the venous pulse.

KEY POINTS:

- Blood flow in the arteries is pulsatile. Pressure at the peak of ejection is called the systolic pressure while the lowest pressure occurs at the end of diastole (the diastolic pressure).

- The difference between the systolic and diastolic pressures is called the pulse pressure. In a healthy young adult, arterial blood pressure at rest will be around 120/80 mmHg.

- The mean arterial pressure is a time-weighted average, which is calculated as the sum of the diastolic pressure plus one-third of the pulse pressure.

The arterioles are the main source of vascular resistance

Measurement of the pressures in the different kinds of blood vessel show that the largest fall in pressure in the systemic circulation occurs as the blood passes through the arterioles (Figure 19.6). They therefore have the greatest resistance to the flow of blood. The majority of arterioles are in a state of tonic constriction due to the activity of the sympathetic nerves that supply them. As a result, their effective cross-sectional area is much less than the total cross-sectional area they would offer if they were all fully dilated.

As the resistance of a vessel depends on its diameter, major changes in blood flow to a particular region can be achieved by adjustment of the calibre of the arterioles.

This adaptation is important in regulating the distribution of the cardiac output between the various vascular beds. The mechanisms by which this regulation is achieved are discussed below (pp. 376–381).

The capillary pressure

One might expect that, as the capillaries have the smallest diameter, they would be the principal site of vascular resistance. However, the overall resistance to blood flow depends both on the diameter of the vessels and on the total cross-sectional area available for the passage of the blood. The cross-sectional area offered by the capillaries is about 25 times that of the arterioles. Moreover, as the capillaries have no smooth muscle in their walls, they cannot be constricted. As a result, they offer relatively little resistance to blood flow.

Venous pressure

The blood volume of a normal adult is about 5 litres but the distribution of this blood is not even throughout

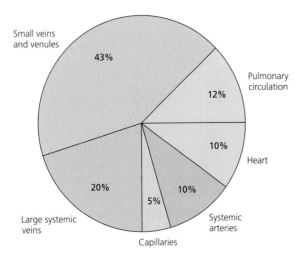

Figure 19.7 A pie chart diagram to show how the blood is distributed between the different parts of the circulation of a man at rest. Note the very high proportion of the blood in the systemic veins (c. 60%).

the circulation (Figure 19.7). The heart and lungs each contain about 500 ml of blood and the systemic arteries account for a further 600 ml while the capillaries have still less (about 250 ml). The bulk of the blood (about 3–3.5 litres) is found in the veins. The veins, particularly the large veins, thus act as a reservoir for blood and are called **capacitance vessels**.

As the walls of the veins are relatively thin and possess little elastic tissue, blood returning to the heart can pool in the veins simply by distending them. The degree of venous pooling is regulated by the tone of the smooth muscle (known as **venomotor tone**), which, in turn, is governed by the activity of the sympathetic nerves supplying the veins. During periods of activity when the cardiac output is high, venomotor tone is increased and the diameter of the veins is correspondingly reduced. Consequently, blood stored in the large veins is mobilized for distribution to exercising tissues.

Although veins hold so much blood, the average venous pressure measured at the level of the heart is only around 2 mmHg (0.27 kPa) compared with the average arterial pressure of about 100 mmHg (13.3 kPa). This pressure difference is sufficient to drive the blood into the central veins and thence into the right side of the heart where pressure is essentially zero (i.e. equal to that of the atmosphere).

The effect of gravity and the skeletal muscle pump on venous pressure

When a person stands up, pressure is increased in all the veins below the heart and reduced in all those above the heart as a result of the effects of gravity (Figure 19.8). In an adult, the pressure in the veins of the foot increases by about 90 mmHg (12 kPa) on standing. Consequently, the veins in the lower limbs become distended and they

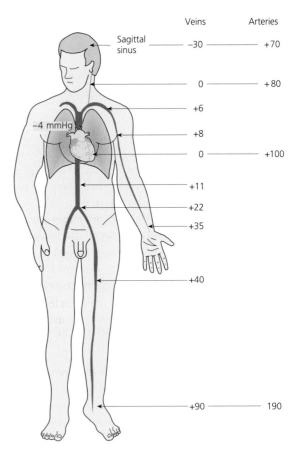

Figure 19.8 The effect of gravity on the pressures in the arteries and veins of an adult male who is standing quietly. The figures are approximate and will depend on the height of the individual. Below the level of the heart, the hydrostatic pressure increases progressively. Nevertheless, the pressure difference between the arteries and veins remains constant as the hydrostatic forces affect both arteries and veins equally. Above the heart, the pressures fall, but the rigidity of the skull prevents collapse of the cerebral veins.

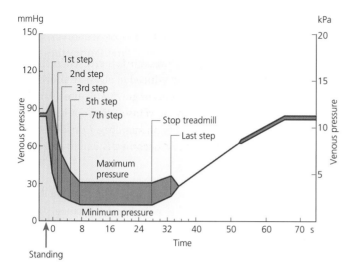

Figure 19.9 The pressure changes in a dorsal vein of the foot when a subject is first standing and then begins to walk on a treadmill. During walking, the active muscles help to 'pump' the blood towards the heart. As a result, venous pressure falls and stabilizes at a lower level, where it remains until exercise ceases. The pressure then progressively rises towards its original level.

accumulate blood (an effect sometimes referred to as **venous pooling**). The additional blood comes mostly from the thorax and abdomen. As a result, there is a transient fall in venous return, which leads to a reduction in stroke volume. This leads to a transient fall in blood pressure known as **postural hypotension**, which is rapidly corrected by the baroreceptor reflex (see p. 379).

When a skeletal muscle contracts, it compresses the veins within it. As the limb veins contain valves that prevent the backward flow of blood, the compression of the veins forces the blood towards the heart. This is known as the **skeletal muscle pump**. As Figure 19.9 shows, the squeezing action of the muscles on the veins leads to a progressive decline in venous pressure measured at the level of the foot. However, once exercise ends, venous pressure begins to rise once more. When the muscle pumps are less active, as in a bedridden subject, blood tends to accumulate in the veins and there is a risk of **deep vein thrombosis** (see Chapter 20, p. 407). A similar situation occurs during prolonged periods of standing or sitting, as experienced on a long-haul flight or similar journey.

KEY POINTS:

- The arterioles are the main sites of vascular resistance; the capillaries offer little resistance to blood flow. The chief determinant of capillary blood flow is the calibre of the arterioles supplying a particular capillary bed.

- The veins offer little resistance to blood flow and contain around two-thirds of the total blood volume.

19.6 Pressure and blood flow in the circulation

The flow of blood through any part of the circulation is driven by the difference in pressure between the arteries that supply the region in question and the veins that drain it. This pressure difference is known as the **perfusion pressure**. The resistance offered by the blood vessels to the flow of blood is known as the **vascular resistance**. The relationship between perfusion pressure, blood flow and vascular resistance is described by the following simple equations:

$$\text{Perfusion pressure} = \text{the arterial pressure } minus \text{ venous pressure} \tag{1}$$

and

Blood flow = the perfusion pressure *divided by* the vascular resistance (2)

Thus, blood flow will increase if the perfusion pressure is increased or if the vascular resistance is decreased. Conversely, if perfusion pressure falls or the vascular resistance increases, the blood flow will fall.

The cardiac output represents a volume of blood flowing round the circulation each minute so it is given as litres per minute. In healthy subjects, the cardiac output is the same in both the systemic and pulmonary circulations. If this were not the case, then blood would accumulate either in the lungs or in the systemic circulation. The cardiac output (CO) is equal to the heart rate (HR) times the stroke volume (SV) (see Chapter 17). In symbols:

$$CO = HR \times SV \qquad (3)$$

As the arterial blood pressure provides the force driving the blood around the body while the central venous pressure is close to zero, we can rearrange Equation 2 above to show the relationship between cardiac output and arterial blood pressure.

The cardiac output is the total amount of blood flowing through the systemic circulation and the perfusion pressure is equal to the mean arterial pressure (MAP). The total peripheral resistance (TPR) is the sum of all the vascular resistances within the systemic circulation, which is determined by the total cross-sectional area of the arterioles that are open at that time. Putting these values into Equation 2:

$$CO = MAP/TPR \qquad (4)$$

Rearranging:

$$MAP = CO \times TPR \qquad (5)$$

Or, in words:

Mean arterial pressure = cardiac output × total peripheral resistance.

> **KEY POINTS:**
> - Blood flows through the systemic circulation from the aorta to the veins because the pressure in the aorta and other arteries is higher than that in the veins.
> - This pressure is known as the arterial blood pressure and it is derived from the pumping activity of the heart.
> - The arterial blood pressure is determined by both the cardiac output and the total peripheral resistance.

19.7 The mechanisms that control the calibre of blood vessels

The smooth muscle of all blood vessels exhibits a degree of resting tension known as 'tone'. Changes in vascular tone alter the calibre of the blood vessels and so alter vascular resistance. If the tone is increased (i.e. if the smooth muscle contracts further), **vasoconstriction** occurs and vascular resistance increases. If tone decreases, there is **vasodilatation** and a fall in vascular resistance. The level of resting or basal tone varies between vascular beds. In areas where it is important to be able to increase blood flow substantially, such as skeletal muscle, basal tone is high while in the large veins basal tone is much lower.

The tone of a blood vessel is controlled by a variety of factors. These fall into two broad categories:

- **Intrinsic (or local) control of blood vessels** is brought about by the response of the smooth muscle to stretch, temperature and locally released chemical factors.
- **Extrinsic control** is exerted by the autonomic nervous system and by circulating hormones.

The diameter of all the major arteries, except the aorta and the principal veins, is regulated by circulating hormones and the autonomic nerves. That of the arterioles and small veins is regulated by hormones, autonomic nerves and by local chemical factors. As the capillaries and postcapillary venules have no smooth muscle, their diameter is not regulated.

Local control of blood vessels

Figure 19.10 shows that, over a certain range, maintained changes in blood pressure have little effect on the flow of blood through a particular vascular bed. If the pressure is raised quickly, blood flow increases at first but then returns close to its original level. Equally, if pressure falls quickly, blood flow also falls before returning towards its previous level. This relative stability of blood flow is known as **autoregulation**. It occurs independently of the nervous system and is the result of direct changes in vascular tone in response to changes in perfusion pressure.

Autoregulation is seen in most vascular beds. Nevertheless, as there is only a certain amount of blood available at any one time, there needs to be some mechanism that ensures that the needs of the body as a whole are met. This requirement is met by the autonomic nerves that innervate the blood vessels and by hormones such as adrenaline and vasopressin. These factors provide what is known as extrinsic control of the blood vessels, which can override the intrinsic or local regulatory mechanisms. For example, the blood flow to the gastrointestinal tract falls during exercise, even if there is food in the stomach. This allows more of the circulating blood to be diverted to the exercising muscles.

Vasodilatation occurs in response to a variety of metabolic by-products

Metabolism within cells gives rise to a number of chemical by-products, such as adenosine, carbon dioxide and lactic acid. When the activity of a tissue is increased, these by-products accumulate locally and cause relaxation

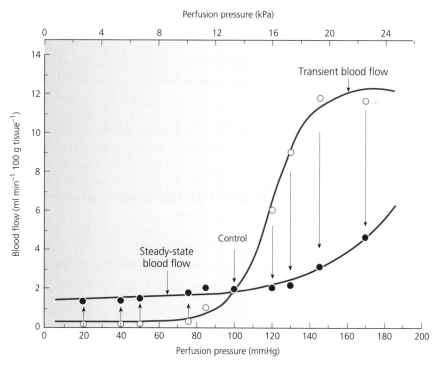

Figure 19.10 A graphical representation of the autoregulation of blood flow in an isolated, perfused skeletal muscle of a dog. The open circles represent the blood flow measured immediately after the perfusion pressure had been raised or lowered from the control level. As perfusion pressure is altered, there is a transient rise or fall in blood flow but autoregulatory mechanisms quickly restore blood flow to levels close to control (here shown by the red symbols). On the left of the diagram the upward pointing arrows show the increase in blood flow seen shortly after the perfusion pressure was reduced. On the right side, the blood flow was initially very high but quickly reverted to values closer to the control level (downward pointing arrows).

of the vascular smooth muscle and vasodilatation. In turn, this results in an increase in the flow of blood through the affected vascular bed. This increase in blood flow is known as **functional** or **metabolic hyperaemia** (it is also called active hyperaemia). Functional hyperaemia has the important effect of facilitating the removal of waste products from the vicinity of the actively metabolizing cells. It is particularly significant in tissues such as exercising muscle and the heart.

If the artery supplying blood to a tissue is compressed, blood flow is interrupted and the tissue becomes **ischaemic**. When the compression is relieved, the blood flow is, for a short time, greater than normal. This response is called **reactive hyperaemia**. It is a response to the accumulation of tissue metabolites during the period of ischaemia. The increased blood flow ensures a fresh supply of nutrients and oxygen to the deprived tissue and the rapid removal of metabolic waste products.

The vasodilatation occuring in response to locally released chemicals may play a role in autoregulation:

- If the blood flow through a tissue is insufficient, there will be an accumulation of local metabolites, which will act to cause vasodilatation and so restore blood flow.
- Conversely, when blood flow is high, the vasodilator chemicals are rapidly removed. Consequently, vascular tone will increase and blood flow will decline to its original level.

Local hormones influence blood flow

A number of so-called local hormones, or **autocoids**, released and acting locally are believed to alter blood flow through their role in processes such as inflammation and blood clotting. Such agents include **histamine**, **prostaglandins** and **platelet-activating factor**.

The endothelial cells of arteries and veins synthesize the inorganic gas **nitric oxide** (**NO**) in response to a wide variety of stimuli. Nitric oxide is a product of the cleavage of arginine by an enzyme present in the endothelial cells called nitric oxide synthase. Nitric oxide synthesis can be inhibited by certain analogues of arginine, and administration of such inhibitors to human subjects brings about vasoconstriction. This suggests that nitric oxide exerts a continuous or tonic vasodilator influence on the vasculature.

KEY POINT:

- The intrinsic regulation of the arterioles by locally released vasoactive substances accounts in large part for the autoregulation of blood flow.

Extrinsic mechanisms of blood vessel control

The mechanisms described above all exert local control over particular vascular beds. Superimposed upon these mechanisms is the overall control of the heart and circulation exerted by the nervous and endocrine systems. The purpose of this extrinsic regulation is to provide for the needs of the body as a whole by diverting blood to where it is most needed.

The nervous control of the blood vessels

Sympathetic vasoconstrictor fibres predominate in most vascular beds

The sympathetic vasoconstrictor fibres are constantly active and continuously regulate the diameter of the arterioles. They release noradrenaline (norepinephrine) that acts on the α-adrenoceptors of the vascular smooth muscle to cause it to contract and this maintains a degree of vasoconstriction in the majority of vascular beds (see Chapter 13). Interruption of the tonic activity of the sympathetic nerves (for example by the administration of α-blockers or by cutting the sympathetic nerves) leads to a significant rise in blood flow in the vessels of many tissues.

Vasodilatation induced by a fall in sympathetic vasoconstrictor fibre activity is also physiologically important. It contributes to the regulation of arterial blood pressure via the baroreceptor reflex (see below) and is partly responsible for producing vasodilatation in the vessels of the skin when it is necessary to lose heat to maintain a normal body temperature (Chapter 33).

Parasympathetic vasodilator nerves

Parasympathetic vasodilator fibres innervate the salivary glands, the exocrine pancreas, the gastrointestinal mucosa, the genital erectile tissue and the cerebral and coronary arteries. They are not tonically active but release acetylcholine when they are stimulated. The acetylcholine then acts on the muscarinic receptors of the vascular smooth muscle to cause it to relax. Blood flow to the tissue will then increase. The role of the parasympathetic vasodilator innervation of the erectile tissue is described in Chapter 29.

The baroreceptor reflex

The arterial blood pressure is very closely regulated by the body. It is monitored by **arterial baroreceptors**, which are stretch receptors found in the walls of the carotid sinuses and the aortic arch. A sharp rise in arterial blood pressure will stretch the walls of the aorta and carotid arteries. This activates the baroreceptors, which relay this information via the carotid sinus and aortic nerves to the brainstem where it elicits an inhibition of the sympathetic activity to the heart and vasculature. The reduced activity of the sympathetic vasoconstrictor fibres results in vasodilatation, which, in its turn, leads to a fall in peripheral resistance. At the same time, the heart rate declines (bradycardia) as a consequence of increased parasympathetic activity in the vagus nerve. The net result is a fall in arterial blood pressure that rapidly offsets the initial rise. The changes in peripheral vascular tone and heart rate that occur in response to activation of the baroreceptors form the **baroreceptor reflex** or **baroreflex**, which is illustrated in Figure 19.11.

A rapid fall in arterial blood pressure brings about a reduction in the baroreceptor discharge that normally occurs during the peak of the arterial pressure wave. In response, there is an increase in heart rate and stroke volume together with constriction of the vessels supplying skeletal muscle, skin, kidneys and gut. As a result, blood pressure is restored to normal levels.

The baroreceptor reflex stabilizes blood pressure following a change in posture

When a person moves quickly from a lying to a standing position, there is a significant fall in venous return to the heart (as blood is transferred to the lower body under the influence of gravity). In turn, this leads to a fall in cardiac output and arterial blood pressure

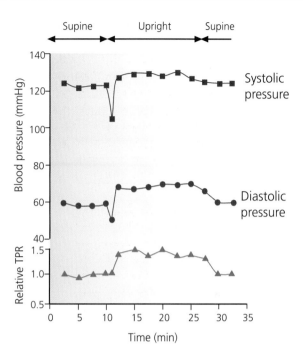

Figure 19.11 The changes in blood pressure and total peripheral resistance (TPR) seen as a subject is tilted from a horizontal to a vertical position and back again. Note the initial drop in pressure that the baroreceptor reflex rapidly corrects by an increase in total peripheral resistance (constriction of the arterioles).

(**postural hypotension**). When arterial blood pressure falls, the baroreceptor reflex is activated: heart rate is increased and there is a rise in total peripheral resistance. These changes normally restore the blood pressure to its normal level. Because the baroreceptor reflex normally takes a few seconds to become fully effective, the blood supply to the brain is briefly reduced when someone stands up quickly, giving rise to a brief episode of dizziness. This sequence of events is summarized in Figure 19.12. Postural hypotension becomes more marked with advancing age and is a common cause of falls in the elderly.

Hormonal control of the blood vessels

Under normal conditions, the circulation is under both nervous and endocrine control. Long-term regulation of blood pressure requires the co-operation of a number of hormonal mechanisms that work alongside the

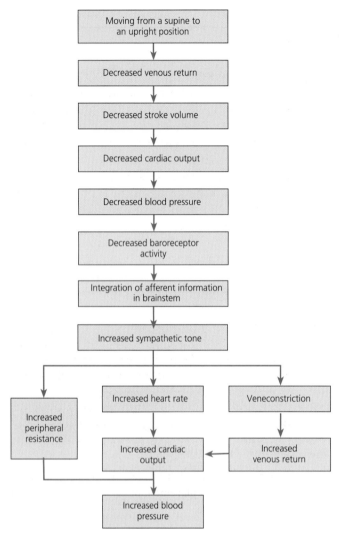

Figure 19.12 A flow chart showing the sequence of cardiovascular changes initiated by the baroreceptors following a fall in arterial blood pressure and resulting in restoration of normal pressure.

autonomic nerves to regulate both plasma volume and vascular tone. The major regulatory hormones are **adrenaline** (epinephrine), **vasopressin** (ADH) and the **renin–angiotensin–aldosterone system**.

The adrenal medullary hormones – adrenaline and noradrenaline

Adrenaline (epinephrine) and noradrenaline (norepinephrine) are secreted by the adrenal medulla during stress of all kinds, including exercise. In humans,

about 80% of the secretion is adrenaline, which acts on the α-adrenoceptors and β-adrenoceptors of the blood vessels (see Chapter 13). When adrenaline activates the α-adrenoceptors, the smooth muscle of the blood vessels contracts and there is a vasoconstriction. When it activates β-adrenoceptors, it causes the vascular smooth muscle to relax and brings about vasodilatation. Noradrenaline has a much greater affinity for α- than for β-adrenoceptors, and will therefore normally cause vasoconstriction.

Adrenaline interacts with the β-adrenoceptors present in the smooth muscle of blood vessels located in skeletal muscle, the heart and the liver. In all these tissues, it elicits a vasodilatation. This adaptation causes the blood vessels of the heart and skeletal muscle to dilate, thereby increasing the blood flow to these tissues during the 'fight or flight' reaction, while vessels elsewhere constrict (e.g. those of the skin and gastrointestinal tract). Blood is thus diverted preferentially to those tissues that require a high blood flow to meet the demands made on them.

Vasopressin (antidiuretic hormone or ADH)

Full details of the secretion and actions of this hormone can be found in Chapter 16. Briefly, although vasopressin is chiefly concerned with the regulation of fluid excretion by the kidneys, it has powerful effects on the vasculature. It is secreted into the circulation from the posterior pituitary gland in response to the fall in blood pressure that follows a substantial haemorrhage (see Box 19.2). At such times, vasopressin causes a powerful vasoconstriction in many tissues, which helps to maintain arterial blood pressure (hence its name). In the cerebral and coronary vessels, however, vasopressin seems to elicit vasodilatation. The net effect is a redistribution of the blood to the essential organs – the heart and brain.

The renin–angiotensin–aldosterone system

Renin is not a hormone but a proteolytic enzyme that is secreted by the kidneys in response to a fall in the sodium concentration in the distal tubule. It acts on an inactive peptide in the blood called angiotensinogen,

Box 19.2 Cardiovascular adjustments in haemorrhage

If there is an acute loss of blood, the cardiovascular system rapidly adjusts to maintain blood pressure and to preserve blood flow to vital organs. However, the exact response of the body to blood loss depends on the amount lost.

Significant cardiovascular adjustments begin as blood loss exceeds about 5% of total blood volume (c. 250 ml). The loss of this amount of blood leads to a diminished venous return and a reduced stimulation of the low-pressure receptors, which results in a vasoconstriction of the cutaneous, muscle and splanchnic vessels. These changes occur before there is any significant fall in blood pressure.

As blood loss increases beyond about 5%, the venous return falls further and there is a fall in both cardiac output and arterial blood pressure. The fall in blood pressure activates the arterial baroreceptor reflex so that there is an increase in heart rate and in arteriolar tone. Together, these changes restore the blood pressure to near normal levels. The increased activity of the sympathetic nerves also increases venous tone, which results in a mobilization of the blood from the large veins (the capacitance vessels). This helps to maintain venous return and so minimizes the fall in cardiac output. These adjustments occur rapidly following blood loss.

If blood loss exceeds 8–10%, there is a further fall in cardiac output and an enhancement of the activity of the sympathetic nerves. The heart rate continues to rise, and the peripheral vasoconstriction increasingly diverts blood away from the skin, muscles and viscera towards the brain and heart. There is an increase in the rate of synthesis of angiotensin II, which has a powerful vasoconstrictor action. Despite the fall in cardiac output, these changes help to maintain blood pressure and they are the principal mechanisms by which the cardiovascular system adjusts to mild haemorrhage (loss of less than 10% of the blood volume). In blood donation, 450 ml of blood (a blood unit) are routinely taken. Normal healthy adults can sustain this level of blood loss without ill effect.

As blood loss approaches 800 ml, there is a further intensification of sympathetic nerve activity but the venous return is no longer sufficient to maintain blood pressure, which begins to fall. During this phase, catecholamine secretion by the adrenal medulla increases, as does the secretion of vasopressin (ADH) by the posterior pituitary gland. These hormones enhance the vasoconstriction. As a consequence of these adjustments, the pulse is weak and rapid, the skin is cold and clammy and the mouth becomes dry as salivary secretion stops.

Following extensive blood loss (greater than 1 litre), there is an urgent need to restore blood volume. Clinically, this can be achieved by blood transfusion. Physiologically, the restoration of the blood volume requires restoration of the plasma volume, which is relatively rapid, and replacement of the lost red cells, which takes several weeks.

to form angiotensin I, which is then converted in the lungs to its active form, **angiotensin II**. This hormone has two important actions: it stimulates the secretion of **aldosterone** from the adrenal cortex and it causes vasoconstriction. Aldosterone acts on the distal tubules of the kidneys to increase the reabsorption of sodium ions. This is particularly important when the blood volume is low – following a haemorrhage for example. The retention of sodium together with its osmotic equivalent of water helps to restore the blood volume. In turn, this will help to restore a normal arterial blood pressure. Further details of this important hormonal system may be found in Chapter 23.

KEY POINTS:

- Nerves and hormones exert extrinsic control over the heart and circulation in response to information arising from cardiovascular receptors of many kinds.

- Sympathetic vasoconstrictor fibres are the most widespread and important of the nerves that alter the calibre of blood vessels.

- A number of hormonal mechanisms provide extrinsic regulation of plasma volume and vascular tone. These include adrenaline, vasopressin (ADH) and the renin–angiotensin–aldosterone system.

19.8 The microcirculation and tissue–fluid exchange

In this section, the processes responsible for the exchange of fluid, nutrients and metabolites between the blood and the tissues are examined in detail. In general terms, blood flows from the arterioles to the venules via the capillaries, which are the main exchange vessels. As part of the process of tissue exchange, a small volume of fluid passes from the plasma to the interstitial space. While part of this fluid is returned directly to the circulation, most is returned to the blood via the afferent lymphatic vessels. For this reason, the basic organization of the lymphatic system and its role in the regulation of the volume of the interstitial fluid will also be discussed here.

The microcirculation is organized in functional units

The arterioles that branch directly from the arteries are known as primary arterioles and they are extensively innervated by sympathetic nerve fibres. The primary arterioles progressively give rise to secondary and tertiary arterioles, which have less smooth muscle and are more sparsely innervated. The final degree of branching gives rise to the terminal arterioles, which have almost no innervation, their calibre being regulated largely by the local concentration of tissue metabolites (see pp. 377–378).

The terminal arterioles are 10–40 μm in diameter and each arteriole gives rise directly to a group of capillaries known as a cluster or module (Figure 19.13). The flow through the capillary cluster is regulated by the terminal arterioles. The capillaries themselves are 5–8 μm in diameter and about 0.5–1 mm in length. They drain into postcapillary venules whose walls do not contain smooth muscle. The postcapillary venules coalesce into larger venules, and the larger ones then merge into the veins.

The structure of the capillaries

The walls of the capillaries consist of a single layer of flattened endothelial cells. This thin layer is partly covered by cells called **pericytes** and the whole structure is surrounded by a basement membrane. The capillary wall is very thin (about 0.5 μm) so the diffusion path between plasma and tissue fluid is extremely short.

As capillaries are the principal exchange vessels, their density determines the total surface area for exchange between tissue and blood. If the density of capillaries is high, there will be a large surface area for exchange and a relatively short distance between capillaries. Consequently, the delivery of oxygen, glucose and other

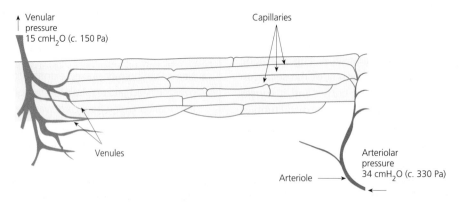

Figure 19.13 A diagrammatic representation of the capillary bed in a relaxed muscle from a rat. On the right, a terminal arteriole feeds a module of capillaries, which eventually coalesce to form postcapillary venules on the left. These in turn drain into the venules and then to the veins.

Levick, J.R. (2003). *An Introduction to Cardiovascular Physiology*, 4th edn. Hodder Arnold: London.

Pocock, G. and Richards, C.D. (2006). *Human Physiology: The Basis of Medicine*, 3rd edn. Oxford University Press: Oxford. Chapter 15, pp. 263–266 and pp. 283–302.

Pharmacology

Grahame-Smith, D.G. and Aronson, J.K. (2002). *Oxford Textbook of Clinical Pharmacology*, 3rd edn. Oxford University Press: Oxford. Chapter 23.

Rang H.P., Dale, M.M., Ritter, J.M. and Flower, R. (2007). *Pharmacology*, 6th edn. Churchill-Livingstone: Edinburgh. Chapters 17 and 18.

Self-assessment questions

Which of the following statements are true and which are false? Answers are given at the end of the book (p.755).

1   a) The arterial blood pressure is normally 120/80 mmHg.
    b) The systemic arterial blood pressure depends solely on the cardiac output.
    c) The mean blood pressure is the arithmetic average of the systolic and diastolic pressures.
    d) The main source of vascular resistance is the capillaries.
    e) The flow of blood in the large veins is pulsatile.

2   a) The diameter of the arterioles is entirely regulated by the sympathetic nervous system.
    b) Autoregulation refers to the nervous control of the blood vessels.
    c) Reactive hyperaemia is due to vasodilatation caused by accumulation of metabolites during a period of occluded blood flow.
    d) Activation of the sympathetic system causes vasoconstriction in the viscera and skin but vasodilatation in skeletal muscle.
    e) Parasympathetic vasodilator fibres innervate the blood vessels of the gastrointestinal tract.

3   a) Solute exchange between the capillaries and tissues occurs mainly by diffusion.
    b) The plasma proteins play an important role in tissue fluid exchange.
    c) The capillaries provide a significant source of vascular resistance.
    d) All of the fluid that passes from the capillaries to the tissues is returned to the blood via the lymphatic circulation.
    e) The lymph has the same ionic composition as the plasma but has a lower oncotic pressure.

4   a) If the arterial pressure suddenly falls, the baroreceptor reflex increases the heart rate.
    b) An increase in arterial pressure elicits a peripheral vasoconstriction that is mediated via the sympathetic nervous system.
    c) The cardiac volume receptors are mainly responsible for the long-term regulation of systemic blood pressure.
    d) The coronary blood flow is mainly regulated by metabolic hyperaemia.

20   The properties of blood

After reading this chapter you should have gained an understanding of:

- The principal functions of the blood
- The physical and chemical properties of plasma and normal values of plasma constituents
- Normal values for the cellular constituents of blood
- The importance of the haematocrit (packed red cell volume)
- The origin of blood cells (haematopoiesis)
- The functions of the red cells, white cells and platelets
- The metabolism of iron and its role in haemoglobin formation
- The carriage of oxygen and carbon dioxide by the blood
- Some important blood disorders including anaemias, leukaemia and thrombocytopenia
- Blood clotting (haemostasis) and prothrombin time
- Blood groups and their importance in blood transfusions
- Anticoagulants and fibrinolytic drugs

20.1   Introduction

Blood consists of a non-cellular fluid matrix (**plasma**) in which are suspended the **erythrocytes** (red blood cells), **leukocytes** (white blood cells) and **thrombocytes** (platelets). Blood is sometimes considered to be a specialized form of connective tissue with the important difference that the plasma is not secreted by the cells. The nature of the cell suspension can be demonstrated by centrifuging a sample of blood in a test tube for a short time at low speed. After centrifugation, the heavier red blood cells are packed at the bottom of the tube, while the plasma may be seen as a clear, pale yellow fluid above the cells, as shown in Figure 20.1. The white blood cells and platelets

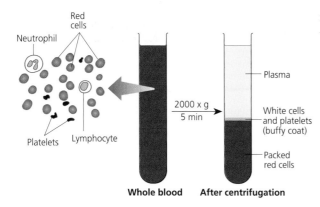

Figure 20.1 Separation of blood into cells and plasma by centrifugation. On the left is a diagrammatic view of the appearance of whole blood showing red cells, leukocytes and platelets.

form a thin layer (the 'buffy coat') between the red cells and the plasma.

The blood provides a vital means of communication between the tissues of large multicellular organisms and is the vehicle of transport around the body for a huge variety of substances. Its functions include the following:

- Absorption of nutrients from the gut and their delivery to the tissues
- Carriage of oxygen from the lungs to the tissues and of carbon dioxide from the tissues to the lungs
- The transport of waste products of metabolism from their sites of production to their sites of disposal
- The carriage of hormones from the endocrine glands to their target tissues
- Defence against disease-producing agents (its immunological role, see Chapter 21).

The circulating blood volume is around 7–8% of body weight, so for a 70 kg man this will be around 5 litres. However, for a newborn baby weighing 3.2 kg, the blood volume will be around 250 ml. Nearly two-thirds of the blood is in the veins and, for an adult, there is around 600 ml each in the heart and lungs. The remainder is in the systemic arteries, arterioles and capillaries (see p. 374).

20.2 The physical and chemical properties of plasma

Normal adults have 35–45 ml of plasma per kg of body weight, i.e. 2.8–3.0 litres in men and around 2.4 litres in women. Plasma is 95% water and 5% solids in solution or suspension. These include mineral ions (electrolytes), small organic molecules and plasma proteins. Typical values for a number of important constituents of plasma are given in Table 20.1.

The ionic constituents of plasma

The principal inorganic cation of the plasma is sodium, which has a concentration of 140–145 mmol l^{-1}. Potassium, calcium, magnesium and hydrogen ions are present in much smaller amounts. Chloride is the principal anion of plasma (around 100 mmol l^{-1}) with bicarbonate, phosphate, sulphate and organic anions present in smaller quantities. In health, the plasma has a pH of 7.35–7.45

and its osmolality is 280–300 mOsmol kg^{-1} water (see Chapter 34). A variety of homeostatic mechanisms normally ensure that the composition of the plasma is kept within biologically safe limits, but the balance may be disturbed by disorders of the kidneys, lungs, liver, cardiovascular system or endocrine organs. For this reason, accurate analysis of plasma composition forms an essential part of diagnosis and treatment.

Plasma proteins

Each litre of plasma contains around 75 g of proteins, which fall into three principal classes: the albumins, the globulins and fibrinogen. **Albumins** are the smallest and the most abundant, accounting for about 60% of the total plasma protein. They act as transport proteins for lipids and steroid hormones and are important in body

Table 20.1 Principal constituents of the plasma

Constituent	Quantity	Units	Remarks
Water	94.5	$g\,l^{-1}$	
Bicarbonate	25	$mmol\,l^{-1}$	Important for the carriage of CO_2 and for H^+ buffering
Chloride	105	$mmol\,l^{-1}$	The principal extracellular anion
Inorganic phosphate	1.1	$mmol\,l^{-1}$	
Calcium	2.5	$mmol\,l^{-1}$	This is total calcium; ionized calcium is about $1.25\,mmol\,l^{-1}$
Magnesium	0.8	$mmol\,l^{-1}$	
Potassium	4	$mmol\,l^{-1}$	
Sodium	144	$mmol\,l^{-1}$	The principal extracellular cation
Hydrogen ions	40	$nmol\,l^{-1}$	This corresponds to a pH value of c. 7.4
Glucose	4.5	$mmol\,l^{-1}$	Major source of metabolic energy, particularly for the CNS
Cholesterol	2.0	$g\,l^{-1}$	
Fatty acids (total)	3.0	$g\,l^{-1}$	
Total protein	70–85	$g\,l^{-1}$	
Albumin	45	$g\,l^{-1}$	Principal protein of the plasma; binds hormones and fatty acids
α-Globulins	7	$g\,l^{-1}$	
β-Globulins	8.5	$g\,l^{-1}$	
γ-Globulins	10.6	$g\,l^{-1}$	Immunoglobulins (antibodies)
Fibrinogen	3	$g\,l^{-1}$	Blood clotting
Prothrombin	1	$g\,l^{-1}$	Blood clotting
Transferrin	2.4	$g\,l^{-1}$	Iron transport

Note that these values are approximate mean values and that even in health there is considerable individual variation.

fluid balance. They provide most of the colloid osmotic pressure (oncotic pressure) that regulates the passage of water and solutes through the capillaries (p. 383). The **globulins** account for around 40% of total plasma protein and are subdivided into alpha (α), beta (β) and gamma (γ) globulins. The α- and β-globulins are made by the liver and act as transport proteins for lipids and fat-soluble vitamins. The γ-globulins are antibodies produced by lymphocytes in response to exposure to specific antigens (see Chapter 21). **Fibrinogen** is an important clotting factor produced by the liver whose role in blood clotting is discussed on pp. 405–406.

KEY POINTS:

- Blood consists of plasma in which are suspended red cells, white cells and platelets. The plasma is about 95% water, the rest consisting of a variety of proteins including albumins, globulins and fibrinogen, mineral ions (chiefly Na^+ and Cl^-) and small organic molecules (e.g. glucose).
- Plasma albumins carry lipids and steroid hormones in the plasma. The α- and β-globulins transport lipids and fat-soluble materials, while the γ-globulins are antibodies and play an essential role in defence against infection.

20.3 The cells of the blood

The blood contains **red cells**, five classes of **white cells** (recognized according to their appearance and staining reactions) and **platelets**. Of these, the red cells are by far the most numerous. Table 20.2 lists the concentrations of the various cell types in whole blood. Mature blood cells have a relatively short lifespan and must therefore be renewed continuously. Replacement is achieved by a process called **haematopoiesis**, which includes erythropoiesis (the formation of red blood cells) and leukopoiesis (the formation of white blood cells).

Blood cells are derived from pluripotent cells of the bone marrow

All the different types of blood cell are formed from a population of cells present within the haematopoietic tissue of bone marrow. These cells are known as **stem cells**, which are **pluripotent**. This means that they have the ability to differentiate into any of the different varieties of mature blood cell through a series of cell divisions. The stages of development of the blood cells are shown diagrammatically in Figure 20. 2.

Infants have haematopoietic tissue in all their bones but, in adults, haematopoietic stem cells are found predominantly in the red bone marrow of the pelvis, vertebrae, clavicles, scapulae and skull and in the proximal ends of the long bones. When samples of bone marrow are required for examination or transplantation, they are usually obtained by aspiration from the pelvic bones.

As may be seen from Figure 20.2, the stem cells proliferate to give rise to more stem cells and two further cell lines – **lymphoid cells** and **myeloid cells**. The lymphoid cells migrate to the lymph nodes, spleen and thymus gland where they differentiate to become lymphocytes. The myeloid cells remain in the bone marrow and differentiate to form the other types of white blood cell, the red blood cells (erythrocytes) and large cells called megakaryocytes, from which platelets are derived.

Erythrocytes (red blood cells)

The red cells are the most abundant cell type in the blood. Each litre of blood normally contains between 4.0 and 6.5×10^{12} red cells, varying according to the age, sex and state of health of the individual. The proportion of the total blood volume that is occupied by the red blood cells is called the **haematocrit ratio** or simply the haematocrit. In males, the haematocrit, measured using a sample of venous blood, is normally between 0.4 and 0.54 while in females the value is slightly lower and lies between 0.37 and 0.47.

The chief function of the red blood cells is to transport the respiratory gases, oxygen and carbon dioxide, around the body. The cells themselves are small, circular, biconcave discs with a diameter of 7–8 μm and a mean volume of $80–95 \times 10^{-15}$ l (80–95 fl). They are thin and flexible and can squeeze through capillaries, which have internal diameters of no more than 5–8 μm. Their shape gives the

Table 20.2 The cellular elements of whole blood

Cell type	Site of production	Typical cell count (l^{-1})	Comments and function
Erythrocytes (red cells)	Bone marrow	5×10^{12} (men)	Transport of O_2 and CO_2
		4.5×10^{12} (women)	
Leukocytes		7×10^9	
Granulocytes: (differential count)			
Neutrophils	Bone marrow	5×10^9 (40–75%)	Phagocytes – engulf bacteria and other foreign particles
Eosinophils	Bone marrow	100×10^6 (1–6%)	Congregate around sites of inflammation – antihistamine properties
			Very short-lived in blood
Basophils	Bone marrow	40×10^6 (<1%)	Circulating mast cells – produce histamine and heparin
Agranulocytes: (differential count)			
Monocytes	Bone marrow	0.4×10^9 (2–10%)	Phagocytes; become macrophages when they migrate to the tissues
Lymphocytes	Bone marrow, lymphoid tissue thymus, spleen	1.5×10^9 (20–45%)	Production of antibodies
Platelets	Bone marrow	250×10^9	Aggregate at sites of injury and initiate haemostatis

Note that, while mean values are given, these are subject to considerable individual variation. The approximate percentage of individual types of leukocyte is given after the number per litre – this is called the **differential white cell count**.

cells a large surface area to volume ratio, which favours efficient gas exchange. The chief constituent of the red cell cytoplasm is haemoglobin (Hb), a protein that binds very strongly to oxygen (see below) and gives the cells their characteristic colour.

Although developing red cells in the bone marrow are nucleated, they lose their nuclei to become reticulocytes just before they are released into the circulation. Within a day or two of entering the bloodstream, the reticulocytes complete their differentiation to become mature erythrocytes. Normally, only 1–3% of the total circulating red cell population consists of reticulocytes at any one time, but their proportion increases during times of accelerated red blood cell production (see below). The proportion of circulating reticulocytes is a useful indicator of the rate of red cell production. In addition to lacking a nucleus, mature red blood cells possess neither mitochondria nor ribosomes. They rely for their energy requirements on glycolysis (the anaerobic breakdown of glucose, see p. 49).

solution is $5.3 \times 0.526 = 2.8 \, \text{ml} \, \text{dl}^{-1}$. Mixed venous blood has a $P\text{co}_2$ of around $6.12 \, \text{kPa}$ ($46 \, \text{mmHg}$) and will therefore contain $6.12 \times 0.526 = 3.2 \, \text{ml} \, \text{CO}_2 \, \text{dl}^{-1}$. Between 5 and 7% of total blood carbon dioxide is carried in physical solution.

Bicarbonate ions

About 90% of the total blood carbon dioxide is carried in the form of bicarbonate ions. Carbon dioxide combines with water to produce carbonic acid by the following reaction:

$$CO_2 + H_2O \rightleftharpoons H_2CO_3 \tag{1}$$

Carbonic acid readily dissociates to form hydrogen ions (H^+) and bicarbonate ions (HCO_3^-):

$$H_2CO_3 \rightleftharpoons H^+ + HCO_3^- \tag{2}$$

Reaction 1 occurs only slowly in plasma but in the red cells there is an enzyme, carbonic anhydrase, that catalyses (speeds up) the reaction. As carbon dioxide diffuses into the red blood cells, carbonic acid is formed, which immediately dissociates to yield bicarbonate and hydrogen ions, as shown in Reaction 2. Much of the bicarbonate moves back out of the cells into the plasma, in exchange for chloride ions. Thus, the plasma returning to the lungs has a higher bicarbonate and lower chloride concentration than the plasma of the arterial blood. This is known as the **chloride shift** or **Hamburger shift**. The hydrogen ions produced by

the dissociation of carbonic acid are buffered by the haemoglobin in the red cells. This buffering is very important because it allows large amounts of carbon dioxide to be carried in the blood (as HCO_3^-), without a significant fall in pH. Indeed, arterial blood contains about $43.9 \, \text{ml} \, \text{CO}_2 \, \text{dl}^{-1}$ as HCO_3^- and has a pH of 7.4 while mixed venous blood contains about $47 \, \text{ml} \, \text{CO}_2 \, \text{dl}^{-1}$ as HCO_3^- and has a pH of 7.35, only 0.05 pH units lower.

Carbamino compounds

A small amount of carbon dioxide (between 5 and 8% of the total) is carried in the blood in the form of carbamino compounds. Some of the carbon dioxide that diffuses into red cells from the tissues combines with amino groups on the haemoglobin molecules to from carbaminohaemoglobin in the reaction:

$$HbNH_2 + CO_2 \rightleftharpoons HbNHCOO^- + H^+$$

In addition, a very small amount of carbon dioxide forms carbamino compounds by combining with amino groups on plasma proteins. Arterial blood contains $2.3 \, \text{ml} \, \text{CO}_2 \, \text{dl}^{-1}$ as carbamino compounds while venous blood contains $3.8 \, \text{ml} \, \text{CO}_2 \, \text{dl}^{-1}$.

Table 20.3 summarizes the ways in which carbon dioxide may be carried in the blood and the amounts carried in each form for arterial and mixed venous blood.

Table 20.3 Normal values for oxygen and carbon dioxide partial pressure and content in arterial and mixed venous blood

	Arterial blood	Mixed venous blood
Partial pressure of oxygen (P_{O_2})	13.3 kPa (100 mmHg)	5.33 kPa (40 mmHg)
O_2 in physical solution	$0.3 \, \text{ml} \, \text{dl}^{-1}$	$0.12 \, \text{ml} \, \text{dl}^{-1}$
O_2 bound to haemoglobin	$19.5 \, \text{ml} \, \text{dl}^{-1}$	$14.1 \, \text{ml} \, \text{dl}^{-1}$
Total O_2	$19.8 \, \text{ml} \, \text{dl}^{-1}$	$14.2 \, \text{ml} \, \text{dl}^{-1}$
Partial pressure of CO_2 (P_{CO_2})	5.3 kPa (40 mmHg)	6.1 kPa (46 mmHg)
CO_2 in physical solution	$2.7 \, \text{ml} \, \text{dl}^{-1}$	$3.1 \, \text{ml} \, \text{dl}^{-1}$
CO_2 as bicarbonate ions	$43.9 \, \text{ml} \, \text{dl}^{-1}$	$47 \, \text{ml} \, \text{dl}^{-1}$
CO_2 as carbamino compounds	$2.3 \, \text{ml} \, \text{dl}^{-1}$	$3.8 \, \text{ml} \, \text{dl}^{-1}$
Total CO_2	$48.8 \, \text{ml} \, \text{dl}^{-1}$	$53.9 \, \text{ml} \, \text{dl}^{-1}$

The values for O_2 content assume that blood haemoglobin is $15 \, \text{g} \, \text{dl}^{-1}$. Calculated figures are given to one decimal place.

The carbon dioxide dissociation curve

The relationship between the P_{CO_2} (in kPa or mmHg) and the total amount of carbon dioxide present in all forms in the blood (in ml dl^{-1}) is called the carbon dioxide dissociation curve. Unlike the oxyhaemoglobin dissociation curve, which shows saturation (Figure 20.5), the carbon dioxide dissociation curve does not become saturated, even at high P_{CO_2}, as shown in Figure 20.7. Nevertheless, the quantity of carbon dioxide carried by the blood depends on the amount of oxygenated Hb present. As the level of oxyhaemoglobin falls, the blood is able to carry more carbon dioxide (Figure 20.6). This is called the **Haldane effect**. Thus, the blood perfusing the tissues is able to take up more carbon dioxide and this facilitates the removal of carbon dioxide from the actively metabolizing tissues. In the lungs, the reverse is true – oxyhaemoglobin levels rise as the blood passes through the capillaries of the lungs and this favours offloading of carbon dioxide from the blood.

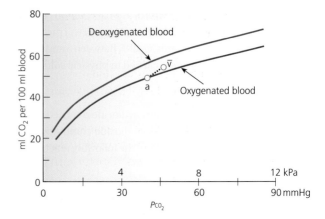

Figure 20.7 The carbon dioxide dissociation curve for whole blood. Point a represents the P_{CO_2} of arterial blood while the point labelled v̄ gives the value for mixed venous blood. Note that mixed venous blood is able to carry more carbon dioxide than arterial blood. This is known as the Haldane effect.

KEY POINTS:

- Carbon dioxide is carried in the blood in three forms: in physical solution, as bicarbonate ion and as carbamino compounds.

- In the red blood cells, carbon dioxide is rapidly converted into carbonic acid, which dissociates to H^+ and HCO_3^-. The H^+ is buffered by haemoglobin while the HCO_3^- diffuses out of the red cells in exchange for Cl^- (the chloride shift).

20.6   Major disorders of red blood cells

Broadly, red blood cell disorders fall into two categories. These are **proliferative disorders**, in which there is an excess of cells, and **deficiency disorders**, in which there are too few cells.

Polycythaemia

This disorder (which is also known as erythrocytosis) is characterized by over-production of red blood cells. This causes an increase in the haematocrit value (to as much as 60–80%) and, as a result, an increase in blood viscosity. Although this is seen as a physiological response to chronic hypoxia in people living at high altitude, it may also arise as a disease process, for example in patients with erythropoietin-secreting tumours of the kidney. The increased viscosity of the blood places an extra load upon the heart and increases the risk of deep vein thrombosis (DVT) and stroke.

Anaemia

This term covers a variety of blood disorders characterized by a reduced number of cells, a reduced haemoglobin concentration, or both. *All types of anaemia result in a fall in the oxygen-carrying capacity of the blood.* Anaemia may result from any of the following causes:

- A reduction in red cell numbers as a result of acute **haemorrhage**. Although plasma volume is quickly restored following blood loss, it takes much longer to replace the lost red cells.

- A reduction in the haemoglobin content of the red cells as a result of **iron deficiency** due to chronic blood loss (a bleeding peptic ulcer for example) or pregnancy.

- A reduction in red cell size. This condition is known as **microcytic anaemia** and is seen in cases of iron deficiency. Normally, mean corpuscular (red cell) volume is generally $80–95 \times 10^{-15}$ l (80–95 fl).

- **Megaloblastic (pernicious) anaemia** is characterized by the presence of abnormal red cells that are larger than normal (macrocytic). It occurs most often in people who lack vitamin B_{12} or folic acid, both of which are essential for normal erythrocyte maturation (see p. 393). Vitamin B_{12} deficiency may be due to an inadequate diet (lacking any foods of animal origin) but is also a consequence of lack of intrinsic factor resulting from total gastrectomy (see Chapter 24 for more details).

- **Aplastic anaemia** occurs when there is failure of bone marrow function. It is associated with a loss of bone marrow stem cells and an increase in the proportion of fat in the marrow cavity. It may be congenital or acquired (due to a viral infection, irradiation or drug exposure) and is associated with a reduction in the number of all blood cell types (**pancytopenia**).

- **Thalassaemia** is the name given to a group of anaemias caused by the hereditary inability to produce one or both of the peptide chains of globin. As a result, Hb is not synthesized normally and red cells appear small and pale. The disorder is found predominantly amongst Mediterranean, African and black American populations.

- **Sickle-cell disease (homozygous sickle-cell anaemia)** is caused by a recessive mutation in one of the genes that control the formation of globin and leads to the production of abnormal haemoglobin called sickle-cell haemoglobin or HbS. It is prevalent in Afro-Caribbean populations with one in five West Africans known to be carriers. In homozygous individuals (those carrying two sickle-cell genes), HbS becomes insoluble when it is deoxygenated. The insoluble HbS crystallizes inside the red cells and causes them to become sickle-shaped. The deformed cells obstruct capillaries and cause a reduction in blood flow to the tissues, which subsequently become hypoxic. Tissue hypoxia is associated with cell damage and intense pain. Virtually every organ is affected but the liver, spleen, heart and kidneys are especially vulnerable to damage because of the increased risk of blood clot formation caused by sluggish blood flow. A small but significant advantage of this disease is that people who carry the HbS gene have a high resistance to malaria as the parasites that cause malaria (protozoa of the genus *Plasmodium*) cannot live in blood cells containing HbS.

> **KEY POINTS:**
>
> - Polycythaemia results from over-stimulation of red blood cell production and leads to a rise in both haematocrit and blood viscosity.
>
> - Anaemia is characterized by a reduction in the oxygen-carrying capacity of the blood. It may arise from reductions in red blood cell number or size, reduction of the haemoglobin content of red cells or abnormalities of haemoglobin structure.

20.7 White blood cells – leukocytes

White blood cells are larger than red blood cells, they possess a nucleus and they are present in much smaller numbers – normal blood contains $4–11 \times 10^9$ white cells per litre. Leukocytes have a vital role in the protection of the body against disease – they are the mobile units of the body's immune system. There are three major categories of white blood cells:

- The **granulocytes** (also called polymorphonuclear leukocytes because their nuclei have a lobed or segmental appearance)

- The **monocytes** (macrophages)
- The **lymphocytes**.

The specific functions of the different types of white blood cell will be discussed in more detail in the next chapter and the following is intended to serve only as a brief introduction.

The monocytes and lymphocytes are also known collectively as **agranulocytes** or mononuclear leukocytes. Granulocytes account for around 70% of the total number of white cells in the blood. Three different types of cell make up the granulocyte population. These are **neutrophils**, **basophils** and **eosinophils** (named according to their staining reactions), and are shown in Figure 20.8.

Neutrophils are the most numerous of the granulocytes, as shown in Table 20.2. They are 12–15 μm in diameter and are phagocytes, that is they are able to engulf and digest bacteria. Neutrophils are able to leave the blood and enter tissues by squeezing between the endothelial cells that form the walls of the postcapillary venules. This process is known as **diapedesis** and is guided by specific homing receptors on the endothelial cells of the postcapillary venules.

Eosinophils normally represent only about 1.5% of the total white blood cell population. They are similar in diameter to neutrophils (12–15 μm) and are important in the body's response to allergic or parasitic disease. A raised eosinophil count (eosinophilia) is often seen in people with conditions such as asthma and hay fever and in children with threadworm infestations. These cells have antihistamine properties and they congregate around sites of inflammation. They are responsible for damage to the bronchiolar epithelium in late-stage asthma.

Basophils are the least numerous of the white blood cells, representing only about 0.5% of the total population. Like the other granulocytes, they are 12–15 μm in diameter. They are closely related to mast cells and are responsible for some of the phenomena associated with local immunological reactions such as local vasodilatation and oedema.

Monocytes are large cells (15–20 μm in diameter) with kidney-shaped nuclei. They often have long cytoplasmic processes. They are derived from bone marrow precursor cells called monoblasts. These differentiate to form promonocytes, large cells that divide twice more to

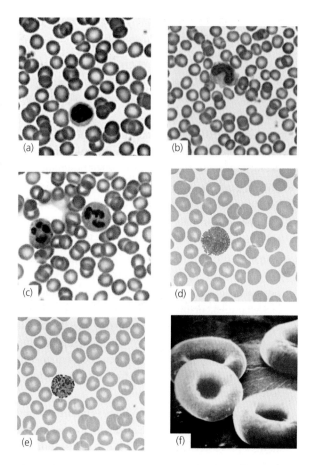

(a)   (b)   (c)   (d)   (e)   (f)

Figure 20.8 The microscopical appearance of the different kinds of white blood cell after staining: (a) a field of red cells with a lymphocyte and several platelets; (b) a monocyte: note the indented nucleus; (c) two neutrophils, with their characteristic lobed nuclei; (d) an eosinophil: note the bilobed nucleus and cytoplasm filled with red-stained granules; (e) a basophil: note the strongly staining blue granules in the cytoplasm; (f) a scanning electron micrograph of a group of red cells (coloured red in the image): note the doughnut-like appearance. The scale of each image can be judged from the red cells, which are 7–8 μm in diameter.

become monocytes. After being released into the bloodstream, they circulate for less than a day before migrating to tissues such as the spleen, liver, lungs, brain and lymph nodes where they mature into macrophages to become part of the **reticuloendothelial system** (**RES**). Macrophages behave in a similar fashion to neutrophils, engulfing and destroying pathogens and other debris.

Furthermore, they participate in immune responses by presenting antigens to the T lymphocytes by means of their cytoplasmic processes (see Chapter 21).

Lymphocytes represent around 25% of the adult white cell population (although in children they are more numerous) and vary in size from 6 to 20 μm in diameter. They are of two types, the T lymphocytes and the B lymphocytes. They are derived from lymphoid precursor cells in the bone marrow and then undergo further differentiation and maturation. B cells mature largely in the bone marrow and T cells in the thymus gland but some also develop within the lymph nodes, spleen and liver. Each has an important role in the protection of the body against infection – B lymphocytes manufacture and secrete specific antibodies and T lymphocytes participate in cell-mediated immune responses. The details of these processes are discussed in the next chapter.

> **KEY POINT:**
>
> - Leukocytes mediate the body's immune responses. They are divided into the granulocytes (basophils, eosinophils and neutrophils) and the agranulocytes (monocytes and lymphocytes).

20.8 Disorders of white blood cells

As for the red blood cells, disorders of the leukocytes fall into two broad categories; proliferative disorders and deficiency disorders.

Proliferative disorders of the white blood cells

Malignant proliferative diseases of the blood include the leukaemias, lymphomas and myelomas. Self-limiting proliferative diseases such as infectious mononucleosis (glandular fever), which is caused by a virus, can also occur.

Leukaemia is characterized by greatly increased numbers of abnormal white blood cells in the bloodstream. Leukaemia may be lymphocytic or myelocytic, depending upon the cell types involved, and it may be acute or chronic. Lymphocytic leukaemia occurs most often in children and involves malignancy of the lymphoid precursors that originate in the bone marrow. Cancerous cells spread to other tissues such as the spleen, lymph nodes and CNS where they continue to proliferate. Myelocytic disease is more common in adults and involves the myeloid stem cells of the bone marrow. The maturation of all the blood cell types including granulocytes, red cells and platelets is affected.

Leukaemic cells are usually non-functional and therefore cannot provide protection against disease. Infections are therefore common in patients with leukaemia. Other consequences of the disease include severe anaemia (due to the suppression of red cell production), an increased tendency to bleed (due to the lack of platelets) and bone pain caused by the infiltration of bone by the rapidly-proliferating leukaemic cells within the bone marrow.

Almost all forms of leukaemia spread eventually to other tissues, particularly those with a rich blood supply such as the spleen, liver and lymph nodes. As they invade these tissues, the growing cancer cells cause extensive damage and place heavy metabolic demands on the body. The energy reserves of the patient are depleted and body protein is broken down. Consequently, weight loss and extreme fatigue are characteristic symptoms of leukaemia.

Leukopenia

This term describes a reduction in white blood cell numbers. It most often involves the neutrophils and in this case the disorder is known as neutropenia. It may arise from a failure of neutrophil production by the bone marrow (due to a genetic disorder, aplastic anaemia or following certain types of chemotherapy or radiotherapy) or may be the result of overgrowth of the bone marrow stem cell population by leukaemic cells. As neutrophils are essential to the normal inflammatory response, infections are common in people with neutropenia.

KEY POINTS:

- Proliferation disorders of the white blood cells include leukaemias, lymphomas and myelomas. In leukaemia, there are high numbers of abnormal white blood cells, which are usually non-functional.
- Leukopenia is defined as an absolute reduction in white blood cell numbers and may be due either to defective production or to accelerated removal of white cells from the circulation.

20.9 Platelets (thrombocytes)

Platelets are irregularly shaped, membrane-bound cell fragments, formed in the bone marrow by budding off from the cytoplasm of giant cells called megakaryocytes, which are derived from haematopoietic stem cells (Figure 20.2). They are 2–4 μm in diameter, rarely possess a nucleus and have a lifespan in the circulation of around 10 days. Normal blood contains 150–400 $\times 10^9$ platelets per litre; although numbers are lower in neonates and in those of Southern European or Middle Eastern racial origin. The factors that control the rate of production of platelets are poorly understood although it is known that the maturation of megakaryocytes and the release of platelets from them is stimulated by a hormone called **thrombopoietin** (**TPO**), which is produced mainly by the liver and kidney.

Platelets play an important role in the maintenance of normal vascular integrity and in the process of blood clotting (haemostasis), described in more detail below.

20.10 Mechanisms of haemostasis

Haemostasis is the process by which excessive blood loss from a damaged blood vessel is prevented while ensuring that coagulation is confined to the site of injury. It involves a series of closely linked events – vasoconstriction, platelet aggregation and coagulation (the formation of a blood clot). Later, blood vessel repair, clot retraction and dissolution complete the healing process.

Vasoconstriction

When the vascular endothelium (the lining of the blood vessels) is damaged, there is a localized contractile response by the vascular smooth muscle that causes the vessel to narrow and restrict blood flow to the damaged region. Arterioles and small arteries may close almost completely. However, this response lasts for only a short time and, to prevent serious blood loss, further haemostatic mechanisms are initiated.

Role of the vessel wall

When the wall of a blood vessel is damaged, a membrane-bound tissue factor, also known as **tissue thromboplastin**, is activated. Tissue factor initiates the process of coagulation (see below). Furthermore, damage to the vascular wall exposes the collagen of the connective tissue beneath the endothelium and stimulates the adhesion of platelets. To prevent inappropriate clotting, the endothelial cells lining an intact vessel secrete a number of substances that help to prevent blood clotting. These include **prostacyclin**, which causes vasodilatation and inhibits platelet aggregation, and **tissue plasminogen activator** (**TPA**), which activates fibrinolysis (the solubilization of fibrin in blood clots).

The role of platelets

Platelets begin to adhere to the endothelium at the site of damage within a few seconds of an injury to a blood vessel. Once begun, adhesion is a self-perpetuating process. Aggregating platelets are activated to synthesize and release ADP, 5-hydroxytryptamine, arachidonic acid and thromboxane A_2, all of which further promote the adhesion of platelets to the walls of the damaged vessel and to each other. As a result, a **platelet plug** is formed at the site of injury, which helps to stem the flow of blood from the damaged vessel.

Blood coagulation

This is the mechanism by which a stable blood clot is formed. To form the clot, a soluble plasma protein called **fibrinogen** must be converted to insoluble threads of **fibrin**

that trap blood cells and plasma. The conversion of fibrinogen to fibrin is catalysed by the enzyme **thrombin**, derived from an inactive precursor in the plasma called **prothrombin**. The activation of prothrombin to thrombin is the result of a complex series of events involving the sequential activation of a number of factors (known as **clotting factors** or coagulation factors) that are normally present in the blood in an inactive form. These factors have been assigned names or Roman numerals throughout the scientific and medical literature. Here, the most common names will be used. Many of the clotting factors are synthesized in the liver and their manufacture is dependent upon vitamin K.

Figure 20.9 illustrates the principal steps of the coagulation pathway (sometimes called the clotting cascade).

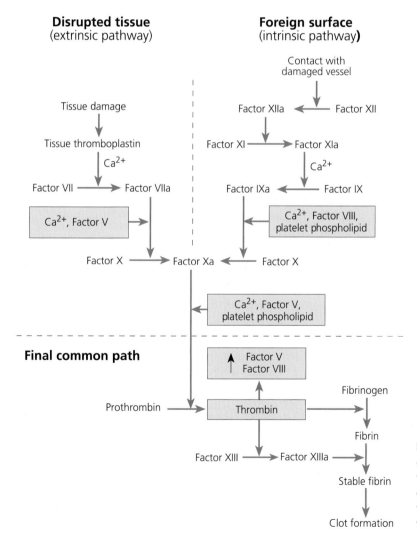

Figure 20.9 A flow chart illustrating the extrinsic and intrinsic pathways leading to the formation of a blood clot. Note the role of Factor Xa in converting prothrombin to thrombin, which is essential for converting fibrinogen to fibrin.

From the diagram, it may be seen that the conversion of prothrombin to thrombin is stimulated by activated Factor X (Factor Xa) and also that there are two pathways by which Factor Xa may be formed. Although it is convenient to separate these pathways into the so-called **intrinsic** and **extrinsic** pathways, in reality the two are closely interlinked and both are activated when blood passes out of the vascular system. All the elements required to activate the intrinsic pathway are present in normal blood. The extrinsic pathway, however, is activated by a factor from outside the blood (tissue factor). Both pathways lead to the formation of activated Factor X (Factor Xa) at the end of the first stage of coagulation. Further steps in the clotting process are common to both pathways and involve the enzymatic conversion of inactive prothrombin to thrombin. Thrombin then initiates the conversion of soluble fibrinogen to insoluble fibrin strands (a process known as polymerization). Platelets, other blood cells and plasma are trapped within the fibrin mesh to form a clot and bind the edges of the damaged vessel together.

The role of calcium in haemostasis

Calcium ions are required for many of the steps involved in the clotting process. Adequate plasma levels of calcium are therefore essential for the normal clotting of blood. In practice, plasma calcium levels never fall low enough to interfere with blood clotting. It is, however, possible to prevent the coagulation of stored blood by reducing the calcium concentration of the plasma. This may be achieved by the addition of substances such as EDTA or citrate, which bind calcium (calcium chelators).

Clot retraction

After coagulation, the clot begins to shrink slowly as serum (a clear fluid similar to plasma but lacking the clotting factors and fibrinogen) is extruded from it. The mechanism of clot retraction is not fully understood but it is thought that thrombin may initiate the process by stimulating the release of stored intracellular calcium in platelets. The increase in intracellular calcium then triggers the contraction of proteins within the platelets (rather like muscle contraction). As the platelets change their shape, they pull the fibrin strands of the clot together, retracting the clot and squeezing out the entrapped fluid.

Dissolution of the clot

Once the wall of the damaged blood vessel has healed, the blood clot is removed by a process called **lysis**. An enzyme called **plasmin** is responsible for digesting the fibrin to disperse the clot. Plasmin is present in the blood at all times, in its inactive form **plasminogen**. It is activated by substances released from damaged cells and by certain drugs often referred to as 'clot busters' (see below).

Prevention of inappropriate clotting

The formation of blood clots in intact blood vessels is normally prevented by a variety of mechanisms. Undamaged vascular endothelial cells prevent clotting by releasing substances that inhibit coagulation (**anticoagulants**). These include:

- **Prostacyclin**: this inhibits platelet aggregation.
- **Heparin**: this inhibits both platelet aggregation and the action of thrombin.
- **Thrombomodulin**: this is a protein that binds to thrombin to form a complex, which, in turn, activates another plasma protein called **protein C**. Protein C inhibits the actions of certain clotting factors. It also stimulates the activation of plasminogen to active plasmin. By dissolving fibrin, plasmin will disperse any small clots that may begin to form.

If the normal anticoagulation mechanisms fail and a blood clot (or **thrombus**) forms within an intact blood vessel, it may block that vessel. One of the most common causes of thrombosis is **atherosclerosis**. This condition is characterized by the formation of fibrous, fatty lesions, called **plaques**, in the inner lining of large or medium-sized arteries such as the aorta, the coronaries and the large vessels supplying the brain. If a clot lodges within a coronary artery (a coronary thrombosis), the region of myocardium (heart muscle) supplied by that artery will become ischaemic and die. This is commonly called a **heart attack** or **myocardial infarction** (**MI**). It is usually serious and may be fatal. If the clot forms or lodges in one of the blood vessels supplying the brain, the neurones in the affected area of the brain will die, giving rise to a **stroke**.

Certain conditions favour the formation of thrombi within blood vessels. These include damage to the vessel wall, sluggish blood flow (stasis) and alterations in certain components of the clotting process. Prolonged immobility (e.g. long periods of immobility in bed after surgery) results in failure of the leg muscles to pump blood back to the heart by the normal rhythmic contractions that accompany walking. The result is sluggish blood flow and pooling of the blood in the deep veins of the legs with an increased risk of clot formation (**deep vein thrombosis** or **DVT**). Sometimes small clots (emboli) break off a DVT and travel to the pulmonary circulation where they may lodge in a small vessel to cause a **pulmonary embolism** (**PE**). Large areas of the lung may experience a loss of blood flow and gas exchange will be severely, perhaps fatally, compromised. To prevent venous pooling and reduce the risk of developing a DVT, it is advisable to perform some gentle exercise or use elastic compression stockings (TED stockings).

Anticoagulant therapies

Dispersion of venous thromboses relies on the use of anticoagulants such as heparin and on fibrinolytic ('clot-busting') agents such as **streptokinase** or, more recently, **recombinant tissue plasminogen activator**. Oral anticoagulants such as **warfarin** are used for the long-term management of patients at risk of developing DVT, PE or systemic embolism. These drugs work by inhibiting the actions of vitamin K, thereby reducing the synthesis of clotting factors by the liver (particularly prothrombin and factors VII, IX and X).

Low-dose **aspirin** treatment is often used to reduce the risk of MI in patients with angina and to reduce the risk of stroke in patients who have suffered transient ischaemic attacks. Aspirin exerts its anticoagulant action by inhibiting platelet aggregation.

The effectiveness of anticoagulation therapy is monitored by measuring the prothrombin clotting time and is expressed as the **international normalized ratio** (**INR**). The INR is the patient's prothrombin time divided by the laboratory's control prothrombin time, which is determined using a commercial thromboplastin preparation. All commercial preparations are compared to a standard international reference preparation. This provides a calibration value called an International Sensitivity Index. A normal INR is between 0.9 and 1.2. If a patient has an INR of 3, this means that their blood takes three times longer to clot than that of a normal subject.

> **KEY POINTS:**
>
> - Following damage to the vascular endothelium, a cascade of events is initiated that ultimately leads to the formation of a blood clot.
>
> - A platelet plug forms at the damaged site. This is followed by the development of a blood clot. Clot retraction and dissolution complete the healing process.
>
> - In the formation of a blood clot, fibrinogen is converted into insoluble threads of fibrin that trap blood cells and plasma. This reaction is catalysed by thrombin.
>
> - A number of clotting factors participate in the events leading to the formation of thrombin. The clotting mechanism requires calcium ions and the phospholipids present in the membranes of the platelets.
>
> - Anticoagulant therapies such as infusion of heparin or tissue plasminogen activator are used to disperse unwanted blood clots. Oral anticoagulants such are warfarin are used to treat patients at risk of developing venous thromboses.

20.11 Failure of the normal clotting mechanisms

Blood may fail to clot normally for a number of reasons. These include thrombocytopenia, thrombocytopathia, structural disorders of the blood vessels, impaired synthesis of clotting factors and hereditary blood clotting disorders.

Thrombocytopenia is defined as a decrease in the number of circulating platelets to less than 150×10^9 per litre. However, problems with blood clotting do not normally appear until the platelet count falls below about 25×10^9 per litre. The condition is characterized by excessive blood loss following injury, spontaneous bleeding from the mucous membranes of the nose, mouth and gastrointestinal tract, and the appearance of bruised areas (ecchymoses) and tiny reddish spots (petechiae) on the arms and legs. Thrombocytopenia may be caused by certain drugs or by certain diseases. The latter include aplastic anaemia (bone marrow failure) and invasion of the bone marrow by malignant cells as in leukaemia.

Thrombocytopathia is the clinical term for disordered platelet function and this may also impair the clotting process. Defects may be inherited or acquired following disease or drug treatments. An example of an inherited disorder of platelet function is **von Willebrand's disease**, in which there is a loss of platelet adhesion. Drugs that interfere with platelet function include aspirin and other non-steroidal anti-inflammatory drugs (NSAIDs), dextran and certain antibiotics (e.g. cephalosporins).

Vascular disorders

Abnormal bleeding may occur from vessels that are structurally weak or that have been damaged by inflammation or immune responses. In scurvy (vitamin C deficiency), for example, blood vessels become fragile due to a loss of adhesion between the cells lining the vascular endothelium. In Cushing's disease (see Chapter 16), high circulating levels of cortisol cause a loss of protein and a reduction in support for the vascular tissue.

Impaired synthesis of clotting factors

Prothrombin, fibrinogen and Factors V, VII, IX, X, XI and XII are all manufactured by the liver. Liver disease is therefore an important cause of coagulation defects. Vitamin K is necessary for the normal synthesis of prothrombin and of Factors VII, IX and X. Deficiency of this vitamin or impaired absorption (e.g. in biliary obstruction) will also disrupt the clotting process and cause abnormal bleeding. Newborn, especially premature, babies are at particular risk of bleeding because their livers are immature and they have very low levels of vitamin K.

Hereditary disorders of blood clotting

Impairment of blood coagulation can result from deficiencies in any one or more of the protein clotting factors involved in the clotting cascade. Although hereditary defects associated with all of the factors are known to occur, most are extremely rare. The most common inherited clotting disorder is haemophilia A in which there is a deficiency of Factor VIII. The Factor VIII gene is located on the X chromosome, so the condition is sex-linked, affecting 1 in 10 000 males. It may also arise as a new mutation. A well-known example is the genetic transmission of haemophilia from Queen Victoria to her descendants, which is illustrated in Figure 20.10. The symptoms of haemophilia A range from mild to very severe spontaneous bleeding, especially into muscles and joints (haemarthrosis). Repeated bleeding into joints can result in long-term joint damage. Infusions of concentrated Factor VIII are necessary to treat most cases of haemophilia A. Prior to 1986, impure Factor VIII was prepared from donated blood but was not adequately screened for viruses. Subsequently, many haemophilia sufferers contracted diseases, including hepatitis and AIDS. Since then, advances in recombinant DNA technology have enabled pure Factor VIII to be produced, thus avoiding the risk of disease transmission.

Haemophilia B (also called Christmas disease) is another sex-linked clotting disorder, which is caused by a deficiency of Factor IX. It occurs less frequently than haemophilia A and its effects are usually milder. Factor IX concentrate is normally given to treat this condition.

Von Willebrand's disease is caused by a genetic defect that affects the production of von Willebrand factor (vWF), which acts as a carrier for Factor VIII in the plasma and mediates platelet adhesion. As it is usually an autosomal dominant condition (i.e. it is not sex linked), it affects males and females equally. Treatment is normally by administration of a preparation that contains both Factor VIII and vWF.

KEY POINTS:

- Failure of the normal clotting reactions occurs in the following situations:
  - Loss of platelets (thrombocytopenia)
  - Structural disorders of the vasculature
  - Hereditary deficiency of clotting factors (e.g. Factor VIII).

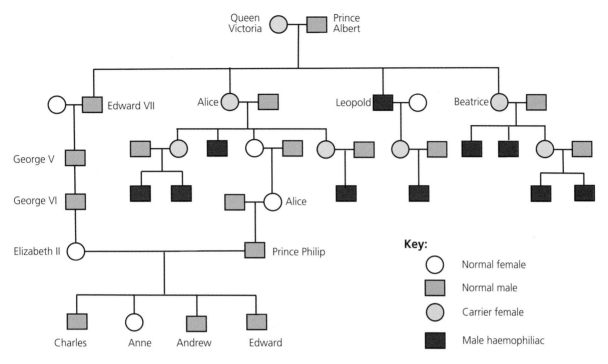

Figure 20.10 The transmission of haemophilia in the descendants of Queen Victoria. In this scheme, only the carriers (all female), the sufferers (all male) and their parents are shown. Note that there are no carriers or haemophiliacs in the present generation of the Royal Family.

20.12 Blood transfusions and the ABO system of blood groups

In modern clinical practice, heavy blood loss caused by trauma or surgery is frequently restored by the transfusion of blood donated by another, unrelated person. Indeed, many healthy people regularly donate blood, which is then stored and later used to provide whole blood, plasma or other blood products for transfusion. Early attempts at blood transfusion, however, were frequently disastrous. The transfused cells aggregated together in large clumps that could block small blood vessels – a process called **agglutination**, illustrated in Figure 20.11. Later, the cell membranes broke down to release haemoglobin into the plasma (**haemolysis**). These adverse reactions were accompanied by fever, hypotension, jaundice and renal failure, and were often fatal. When such clinical signs follow the transfusion of blood, the transfused blood is said to be **incompatible** with that of the recipient.

Figure 20.11 An example of blood agglutination following the mixing of incompatible blood groups. Note the pronounced clumping of red cells on the edge of the droplet of blood.

What is the basis of this incompatibility and why is some blood compatible while other blood is not? It is now known that agglutination results from an interaction between proteins known as **antigens** and **antibodies** present in the blood. Two kinds of antigen, called **agglutinogens**, may be found on the membranes of human red blood cells. They are called A and B and they may be present separately, together, or not at all. Hence, there are four different blood groups that constitute the so-called ABO system. They are:

- Group A (A agglutinogen only)
- Group B (B agglutinogen only)
- Group AB (both A and B agglutinogens)
- Group O (neither agglutinogen present).

Normal human plasma may contain antibodies that cause red cells to stick together to form clumps (agglutinate). These antibodies are called **agglutinins** and, unlike most other antibodies (see Chapter 21), they occur naturally and are inherited according to Mendelian Laws, as described in Chapter 30. There are two kinds of agglutinins known as anti-A and anti-B (or as agglutinins α and β). If a person's red blood cells possess agglutinogen A (i.e. they have blood group A), then their plasma will contain antibodies to agglutinogen B (anti-B). If the cells possess agglutinogen B (i.e the person has blood group B), the plasma will contain anti-A antibodies. It follows, therefore, that people with blood group AB have no agglutinins in their plasma (they do not agglutinate their own blood!) while those with group O blood have both anti-A and anti-B antibodies in their plasma. Table 20.4 gives the relationships between the different groups, and their approximate frequency in the general population of the UK and USA.

Rhesus blood group system

In addition to A and B antigens, red blood cells may possess another antigen, the Rhesus (Rh)-antigen (also known as the Rh-factor or the D-antigen). Roughly 85% of people are Rh-positive, that is their red cells possess the D-antigen. The remainder are said to be Rh-negative. Unlike the anti-A and anti-B antibodies that arise naturally in the plasma, anti-Rh antibodies only develop in the blood of Rh-negative individuals if they come into contact with cells that possess the Rh-antigen. This may occur if a Rh-negative individual is transfused with Rh-positive blood, or in pregnancy when a Rh-negative mother is carrying a Rh-positive fetus (this can happen because the D-antigen is inherited by simple Mendelian principles).

If Rh-positive fetal red blood cells leak into the maternal circulation, the mother may develop anti-Rh antibodies. This immunization of the mother by the baby's red blood cells may happen at any time during pregnancy but is most likely to occur when the placenta separates from the wall of the uterus while the mother is giving birth. This means that, although the newly delivered baby is unlikely to be affected, any subsequent Rh-positive

Table 20.4 Characteristics of the ABO blood groups

Blood group	% of population	Agglutinogen on red cells	Agglutinin in plasma	Compatible blood groups	Notes
A	41	A	Anti-B (b)	Group A, group O	
B	10	B	Anti-A (a)	Group B, group O	
AB	3	A & B	None	All groups	Universal recipient
O	46	None	Anti-A and anti-B (α and β)	Group O	Universal donor

Note that for all blood transfusions cross-matching is essential as it establishes whether there will be any incompatibility between donor blood and the recipient and not just those due to the ABO system.

fetus may be exposed to the anti-Rh antibodies that developed in the mother's blood during the previous pregnancy. These antibodies are small enough to pass across the placenta into the fetal circulation where they may cause a severe agglutination reaction – a condition called **haemolytic disease of the newborn**. Around half of the babies affected by this condition will require at least partial replacement of their blood by transfusion. In the absence of appropriate preventative measures, the condition will occur roughly once in every 160 births. However, in many countries, this problem is now largely avoided by giving Rh-negative mothers injections of anti-D immunoglobulin (IgG), which coats any fetal cells present and thus prevents the formation of antibodies. This may be done either after delivery or, more often prophylactically, at weeks 28 and 34 of gestation.

Although ABO incompatibility between mother and fetus is common, there are rarely problems with haemolysis. This is because the A and B agglutinins are much larger proteins than the anti-Rh antibodies and therefore do not readily cross the placenta. Furthermore, the A and B antigens do not develop fully until after birth.

Other blood groups

Soon after the original description of the ABO system of blood groups, it was discovered that group A could be further subdivided into two groups, A1 and A2. Other groups, such as the M, N, P and Lewis groups, are also recognized but are not generally of significance in blood transfusions. Other red cell antigens, such as Kell, Duffy and Kidd, occur in certain individuals and can stimulate the development of antibodies if transfused into people lacking the antigen.

Cross-matching of blood for transfusion

To avoid the problems of incompatibility described above, blood for transfusion is normally **cross-matched** to that of the recipient. In this process, red blood cells from the donor are mixed with some serum (plasma) from the recipient. The sample is subsequently examined to check for agglutination reactions that would indicate incompatibility. This test screens for all the serum agglutinins present in the recipient's blood and not just those of the ABO system. An example is shown in Figure 20.12.

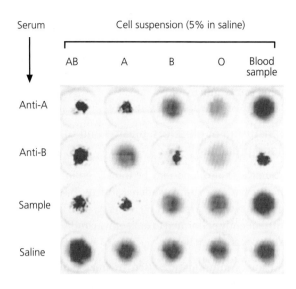

Figure 20.12 Cross-matching of blood. Drops of anti-A and anti-B serum are placed in shallow wells on a porcelain plate, as shown in the figure, together with plasma from the blood sample and a saline control. Drops of blood with known blood groups and the test sample are added to the wells, as shown along the top of the figure, and mixed with the serum. If the blood samples are compatible, the mixed blood sample appears uniform, but if the blood is incompatible with the serum, the red cells agglutinate and precipitate as shown. In this case, the blood sample (on the right) is agglutinated by anti-B serum and its serum agglutinates the red cells of groups A and AB. So this blood sample is group B as both group A and group AB have group A agglutinogen on their red cells.

Although cross-matching of the donor and recipient blood is preferable, in emergencies it is generally safe to transfuse group O blood into people of other blood groups because group O red cells have neither A nor B antigen and cannot, therefore, be agglutinated by the antibodies found in the plasma of other groups. For this reason, group O is often called the **universal donor** group. Group AB, by contrast, is known as the **universal recipient**. This is because the plasma of group AB blood contains neither anti-A nor anti-B antibodies and cannot therefore agglutinate the red cells of any blood group. The plasma agglutinins present in the donor blood do not generally cause adverse reactions because they become diluted in the recipient's circulation.

For some patients, autologous blood transfusions are an option. Here, the patient donates their own blood prior to elective surgery. Normally, several units of blood are taken over a relatively short period during which the patient is given iron supplements. As for other donations, the donated blood is screened for infectious agents and stored at 4°C until required.

Transfusion of platelets, plasma and other blood products

Although whole blood may be transfused into a recipient who has, for example, suffered an acute haemorrhage with hypovolaemia (see p. 747), other types of transfusion are frequently performed to treat other conditions. Some examples are given below:

- Platelet infusions are used to treat patients with thrombocytopenia or defects of platelet function.
- Fresh frozen plasma (which contains all the plasma proteins and clotting factors) may be administered to patients with coagulation abnormalities or liver disease or to reverse the effects of oral anticoagulant or thrombolytic ('clot-busting') therapies.
- Highly purified concentrates of specific coagulation factors may be prepared in freeze-dried powder form

and used to treat clotting disorders such as haemophilia and von Willebrand's disease.

- Immunoglobulins (Igs) prepared from pooled donated plasma are used to provide prophylaxis against a variety of diseases including hepatitis B, tetanus and rabies.

> **KEY POINTS:**
>
> - In the ABO system, two kinds of antigen occur on human red cells: A and B. They may be present separately, together, or be completely absent so giving rise to four groups: A, B, AB and O. Human plasma may contain anti-A and anti-B antibodies.
> - When plasma containing an antibody (e.g. anti-A) is mixed with red cells possessing an antigen with which it can react (in this case A), the red cells agglutinate.
> - For successful blood transfusion, the blood of the donor must be compatible with that of the recipient. This is established by cross-matching of blood.

Recommended reading

Physiology of gas transport

Hlastala, M.P. and Berger, A.J. (2001). *Physiology of Respiration*, 2nd edn. Oxford University Press: New York. Chapter 6.

Levitzky, M.G. (2003). *Pulmonary Physiology*, 6th edn. McGraw-Hill: New York. Chapter 7.

Biochemistry of haemoglobin

Berg, J.M., Tymoczko, J.L. and Stryer, L. (2002). *Biochemistry*, 5th edn. Freeman: New York. pp. 269–274.

Elliott, W.H. and Elliott, D.C (2008). *Biochemistry and Molecular Biology*, 4th edn. Oxford University Press: Oxford. Chapter 4.

Haematopoiesis

Alberts, B., Johnson, A., Lewis, J., Raff, M., Roberts, K. and Walter, P. (2008). *Molecular Biology of the Cell*, 5th edn. Garland: New York. Chapter 23, pp. 1450–1462.

Junquieira, L.C. and Carneiro, J. (2005). *Basic Histology*, 11th edn. McGraw-Hill: New York. Chapters 12 and 13.

Pharmacology of the blood

Rang H.P., Dale, M.M., Ritter, J.M. and Flower, R. (2007). *Pharmacology*, 5th edn. Churchill-Livingstone: Edinburgh. Chapters 20 and 21.

Haematology

Hoffbrand, A.V., Pettit, J.E. and Moss, P.A.H. (2001). *Essential Hematology*, 4th edn. Blackwell Science: Oxford.

Self-assessment questions

Which of the following statements are true and which are false? Answers are given at the end of the book (p. 755).

1 Regarding the plasma:
  a) It normally accounts for about 4% of body weight.
  b) It has an osmolality of about $140 \, mOsmol \, kg^{-1}$.
  c) It has about $140 \, mmol \, l^{-1}$ of sodium.
  d) It contains about 5% albumin by weight.
  e) Plasma is about 60% water.

2 For a normal healthy adult male of 60–70 kg body weight:
  a) The blood volume would be about 5 litre.
  b) The blood pH is 7.0.
  c) There are about 4.5×10^{12} red cells per litre.
  d) There are about 1×10^9 leukocytes per litre.
  e) The blood contains about 20% of total body iron.

3 Normal human red cells:
  a) are about $7 \, \mu m$ in diameter.
  b) are produced in the bone marrow from erythroblasts.
  c) do not have a nucleus.
  d) are released into the circulation as mature cells.
  e) live in the circulation for about 10 days.

4 The following statements refer to gas carriage by the blood:
  a) After leaving the lungs, each dl of blood contains about 20 ml of oxygen.
  b) Each dl of mixed venous blood contains about 5 ml of carbon dioxide.
  c) Most of the carbon dioxide in blood is carried as bicarbonate.
  d) As the P_{CO_2} rises, the affinity of haemoglobin for oxygen is increased.
  e) The affinity of haemoglobin for carbon monoxide is lower than its affinity for oxygen.

5 Platelets:
  a) are produced in the bone marrow from megakaryocytes.
  b) are present in greater number than red cells.
  c) will adhere to the walls of damaged blood vessels.
  d) secrete the main clotting factors.

6 The following statements refer to the mechanisms of haemostasis:
  a) Only the extrinsic clotting cascade leads to the activation of Factor X.
  b) Coagulation can be prevented by adding EDTA or citrate to a sample of blood.
  c) Haemostasis is inhibited in healthy blood vessels by warfarin secreted by the vessel wall.
  d) Haemostasis is initiated when tissue factor comes into contact with blood.
  e) Failure of haemostasis is always caused by a deficiency of Factor VIII.

21 Defence against infection: inflammation and immunity

After reading this chapter you should have gained an understanding of:

- The passive mechanisms by which the body resists infection
- How the body recognizes invading organisms
- The natural immune system
- The inflammatory response
- The adaptive immune system and the role of the lymphocytes
- Disorders of the immune system
- The need for tissue matching in transplantation

21.1 Introduction

To defend themselves against infections, animals have two basic strategies: they use passive barriers to prevent disease-producing organisms (pathogens) entering the body and they actively attack those organisms that have become lodged in the tissues. To eliminate an invading organism, the host must first be able to distinguish it from its own cells. Secondly, it must neutralize or kill it. Finally, it must dispose of the remains in such a way that does no further harm. These functions are performed by the **immune system**, which may be conveniently divided into the **innate** or **natural immune system** and the **adaptive immune system**.

The immune system is a complex network of organs, cells and circulating proteins. The principal organs of the immune system are the bone marrow (particularly that of the flat bones of the pelvis and sternum), the thymus, the spleen, the lymph nodes and the lymphoid tissues associated with the epithelia that line the gut and airways (Figure 21.1). The cells of the immune system include the leukocytes of the blood (see Chapter 20), mast cells and various accessory cells that are scattered throughout the body. The proteins of the immune system are **antibodies** and **complement**. The cells of the immune system recognize foreign materials by their surface molecules. Those molecules that generate an immune response are called **antigens**.

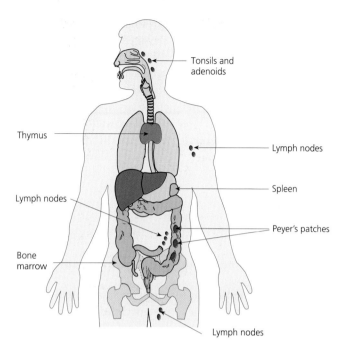

Figure 21.1 The location of the major lymphoid organs. The thymus and bone marrow are the primary lymphoid tissues; the remainder are secondary lymphoid tissues.

21.2   Passive barriers to infection

The first barrier encountered by most invading organisms is the skin, which prevents them from infiltrating the tissues. Bacterial growth on the skin surface is also inhibited by secretions from the sweat glands and sebaceous glands. When the skin is broken, either by abrasion or by burns, infection may become a significant problem.

The skin forms a continuous sheet with the membranes that line the airways, gut and urogenital tract. The epithelia of these membranes are less rugged than that of the skin but they still provide an effective barrier to invasion by micro-organisms. For example, the epithelia that line the airways are protected by a thick layer of mucus, which traps many bacteria and viruses and prevents them adhering to the underlying cells. The mucus is then eliminated via the mucociliary escalator and coughed up (see Chapter 22, p. 457). Other regions that are vulnerable to infection are regularly flushed by sterile fluid (e.g. the urinary tract) or by fluids that contain bactericidal

agents. For example, the external surface of the eye is washed by fluid from the tear glands, which both flushes the surface to remove foreign materials and contains the bactericidal enzyme lysozyme. Other body secretions, such as the semen and breast milk, also contain antibodies and bactericidal substances.

The food we eat is inevitably contaminated by bacteria and other micro-organisms. The gut has several stratagems to combat infection arising from this source. The mucous membranes of the mouth and upper gastrointestinal tract are protected by lysozyme and antibodies of the IgA class (see below) secreted by the salivary glands. Many bacteria are killed by the low pH of the gastric juice. The mucosal surface of the gut also possesses mucus glands that secrete a layer of mucus to both lubricate the passage of food and protect the surface epithelium from infection. Despite these barriers, the intestine contains a healthy bacterial population. These are commensal organisms, which provide the body with a further line of defence. The normal bacterial flora of the gut both competes with potential pathogens for essential nutrients and secretes inhibitory factors to kill them. If the balance is upset, for example by the administration of antibiotics, then pathogens that are normally present in the gut such as *Clostridium difficile* can multiply.

Clearly, these passive mechanisms do not always prevent the ingress of pathogens: the skin can be penetrated by ectoparasites such as ticks and mosquitoes, which may themselves be infected with micro-organisms such as *Plasmodium* (the organism that causes malaria). Small pathogens such as bacteria and viruses may penetrate the body's defences via the internal epithelia such as those of the airways. Those that enter the gut may overwhelm the defences afforded by the natural commensal bacteria, as happens in typhoid fever, for example. When infection occurs, the active processes of immunity come into play.

21.3   Self and non-self

Before the body can mount a defence against infection, it first needs to know the difference between the normal cells of the body and those of invading organisms. How does the immune system recognize 'self' from 'non-self'? It is now known that, just as red cells possess surface proteins that determine particular blood groups, other cells possess integral membrane proteins that identify them as being host cells. By using these markers, the immune system can distinguish the cells of the body from those of invading organisms. The molecules of the immune system that detect a general 'non-self' characteristic are said to be **non-specific** while those that can detect a particular invading organism among the thousands of possible candidates are called **specific recognition molecules**. As we shall see, non-specific recognition is characteristic of the natural immune system while the adaptive immune system can identify and destroy a specific type of invading organism.

The proteins that identify host cells are known as the **major histocompatability complex** or **MHC**. Their rather unfortunate name arises from the history of their discovery. They were first detected as the proteins responsible for the rejection of tissue grafts between a donor and a recipient animal. In human immunology, the MHC is known as the **HLA complex** (for human leukocyte antigen). The MHC proteins incorporate parts of foreign proteins and expose them to cells of the immune system known as T lymphocytes (see p. 420). The MHC of each person is unique to them (except for identical twins).

21.4   The natural immune system

The natural immune system consists of the innate defence mechanisms that do not change very much either with age or following infections. It consists of four kinds of cells and three different classes of proteins.

The cells of the natural immune system are:

- phagocytes (neutrophils and macrophages)
- natural killer cells
- mast cells
- eosinophils.

The classes of protein are:

- complement
- interferons
- acute-phase proteins.

The cells of the natural immune system

The **neutrophils** are the most common type of white cell in the blood (see Chapter 20, pp. 401–403). They are able to pass from the blood into the intercellular spaces by diapedesis (see below) and actively phagocytose disease-producing bacteria. The enzymes within the cytoplasmic granules then kill the invading organisms and digest them. As a result of this action, the neutrophils form the first line of defence against infection.

The **macrophages** are formed in the bone marrow and released into the blood as monocytes. Within 2 days, they migrate to tissues such as the spleen, lungs and lymph nodes where they mature. Macrophages contain a large number of lysosomes and phagocytic vesicles, containing the remains of ingested materials. They are found in all tissues, even in the brain where they are known as **microglia** (Figure 21.2). Macrophages are situated around the basement membrane of small blood vessels. They also line both the spleen sinusoids and the medullary sinuses of the lymph nodes where they are able to remove particulate matter from the circulation. In the liver, they are known as **Kupffer cells**.

Viruses lack the ability to replicate by themselves. Instead, they subvert the genetic machinery of host cells to make copies of themselves. For this reason, it is important that those cells that become infected are destroyed before the virus has time to replicate and infect neighbouring cells. The cells that perform this vital function are known as **natural killer** (**NK**) **cells**. They are large, granular lymphocytes that recognize virus-infected cells from MHC cell-surface markers that have incorporated part of the viral protein. When a natural killer cell has recognized a target, it is activated and releases cytotoxic proteins onto the target cell, which responds by undergoing a preprogrammed cell death (apoptosis, see p. 47) so preventing viral replication by the infected cell.

Mast cells are found in the skin and gut and around blood vessels. They play an important role in the acute inflammatory response by releasing histamine and other substances to increase local blood flow. Histamine also increases the permeability of the capillary cells to complement and antibodies.

Eosinophils are the least numerous of the white cells of the blood. They contain granules that take up the dye eosin (see Figure 20.8). Eosinophils appear to play an important role in combating worm infections. Worms are too large to be phagocytosed by a single cell so they must be attacked extracellularly. Eosinophils are particularly attracted to parasites whose outer membranes have been coated with antibody of the IgE class (see below) and attack the outer membrane of the parasites to inactivate or kill them.

The proteins of the natural immune system

Complement is the name given to a group of about 20 plasma proteins that play an important part in the control

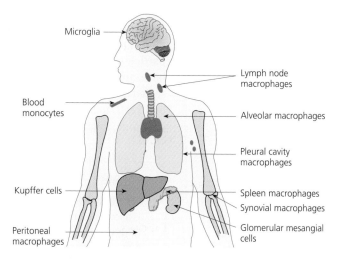

Figure 21.2 The mononuclear phagocyte system. Monocytes are formed in the bone marrow where they mature before being released into the circulation. They migrate to tissues such as the spleen, liver, lungs and lymph nodes where they take up residence as mature macrophages. The neutrophils (the other main class of phagocyte) remain in the circulation until they participate in an inflammatory reaction.

of infections, particularly those caused by bacteria and fungi. Like the clotting factors, the **complement proteins** are enzymes that can be sequentially activated.

Activation of the complement system occurs by the breakdown of a component known as C3. This results in the formation of two smaller proteins known as C3a and C3b: C3a activates phagocytes and C3b binds to the surface of microbes and facilitates their uptake by phagocytes. Both fragments, together with another complement fraction known as C5, play an important role in the initiation of the inflammatory response (see below).

When cells are infected by a virus, they make **interferons** (of which there are many different kinds) and secrete them into the extracellular fluid. The interferons then bind to receptors on neighbouring cells, which respond by reducing their rate of mRNA translation. This results in the infected cell being surrounded by a layer of cells that cannot replicate the virus, so forming a barrier to prevent the spread of the infection. Finally, the natural killer cells seek out and destroy any infected cells.

The **acute-phase proteins** are a group of plasma proteins synthesized by the liver that show a huge increase in concentration during an infection. They include C-reactive protein and complement components – including C3. As described above, complement C3b binds to the surface of invading organisms. This surface coating is known as **opsonization**. Both C-reactive protein and antibodies also opsonize foreign organisms. As the phagocytes have receptors for the coating proteins, they are able to recognize opsonized particles and engulf and destroy them.

KEY POINTS:

- The natural immune system provides the innate defence mechanisms.

- It consists of four kinds of cells, the phagocytes, natural killer cells, mast cells and eosinophils, plus three different classes of proteins: complement, interferons and acute-phase proteins.

The acute inflammatory response

When the body becomes injured or infected, a number of physiological changes occur in the affected area. There is a local vasodilatation, increased permeability of the capillaries and infiltration of the damaged tissues by white cells. These changes constitute the inflammatory response, which brings plasma proteins and leukocytes to the point of injury. The processes involved in the inflammatory response are summarized in Figure 21.3. Inflammation can be caused by a variety of stimuli including traumatic injury, infection and cellular necrosis.

For an injury to the skin, the stages of the inflammatory response are as follows. There is a reddening of the skin at the site of injury, which results from vasodilatation. This is rapidly followed by local tissue swelling due to the accumulation of fluid by the affected tissues. The skin of the surrounding area becomes flushed (the flare). These three components of the inflammatory response constitute the **triple response**. Injury to internal organs is accompanied by a similar sequence of events.

If the infection or trauma is sufficiently extensive, the injured tissues become infiltrated by polymorphonuclear leukocytes, particularly neutrophils. The vascular endothelium in the injured area becomes modified, and the neutrophils attach themselves to the wall of the postcapillary venules and squeeze between the endothelial cells to pass into the tissues (this is known as **diapedesis**). Once in the tissues, the leukocytes phagocytose any invading organisms or cell debris. They also seem to play an important role in the healing process. Finally, the inflammatory response declines as the damaged tissue becomes healed. This phase is known as resolution. If the invading organism or any particles that triggered the inflammatory response have not been eliminated, the offending material is sealed off by a layer of macrophages, lymphocytes and other cells to form a **granuloma**, which is characteristic of chronic inflammation.

KEY POINT:

- When the body becomes injured or infected, there is a local vasodilatation, increased permeability of the capillaries leading to local oedema and infiltration of the damaged tissues by white cells. These changes constitute the inflammatory response.

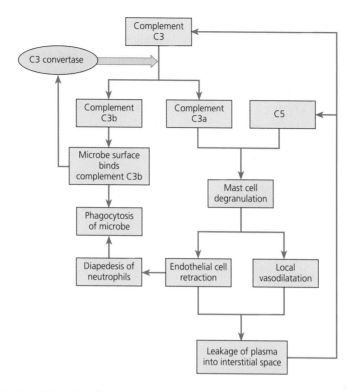

Figure 21.3 The mechanism by which an invading organism triggers an inflammatory response. The initiation of this series of events begins with the activation of an enzyme known as C3 convertase, which cleaves complement C3 into two fragments and so promotes the inflammatory reaction by different pathways.

21.5 The adaptive immune system

It is well known that, while exposure to certain disease-producing organisms (e.g. the chicken pox virus) will cause disease on the first exposure, a subsequent exposure will not generally result in infection. Nevertheless, the resistance to that infection does not extend to other diseases. Experience of chicken pox does not prevent infection by the measles virus. These facts highlight two important features of our immune system:

- Resistance is acquired by one exposure to an invading organism and then lasts for many years – sometimes even for a whole lifetime.

- The resistance is specific for that organism.

In immunological terminology, the response is *specific* and has *memory*. These characteristics distinguish the response of the adaptive immune system from that of the natural immune system.

The cells of the adaptive immune system are the **lymphocytes**. The lymphoid system consists of the total mass of tissue associated with the lymphocytes and their function. It is disseminated throughout the body as shown in Figure 21.1. The tissues in which the lymphocytes mature (the bone marrow and thymus)

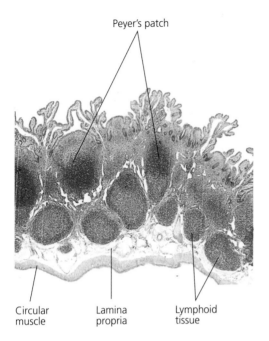

Peyer's patch

Circular muscle

Lamina propria

Lymphoid tissue

Figure 21.4 A section through a Peyer's patch of the small intestine. Note the extensive lymphoid tissue.

are known as **primary lymphoid tissue** while the lymph nodes, spleen and other lymphoid tissues are **secondary lymphoid tissue**. The lymphoid tissues of the airways and gut are known as mucosa-associated lymphoid tissue (MALT) while those of the gut are known as gut-associated lymphoid tissue (GALT). An example is shown in Figure 21.4.

> **KEY POINTS:**
>
> • The adaptive immune system provides a mechanism for defending the body against an extraordinarily wide range of organisms.
>
> • Unlike the response of the natural immune system, that of the adaptive immune system is specific and has memory. Indeed, once resistance has been acquired, it usually lasts for many years.

There are two principal **classes of lymphocyte**, which are known as **B cells** and **T cells**. The B cells mature in the bone marrow and secrete antibody while the T cells mature in the thymus gland and secrete signalling molecules known as **cytokines** or proteins that kill affected cells, known as **cytotoxins**. Lymphocytes are stimulated by antigens that bind to their surface receptors. Individual lymphocytes respond only to one antigen and, when they are stimulated, they proliferate by mitosis to form a population of cells with an identical specificity, called a **clone**. As shown in Figure 21.5, some of the cells continue to proliferate and carry out their specific immunological function (see below) while others remain in the lymphoid tissue as **memory cells** that are able to respond quickly to the same antigen should the infection recur.

Lymphocytes continuously circulate through the tissues

To monitor the tissues of the body for invading organisms, the lymphocytes continuously circulate throughout the tissues. Normally, like the red cells, the lymphocytes remain in the centre of a blood vessel but, when they reach a target tissue, some of them become attached to the vessel wall. This process is guided by homing receptors that are specific for particular tissues. The lymphocytes then flatten and squeeze between the endothelial cells (diapedesis) and migrate through the surrounding tissues, which they scrutinize for the presence of foreign antigens. From the tissues, they pass into the lymph nodes, which they enter by way of the afferent lymphatic vessels (see p. 383). From the lymph nodes, they pass into the efferent lymphatics and return to the blood via the thoracic duct.

Lymph nodes

The lymph nodes are swellings that occur at various points along the lymphatic vessels. Humans have around 500 lymph nodes scattered throughout the body. Each lymph node consists of an outer capsule beneath which lies the subcapsular sinus, fed by the afferent lymphatics (Figure 21.6). Below the

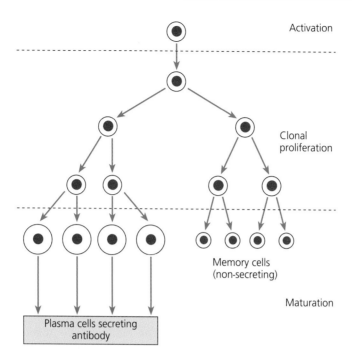

Activation

Clonal
proliferation

Memory cells
(non-secreting)

Maturation

Plasma cells secreting
antibody

Figure 21.5 The generation of a B-lymphocyte clone. An antigen activates a naïve lymphocyte, which then proliferates. Some of the clonal cells mature and secrete antibody of the same specificity as that of the receptor (i.e. it is capable of binding to the initiating antigen) while other cells remain in the lymphoid tissues as memory cells. Activated T cells also proliferate in response to a foreign antigen.

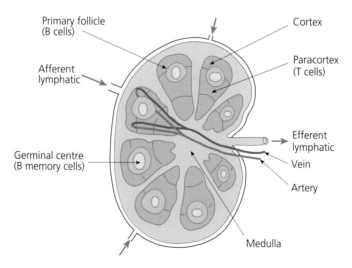

Primary follicle
(B cells)

Cortex

Afferent
lymphatic

Paracortex
(T cells)

Efferent
lymphatic

Germinal centre
(B memory cells)

Vein

Artery

Medulla

Figure 21.6 The structure of a typical lymph node. Note the compartmentation of the B and T lymphocytes.

subcapsular sinus lies the cortex, which is organized into primary follicles, containing B cells, and secondary follicles, also known as **germinal centres**, which contain the memory cells. The space between the follicles is called the paracortex and is populated by T cells. At the centre of the gland lies the medulla, which contains antibody-secreting B cells and macrophages. The arrangement of the lymph nodes allows the afferent lymph to percolate through the tissue to the efferent lymphatic. In this way, any antigens that may be present are exposed to cells capable of mounting an appropriate immune response.

KEY POINTS:

• The lymphocytes are the cells of the adaptive immune system. There are two principal classes of lymphocyte: B cells and T cells, which mature in the bone marrow and thymus, respectively.

• To scan the body for invading organisms, the lymphocytes pass from the blood vessels, through the tissues and re-enter the venous blood by way of the lymph nodes and the thoracic duct.

• When lymphocytes are stimulated by antigens, they proliferate to form a population of cells with an identical specificity called a clone.

• Some of the clonal cells continue to proliferate and carry out their specific immunological function, such as antibody production, while others remain in the lymphoid tissue as memory cells, able to respond to a similar challenge in the future.

Antibodies

As mentioned above, the B cells secrete antibodies in response to an infection. There are five main types of antibody or immunoglobulin (Ig) known as IgA, IgD, IgE, IgG and IgM:

1. IgA is the most common immunoglobulin in secretions such as saliva and bile. It is also found in colostrum (the milk secreted during the first week of lactation, see Chapter 29, p. 640).

2. IgD acts as a surface receptor on B cells together with IgM.

3. IgE binds to mast cells to facilitate their inflammatory response to an antigen. It is implicated in allergic reactions.

4. IgG is the most abundant class of immunoglobulin in plasma. As IgG molecules can cross the placental membrane, they provide the fetus with ready-made antibodies that protect it both *in utero* and for some months after birth.

5. IgM is the first antibody to be produced during the primary immune response. It activates complement.

Antibodies have two main functions:

1. To bind an antigen.

2. To elicit a response that results in the removal of the antigen from the body.

The antibody acts together with complement to stimulate the phagocytes to kill and digest any infecting organism. The combination of antibody, complement and phagocyte is very effective but a lack of any one component seriously compromises the ability of the body to mount an adequate immune response. In some cases, the binding of the antibody to its antigen is sufficient to inactivate the invading organism. Thus, when viruses are coated with antibody, they cannot infect the host cells and so are prevented from proliferating.

Antibody diversity

Although antibodies can be grouped into five main classes, the number of different antibody molecules is immense. Estimates suggest that there may be more than a thousand million million (10^{15}) different kinds of antibody molecule in a single individual, each with its own specificity. These antibodies constitute the pre-immune antibody repertoire. How is such a large number of different proteins generated when human DNA has only about 20 000 (2×10^4) genes? The answer to this problem lies in the way in which antibody molecules are encoded by DNA. Each antibody consists of four polypeptide chains: two light chains: and two heavy chains. The DNA encodes each of these polypeptides by combining different gene segments. There are, therefore, many different DNA sequences from which the polypeptide chains of the antibodies can ultimately be synthesized. Moreover, during the proliferation of a B-cell clone, point mutations occur in the DNA, which generates still further antibody diversity. This explains why the affinity of antibody for a particular antigen frequently increases with time after the initial exposure.

> **KEY POINTS:**
> - Antibodies have two main functions: (1) to bind an antigen and (2) to elicit a response that results in the removal of the antigen from the body.
> - The antibody acts together with complement to stimulate the phagocytes, with the result that the organism carrying the antigen is killed and digested.

T lymphocytes and cell-mediated immunity

Like all lymphocytes, the T lymphocytes (or T cells) circulate between the blood and the tissues. When they encounter a cell with a foreign antigen on its surface, T cells secrete cytokines or cytotoxic molecules. As these substances act in a paracrine fashion, the effects of T-cell activation are local and usually only one cell responds (hence the term cell-mediated immunity). **Helper T cells** secrete cytokines to stimulate B cells to proliferate and secrete antibody. **Cytotoxic T cells** secrete cytotoxic substances to kill a target cell. Each type has a different role to play in combating infection.

Antigen processing and presentation

While B cells respond to circulating antigens, T cells respond to cells whose MHC molecules have bound a foreign peptide – such as a cell infected with a virus. An individual T cell recognizes an MHC molecule that incorporates a particular foreign peptide. In effect, the MHC molecules are exposing the foreign peptide to the T cell, that is, they are presenting it to the T cell. This is known as **antigen presentation**.

The sequence of events leading to antigen presentation is as follows. When cells become infected by viruses or intracellular bacteria, foreign proteins are expressed within their cytoplasm. These foreign proteins are broken down

into peptides by the cell and become associated with MHC proteins before being inserted in the plasma membrane, where they can be recognized by circulating T cells.

T-cell receptors

The T-cell receptor has two components, a part that recognizes specific host-cell markers and a part that investigates the MHC for the presence of foreign antigen. Helper T cells identify B cells bearing a particular foreign peptide and secrete cytokines to stimulate them to proliferate. This increases the size of the particular clone and the amount of antibody secreted (Figure 21.7). As a result, the antibody response to the specific antigen detected by the T cell is enhanced.

T cells that respond to the MHC on macrophages do so by proliferating and stimulating the affected macrophages to activate their normal killing mechanisms, as shown in Figure 21.8. This process is important for those cells that have become infected with intracellular bacteria (e.g. the bacillus that causes tuberculosis). Viral infections are combated in a similar way. The MHC of infected cells incorporates part of the viral protein, which is recognized by the receptors on the cytotoxic T cells. The T cell then secretes cytotoxins onto an infected cell to cause it to undergo apoptosis (Figure 21.9).

The antibody response to infection

When people are first exposed to an infectious agent, they rely on the natural immune system for their defence. Indeed, the complement system is particularly important in young children who have not acquired a wide antibody repertoire. If the infection is severe, B cells begin to make antibody but this only occurs after a lag phase during which B and T cells are activated. Initially, the B cells secrete mostly IgM, which reaches a peak after about a week (Figure 21.10). This antibody is able to activate complement and so enhances the ability of the natural immune system to combat the infection. As the disease proceeds, some B cells switch to the production of IgG and the level of this antibody in plasma continues to rise after the IgM level begins to decline. IgG levels peak after about 2 weeks and then slowly decline as the infection wanes. This pattern of antibody secretion is known as the **primary response**.

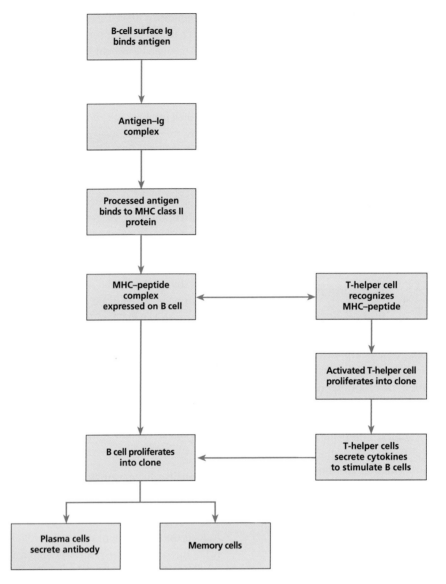

Figure 21.7 The cellular basis of the antibody response. T helper cells are activated by B cells and proliferate. They then use chemical signals (cytokines) to stimulate B cells to proliferate into a clone. Some of these clonal cells develop into antibody-secreting plasma cells while others remain as memory cells to be activated should the infection recur.

A subsequent infection by the same organism is met by a much more prompt response – the **secondary response**. IgG levels rapidly rise above those seen during the primary response and remain elevated for a longer period (Figure 21.10). In contrast, the IgM levels follow approximately the same time course as the primary response. The rapid and augmented secondary response is due to the recruitment of specific memory cells – cells that arose during the development of the initial clone but which did not mature into antibody-secreting plasma cells. The B-lymphocyte memory cells, in particular, respond to very small levels of antigen once they have been primed. This memory function of the adaptive immune system makes possible artificial immunization against specific diseases such as poliomyelitis (see Box 21.1) and smallpox ('vaccination').

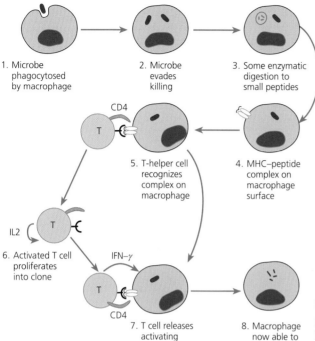

Figure 21.8 The processes by which T cells activate macrophages. IFN-γ is gamma interferon and IL2 is interleukin 2; CD4 is a protein on the surface T cells that can recognize the host part of the MHC protein.

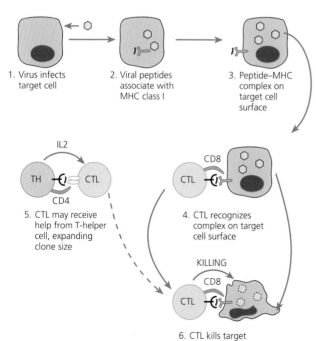

Figure 21.9 The mechanism by which cytotoxic T lymphocytes (CTLs) kill cells infected with a virus. CD4 and CD8 are proteins on the surface of T cells that can recognize the host part of the MHC protein. TH, helper T cell.

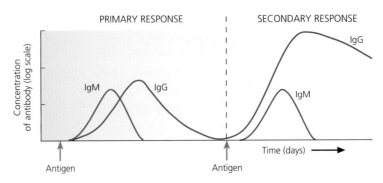

Figure 21.10 Primary and secondary antibody responses to an infection. Note the relatively slow antibody response to the first exposure and the early secretion of IgM. Compare this with the very rapid IgG response to the second challenge.

Box 21.1 Peyer's patches and the oral polio vaccine

One component of the mucosal-associated lymphoid tissue (MALT) is the organized gut-associated lymphoid tissue (GALT), which includes the Peyer's patches. These are lymphoid aggregates in the gut mucosa, found predominantly in the ileum. They house large numbers of naïve T and B lymphocytes and are covered by an epithelial layer containing cuboidal cells, known as M cells. These are specialized antigen-transporting cells that do not express one type of MHC molecule (MHC II). They lie over the phagocytic cells that take up the antigens before migrating to the local lymphoid tissue, where they activate the lymphocytes. Once exposed to the antigen, lymphocytes within the Peyer's patch become sensitized and begin to secrete IgA onto the epithelial surface that binds to, and inactivates,

the antigen. In this way, the Peyer's patches play an important part in the immunological homeostasis of the gastrointestinal tract.

A further example of the importance of the GALT is provided by the effectiveness of the oral polio vaccine. The polio virus is normally transmitted via the faecal–oral route and may cause paralysis as a result of damage to neurones of the anterior horn of the spinal cord. The polio vaccine is administered routinely to children in three doses during the first year of life (with boosters in childhood and adolescence). An attenuated form of the virus is given to the child orally. Once administered, the attenuated virus colonizes the intestine where it triggers an immune response mediated by cells of the Peyer's patches and other GALT cells.

KEY POINTS:

- T cells may be cytotoxic or helper cells. The effects of T-cell activation are local and usually only one target cell responds. T cells respond to those cells with MHC molecules that have bound a foreign peptide. Thus, T cells are able to respond to cells infected with a virus.

- On initial exposure to an antigen, the B cells secrete IgM and IgG. Plasma levels of these antibodies decline after 1 to 2 weeks. A subsequent infection by the same organism is met by a much more prompt increase in the plasma levels of the appropriate IgG antibody.

21.6 Disorders of the immune system

Like other organ systems, the function of the immune system may become disturbed. Broadly speaking, the disorders are due to inappropriate immune activity (**hypersensitivity**) or to a failure of the immune system (**immunodeficiency**). Each will be discussed in turn.

Hypersensitivity

The adaptive immune system is very powerful and capable of responding to a wide variety of antigens. In normal people, the response is appropriate and correctly targeted against the invading organism. Sometimes, however, the activity of the immune system leads to pathological changes in the host tissues. Such reactions are called hypersensitive reactions and may be grouped under one of four headings:

- Type I are allergic reactions, which are mediated by IgE antibodies and mast cells. Examples are hay fever and asthma.

- Type II is antibody-dependent cytotoxic hypersensitivity. In this case, IgG reacts with host cells so initiating the processes that lead to their destruction. This unwanted reaction may follow skin or organ transplants, leading to the rejection of the graft or transplant (see below). Occasionally, the lymphocytes attack the host cells themselves (i.e. they respond to the normal host antigens). This is known as **autoimmunity**.

- Type III is immune complex-mediated hypersensitivity. When an antibody reacts with an antigen, it forms an immune complex that may precipitate. If these complexes are not phagocytosed rapidly (their usual fate), they may accumulate in the small blood vessels. When this happens, they are attacked by complement and neutrophils. The ensuing reactions may cause damage to the endothelium at a critical point in the circulation and compromise the function of the affected organ. One example is **glomerulonephritis** (inflammation of the renal glomeruli), which results from the deposition of immune complexes in the glomerular capillaries and is a common cause of renal failure.

- Type IV or cell-mediated delayed hypersensitivity can manifest itself as an allergic reaction to certain parasites (e.g. bacteria and fungi), as a contact dermatitis resulting from sensitization to a chemical agent or by the rejection of a tissue graft or transplant. This type of hypersensitivity is mediated principally by the T cells.

Defects of the complement system

These appear to be mainly due to deficiency of particular components of the complement system. Low levels of certain of the complement proteins lead to difficulty in clearing antigen–antibody complexes from the plasma. Such complexes are a common feature of **systemic lupus erythematosus**, in which deposition of immune complexes in the renal glomeruli leads to acute renal failure as discussed above. If other complement components are low, or even absent, there is a predisposition to serious infections by organisms such as those that cause disseminated gonorrhea and menigococcal meningitis.

Self-tolerance

In autoimmunity, the lymphocytes of the immune system mount an attack on normal host cells. Although these reactions are frequently triggered by an infection, their very existence raises an important question: why are the cells of the immune system normally tolerant of the host cells? The explanation is as follows. Only those T cells that react to self-MHC molecules that incorporate a foreign peptide are released into the circulation. If a T cell reacts with unaltered MHC molecules, it will not be released from the thymus gland but will be destroyed. It is thought that many B cells do express antibody receptors that react with host cells. Those with a high affinity for the host cell-surface markers are deleted in the bone marrow and never reach the circulation (this is comparable with the fate of T cells that react with self-MHC molecules). Any remaining B cells that react with host antigens will do so only weakly and even these are ineffective because they do not get help from the T cells.

Immunodeficiency and human immunodeficiency virus

Immunodeficiency of genetic origin is called **primary immunodeficiency**. If it is due to some other cause, it is **secondary immunodeficiency**. Primary immunodeficiencies are rare and are usually due to a defect in a single gene. In all cases, there is an increased susceptibility to infection. Secondary immunodeficiency is far more common, particularly in adults. The commonest cause is malnutrition, but damage to the immune system by certain infections (notably HIV; see below), tumours, traumas and some medical interventions also impairs the function of the immune system (e.g. treatment with immunosuppressive drugs or exposure of the bone marrow to high levels of X-irradiation).

The human immunodeficiency virus (HIV) is perhaps the best-known infectious agent that compromises the function of the immune system. Unlike other agents, the virus attacks the cells of the immune system, particularly the T cells. HIV is a retrovirus and can insert its genetic material into the host-cell DNA. The stimulation of an infected T cell therefore results in replication of the virus. This leads to a slow depletion of the T-cell population and an increased susceptibility to infection. The resulting disease is known as acquired immunodeficiency syndrome or AIDS.

KEY POINTS:

- The immune system may react powerfully to an antigen (hypersensitivity) or it may fail to mount an adequate immune response (immunodeficiency).

- Hypersensitive reactions may be grouped under one of four headings: allergic reactions, e.g. hay fever and asthma; antibody-dependent cytotoxic hypersensitivity; immune complex-mediated hypersensitivity; and cell-mediated delayed hypersensitivity.

- Immunodeficiency may be of genetic origin (primary immunodeficiency) or due to some other cause (secondary immunodeficiency).

21.7   Transplantation and the immune system

Tissue transplantation from one person to another to treat organ failure has long been a major goal of medicine. The first such procedure to be wholly successful was the transfusion of blood. This depended on the correct identification of the antibodies (agglutinins) and antigens (agglutinogens) present in both donor and recipient blood, as described in Chapter 20, p. 409. In the same way, successful skin grafts or organ transplants require a close match between the specific cell markers of both donor and recipient.

The problems associated with transplantation are very well illustrated by the difficulties experienced in successfully grafting skin. If the skin becomes severely damaged, for example as a result of extensive burns, the healing process cannot make good the lost germinal tissue and the wound contracts as it becomes infiltrated by connective tissue. This results in disfigurement and distortion of the neighbouring tissue. If the affected area is large, there may be a continued loss of fluid from the damaged area, which will also be susceptible to infection. For these reasons, it is sometimes desirable to graft some healthy skin from another part of the body onto the site of injury. Such grafts (which are known as **autografts**) are usually successful. The transplanted tissue is quickly infiltrated by blood vessels and heals into place. The donor area also heals rapidly. If the damaged area is very extensive, however, it may be impossible to find sufficient undamaged skin to act as a source for the grafts. In this case, it is necessary to consider grafting skin from someone else – to use an **allograft**.

A skin graft from another individual will initially take quite well, but after about a week it will be rejected. The solution to graft rejection is to match the tissue of the donor as closely as possible with that of the recipient by tissue typing and to inhibit the activity of the immune system with immunosuppressant drugs. This approach has proved very successful with skin grafts and with kidney and heart transplants, provided the HLA antigens are well matched. Close tissue matching is not required for corneal grafts because the cornea does not have blood vessels and so is not subject to attack by the lymphocytes.

21.8   Anti-inflammatory drugs

Inflammation is often painful and disabling. For these reasons, it is desirable to reduce an inflammatory reaction using suitable drugs. The main anti-inflammatory drugs are the non-steroidal anti-inflammatory drugs (**NSAIDs**) and the **glucocorticoids**. The NSAIDs are very widely used both by medical professionals and by the public at large. The most commonly used are **aspirin**, **ibuprofen** and **paracetamol** (known as acetaminophen in the USA). All are available as over-the-counter medications. Other commonly used NSAIDs are **naproxen** and **diclofenac**. The NSAIDs inhibit the conversion of arachidonic acid to prostaglandins and thromboxanes (see Chapter 7, p. 91) and have three main actions: they provide relief from some sorts of pain (analgesia); they reduce the rise in body temperature (fever) caused by infections (an antipyretic effect); they reduce inflammation (their anti-inflammatory effect). Paracetamol has only a weak anti-inflammatory effect but is useful as an analgesic and antipyretic.

In addition to their role in regulating metabolism (see Chapter 16, p. 305), glucocorticoids have powerful anti-inflammatory and immunosuppressive actions. They affect all inflammatory reactions and depress the activity of the neutrophils and macrophages by suppressing the expression of the genes responsible for synthesizing the proteins that recruit these cells to the site of inflammation. They inhibit the synthesis of prostaglandins and decrease the production of IgG antibodies. They also decrease the plasma levels of complement proteins. A number of different glucocorticoids are used as anti-inflammatory agents, including **hydrocortisone** (cortisol), **prednisolone** and **dexamethasone** (see also Chapter 10, p. 167).

Glucocorticoids act as powerful immunosuppressants because they inhibit the clonal proliferation of helper T cells. Other immunosuppressants, such as **cyclosporine** and **tacrolimus**, have a similar action. Various cytotoxic drugs are also used as immunosuppressants, including **cyclophosphamide**, **chlorambucil** and **azathioprine**.

 ## Recommended reading

Cell biology

Alberts, B., Johnson, A., Lewis, J., Raff, M., Roberts, K. and Walter, P. (2008). *Molecular Biology of the Cell*, 5th edn. Garland: New York. Chapters 24 and 25.

Immunology

Playfair, J.H.L. and Bancroft, G.J. (2004). *Infection and Immunity*, 2nd edn. Oxford University Press: Oxford.

Pharmacology

Rang, H.P., Dale, M.M., Ritter, J.M. and Moore, P. (2007). *Pharmacology*, 6th edn. Churchill-Livingstone: Edinburgh. Chapters 15 and 16.

Self-assessment questions

Which of the following statements are true and which are false? Answers are given at the end of the book (p. 755).

The following statements relate to the natural immune system:

1 a) The monocytes of the blood are the precursors of tissue macrophages.
b) Neutrophils are only able to destroy bacteria circulating in the blood.
c) Macrophages use nitrous oxide to kill bacteria they have engulfed.
d) Complement facilitates the uptake of bacteria by phagocytes.
e) Natural killer cells act by destroying cells infected with a virus.

2 a) Tissue swelling is the first stage of the inflammatory response.
b) The capillaries at a site of inflammation exude plasma.
c) Mast cells secrete chemotactic agents that attract lymphocytes.
d) Macrophages secrete vasoactive materials during the inflammatory response.
e) The inflammatory response can be triggered by complement binding to the surface of micro-organisms.

The following statements relate to the adaptive immune system:

3 a) The lymphocytes are the most abundant white cells of the blood.
b) Lymphocytes only leave the blood at sites of inflammation.
c) B lymphocytes mature in the bone marrow.
d) B lymphocytes respond to antigens in the extracellular fluid.
e) All B lymphocytes become plasma cells after they have been stimulated.

4 a) B lymphocytes secrete IgG when they are activated.
b) T cells respond to cells infected with a virus.
c) T lymphocytes mature in the thyroid gland.
d) All T cells secrete cytotoxic materials.
e) T cells respond to proteins on the host cells.

5 a) The antigen receptor on a lymphocyte has the same specificity as the antibody it secretes.
b) Antibody molecules have two identical antigen-binding sites.
c) B lymphocytes secrete IgM and IgG in response to an infection.
d) IgM crosses the placenta and provides a fetus with antibody protection.
e) Mast cells play an essential role in allergy.

Respiration

22

After reading this chapter you should have gained an understanding of:

- The anatomy of the respiratory system
- The gas laws as they apply to respiration
- The mechanics of ventilation
- The principles of gas exchange in the alveoli
- The role of pulmonary surfactant
- The pulmonary circulation
- The factors that determine the ratio of ventilation to blood flow in different parts of the lung
- The origin of the respiratory rhythm
- Chemoreceptors and their role in the control of respiration
- The lung defence systems
- The pharmacology of the respiratory system

22.1 Introduction

The energy that people need for their normal activities is mainly derived from the oxidative breakdown of foodstuffs – particularly carbohydrates and fats. During this process, which is called **internal** or **cellular respiration**, oxygen is utilized by the mitochondria and carbon dioxide is produced. The oxygen required for this process is ultimately derived from the atmosphere by the process of **external respiration** or breathing, which also serves to eliminate the carbon dioxide pro-

duced by the cells. This chapter is concerned with the process of external respiration and the factors that regulate it.

Although the lungs are primarily concerned with gas exchange between the air deep in the lungs and the blood that perfuses them, they also have a variety of non-respiratory functions including trapping blood-borne particles (e.g. small fragments of blood clots) and the metabolism of a variety of vasoactive substances.

To fully understand the process of breathing and gas exchange, the following questions need to be addressed:

1. What mechanisms cause air to move into and out of the lungs?

2. How is oxygen taken up in the lungs and how is it carried in the blood?

3. How is carbon dioxide transferred from the cells to the lungs?

4. How efficient are the lungs in matching their ventilation to their blood flow?

5. How is the respiratory rhythm generated?

6. How does the body adjust the rate and depth of respiration to suit the varying demands of exercise?

7. What mechanisms prevent the lungs from becoming clogged with particles from the air?

Most of these questions are discussed in this chapter but gas transport via the blood is discussed in Chapter 20 and the physiology of exercise is discussed in Chapter 32. To understand the factors that determine the uptake and loss of the respiratory gases, it is useful to have a basic understanding of their physical properties. For this reason, a brief account is given in Box 22.1.

Box 22.1 The gas laws and their applicability to respiration

The air we breathe and that in our lungs is a mixture of gases consisting mainly of nitrogen, oxygen, carbon dioxide and water vapour. **Dalton's law of partial pressures** states that the total pressure of a gas mixture is the sum of the pressures that each gas would exert if it were present on its own in the same volume. In symbols, it is written as follows:

$$P_T = P_{N_2} + P_{O_2} + P_{CO_2} + P_{H_2O}$$

where P_T is the total pressure of the gas mixture and P_{N_2}, P_{O_2}, P_{CO_2}, and P_{H_2O}, are the partial pressures of nitrogen, oxygen, carbon dioxide and water vapour, respectively. In respiratory physiology, the concentration of dissolved gases is usually given as their partial pressure, even when they are present in a solution with no gas phase (e.g. in the arterial blood).

The amount of gas that is dissolved is proportional to its partial pressure in the gas phase. This is **Henry's law**, which can be written as:

$$V = sP$$

where s is the solubility coefficient ($ml\,l^{-1}\,kPa^{-1}$ or $ml\,l^{-1}\,mmHg^{-1}$), V is the volume of dissolved gas in a litre of the liquid phase and P is the partial pressure of the gas under consideration. For oxygen at body temperature (37°C), s is $0.225\,ml\,l^{-1}\,kPa^{-1}$ ($0.03\,ml\,l^{-1}\,mmHg^{-1}$). A simple calculation can be used to show the volume of oxygen

dissolved in one litre of water or, of more importance here, blood plasma. When P_{O_2} is 13.33 kPa (100 mmHg), the volume of oxygen dissolved (V) is:

$$V = 0.225 \times 13.33 = 3\,ml\ per\ litre$$

Similar calculations can be made for carbon dioxide, where $s = 5.1\,ml\,l^{-1}\,kPa^{-1}$ ($0.68\,ml\,l^{-1}\,mmHg^{-1}$), and for nitrogen, where $s = 0.112\,ml\,l^{-1}\,kPa^{-1}$ ($0.015\,ml\,l^{-1}\,mmHg^{-1}$).

Note that this relationship only applies to dissolved gas. *Where the gas enters into chemical combination, the total amount in the liquid phase is the sum of that chemically bound plus that in physical solution.*

The partial pressure of a gas can readily be converted to the equivalent molar concentration using Avogadro's law. For example, when carbon dioxide has a partial pressure of 5.33 kPa (40 mmHg), each litre of plasma will dissolve:

$$5.33 \times 5.1 = 27.2\,ml$$

Since at standard temperature and pressure (STP) 1 mole of CO_2 occupies 22.4 l, this corresponds to:

$$27.2 \times 10^{-3} / 22.4 = 1.2 \times 10^{-3}\,mol\,l^{-1}\ (1.2\,mmol\,l^{-1}).$$

Boyle's law states that the pressure exerted by a gas is inversely proportional to its volume. In symbols, this is written as:

$$P \propto 1/V$$

Charles' law states that the volume occupied by a gas is directly related to the absolute temperature (T). In symbols:

$$V \propto T$$

Boyle's Law and Charles' Law are combined in the **ideal gas law**, which is written as follows:

$$PV = nRT$$

where n is the number of moles of gas and R is the gas constant (8.31 joules K^{-1} mol^{-1}). Each mole of gas occupies 22.4 litres at STP.

As the volume of a gas or gas mixture increases with rising temperature and decreases with rising pressure, it is essential to state the conditions under which a gas sample is obtained. So for the volume V_1 of a gas sample at temperature T_1 and pressure P_1:

$$P_1 V_1 = nRT_1 \qquad (1)$$

and for the volume V_2 at temperature T_2 and pressure P_2:

$$P_2 V_2 = nRT_2 \qquad (2)$$

Dividing Equation 1 by Equation 2:

$$\frac{P_1 V_1}{P_2 V_2} = \frac{nRT_1}{nRT_2}$$

rearranging and cancelling terms:

$$\frac{P_1 V_1}{T_1} = \frac{P_2 V_2}{T_2}$$

gives:

$$V_2 = \frac{V_1 P_1}{P_2} \times \frac{T_2}{T_1}$$

Sample calculation

A 1 litre gas sample taken at a room temperature of 22°C (295K) when the atmospheric pressure was 750 mmHg, would have a volume at STP (273K and 760 mmHg) of:

$$V_{STP} = \frac{1 \times 750 \times 273}{760 \times 295} = 0.91 \text{ litres}$$

22.2 The composition of expired air

Expired air contains less oxygen and more carbon dioxide than inspired air. Standard values for the partial pressures of the gases present in expired and alveolar air are given in Table 22.1. Note that, although atmospheric nitrogen is not involved in metabolism, its partial pressure changes as it becomes diluted by the water vapour and carbon dioxide from the lungs.

The ratio of the carbon dioxide produced divided by the oxygen uptake is called the **respiratory exchange ratio** (**R**), which, under steady-state conditions, is equal to the **metabolic respiratory quotient** (**RQ**) (see Chapter 32 for further details of the RQ).

$$\text{Respiratory exchange ratio} = \frac{[\text{Volume of } CO_2 \text{ produced}]}{[\text{Volume of } O_2 \text{ taken up}]}$$

The respiratory exchange ratio varies according to the type of food being metabolized to produce ATP. It ranges from 0.7 when fats are the principal source of energy to 1.0 for carbohydrates. Usually, R is around 0.75–0.8 as both carbohydrates and fats are metabolized at the same time.

KEY POINTS:

- Expired air has less oxygen and more carbon dioxide than room air.

- The ratio of the amount of carbon dioxide expired to the amount of oxygen taken up is known as the respiratory exchange ratio.

- The respiratory exchange ratio depends on the nature of the foodstuffs being metabolized.

Table 22.1 Standard values for respiratory gases

	N_2	O_2	CO_2	H_2O
Inspired air				
(kPa)	79.6	21.2	0.04	0.5
(mmHg)	597	159	0.3	3.7
% total	78.5	20.9	0.04	0.5
Expired air				
(kPa)	75.5	16	3.6	6.3
(mmHg)	566	120	27	47
% total	74.5	15.8	3.5	6.2
Alveolar air				
(kPa)	75.9	13.9	5.3	6.3
(mmHg)	569	104	40	47
% total	74.9	13.7	5.2	6.2

22.3 The structure of the respiratory tree

The lungs are the principal organs of the respiratory system and provide the surface over which oxygen is absorbed and carbon dioxide is excreted. As the lungs are situated in the chest, air from the atmosphere must pass through the nose or mouth and enter the trachea before it can reach the respiratory surface of the lungs, where gas exchange occurs. The lungs are supplied with blood via the pulmonary and bronchial circulations, which are discussed in detail on p. 448.

During quiet breathing, air is normally taken in via the nose, but during heavy exercise, air is taken in via the mouth, which offers much less resistance to air flow. After entering the nose or mouth, the air passes through the pharynx to the larynx. The **trachea** links the larynx to the lungs. In an adult, it is about 1.8 cm in diameter and 10–12 cm in length. It is the first component of the respiratory tree – the branching set of tubes that link the respiratory surface to the atmosphere.

Within the chest, the trachea branches to form the two main primary **bronchi** – one for each lung. In turn, the bronchi branch to give rise to two smaller branches on the left and three on the right corresponding to the lobes of the lung. (The left lung has two lobes while the right has three.) Within each lobe, the bronchi divide into two smaller branches and these smaller branches also divide into two, and so on until the final branches reach the respiratory surface. The nose, mouth and upper trachea lie outside the chest (thorax) while the lower trachea and the remainder of the respiratory tree lie within the chest. A diagram of the arrangement of the trachea and lungs within the chest is shown in Figure 22.1.

In all, there are 23 generations of airways between the atmosphere and the alveoli. The trachea is generation 0, while the two main bronchi form generation 1. The bronchi form the first 11 generations of airways. From the 12th to the 19th generations, the airways are

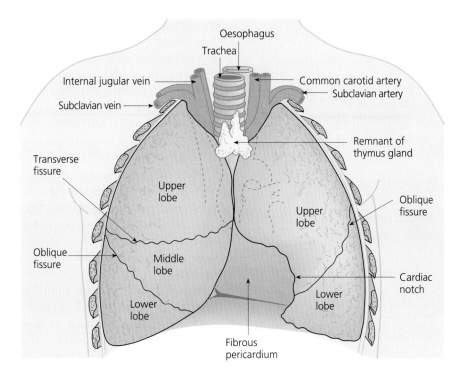

Figure 22.1 Diagrammatic representation of the arrangement of the lungs within the thorax. Note that there are the three lobes on right and two on the left. The heart lies on the left side within the pericardium.

known as **bronchioles**. The 16th generation of airways are the **terminal bronchioles**. The airways as far as the terminal bronchioles are concerned with warming and moistening the air on its way to the respiratory surface. They are known as **conducting airways** and play no significant part in gas exchange. From generation 17 to generation 19, the airways begin to participate in gas exchange. These transitional airways are called the **respiratory bronchioles**, which give rise to the principal gas exchange structures: the **alveolar ducts** and **alveolar sacs** (generations 20–23). The alveolar sacs consist of two or more **alveoli**. The respiratory bronchioles, alveolar ducts and alveoli provide a total area for gas exchange of about 60–80 m² in an adult.

> **KEY POINTS:**
>
> - The airways consist of the nose, pharynx, larynx, trachea and a branching tree of bronchi and bronchioles.
>
> - The first 16 generations of bronchi and bronchioles form the conducting airways.
>
> - The remaining seven generations (the respiratory bronchioles, alveolar ducts and alveoli) comprise the transitional and respiratory airways.

22.4 The structure of the airways

The trachea and the primary bronchi in humans consist of three main layers: a lamina propria, a submucosa and an adventitia. Overlying the lamina propria is the respira-

tory epithelium, which is discussed in more detail below. The main features of the trachea in cross-section can clearly be seen in Figure 22.2. The trachea and bronchi

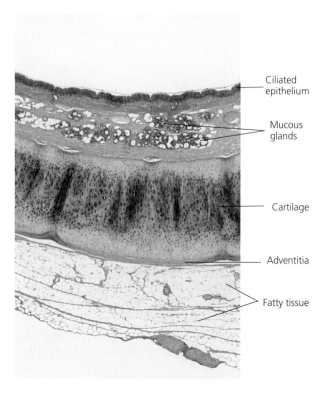

Ciliated epithelium

Mucous glands

Cartilage

Adventitia

Fatty tissue

Figure 22.2 A section through dog trachea to show the principal tissue layers. Note the green-stained cartilage and the large number of secretory glands just beneath the epithelium.

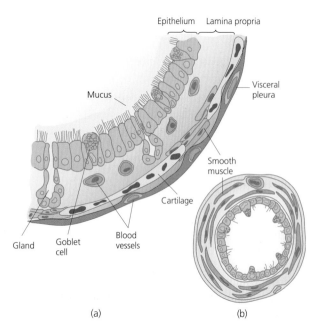

Epithelium   Lamina propria

Mucus

Visceral pleura

Smooth muscle

Cartilage

Gland   Goblet cell   Blood vessels

(a)   (b)

are stiffened and held open by C-shaped rings of cartilage, the open part of which is on the posterior (or dorsal) aspect. The ends of the cartilage rings are held together by ligaments and bundles of smooth muscle. In the smaller bronchi, the structural role of the cartilage rings is taken by overlapping plates of cartilage. The bronchioles, which are less than 1 mm diameter, have no cartilage and are easily collapsed when the pressure outside the lung exceeds the pressure in the airways; this happens during a forced expiration. Smooth muscle is found in the walls of all the airways, including the alveolar ducts, but not in the walls of the alveoli themselves. The outermost part of the bronchiolar wall – the adventitial layer – is composed of dense connective tissue, including elastic fibres. The structure of the bronchi and bronchioles is illustrated in Figure 22.3.

From the nasal passages to the small bronchi, the airways are lined with a ciliated pseudostratified columnar epithelium (see p. 74 for the classification of epithelia). This epithelium progressively changes in character as the airways branch. In the upper airways, it consists mainly of ciliated columnar epithelial cells and goblet cells (which are easily distinguished in the beautiful scanning electron micrograph shown in Figure 22.4). Basal cells are small, rounded cells that lie on the basement membrane but do not reach the surface of the epithelium. Several other cell types are also present but are not easily distinguished by light microscopy. The upper airways have the highest proportion of mucus-secreting goblet cells. However, as the bronchi divide, the proportion of goblet cells declines progressively. Goblet cells are not found in the epithelium of the terminal bronchioles and later generations of airways. In the upper airways, the cilia of the epithelial cells beat continuously and slowly move the secretions of the goblet cells and submucosal glands towards the mouth.

Figure 22.3 Simplified diagrams to show the structure of (a) the bronchi and (b) the bronchioles. Note that the bronchus has cartilage, a ciliated epithelium and a thicker lamina propria. In the bronchiole, there is no cartilage and the epithelium has few ciliated cells. The bronchiole has a higher proportion of smooth muscle than the bronchus.

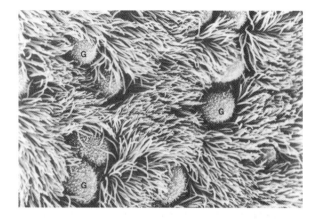

Figure 22.4 A scanning electron micrograph of the ciliated epithelium of rat trachea. Note the abundant cilia and the interspersed goblet cells (labelled G).

This arrangement is known as the **mucociliary escalator**, which plays an important role in the removal of inhaled particles (see Figure 22.5 and p. 457).

Within the lamina propria, submucosal glands are plentiful and they discharge their secretions onto the surface of the epithelium. In addition, the lamina propria has both collagen and elastic fibres that provide flexible support for the walls of the conducting airways. Elastic fibres are especially abundant in the bronchi and bronchioles. Bundles of smooth muscle surround the airways as far as the alveolar ducts. This muscle serves to adjust the diameter of the airways – for example during the bronchoconstriction caused by inhalation of smoke and the bronchodilatation that occurs during exercise. The lamina propria also has numerous blood vessels.

By the terminal bronchioles, the respiratory epithelium has become cuboidal in type. It progressively becomes flattened as the respiratory bronchioles give rise to the alveolar ducts and alveoli. The walls of the alveoli consist of a thin epithelial layer comprising two types of cells, called the alveolar type I and type II cells. The type I cells

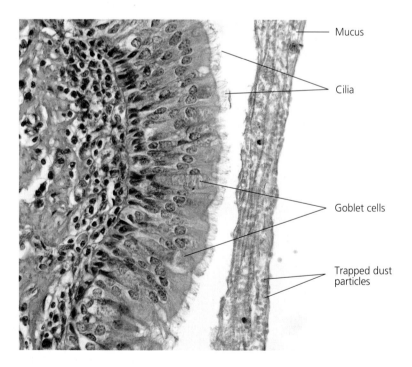

Mucus

Cilia

Goblet cells

Trapped dust particles

Figure 22.5 Enlarged view of the ciliated epithelium of the trachea. Note the large number of mucus-secreting goblet cells and the prominent ciliated epithelium. Trapped dust particles are clearly visible in the overlying mucus layer.

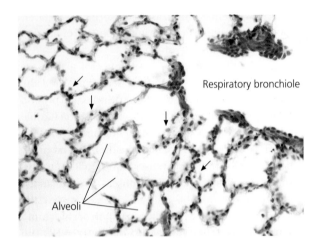

Figure 22.6 A section of lung showing part of a respiratory bronchiole and alveoli. Note the extremely thin alveolar walls, which are shared between adjacent alveoli (the interalveolar septa). The arrows point to type II alveolar cells.

are squamous epithelial cells that form most of the alveolar wall while the type II cells are thicker (Figure 22.6) and produce the fluid layer that lines the alveoli. The type II cells also synthesize and secrete **pulmonary surfactant** (see p. 445).

There are about 300 million alveoli in the adult lung and each is almost completely enveloped by pulmonary capillaries (see Figure 22.16). Estimates suggest that there are about 1000 pulmonary capillaries for each alveolus. This provides a huge area for gas exchange by diffusion. The alveolar epithelial cells and pulmonary capillary endothelial cells are very thin so that the pulmonary blood is separated from the alveolar air by as little as 0.5 μm, as shown in Figure 22.7. Interspersed between the capillaries in the walls of the alveoli are the elastic and collagen fibres that form the connective tissue of the lung. This connective tissue links the alveoli together to form the lung **parenchyma**, which is sponge-like in appearance (Figure 22.6).

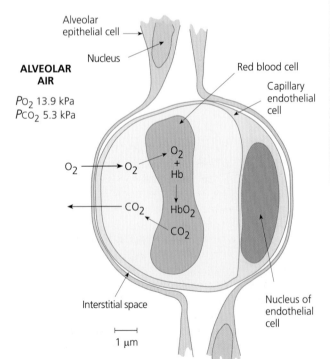

ALVEOLAR AIR

PO_2 13.9 kPa
PCO_2 5.3 kPa

KEY POINTS:

- The airways are lined by a ciliated epithelium that contains many mucus-secreting cells.

- The trachea and bronchi are kept open by rings or plates of cartilage. The bronchioles have no cartilage and their walls consist mainly of smooth muscle.

- The alveoli are the principal sites of gas exchange. They are blind sacs whose walls consist of a very thin epithelium beneath which lies a dense network of pulmonary capillaries.

Figure 22.7 A diagrammatic representation of the alveolar cells that separate the alveolar air space from the blood in the pulmonary capillaries.

22.5 The structure of the chest wall

The lungs are not capable of inflating themselves; inflation is achieved by changing the dimensions of the chest wall by means of the respiratory muscles. The principal respiratory muscles are the **diaphragm** and the internal and external intercostal muscles. The **external intercostal muscles** are arranged in such a way that they lift the ribs upwards and outwards as they contract while the **internal intercostal muscles** pull the ribs downwards, in opposition to the external intercostal muscles. In addition, some other muscles that are not involved during normal quiet breathing may be called upon during heavy exercise. These are known as the accessory muscles, which are shown in Figure 22.8. The scalenes and sternocleidomastoids assist in inspiration, while the abdominal muscles assist in expiration.

The chest wall is lined by a membrane called the **parietal pleura** (Figure 22.9). This is separated from the **visceral pleura** (the membrane that covers the lungs) by a thin layer of liquid, which serves to lubricate the surfaces of the pleural membranes as they move during respiration.

The total volume of intrapleural fluid is about 10 ml. It is an ultrafiltrate of plasma and is normally drained by the lymphatic system that lies beneath the visceral pleura. The pleural membranes themselves are joined at the top and bottom of the lungs. They consist of two layers of collagenous and elastic connective tissue. Beneath the visceral pleura lies the limiting membrane of the lung itself that, together with the visceral pleura, limits the expansion of the lungs. The lungs are separated from the chest wall only by the pleural membranes and, in health, they occupy the entire cavity of the chest except for the region that contains the heart and great vessels (the mediastinum).

> **KEY POINT:**
>
> - The chest wall is formed by the rib cage, the intercostal muscles and the diaphragm. It is lined by the pleural membranes and forms a large, gas-tight compartment that contains the lungs.

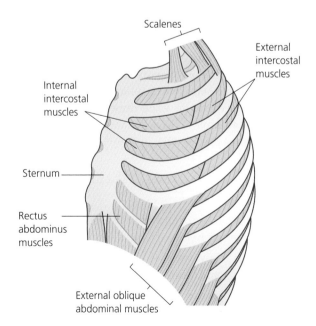

Figure 22.8 The principal respiratory muscles of the human chest excluding the diaphragm. Between each pair of ribs there are two layers of muscle, the external intercostal muscles and the internal intercostal muscles. The figure illustrates the orientation of the muscle fibres in these muscle groups. Note that the angle of the external intercostal muscles allows them to lift the rib cage when they shorten, so expanding the chest. The internal intercostal muscles act to lower the rib cage. The contraction of the scalene muscles and sternocleidomastoid muscles (see Figure 10.24a) also acts to lift the rib cage while contraction of the abdominal muscles tends to compress the abdominal contents and force the diaphragm upwards into the chest, assisting expiration.

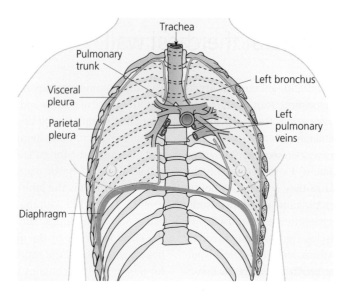

Figure 22.9 The arrangement of the pleural membranes. The parietal pleura lines the chest cavity while the visceral pleura covers the lungs.

22.6 Innervation of the respiratory system

The respiratory muscles do not contract spontaneously. Rhythmical breathing depends on nerve impulses in the **phrenic** and **intercostal nerves**, which are the motor nerves serving the diaphragm and intercostal muscles. The rhythmical discharge of these nerves is governed by the activity of specific groups of nerve cells that are located in the brainstem. This aspect of respiration will be discussed in Section 22.16 (p. 452).

The smooth muscle of the bronchi and bronchioles is innervated by cholinergic parasympathetic fibres, which reach the lungs via the vagus nerves. Activation of these nerve fibres causes a narrowing of the airways (**bronchoconstriction**). Sympathetic nerves innervate the blood vessels of the bronchial circulation but there appears to be no direct sympathetic innervation of the bronchiolar smooth muscle itself. Bronchodilatation occurs in response to circulating adrenaline and noradrenaline (epinephrine and norepinephrine), which act on β-adrenergic receptors to cause relaxation of the smooth muscle. Inhalation of β-adrenergic drugs, such as salbutamol, is used to overcome the bronchospasm that occurs during asthmatic

attacks (see p. 462). In addition to their cholinergic and adrenergic innervation, the tone of the smooth muscle of the airways is also regulated by autonomic fibres that secrete nitric oxide.

The lungs themselves contain slowly adapting stretch receptors, irritant receptors and pulmonary C-fibre endings, which send information to the CNS via visceral afferent fibres in the vagus nerves. These receptors play an essential role in the respiratory reflexes, as discussed below (see pp. 456–457).

KEY POINTS:

- The muscles of respiration receive their motor innervation via the phrenic and intercostal nerves.

- The airways are innervated by parasympathetic autonomic nerves.

- Sensory nerve fibres travel from the lungs to the CNS via the vagus nerves.

22.7 The mechanics of breathing

It is common knowledge that breathing is associated with changes in the volume of the chest. During inspiration, the chest is expanded and air enters the lungs. During expiration, the volume of the chest decreases and air is expelled from the lungs. In this section, the mechanisms responsible for the changes in the dimensions of the chest will be examined, followed by a discussion of the factors governing the flow of air in the airways.

The lung volumes

The movement of air into and out of the lungs is known as **ventilation** and the volume of air that moves in and out of the chest during breathing can be measured with the aid of an instrument called a **spirometer**, which, in its simplest form, consists of an inverted bell with a water seal to form an airtight chamber. The bell is free to move in the vertical direction and the movements can be recorded onto a chart or logged by a computer (Figure 22.10). Portable, hand-held electronic devices are now available for routine clinical examination.

The relationship between the various lung volumes is shown in Figure 22.11. When the chest is expanded to its fullest extent and the lungs allowed time to inflate fully, the amount of air they contain is at its maximum. This

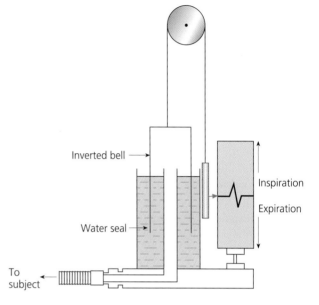

Figure 22.10 A simplified diagram of a recording spirometer. The subject breathes in and out through the mouth via the flexible tube shown bottom left. Inspiration draws air from the bell and the volume of air trapped within the bell decreases. It increases during expiration. These changes in volume are recorded on a calibrated chart as shown or by a computer. In use, the spirometer would normally be connected to a more elaborate gas circuit with a soda-lime canister to absorb exhaled carbon dioxide.

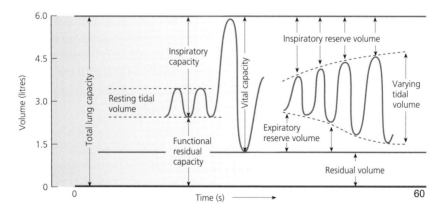

Figure 22.11 The subdivisions of the lung volumes. The figure shows an idealized spirometry record of the changes in lung volume during normal breathing at rest (resting tidal volume) followed by a large inspiration to total lung capacity followed by a full expiration to the residual volume. This measures the vital capacity. Note that the residual volume and functional residual capacity cannot be measured by a spirometer. Instead, they are measured by the helium dilution method discussed in Box 22.2. As shown on the right-hand side of the figure, the tidal volume is highly variable – in exercise for example – and the inspiratory and expiratory reserve volumes become smaller as tidal volume increases.

is the **total lung capacity**. If this is followed by a maximal expiration, the lungs will still contain a volume of air that cannot be expelled. This is called the **residual volume** (**RV**). The amount of air breathed out during a maximal expiration following a maximal inspiration is called the **vital capacity** (**VC**).

The air taken in and exhaled with each breath is known as the **tidal volume** and is about 500 ml in an adult male at rest. The tidal volume is normally much less than the vital capacity (which is about 5–6 litres). The difference in lung volume at the end of a normal inspiration and the total lung capacity is known as the **inspiratory reserve volume** (**IRV**) and the amount of air that can be forced from the lung after a normal expiration is called the **expiratory reserve volume** (**ERV**). The **functional residual capacity** (**FRC**) is the volume of air left in the lungs at the end of a normal expiration. The various lung volumes depend on height (they are larger in tall people), age, sex (the volumes tend to be smaller in women than in men of similar body size) and training. Typical values for healthy, young, adult males are given in Table 22.2.

Although the vital capacity and residual volume are relatively fixed in any one person, the tidal volume varies according to the requirements of the body for oxygen. Consequently, the inspiratory and expiratory reserve volumes are also variable – the larger the tidal volume, the smaller the inspiratory and expiratory reserve volumes

(i.e. there is a smaller margin remaining for increasing the tidal volume). With the exception of the residual volume and functional residual capacity, all the lung volumes can be directly measured by spirometry. (The functional residual capacity and residual volume can be measured by the helium dilution method, explained in Box 22.2.) The total lung volume can be obtained by adding the residual volume to the vital capacity.

Movements of the chest during breathing

The various muscles of respiration are called on at different times. The diaphragm is the principal respiratory muscle. It forms a continuous sheet that separates the thorax from the abdomen. At rest, it assumes a dome-like shape. When it contracts during inspiration, the crown of the diaphragm descends, thereby increasing the volume of the chest. The pressure within the lungs falls and this draws air in from the atmosphere. During expiration, the diaphragm smoothly relaxes and the volume of the chest decreases. This increases the pressure within the lungs and air is expelled.

When the demand for oxygen increases, the other muscles of inspiration are called into play. The chest wall is lifted upward and outward by the activity of the external intercostal muscles and the diaphragm contracts more strongly so that the volume of the chest is increased further.

At rest, expiration is largely passive as the chest wall recoils from being stretched, but in exercise, the internal intercostal muscles also contract to speed up the decrease in the volume of the chest, so that expiration under these conditions becomes an active process. Powerful expiration may also be assisted by contraction of the abdominal muscles that force the abdominal contents against the diaphragm, pushing it upwards and reducing the volume of the chest (for example during singing or playing a wind instrument).

Table 22.2 Typical values for respiratory variables in a healthy young adult male at rest

Total lung volume (litres)	6.0
Vital capacity (litres)	4.8
Residual volume (litres)	1.2
Tidal volume (litres)	0.5
Respiratory frequency (breaths min⁻¹)	12
Minute ventilation (litres min⁻¹)	7.2
Functional residual capacity (litres)	2.2
Inspiratory capacity (litres)	3.8
Inspiratory reserve volume (litres)	2.5
Expiratory reserve volume (litres)	1.0

KEY POINTS:

- Ventilation is the volume of air moved into and out of the lungs. It is driven by changes in the dimensions of the chest arising from the contraction and relaxation of the muscles of respiration.

- Inspiration is an active process that depends on the contraction of the diaphragm and external intercostal muscles. Expiration is largely passive and is due to the elastic recoil of the lungs and chest wall.

Box 22.2 Determination of functional residual capacity and residual volume

The various subdivisions of the lung volume are of interest in a number of respiratory diseases. It is therefore desirable that they can be measured with some accuracy. While the *change* in the volume of the lungs during various breathing manoeuvres can be directly measured by a spirometer, the volume of air left in the lungs at the end of a normal expiration (the functional residual capacity or FRC) and the residual volume (RV) cannot be measured in this way. Instead, they are measured by the helium dilution method.

To determine the FRC, a subject is asked to breathe out normally and then is asked to inspire from a spirometer filled with a known volume of air containing a known concentration of the inert gas helium. As the gas containing helium is breathed in and out, the helium is diluted by the volume of air left in the lungs. A subsequent measurement of the helium concentration permits this

additional volume to be calculated from the following formula:

$$FRC = \left(\frac{\text{initial helium concentration}}{\text{final helium concentration}} - 1\right)$$
$$\times \text{volume of spirometer}$$

To measure the RV, a similar procedure is used, but the subject is asked to perform a maximal expiration before breathing from the spirometer containing the helium.

Total lung capacity = functional residual capacity
+ inspiratory capacity

Or:

Total lung capacity = residual volume + vital capacity

If either the FRC or the RV is known, all the other subdivisions of the lung volume (including the total lung volume) can be determined with the aid of a spirometer.

22.8 The intrapleural pressure

In health, the lungs are expanded to fill the thoracic cavity because the pressure outside the lungs is less than that of the air in the alveoli. This pressure is known as the **intrapleural pressure** or **intrathoracic pressure**. At the end of a quiet expiration, it is found to be about 0.5 kPa (5 cmH$_2$O) below that of the atmosphere. By convention, pressures less than that of the atmosphere are called negative pressures and those above atmospheric pressure are called positive pressures. Thus, the intrapleural pressure at the end of a quiet expiration is –0.5 kPa or –5 cmH$_2$O. During inspiration, the chest wall expands and the intrapleural pressure becomes more negative reaching a maximum of about –1 kPa (–10 cmH$_2$O). The fall in intrapleural pressure expands the alveoli and the alveolar pressure falls below that of the atmosphere. This pressure difference draws air into the lungs. During expiration, the sequence is reversed: intrapleural pressure falls, alveolar pressure rises above that of the atmosphere and air is expelled from the lungs. These events are summarized in Figure 22.12.

If the chest wall is punctured, there is an inrush of air from the atmosphere into the cavity of the chest (**pneumothorax**). Under these conditions, the intrapleural pressure becomes equal to that of the atmosphere and the lungs will collapse. This shows that the negative value of the intrapleural pressure is due to the elastic recoil of the lungs.

The change in the volume of the chest that results from a given change in intrapleural pressure is called the **compliance**. It is a measure of the ease with which the chest volume can be changed and is determined when there is no movement of air into or out of the lungs (**static compliance**). If compliance is high, there is little resistance to expansion of the chest; conversely, if it is low, the chest is expanded only with difficulty. For healthy young subjects, the static compliance has a typical value of 1.0 l kPa^{-1} (0.1 l cmH$_2$O^{-1}), as shown by the straight black line in Figure 22.13.

During normal breathing, a larger change in pressure is required to move a given volume of air into the chest

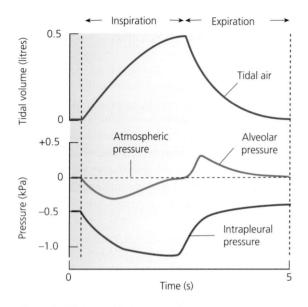

Figure 22.12 A simplified diagram of the changes in intrapleural and alveolar pressure during a single respiratory cycle. Note that the changes in intrapleural pressure occur before the change in alveolar pressure and are more prolonged.

than would be expected from the static compliance of the chest. This is shown in Figure 22.13. The additional pressure is required to overcome additional **non-elastic resistances**, which are:

1. The resistance of the airways to the movement of air (the **airways resistance**)

2. The frictional forces arising from the viscosity of the lungs and chest wall (tissue resistance)

3. The inertia of the air and tissues.

Of these, the airways resistance is by far the most important. The pressure–volume relationship during a single respiratory cycle is thus a closed loop, as shown in Figure 22.13. The physiological reason for this property (known as **hysteresis**) will be discussed below.

The compliance of the intact chest is determined by the elasticity of the chest wall and that of the lungs. (An elastic body is one that resumes its original dimensions after the removal of an external force by which it has become deformed.) The elasticity of the chest wall (and therefore its compliance) is determined mainly by that of its muscles, ligaments and tendons. The elasticity of the lungs is determined by two major factors: the elastic

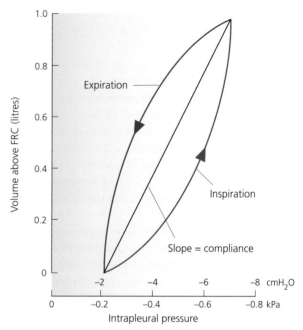

Figure 22.13 The pressure–volume relationship for a single respiratory cycle. The compliance of the respiratory system is given by the slope of the line. This would represent the change in volume if the work of respiration was against purely elastic resistances. However, during inspiration, additional pressure is needed to overcome the airways resistance and other resistive forces. This is shown by the curve to the right of the compliance line. The curve to the left of the compliance line shows the resistive work done during passive expiration.

fibres of the lung parenchyma and the surface tension of the liquid film that lines the alveoli.

If the chest wall is cut open when it is at its relaxation volume, the ribs spring outwards and the lungs collapse inwards. The volume assumed by the chest wall is greater, and that of the lungs is smaller, than that of the intact chest. This shows that the dimensions of the chest at rest reflect the balance of the forces acting on the chest wall and the lungs.

> **KEY POINT:**
>
> • As the chest expands during inspiration, the pressure in the alveoli falls below that of the atmosphere causing air to enter the lungs. As the chest volume falls during expiration, the pressure in the alveoli rises above that of the atmosphere and air is expelled from the lungs.

22.9 Surface tension and the elasticity of the lungs

During the initial stages of inflation with air, collapsed lungs require a considerable pressure before they begin to increase in volume (about 0.8–1.0 kPa; phase 1 in Figure 22.14). The lungs then expand roughly in proportion to the increase in pressure (phase 2) until they approach their maximum capacity (phase 3). Once they have been fully expanded, the volume of the lungs changes slowly during deflation until the pressure holding them open decreases to about 0.8 kPa (phase 4), whereupon their volume declines more steeply as the pressure falls. The unequal pressure required to maintain a given lung volume during inflation with air as opposed to deflation accounts for the hysteresis in the pressure–volume relationship seen during the respiratory cycle (Figure 22.13). If the lungs are inflated with isotonic saline (0.9% NaCl), however, the pressures required to expand the lung to a given volume are much reduced and there is little or no hysteresis.

Why is it more difficult to inflate the lungs with air than with isotonic saline? When the lungs are inflated with saline, the only force opposing expansion is the tension in the elastic elements of the parenchyma that become stretched as the lungs expand. When the lungs are inflated by air, however, the surface tension at the air–liquid interface in the alveoli also opposes their expansion. To minimize the effort required to expand the lungs, the type II cells of the alveoli secrete **lung surfactant** (also called **pulmonary surfactant**), which lowers the surface tension of the liquid lining the air spaces of the lung. This has the effect of reducing the pressures needed to hold the alveoli open. In addition, as surfactant lowers surface tension, it helps to prevent fluid accumulating in the alveoli and so plays an important role in keeping the alveolar air space dry. By lowering surface tension, lung surfactant reduces the work of breathing. This effect of lung surfactant on surface tension is particularly important when the lungs first expand at birth (see p. 631).

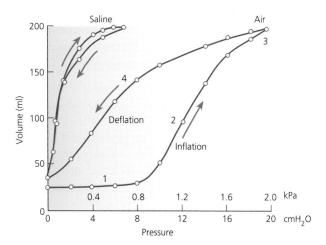

Figure 22.14 Pressure–volume relationships for isolated cat lungs when inflated by air or with isotonic saline. Note the low pressures required to expand the saline-filled lungs and that the curve for inflation is virtually the same as for deflation. For air-filled lungs, much greater pressures are required for a given volume change. The curve shows hysteresis, a greater pressure being required to inflate the lungs to a given volume compared with the pressure required to hold that volume during deflation. See text for further explanation.

22.10 Airways resistance

Like the flow of blood in the circulatory system, the flow of air through the airways may be either laminar or turbulent. **Laminar flow** occurs at low linear flow rates (flow rate in volume per second divided by the cross-sectional area), but when the linear flow rate increases beyond a critical velocity, the orderly pattern of flow breaks down, eddies form and the flow becomes turbulent. **Turbulent flow** is more likely to occur in large-diameter, irregularly branched tubes when the flow rate is high. This is the situation in the nose, pharynx and larynx, and accounts for about a third of the total airways resistance. This can be significantly reduced by breathing through the mouth – a fact that is widely exploited during heavy exercise. The remaining two-thirds of the airways resistance is located within the upper part of the tracheobronchial tree. This is called the **upper airways resistance**. By the time air reaches the smallest airways, the total cross-sectional area of the airways is high and the airflow is laminar, so that the resistance becomes very small.

> **KEY POINTS:**
> - During breathing, the total pressure required to inflate the chest is the pressure required to expand the elastic elements of the chest (measured by the compliance) plus the pressure required to overcome the airways resistance.
> - The compliance of the lungs is determined both by the elastic elements in the lung parenchyma and by the surface tension of the air–liquid interface of the alveoli. The surface tension is reduced below that of water by pulmonary surfactant secreted by the type II alveolar cells.

22.11 Alveolar ventilation and dead space

Broadly, the respiratory system can be considered to consist of two parts: the conducting airways and the area of gas exchange. In dividing the respiratory system in this way, it becomes obvious that not all the air taken in during a breath reaches the alveolar surface. Some of it must occupy the airways that connect the respiratory surface to the atmosphere. This air does not take part in gas exchange and is known as the **dead space**. The remaining fraction of the tidal volume enters the alveoli. Thus:

Tidal volume = dead space + volume of air entering the alveoli

As not all of the air that enters the alveoli takes part in gas exchange, two different types of dead space are distinguished:

1. The **anatomical dead space**, which is strictly the volume of air taken in during a breath that does not mix with the air in the alveoli. It is the volume of air in the conducting airways.

2. The **physiological dead space**, which is the volume of air taken in during a breath that does not take part in gas exchange.

As with other lung volumes, such as vital capacity, the anatomical and physiological dead spaces depend on the body size, the age and the sex of the individual. The physiological dead space is equal to the volume of the non-respiratory airways plus the volume of air that enters those alveoli that are not perfused with blood, as these alveoli cannot participate in gas exchange. The physiological dead space can be estimated by measuring the carbon dioxide content of the expired and alveolar air using the Bohr equation (see Box 22.3).

In a normal healthy adult, the anatomical and physiological dead space are about the same – 150 ml for a tidal volume of 500 ml. In some diseases of the lung, such as emphysema, the physiological dead space can greatly exceed the anatomical dead space.

Box 22.3 The Bohr equation for calculating the physiological dead space

The Bohr equation is:

$$V_D = V_E \left(1 - \frac{F_E}{F_A}\right)$$

Where V_D is the physiological dead space and V_E is the volume of expired air. F_E is the fraction of CO_2 in the expired air and F_A is the fraction of CO_2 in the alveolar air.

The volume and CO_2 content of the expired air can be readily measured and a sample of alveolar air can be obtained by asking a subject to expire fully through a long thin tube. The last part of the expired volume will have the same composition as the alveolar air. This gas can be sampled and its F_{CO_2} determined. This will give an average for the composition of the alveolar air.

A worked example

If the fractional content of CO_2 in the expired and alveolar air was 0.036 (3.6%) and 0.052 (5.2%) and the tidal volume was 500 ml what is the physiological dead space?

Using the Bohr equation:

$$V_D = 500 \times (1 - (0.036 / 0.052))$$
$$= 500 \times (1 - 0.69)$$
$$= 155 \text{ ml}$$

The amount of air that is taken in during each breath (the tidal volume, V_T) multiplied by the frequency of breathing (f) is known as the **minute volume**. The fraction of the minute volume that ventilates the alveoli is known as the **alveolar ventilation**. So:

$$fV_T = fV_D + fV_A$$

In the case of a subject breathing a tidal volume of 0.5 litres 12 times a minute:

Minute volume $= 12 \times 0.5 = 6 \, l \, min^{-1}$

If the dead space is 0.15 l (V_D=150 ml), the alveolar ventilation is:

$12 \times (0.50 - 0.15) = 12 \times 0.35$
$= 4.2 \, l \, min^{-1}$

The alveolar ventilation is not uniform throughout the lung

The ventilation of the lung is its change in volume relative to its resting volume during a single respiratory cycle: the greater the relative change in volume, the greater the ventilation. Measurements with tracer gases show that the inspired air is not distributed evenly to all parts of the lung. The pattern of ventilation depends on posture (i.e. whether the subject is upright or lying down), on the amount of air inspired and on the rate of inspiration. In an upright subject,

during a slow inspiration following a normal expiration, the base of each lung is ventilated about 50% more than the apex (Figure 22.15). This difference is reduced if the subject lies down.

What are the causes of uneven pulmonary ventilation? Firstly, during inspiration, the volume of the lower part of the chest increases significantly more than the upper part. This situation arises because the lower ribs are more curved and more mobile than the upper ribs. Secondly, the descent of the diaphragm expands the lower lobes of

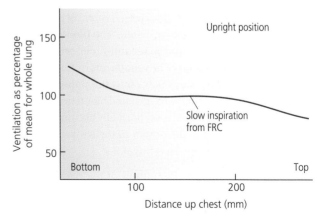

Figure 22.15 The distribution of ventilation in the normal upright human lung. The data shown are for a slow inspiration following a normal expiration. The base of the lung is better ventilated than the top.

the lungs more than the upper ones, which are attached to the main bronchi. Thirdly, the compliance of the lung is not uniform. The peripheral lung tissue is more compliant than the tissue that is attached to the stiffer airways. The combination of these factors results in differing regions of the lung exhibiting differing amounts of ventilation.

KEY POINTS:

- The anatomical dead space is the volume of air taken in during a breath that does not mix with the air in the alveoli. It is a measure of the volume of the conducting airways.
- The physiological dead space is the volume of air taken in during a breath that does not take part in gas exchange.
- The minute ventilation is the tidal volume multiplied by the frequency of breathing.
- The alveolar ventilation is the volume of air entering the alveoli per minute.
- The ventilation of the lung is not uniform, being somewhat greater at the base than at the apex. This situation arises in part because the base of the lungs expands proportionately more than the apex.

22.12   The bronchial and pulmonary circulations

The lungs receive blood from two sources: the bronchial circulation and the pulmonary circulation (Figure 22.16).

The bronchial circulation. The bronchial arteries arise from the aortic arch, the thoracic aorta or their branches (mainly the intercostal arteries). These arteries supply oxygenated blood to the smooth muscle of the principal airways (the trachea, bronchi and bronchioles as far as the respiratory bronchioles), the intrapulmonary nerves, nerve ganglia and the interstitial lung tissue. The blood draining the airways is deoxygenated. The blood from the upper airways (as far as the second-order bronchi) drains into the right atrium. The venous return from the later generations of airways flows into the pulmonary veins where it mixes with the oxygenated blood from the alveoli. The bronchial circulation normally accounts for only a very small part of the output of the left ventricle (about 2%).

The pulmonary circulation. The output of the right ventricle passes into the pulmonary artery, which subsequently branches to supply the individual lobes of the lung. The output of the right ventricle is equal to that of the left ventricle so that, at rest, about 5 litres of blood pass through the pulmonary vessels each minute. The pulmonary arteries branch along with the bronchial tree until they reach the respiratory bronchioles. Here they form a dense capillary network to provide a vast area for gas exchange that is similar in extent to that of the alveolar surface (Figure 22.16). The capillaries drain into pulmonary venules, which arise in the septa of the alveoli. Small veins merge with larger veins that have a segmental arrangement. Finally, two large pulmonary veins emerge from each lung to empty into the left atrium.

Blood flow through the upright lung is greatest at the base and least at the apex

As in other vascular beds, the flow of blood through the lungs is determined by the perfusion pressure and the vascular resistance. Compared with the systemic circulation, however, the pressures in the pulmonary arteries are rather low, the systolic and diastolic pressures being about 3.3 and 1.0 kPa respectively (*c*. 25/8 mmHg). As a column of blood 1 cm high will exert a pressure of about 0.1 kPa, the pressure in the pulmonary artery during systole is sufficient to support a column of blood 33 cm high. During diastole, however, the pressure is sufficient only to support a column about 10 cm high. As a result, differences in the hydrostatic pressures of the blood in various parts of the pulmonary circulation will exert a considerable influence on the distribution of pulmonary

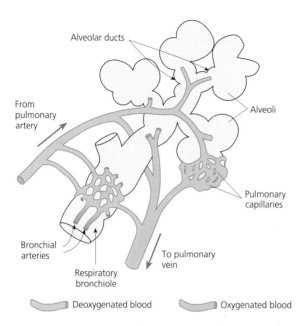

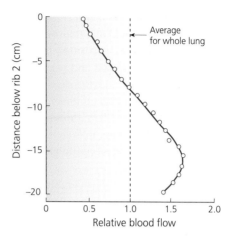

Figure 22.17 The distribution of blood flow in the upright lung. Note the relatively low blood flow at the top of the lungs compared with the lower regions.

Figure 22.16 A schematic drawing showing the arrangement of the pulmonary and bronchial circulations in relation to the alveoli. Note that the bronchial circulation does not supply the alveoli. The blood returning from lower airways drains into the pulmonary veins having bypassed the alveoli. This lowers the saturation of the haemoglobin in the arterial blood by a small amount.

occurs only during systole and not during diastole. The distribution of blood flow in the upright lung is shown in Figure 22.17.

blood flow when the body is upright. Moreover, because the pressures in the pulmonary circulation are low, the pressure of the air in the alveoli has a marked effect on vascular resistance and hence on blood flow.

When the body is upright, the base of the lung lies below the origin of the pulmonary artery and the blood flow is high. Above the origin of the pulmonary artery, the pressure is diminished by the hydrostatic pressure. Consequently, the blood flow falls with distance up the lung until, at the apex of the lung, which is about 15 cm above the origin of the pulmonary artery, blood flow

KEY POINTS:

• The lungs receive their blood supply via the bronchial and the pulmonary circulations.

• The bronchial circulation is part of the systemic circulation and supplies the trachea, bronchi and bronchioles as far as the respiratory bronchioles.

• The pulmonary circulation is supplied by the output of the right ventricle and the blood in this vascular bed participates in gas exchange.

• The pressures in the pulmonary artery are much lower than those in the aorta.

22.13 The matching of pulmonary blood flow to alveolar ventilation

In the upright lung, both ventilation and perfusion fall with height above the base. As the local blood flow falls more rapidly than ventilation with distance up the lung, the ratio of alveolar ventilation to blood flow ($\dot{V}_A/\dot{Q}$ ratio) will vary. As the physiological purpose of ventilating the lung is to promote gas exchange between the blood and the alveolar air, this variation has considerable physiological significance. For the lungs as a whole, the alveolar ventilation is about 4.2 l min^{-1} while the resting cardiac output is about 5.0 l min^{-1} so that the average value for the $\dot{V}_A/\dot{Q}$ ratio is 4.2/5.0 = 0.84. The base of the lung is relatively well perfused and ventilated with a $\dot{V}_A/\dot{Q}$ ratio of about 0.6. The ratio rises slowly with distance from the base (measured by rib number in Figure 22.18). These are average figures for the various segments of the lung but they are not constant. In exercise, for example, the ventilation increases disproportionately more than the pulmonary blood flow.

The $\dot{V}_A/\dot{Q}$ ratio can vary considerably from infinity (ventilated alveoli that are not perfused) to zero (for blood that passes through the lung without coming into contact with the alveolar air). The P_{O_2} of blood leaving poorly ventilated parts of the lungs will not be compensated by well-oxygenated blood leaving relatively over-ventilated parts. This situation arises because the oxygen content of blood from the over-ventilated alveoli is not significantly higher than that from well-matched alveoli while that from poorly ventilated areas will be substantially below normal. Thus, when blood from well-ventilated and poorly ventilated regions becomes mixed in the left side of the heart, *the oxygen content of the two streams of blood are averaged* (*not* their partial pressures). The oxygen content of the blood from each region will depend on the haemoglobin concentration, the partial pressure of oxygen and the shape of the oxygen dissociation curve.

To take a specific example, if equal volumes of blood with a haemoglobin concentration of 15 g dl^{-1} from well-ventilated and poorly ventilated parts of the lungs are mixed when the P_{O_2} of the blood draining the poorly ventilated region is 5.4 kPa and that of the normally ventilated region is 13.3 kPa, the P_{O_2} of the mixed blood will be approximately 7.6 kPa rather than the average of the two partial pressures, which is 9.3 kPa. This is evident from the following calculation:

Each gram of Hb carries 1.34 ml of oxygen when fully saturated. Therefore, for the sample with a P_{O_2} of 13.3 kPa (haemoglobin 97% saturated):

The oxygen bound to haemoglobin is
$$15 \times 1.34 \times 0.97 = 19.5 \, \text{ml dl}^{-1}$$
The amount in solution is 13.3 $\times$ 0.0225 = 0.3 ml
The total O_2 content of the sample is therefore
$$19.5 + 0.3 = 19.8 \, \text{ml dl}^{-1}$$

For the sample with a P_{O_2} of 5.4 kPa (haemoglobin 75% saturated):

The oxygen bound to haemoglobin is
$$15 \times 1.34 \times 0.75 = 15 \, \text{ml dl}^{-1}$$
The amount in solution is 5.4 $\times$ 0.0225 = 0.12 ml
The total O_2 content of the sample is therefore
$$15 + 0.12 = 15.12 \, \text{ml dl}^{-1}$$
The average oxygen content is therefore
$$(15.12 + 19.8) / 2 = 17.5 \, \text{ml dl}^{-1}$$

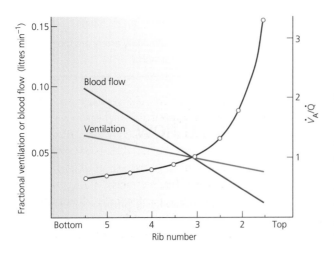

Figure 22.18 The distribution of ventilation, blood flow and the ventilation–perfusion ratio in the normal upright lung for a subject at rest. Straight lines have been drawn through the data for ventilation and blood flow (left vertical axis). Note that the ventilation–perfusion ratio (right vertical axis) rises slowly at first, then rapidly towards the top of the lung. The best matching of blood flow to ventilation occurs at the level of the third rib. Above that level, the ventilation is much more efficient than the perfusion.

A sample with $17.5\,ml\,dl^{-1}$ is approximately 87% saturated and would have a partial pressure of around 7.6 kPa.

The mixing of venous blood with oxygenated blood is known as **venous admixture**. It occurs naturally when blood from the bronchial circulation drains into the pulmonary veins. When venous blood completely bypasses the lungs, it is called a right–left shunt and is commonly seen in congenital heart disease where deoxygenated blood from the right side of the heart mixes with oxygenated blood from the pulmonary veins. As for the case when blood from poorly ventilated alveoli mixes in significant quantities with arterialized blood from well-ventilated alveoli, a right–left shunt will reduce the P_{O_2} and the oxygen content of the blood reaching the systemic circulation.

> **KEY POINTS:**
>
> - Because the systolic and diastolic pressures in the pulmonary arteries are low, the effects of gravity on regional blood flow are very significant. As a result, there is considerable variation in blood flow in the upright lung. The base of the lung is relatively well perfused compared with the apex.
>
> - In the upright lung, both ventilation and perfusion fall with height above the base but the blood flow falls significantly faster than ventilation.

22.14 Gas exchange and diffusion capacity of the lungs

As the inspired air passes through the passages of the lung, it is warmed, moistened and progressively cleared of particles. When it reaches the respiratory bronchioles and alveoli, it equilibrates with the gas in the alveoli by simple diffusion. To be able to oxygenate the blood, a molecule of O_2 must first dissolve in the aqueous layer covering the alveolar epithelium before diffusing across the thin membranes that separate the alveolar air spaces from the blood. Once in the plasma, the oxygen diffuses into the red cells where it binds to haemoglobin before being transported around the body (see Figure 22.7).

When a subject is resting, the blood takes about one second to pass through the pulmonary capillaries but during severe exercise it takes as little as 0.3 seconds. Despite the short time available, in healthy subjects the blood still becomes fully equilibrated with the alveolar air during its transit through the pulmonary capillaries. This ability of the lungs to ensure equilibration between the blood of the pulmonary capillaries and the alveolar air is measured by its **diffusing capacity** (sometimes called its **transfer factor**). Diffusing capacity is impaired if the alveolar membranes become thickened by disease, as in emphysema or fibrosis. If the alveoli become filled with fluid (as in pulmonary oedema), the diffusing capacity is also significantly reduced.

22.15 Fluid exchange in the lungs

In health, the alveoli are dry. What prevents fluid leaking from the pulmonary capillaries into the alveoli? Like other vascular beds, the exchange of fluid between the capillaries and the interstitial fluid is governed by Starling forces (Chapter 19, p. 383). Normally, any filtered fluid is returned to the circulation via the pulmonary lymphatics and the alveoli are free of fluid.

As the capillary pressure in the upper part of the lungs is low, there is little fluid formation in this part of the lung.

However, as the Starling forces favour filtration in the lower regions, a small increase in the pressure within the pulmonary capillaries will lead to a greater filtration of fluid. If this exceeds the drainage capacity of the lymphatics, fluid will accumulate in the interstitium, resulting in **pulmonary oedema**. This may occur during left-sided heart failure and following mechanical or chemical damage to the lining of the alveoli. The fluid first accumulates within the pulmonary interstitium and the lymphatic vessels. Above a critical pressure, fluid will enter the alveoli themselves, flooding them and seriously compromising their ability to participate in gas exchange.

Fluid accumulation in the space between the pleural membranes is known as a **pleural effusion**. Normally only about 10 ml of fluid is present, but in certain diseases (e.g. bronchial carcinoma) more than a litre of fluid may accumulate. While this fluid does not directly affect gas exchange, it will displace air from the lungs and reduce the vital capacity.

22.16 The origin of the respiratory rhythm

No normal person has to think about when, or how deeply, to breathe. Breathing is an automatic, rhythmical process that is constantly adjusted to meet the everyday requirements of life such as exercise and speech. To account for this remarkable fact, it is necessary to consider three important questions:

1. Where does this rhythmical activity originate?

2. How is it generated?

3. How is the rate and depth of respiration controlled?

The basic respiratory rhythm is maintained even if all the afferent nerves from the lungs and chest are cut. If the spinal cord and phrenic nerve are cut, breathing ceases. From these and other experiments it is clear that:

1. The respiratory muscles themselves have no intrinsic rhythmic activity.

2. The medulla has all of the neuronal mechanisms required to generate and maintain a basic respiratory rhythm.

How is the respiratory rhythm generated? There are two groups of neurones in the medulla that discharge action potentials with an intrinsic rhythm that corresponds to that of the respiratory cycle. These are known as the **dorsal respiratory group** and the **ventral respiratory group**. The neurones of the dorsal respiratory group mainly discharge their action potentials just prior to and during inspiration and are therefore mainly inspiratory neurones. The neurones of the ventral respiratory group are both inspiratory and expiratory. They also receive inputs from the dorsal respiratory group. The dorsal and ventral respiratory groups receive a variety of inputs from higher centres in the brain including the cerebral cortex and pons. They also receive inputs from the carotid bodies (which are the peripheral chemoreceptors that sense the P_{O_2}, P_{CO_2} and pH of the arterial blood; see below) and the vagus nerve (which carries afferent nerve fibres from the lungs). It appears that inspiration is initiated by neurones of the dorsal respiratory group. This intrinsic activity subsequently sums with afferent activity coming from lung stretch receptors to switch off inspiration and commence expiration. A simple diagram of the arrangement of the respiratory control pathway is shown in Figure 22.19.

Throughout the respiratory cycle, the respiratory muscles are active. During inspiration, the activity of the inspiratory muscles (the diaphragm and external intercostal muscles) progressively increases, additional motor units are recruited and the muscles shorten progressively, thereby expanding the volume of the chest (see Section 22.7). During expiration, the activity of the inspiratory muscles gradually declines allowing the chest to return to its resting volume. The expiratory muscles show a reciprocal pattern, with increasing activity during expiration and falling activity during inspiration. The progressive modulation of the tone of the respiratory muscles provides a smooth transition from expiration to inspiration.

Voluntary control of respiration

Normal regular breathing (or **eupnoea**) is an automatic process, although the rate and depth of breathing can be readily adjusted by voluntary means. For example, it is possible to suspend breathing for a short period. This breath holding is known as **voluntary apnoea** and its duration is normally limited by the rise in arterial P_{CO_2}. Equally, it is possible to increase the rate and depth of breathing deliberately during **voluntary hyperventilation** (also known as **voluntary hyperpnoea**). The pathways involved in voluntary regulation are not known with any certainty but presumably have their origin in the motor cortex. A fine degree of control over the muscles of respiration is possible. This is important during speech, singing or the playing of wind instruments.

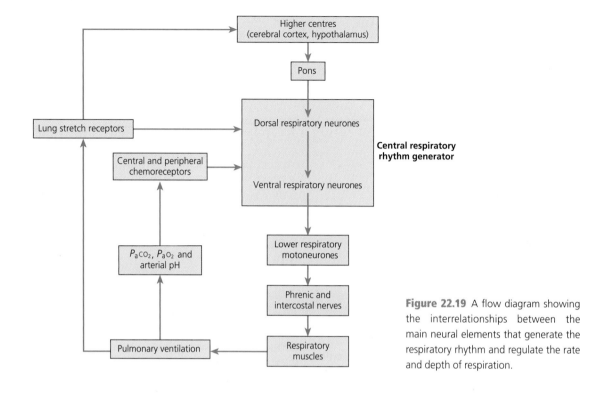

Figure 22.19 A flow diagram showing the interrelationships between the main neural elements that generate the respiratory rhythm and regulate the rate and depth of respiration.

KEY POINTS:

- The diaphragm and intercostal muscles have no inherent rhythmic activity themselves but contract in response to efferent activity in the phrenic and intercostal nerves.
- The basic respiratory rhythm originates in the medulla.

22.17 The chemical regulation of respiration

The purpose of respiration is to provide the tissues with oxygen and to remove the carbon dioxide derived from oxidative metabolism. This is achieved by close regulation of the partial pressures of carbon dioxide and oxygen in the arterial blood (i.e. the P_aCO_2 and P_aO_2), which are maintained within very close limits throughout life. Indeed, the P_aCO_2 and P_aO_2 vary little between deep sleep and severe exercise, where the oxygen consumption and carbon dioxide output of the body may increase more than tenfold. Clearly, to achieve such remarkable stability the body needs some means of sensing the P_aCO_2 and P_aO_2 and

relaying that information to the neurones that determine the rate and depth of ventilation. This role is performed by the peripheral and central chemoreceptors.

The **peripheral arterial chemoreceptors** – the carotid bodies – are small organs about $7 \times 5\,mm$ in size which are located just above the point where the coronary artery branches to give rise to the left and right carotid arteries. The carotid bodies are anatomically and functionally separate from the arterial baroreceptors, which are located in the wall of the carotid sinus (Figure 22.20). Nevertheless, afferent fibres from the carotid body and

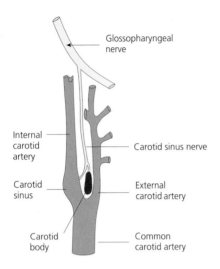

Figure 22.20 A simplified diagram to show the relative positions of the carotid body and carotid sinus on the left side as seen from the front.

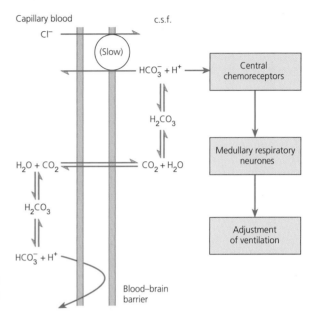

Figure 22.21 A schematic diagram to illustrate how the P_{CO_2} of the capillary blood in the brain stimulates the central chemoreceptors. A rise in plasma carbon dioxide leads to increased carbon dioxide uptake into the brain where it is converted to bicarbonate and hydrogen ions by carbonic anhydrase. The hydrogen ions stimulate the central chemoreceptors and this increases the rate and depth of respiration. A fall in plasma carbon dioxide has the opposite effect.

the carotid sinus run in the same nerve, the carotid sinus nerve, which is a branch of the glossopharyngeal (IX cranial) nerve. The **aortic bodies** are diffuse islets of tissue scattered around the arch of the aorta, which have a similar microscopic structure to the carotid bodies. There is, however, no evidence to suggest that they act as chemoreceptors in man, although they may do so in other species.

The carotid bodies respond to changes in the P_aO_2, P_aCO_2 and pH of the arterial blood and are the only receptors that are able to elicit a ventilatory response to low oxygen partial pressure (hypoxia). Thus, after they have been surgically removed for therapeutic reasons, the ventilatory response to hypoxia is lost – even though the aortic bodies remain intact. When breathing normal room air, the influence of the carotid bodies on the rate of ventilation is small. For example, if a subject suddenly switches from breathing room air to breathing 100% oxygen, the minute volume falls by about 10% for a brief period before returning to its previous level. During hypoxia, however, the carotid bodies play an important role in stimulating ventilation.

The **central chemoreceptors** respond to changes in the pH of the cerebrospinal fluid (CSF) that result from alterations in P_aCO_2. They are located on or close to the ventral surface of the medulla near to the origin of the

glossopharyngeal and vagus nerves, and provide most of the chemical stimulus to respiration under normal resting conditions. The mechanism by which they sense the P_aCO_2 is illustrated in Figure 22.21. Increased P_aCO_2 results in an increase in the P_{CO_2} of the CSF and the hydration reaction for CO_2 is driven to the right, leading to the increased liberation of hydrogen ions:

$$CO_2 + H_2O \rightleftharpoons H_2CO_3 \rightleftharpoons H^+ + HCO_3^-$$

Unlike the blood, the CSF has little protein so the hydrogen ions produced by this reaction are not buffered to any great extent. As a result, the pH falls in proportion to the rise in P_{CO_2} and the rise in hydrogen ion concentration stimulates the chemoreceptors.

The effects of breathing different gas mixtures

When air containing a significant amount of carbon dioxide is inhaled, its partial pressure in the alveoli and

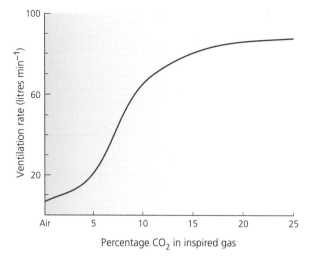

Figure 22.22 The effect of breathing carbon dioxide on ventilation. The figure shows the relationship between the concentration of carbon dioxide in the inspired air and the total ventilation for a normal subject. Note the steep rise in ventilation as the carbon dioxide concentration increases from 5% to 10%.

arterial blood rises. This is known as **hypercapnia**. If a subject deliberately hyperventilates for a brief period, the partial pressure of carbon dioxide in the alveoli and arterial blood falls, as it is lost from the lungs faster than it is being generated in the tissues. This fall in the partial pressure of carbon dioxide is known as **hypocapnia**. If subjects breathe a gas mixture that has a P_{O_2} lower than the normal 21.2 kPa (159 mmHg), their arterial P_{O_2} will fall. This is known as **hypoxaemia**. If the oxygen content is insufficient for the needs of the body, the subject is said to be **hypoxic**. The total absence of oxygen is **anoxia**.

The relative importance of carbon dioxide and oxygen in determining the ventilatory volume is readily investigated by asking subjects to breathe different gas mixtures. If a normal healthy subject breathes a gas mixture containing 21% oxygen, 5% carbon dioxide and 74% nitrogen for a few minutes, their ventilation increases about threefold. A higher fraction of carbon dioxide level in the inhaled gas mixture will stimulate breathing even more. Even a single breath of air containing an elevated carbon dioxide level is sufficient to increase ventilation for a short time. Conversely, if a subject hyperventilates for a brief period, the subsequent ventilation is temporarily decreased. Thus,

any manoeuvre that alters the partial pressure of carbon dioxide in the alveolar air ($P_{A_{CO_2}}$) results in a change in ventilation that tends to restore the $P_{A_{CO_2}}$ to its normal value (5.3 kPa or 40 mmHg). The relationship between the concentration of carbon dioxide in the inspired air and total ventilation is shown in Figure 22.22.

By contrast, if the same subject breathes a mixture of 15% oxygen and 85% nitrogen, there is little change in the rate of ventilation at normal barometric pressure. Indeed, hypoxia only tends to stimulate ventilation strongly when the alveolar P_{O_2} falls below about 8 kPa (60 mmHg) (see Figure 22.23). As the alveolar P_{O_2} falls further, ventilation increases steeply.

From these observations, it appears that the principal chemical stimulus to respiration is the P_{CO_2} of the alveolar air rather than the P_{O_2}. At first, this may appear strange, as the main purpose of gas exchange is to maintain the oxygenation of the tissues. The reason for the relatively small ventilatory effect of mild hypoxia can be understood by looking at the oxyhaemoglobin dissociation curve, which shows that at a P_{O_2} of 8 kPa (60 mmHg) the haemoglobin is still about 90% saturated. Below this value, the percentage saturation rapidly falls. Consequently, at normal atmospheric pressure (101 kPa or 760 mmHg), hypoxia would be a relatively weak stimulus to ventilation.

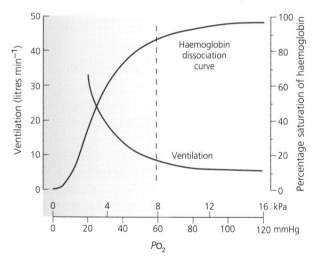

Figure 22.23 The effect of acute hypoxia on pulmonary ventilation compared with the oxyhaemoglobin dissociation curve. The sensitivity of ventilation to the inspired P_{O_2} becomes much steeper below about 8 kPa (60 mmHg), which is the point at which the oxyhaemoglobin dissociation curve also becomes very steep.

22.18   Other respiratory reflexes

The upper airways have slowly adapting stretch receptors. When the lungs are inflated, these receptors send impulses to the dorsal respiratory group of neurones via the vagus nerves. This afferent information inhibits respiratory activity and so acts to limit inspiration. This is known as the **Hering–Breuer lung inflation reflex**. If the lungs are inflated by positive pressure, the frequency of respiratory movements falls and may cease altogether (**apnoea**). In man, the Hering–Breuer reflex is not activated at normal tidal volumes. It is, however, activated when tidal volumes exceed about 0.8–1 litres. For this reason, it is thought that the Hering–Breuer reflex may play a role in regulating inspiration during exercise.

Cough and sneeze. In addition to their stretch receptors, the airways possess receptors that respond to irritants. Activation of these receptors elicits a range of reflexes. When the irritant receptors of the upper airways are stimulated, they elicit a cough or, in the case of irritants on the nasal mucosa, a sneeze. The initial phase of either response is a deep inspiration followed by a forced expiration against a closed glottis. As the pressure in the airways rises, the glottis suddenly opens and the trapped air is expelled at high speed. This dislodges some of the mucus covering the epithelium of the airways and helps to carry the irritant away with it via the mouth or nose.

Swallowing. During swallowing, respiration is inhibited. This is part of a complex reflex pattern: as food or drink passes into the oropharynx, the nasopharynx is closed by the upward movement of the soft palate and the contraction of the upper pharyngeal muscles. Respiration is inhibited at the same time and the laryngeal muscles contract, closing the glottis. This prevents aspiration of food into the airways.

Pulmonary chemoreflex. Inhalation of smoke and noxious gases, such as sulphur dioxide and ammonia, stimulates irritant receptors (also known as rapidly adapting receptors) within the tracheobronchial tree and elicits a powerful pulmonary chemoreflex in which there is a constriction of the larynx and bronchi, and an increase in mucus secretion. If the lungs become congested, breathing becomes shallow and rapid (called **pulmonary tachypnoea**). The receptors that mediate this response are C-fibre endings located in the interstitial space of the alveolar walls. These receptors were previously known as J-receptors (for juxtapulmonary capillary receptors). The role they play in normal breathing, if any, is not known.

Other reflex modulations of respiration. The normal pattern of breathing is modified by many other factors. For example, passive movement of the limbs results in an increase in ventilation that is believed to occur as a result of stimulation of proprioceptors in the muscles and joints. This reflex may play an important role in the increase in ventilation during exercise (see pp. 689–691). Pain results in alterations to the normal pattern of respiration. Abdominal pain (e.g.

postoperative pain) can be so sharp that it causes a reflex inhibition of inspiration and apnoea. Prolonged, severe pain is associated with fast, shallow breathing. Immersion of the face in cold water elicits the **diving response** in which there is apnoea, bradycardia and peripheral vasoconstriction.

> **KEY POINT:**
>
> - A number of reflexes directly influence the pattern of breathing. These include the cough reflex, the Hering–Breuer lung inflation reflex and swallowing.

22.19 Pulmonary defence mechanisms

As all city dwellers know, the air we breathe is full of particulate matter, some of which is inhaled with each breath. A respiratory minute volume of $6 \, \mathrm{l\,min^{-1}}$ results in the intake of over 8500 litres of air each day. If the concentration of particles in the air is only 0.001% (10 parts per million), this volume of air would include 85 ml of particulate matter. Clearly, unless some mechanism existed for the removal of this material, our lungs would rapidly become clogged with dust and debris. Moreover, not all of the inhaled material is biologically inert. Some of it will be infectious agents (bacteria and viral particles) and some will be allergenic (e.g. pollen). The lungs therefore need to remove the inert material and inactivate the infectious and allergenic agents.

As mentioned earlier, the airflow through the nose and upper airways is rapid and turbulent. As a result, large particles (>10–15 μm) are brought into contact with the mucus lining these passages and are entrapped by it. This results in filtration of the air and removal of most of the large particles before they reach the trachea. In addition to filtering the incoming air, the upper airways also warm and moisten it. As in the upper airways, the airflow in the trachea and bronchi is turbulent and this turbulence brings the incoming air stream into contact with the wall of the airways. As a result, most of the remaining large particles (5–10 μm) become lodged in the mucous lining of the upper respiratory tree, as shown in Figure 22.5. Further down the airways, the airflow becomes slow and laminar. In these regions of the lung, smaller particles (0.2–5 μm) settle on the walls of the airways under the influence of gravity. Only the smallest particles reach the alveoli where most remain suspended as aerosols and are subsequently exhaled. Nevertheless, about one-fifth of these small particles become deposited in the alveolar ducts or in the alveoli themselves where they are engulfed by the alveolar macrophages.

As described earlier (p. 436), the respiratory tract from the upper airways to the terminal bronchioles is lined with a ciliated epithelium covered by a layer of mucus. The cilia beat steadily towards the pharynx and this slowly moves the mucus upwards. This is known as the **mucociliary escalator**. As it reaches the pharynx, the mucus is swallowed or coughed up and expectorated. The mucociliary escalator is very efficient in removing those inhaled particles that became trapped in the mucous layer during the passage of air to the alveoli. By using small particles labelled with a radioactive tracer, it has been shown that most of the trapped particles are removed within 24 hours.

> **KEY POINTS:**
>
> - Particulate matter entering the airways becomes lodged in the mucous lining of the respiratory tree. Most of this material is removed by the mucociliary escalator.
>
> - Material that becomes deposited in the alveolar ducts or in the alveoli is ingested by alveolar macrophages.

22.20 Tests of ventilatory function

In the diagnosis and treatment of respiratory diseases, assessment of pulmonary function is of considerable importance. Key tests of ventilatory function are:

- The vital capacity
- The forced vital capacity
- The peak expiratory flow rate
- Maximal ventilatory volume.

To measure the **vital capacity**, a subject is asked to make a maximal inspiration and then to breathe out as much air as possible. The volume of air exhaled is measured by spirometry, as described earlier. Note that, in this test, the time taken to expel the air is not taken into account and it is usual to estimate the vital capacity during expiration, rather than inspiration. The normal values depend on age, sex and height but, for a healthy adult male of average height and 30 years of age, vital capacity is about 5 litres. For women of the same age, the vital capacity is about 3.5 litres.

In the **forced vital capacity** (**FVC**) test, the subject is asked to make a maximal inspiration and then to breathe out fully as fast as possible. The air forced from the lungs is measured as a function of time (with an instrument called a **pneumotachograph**). After the final quantity of air is forced from the lungs, only the residual volume (RV) is

left. In healthy young subjects, around 85% of the vital capacity is forced from the lungs within the first second. This is known as the **forced expiratory volume** at 1 second or **FEV_1**. The remaining volume is expired over the next few seconds (Figure 22.24). The FEV_1 declines with increasing age. Even so, a healthy 60-year-old man should have a value of around 70%. By way of contrast, a person with an obstruction of the airways (e.g. during an asthmatic attack) would have a much lower FEV_1. In severe cases, the FEV_1 can be less than 40%. The vital capacity and FVC test are also useful in the diagnosis of restrictive lung diseases such as fibrosis of the lung. In restrictive disorders, the ability of the lung to expand normally is compromised. As a result, the FEV_1 may be normal but the vital capacity will be much reduced.

The **peak expiratory flow rate** (also known as the **maximal expiratory flow**) is also used to distinguish between obstructive and restrictive diseases. The peak airflow (in litres per second) is measured with a pneumotachograph or a Wright peak flow meter during a forced expiration following a full inspiration. The maximum flow rate is normally reached in the first tenth of a second of the forced expiration and is measured in litres per second. Healthy young adults are able to achieve flow rates of $8-10 \text{ls}^{-1}$. As with the FEV_1, obstructive airway disease results in a reduced peak flow (Figure 22.25).

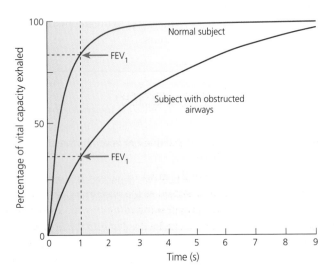

Figure 22.24 Forced vital capacity test for a normal subject and for a subject with obstructed airways. Note the marked difference in the FEV_1 values and the time required for the subject with obstructed airways to expel the air taken in during inspiration to total lung capacity.

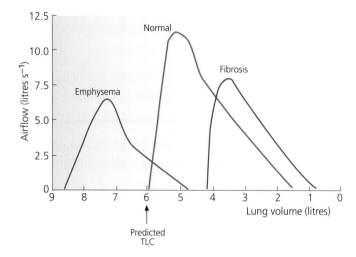

Figure 22.25 The relationship between maximal airflow and lung volume for a normal subject and subjects of the same stature suffering from emphysema (red curve) and pulmonary fibrosis (purple curve). For all three subjects, the predicted total lung capacity (TLC) is approximately 6 litres and residual volume is around 1.5 litres as shown by the normal blue curve. In emphysema, total lung capacity is nearly 9 litres and residual volume is much higher than normal at about 4.8 litres. In the subject suffering from fibrosis, total lung capacity and residual volume are smaller than normal.

The maximum minute volume attainable by voluntary hyperventilation is known as the **maximal ventilatory volume** (**MVV**) or **maximum breathing capacity** (**MBC**). The subject is asked to breathe in and out as fast and as deeply as possible through a low-resistance circuit for 15–30 s. This test involves the whole respiratory system during inspiration and expiration. As for other respiratory variables, the MVV varies with the age and sex of the subject. Healthy young men of 20 years of age can attain a MVV equal to 150 l min^{-1}. By the age of 60 years, however, the MVV for normal men has fallen by about one-third to about 100 l min^{-1}. For women, the equivalent values are 100 l min^{-1} at age 20 falling to about 75 l min^{-1} by age 60. The MVV is dependent on airways resistance, the compliance of the lungs and chest wall and on the activity of the muscles of respiration. As a result, it is a sensitive measure of ventilatory function. It is profoundly reduced in patients with obstructed airways (e.g. asthma) and in those with decreased compliance (e.g. pulmonary fibrosis).

> **KEY POINT:**
>
> - Clinical assessment of ventilatory function can be made using a variety of tests. These include measurement of the vital capacity, the FEV$_1$, peak expiratory flow rate and the maximal ventilatory volume.

22.21 Some common disorders of respiration

Normally, respiration continues unnoticed and uneventfully. It is only when things go wrong that we become aware of our breathing. Difficulty in breathing causes distress known as **dyspnoea**. It is a subjective phenomenon during which a patient may report being breathless. It may be quite normal – as in a subject who has rapidly climbed to a high altitude (where the atmospheric P_{O_2} is low) or it may reflect some organic disease. In both cases, it is the sense of breathlessness that limits the ability to undertake exercise.

The ability of the lungs to provide enough air for the body's needs is known as the **ventilatory capacity**. If this is less than normal, some form of respiratory disorder is present. These can be divided into those in which the airways are obstructed, those in which the expansion of the lungs is restricted and those in which the respiratory muscles are weakened and unable to expand the chest fully. Some of the commonly occurring respiratory disorders are discussed here.

Asthma is a condition in which a person has difficulty in breathing, particularly expiration. Asthmatic attacks are characterized by the sudden onset of dyspnoea. This is the result of bronchospasm, which usually occurs in response to an allergen that is present in the environment, although it may also occur in response to exercise or to cold. The spasm of the smooth muscle of the bronchi leads to a reduction in the diameter of the airways. During an asthmatic attack, the FRC and RV are increased although vital capacity is normal. The FEV_1 and peak flow are markedly reduced, in severe cases by more than half (Figure 22.24). This limits the ventilatory capacity and results in the dyspnoea. In chronic asthma, there is destruction of the bronchial epithelium, and FRC and RV become permanently elevated.

Emphysema is a condition in which the alveoli are increased in size due to destruction of the lung parenchyma. It is frequently (but not only) caused by cigarette smoking. The lungs of smokers become invaded by neutrophils, which secrete proteolytic enzymes that damage the parenchyma of the lungs. In addition, the smoke inhibits the movement of the bronchial cilia, slowing the removal of particulate matter from the airways. The irritant effect of the smoke is the probable cause of the increased secretion of mucus in the larger airways. These effects combine to increase the chances of infection, which results in chronic inflammation of the bronchiolar epithelium. As a result, the diameter of the airways is reduced and, as in asthma, it becomes difficult to exhale, leading to the entrapment of air (from which the disease takes its name). As a result of the loss of parenchymal tissue, the traction on the airways is reduced and their resistance is increased.

In **pulmonary fibrosis**, the alveolar wall becomes thickened and this decreases the diffusing capacity of the lung. The diffusing capacity is also reduced in **pneumonia**. In this case, bacterial infection leads to fluid accumulation in the alveoli. In severe cases, a pulmonary lobe may become filled with fluid containing bacterial toxins ('consolidated'). In this situation, although there is a local increase in blood flow, there is no significant oxygenation of the blood. As a result, there may be a significant reduction in the amount of oxygen carried in the blood leaving the lungs (hypoxic hypoxia; see below).

Cystic fibrosis is the most common inherited disorder of people of European ancestry. It is a recessive disorder caused by a mutation of the gene encodes an epithelial chloride channel known as CFTR (cystic fibrosis transmembrane conductance regulator). When the gene is defective, the epithelial chloride channel of the submucosal glands in the upper airways fails to open normally in response to cyclic AMP. This results in a decrease in fluid secretion so that the mucus of the airways becomes abnormally thick and difficult to dislodge. Consequently, the normal clearance of mucus via the mucociliary escalator is reduced, resulting in obstruction of the small airways and frequent bronchial infections. The disease also affects the secretions of the gastrointestinal tract and sweat glands. It is normally diagnosed by the sweat test – this measures the amount of sodium and chloride in skin sweat, which is abnormally high in those with the affected gene.

Sleep apnoea describes a number of conditions in which breathing temporarily stops. Two principal types can be recognized: obstructive sleep apnoea and central sleep apnoea. **Obstructive sleep apnoea** results from physical obstruction of the upper airway. During deep sleep, the muscles of the mouth, larynx and pharynx relax. As a result, there is a tendency for the pharynx to collapse, so obstructing the airway. Periods of increasing obstruction are accompanied by loud snoring followed by a period of silence when obstruction is complete. $P_{a}O_2$ falls and $P_{a}CO_2$ rises – both stimulating respiratory effort. Breathing movements then increase in intensity until the intrathoracic pressure overcomes the obstruction and ventilation resumes – often with a loud snore. **Central sleep apnoea** occurs when the respiratory drive is not able to initiate breathing movements. During the apneustic phase, the $P_{a}CO_2$ rises and the patient wakes. Once awake, the patient resumes breathing, their $P_{a}CO_2$ returns to normal and they fall asleep once again. If this cycle is repeated many times during the night, sleep may become severely disturbed.

Periodic breathing: the importance of the chemical control of breathing is revealed by a pattern known as **Cheyne–Stokes breathing**. This consists of a periodic change in the frequency and tidal volume. The pattern

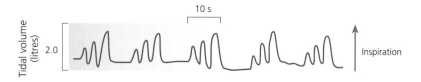

Figure 22.26 A record of respiratory movements to illustrate the pattern of Cheyne–Stokes breathing. In this example, the record was obtained from a healthy subject who had recently moved to a high altitude.

of breathing alternates between apnoea and mild hyperventilation (Figure 22.26). Breathing begins with slow, shallow breaths; the frequency and tidal volume gradually increase to a maximum before slowly subsiding into apnoea. This cyclical pattern is seen in patients who are terminally ill or who have suffered brain damage. It is sometimes seen in normal people during sleep, especially at high altitude.

22.22 Insufficient oxygen supply to the tissues – hypoxia and its causes

The respiratory system is principally concerned with gas exchange – with the uptake of oxygen from the air and the elimination of carbon dioxide from the pulmonary blood. If the oxygen content of the blood is reduced, there may be insufficient oxygen to support the aerobic metabolism of the tissues. This condition is known as **hypoxia**. There are four principal types of hypoxia: hypoxic hypoxia, anaemic hypoxia, stagnant hypoxia and histotoxic hypoxia. Each will be discussed in turn.

Hypoxic hypoxia. This refers to hypoxia that results from a low arterial P_{O_2}. There are a number of causes but in each case there is a lowering of the oxygen content of the systemic arterial blood. If the alveolar P_{O_2} is low, the arterial P_{O_2} will inevitably follow and so will the oxygen content. This is normal at high altitude. Other causes include reduced ventilation (e.g. as a result of a drug overdose); severe weakness of the muscles that support respiration as in poliomyelitis and myasthenia gravis; airway obstruction; and ventilation–perfusion inequality.

Anaemic hypoxia is caused by a decrease in the amount of haemoglobin available for binding oxygen so that the oxygen content of the arterial blood is abnormally low. It may be due to blood loss, reduced red cell production or to the synthesis of abnormal haemoglobin because of a genetic defect.

Stagnant hypoxia is the result of a low blood flow. It may occur peripherally due to local vasoconstriction (e.g. exposure of the extremities to cold) or it may result from a reduced cardiac output. As the blood flow through the metabolizing tissues is very slow, excessive extraction of the available oxygen occurs and venous P_{O_2} is very low. This gives rise to peripheral cyanosis (blue lips and mucous membranes).

Histotoxic hypoxia refers to poisoning of the oxidative enzymes of the cells, by cyanide for example. In this situation, the supply of oxygen to the tissues is normal but they are unable to make full use of it.

Oxygen therapy in hypoxia

The administration of a high partial pressure of oxygen can be beneficial in the treatment of hypoxic hypoxia. By increasing the partial pressure of oxygen in the alveoli, the oxygen content of the blood leaving the lungs is raised. This will lessen any central cyanosis and alleviate any dyspnoea. Oxygen therapy will be of less value in other forms of hypoxia.

Note that newborn infants should not be exposed to partial pressures of oxygen greater than about 40 kPa (*c.* 300 mmHg) as they are particularly sensitive to its toxic effects (see Chapter 30 of Pocock and Richards (2006), *Human Physiology: The Basis of Medicine*).

Respiratory failure

The term respiratory failure applies to the situation in which the respiratory system is unable to maintain normal values of arterial P_{O_2} and P_{CO_2}.

In **type I respiratory failure**, P_aO_2 is low while P_aCO_2 is normal or low. This occurs when there is a major right to left shunt of deoxygenated blood, or when the $\dot{V}_A/\dot{Q}$ ratio is abnormal. Such a situation may arise during pneumonia, pulmonary oedema and during adult respiratory distress syndrome.

In **type II respiratory failure** P_aO_2 is low while P_aCO_2 is elevated. This situation occurs when alveolar ventilation is not sufficient to excrete the carbon dioxide produced by the normal metabolism of the body. This is known as **ventilatory failure**, which may be caused by many different factors (see below). The commonest cause is chronic obstructive pulmonary disease.

> **KEY POINTS:**
>
> - Hypoxia is a condition in which the metabolic demand for oxygen cannot be met by the circulating blood.
>
> - Four principal types of hypoxia are recognized: hypoxic hypoxia, anaemic hypoxia, stagnant hypoxia and histotoxic hypoxia. Hypoxic hypoxia can be treated by administration of oxygen but this is of less value in other forms of hypoxia.

22.23   Drugs used to treat disorders of the respiratory system

Asthma

Three factors contribute to the increase in airways resistance seen in bronchial asthma. These are:

- Excessive constriction of the bronchi (bronchoconstriction)
- Bronchial oedema (swelling of the bronchial mucosa)
- Secretion of excessive amounts of mucus by the bronchi.

The drugs used to treat the condition act to reduce one or more of these factors. The best-known drugs for treating asthma are β_2 **adrenoceptor agonists**, such as **salbutamol** and **terbutaline**. They act directly on the bronchial smooth muscle to relax it and have little effect on the heart. **Salmeterol** and **formoteral** are longer-acting drugs with a similar mode of action. Atropine-derived **anticholinergic drugs**, such as **ipratropium** and **oxitropium**, inhibit the bronchoconstricting action of vagal nerves and have the additional benefit of reducing mucus secretion. Inflammation of the bronchi can be treated by **corticosteroids**, such as **beclomethazone** and **budesonide**. Other drugs that have been used are **xanthene derivatives** (e.g. theophilline) and cromones (e.g. sodium cromoglicate). Most of these drugs can be given by inhalation, which permits their delivery directly to the airways. This mode of administration helps to confine their action to the lungs and avoid the involvement of other organ systems, particularly the heart. Of those listed, only the xanthenes cannot be administered in this way.

Chronic obstructive lung disease and lung infections

In chronic obstructive lung disease (COLD – also known as chronic obstructive pulmonary disease or COPD), the principal aims of treatment are to prevent infections and relieve bronchoconstriction and mucus accumulation in the airways. To minimize the risk of encouraging the emergence of resistant bacteria, antibiotics are only given when necessary and bronchoconstriction is treated as in asthma, ipratropium being particularly helpful in such cases. Pulmonary infections such as pneumonia and tuberculosis are treated with antibiotics. Pneumonia may be caused by a number of bacteria or it may be of viral origin. For pneumonia of bacterial origin, the choice of antibiotic should be guided by the microbiology. A wide range of antibiotics may be used including drugs derived from penicillin (**benzylpenicillin**, **ampicillin**, **amoxicillin**), **erythromycin** and **gentamicin**. Tuberculosis is generally treated with a combination of **isoniazid** and **rifampicin** (also known as rifampin).

Cough

The cough reflex is a normal response to irritants in the airways but it may become so persistent and unproductive that a cough suppressant is called for. For a dry unproductive cough, **codeine phosphate** may be helpful. In patients with a terminal illness and an intractable dry cough (e.g. those with a bronchial carcinoma), the cough suppressant action of more powerful opiates such as **morphine** are helpful. Expectorants are of dubious clinical effectiveness, steam inhalation being as effective.

Recommended reading

Anatomy

MacKinnon, P.C.B. and Morris, J.F. (2005). *Oxford Textbook of Functional Anatomy*, 2nd edn, Vol. 2 . *Thorax and Abdomen*, 2nd edn. Oxford University Press: Oxford. pp. 35–63.

Histology

Junquieira, L.C. and Carneiro, J. (2005). *Basic Histology*, 11th edn. McGraw-Hill: New York. Chapters 12 and 13.

Pharmacology of the respiratory system

Grahame-Smith, D.G. and Aronson, J.K. (2002). *Clinical Pharmacology and Drug Therapy*, 3rd edn. Oxford University Press: Oxford. Chapter 24.

Rang, H.P., Dale, M.M., Ritter, J.M. and Flower, R. (2007). *Pharmacology*, 6th edn. Churchill-Livingstone: Edinburgh. Chapter 22.

Physiology of the respiratory system

Hlastala, M.P. and Berger, A.J. (2001). *Physiology of Respiration*, 2nd edn. Oxford University Press: New York.

Levitzky, M.G. (2003). *Pulmonary Physiology*, 6th edn. McGraw-Hill: New York.

Pocock, G. and Richards, C.D. (2006). *Human Physiology: The Basis of Medicine*, 3rd edn. Oxford University Press: Oxford. Chapter 16.

Self-assessment questions

In Questions 1 to 6, which of the statements are true and which are false? Answers are given at the end of the book (p.755).

1 The following relate to the circulation through the lungs:

a) All the output from the right side of the heart passes through the lungs.

b) The pressures in the pulmonary arteries are similar to those in the systemic arteries.

c) The distribution of pulmonary blood flow is dependent on posture.

d) In an upright man, pulmonary blood flow is greatest at the base of lung.

2 The following relate to the airways and alveoli:

a) The bronchioles are prevented from collapsing by the cartilage in their walls.

b) The upper airways play an important role in protecting the lungs from airborne particles.

c) The airways resistance is greatest in the terminal bronchioles.

d) The alveoli are the only site of gas exchange.

e) In a healthy lung, the distance between the alveolar air and the blood in the pulmonary capillaries is less than 1 μm.

3 In a normal healthy individual with a total lung capacity of 6 litres:

a) The tidal volume at rest is about 0.6 litres.

b) The vital capacity is equal to the total lung capacity.

c) The functional residual capacity is about 4 litres.

d) The expiratory reserve volume at rest would be about 1 litre.

e) The FEV_1 would be about 3.5 litres.

4 The following relate to the mechanics of ventilation:

a) The compliance is the change in the volume of the lungs with pressure.

b) The total compliance of the chest is equal to the compliance of the lungs.

c) The recoil of the lungs assists inspiration.

d) The compliance of the lungs is determined by surface tension forces in the alveoli and by the elastic tissues of the lung parenchyma.

e) Pulmonary surfactant reduces the work of breathing.

5 The following relate to ventilation and gas exchange:

a) The physiological dead space is the volume of air taken in during a breath that does not enter the alveoli.

b) The physiological dead space is always greater than the anatomical dead space.

c) In an upright lung, the alveolar ventilation is higher at the base than it is at the apex.

d) The $\dot{V}_A/\dot{Q}$ ratio for a subject in a sitting position increases with distance above the base of the lung.

6 The following are concerned with the control of breathing:

a) The principal muscle of respiration is the diaphragm.

b) The depth and rate of ventilation is increased when a subject breathes air containing 5% carbon dioxide.

c) The central chemoreceptors sense the carbon dioxide tension of the arterial blood.

d) The peripheral chemoreceptors are located in the carotid sinus and aortic arch.

e) The peripheral chemoreceptors only respond to changes in the partial pressure of oxygen in the arterial blood.

Quantitative problems:

7 A sample of gas taken at room temperature of 20°C when the atmospheric pressure was 765 mmHg has a volume of 11.5 litres. What is its volume at STP (i.e. at 273 K and 760 mmHg)?

8 A subject breathes out 25 litres of air in 5 minutes, which contains 15.8% oxygen and 4.4% carbon dioxide. What is the respiratory exchange ratio? (Oxygen in the atmospheric air is 20.9%; there is a negligible amount of carbon dioxide.)

9 In an estimate of residual volume, a subject breathes an air–helium mixture from a spirometer. The spirometer has a volume of 6 litres and the concentration of helium in the air in the spirometer before the subject breathes the gas mixture is 4.5%. After the subject has breathed from the spirometer, the fractional concentration of helium has fallen to 3.8%. What is the functional residual volume?

10 What is the minute volume of a subject taking 12 breaths a minute with a tidal volume of 550 ml?

11 If the percentage of carbon dioxide in the expired and alveolar air was 3.3% and 5.25% and the tidal volume was 450 ml, what is the physiological dead space for this subject?

12 Using the data from Question 11, calculate the alveolar ventilation of the subject if they are taking 15 breaths a minute with a tidal volume of 520 ml.

23 The kidney and the urinary tract

After reading this chapter you should have gained an understanding of:

- The anatomy of the kidney and the urinary tract
- The structure of the nephron and the organization of its blood supply
- The regulation of renal blood flow
- The formation of the urine: filtration, secretion and absorption
- The concept of renal clearance
- The role of the distal tubule in the regulation of the ionic balance of the body
- How antidiuretic hormone (ADH) regulates plasma osmolality
- Bladder function
- Drugs that act on the kidneys

23.1 Introduction

All animals feed on other organisms, both to provide material for tissue growth and maintenance, and to provide themselves with the energy for other activities. As with other animals, humans take in varying amounts of many different substances in their diet. However, to maintain normal body function they also need to keep the composition of the extracellular fluid stable. The kidneys play an essential role in this process by excreting those substances that are present in excess while conserving those that are scarce. The kidneys also excrete the bulk of non-volatile waste products.

As well as their primary regulatory and excretory roles, the kidneys also produce the hormone **erythropoietin**, which regulates the production of red blood cells (see Chapter 20), and the enzyme **renin**, which is important in the regulation of sodium balance via aldosterone secretion. The kidneys synthesize **1,25-dihydroxycholecalciferol** (also called calcitriol) from vitamin D, which stimulates the absorption of calcium from the gut and promotes the calcification of bone. Finally, together with the liver, the kidneys synthesize glucose from amino acids during fasting (gluconeogenesis; see Chapter 4, p. 52).

23.2 The anatomy of the kidney and urinary tract

The kidneys lie high in the abdomen on its posterior wall, either side of the vertebral column. They form the upper part of the urinary tract and the urine they produce is delivered to the bladder by a pair of thin muscular tubes called the **ureters**. The urine continuously accumulates in the bladder, which periodically empties its contents via the **urethra** under the control of an external sphincter – a process known as **micturition**.

To ensure efficient regulation of the internal environment, the kidneys receive around a quarter of the resting cardiac output via the renal arteries. These arise directly from the abdominal aorta. Blood leaves the kidneys via the renal veins, which drain directly into the inferior vena cava. The kidneys and lower urinary tract are supplied by nerves from both divisions of the autonomic nervous system (see below). The arrangement of the principal structures of the urinary tract is shown in Figure 23.1.

In adult humans, each kidney is about 10 cm long and weighs about 140 g. A simple diagram of the gross structure of the kidney is shown in Figure 23.2. Facing the midline of the body is an indentation called the **hilus** through which the renal artery enters and the renal vein and ureter leave. If a kidney is cut open, two regions are easily recognized – a dark brown **cortex** and a paler inner region, which consists of the **renal pelvis** and the

renal **medulla**. In humans, the medulla is located in large, conical masses known as the **renal pyramids**. The apex or **papilla** of each pyramid lies in the central space of the renal pelvis, which collects the urine prior to its passage to the bladder via the ureter. The central space itself is divided into areas known as the **renal calyxes** (or **calyces**).

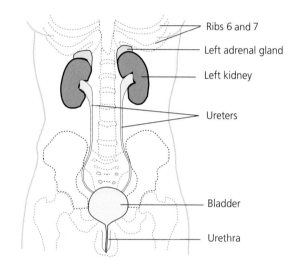

Figure 23.1 labels:
Ribs 6 and 7
Left adrenal gland
Left kidney
Ureters
Bladder
Urethra

Figure 23.1 The principal structures of the male renal system and their position relative to the chest and pelvis.

23.3 The structure of the nephron

Each human kidney contains about 1.25 million **nephrons**, which are the functional units of the kidney. Each nephron originates in the renal cortex and consists of a **renal corpuscle** attached to a long, thin tube as shown in Figures 23.2 and 23.3. The renal corpuscle consists of a capsule of epithelial cells (**Bowman's capsule**) that envelops a tuft of capillaries known as a **glomerulus**. The glomerular capillaries originate from an afferent arteriole and

recombine to form an efferent arteriole (Figure 23.4). They are separated from the fluid-filled space of Bowman's capsule by an epithelial layer consisting of cells known as podocytes. Between the glomerular capillaries are clusters of phagocytes, called mesangeal cells.

The first segment of the tubule is called the **proximal tubule**. It arises directly from Bowman's capsule. The epithelial cells of the proximal tubule are cuboidal

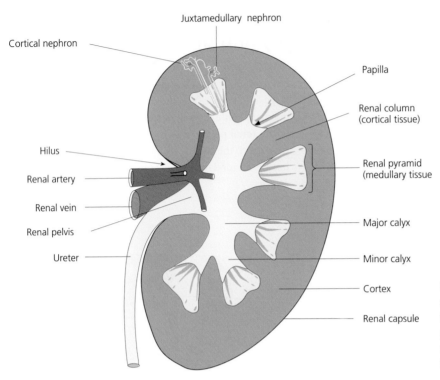

Cortical nephron

Juxtamedullary nephron

Papilla

Renal column
(cortical tissue)

Hilus

Renal artery

Renal vein

Renal pelvis

Ureter

Renal pyramid
(medullary tissue

Major calyx

Minor calyx

Cortex

Renal capsule

Figure 23.2 Diagrammatic representation of the internal organization of the left kidney showing the principal anatomical features. Two nephrons are shown in relation to the renal cortex and medulla.

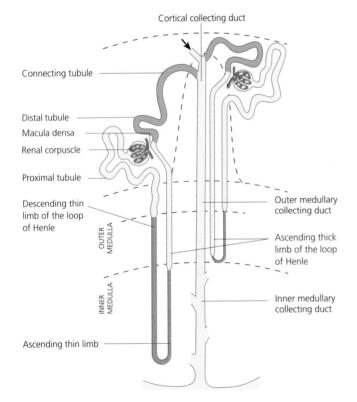

Cortical collecting duct

Connecting tubule

Distal tubule

Macula densa

Renal corpuscle

Proximal tubule

Descending thin limb of the loop of Henle

OUTER MEDULLA

INNER MEDULLA

Ascending thin limb

Outer medullary collecting duct

Ascending thick limb of the loop of Henle

Inner medullary collecting duct

Figure 23.3 A diagram of a short-looped (cortical) and a long-looped (juxtamedullary) nephron to show their basic organization. Note that the early distal tubule of each type of nephron is in contact with the afferent arteriole of its own glomerulus.

Macula densa   Capillary tuft   Bowman's capsule   Proximal tubule

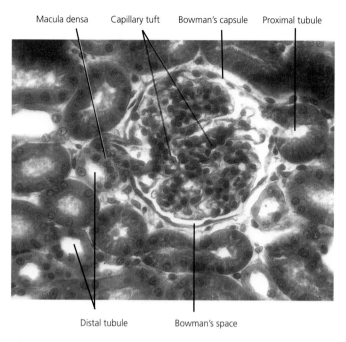

Distal tubule   Bowman's space

(a)

Macula densa cells

Distal tubule

Afferent arteriole

Efferent arteriole

Renal nerves

Granular cells

Mesangeal cells

Capillary tuft

Podocytes

Basement membrane

Bowman's capsule

Bowman's space

Capillary endothelial cells

Proximal tubule

(b)

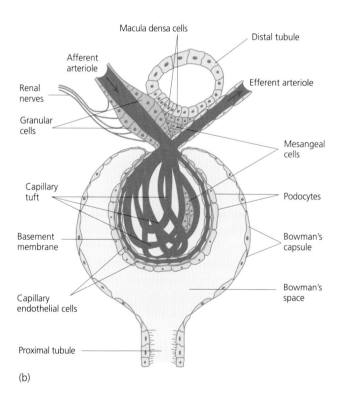

Figure 23.4 The principal features of a renal glomerulus and the juxtaglomerular apparatus. (a) Cross-section of a renal corpuscle. The plane of the section fortuitously shows both the origin of the proximal tubule and, at the opposite end, the macula densa of the distal tubule. (b) A diagrammatic representation of a renal corpuscle. The wall of the afferent arteriole is shown as thickened close to the point of contact with the distal tubule where the juxta-glomerular cells are located. These granular cells secrete the enzyme renin in response to low sodium in the distal tubule.

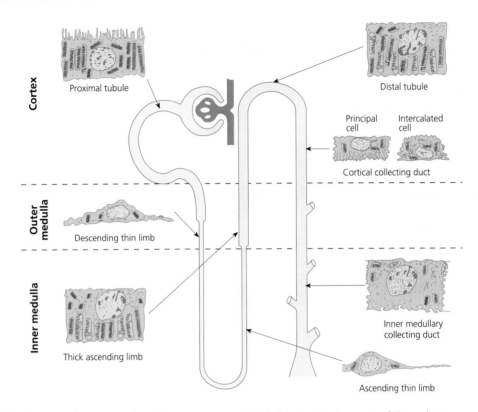

Figure 23.5 A diagrammatic representation of the appearance of the tubular cells at various parts of the nephron.

in appearance and are rich in mitochondria. They are closely fused with one another via tight junctions near their apical surface, which is densely covered by microvilli giving rise to a prominent brush border (Figure 23.5).

As the proximal tubule passes deeper into the kidney, it enters the medulla. Here, the character of the tubule wall changes: the epithelial cells are thin, flattened and have few mitochondria. This segment is known as the descending thin limb of the **loop of Henle** (Figure 23.3). The tubule turns back towards the renal cortex forming the ascending thin limb, which merges with the ascending thick limb of the loop of Henle. The cells of the thick segment are cuboidal in appearance and, like those of the proximal tubule, are rich in mitochondria. The **distal tubule** arises from the thick limb of the loop of Henle and its wall is similar in appearance.

The distal tubule contacts the afferent arteriole close to the glomerulus from which the tubule originated. Here, the tubular epithelium is modified to form the **macula densa** and the wall of the afferent arteriole is thicker due to the presence of **juxtaglomerular** or **granular cells**. These are modified smooth muscle cells that contain secretory granules. The juxtaglomerular cells, macula densa and associated cells form the **juxtaglomerular apparatus** (Figure 23.4), which plays an important role in the regulation of the body's sodium balance.

The distal tubules of a number of nephrons merge to form **collecting ducts**, which pass through the cortex and medulla to the renal pelvis. The epithelium of the collecting ducts consists of two cell types, principal cells or P cells, which play an important role in the regulation of sodium balance, and intercalated cells or I cells, which are important in regulating acid–base balance.

23.4 The organization of the renal circulation

The renal artery enters the hilus and branches to supply the different parts of the kidney. Arcuate arteries course around the outer part of the medulla and give rise to cortical radial arteries, which ascend towards the renal capsule. These arteries give rise to the **afferent arterioles** of the renal corpuscle, as shown in Figure 23.4. Each afferent arteriole gives rise to a network of capillaries (the glomerular tuft) within Bowman's capsule. The capillaries recombine to form the **efferent arterioles**. In the outer cortex, the efferent arterioles give rise to a rich supply of capillaries that covers the renal tubules (the **peritubular capillaries**; Figure 23.6). Blood from the peritubular capillaries first drains into **stellate veins** and thence into the cortical radial veins and arcuate veins. In contrast, the efferent arterioles of nephrons close to the medulla (the juxtamedullary nephrons) give rise to a series of straight vessels that pass into the medulla. These are known as the descending vasa recta (from the Latin for straight vessel), which provide the blood supply of the outer and inner medullary regions as shown in Figure 23.6. Blood from the ascending vasa recta drains into the arcuate veins.

The renal blood vessels are innervated both by postganglionic fibres from the sympathetic paravertebral chain and by efferent fibres from the parasympathetic nervous system via the vagus nerves. This innervation provides nervous control of the renal circulation that can override the intrinsic autoregulation of blood flow (see the next section).

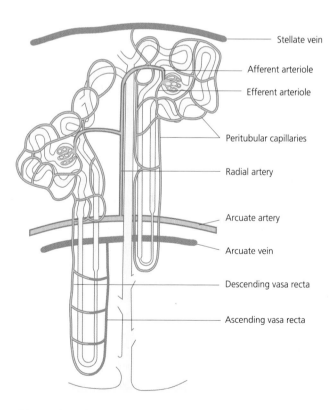

Stellate vein

Afferent arteriole

Efferent arteriole

Peritubular capillaries

Radial artery

Arcuate artery

Arcuate vein

Descending vasa recta

Ascending vasa recta

Figure 23.6 The organization of the renal circulation. Arterial blood is shown in red, venous blood in blue. Note that the efferent arterioles of the superficial nephrons give rise to the peritubular capillaries while those of the deeper juxtaglomerular nephrons give rise to straight vessels that pass deep into the renal medulla (the descending vasa recta).

23.5 Renal blood flow is kept constant by autoregulation

In normal adult humans, the total renal blood flow (i.e. the blood flow for both kidneys) is about $1.25 \, l \, min^{-1}$ or about a quarter of the resting cardiac output. If arterial pressure is altered over the range 10–26 kPa (80–200 mmHg), the renal blood flow remains remarkably constant (see the red line in Figure 23.7). The stability of renal blood flow persists even after the renal nerves have been cut and can be observed in isolated perfused kidneys. It is therefore due to mechanisms intrinsic to the renal circulation and is called **autoregulation** (see also Chapter 19, p. 377).

What are the mechanisms that underlie autoregulation? Two hypotheses have been advanced, which are not mutually exclusive:

- The **myogenic hypothesis**, which proposes that autoregulation is due to the response of the renal arterioles to stretch. An increase in pressure will distend the arteriolar wall and stretch the smooth muscle fibres, which then contract after a short delay. The resulting vasoconstriction will increase vascular resistance and decrease blood flow.

- The **metabolic hypothesis**, which proposes that humoral factors secreted by the renal tissue maintain a degree of vasodilatation. An increase in perfusion pressure will lead to an increased blood flow, which, in turn, will leach out more humoral factors and so decrease the vasodilatation. Conversely, a decrease in blood flow will allow the humoral factors to accumulate, causing relaxation of the vascular smooth muscle and vasodilatation. These humoral factors have not been unambiguously identified but likely candidates include adenosine, prostaglandins and the inorganic gas nitric oxide.

In addition to autoregulation, renal blood flow can also be controlled by the activity of the sympathetic nerves supplying the afferent arterioles and by hormones circulating in the blood. Sympathetic stimulation causes vasoconstriction of the afferent arterioles and thus reduces renal blood flow. Circulating adrenaline and noradrenaline (epinephrine and norepinephrine) also cause vasoconstriction in the renal circulation. These factors cause a fall in renal blood flow during exercise. Angiotensin II and antidiuretic hormone (vasopressin) are other hormones that act as powerful vasoconstrictors, particularly following severe haemorrhage (see Chapter 19, Box 19.2).

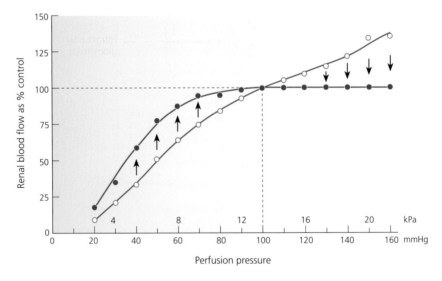

Figure 23.7 The autoregulation of renal blood flow. The renal blood flow for an isolated dog kidney was first allowed to stabilize at a perfusion pressure of 13.3 kPa (100 mmHg). The perfusion pressure was then abruptly altered to a new value and the blood flow measured immediately after the change in pressure (open circles and blue line). After a short period, the blood flow stabilized at a new level (shown by the filled circles and red line). The data show that the steady-state renal blood flow remains essentially constant once the arterial pressure rises above about 10 kPa (75 mmHg).

23.6 The kidneys regulate the plasma composition by ultrafiltration followed by selective modification of the filtrate

The kidneys regulate the plasma composition by three key processes: filtration, reabsorption and secretion (Figure 23.8). Firstly, some of the plasma flowing through the glomerular capillaries is forced through the capillary wall into Bowman's space by the hydrostatic pressure of the blood. This is filtration or, more accurately, **ultrafiltration**. Then, as this fluid passes along the renal tubules, its composition is modified both by the reabsorption of some substances and by the secretion of others. **Reabsorption** is defined as movement of a substance from the tubular fluid to the blood. **Secretion** is defined as movement of a substance from the blood into the tubular fluid.

The main stages of urine formation may be summarized as follows:

- Firstly, about a fifth of the plasma is filtered into Bowman's space from where it passes along the proximal tubule (Step 1 of Figure 23.8).

- In the proximal tubule, many substances are reabsorbed while others are secreted. Normally, all the filtered glucose and amino acids as well as most of the sodium, chloride and bicarbonate are reabsorbed here together with an osmotic equivalent of water so that, by the end of the proximal tubule, about two-thirds of

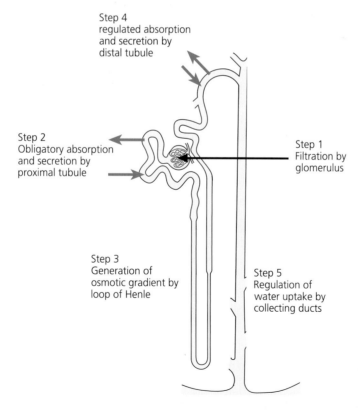

Step 4
regulated absorption
and secretion by
distal tubule

Step 2
Obligatory absorption
and secretion by
proximal tubule

Step 1
Filtration by
glomerulus

Step 3
Generation of
osmotic gradient by
loop of Henle

Step 5
Regulation of
water uptake by
collecting ducts

Figure 23.8 The main stages of urine formation by the kidney. Fluid is first filtered by the glomerulus. As it passes along the proximal tubule, it is modified by selective reabsorption and secretion. The reabsorption of ions by the loop of Henle without an osmotic equivalent of water leads to the generation of an osmotic gradient in the medulla, which is exploited to regulate the uptake of water by the collecting ducts.

the filtered fluid has been reabsorbed. As this phase of reabsorption is not closely linked to the ionic balance of the body, it is often called the **obligatory phase** of reabsorption (Step 2 of Figure 23.8).

- The loop of Henle is concerned with establishing and maintaining an osmotic gradient in the renal medulla. It does this by transporting sodium chloride from the tubular fluid to the space surrounding the tubules (the interstitium) without permitting the osmotic uptake of water. As a result, the osmolality of the fluid leaving the loop of Henle is lower than the plasma, while that of the interstitium is higher (Step 3 of Figure 23.8).

- The distal tubule regulates the ionic balance of the body by adjusting the amount of sodium and other ions it reabsorbs according to the requirements of the body. It also secretes hydrogen ions, which leads to the acidification of the urine. The fluid leaving the distal tubule is relatively dilute (Step 4 of Figure 23.8).

- As the urine passes through the collecting ducts, if the osmolality of the body fluids is relatively high (>290 mOsmol kg^{-1}), water is absorbed under the influence of antidiuretic hormone (ADH) and concentrated urine is excreted. If the osmolality of the body fluids is relatively low (<285 mOsmol kg^{-1}), little ADH is secreted and dilute urine is excreted (Step 5 of Figure 23.8). (See Chapter 2, p. 22 for an explanation of osmoles (Osmol).)

The formation of the glomerular filtrate

Analysis of samples of fluid taken from Bowman's capsule by micropuncture has shown that the glomerular filtrate has the same composition as plasma except that there is very little protein. The simplest explanation of this key observation is that the capsular fluid is formed by allowing the free passage of ions and small molecules such as glucose into the capsular space while retaining the plasma proteins in the blood. The volume of plasma that is filtered by the kidneys in a given amount of time is determined by the Starling forces exerted on the glomerular capillaries and by the ease with which fluid can pass across their walls (see Chapter 19, p. 383).

Measurement of the glomerular filtration rate

The rate at which the two kidneys form the ultrafiltrate is known as the **glomerular filtration rate** or **GFR** and has units of millilitres per minute. Although measurement of the GFR is important for an understanding of renal physiology, it cannot be measured directly. It can, however, be estimated by measuring the rate of excretion of substances that are freely filtered but are then neither absorbed nor secreted by the renal tubules. For accurate measurement, such substances should have no influence on any physiological parameter that may alter renal function, such as blood pressure or renal blood flow. These criteria are met by the plant polysaccharide inulin, which is excreted by the kidneys in direct proportion to its plasma concentration over a very wide range (Figure 23.9). The GFR measured by the rate of inulin excretion is called the inulin clearance and is generally about 125 ml min^{-1} for adult humans.

Despite its theoretical advantages, the use of inulin is not very convenient for clinical purposes, as a steady concentration needs to be maintained in the plasma for accurate measurement. For this reason, it

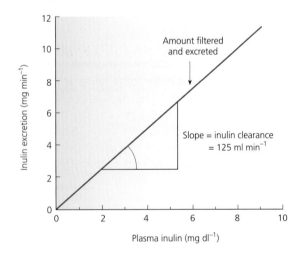

Figure 23.9 The relationship between the plasma inulin concentration and the amount excreted in the urine. The amount of inulin excreted rises in direct proportion to the plasma inulin concentration as expected for a substance that is filtered but not secreted or reabsorbed. The slope of the line is equal to the inulin clearance.

is desirable to use a substance that is normally present in the plasma that is filtered, but is neither secreted nor reabsorbed by the renal tubules. In addition, the plasma concentration of such a substance should not fluctuate rapidly. These criteria are largely met by creatinine (a metabolite of creatine) and **creatinine clearance** is generally used to measure GFR in clinical practice.

> **KEY POINTS:**
>
> - The amount of fluid passing into Bowman's capsule is determined by the balance of Starling forces acting on the glomerular capillaries.
> - The rate at which the kidneys filter the plasma is known as the glomerular filtration rate or GFR.

23.7 How can we tell whether a substance is reabsorbed or secreted by the kidneys?

When a substance is simply filtered and excreted unchanged, the amount excreted is directly proportional to the plasma concentration (Figure. 23.9). It represents the **filtered load** and is equal to the plasma concentration of the substance multiplied by the GFR (see Box 23.1). If a substance is reabsorbed, the amount excreted will be less than the filtered load (Figure 23.10). For example, glucose and amino acids are freely filtered but healthy people have virtually no glucose or free amino acids in their urine so there must be tubular mechanisms for the reabsorption of these substances. Conversely, if filtration is followed by tubular secretion, the amount appearing in the urine will be greater than the filtered load (Figure 23.11). As much as 70% of the dye phenolsulphonphthalein is removed from the blood in a single pass through the kidneys. As 75% is bound to plasma proteins, only 5% of the dye could have appeared in the urine by filtration. The remainder must have been secreted into the tubule.

The transport maximum

When polar substances such as glucose are reabsorbed, carrier molecules are required to permit their movement across the apical and basal membranes of the tubular cells. As each cell has a limited number of carriers, it is possible to saturate the transport capacity of the tubule if the plasma concentration rises too high. The maximum amount of a solute that can be transported by the renal

tubules each minute is called the transfer or **transport maximum (Tm)**.

Glucose excretion in uncontrolled diabetes mellitus (see Chapter 16) provides a clear example of the importance of the transport maximum in tubular transport. If glucose in the plasma rises above its normal level of about $5\,mmol\,l^{-1}$, no glucose appears in the urine until the plasma concentration exceeds $10-12\,mmol\,l^{-1}$. This value is known as the **renal threshold**. Above the renal threshold, the amount of glucose appearing in the urine increases slowly at first, then, as the transport process responsible for glucose reabsorption becomes fully saturated (above about $17\,mmol\,l^{-1}$), the urinary glucose rises in line with the increase in the plasma concentration, as shown in Figure 23.10. The appearance of glucose in the urine is known as **glycosuria**. As the excreted glucose is accompanied by an osmotic equivalent of water, glycosuria is accompanied by an increase in urine production known as an **osmotic diuresis**.

Active tubular secretion occurs by carrier-mediated processes analogous to those discussed above for glucose reabsorption, but which operate in the opposite direction. It is known that many organic substances are secreted by the tubules including *p*-aminohippurate (PAH), penicillin, and other organic anions and cations. The relationships between filtration, secretion and excretion of PAH are illustrated in Figure 23.11.

As PAH is almost totally removed from the plasma as it passes through the renal circulation, its clearance

Box 23.1 Renal clearance and its significance

The rate at which a substance is filtered is equal to the glomerular filtration rate (GFR) multiplied by the plasma concentration (P_c):

Rate of filtration $= P_c \times$ GFR $\mathrm{mg\,min^{-1}}$

This is known as the **filtered load** as it represents the quantity of a given substance delivered to the renal tubules each minute.

The rate at which the same substance is excreted is simply its concentration in the urine (U_c) multiplied by the amount of urine produced per minute ($\dot{V}$) so:

Rate of excretion $= U_c \times \dot{V}$ $\mathrm{mg\,min^{-1}}$

For a substance that is neither reabsorbed nor secreted (such as the plant polysaccharide inulin), the rate of excretion must be the same as the filtered load:

$P_c \times$ GFR $= U_c \times \dot{V}$

Rearranging:

GFR $= \dfrac{U_c \times \dot{V}}{P_c}$ $\mathrm{ml\,min^{-1}}$

Thus, the glomerular filtration rate can be calculated if the urine concentration, urine flow rate and plasma concentration of inulin are known. The GFR measured in this way is called the inulin clearance and has units of $\mathrm{ml\,min^{-1}}$. (It represents the volume of plasma filtered in each minute.) For normal healthy adults, it is usually around $125\,\mathrm{ml\,min^{-1}}$.

The concept of clearance can be extended to other substances that are secreted or reabsorbed by the renal tubules. Using the same basic formula but substituting C_s (for clearance of substance s) for GFR:

$C_s = \dfrac{U_c \times \dot{V}}{P_c}$ $\mathrm{ml\,min^{-1}}$

C_s is known as the renal clearance of substance s and represents the volume of plasma from which a given substance has been removed in 1 minute. In other words, it represents the minimum volume of plasma from which the excreted substance could have been derived. If a substance has a clearance smaller than the GFR, either the kidneys must reabsorb it (e.g. glucose) or it is not freely filtered (e.g. plasma proteins). If a substance has a clearance larger than the GFR, it must be secreted by the renal tubules (e.g. PAH).

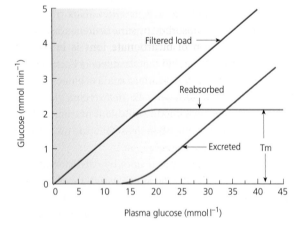

Figure 23.10 The quantities of glucose reabsorbed and excreted plotted as a function of the plasma concentration. The blue line shows that glucose appears in the urine when the plasma concentration exceeds about $12\,\mathrm{mmol\,l^{-1}}$. Above $20\,\mathrm{mmol\,l^{-1}}$, the extra amount of glucose excreted rises in proportion to the plasma concentration. The red line shows the filtered load in $\mathrm{mmol\,min^{-1}}$ and this is calculated from the GFR (here taken as $120\,\mathrm{ml\,min^{-1}}$) and the plasma concentration. The green line indicates the amount of glucose reabsorbed by the tubules, which is the difference between the filtered load and the amount excreted. Reabsorption reaches a maximum when the glucose load exceeds $2\,\mathrm{mmol}$ ($360\,\mathrm{mg}$) $\mathrm{min^{-1}}$. This is the transport maximum for glucose or $\mathrm{Tm_g}$.

can be used to estimate the renal plasma flow (RPF) in a relatively non-invasive way using the Fick principle discussed in Chapter 17 (p. 342). From estimates of GFR and renal plasma flow, it is possible to determine the proportion of plasma filtered as it passes through the kidneys. This is called the **filtration fraction**. Thus, if the GFR is $125\,\mathrm{ml\,min^{-1}}$ and renal plasma flow is $625\,\mathrm{ml\,min^{-1}}$, the filtration fraction is $125/625 = 0.20$ or 20%.

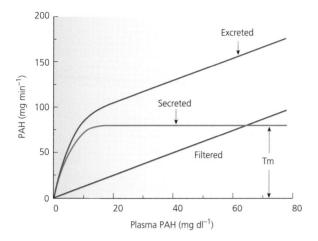

Figure 23.11 The amount of *p*-aminohippurate (PAH) excreted plotted as a function of the plasma concentration. The blue line indicates the amount excreted (in mg min^{-1}) while the red line indicates the filtered load. The amount secreted by the tubules (indicated by the green line) is given by the difference between the filtered load and the amount excreted. Tubular secretion of PAH reaches a maximum when the plasma concentration is about 18 mg dl^{-1} and the Tm$_{PAH}$ is about 80 mg min^{-1}.

23.8 The functions of the proximal tubule

In the course of a day, around 180 litres of plasma are filtered by the renal glomeruli (about 125 ml of plasma are filtered each minute and there are 1440 minutes in a day; so 125 × 1440 = 180 000 ml (or 180 litres) of plasma are filtered daily). Direct measurements have shown that virtually all the filtered glucose, lactate and amino acids have been reabsorbed as the tubular fluid traverses the proximal tubule. This represents a major challenge to the transport systems of the proximal tubules.

To take a specific example, glucose has a plasma concentration of about 5 mmol l^{-1} and about 180 litres of plasma are filtered each day. This means that about 160 g of glucose are filtered each day (5 mmol l^{-1} is equal to 0.9 g l^{-1} so 180 l × 0.9 g l^{-1} = 162 g). All of this glucose is reabsorbed by the proximal tubules. Similar calculations can be made for many other substances. However, as glucose has the same concentration in the plasma and the glomerular filtrate, it is clear that its uptake is not favoured by a concentration gradient across the wall of the proximal tubule. The same applies to all the other small molecules found in the plasma.

How does the kidney solve this problem? A specific carrier molecule on the brush border of the proximal tubule cells binds both glucose and sodium. The inward movement of sodium down its electrochemical gradient provides the energy for glucose uptake, which then accumulates within the cell until its concentration exceeds that of the plasma. The glucose can then diffuse out of the cell via the basolateral surface by the use of another carrier protein that is not sodium dependent. The arrangement of these two carriers on the apical and basolateral surfaces of the cells therefore permits the **secondary active transport** of glucose. This process is summarized in Figure 23.12. Amino acids and some other solutes are reabsorbed by similar mechanisms.

The reabsorption of bicarbonate ions is important for maintaining the stability of the blood pH. As shown in Figure 23.13, the reabsorption of bicarbonate is also linked to that of sodium. The cells of the proximal tubule secrete hydrogen ions into the tubular fluid, which leads to the conversion of bicarbonate ions to carbon dioxide and water. This process is helped by an enzyme called carbonic anhydrase, which is located on the brush border of the proximal tubule cells. The carbon dioxide diffuses into the tubular cells, where it is quickly reformed into carbonic acid by intracellular carbonic anhydrase. The bicarbonate formed by this reaction leaves the cell via the basolateral membrane in exchange for chloride ions and passes into the circulation. The processes involved are summarized in Figure 23.13.

Under normal circumstances, virtually no protein is found in the urine. If, however, the glomeruli become

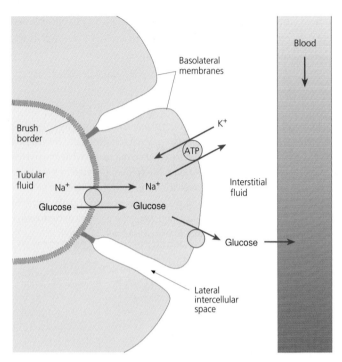

Figure 23.12 A schematic diagram illustrating the processes responsible for the reabsorption of glucose in the proximal tubule. Glucose and sodium bind to a carrier protein on the brush border, which allows them to cross the apical membrane. The sodium gradient provides the energy for the uptake process. Glucose leaves the cell via another carrier protein, which does not require sodium, located on the basolateral membrane. The sodium gradient across the plasma membrane of the proximal tubule cells is maintained by the activity of the sodium pump (Na^+, K^+ ATPase) of the basolateral membrane. Similar processes are responsible for the reabsorption of the amino acids and organic acids such as lactate.

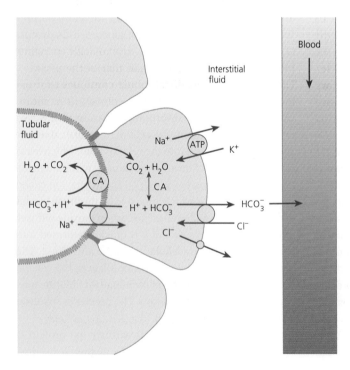

Figure 23.13 A schematic representation of bicarbonate reabsorption in the proximal tubule. H^+ secreted into the lumen lowers the pH of the tubular fluid and this favours the conversion of HCO_3^- to carbonic acid, which is converted to carbon dioxide and water by the carbonic anhydrase (CA) of the brush border membrane. The carbon dioxide diffuses down its concentration gradient into the tubular cell where it is reconverted to carbonic acid by intracellular carbonic anhydrase. It ionizes to form HCO_3^-, which leaves the basolateral surface of the cell in exchange for chloride. The H^+ formed is secreted into the lumen via Na^+–H^+ exchange to promote further HCO_3^- reabsorption.

diseased, significant amounts of protein may pass into the filtrate. As the proximal tubule has a very limited capacity for reabsorbing protein, most will be lost to the urine, which may have a frothy appearance. The presence of protein in the urine is called **proteinuria** (see Section 23.13).

Water absorption in the proximal tubule is directly linked to solute uptake

The proximal tubule reabsorbs about two-thirds of the filtered sodium, potassium, chloride, bicarbonate and other solutes. The uptake of sodium, chloride, glucose and other solutes by the tubular cells results in a transfer of osmotically active particles from the tubular lumen to the extracellular space. This transport is accompanied by an osmotic equivalent of water. This is known as the obligatory phase of water reabsorption.

> **KEY POINTS:**
> - As the filtrate passes along the proximal tubule, all the filtered amino acids, glucose and most of the bicarbonate are reabsorbed.
> - The absorption of these solutes is accompanied by an osmotic equivalent of water.
> - The proximal tubule also secretes some organic anions and cations into the tubular fluid.

23.9 Tubular transport in the loop of Henle

On entering the descending thin limb of the loop of Henle, the tubular fluid is still isotonic with the plasma, but, as it passes down the descending limb, the tubular fluid becomes increasingly hypertonic. This change in osmolality occurs as the tubular fluid equilibrates with the fluid of the medullary interstitium, which is increasingly hypertonic with respect to the plasma the closer it is to the apex of the papilla. As the tubular fluid flows round the hairpin bend, it enters the thin ascending limb of the loop of Henle. Here, the wall is impermeable to water so the tubular fluid does not equilibrate with the interstitium. Consequently, the tubular fluid entering the thick limb of the loop of Henle is strongly hypertonic with respect to the plasma.

The cells of the thick limb of the loop possess an electroneutral symporter (see p. 58) that transports sodium, potassium and chloride ions from the tubular fluid into the cell without the transfer of an osmotic equivalent of water (Figure 23.14). Hence, as the tubular fluid flows along the thick segment, the transfer of these ions from the fluid of the tubular lumen results in a fall in its osmolality and a rise in the osmolality of the interstitium. This establishes an osmotic gradient in the inner medulla, a process that is aided by the countercurrent flow in the descending and ascending limbs of the loop. When the body is dehydrated, this osmotic gradient is exploited to reabsorb water from the tubular fluid as it passes through the collecting ducts.

By the time the tubular fluid has reached the beginning of the distal tubule, it has an osmolality of about $150\,\text{mOsmol\,kg}^{-1}$ (i.e. about half that of the plasma). The early part of the distal tubule continues to reabsorb sodium and chloride ions without an osmotic equivalent of water, so that the tubular fluid becomes progressively more dilute as it flows towards the collecting ducts.

> **KEY POINTS:**
> - Sodium, potassium and chloride ions are transported from the tubular fluid by a symporter located in the cells of the thick ascending limb.
> - This transport is not accompanied by an osmotic equivalent of water.
> - The fluid leaving the thick limb is hypotonic with respect to the plasma.

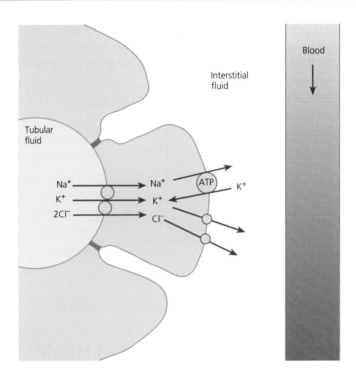

Figure 23.14 The transport processes responsible for the uptake of sodium and chloride in the thick ascending limb of the loop of Henle. Sodium, potassium and chloride are transported into the tubular cells by an electroneutral cotransporter. N.B. In this section of the nephron the ion movements occur without the osmotically driven uptake of water.

23.10   Ionic regulation by the distal tubules

The reabsorption of sodium, potassium, calcium and water by the proximal tubule and ascending loop of Henle largely takes place regardless of the ionic balance of the body. In the later distal tubule and collecting ducts, however, the uptake and secretion of ions is closely regulated. In addition, the distal tubule and collecting ducts play an important role in both acid–base balance and water balance.

While about 12% of the filtered load of sodium is reabsorbed by the **principal cells** (**P cells**) of the distal tubule and collecting ducts, their capacity to reabsorb sodium is regulated by the activity of the juxtaglomerular apparatus. When the sodium of the fluid in the distal tubule is low, the cells of the macula densa cause the granular cells of the afferent arteriole to secrete the proteolytic enzyme **renin**

into the blood. (N.B. renin should not be confused with rennin, which is an enzyme found in the stomach of young mammals and is responsible for curdling milk.) Renin converts a plasma peptide called angiotensinogen into a smaller peptide known as angiotensin I. This, in turn, is converted to angiotensin II by **converting enzyme**, which is found on the capillary endothelium of the lungs and some other vascular beds. (Angiotensin-converting enzyme is often abbreviated as ACE and is the target for some antihypertensive drugs, e.g. captopril.) Angiotensin II acts on the zona glomerulosa cells of the adrenal cortex to stimulate the secretion of the hormone aldosterone (Figure 23.15).

Aldosterone stimulates the production of sodium channels, which become inserted in the apical membranes of the principal cells of the distal tubule

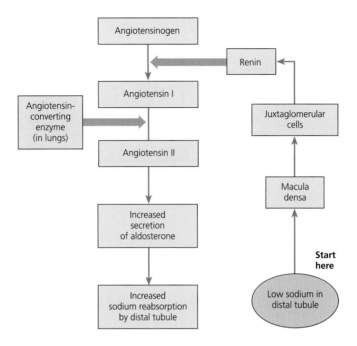

Figure 23.15 The regulation of aldosterone secretion by the sodium load in the distal tubule. The cells of the macula densa sense the sodium load delivered to the distal tubule and regulate the level of circulating renin according to the body's sodium requirements. If the sodium concentration in the tubular fluid is low, the juxtaglomerular cells increase their secretion of renin. This ultimately results in increased sodium reabsorption.

and collecting ducts. It also stimulates the synthesis of Na$^+$, K$^+$ ATPase molecules, which are inserted in the basolateral membrane. By increasing the ability of the principal cells to transport sodium, aldosterone promotes sodium reabsorption from the tubular fluid. Increasing the activity of the sodium pump also raises intracellular potassium and this can pass into the tubular fluid down its concentration gradient. Thus, aldosterone increases the ability of the nephron to reabsorb sodium and to secrete potassium. As the action of aldosterone requires the synthesis of new proteins, its effect is not immediate but is delayed by an hour or so and reaches its maximum after about a day.

The intercalated cells regulate acid–base balance by secreting hydrogen ions

Although most of the filtered bicarbonate is reabsorbed in the proximal tubule, there is 1–2 mmol l^{-1} in the fluid entering the distal tubule. Under normal circumstances, all of this bicarbonate is reabsorbed and acid urine is excreted. The reabsorption of bicarbonate by the intercalated or **intercalated cells** (**I cells**) differs from that of the proximal tubule. The I cells actively secrete hydrogen ions into the lumen via an ATP-dependent pump, as shown in Figure 23.16. As in the proximal tubule, the decrease in the pH of the tubular fluid favours the conversion of bicarbonate ions to carbon dioxide and water, and the liberated carbon dioxide diffuses down its concentration gradient into the tubular cells where carbonic anhydrase catalyses the reformation of carbonic acid within the I cells. The carbonic acid dissociates into hydrogen ions and bicarbonate ions. The hydrogen ions are secreted into the lumen and the bicarbonate ions exit the tubular cells via a bicarbonate–chloride exchanger located in the basolateral membrane.

This active secretion of hydrogen ions into the lumen of the distal tubule and collecting ducts results in a fall in the pH of the tubular fluid, which can reach values as low as 4–4.5, much lower than elsewhere in the

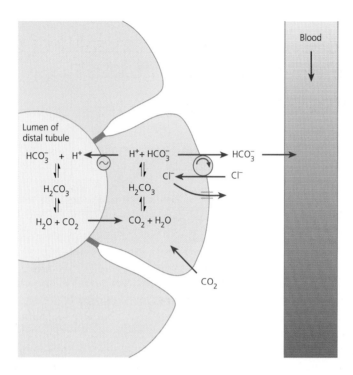

Figure 23.16 The mechanism by which the intercalated cells of the late distal tubule and cortical collecting ducts secrete hydrogen ions.

nephron. As a pH of 4 corresponds to a free hydrogen ion concentration of $0.1\,mmol\,l^{-1}$, only 0.15 mmol of free hydrogen ion can be excreted each day for a normal daily urine output of 1.5 litres. Nevertheless, the body produces about 50 mmol of non-volatile acid a day (mostly sulphuric acid) that needs to be excreted. By buffering the non-volatile acids in the tubular fluid with phosphate, around 25 mmol of free acids can be excreted. The remaining non-volatile acids are excreted as their ammonium salts.

Calcium ions are absorbed from the distal tubule by an active process that can be stimulated by parathyroid hormone

Around 70% of the filtered load of calcium is reabsorbed in the proximal tubules, 20% in the ascending loop of Henle and most of the remainder in the distal tubules and cortical collecting ducts. Only about 1% of the filtered calcium is normally excreted. Calcium

reabsorption by the distal tubule and cortical collecting ducts is stimulated by parathyroid hormone (PTH), which plays a major role in calcium homeostasis. Conversely, an increase in PTH secretion decreases the reabsorption of phosphate by the proximal tubule. Further details of the regulation of calcium balance can be found in Chapter 16, pp. 310–315.

KEY POINTS:

- In the second part of the distal tubule and in the collecting duct, the principal cells absorb sodium and secrete potassium.

- The efficacy of sodium uptake and potassium secretion is regulated largely by the hormone aldosterone.

- The intercalated cells secrete hydrogen ions and reabsorb bicarbonate.

23.11 The kidney regulates the osmolality of the plasma by adjusting the amount of water reabsorbed by the collecting ducts

In a normal individual, water intake varies widely according to circumstances. As a result, the osmolality of urine can range from as little as $50\,\mathrm{mOsmol\,kg^{-1}}$ following a large water load to around $1200\,\mathrm{mOsmol\,kg^{-1}}$ in severe dehydration. How do the kidneys produce urine with such a wide range of osmolality?

In essence, the kidney generates an osmotic gradient within the medulla by transporting sodium, potassium and chloride ions from the lumen of the ascending loop of Henle into the interstitial space without the transfer of an osmotic equivalent of water. This osmotic gradient is then used to reabsorb water from the urine as it passes through the medulla. The amount of water reabsorbed is regulated by the level of antidiuretic hormone (ADH or vasopressin) circulating in the blood. When the plasma osmolality is low, little ADH is secreted and copious dilute urine is produced. Conversely, when plasma osmolality is high ADH secretion is increased. Under such circumstances, water reabsorption in the distal nephron is increased with the result that a small volume of concentrated urine is produced.

Antidiuretic hormone regulates the absorption of water from the collecting ducts

The absorption of ions by the ascending limb of the loop of Henle results in the tubular fluid becoming hypo-osmotic as it approaches the distal tubule. During its passage along the first third of the distal tubule, sodium and chloride continue to be transported from the lumen to the interstitium. Moreover, the tubular epithelium is still relatively impermeable to water, so the tubular fluid becomes progressively more dilute. By the time the fluid reaches the last third of the distal tubule, there is a substantial osmotic gradient in favour of water reabsorption. However, both this part of the distal tubule and the collecting ducts are impermeable to water unless ADH is present.

ADH secretion is regulated by neurones in the hypothalamus known as **osmoreceptors**, which monitor the osmolality of the plasma. When plasma osmolality is below $285\,\mathrm{mOsmol\,kg^{-1}}$, the osmoreceptors are inactive and ADH secretion by the posterior pituitary is very low. A small increase in plasma osmolality above the threshold of $285\,\mathrm{mOsmol\,kg^{-1}}$ will activate the osmoreceptors, which then stimulate ADH secretion. The degree of stimulation depends on the increase in osmolality above the threshold (Figure 23.17). This osmotic regulation of ADH secretion is central to the control of plasma osmolality.

ADH increases the permeability of the last third of the distal tubule and that of the whole of the collecting duct to water. The result is a movement of water down its osmotic gradient into the tubular cells and thence into the interstitial fluid and the plasma. *This water movement is independent of solute uptake* and therefore results in an

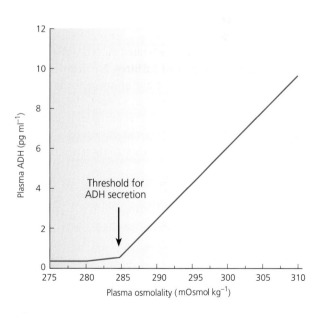

Figure 23.17 The effect of changes in plasma osmolality on circulating levels of ADH.

increase in urine osmolality and a fall in plasma osmolality that is directly related to the amount of water reabsorbed.

When the body has excess water and plasma osmolality is less than 285 mOsmol kg^{-1}, ADH secretion will be suppressed. Under these circumstances, water will not be reabsorbed during its passage through the collecting ducts and a large volume of dilute urine will be produced (giving rise to a **diuresis**). In contrast, during dehydration, ADH secretion is stimulated and acts on the collecting ducts to increase their permeability to water. This results in the production of concentrated urine.

Only some 10–15% of the total filtered load of water is subject to regulation by ADH as the remainder has been reabsorbed along the earlier parts of the nephron together with ions and other solutes. However, if the posterior pituitary is unable to secrete ADH, or if the collecting ducts are unable to respond to it, a large volume of dilute urine is produced. This condition is known as **diabetes insipidus**. In excess of 15 litres of urine can be excreted per day, which must be made up by increased water intake if life-threatening dehydration is to be avoided.

ADH controls the insertion of water channels into the apical membrane of the P cells of the collecting ducts

How does ADH regulate the permeability of the collecting ducts to water? It now appears that ADH circulating in the blood binds to receptors on the basolateral surface of the P cells of the collecting ducts. These receptors are coupled to the enzyme adenylyl cyclase so that activation of these receptors results in an increase in intracellular cyclic AMP levels (see p. 89). This initiates a process that leads to the fusion of vesicles containing water channels with the apical membrane. When ADH levels fall, so do the levels of cyclic AMP in the P cells and the water channels are taken back into the cells to await recycling. The permeability of the apical membrane of the P cells to water is therefore regulated by the insertion and removal of specific water channels, as shown in Figure 23.18. The water that enters the cell across the apical membrane passes freely into the lateral intercellular space from where it enters the plasma.

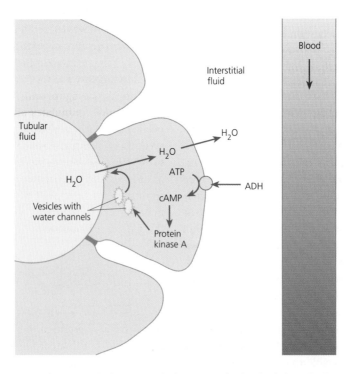

Figure 23.18 A schematic drawing of the control of water uptake by ADH in the distal tubule and collecting duct.

KEY POINTS:

- The operation of the loop of Henle gives rise to an osmotic gradient between the outer border of the renal medulla and the papilla of the renal pyramids.

- This gradient arises from the transport of sodium, potassium and chloride by the thick limb of the loop of Henle and the first part of the distal tubule, which occurs without the movement of an osmotic equivalent of water.

- The flow of fluid in different parts of the nephron runs in opposite directions (there is a countercurrent arrangement). This concentrates sodium, chloride and urea in the renal medulla.

- The collecting ducts are impermeable to water unless antidiuretic hormone (ADH) is present.

23.12 The collection and voiding of urine

The renal calyces, ureters, urinary bladder and urethra comprise the **urinary tract**, which is concerned with collecting the urine formed by the kidneys and storing it until a convenient time occurs for the bladder to be emptied. The epithelia lining the urinary tract are impermeable to water and solutes and so do not modify the composition of the urine.

Urine passes from the kidney to the bladder by the peristaltic action of the muscle in the wall of the ureters

The ureters are tubes about 30 cm long, which consist of an epithelial layer surrounded by circular and longitudinal bundles of smooth muscle. In addition, some muscle fibres are disposed in a spiral arrangement around the ureter. When the renal calyces and upper regions of the ureters become distended due to the accumulation of urine, peristaltic contractions occur in the ureters propelling the urine towards the bladder. The ureters pass for 2–3 cm obliquely through the bladder wall in a region known as the **trigone** before they finally empty into the bladder just above the neck. This arrangement prevents the reflux of urine when the pressure within the bladder rises above that in the ureters.

The bladder itself consists of two principal parts: the **body** or **fundus**, which serves to collect the urine, and the **bladder neck** or **posterior urethra**, which is 2–3 cm in length. The bladder is lined by a mucosal layer, which becomes greatly folded when the bladder is empty. The bladder wall consists of smooth muscle and elastic tissue. The smooth muscle is of the single unit type and is known as the **detrusor muscle**. The wall of the bladder neck has a higher proportion of elastic tissue interlaced with the smooth muscle. The tension in the wall of the bladder neck keeps this part of the bladder empty of urine during normal filling and so behaves as an internal sphincter. The urethra passes through the urogenital diaphragm, which contains a layer of striated muscle called the external sphincter that is under voluntary control via the pudendal nerves.

During the storage of urine, the progressive distension of the bladder stimulates afferent nerve fibres in the pelvic nerves. The activity in these afferents excites the sympathetic fibres of the hypogastric nerve and this leads to inhibition of the detrusor muscle and constriction of the neck of the urethra. In addition, the external sphincter is held closed by activity in the pudendal nerves. These responses are 'guarding reflexes', which act to promote continence.

The micturition reflex is responsible for the voiding of urine

The process by which the bladder normally empties is known as micturition. It is controlled by the sacral segments of the spinal cord. As the bladder fills, it becomes distended and the detrusor muscle becomes stretched and contracts slightly. This results in a rise in pressure

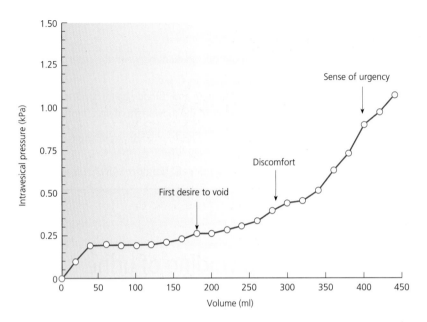

Figure 23.19 Pressure–volume relations for the normal human bladder. Note that, after the initial phase of filling, the volume increases three- to fourfold with little increase in intravesical pressure. As the volume increases further the pressure rises more and more steeply. It is this increase in pressure that stimulates the desire to void.

within the bladder (the **intravesical pressure**) of about 300 Pa (3 cmH$_2$O; Figure 23.19). Further filling results in little change in pressure until the bladder volume reaches 200–300 ml when the first desire to void is experienced. Further increases in volume lead to increased intravesical pressure until at about 400–450 ml the pressure begins to rise steeply as more and more urine accumulates in the bladder. At these volumes, the bladder undergoes periodic reflex contractions and there is an urgent need to urinate.

Normally, micturition is a voluntary act controlled by corticospinal impulses sent to the lumbrosacral region of the spinal cord. During urination, the detrusor muscle contracts fully. Urination is further aided by contraction of the abdominal muscles, which raise intra-abdominal pressure, and by relaxation of the muscles of the urogenital diaphragm, which permit the dilatation of the urethra. Once urination occurs, the bladder is normally almost fully emptied, with only a few millilitres of urine remaining.

The inappropriate voiding of urine is known as **urinary incontinence**. It has many causes. In the elderly, it is very common and may reflect nothing but the fact that impaired mobility prevents the individual from reaching the toilet in time. In obstruction of the lower urinary tract, the sufferer cannot fully empty their bladder, which becomes very full and continually leaks small amounts of urine. Incontinence of such small volumes is called overflow incontinence and is frequently seen in elderly men with prostate enlargement.

Hyperexcitability of the detrusor muscle of the bladder wall may give rise to urge incontinence. The involuntary loss of urine during coughing, sneezing or strenuous activity is known as stress incontinence and results from weakness of the pelvic floor muscles, which are not able to prevent urine leakage under physical stress. Those suffering from dementia often become incontinent as their ability to respond to the normal signals from the bladder declines with the loss of cognitive function.

If the spinal cord is damaged above the sacral segments (e.g. through traumatic injury to the spine in the thoracic region), there is a complete suppression of the micturition reflex. Over time, the bladder reflexes gradually return. Despite this recovery, such patients have no control over when they urinate and micturition is triggered when the pressure within the bladder reaches a threshold level. This is known as **automatic bladder**. In this situation, the urine does not dribble constantly from the urethra but is periodically voided as in a normal subject. A similar situation exists in infants before they have learnt voluntary control.

23.13   Urine composition

Normal, fresh urine has a slight aromatic odour that can readily be masked by aromatic compounds from certain foodstuffs (e.g. asparagus, coffee or garlic). Its characteristic pale yellow colour is principally due to the presence of pigments known as urochromes. The unpleasant, fetid odour often associated with urine is due to the production of ammonia by bacterial decomposition of urea.

Subjects living in a temperate climate who are drinking normally produce about 1.5 litres of urine each day containing about 50–70 g of solids. However, the precise chemical composition and osmolality of the urine is very variable (Table 23.1). It changes with the diet and the state of water balance of the individual. The main electrolytes found in urine are sodium, chloride, potassium, calcium and phosphate while the main non-electrolytes are urea and creatinine. The urine is normally somewhat acid compared with the plasma and its pH may range between 4.8 and 7.5, but for people eating a normal mixed diet, it usually lies between 5 and 6.

As the urine is derived from an ultrafiltrate of the plasma, it is not surprising to find that it contains traces of most of the plasma constituents. Nevertheless, significant amounts of some plasma constituents, such as proteins, glucose and amino acids, are not normally found in the urine of healthy subjects. However, important diagnostic information can be obtained by analysing the chemical composition of the urine (see below). In addition, urine colour also gives information regarding the health of an individual. For example, it is often strongly coloured during infections and has a brownish-red colour when haemoglobin is present.

The analysis of urine and its role in diagnosis

The analysis of urine (urinalysis) is a diagnostic tool used to detect a variety of metabolic and renal disorders. Routine analysis of urine by a simple 'dipstick' method is performed on admission to hospital, at antenatal

Table 23.1 Typical values for the constituents of plasma and urine

	Plasma	Urine	Units
Na^+	140–150	50–130	mmol l^{-1}
K^+	3.5–5	20–70	mmol l^{-1}
Ca^{2+}	1.35–1.50	10–24	mmol l^{-1}
Phosphate	0.8–1.25	25–60	mmol l^{-1}
Cl^2	100–110	50–130	mmol l^{-1}
Creatinine	0.06–0.12	6–20	mmol l^{-1}
Urea	4–7	200–400	mmol l^{-1}
Glucose	3.9–5.2	0	mmol l^{-1}
pH	7.35–7.4	4.8–7.5	$(-\log_{10}[H^+])$
Osmolality	281–297	50–1300	mOsmol kg^{-1}

Note that the composition of the urine is subject to considerable variation. Moreover, some important constituents of plasma such as protein, glucose and bicarbonate are absent from normal urine.

appointments, in GP surgeries and in outpatient clinics. The test is simple and takes only a few minutes to complete. A sample of urine is collected from the individual to be tested. This is usually a midstream urine sample (MSU) so that contamination of the urine by urethral debris is minimized. The container must be clean but if the sample is to be tested for the presence of bacteria, it should also be sterile. The urine should be tested as soon as possible after collection, as prolonged exposure to the air may cause changes to the sample that affect the analysis. It is important to note the colour, clarity and odour of the sample prior to testing. A fishy odour or a cloudy appearance, for example, strongly suggests the presence of a urinary tract infection, while certain foods and drugs can alter the colour of the urine.

A variety of different types of disposable testing strips are available for urinalysis, the most common of which consists of a thin strip of plastic housing ten absorbent cellulose patches impregnated with reagents that indicate the presence of specific constituents of urine by showing a colour change when dipped into the sample. Strips are normally supplied in screw-topped containers with a manufacturer's colour comparison chart. The substances detected and the sensitivity of the tests are as follows:

- Glucose (between 0 and 112 mmol l^{-1})
- Bilirubin (negative, small, moderate or large)
- Ketones (acetoacetic acid) (between 0 and 16 mmol l^{-1})
- Specific gravity (between 1.005 and 1.030)
- Blood (negative, trace, small, moderate or large)
- pH (between 5.0 and 8.5)
- Protein (between 0 and 2.4 g l^{-1})
- Urobilinogen (3–132 mmol l^{-1})
- Nitrite (any degree of pink colour indicates the presence of nitrites)
- Leucocytes (white blood cells) (negative, trace, small, moderate or large).

The strip is dipped briefly into the urine sample and then removed so that the patches are covered but the strip is not saturated. The test strip is then shaken or tapped gently to remove excess urine and held horizontally alongside the colour comparison chart so that the colour changes can be read and noted. The degree of colour change provides a semiquantitative assessment of the amount of substance present. For the most accurate results, the patches must be read at the times indicated on the container. Glucose, for example, should be read 30 seconds after immersion and so on. Automatic strip readers are also available in some clinical environments.

Interpretation of urinalysis results

Glucose is not detectable in the urine of a healthy person because all the glucose filtered at the glomerulus is reabsorbed in the proximal tubule (see Section 23.8). Glucose is only excreted in the urine if the quantity of filtered glucose exceeds the transport maximum for glucose reabsorption. The presence of glucose in the urine (glycosuria) is strongly suggestive of the condition diabetes mellitus, although blood tests and a glucose tolerance test will be required to confirm the diagnosis (see Chapter 16).

Bilirubin is a yellowish pigment derived from the breakdown of senescent red blood cells. It is normally conjugated with glucuronic acid in the liver and excreted in the bile (see Chapter 25). It does not normally appear in the urine. The presence of bilirubin in the urine may indicate a problem associated with liver function or with the normal delivery of bile to the small intestine (such as the presence of gallstones in the bile ducts).

Ketones are derived from the metabolism of fats. They only appear in the urine in measurable quantities when fats, rather than carbohydrates, are being used by the body as the principal source of energy. Ketonuria may occur in a healthy fasting individual or in someone who is following a low-carbohydrate diet (such as the Atkins diet). It may also occur in the later stages of pregnancy when large amounts of glucose are being used up by the fetus and the maternal energy supply has to come from fats. While these conditions represent normal physiological adaptations, prolonged or excessive ketonuria is more likely to be associated with diabetes mellitus, in which glucose cannot be utilized normally. In such cases, the level of ketones in the blood may become high enough to give rise to the serious and potentially life-threatening condition of ketoacidosis.

Specific gravity is a measure of the number of dissolved particles in the urine. Strongly coloured urine with a high specific gravity indicates dehydration, while dilute urine, which is pale in colour and has a low specific gravity, indicates that the individual is well hydrated. Low specific gravity coupled with symptoms of dehydration may indicate diabetes insipidus in which there is a failure of ADH secretion by the posterior pituitary gland (see Chapter 16).

Blood in the urine (haematuria) may have a number of origins. Menstruation, catheterization or any minor lesion of the urethra or surrounding tissue may result in haematuria. Heavy exercise may also lead to the transient appearance of blood in the urine (particularly in young males) for reasons that are not understood. More serious causes of haematuria include infections, tumours of the kidney or genitourinary tract, or renal calculi (stones). It is important to remember, however, that a number of foods, such as beetroot and blackberries, can cause a red discoloration of the urine that may be mistaken for blood. Certain medications, including phenothiazines (used in the treatment of tuberculosis), may also cause the urine to appear reddish in colour.

The hydrogen ion concentration of the urine (i.e. its acidity) is measured on the pH scale. Normal urine shows a wide range of pH (4.5–8.0) although it is most commonly between 5 and 7 as excess acid produced by metabolic reactions is excreted in the urine (see p. 481). Certain disease states are associated with urine of extremely high or low pH. A very alkaline pH, for example, may indicate a failure of the renal tubules to secrete acid normally (renal tubular acidosis). Prolonged vomiting will also result in urine of high pH, as the kidneys compensate for the loss of gastric acid by secreting fewer hydrogen ions (see pp. 720–724). The presence of ketones in the urine, however, will lower urinary pH as these metabolites are acidic.

Protein is not normally present in the urine in more than trace amounts. As explained in Section 23.8, those proteins that are filtered at the glomerulus are normally reabsorbed in the proximal tubule by pinocytosis. Proteinuria may indicate the presence of a urinary tract infection (UTI), damage to the glomerulus, hypertension or, in pregnancy, pre-eclampsia.

Urobilinogen appears in normal urine at concentrations below $17\,\mathrm{mmol\,l^{-1}}$. Urobilinogen is derived from the actions of gut flora on bilirubin glucuronide and most of it is eliminated from the body via the faeces. An increased urinary urobilinogen concentration may indicate restricted liver function or excessive breakdown of red blood cells. However, eating large numbers of bananas may result in false positives, as may taking drugs such as chlorpromazine, an antipsychotic agent that is often used as an anti-emetic.

Nitrites are absent from the urine of a healthy person. Their presence suggests a urinary tract infection. Normal urine contains nitrates as waste products. Some strains of bacteria convert the nitrates to nitrites while the urine is stored in the bladder.

Leucocytes are absent from the urine of a healthy person. They will be detected in the urine in the presence of a urinary tract infection as a result of the inflammatory response.

Although urinalysis is undoubtedly a very useful clinical tool, it is important to remember that, on its own, it is unable to confirm any diagnosis. Further laboratory urine and blood tests, as well as more detailed investigation of the kidneys, urinary tract and other organs, will be required.

23.14 Drugs acting on the kidneys

Many different clinical conditions lead to oedema – the abnormal accumulation of fluid in the interstitial space. For example, in chronic heart failure the plasma volume is expanded because of the retention of sodium and water. This lowers the oncotic pressure and increases the capillary pressure so that the Starling forces increasingly favour filtration. Fluid accumulates in the tissues and oedema results. In right-sided heart failure, the increased capillary pressure leads to oedema in the periphery, which is most noticeable in the ankles. In left-sided heart failure (congestive heart failure), there is raised pressure in the left atrium. This results in an increased back pressure on the pulmonary capillaries causing an accumulation of fluid in the pulmonary interstitium, pulmonary oedema and impaired gas exchange. As part of the process of restoring the situation to normal, it is necessary to eliminate the excess sodium and water. This can be achieved by treating the patient with drugs that promote the loss of both sodium and water in the urine. These are called **diuretics**.

Diuretics may act indirectly, by exerting an osmotic pressure that is sufficient to inhibit the reabsorption of water and sodium chloride from the renal tubules, or they may act directly by inhibiting active transport in various parts of the nephron, as shown in Figure 23.20.

The commonly used diuretics are:

- Osmotic diuretics
- Loop diuretics
- Thiazides
- Potassium-sparing diuretics
- Aldosterone antagonists.

Osmotic diuretics such as the sugar mannitol are filtered at the glomerulus but they are not transported by the cells of the proximal tubule. In consequence, as other substances are transported and the proportion of

the original filtered volume falls, these substances accumulate and exert sufficient osmotic pressure to inhibit tubular reabsorption of water. As absorption by the proximal tubule is iso-osmotic, a decrease in water reabsorption allows more sodium to reach the distal nephron, with the result that more sodium is excreted. There is, therefore, a loss of both sodium and water. Nevertheless, the osmotic diuretics are more effective in increasing water excretion than they are in increasing sodium excretion. Their main use is in the treatment of acutely raised intracranial pressure and to prevent acute renal failure. They are normally given intravenously.

Diuretics that act by inhibiting active transport are exemplified by loop diuretics such as **furosemide** and **bumetanide**. These drugs inhibit the cotransport of sodium, potassium and chloride by the ascending thick limb of the loop of Henle (Figure 23.14). As inhibition of this transport decreases the ability of the nephron to concentrate urine, the effects of the loop diuretics are twofold: they increase sodium excretion by inhibiting sodium chloride transport and increase water loss through impairment of the countercurrent mechanism responsible for establishing the osmotic gradient in the renal medulla. Loop diuretics are used to treat congestive heart failure and chronic renal failure. They are the most potent diuretics in current clinical use and produce a very pronounced diuresis. Unfortunately, they also promote a marked loss of potassium in the urine. The loop diuretics may be given orally or intravenously.

Another group of clinically useful diuretics are the **thiazides** and their relatives, which inhibit the cotransport of sodium and chloride by the cells of the early distal tubule. They have little or no effect on the ion transport of the thick segment of the loop of Henle. Examples are **bendroflumethiazide**, **hydrochlorothiazide** and **metolazone**. These drugs are normally given orally and are used to treat hypertension, and congestive heart failure. The thiazides are less potent than the loop diuretics. Nevertheless, they also increase potassium loss in the urine and are therefore potentially capable of upsetting normal potassium balance, which may lead to cardiac arrhythmias. Potassium-sparing diuretics, such as **amiloride** and **triamterine**, minimize these problems. These drugs act on the distal tubule, connecting tubules and collecting ducts to inhibit both sodium absorption and potassium excretion by the principal cells. They are not very potent as diuretics but are often given with thiazides to minimize potassium loss.

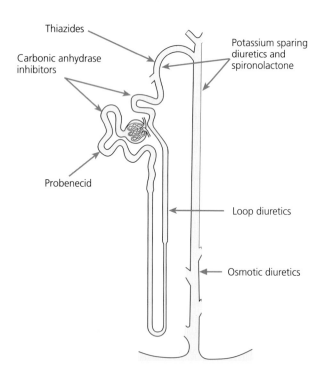

Figure 23.20 The sites at which various classes of drug act on the kidneys.

The mineralocorticoid aldosterone acts as a sodium-conserving hormone. The excessive retention of sodium leads to expansion of the extracellular volume and hypertension, which may exacerbate the problems of heart failure. Drugs that antagonize the action of aldosterone could potentially be used to reverse these conditions. The search for drugs with these properties led to the discovery of **spironolactone**. Spironolactone acts on the distal tubule by antagonizing the sodium-retaining action of aldosterone. It is administered orally but the severity of its side-effects limits its clinical use to the treatment of oedema caused by primary hyperaldosteronism (Conn's syndrome) or secondary hyperaldosteronism resulting from cirrhosis of the liver. Other drugs with a diuretic action are carbonic anhydrase inhibitors such as **acetazolamide**. These drugs are now rarely used clinically as diuretics but are used to treat glaucoma.

Probenecid

Uric acid passes freely into the glomerular filtrate. It is largely reabsorbed by the proximal tubule but around 10% of the amount filtered is excreted. The balance between filtration and reabsorption maintains plasma uric acid concentration at around $0.3\,\mathrm{mmol\,l^{-1}}$. However, in some unfortunate individuals, plasma uric acid concentrations are much higher, leading to the deposition of uric acid crystals in the joints. These crystals cause severe inflammation and acute pain, a condition known as gout. Probenecid inhibits reabsorption of uric acid by the proximal tubule and thereby lowers plasma uric acid, relieving the pain and inflammation. However, as uric acid is the end product of purine metabolism, allopurinol is now usually given to inhibit the synthesis of uric acid by xanthine oxidase. A summary of the drugs acting on the kidney is given in Table 23.2.

In common with the majority of orally administered drugs, antibiotics such as penicillin are rapidly excreted unchanged by the kidneys. This fact is exploited to treat bladder infections. For symptomatic relief of inflammatory conditions in the lower urinary tract, potassium citrate is occasionally given orally. The citrate is metabolized and the increased potassium excretion renders the urine alkaline, and less irritating to the urinary tract.

Table 23.2 Drugs acting on the kidneys

Drug class	Specific example	Principal action	Clinical uses
Osmotic diuretics	Mannitol	Maintains high osmolality in renal tubules to promote urine flow	Prevention of acute renal failure; treatment of acutely raised intracranial pressure
Loop diuretics	Furosemide (frusemide)	Inhibits active transport of ions by the thick segment of the loop of Henle	Treatment of congestive heart failure and chronic renal failure
Thiazides	Bendroflumethiazide	Inhibits active transport of sodium and chloride in the early distal tubule	Treatment of hypertension
Potassium-sparing diuretics	Amiloride	Inhibits sodium uptake in the distal tubule and collecting ducts	Prevention of potassium deficiency when given with thiazide diuretics
Aldosterone antagonists	Spironolactone	Inhibits the action of aldosterone on the distal tubule and collecting ducts	Treatment of hyperaldosteronism
Probenecid		Inhibits reabsorption of uric acid by the proximal tubule	Treatment of gout

Recommended reading

Anatomy of the urinary tract

MacKinnon, P.C.B. and Morris, J.F. (2005). *Oxford Textbook of Functional Anatomy,* 2nd edn, Vol. 2. *Thorax and Abdomen*. Oxford University Press: Oxford. pp. 91–98.

Pharmacology and the kidney

Rang, H.P., Dale, M.M., Ritter, J.M. and Flower, R. (2007). *Pharmacology*, 6th edn. Churchill-Livingstone: Edinburgh. Chapter 23.

Renal physiology

Koeppen, B.M. and Stanton, B.A. (2001). *Renal Physiology*, 3rd edn. Mosby: St Louis.

Valtin, H. and Schafer, J.A. (1995). *Renal Function*, 3rd edn. Little, Brown & Co: Boston.

Self-assessment questions

Which of the following statements are true and which are false? Answers are given at the end of the book (p.755).

1. a) The urine concentration of potassium is greater than that of plasma.
   b) The concentration of calcium in the plasma is greater than that in the urine.
   c) Urine normally contains about $2\,mmol\,l^{-1}$ of bicarbonate.
   d) Urine pH is normally greater than that of plasma.
   e) Urine normally contains no measurable quantity of protein.

2. a) Renal plasma flow can be measured by the clearance of *p*-aminohippurate.
   b) The kidneys receive about a fifth of the resting cardiac output.
   c) As cardiac output increases during exercise, renal blood flow rises.
   d) Renal blood flow is maintained within narrow limits by autoregulation.
   e) Increased activity in the renal sympathetic nerves results in increased blood flow to the kidneys.

3. a) The glomerular filtrate has the same composition as plasma.
   b) The glomerular filtration rate can be determined by measuring the clearance of creatinine.
   c) A substance that has a clearance less than that of creatinine must have been reabsorbed by the renal tubules.
   d) The glomerular filtration rate is about $125\,ml\,min^{-1}$ in a healthy adult male.
   e) The glomerular filtration rate depends on the pressure in the afferent arterioles.

4. a) The appearance of glucose in the urine reflects a saturation of the glucose carriers of the proximal tubule.
   b) Glucose absorption by the renal tubules is independent of sodium uptake.
   c) The transport maximum for glucose is $30\,mg\,min^{-1}$.
   d) In a healthy person, the proximal tubules reabsorb all of the filtered glucose.

5. a) The uptake of sodium ions is regulated by the proximal tubule.
   b) Macula densa cells secrete renin when plasma sodium is low.
   c) Angiotensin II is formed from renin by the action of an enzyme found on the endothelium of the pulmonary blood vessels.
   d) Angiotensin II acts on the adrenal medulla.
   e) The adrenal glands secrete aldosterone in response to stimulation by angiotensin II.

The digestive system 24

After reading this chapter you should have gained an understanding of:

- The basic anatomical organization of the gastrointestinal system, including its nerve and blood supplies
- The motility of the gut and its regulation by nerves and hormones
- Chewing and swallowing
- The secretion and functions of saliva
- The functions of the stomach and the regulation of gastric secretion
- The digestion of food and the absorption of nutrients
- The functions of the large intestine
- Defecation, diarrhoea and constipation
- The actions of drugs on gastrointestinal secretion and motility

24.1 Introduction

The body requires food to provide the raw materials for the production of energy, and for the growth and repair of tissues. The provision of these materials is the role of the gastrointestinal tract, which breaks down the fats, proteins and carbohydrates of the diet to smaller molecules that can be absorbed across the gut wall into the bloodstream. The circulation then distributes the products of digestion to the tissues that require them.

The processing of food first requires it to be stored in the stomach. Subsequently, it is subjected to mechanical and chemical (enzymatic) digestion before being absorbed. To permit optimal digestion and absorption, the food and digestion products must move along the tract at an appropriate rate. Moreover, in performing its digestive function, the gut must protect itself against self-digestion. Finally, the indigestible remains, together with water, bile pigments, mucus and bacteria, are eliminated from the anus as faeces.

24.2   Anatomical organization of the gastrointestinal tract

The gastrointestinal (or GI) tract and its accessory structures are illustrated in Figure 24.1. It is essentially a muscular, hollow tube that communicates with the outside world at the mouth and anus. The gut consists of the oral cavity, pharynx, oesophagus, stomach, small intestine (duodenum, jejunum and ileum), large intestine (caecum, colon and rectum) and anus. Accessory structures include the teeth, tongue, salivary glands, liver, gall bladder and pancreas. Each component of the tract is adapted to carry out one or more specific functions.

KEY POINTS:

- The GI tract consists of the oral cavity, pharynx, oesophagus, stomach, small intestine, large intestine and anal canal.

- Accessory organs include the teeth, tongue, salivary glands, exocrine pancreas, liver and gall bladder.

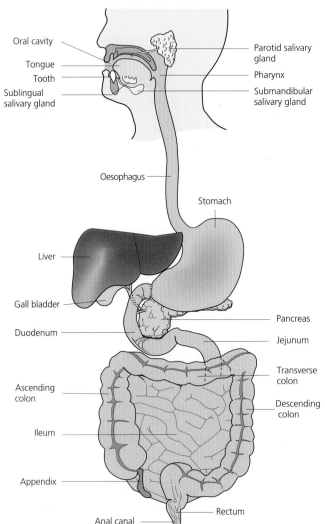

Oral cavity
Tongue
Tooth
Sublingual salivary gland
Parotid salivary gland
Pharynx
Submandibular salivary gland
Oesophagus
Stomach
Liver
Gall bladder
Duodenum
Pancreas
Jejunum
Transverse colon
Ascending colon
Descending colon
Ileum
Appendix
Rectum
Anal canal

Figure 24.1 A schematic diagram showing the GI tract and its principal accessory organs. The different parts are not drawn to scale.

24.3 The principal structural features of the gut wall

Figure 24.2 illustrates the basic plan of the layers of the gut wall. Although specific adaptations are seen along the length of the gut, the main features of the wall of the GI tract are similar throughout. The wall consists of an outer layer known as the serosa, two layers of smooth muscle, the submucosa and the mucosa. Linking the gut to the abdominal wall is a thin sheet of connective tissue called the mesentery, through which blood vessels, lymphatics and nerves reach the GI tract.

The outermost layer of the GI tract, the serosa, forms the visceral peritoneum, which is continuous with the parietal peritoneum that lines the abdominal cavity.

The peritoneum consists of a layer of simple squamous epithelium (see p. 74). The mesentery consists of a layer of loose connective tissue covered by two layers of peritoneum.

The smooth muscle of the gut wall allows the gut both to move food and digestion products along the tube at an appropriate rate and to mix food with digestive secretions (see next section). Throughout most of the gut, the smooth muscle is arranged in two distinct layers (together known as the muscularis externa), one in which the muscle fibres are circularly arranged and a thinner outer layer in which the fibres are arranged

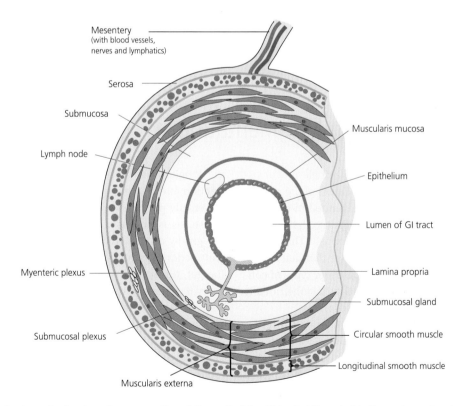

Figure 24.2 A schematic drawing of a cross section of the wall of the GI tract. Although this figure shows the structural features common to all parts of the gut, there are considerable variations in the organization of the main layers of the different regions. The epithelium, lamina propria and muscularis mucosa form the mucosa.

longitudinally. The stomach possesses a third, obliquely arranged smooth muscle layer inside the circular layer, which allows vigorous churning and mixing contractions that help to break down food into small pieces. At intervals throughout the tract, the circular smooth muscle is thickened and modified to form muscular rings called sphincters. Sphincters control the passage of material from one part of the gut to another. They are found at the junctions between the oesophagus and stomach, the stomach and duodenum, the ileum and caecum, and in the anus where they control the process of defecation.

The submucosa consists of loose connective tissue with collagen and elastin fibrils, blood vessels, lymphatics and, in some regions, exocrine glands. The mucosa is the innermost layer of the gut wall. It is further subdivided into three regions: a layer of epithelial cells, a basement membrane (the lamina propria) and a thin layer of smooth muscle – the muscularis mucosa. The characteristics of the epithelium vary greatly from one region of the GI tract to another.

For example, an abrasion-resistant stratified squamous epithelium lines the oesophagus, but in the small intestine the absorptive surface is covered with a simple columnar epithelium.

> **KEY POINTS:**
>
> - The wall of the GI tract consists of the following principal layers beginning from the outside: the serosa, a layer of longitudinal smooth muscle, a layer of circular smooth muscle, the submucosa and the mucosa.
>
> - The passage of material from one part of the gut to another is controlled by rings of smooth muscle called sphincters. They are found at the junctions between the oesophagus and stomach, the stomach and duodenum, the ileum and caecum, and the anus.

24.4   Peristalsis and segmentation

The smooth muscle of the gut is of the single-unit or visceral type (see Chapter 9 for more details of the properties of smooth muscle). Signals generated in one fibre are propagated to neighbouring fibres so that sections of smooth muscle contract synchronously. The smooth muscle fibres maintain a level of tone that determines the length and diameter of the gut. Other types of contraction are superimposed upon this basal tone, notably peristaltic and segmentation contractions.

Peristaltic contractions are concerned with the propulsion of GI contents along the tract and consist of successive waves of contraction and relaxation of the smooth muscle, as shown in Figure 24.3. The longitudinal muscle contracts first and halfway through its contraction the circular muscle begins to contract. During the latter part of the circular muscle contraction,

the longitudinal muscle relaxes. This alternate contraction and relaxation of the muscle layers results in the steady progression of material along the GI tract. Peristalsis is normally triggered by the distension of the gut by a food bolus. The frequency of peristaltic contractions gradually decreases with distance along the gut, for example contractions occur roughly once every 5 seconds in the duodenum but only once every 15 seconds or so in the colon.

Segmentation, which is the main motility pattern in the small intestine, allows mixing of the GI contents and facilitates digestion. Segmentation contractions occur at around 12 contractions per minute in the duodenum but less frequently in the jejunum and ileum. Their frequency is increased by the arrival of partially digested food (known as chyme) in the duodenum.

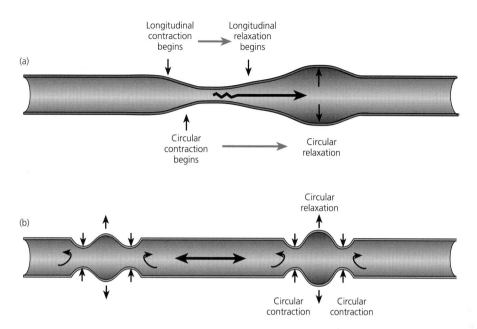

Figure 24.3 The two main types of contractile activity in the smooth muscle of the GI tract: (a) peristaltic contractions, which propel the food thorough the GI tract; (b) segmentation movements, which promote both the mixing of food with the intestinal secretions and the presentation of the products of digestion to the mucosal surface for absorption.

24.5 The blood supply of the gastrointestinal tract

The gut receives a rich blood supply. Within the abdominal cavity, the gut can be divided into the foregut (the lower oesophagus, stomach and upper duodenum), the midgut (the last part of the duodenum, jejunum, ileum, and ascending and transverse colon) and the hindgut (the last part of the transverse colon to the anal canal). The foregut is supplied by the coeliac artery via the dorsal mesentery. The midgut is supplied by the superior mesenteric artery, as shown diagrammatically in Figure 24.4, and the hindgut is supplied by the inferior mesenteric artery (see Figure 24.32 below).

The combined circulation to the stomach, liver, pancreas, intestine and spleen (which has no digestive function) is called the splanchnic circulation and accounts for 20–25% of the cardiac output at rest. Following a meal, blood flow is even greater as it is stimulated by products of digestion and hormones including gastrin and cholecystokinin (CCK) (see below). Note that

venous blood from the stomach, pancreas, intestine and spleen ultimately flows into the portal vein, which provides about 70% of the blood supply to the liver. The remainder of the hepatic blood supply is provided by the hepatic artery, which supplies most of the oxygen required by the liver. The portal circulation allows the rapid delivery of digestion products from the GI tract to the liver where they will undergo further processing. The organization of the splanchnic circulation is illustrated in Figure 24.5.

> **KEY POINT:**
>
> • The gut receives a rich blood supply. The combined circulation to the stomach, liver, pancreas, intestine and spleen is called the splanchnic circulation. At rest, it accounts for 20–25% of the cardiac output.

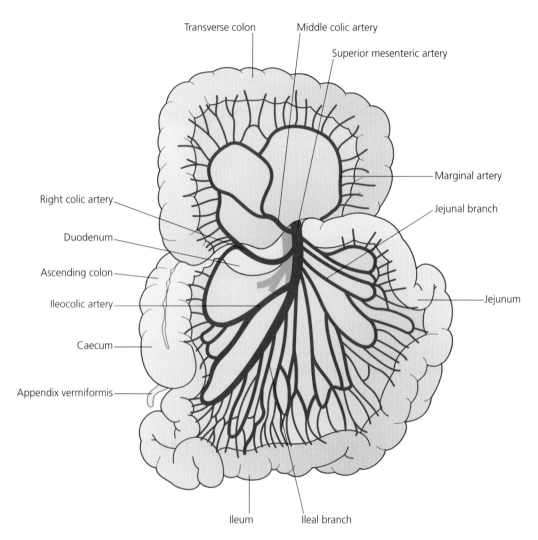

Figure 24.4 The arterial blood supply to the small intestine and upper colon. The superior mesenteric artery gives rise to a number of smaller arteries, which then give rise to a series of loops and from these small arteries branch off to supply different segments of the wall of the intestine. The venous drainage follows a similar pattern.

24.6 Regulation of the gastrointestinal tract by nerves and hormones

The gut has an extensive afferent and efferent innervation, which allows for fine control of secretory and motor activity. Furthermore, a number of peptide hormones influence both the motility and secretory function of parts of the GI tract. The innervation of the gut consists of intrinsic pathways (the enteric nervous system) and autonomic extrinsic pathways.

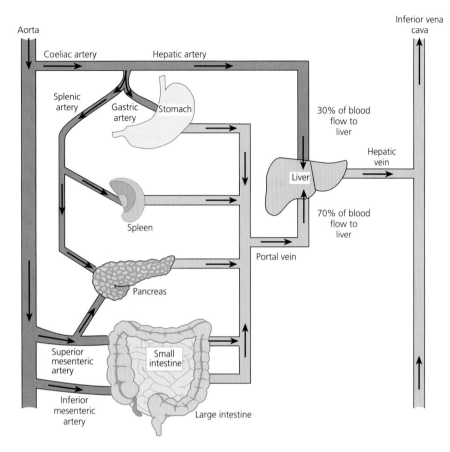

Figure 24.5 A schematic representation of the splanchnic circulation. Note the dual blood supply to the liver.

The enteric nervous system

The enteric nervous system is made up of two networks of nerve fibres and ganglion cell bodies situated along the length of the wall of the GI tract. These nerve networks are called intramural plexuses and their location is illustrated in Figure 24.2. The myenteric plexus (also called Auerbach's plexus) lies between the circular and longitudinal smooth muscle layers while the submucosal plexus (also called Meissner's plexus) lies within the submucosa. There are neuronal connections between the two plexuses and between the enteric nervous system and the extrinsic nerve supply. The myenteric plexus regulates the motility of the GI tract. The submucosal plexus is concerned chiefly with the regulation of the secretory activity and blood flow to the gut.

In an inherited disorder known as Hirschsprung's disease, those nerve cells that normally migrate in early development to form the ganglia of the myenteric plexus of the lower parts of the GI tract fail to do so. Consequently, the affected regions do not receive the normal nerve signals that propel the intestinal contents onwards with the result that the motility of the gut is reduced. This in turn leads to severe constipation or even intestinal obstruction.

Afferent innervation of the gastrointestinal tract

Chemoreceptors and mechanoreceptors are present in the mucosa and outer muscle layers of the gut wall. Chemoreceptors are stimulated by substances present in foods while mechanoreceptors respond to distension or irritation of the gut wall. Afferents arising from these receptors mediate secretory or motor responses either through local reflexes or via the CNS. Many of the afferents travel to the CNS in the vagus nerve.

The extrinsic innervation of the gastrointestinal tract

The nerve plexuses of the enteric nervous system are linked to the CNS via afferent fibres and receive efferent input from both the sympathetic and parasympathetic divisions of the autonomic nervous system, as shown in Figure 24.6. (The organization of the autonomic nervous system is discussed in Chapter 13.)

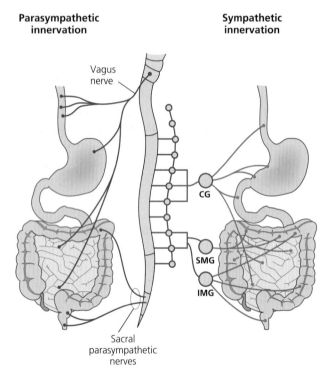

Figure 24.6 The innervation of the gut by the parasympathetic and sympathetic divisions of the autonomic nervous system. Preganglionic fibres are shown in blue and postganglionic sympathetic fibres in green. Postganglionic parasympathetic fibres are embedded within the organs served by the parasympathetic preganglionic fibres and are omitted for clarity. CG, coeliac ganglion; IMG, inferior mesenteric ganglion; SMG, superior mesenteric ganglion.

Sympathetic innervation. The preganglionic fibres of the GI sympathetic innervation arise from the thoracic and upper lumbar regions of the spinal cord (segments T8–L2). The cell bodies of the postganglionic fibres lie within the coeliac, superior and inferior mesenteric and hypogastric plexuses. Some of the sympathetic postganglionic fibres innervate secretory cells or circular smooth muscle fibres via the enteric nervous system, while others directly innervate the smooth muscle of arterioles within the GI tract. In general, sympathetic stimulation reduces the secretion and motor activity of the gut. In contrast, the circular smooth muscle of the GI sphincters shows increased constriction in response to sympathetic stimulation. Constriction of the arterioles following sympathetic activation directs blood away from the splanchnic blood vessels, and is part of the response to exercise or haemorrhage.

Parasympathetic innervation. The vagus nerve relays parasympathetic innervation to the oesophagus, stomach, small intestine, liver, pancreas, caecum, appendix, ascending colon and transverse colon. The remainder of the colon receives parasympathetic innervation via the hypogastric plexus. All of the parasympathetic fibres terminate within the myenteric plexus. Increased activity in the parasympathetic nerves stimulates both the motility and the secretory activity of the gut. The parasympathetic innervation therefore opposes the effects of the sympathetic innervation.

Hormonal regulation of the gut

The effects of the GI innervation on secretion and motility are supplemented by those of a variety of peptide hormones acting through endocrine or paracrine pathways. The relative importance of nervous and hormonal influences differs throughout the tract. Salivary secretion, for example, is controlled almost exclusively by the autonomic nerves, while pancreatic secretion is regulated principally by hormones. Three major hormones that exert an endocrine influence on the activity of the gut are:

- Gastrin (secreted by the stomach)
- Secretin (secreted by the duodenum)
- Cholecystokinin (secreted by the duodenum).

> **KEY POINTS:**
>
> - The activity of the GI tract is regulated by the enteric nervous system, by hormones and by extrinsic nerves.
>
> - The enteric nervous system consists of networks of nerve cells in the wall of the GI tract. These nerve networks mediate a number of intrinsic reflexes that control secretory and contractile activity.
>
> - Afferent fibres send information to the brain mainly via the vagus nerve. Both sympathetic and parasympathetic nerves supply the GI tract. The parasympathetic innervation acts to increase gut motility and secretory activity.
>
> - Both endocrine and paracrine hormones play an important role in regulating the secretory and motor activity of the GI tract.

24.7 Ingestion of food, chewing and salivary secretion

Food is ingested via the mouth (also known as the oral or buccal cavity) where it stimulates the taste buds, is chewed and mixes with saliva. As a result, food is tasted, broken into smaller pieces and lubricated in readiness for swallowing. The presence of food in the mouth plays an important part in the control of gastric secretion (see below, p. 509).

The mouth

The external opening of the mouth is bounded by the upper and lower lips. Within the mouth lies the tongue, which is attached to the hyoid bone and the lower jaw (or mandible). The cheeks form the walls of the mouth and act together with the tongue to position the food between the teeth when chewing. The hard palate forms the anterior roof of the mouth while the soft palate forms the posterior roof. A fold of tissue known as the uvula hangs down from the soft palate in front of the oropharynx, the passage by which food leaves the mouth and enters the oesophagus. At the back of the mouth lies a mass of lymphoid tissue, the palatine tonsils, which form part of the gut-associated lymphoid tissue (GALT; see Chapter 21). The whole of the buccal cavity is lined by moist mucous membranes and is covered by a stratified epithelium, 15–20 layers of cells thick, which is adapted to withstand considerable mechanical shearing forces.

The teeth

The teeth are embedded in sockets within the bones of the upper and lower jaws (the maxilla and mandible,

respectively). During life, the mouth has two sets of teeth, the deciduous or milk teeth and the permanent teeth of adult life. The deciduous teeth normally erupt between the ages of around 6 to 30 months with the incisors erupting first, followed by the first molars, the canines and finally the second molars. In all, there are 20 deciduous teeth. The permanent teeth erupt over a period of 10–15 years from the age of 6 or so and replace the deciduous teeth, which are gradually shed. There are 32 permanent teeth: eight incisors, four canines, eight premolars and 12 molars. Each tooth has a characteristic morphology related to its function. The incisors are adapted for biting and cutting food while the premolars and molars are flatter and broader, with several cusps that enable them to grind food during mastication (chewing). Figure 24.7 illustrates the structural features of an incisor. The tooth is subdivided into the crown, which protrudes from the gum (or gingiva), and the root, which is embedded in the bone of the jaw and held in the socket by cement and the periodontal ligament. The pulp cavity, in the centre of the tooth, contains blood vessels, lymphatics and nerves. The pulp cavity is surrounded by a hard layer of dentine. The crown of the tooth is covered by a layer of even harder material called enamel. The teeth are innervated by branches of the trigeminal nerve (CN V).

The tongue

The tongue is composed of skeletal (voluntary) muscle and occupies the floor of the mouth. Its upper surface

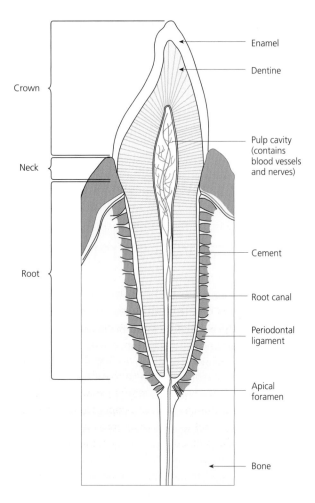

Figure 24.7 A diagrammatic representation of a longitudinal section through a canine tooth to illustrate the major structural features.

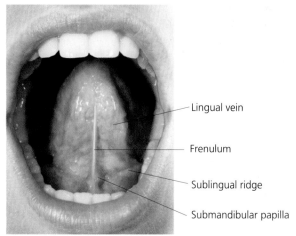

Figure 24.8 A view of the underside of the tongue showing the frenulum, openings of the submandibular glands (the submandibular papillae) and the lingual veins.

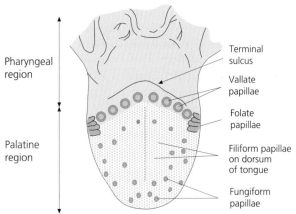

Figure 24.9 A view of the upper surface (or dorsum) of the tongue, showing the location of the three types of papillae concerned with the sense of taste.

is covered with stratified squamous epithelium while its underside is covered with smooth, thin mucous membrane. As mentioned above, it is attached to the hyoid bone and the lower jaw at the back of the mouth. A fold of mucous membrane called the frenulum forms the anterior attachment to the floor of the mouth, which can easily be seen by raising the tongue (Figure 24.8). Occasionally the frenulum is very short and interferes with the ability of the tongue to facilitate speech ('tongue-tie'). In such cases, the frenulum is cut to free the tongue. In addition to its role in speech, the tongue is necessary for chewing, swallowing and the perception of taste (see Chapter 14).

The upper surface (or dorsum) of the tongue is divided into two main regions by a V-shaped depression called the terminal sulcus: an anterior or palatine region and a posterior or pharyngeal region (Figure 24.9). All over its upper surface there are small projections called papillae, which give the tongue its roughness. Four different types can be identified: filiform, folate, fungiform and vallate. The large, round vallate papillae are located just in front of the terminal sulcus. The filiform papillae are minute, pointed projections that cover most of the anterior part of the or dorsum of the tongue. The folate papillae are located on the sides of the tongue while the fungiform papillae are scattered over the upper surface. The taste buds are associated with the folate, fungiform and vallate papillae.

Mastication (chewing)

Chewing involves movements of the jaw and tongue. The muscles of the cheeks also participate in the process by holding the food within the buccal cavity. Considerable forces can be generated on the molars during chewing, although this is significantly reduced in people who wear dentures. As a result of chewing, food is broken into smaller pieces and mixed with saliva, which lubricates it and prepares it for swallowing.

The secretion of saliva

Saliva is secreted into the mouth by three pairs of salivary glands at a rate of around 1500 ml day^{-1}. These are the submandibular, sublingual and parotid salivary glands. Each gland consists of a number of lobules surrounded by a fibrous capsule. Each lobule is made up of a ball of cells called acinar cells, which is drained by a small duct. These combine to form larger ducts that eventually empty into the mouth. During active secretion, about 70% of the saliva produced by the glands comes from the submandibular glands, which, as their name implies, are situated below the mandible (Figure 24.1). A further 25% is contributed by the large parotid glands, which lie at the angle of the jaw and whose ducts drain into the mouth opposite the second molars on each side. The remaining 5% is secreted by the sublingual glands situated below the tongue in the floor of the mouth (the sublingual pockets).

Saliva contains water, electrolytes, mucus, an antibacterial substance called lysozyme and an enzyme called salivary α-amylase (ptyalin). It performs several important functions: saliva softens and lubricates food to facilitate swallowing; it dissolves substances in foods to make them available to the taste cells of the tongue; and it initiates the digestion of starch to smaller polysaccharides (by the enzyme salivary amylase). Saliva also maintains oral comfort, promotes wound healing in the mouth and lowers the risk of oral infection. Finally, by maintaining the mucous membranes in a moist state, it facilitates speech.

The regulation of salivary secretion

The rate at which saliva is produced is controlled primarily by reflexes mediated by the autonomic nervous system. The rate of secretion may increase from the resting value of around 0.5 ml min^{-1} to as much as 7 ml min^{-1} during maximal stimulation by the taste and smell of food. Sensory receptors in the mouth, pharynx and olfactory area relay information about the presence of food in the mouth, its taste and smell, to the salivatory nuclei, which are situated in the medulla of the brain. This area also receives information from the hypothalamic appetite area and regions of the cerebral cortex associated with the perception of taste and smell. Activity in the parasympathetic nerves supplying the salivary glands promotes an increase in both blood flow to the glands and the rate of secretion of watery saliva rich in amylase and mucus. The acetylcholine secreted by the parasympathetic nerve endings acts on muscarinic receptors. For this reason, administration of muscarinic antagonists such as atropine and hyoscine (scopolamine) inhibits salivary secretion and these drugs are used in premedication for general anaesthesia to reduce salivary secretion.

The response of the salivary glands to sympathetic stimulation is variable. Sympathetic stimulation tends to enhance the output of amylase by the acinar cells but at the same time significantly reduces blood flow to the glands. As a result, the net effect is a fall in the rate of salivary secretion. Indeed, a dry mouth is a characteristic feature of the sympathetic response to stress and fear. A summary of the mechanisms involved in the control of salivary secretion is shown in Figure 24.10.

Xerostomia is the name given to the condition in which the mouth becomes very dry as a result of a significant reduction in salivary secretion. It is relatively common, and has important consequences and many causes. Chewing, tasting and swallowing are all compromised in this condition and there is a predisposition to dental caries (tooth decay) and mouth infections after dental or oral surgery. The most common causes include irradiation of the head and neck, and medications such as tricyclic antidepressants and drugs that mimic the effects of activation of the

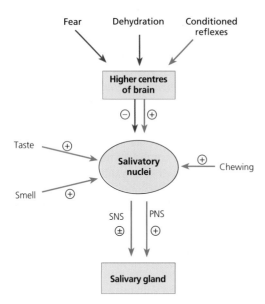

Fear   Dehydration   Conditioned
reflexes

**Higher centres
of brain**

Taste

Smell

**Salivatory
nuclei**

Chewing

SNS   PNS

Salivary gland

Figure 24.10 A summary of the major factors that influence the secretion of saliva. Factors that stimulate secretion are shown in green while those that inhibit secretion are shown in red. PNS, parasymapthetic nervous system; SNS, sympathetic nervous system.

sympathetic nervous system (sympathomimetic drugs). Certain autoimmune inflammatory conditions can also cause xerostomia. Xerostomia is sometimes treated with **pilocarpine**, a drug that mimics the effects of stimulation of the parasympathetic nerves.

The digestive actions of saliva

Although the principal digestive function of saliva is to lubricate food and facilitate swallowing, saliva contains salivary amylase, an enzyme that is able to degrade complex polysaccharides such as starch and glycogen to maltose, maltriose and dextrins. As this enzyme works optimally at pH 6.9, it becomes inactivated once the food is completely mixed with gastric juice in the stomach.

> **KEY POINTS:**
>
> - In the mouth, food is mixed with saliva as it is chewed. The salivary glands (parotid, submandibular and sublingual) secrete about 1500 ml of saliva each day, which contains mucus, and an α-amylase, which initiates the breakdown of starch.
>
> - Salivary secretion is controlled by reflexes mediated by the autonomic nervous system. Parasympathetic stimulation promotes an abundant secretion of watery saliva rich in amylase and mucus. Blood flow to the salivary glands is also enhanced. Sympathetic stimulation promotes the output of amylase but reduces blood flow to the glands. The overall effect of sympathetic stimulation is generally a reduction in the rate of salivary secretion.

24.8   Swallowing (deglutition)

As a result of chewing, the food is formed into a softened ball or bolus, which is then swallowed. From the mouth, the bolus moves into the oropharynx (the part of the pharynx that lies just behind the mouth) and then into the laryngopharynx, both of which are common passages for food, fluids and air. Swallowing occurs in three phases. The first (oral) phase is voluntary while the subsequent pharyngeal and oesophageal phases are involuntary and are mediated by the autonomic nervous system.

As the involuntary phase of the swallowing reflex begins, the soft palate is raised towards the posterior wall of the pharynx to prevent food entering the nasal passages. Swallowing also initiates a wave of peristaltic contraction that propels the food bolus through the relaxed upper oesophageal sphincter (Figure 24.11) into the oesophagus. The larynx rises, the glottis closes, the epiglottis tilts to cover the closed glottis and respiration is inhibited (deglutition apnoea). In this way, food is prevented from

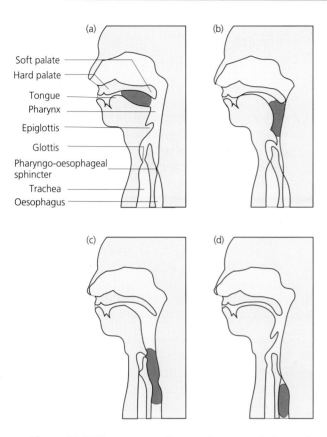

Soft palate
Hard palate
Tongue
Pharynx
Epiglottis
Glottis
Pharyngo-oesophageal sphincter
Trachea
Oesophagus

(a) (b) (c) (d)

Figure 24.11 The sequence of events that occurs during swallowing. The food bolus is shown in brown. Note how the tongue first moves the bolus from the mouth to the pharynx to initiate the swallowing reflex proper (compare (a) and (b)). As the bolus reaches the larynx, the epiglottis closes off the airway to prevent food particles entering the trachea.

entering the trachea. In the final (oesophageal) phase of swallowing, the wave of contraction that was initiated in the pharynx continues along the length of the oesophagus and carries the food towards the stomach. The wave takes 7–9 seconds to travel the length of the oesophagus and is normally sufficient to transport the bolus all the way to the stomach. Mucus, secreted by glands of the oesophageal submucosa, facilitates the movement of the food bolus. Fluids normally reach the stomach much more quickly (within 1–2 seconds).

At the junction between the oesophagus and the stomach there is a modified ring of circular smooth muscle, the lower oesophageal (or cardiac) sphincter. At rest, this remains closed to prevent stomach contents entering the oesophagus, but as the wave of peristalsis reaches the lower part of the oesophagus, the sphincter relaxes to allow the food bolus to enter the stomach.

Reflux of gastric contents into the oesophagus may occur as a result of a number of conditions. Gastro-oesophageal reflux occurs when the abdominal pressure exceeds the thoracic pressure, or when the lower oesophageal sphincter remains relaxed for long periods. The former may occur following a heavy meal, particularly on lying down, during bending and stooping, or during pregnancy, in which the growing baby causes displacement of the abdominal organs and a rise in abdominal pressure. The presence of gastric acid in the oesophagus (and occasionally even the pharynx) causes retrosternal pain and discomfort known as 'heartburn' or 'indigestion'.

Gastro-oesophageal reflux may also be caused by **hiatus hernia**. There are two main forms of this condition, the sliding hernia and the rolling hiatus hernia (or para-oesophageal hernia). In the sliding hernia, the most common type of hiatus hernia, the gastro-oesophageal junction slides through the oesophageal hiatus (aperture) in the diaphragm and lies above the diaphragm. In a rolling hiatus hernia, the sphincter remains below the diaphragm but part of the stomach rolls up through the hiatus to lie alongside the oesophagus (Figure 24.12). Treatments for gastro-oesophageal reflux disorder (GORD) include lifestyle advice, antacids such as sodium bicarbonate, calcium carbonate, magnesium and aluminium salts, which neutralize the acidic chyme, and drugs such as ranitidine and omeprazole that inhibit gastric acid secretion (see p. 513).

Dysphagia

Dysphagia, or difficulty in swallowing, affects many people and its symptoms range from mild discomfort of the throat during swallowing to an inability to swallow either liquid or solid foods comfortably. Common causes of dysphagia include dry mouth, chronic heartburn, lack of co-ordination of the muscles of swallowing, oesophageal strictures and tumours, and CNS disorders such as stroke, multiple sclerosis and Parkinson's disease.

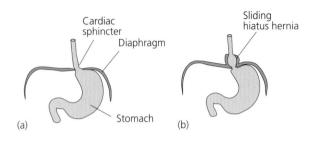

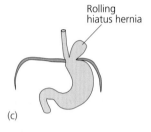

Figure 24.12 Diagrammatic representation of a sliding hiatus hernia (b) and a rolling hiatus hernia (c) compared with the normal arrangement (a).

KEY POINTS:

• The first stage of swallowing is voluntary but subsequent stages are under reflex autonomic control.

• As food is pushed to the back of the mouth by the tongue, a wave of peristalsis is initiated, which propels the food bolus through the upper oesophageal sphincter into the oesophagus and thence to the stomach.

24.9 The stomach

Below the oesophagus, the GI tract expands to form the stomach, which lies on the left side of the upper abdominal cavity. The stomach is continuous with the oesophagus at the lower oesophageal (cardiac) sphincter and the duodenum at the pyloric sphincter. Although its size and shape varies between individuals and with its degree of fullness, the stomach is about 25 cm long and roughly J-shaped. When empty, the stomach has a volume of around 50 ml but this can increase to 4 litres or so after a large meal. As the stomach fills, longitudinal folds (rugae) on the inner surface of the empty stomach flatten to allow distension.

Figure 24.13 illustrates the gross anatomy of the stomach. The cardiac region (or cardia) surrounds the cardiac orifice through which food enters the stomach. The expanded part lateral to the cardiac region is the **fundus**. The mid-region is the body, which merges with the funnel-shaped pyloric region. The widest part of the pyloric region is called the **antrum**, which narrows to form the pyloric canal terminating in the pylorus itself. The convex lateral surface of the stomach is called the greater curvature while the concave medial surface is the lesser curvature. A double layer of peritoneum (the lesser omentum) extends from the liver to the lesser curvature. The greater omentum, another layer of the peritoneum, covers the abdominal organs and attaches to the transverse colon and the posterior body wall (Figure 24.14). This layer contains fatty tissue, which helps to protect and insulate the abdominal organs, and collections of lymph nodes, which house cells of the immune system.

The functions of the stomach include:

• The temporary storage of food

• Mechanical breakdown of food to form a semiliquid mixture of gastric juice and small food particles (chyme)

• The secretion of acid by the gastric glands

• Chemical digestion of proteins to polypeptides by pepsins (proteolytic enzymes)

• The controlled release of chyme into the duodenum

• The secretion of intrinsic factor, required for the absorption of vitamin B_{12} by the terminal ileum.

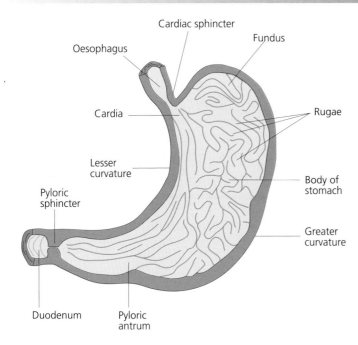

Figure 24.13 A longitudinal view to show the major anatomical regions of the stomach. Note the prominent folds of the stomach mucosa known as rugae.

Gastric secretion and motility are regulated by both hormonal and nervous pathways. Mucus secreted by the gastric glands also plays an important part in protecting the stomach from erosion by the acidic, proteolytic contents of the gastric juice.

Blood supply of the stomach

The arterial blood supply to the stomach is provided by the gastric arteries, which arise from the coeliac artery to form a plexus (network) of vessels within the submucosal layer. From here, the vessels branch extensively to provide the mucosal layer with a rich vascular network. Venous drainage is via the gastric veins. These empty into the portal vein that carries blood to the liver. Blood flow to the stomach increases significantly in response to a meal. The lymphatics of the stomach lie along the arteries and drain into lymph nodes around the coeliac artery.

Structure of the gastric mucosa

The general organization of the stomach wall resembles that illustrated in Figure 24.2. However, as well as the longitudinal and circular smooth muscle layers, the muscularis

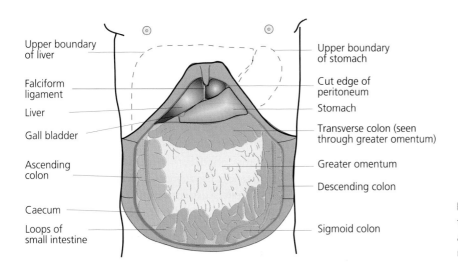

Figure 24.14 A diagram showing the anterior abdominal wall cut away to reveal the viscera in their normal positions.

externa of the stomach possesses a third layer of smooth muscle between the circular layer and the submucosa. In this layer, the smooth muscle fibres are arranged obliquely and this orientation may contribute to the grinding, churning and mixing movements that are characteristic of the stomach. Furthermore, the circular layer varies in thickness in different parts of the stomach, being thickest in the antral region where the most vigorous contractile activity occurs. In the pylorus, the circular muscle is modified to form the pyloric sphincter, which regulates gastric emptying.

The stomach is lined by a layer of simple columnar epithelial cells that secrete an alkaline fluid rich in mucus. This fluid provides a protective layer over the stomach lining, which helps to prevent autodigestion. In the fundus and body of the stomach, the epithelial layer contains many small openings ('gastric pits') into which the secretions of the gastric glands empty. The gastric pits occupy about half of the surface area of the mucosa.

Figure 24.15 is a diagrammatic representation of a gastric gland to illustrate the different cell types. The gastric glands are mainly simple branched tubes, lined with a single layer of cells, that changes in its characteristics with the distance from the surface epithelium. Each cell type has specific morphological characteristics that adapt them to their particular secretory role. The main cell types are:

- **Mucous neck cells** situated near to the opening of the gland. These cells secrete mucus.
- **Chief cells**, located in the deeper parts of the glands, which secrete pepsinogen, the inactive form of the protein-digesting enzyme pepsin.
- **Parietal cells** (also known as **oxyntic cells**), which are scattered between the chief cells and secrete hydrochloric acid and intrinsic factor.
- **Enteroendocrine cells** or G cells (also known as argentaffin cells), which are also found amongst the chief cells. These secrete the hormone gastrin into the bloodstream.

Glands in the fundus and body of the stomach are rich in both parietal and chief cells while those in the antral and pyloric regions secrete predominantly pepsinogen, mucus and gastrin. In the cardiac region, the gastric glands consist almost entirely of mucus-secreting cells. These regional variations mean that the consequences of partial gastrectomy will depend upon the particular part of the stomach that has been removed.

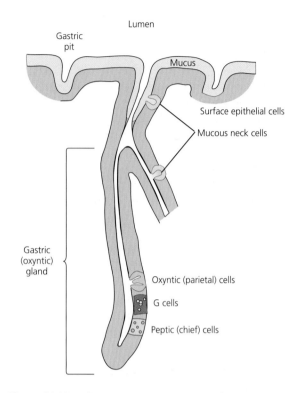

Figure 24.15 A diagrammatic representation of a gastric gland to show the various secretory cell types present. The parietal cells secrete hydrochloric acid, the chief cells secrete pepsinogen and the G cells secrete the hormone gastrin.

KEY POINTS:

- The functions of the stomach include storage of food, mixing, churning and kneading the food to produce chyme, and the secretion of acid, enzymes, mucus and intrinsic factor.
- In addition to the circular and longitudinal smooth muscle layers, the stomach wall possesses a third, obliquely arranged muscle layer that promotes churning movements.
- The surface epithelium of the gastric mucosa is composed almost entirely of cells that secrete an alkaline fluid containing mucus. Gastric acid is secreted by the gastric glands.
- The gastric glands contain mucous cells, chief cells, which secrete pepsinogens, and parietal cells, which produce gastric acid and intrinsic factor. A variety of enteroendocrine cells is also present, for example the G cells, which secrete gastrin.

Secretion by the gastric glands

The gastric glands produce around 2–3 litres of gastric juice each day. Secretion is minimal during fasting but increases significantly following the ingestion of a meal. Gastric juice contains water, electrolytes (salts), hydrochloric acid, pepsinogens and intrinsic factor. The electrolyte composition depends upon the rate of secretion but the potassium content of gastric juice is always higher than that of the plasma. For this reason, prolonged vomiting may cause not only metabolic alkalosis (as stomach acid is lost) but also hypokalaemia (low plasma potassium).

The secretion of hydrochloric acid

The ingestion of a meal stimulates the secretion of hydrochloric acid by the parietal cells of the gastric glands. As a result, the pH of gastric juice falls to around 2. This highly acidic environment is important for several reasons:

- It helps to break down the connective tissue and muscle fibres of meat.
- It activates pepsinogens to active pepsins and provides optimal conditions for their enzymic activity.
- It aids in the absorption of calcium and iron by allowing them to form soluble salts.
- It kills many of the micro-organisms that may cause GI infection, e.g. salmonella, cholera, dysentery and typhoid.

Parietal calls are structurally adapted for the secretion of hydrochloric acid. Their cytoplasm contains numerous mitochondria and they are filled with a branching system of tubules derived from the endoplasmic reticulum. These tubules are lined by microvilli that possess the apparatus for hydrogen ion secretion. When a meal is eaten and the gastric glands are stimulated to secrete, the tubular structures fuse to form deep invaginations of the apical membrane known as secretory canaliculi. In this way, the surface area of the apical surface of a parietal cell is significantly increased.

The steps involved in the secretion of hydrochloric acid are summarized in Figure 24.16. Both hydrogen and chloride ions are secreted by active transport against concentration and electrochemical gradients, respectively, so that acid secretion is highly energy dependent. Hydrogen ion secretion is achieved by proton pumps in the secretory membranes of the canaliculi that exchange hydrogen ions for potassium ions (H^+, K^+ ATPase). The hydrogen ions are derived from the dissociation of carbonic acid within the parietal cells. The bicarbonate ions generated by this ionization leave the parietal cells in exchange for chloride ions at the basolateral membrane. As a result of this movement of bicarbonate ions out of the cell, the venous blood leaving the stomach after a meal is slightly more alkaline than the arterial blood (the 'alkaline tide').

The secretion of enzymes by the gastric glands

A number of proteolytic enzymes are secreted by the chief cells of the gastric glands. These are known collectively as pepsin and are stored and secreted as inactive precursors called **pepsinogens**. In the acidic environment of the stomach, the pepsinogens are activated to pepsin, which is an endopeptidase. This means that it hydrolyses peptide bonds within the protein molecule to liberate polypeptides and a few free amino acids. A lipase is also secreted by the gastric glands. It shows the greatest activity against the triglycerides found in milk so it is probably more significant in children than in adults.

The secretion of intrinsic factor by the gastric glands

The parietal cells of the stomach glands secrete a glycoprotein called intrinsic factor, which is essential for the absorption of vitamin B_{12}. Intrinsic factor binds to the vitamin in the duodenum and protects it from the actions of the enzymes of the small intestine. The complex of vitamin B_{12} and intrinsic factor is absorbed into the blood as chyme passes through the lower part of the ileum. Vitamin B_{12} is needed for the production of mature red blood cells (see Chapter 20) and, in its absence, a condition called pernicious anaemia develops in which abnormally large, immature red blood cells are produced by the bone marrow. People who have had part or all of their stomach removed (gastrectomy) require regular intramuscular injections of vitamin B_{12}.

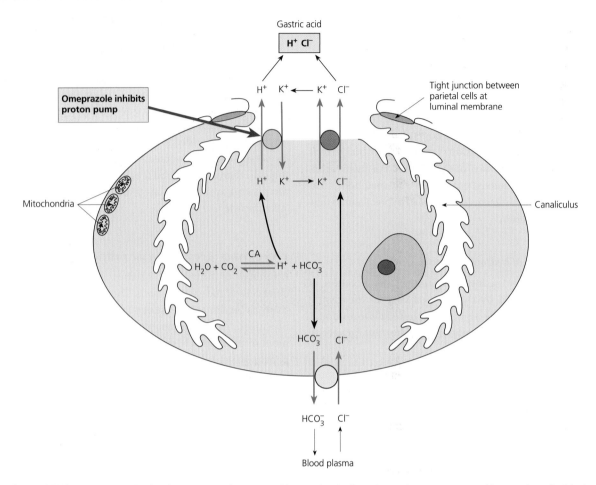

Figure 24.16 The steps involved in the secretion of gastric acid by a parietal cell. Hydrogen ions are generated from carbon dioxide via the action of carbonic anhydrase (CA) and the subsequent ionization of carbonic acid. The hydrogen ions are pumped into the lumen of the stomach in exchange for potassium ions. Chloride ions enter the cell in exchange for bicarbonate ions and are transported into the stomach lumen together with potassium ions. The widely used drug omeprazole reduces acid secretion by direct inhibition of the proton pump, which exchanges intracellular hydrogen ions for potassium ions.

KEY POINTS:

- Gastric juice contains salts, water, hydrochloric acid, pepsinogen and intrinsic factor.
- The parietal cells secrete hydrochloric acid and a glycoprotein called intrinsic factor, which is essential for the absorption of vitamin B_{12}.
- A number of proteolytic enzymes are secreted by the chief cells of the gastric glands. They are released as inactive pepsinogens and activated by the acidic environment in the stomach.

Why doesn't the stomach digest itself?

Gastric juice is extremely acidic and contains powerful proteolytic enzymes. The gastric mucosa is exposed to these corrosive conditions and must protect itself from autodigestion. It does this in several ways:

- Tight junctions between the mucosal epithelial cells help to prevent gastric juice from leaking into the underlying layers of tissue.
- Mucus, secreted by the neck cells and the surface epithelial cells, together with alkaline fluid from the

surface epithelial cells, forms a protective layer up to 200 μm thick over the mucosa.

- Prostaglandins of the E series increase the thickness of the mucous barrier, enhance the secretion of alkaline fluid and stimulate blood flow to the mucosa. In this way, an improved supply of nutrients is brought to any damaged area of mucosa.

- The gastric epithelium is renewed continuously through growth, migration and shedding (desquamation) of the epithelial cells.

The formation of gastric ulcers

If gastric juice is allowed to gain access to the mucosal cells, gastric irritation (gastritis) occurs, and eventually a gastric ulcer may develop in which the mucosa becomes eroded, exposing the submucosa to attack by the gastric contents. Such ulcers may bleed (as in the example shown in Figure 24.17) and may result in severe blood loss. Many drugs promote ulcer formation by altering the rates of acid and mucus production. They include caffeine, nicotine and non-steroidal anti-inflammatory drugs (NSAIDs) such as aspirin and ibuprofen. NSAIDs act by inhibiting synthesis of the prostaglandins that are necessary for normal mucus secretion. Stress is associated with an increased incidence of gastric ulcer formation. Infection of the gastric mucosa by the bacterium *Helicobacter pylori* (*H. pylori*) is the cause of over 80% of gastric and duodenal ulcers. This organism stimulates the production of gastrin, which, in turn, causes an increase in the rate of hydrochloric acid secretion (see below). It also interferes with the function of the protective mucosal barrier. Eradication of the infection by treatment with antibacterial drugs, such as clarithromycin (usually in combination with drugs that inhibit acid secretion directly), brings about a fall in acid secretion and allows the mucosa to heal.

The regulation of gastric secretion

Nervous and endocrine mechanisms interact to regulate the secretion of hydrochloric acid and pepsinogens by the gastric glands. Figure 24.18 summarizes these mechanisms. Gastric secretion is normally considered to occur in three overlapping phases: the cephalic, gastric and intestinal phases.

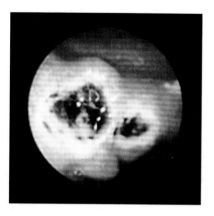

Figure 24.17 A picture of a gastric ulcer viewed through an endoscope. Note the unhealthy white areas and the dark regions, which indicate that the ulcer has been bleeding.

The cephalic phase begins before food arrives in the stomach and may account for up to 35% of the total secretion seen in response to a meal. This phase of secretion is stimulated by the anticipation of food, its smell, taste and its presence in the mouth. Nervous signals originating in the cerebral cortex or appetite-regulating areas of the brain travel to the stomach via the vagus nerve. This increases gastric secretion both directly and indirectly via the stimulation of gastrin secretion by G cells within the antral gastric glands. Furthermore, both activity in the vagus nerves and gastrin stimulate the secretion of histamine from mast cells in the gastric mucosa. Histamine acts on the H_2 receptors of parietal cells to stimulate the secretion of H^+ ions.

The gastric phase accounts for around 60% of total gastric secretion in response to a meal and begins once food has started to arrive in the stomach. Distension of the stomach activates mechanoreceptors in the stomach wall that activate both local reflexes (i.e. those reflexes controlled by nerve networks within the gut wall) and vagovagal reflexes (those in which receptors in the gut wall send information to the brainstem via the vagus nerves, which also carries the appropriate motor instructions to the gut). The vagovagal reflexes directly stimulate gastric secretion and increase the secretion of gastrin, which is a powerful stimulus for acid secretion by the parietal cells. Gastrin also enhances the release of enzymes and mucus. Certain amino acids derived from the breakdown of proteins by pepsin also exert a

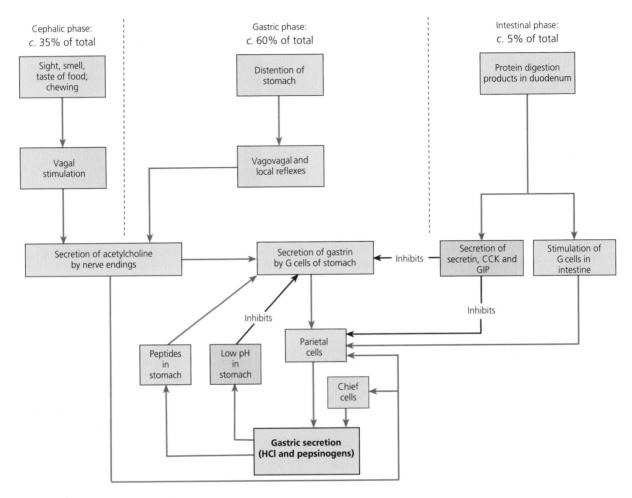

Figure 24.18 A flow chart illustrating the regulation of gastric secretion and the major factors involved. Secretin, cholecystokinin (CCK) and gastric inhibitory peptide (GIP) are secreted by enteroendocrine cells in the epithelium of the upper small intestine and have an inhibitory action on gastrin secretion, as does a low pH in the lumen of the stomach. The stimulatory action of gastrin on mucus and enzyme secretion is omitted for clarity.

direct stimulatory effect on the G cells, thus indirectly increasing gastric secretion. The gastric phase of secretion is self-limiting because gastrin secretion is inhibited when the pH of the stomach contents falls below 3. This inhibition is mediated by somatostatin secreted by the D cells of the gastric mucosa.

The intestinal phase of gastric secretion begins as partially digested food starts to enter the duodenum. This stimulates the secretion of a small amount of gastrin by G cells in the duodenal mucosa, which stimulates the gastric glands to secrete a further quantity of gastric juice. This phase of secretion makes only a small (5%) contribution to the total gastric secretion and is quickly inhibited by hormones of the upper small intestine, notably secretin, cholecystokinin (CCK) and gastric inhibitory peptide (GIP).

Disorders of gastric secretion and their treatment

Reduced gastric secretion. This is relatively rare and usually only affects the elderly or those who have atrophy (degeneration) of the gastric mucosa. Achlorhydria is a condition in which gastric acid secretion is reduced so that gastric pH never falls below pH 4. It is caused by a loss of parietal cells. Any condition that produces chronic gastritis can lead to parietal cell loss, but the most common is chronic infection with *H. pylori* (see above). Prolonged treatment with drugs that block the secretion of hydrogen ions may also lead to the development of achlorhydria. People with achlorhydria usually have normal digestive function but show an elevated level of circulating gastrin because the gastric contents are never acidic enough to inhibit gastrin secretion.

Increased gastric secretion. This is much more common and has many causes. Stress and a variety of drugs and constituents of foods are known to increase the rate at which HCl is produced by the gastric glands. A rare gastrin-secreting tumour of the pancreas causes Zollinger–Ellison syndrome in which gastric acid secretion is so high that erosion of the mucosa occurs, resulting in ulceration of the stomach wall.

Drug therapy for excess production of gastric acid

Several types of drugs may be used to treat excessive acid production and allow healing of the gastric mucosa. These include:

- **Antacids** such as sodium bicarbonate, magnesium hydroxide and aluminium hydroxide, which are often the first course of treatment. These work by raising the pH of the stomach contents but they also hasten gastric emptying and tend to stimulate the secretion of gastrin. Both of these effects tend to be counterproductive.

- **Mucosal strengtheners** such as **sucralphate** and **bismuth chelate**. These polymerize below pH 4 to form a sticky gel that adheres to the base of ulcers forming a coat that protects against further erosion and promotes healing.

- **Histamine H$_2$-receptor antagonists** such as **cimetidine**, **ranitidine**, **famotidine** and **nizaditine**. These block the actions of histamine on the H$_2$ receptors of the parietal cells, thereby reducing HCl secretion.

- **Proton pump inhibitors** such as **omeprazole** and **lansoprazole**. These drugs are inactive at neutral pH but become active in acidic environments and inhibit the H$^+$, K$^+$ ATPase (proton pump) that is responsible for transporting H$^+$ ions out of the parietal cells (Figure 24.16). Currently, these drugs are the most powerful agents available for suppressing the secretion of gastric acid.

The motor functions of the stomach

When the stomach is empty, it shows only weak contractile activity, but after a meal, weak peristaltic contractions begin in the body of the stomach. As the wave of contraction approaches the pyloric region, where the muscle is thicker, the contractile activity becomes more powerful, reaching maximum strength close to the gastroduodenal junction. These peristaltic waves occur roughly once every 20 seconds and serve to push the gastric contents forwards, towards the pyloric sphincter. As pressure in the antral region rises, the sphincter opens and a small amount of chyme (normally only a few millilitres) is delivered into the first part of the duodenum (the duodenal bulb). The sphincter closes again almost immediately so that no more chyme can leave the stomach. Instead, because of the high pressure in the antrum, chyme is forced backwards into the body of the stomach. This retropulsion helps to break down food particles and mix them with the gastric juice. Gastric motility is enhanced by many of the nervous and hormonal factors that also stimulate gastric secretion. These include distension, gastrin and activity in the parasympathetic nerves.

The regulation of gastric emptying

It is essential that gastric contents are delivered to the small intestine at a rate that is compatible with complete digestion and absorption of nutrients. As the gastric and duodenal environments are very different, it is also important that the stomach does not empty too quickly and that the duodenal contents are not regurgitated into the stomach. The gastric mucosa is resistant to low pH but may be damaged by the detergent nature of bile. The duodenum, on the other hand, is resistant to the detergent effects of bile but cannot tolerate low pH. Regurgitation of duodenal contents may therefore cause a gastric ulcer, while emptying the stomach too quickly may cause ulcers to form in the duodenum.

Many factors contribute to the regulation of gastric emptying and the process is still not fully understood. Receptors of various kinds are present within the duodenum and contribute to the control of gastric emptying via the so-called **enterogastric reflex** (a term used to describe all of the nervous and hormonal mechanisms that mediate intestinal control of gastric emptying). In general, the rate at which the stomach empties is proportional to its degree of distension; thus, the fuller it is, the more rapidly it empties. The physical and chemical nature of the foods eaten also influences gastric emptying. Foods that are high in fat and/or protein tend to slow the rate of emptying, as do mixtures of foods and gastric secretions that are highly acidic or hypertonic in nature. The mechanisms that account for this slowing are not clear, but the hormones CCK and glucagon-like peptide 1 (GLP-1), secreted by the small intestine in response to the presence of the digestion products of fats in the duodenum, are both known to delay gastric emptying. The effects of protein digestion products may be mediated by gastrin while those of acidic chyme are mediated by secretin, which is released by the duodenum in response to the arrival of acidic gastric contents.

In those patients who have had part or all of their stomach removed (partial or total gastrectomy), food enters the small intestine in an uncontrolled fashion so that relatively large amounts of hypertonic food enter the duodenum. This phenomenon is known as 'dumping' and gives rise to a number of unpleasant symptoms. These include nausea and pain associated with the distension of the upper small intestine. At the same time, large amounts of fluid are secreted into the small intestine in response to the hypertonic contents and this may cause a fall in the circulating blood volume. This, in turn, may be associated with sweating, faintness and palpitations as the sympathetic nervous system is activated. The symptoms of 'dumping' can be alleviated by reducing the size of meals so that the entry of food into the small intestine is restricted.

Absorption by the stomach

Very little absorption takes place in the stomach. Alcohol is absorbed here due to its high lipid solubility. Aspirin is a weak acid, which is converted to its non-ionized form by the low pH of the stomach contents. This renders it more lipid-soluble so that some of the ingested aspirin can be absorbed by the stomach, although most is absorbed in the upper small intestine. The stomach is virtually impermeable to water.

> **KEY POINTS:**
>
> - The stomach stores food, mixes it with gastric juice and breaks it into smaller pieces to form a semiliquid chyme. The stomach then delivers chyme to the duodenum in a controlled fashion, at a rate compatible with full digestion and absorption by the small intestine.
>
> - The emptying of the stomach is regulated by neural reflexes and by hormones.

Vomiting (emesis)

Vomiting is the sudden and forceful expulsion of the gastric contents from the mouth (and sometimes also the nose). It is generally a protective mechanism that rapidly removes ingested toxic substances from the GI tract, but may be triggered by many other stimuli. These include pain, endocrine factors (e.g. the sickness of early pregnancy), disturbances of the vestibular apparatus (motion sickness, Ménière's disease), certain drugs, radiation damage and a rise in intracranial pressure, as well as emotions mediated by higher centres of the brain in response to unpleasant sights or smells. Vomiting is usually accompanied by nausea and is often preceded by retching (heaving) and a number of other characteristic physiological changes such as sweating, pallor and salivation.

The vomiting reflex is co-ordinated by neurones in the dorsal portion of the reticular formation of the medulla (near to the areas that regulate cardiovascular and respiratory function). It receives afferent information from many areas of the body including the GI tract, liver, gall bladder, urinary bladder, uterus and kidneys, the cerebral cortex and the vestibular apparatus. It also receives information from an area known as the chemoreceptor trigger zone, which lies in the floor of the fourth ventricle and may mediate the emetic effects of anaesthetic agents and certain other drugs. The motor impulses responsible for the action of vomiting travel via the trigeminal, facial, glossopharyngeal, vagus and hypoglossal nerves (CN V, VII, IX, X and XII).

Vomiting begins with a deep inspiration followed by closure of the glottis, which prevents any vomited material (known as vomitus) from entering the trachea. The soft palate rises to close off the nasopharynx. The stomach and pyloric sphincter relax and the duodenum contracts. This reverses the normal pressure gradient in this part of the gut and intestinal contents are allowed to enter the stomach ('reverse peristalsis'). The diaphragm and abdominal wall then contract powerfully, the gastro-oesophageal sphincter opens and the pyloric sphincter closes. The resulting rise in intragastric pressure leads to expulsion of the gastric contents from the mouth. Prolonged vomiting can lead to metabolic alkalosis (a rise in plasma pH) as a result of the loss of acid from the stomach (see Chapter 34).

Anti-emetic drugs

Although vomiting often occurs as a protective mechanism, prolonged or repeated vomiting is undesirable and may require treatment with anti-emetic drugs. Motion sickness may be prevented by anticholinergic agents such as **hyoscine**, or by antihistamines such as **cinnarizine**. These drugs act directly on the vomiting centre of the medulla. Ménière's disease is often treated with **betahistine**, which is thought to act by reducing the secretion of endolymph in the inner ear (see Chapter 14). The chemoreceptor trigger zone possesses many receptors for both dopamine and 5-hydroxytryptamine (5-HT$_3$ receptors in particular) and drugs that are antagonists for these receptors have anti-emetic effects. Dopamine antagonists include **prochlorperazine**, **metaclopramide** and **domperidone**, while 5-HT$_3$ antagonists include **ondansetron** and **granisetron**.

KEY POINTS:

- Vomiting is a protective mechanism whereby noxious or potentially toxic substances are expelled from the GI tract. The vomiting reflex is co-ordinated in the medulla of the brain.

- Prolonged vomiting can cause metabolic alkalosis through the loss of gastric acid and hypokalaemia. Consequently, prolonged or repeated vomiting is undesirable and may require treatment with anti-emetic drugs.

24.10 The small intestine

In life, the small intestine is about 4 metres in length with a diameter of about 2.5 cm. It is divided into three sections: the duodenum, the jejunum and the ileum (see Figure 24.1). As discussed above, the digestive processes occurring in the stomach result in the formation of chyme, which empties into the first segment of the small intestine, the duodenum. Here the chyme is mixed with pancreatic juice and with bile from the liver and the final stage of digestion begins. Digestion is completed in the jejunum and ileum where the resulting nutrients are absorbed.

The word duodenum literally means 12 finger-widths and this part of the small intestine is about 25 cm long. As Figure 24.19 illustrates, it forms an arc in which the head of the pancreas lies. Pancreatic juice and bile are delivered into the duodenum by ducts that unite close to the duodenum at the hepatopancreatic ampulla. The sphincter of Oddi controls the entry of bile and pancreatic juice into the duodenum.

A sharp flexure (bend) forms the junction between the duodenum and the jejunum. The jejunum is about

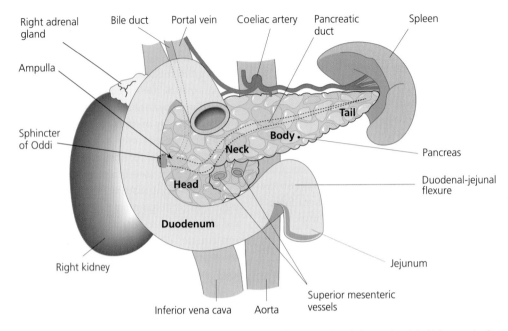

Figure 24.19 The anatomical arrangement of the duodenum, jejunum and pancreas in relation to the right kidney and spleen. The left kidney, which lies beneath the tail of the pancreas, is not shown.

1.5 metres in length and extends to the ileum, which is a coiled tube about 2.5 metres long. Both the jejunum and the ileum are supported by a mesentery that contains branches of the superior mesenteric artery and the venous and lymphatic drainage vessels of the small intestine. At rest, about 10% of the cardiac output flows to the intestine and this increases substantially in response to a meal.

Specific structural adaptations of the small intestine

The small intestine is highly adapted to carry out its digestive and absorptive functions. Most importantly, it has a huge surface area, estimated at around 200 m². This is due not only to its length but also to certain characteristic structural modifications of its wall. The mucosal and submucosal layers of the small intestine (and in particular the jejunum) are thrown into deep folds called circular folds (Figure 24.20a), which force the chyme into a spiral motion as it passes through the lumen of the small intestine. This spiral motion slows the progress of the chyme and allows more time for the intestinal juices to act on the food particles.

The folded surface of the small intestine is covered with finger-like projections, between 0.5 and 1.5 mm high,

called villi (Figure 24.20b). The exact size and shape of the villi differ throughout the gut but they all show similar histological characteristics. The surface of each villus is formed mainly by columnar epithelial cells (known as enterocytes) bound by tight junctions. The mucosal surfaces of these cells consist of tiny processes or microvilli, which form the so-called brush border and contribute further to the vast surface area available for absorption. In addition to the enterocytes, other cell types lie within the surface epithelium. These include enteroendocrine cells that secrete secretin, somatostatin, CCK and 5-HT.

Each villus contains connective tissue, a little smooth muscle, a capillary network and a modified lymph vessel (lacteal) that opens into the local lymphatic circulation. Between the villi are simple tubular glands called the **crypts of Lieberkuhn**. These are particularly numerous in the duodenum and jejunum but are also present in the ileum. They secrete around 1.5 litres of isotonic fluid each day. The submucosal layer of the duodenum also contains **Brunner's glands**, which secrete a mucus-rich alkaline fluid. Local aggregations of lymphoid tissue (**Peyer's patches**), as shown in Figure 24.20a, are found scattered throughout the submucosa of the small intestine but are most abundant in the ileum.

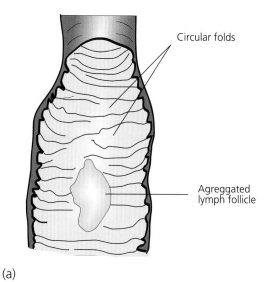

(a)

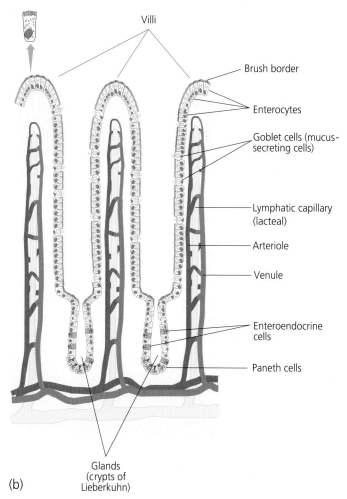

Villi

Brush border

Enterocytes

Goblet cells (mucus-secreting cells)

Lymphatic capillary (lacteal)

Arteriole

Venule

Enteroendocrine cells

Paneth cells

Glands (crypts of Lieberkuhn)

(b)

Figure 24.20 Key structural characteristics of the small intestine. (a) A section of ileum cut open to reveal circular folds and aggregated lymphoid tissue (a Peyer's patch). (b) A schematic view of intestinal villi in longitudinal section. An enlarged view of one of the cells lining the small intestine is shown above. Note the microvilli on the apical surface, which form the brush border. Mucus-secreting goblet cells are shown in yellow. Enteroendocrine cells are located in the intestinal glands (crypts of Lieberkuhn). These secrete a variety of hormones that regulate both the motility and secretory activity of the small intestine.

The epithelium of the small intestine shows a very rapid rate of cell proliferation, and renews itself every 6 days or so. Senescent (old or dying) cells are replaced by new cells that migrate towards the top of the villus from a population of undifferentiated cells at its base. The replaced cells are shed into the intestinal lumen. The rate at which the cells proliferate is reduced by irradiation, starvation, prolonged parenteral feeding and certain drugs used in the treatment of cancer, such as methotrexate.

> **KEY POINTS:**
>
> - The small intestine provides a huge surface area for nutrient absorption. The mucosal surface is covered with projections called villi, the surface of which is covered by brush border membranes.
>
> - Simple tubular glands called the crypts of Lieberkuhn lie between the villi. These contain many different cell types including enteroendocrine cells.
>
> - The intestinal epithelium is self-renewing, replacing itself completely every 6 days or so.

Motility of the small intestine

Peristaltic and segmentation contractions, the important propulsive and mixing movements described on p. 497, occur independently of extrinsic innervation. They are mediated by the enteric nervous system. Segmentation is characterized by closely spaced contractions that serve to mix chyme with the intestinal secretions and to expose digestion products to the absorptive surfaces. Peristaltic contractions occur less frequently and appear to move the chyme only a short distance (short-range contractions). It normally takes between 3 and 5 hours for chyme to move through the small intestine – although it can take as long as 10 hours. Although peristalsis and segmentation are mediated by the enteric nervous system, intestinal motility is also modified by the autonomic nerves that supply the gut and by a number of hormones acting on the intramural plexuses. Parasympathetic activity increases the excitability of the intestinal smooth muscle while sympathetic activity reduces it.

Extrinsic nerves also have a role in certain long-range intestinal reflexes. These include the ileogastric and gastroileal reflexes that reflect interactions between the stomach and the terminal ileum and serve to match emptying of the small intestine to the arrival of chyme in the duodenum. The ileogastric reflex refers to the reduction in gastric motility that occurs in response to distension of the ileum. The gastroileal reflex describes the increase in motility of the terminal ileum that occurs whenever there is an increase in secretory and/or motor activity of the stomach.

During periods of fasting, or once a meal has been processed, the pattern of contractile activity in the small intestine shows interesting changes. Segmentation contractions wane and periodic bursts of peristaltic activity are seen in which the contents of the gut are moved long distances (up to 70 cm) along the gut. These are called 'housekeeper' contractions and serve to sweep the last vestiges of the previous meal into the large intestine.

In addition to the motor activity of the intestinal smooth muscle, the villi themselves also show both piston-like contractions and swaying pendular movements that contribute to the mixing of chyme with intestinal secretions. The piston-like contractions also probably help to empty the lacteals into the lymphatics.

Emptying of the small intestine

From the terminal ileum, chyme moves into the first part of the large intestine, the caecum. The ileocaecal sphincter controls the rate at which the ileum empties to ensure that water and electrolytes are fully absorbed from the chyme by the colon. The sphincter normally remains closed until a short-range peristaltic contraction arrives at the terminal ileum. The sphincter then relaxes to allow a small amount of chyme to pass into the large intestine. The gastroileal and ileogastric reflexes ensure that the filling of the duodenum and emptying of the ileum are matched. For example, after a meal, ileal emptying is hastened through the operation of the gastroileal reflex.

> **KEY POINTS:**
>
> - During digestion, the small intestine exhibits two types of movement: segmentation and peristalsis.
>
> - The intestinal villi exhibit both piston-like contractions and swaying pendular movements.

24.11 Digestion

During its passage through the small intestine, the chyme leaving the stomach is first neutralized by alkaline juice, secreted by the pancreas, and by bile, secreted by the liver. Subsequently, the carbohydrates, fats and proteins are completely broken down into small molecules that can be absorbed into the bloodstream or lymphatic system and subsequently utilized by cells throughout the body. Through the actions of a wide variety of enzymes, carbohydrates are digested to monosaccharides such as glucose; fats are digested to fatty acids and glycerol; and proteins are reduced to small peptides and their constituent amino acids.

The pancreas secretes between 1000 and 1500 ml of fluid each day, which consists of a watery component, secreted almost entirely by the cells lining the ducts, and enzymes, secreted by the acinar cells. A list of the principal enzymes present in the pancreatic juice is given in Table 26.1 (p. 550). Many of them, including the proteolytic and some of the lipolytic enzymes, are stored and secreted in an inactive form (zymogen), which is subsequently activated within the lumen of the small intestine. In this way, the pancreas, like the stomach, avoids self-digestion.

The principal proteolytic enzyme of pancreatic juice is **trypsin**. It is secreted in the form of inactive trypsinogen, which is converted to trypsin in the duodenum. Trypsin then activates other proteolytic enzymes such as chymo-trypsin. Trypsin and chymotrypsins are endopeptidases, which hydrolyse peptide bonds within the protein molecules to liberate polypeptides of various sizes as well as free amino acids. Other enzymes, such as carboxypeptidases and aminopeptidase, digest the polypeptides further to release small peptides and free amino acids, as shown in Figure 24.21.

The digestion of starch that was initiated by salivary α-amylase in the mouth is completed by pancreatic α-**amylase**, which is secreted in its active form. This enzyme converts starch to maltose and maltriose. Enzymes on the intestinal brush border then complete the breakdown of starch to glucose, as shown in Figure 24.22. In response to fatty chyme entering the duodenum, the gall bladder contracts and forces bile into the duodenum, where it emulsifies fats prior to their digestion by **pancreatic lipase** and colipase. Fats are hydrolysed by lipase to release free fatty acids, monoglycerides and glycerol (Figure 24.23), which are then absorbed by the small intestine. Further details of the control of the secretion of bile and pancreatic juice are given in the following two chapters.

The intestinal juice (or succus entericus) contains very few enzymes that are derived specifically from the secretions of the small intestine itself. The majority of the enzymes in it are of pancreatic origin. However,

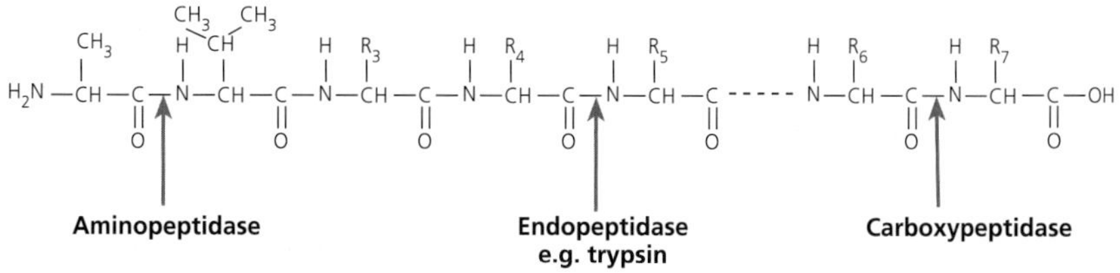

Figure 24.21 The major enzymes involved in the digestion of proteins. Trypsin and chymotrypsin split peptide bonds within a protein molecule. Such enzymes are called endopeptidases. Aminopeptidase and carboxypeptidase attack peptides at their amino and carboxyl terminals, respectively. After a protein has been subjected to attack by these enzymes, both free amino acids and small peptides are liberated, which are then absorbed by the small intestine by the processes illustrated in Figure 24.25.

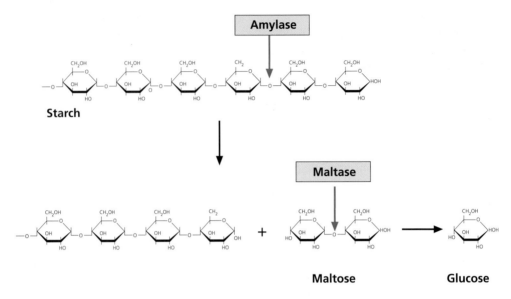

Starch

Maltose   **Glucose**

Figure 24.22 The processes by which starch is digested in the small intestine. Starch is a large polysaccharide consisting of glucose molecules linked together to form long chains. Pancreatic amylase splits the bonds between the glucose residues to release maltose, which is subsequently converted to glucose by the enzyme maltase located on the brush border of the enterocytes.

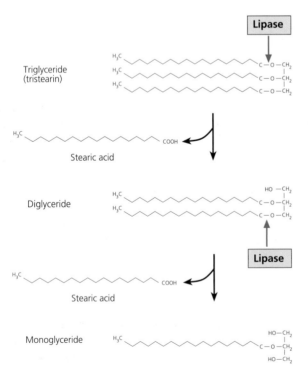

Figure 24.23 The steps by which fats are digested in the small intestine. The fats are first emulsified by bile secreted from the gall bladder to allow them to mix with the watery contents of the small intestine. The triglycerides are then attacked by pancreatic lipase to split off the two outer fatty acid residues as shown. The monoglycerides and the fatty acids are then absorbed.

some enzymes are located at the brush border of the enterocytes and carry out digestion at the surface of the epithelium. These enzymes include disaccharidases (such as maltase and sucrase), peptidases and phosphatases. A key brush border enzyme, called enteropeptidase, activates pancreatic trypsinogen to release active trypsin.

Secretion of fluid by the small intestine

The cells of the intestinal crypts secrete around 1.5 litres of isotonic fluid (succus entericus) each day. In addition, Brunner's glands in the duodenum secrete an alkaline fluid rich in bicarbonate and mucus. Together, these secretions help to protect the duodenal mucosa against corrosion by the acidic chyme. Although the principal stimulant for the secretion of intestinal juice is distension of the intestine by acidic or hypertonic chyme, secretion may be stimulated further by vagal (parasympathetic) activity, and the hormones gastrin, secretin and CCK. Sympathetic stimulation, however, causes a significant reduction in the secretion of mucus by Brunner's glands,

leaving the duodenum vulnerable to erosion. This probably explains why many cases of duodenal ulcer are related to stress, in which there is a generalized increase in sympathetic activity.

KEY POINTS:

- All of the major enzymes required for the digestion of fats, carbohydrates and proteins are contained within the pancreatic juice.

- Pancreatic α-amylase is responsible for the digestion of starch to oligosaccharides in the duodenum.

- The proteolytic enzymes are secreted as inactive precursors and are activated in the duodenum. They break down proteins into small peptides and amino acids.

- Pancreatic lipases hydrolyse water-insoluble triglycerides to release free fatty acids and monoglycerides.

24.12 The absorption of digestion products by the small intestine

Each day around 8–10 litres of water and up to 1 kg of nutrients pass across the gut wall into the blood or lymph draining the GI tract. Although bile salts and vitamin B_{12} are absorbed in the ileum, virtually all of the other nutrients have been absorbed by the time the chyme reaches the middle of the jejunum. Water is absorbed along the length of the intestine, leaving 400–1000 ml to be absorbed in the large bowel.

The small intestine has an enormous surface area for absorption by virtue of the mucosal folds, intestinal villi and the microvilli that form the brush border of the enterocytes. Furthermore, the villi have a rich supply of both blood and lymph, as shown in Figure 24.20. However, because the epithelial cells of the intestinal mucosa are joined at their apical surfaces by tight junctions, the products of digestion cannot move between them. Instead, they

must move through the cells and into the interstitial fluid abutting their basal membranes before they can enter the capillary blood. The specific processes by which the digestion products of the principal nutrients (carbohydrates, proteins and fats) are absorbed will be described below. Further details of plasma membrane transport and epithelial transport are given in Chapter 5, while the mechanisms of absorption of specific micronutrients such as calcium and iron are described elsewhere.

The absorption of monosaccharides (the digestion products of carbohydrates)

Carbohydrates are broken down by the gut to simple sugars (monosaccharides) including glucose, galactose and fructose. They are absorbed in the upper small

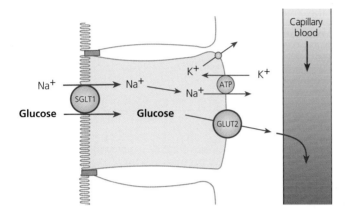

Figure 24.24 A simplified diagram showing the steps by which glucose is absorbed in the small intestine. Glucose is transported into the enterocytes in association with sodium ions via specific carrier molecules (SGLT1). Once it has accumulated in the cells, it diffuses down its concentration gradient by a separate carrier (GLUT2) on the basolateral surface of the cells. From there, it diffuses into the blood and is transported to the liver.

intestine and enter the blood of the portal vein, which transports them directly to the liver. Glucose and galactose are transported into the epithelial cells against their concentration gradients by a sodium-dependent co-transport mechanism similar to that found in the proximal tubule of the nephron (Figure 24.24). They leave the cells at the basolateral membrane by facilitated diffusion. Fructose is absorbed by sodium-independent facilitated diffusion. It does not need to be transported against a concentration gradient as the plasma concentration is always very low.

The absorption of peptides and amino acids (the digestion products of proteins)

The amino acids liberated by the digestion of proteins are absorbed at the brush border of the enterocytes by a sodium-dependent cotransport mechanism similar to that utilized by glucose (Figure 24.25). Once their con-

centration within the enterocytes rises sufficiently, they cross the basolateral surface by facilitated diffusion. The amino acids then enter the capillary blood of the villi, from where they enter the portal vein and are delivered to the liver. Some small peptides are also absorbed, but they are transported into enterocytes by a carrier that is linked not to sodium but to the influx of hydrogen ions, as shown in Figure 24.25. This carrier is also believed to be responsible for the rapid absorption across the gut of certain drugs such as the antihypertensive agent **captopril**. Once inside the intestinal cell, the peptides are broken down into their constituent amino acids, which are transported across the basolateral membrane by the amino acid transporters. Large peptides and intact proteins are not generally absorbed across the gut, although the immunoglobulins present in colostrum appear to pass intact across the intestinal epithelium of the newborn. (Colostrum is the sticky yellowish fluid secreted by the breast in the first week or so after giving birth; see Chapter 29.)

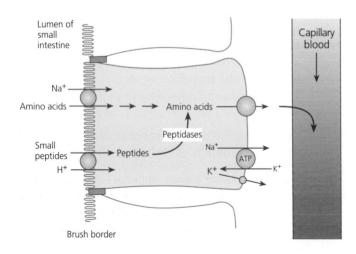

Figure 24.25 The mechanisms by which amino acids and small peptides are absorbed in the small intestine. Amino acids are transported into the enterocytes, together with sodium ions, by specific carriers present in the brush border. Once they have accumulated in the enterocytes they exit the cell by diffusion down their concentration gradients via the basolateral surface and specific carrier molecules. Small peptides are absorbed from the lumen in association with hydrogen ions. Once inside the cell they are broken down into their constituent amino acids before leaving the cell by the amino acid carriers of the basolateral membrane.

The absorption of monoglycerides and free fatty acids

Each day around 80 g of fat are absorbed from the small intestine, largely in the jejunum. The monoglycerides and free fatty acids liberated by the actions of the pancreatic lipases become associated with bile salts and lecithin to form micelles. These have a hydrophilic outer region and a non-polar core, which also contains cholesterol and fat-soluble vitamins (see Chapter 25, p. 537). The hydrophilic outer shell enables the micelle to enter the aqueous layer surrounding the microvilli that form the brush border of the enterocytes. The fat-soluble contents of the core then diffuse passively into the cells lining the intestine while the polar region of the micelle (containing bile salts) remains within the lumen of the gut. Absorption of the bile salts occurs in the terminal ileum.

Once inside the enterocytes, the products of fat digestion undergo further chemical processing. In the smooth endoplasmic reticulum (see Chapter 4), triglycerides are reformed and phospholipids are resynthesized. The lipids accumulate in the vesicles of the smooth endoplasmic reticulum to form **chylomicrons**, which are released from the cell by exocytosis at the basolateral membrane (Figure 24.26). From here, they enter the lacteals that run through the core of the villi and leave the intestine via the lymph. Unlike the sugars and amino acids, lipids are not absorbed into the portal vein and so bypass the liver initially. In the absence of normal levels of bile in the duodenum, fat digestion and absorption are impaired and fat appears in the faeces in abnormally high amounts. This is known as steatorrhoea.

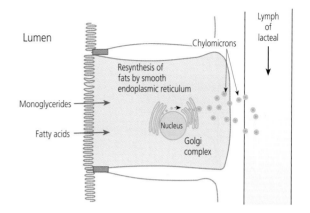

Figure 24.26 The key steps involved in the absorption of fats by the small intestine. Fatty acids and monoglycerides pass across the brush border and are resynthesized into triglycerides by the endoplasmic reticulum. They are then secreted across the basolateral membrane as small fat droplets (chylomicrons) before being absorbed by the lacteals and entering the lymphatic system draining the GI tract.

The absorption of vitamins

The water-soluble vitamins required by the body include vitamin C and the vitamins of the B 'complex' (see Chapter 31). They are absorbed by specific transport mechanisms. Vitamin C, for example, is absorbed in the jejunum by sodium-dependent active transport while folic acid is absorbed (also in the jejunum) by carrier-mediated facilitated diffusion. Vitamin B_{12} is absorbed in the lower ileum by a specific mechanism requiring intrinsic factor, which is secreted by the parietal cells of the gastric glands (Figure 24.27). The fat-soluble vitamins (A, D, E and K) are absorbed in the same way as the products of fat digestion as described earlier.

The absorption of fluid and electrolytes

The gut plays an extremely important role in the overall fluid and electrolyte balance of the body. A summary of its contribution is shown in Figure 24.28. Each day, in addition to the 2 litres or so of fluid that is taken into the body in the diet, around 7 litres of additional fluid is added to the lumen of the gut as a result of the various secretory processes occurring along its length – 2 litres of gastric juice, 1.5 litres of saliva and so on. If all of this fluid were to leave the body in the faeces, dehydration would be

---

KEY POINTS:

- Absorption is the process by which the products of digestion are transported from the GI tract into the blood or lymph draining the gut. Almost all of the absorption of water, electrolytes and nutrients occurs in the small intestine.

- Monosaccharides are absorbed by sodium-dependent cotransport. Amino acids utilize similar mechanisms.

- The products of fat digestion diffuse into the cells and are reprocessed by the smooth endoplasmic reticulum to form chylomicrons, which enter the lacteals of the villi.

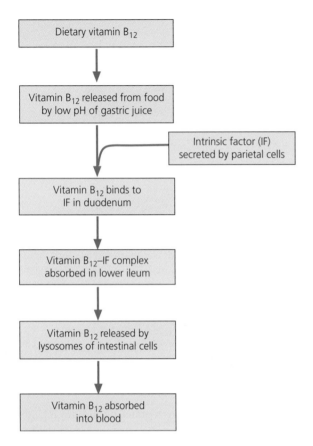

Figure 24.27 The main stages by which vitamin B_{12} is absorbed from the GI tract. The vitamin is released from food by the low pH of the gastric juice. The parietal cells of the gastric glands secrete a glycoprotein known as intrinsic factor, which binds to vitamin B_{12} in the duodenum, protecting it from attack by the digestive enzymes. The complex of vitamin B_{12} and intrinsic factor is then absorbed in the lower ileum by endocytosis.

swift and fatal. In reality, almost all of it is returned to the blood from the chyme as it moves through the intestine. Indeed, in health, the faeces account for only 100–200 ml of water loss from the body each day. About 5–6 litres of fluid are absorbed from the jejunum, 2–3 litres from the ileum and between 400 and 1000 ml from the colon.

Electrolytes from both ingested foods and GI secretions are absorbed along the length of the small intestine (except for calcium and iron whose absorption is mainly restricted to the duodenum). Water follows rapidly along the osmotic gradient generated by the absorption of electrolytes.

Disorders of absorption (malabsorptive states)

A number of disease states may reduce the absorption of nutrients by the small intestine and give rise to a variety of symptoms. These include weight loss, abdominal discomfort and bloating, often accompanied by flatulence, and glossitis (painful loss of the normal epithelium covering the tongue). More specific symptoms are associated with particular malabsorption states. For example, poor fat absorption results in stools that are greasy, soft and malodorous while anaemia is often seen if the absorption of iron or folic acid is impaired. Failure to absorb vitamin K may lead to an increased tendency to bleed. Among the most important of the malabsorption states are:

- **Carbohydrate intolerance** in which there is a deficiency in one or more of the enzymes that digest carbohydrates. The most common of these is lactase deficiency, which results in lactose intolerance.

- **Coeliac disease** (**non-tropical sprue**) in which the patient has an intolerance to gluten, a cereal protein found in wheat. Part of the gluten molecule forms an immune complex within the small intestinal mucosa, promoting the aggregation of natural killer cells (see Section 21.4), which secrete toxins that destroy the enterocytes and cause progressive atrophy of the villi.

- **Crohn's disease**, a chronic inflammatory condition of the gut.

> **KEY POINTS:**
> - Fat-soluble vitamins are absorbed along with the products of fat digestion. Most water-soluble vitamins are absorbed by facilitated transport.
> - The GI tract absorbs 8–10 litres of fluid and electrolytes each day. Failure to absorb fluid results in potentially life-threatening diarrhoea.
> - A number of conditions result from malabsorption of nutrients. Specific syndromes include carbohydrate intolerance, coeliac disease and Crohn's disease.

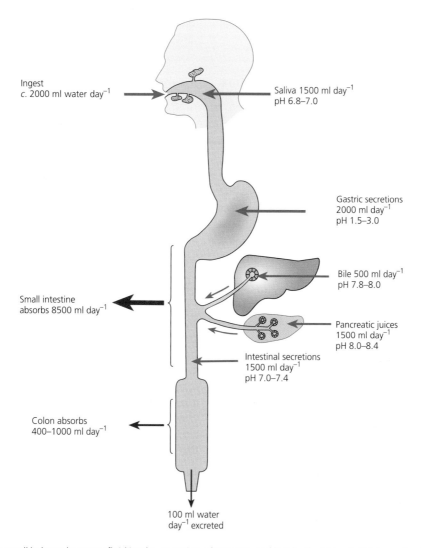

Ingest
c. 2000 ml water day^{-1}

Saliva 1500 ml day^{-1}
pH 6.8–7.0

Gastric secretions
2000 ml day^{-1}
pH 1.5–3.0

Bile 500 ml day^{-1}
pH 7.8–8.0

Small intestine
absorbs 8500 ml day^{-1}

Pancreatic juices
1500 ml day^{-1}
pH 8.0–8.4

Intestinal secretions
1500 ml day^{-1}
pH 7.0–7.4

Colon absorbs
400–1000 ml day^{-1}

100 ml water
day^{-1} excreted

Figure 24.28 The overall balance between fluid intake, secretion, absorption and loss by the GI tract. Fluid secretion is shown by green arrows and fluid absorption by red arrows.

24.13   The large intestine

The large bowel is about 1.3 m in length in an adult and extends from the ileocaecal valve to the anus. It is attached to the posterior abdominal wall by a double layer of peritoneum, the mesocolon. The large intestine is subdivided into four principal regions, the caecum, colon, rectum and anal canal. Each day around 500–1000 ml of chyme enter the caecum and 100–150 g of semisolid waste material (faeces) is eliminated from the anal canal.

The main functions of the large intestine include the storage of food residues prior to elimination, the secretion of mucus and the absorption of water and electrolytes (see above). The colon also accommodates a colony of bacteria (intestinal flora) that synthesize small amounts of vitamin K and some B vitamins.

Histology and innervation of the large intestine and anal canal

The wall of the large intestine differs from that of the small intestine in several respects. There are no villi but large numbers of intestinal glands (crypts) are present. These are populated with columnar absorptive cells and many mucus-secreting goblet cells. The muscularis mucosa is well developed and there is a thick circular layer of smooth muscle. Furthermore, the longitudinal layer of smooth muscle within the muscularis externa is thickened to form three bands, the taeniae coli. Between the thickened regions, the muscle layer is relatively thin, as shown in Figure 24.29. In the colon, tonic contractions of the taeniae pull the wall into pouches (haustra) that give it a puckered appearance (Figure 24.30). The final part of the colon is known as the sigmoid colon, which merges with the rectum, a muscular tube about 15 cm in length. Here the longitudinal muscle forms a complete layer, as in the small intestine. The rectum passes through the floor of the pelvis to merge with the anal canal at the anorectal junction.

The anal canal forms the final part of the GI tract. It is about 3 cm long and lies entirely outside the abdominal cavity. It possesses internal and external sphincters that remain tightly closed except during defecation. The epithelium of the upper portion of the anal canal is similar to that found in the colon but in the final third this changes to the stratified squamous type, reflecting the abrasion received by this area by the passage of the faeces. In the upper part of the canal, the mucous membrane hangs in folds (the anal columns), which are joined at their inferior (lower) ends to form the anal valves. Between the columns are the anal sinuses (Figure 24.31). Two superficial venous plexuses are associated with the anal canal. If these become inflamed, itchy varicosities called haemorrhoids develop.

The large intestine receives both parasympathetic and sympathetic innervation, as shown in Figure 24.6 at the beginning of this chapter. The parasympathetic fibres end chiefly on neurones of the enteric nervous system (the intramural plexuses). The skeletal muscle of the external anal sphincter receives branches of somatic nerves arising from the sacral region of the spinal cord.

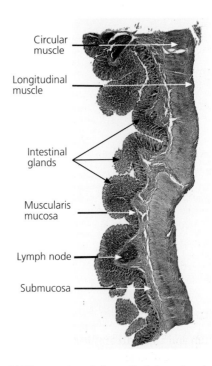

Figure 24.29 A section of the wall of the colon showing the main layers. Note the thin layer of longitudinal muscle and the darkly staining mucosa, which contains the intestinal glands. These glands absorb fluid from the food residues and secrete mucus to aid the movement of faeces through the remainder of the GI tract.

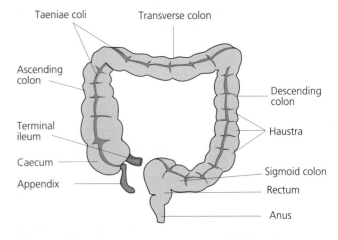

Figure 24.30 A diagrammatic representation of the large intestine showing the bands of longitudinal muscle (the taeniae coli) and haustra of the colon.

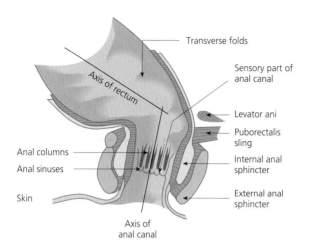

Figure 24.31 A longitudinal section through the rectum and anal canal.

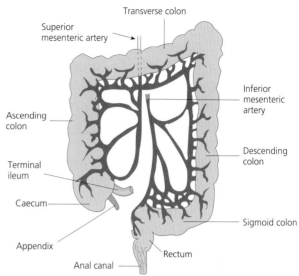

Figure 24.32 The major divisions of the large intestine and their arterial blood supply.

The caecum and appendix

The caecum and appendix are situated in the right iliac fossa. A loose fold of mucosa protrudes into the caecum forming a flap, called the ileocaecal valve, which prevents the regurgitation of chyme from the caecum back into the ileum. The caecum itself is a blind-ended pouch with little or no digestive function in humans. The appendix is a coiled tube about 7 cm long attached to the caecum. It contains a large mass of lymphoid tissue but makes no essential contribution to the immune system of the body. Occasionally, the appendix may become inflamed, a condition known as **appendicitis**, and surgical removal of the inflamed tissue is needed to prevent rupture. If the appendix does rupture, faecal material containing bacteria will enter the abdominal cavity causing a more serious condition called **peritonitis**.

The colon

The total length of the colon is around 1.2 m in an adult. It is divided into four sections: the ascending, transverse, descending and sigmoid colon. The ascending and descending colon are closely attached to the posterior abdominal wall behind the parietal peritoneum (such an attachment is said to be retroperitoneal). The transverse colon is attached to the posterior abdominal wall by a short mesentery. Suspended from the lower border of the transverse colon is the greater omentum, a large, apron-like sheet of mesentery that contains fat (Figure 24.14).

The ascending colon and the first part of the transverse colon receive blood via the superior mesenteric artery while the rest of the colon is supplied by the inferior mesenteric artery, as shown in Figure 24.32. Venous blood enters the superior and inferior mesenteric veins before draining into the hepatic portal vein. Blood vessels supplying the colon penetrate the circular smooth muscle layer creating areas of potential mechanical weakness. Occasionally, herniation of the mucosa in these areas leads to the formation of pouches or **diverticulae**.

Functions of the colon

One of the most important functions of the colon is to absorb a large proportion of the water and salts that remain in the chyme leaving the ileum; typically, this amounts to 400–1000 ml each day. Severe diarrhoea will occur if the colon fails to absorb this fluid, resulting in dehydration (see below).

Many different species of bacteria live in the colon, in symbiosis with their host, performing a number of functions. These include:

- The fermentation of indigestible carbohydrates (cellulose of plant cell walls, etc.) and lipids. These reactions produce short-chain fatty acids along with 500 ml or more of gas (flatus) each day. The short-chain fatty acids are used as a source of energy by the colonocytes.

- The conversion of bilirubin to urobilinogens (see p. 539).
- The degradation of cholesterol and certain drugs.
- The synthesis of vitamins K, B_{12}, thiamine and riboflavin.

The balance of different kinds of bacteria within the colon helps to maintain normal health, and disruption to this balance may lead to illness. This is illustrated by the changes in bowel function (notably diarrhoea) that often accompany the use of antibiotics to treat infection.

Movements of the colon

As in the small intestine, both mixing and propulsive movements occur in the colon. To allow time for the colon to absorb water and electrolytes from the food residues, propulsive movements are relatively slow. Typically, material moves along the colon at a rate of about 5–10 cm h^{-1} and remains within the colon for 16–20 hours. In addition to the short-range peristaltic contractions that move semisolid faecal material slowly towards the rectum, more vigorous propulsive movements of the colon occur several times a day (often after meals). These are called **mass movements** and they cause large segments of the colon to be emptied. Those occurring in the transverse and descending regions serve to move faeces into the rectum and trigger the need to defecate. Analgesics such as **morphine**, **codeine** and **pethidine**, as well as aluminium-based antacids, decrease the frequency of colonic mass movements and can cause constipation. A diet rich in fibre (roughage), which consists of undigested plant cell walls, facilitates the movement of food residues through the large intestine. The fibre adds bulk to the stool and also attracts water to the stool, making it softer and easier to expel.

The segmental mixing movements that occur in the colon are caused by contraction of the circular smooth muscle and serve to constrict the lumen. This type of contraction is called **haustration** because the segments correspond to the thickenings or haustra (see earlier). The haustrations squeeze and roll the faecal material around, exposing it to the absorptive surfaces of the colonic mucosa so that water and electrolyte absorption is optimized.

The rectum and defecation

The rectum is a muscular tube 12–15 cm in length, extending from the sigmoid colon to the anus. It lies in front of the sacrum, behind the prostate gland in men and behind the uterus and vagina in women. The anal canal is the final part of the GI tract, forming an angle with the rectum (Figure 24.31). The anal canal possesses two sphincters, both of which are maintained in a tonic state of contraction. The internal (involuntary) sphincter is supplied by both sympathetic and parasympathetic fibres. Parasympathetic stimulation brings about relaxation of the internal sphincter while sympathetic stimulation causes contraction. The external sphincter is made of skeletal muscle and is innervated by the pudendal nerve. It is under voluntary control, which is learned during the first few years of life. Muscle fibres of the levator ani and the puborectalis muscles, which form part of the pelvic floor, encircle the anus.

The rectum is normally empty, but when a mass movement brings faeces into the rectum, causing distension, stretch receptors in the rectal wall are stimulated and trigger the urge to defecate. Defecation is a complex process involving both reflex and voluntary actions. Although the urge to defecate can be inhibited consciously, if the conditions are not suitable, in a healthy individual, it will eventually become overwhelming and the reflex will proceed. Activity in the parasympathetic nerves stimulates the walls of the sigmoid colon and the rectum to contract, thus moving faeces towards the anus. The anal sphincters relax, allowing faeces to move through the anal canal. Voluntary contractions of the abdominal muscles and closure of the glottis aid in the expulsion of faeces by increasing the intra-abdominal pressure. The neural pathways involved in the defecation reflex are shown in Figure 24.33.

Around 100–150 g of faeces, containing 30–50 g of solids and 70–100 g of water, are normally eliminated each day. The solid portion is made up largely of cellulose, epithelial cells shed from the GI lining, bacteria, some salts and the brown pigment stercobilin. The characteristic odour of faeces is due to the presence of hydrogen sulphide and organic sulphides.

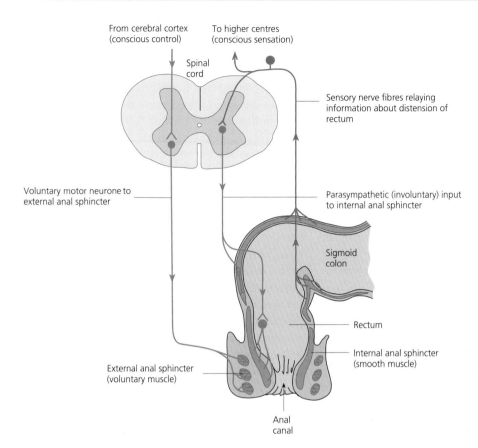

Figure 24.33 The neural pathways involved in the defecation reflex. Afferent nerve fibres are shown in red and motor axons in blue. Sensory information ascends to the brain via the spinothalamic tract, which is not shown in this figure.

KEY POINTS:

- The large intestine consists of the caecum, colon, rectum and anal canal. Its main functions are to store food residues, secrete mucus and absorb water and electrolytes from the food residues prior to their elimination as faeces via the anus.

- The bacteria present in the large intestine perform fermentation reactions that produce short-chain fatty acids and flatus that are absorbed by the colonocytes. Intestinal bacteria also synthesize certain vitamins such as vitamin K.

- The colon exhibits mixing movements (haustrations) and sluggish propulsive movements. Mass movement contractions occur several times daily serving to move faecal material towards the rectum.

Diarrhoea and constipation

Diarrhoea and constipation affect most people at some time in their lives and, for some, can become chronic and debilitating. Diarrhoea is defined as the passage of an excess quantity of stool ($>300\,g\ day^{-1}$). It is usually accompanied by increased frequency of defecation and increased fluidity of the stool. Most cases of diarrhoea are acute and self-limiting, but the loss of water and electrolytes that occurs in severe diarrhoea can lead to dehydration and, if untreated, death. Very young children and the elderly are particularly at risk.

Specific causes of diarrhoea include:

- The presence of non-absorbable substances, such as lactose or sorbitol, in the gut, which reduce the absorption of water by osmosis from the gut lumen

- The presence of bacterial toxins within the gut, as in cholera

- Viral infections
- Loss of autonomic control (e.g. in diabetic neuropathy)
- Inflammatory disorders such as Crohn's disease or ulcerative colitis
- Drugs that increase GI motility, e.g. acetylcholinesterase inhibitors used to treat myasthenia gravis
- Parasites.

Treatments for diarrhoea depend upon the initial cause of the condition but in most cases the primary concern is to restore fluid and electrolyte balance. Hydration can be maintained by administering an **oral rehydration** solution. Typically, these are slightly hypotonic, alkaline and contain sodium and glucose in the correct ratio to encourage uptake of sodium and glucose via the co-transporter of the enterocyte membrane (see p. 522). This will be followed by the movement of water across the gut wall by osmosis. Fluid absorbents and adsorbents such as **kaolin** may also be helpful. Drugs that inhibit GI motility are frequently used in the treatment of diarrhoea. Such agents include the opiates **codeine** and **loperamide**, which increase the time available for intestinal fluid absorption. Specific drug treatments may also be used in diarrhoea of known cause, for example antibiotics for specific infections or anti-inflammatory drugs in Crohn's disease or ulcerative colitis.

Constipation may affect up to 14% of the population at any one time. It must be defined individually but usually means a decrease in the number of bowel movements per week, hard stools and difficult evacuation. There are many causes of constipation including:

- Diet and lifestyle factors
- Emotional and psychological factors
- Functional and mechanical conditions such as inflamed haemorrhoids or anal fissure
- Abdominal surgery
- Neurogenic disorders of the large intestine (e.g. Hirschsprung's, disease)
- Blockage of the gut by a tumour
- Neurological disorders affecting the defecation reflex
- Certain drug treatments (e.g. morphine, codeine, anticholinergic antidepressants, antacids contain-

ing aluminium hydroxide or calcium carbonate, iron salts, diuretics and chemotherapy).

If initial lifestyle and dietary changes are without success, constipation is often treated using a variety of **laxative drugs**, including faecal softeners and lubricants, bulk-forming agents, GI stimulants and osmotic laxatives. Softeners and lubricants include vegetable and mineral oils such as arachis oil, glycerol and di-octyl sodium sulphosuccinate (DSS). The latter is usually administered rectally because it can inhibit bile production and is also a gastric irritant. Bulk-forming agents are hydrophilic compounds that act by absorbing water and swelling, there by increasing the stool volume. The increased bulk stimulates stretch receptors in the rectal wall and promotes defecation. Examples include bran, ispaghula husk, methylcellulose and sterculia.

Laxatives that stimulate GI motility act on the nerve plexuses of the colon to enhance peristalsis. They are used when rapid evacuation of the bowel is needed, for example before surgery on the large bowel. Examples of such drugs include **senna**, **metoclopramide**, **domperidone**, **bisacodyl** and **sodium picasulphate**. Given orally, they usually work within 6–12 hours but bisacodyl acts more rapidly (within 20–60 minutes) if administered rectally.

Osmotic laxatives act by reducing water absorption in the intestine. Examples include **lactulose** and salts of magnesium or sodium (saline purgatives). The latter work very rapidly and dramatically so rectal administration is usual and fluid balance must be maintained. Lactulose is a disaccharide of galactose and fructose. It is hydrolysed to its component sugars by the bacteria of the colon. The sugars are then fermented to acetic acid and lactic acid, which act as osmotic laxatives.

Threadworm infestation of the intestine

Although a variety of multicellular worms and parasites may infest the intestine, causing chronic diarrhoea, malabsorption and anaemia, the most common type of infestation is threadworm, which is thought to affect up to 40% of children at some time between the ages of 3 and 10 years. These tiny worms inhabit the lower small intestine, caecum and proximal colon. They may be successfully treated by a single dose of **mebendazole**, a drug that kills the organisms by blocking glucose uptake and inhibiting microtubule formation.

Recommended reading

Histology

Junquieira, L.C. and Carneiro, J. (2005). *Basic Histology*, 11th edn. McGraw-Hill: New York. Chapter 15.

Physiology of the digestive tract

Johnson, L.R. (2007). *Gastrointestinal Physiology*, 7th edn. Mosby: Philadelphia. [A clear account of the physiology of the digestive tract that is not too detailed.]

Gastrenterology

Kumar, P. and Clarke, M. (2005). *Clinical Medicine*, 6th edn. Saunders: Edinburgh. Chapter 4. [A detailed account of the diseases of the GI tract including clinical investigations and the basics of treatment.]

Self-assessment questions

Which of the following statements are true and which are false? Answers are given at the end of the book (p. 755).

1. a) Swallowing is a purely voluntary activity.
   b) The food bolus is propelled down the oesophagus by segmentation movements.
   c) The saliva contains an enzyme that digests starch.
   d) Activation of the salivary parasympathetic nerves inhibits salivary secretion.
   e) Salivary secretion is inhibited by atropine.

2. a) Blood flow to the stomach increases during a meal.
   b) The gastric mucosa secretes an alkaline fluid containing mucus.
   c) The parietal cells of the gastric glands secrete pepsinogens.
   d) The secretion of gastric acid is inhibited by H_2 receptor blockers such as ranitidine.
   e) Venous blood draining the stomach immediately after a meal has a lower pH than blood in the right atrium.

3. a) Gastric secretion does not begin until food enters the stomach.
   b) Gastric secretion is inhibited by secretin.
   c) Gastric emptying is slower when a fatty meal has been consumed.
   d) The stomach absorbs glucose and amino acids.
   e) Vitamin B_{12} is absorbed in the stomach.

4. a) The plasma bicarbonate concentration will be lower than normal following prolonged vomiting.
   b) Persistent vomiting often leads to metabolic alkalosis.
   c) Gastric emptying is controlled by the pyloric sphincter.
   d) The chyme entering the duodenum stimulates the secretion of pancreatic juice.
   e) The bile pigments are essential for the digestion of fats.

5 a) The innermost layer of the GI wall is known as the mucosa.

b) To permit churning movements, the stomach has two main layers of smooth muscle.

c) The myenteric plexus lies between the longitudinal and circular smooth muscle layers in the gut wall.

d) The villi increase the surface area for absorbing nutrients.

e) The taeniae coli of the colon are formed from its layer of circular smooth muscle.

6 a) About half of the digested carbohydrate is absorbed in the small intestine.

b) Intestinal digestive enzymes are secreted by cells of the crypts of Lieberkuhn.

c) The GI tract absorbs about 8–10 litres of fluid each day.

d) Amino acids are absorbed in the small intestine by cotransport with sodium.

e) Cellulose cannot be digested or absorbed by the human small intestine.

7 a) Parasympathetic activity inhibits intestinal motility.

b) Emptying of the ileum is facilitated by the gastroileal reflex.

c) The colon has a specific type of mixing movement called haustration.

d) Vitamin K can be synthesized by the intestinal flora

e) The mucosa of the anal canal is covered by a simple squamous epithelium.

The liver and gall bladder

25

After reading this chapter you should have gained an understanding of:

- The anatomy of the liver and gall bladder
- The hepatic circulation
- The formation and secretion of bile
- The role of the liver in digestion
- The formation and excretion of bile pigments
- Jaundice
- The role of the liver in whole-body energy metabolism
- The synthesis of plasma proteins
- The endocrine functions of the liver
- Detoxification by the liver
- Liver failure

25.1 Introduction

The liver is the largest internal organ, accounting for around 2% of body weight in an adult and as much as 5% in infants. It has important roles in digestion, excretion, energy metabolism and protein synthesis. The liver also acts as a store for glucose (as glycogen), fat-soluble vitamins, vitamin B_{12} and iron. It also plays a key role in the detoxification and elimination of organic molecules

(including drugs) taken into the blood during the digestion of food. The liver has important endocrine functions. It secretes a number of hormones, including IGF-1 and IGF-2, both of which are important in growth, and hepcidin, which regulates iron metabolism. It also inactivates circulating hormones. These functions are summarized in Table 25.1.

Table 25.1 The normal functions of the liver

Function	Comments
Synthesis and secretion of bile	Bile is important for fat absorption and for the excretion of bile pigments, which are derived from the breakdown of haemoglobin.
Carbohydrate metabolism	The liver contains the body's reserve of carbohydrate in the form of glycogen. It is also able to form glucose by gluconeogenesis.
Fat metabolism	Absorption of fats and fat-soluble vitamins (see p. 523); the storage of fat-soluble vitamins; synthesis of lipoproteins.
Protein synthesis	The liver is the major source of plasma proteins including albumin and the clotting factors.
Endocrine function	The liver synthesizes and secretes IGF-1 and IGF-2 in response to the secretion of growth hormone by the anterior pituitary. It also secretes hepcidin, which regulates iron absorption from the small intestine, and thrombopoietin, which regulates the number of platelets in the blood.
Detoxification	Inactivation of hormones; conjugation of various drugs and toxins for excretion; conversion of ammonia to urea.
Iron storage	Essential for erythropoiesis.
Storage of vitamin B_{12}	Required for normal erythropoiesis.
General	As the liver is situated between the gut and the general circulation, it is able to protect the body by inactivating toxic materials absorbed from the gut.

25.2 The structure of the liver

The liver is situated in the upper right quadrant of the abdominal cavity, as shown in Figure 24.1. It consists of four lobes surrounded by a tough capsule of connective tissue, called **Glisson's capsule**. A thin layer of mesentery, the **lesser omentum**, links the liver to the lesser curvature of the stomach, while the **falciform ligament**, which also separates the large right and left lobes, attaches the liver to the diaphragm and anterior abdominal wall. The right lobe lies over the right kidney and is significantly larger than the left lobe. Two smaller lobes, the caudate and quadrate lobes, lie on the posterior surface of the right lobe, as shown in Figure 25.1b.

The **gall bladder** rests in a recess on the inferior surface of the right lobe, alongside the quadrate lobe. It is a pear-shaped sac, 7–10 cm in length, which stores and concentrates bile until it is needed for the processing of fats in the small intestine. Bile leaves the liver through a series of ducts that finally join together to form the common hepatic duct. As the hepatic duct passes towards the duodenum, it fuses with the cystic duct that drains the gall bladder, to form the bile duct (Figure 25.2). Finally, just before it enters the duodenum, the bile duct fuses with the main pancreatic duct. The entry of bile and pancreatic juice into the duodenum is regulated by the sphincter of Oddi, which also prevents the reflux of bile.

Microscopically, the liver consists of between 50 000 and 100 000 structural and functional units called **lobules**. These are irregularly shaped, polyhedral structures. Each lobule is bounded by a thin layer of connective tissue, which is not especially well defined in man. When cut across, the lobules are roughly hexagonal in shape and are 1–2 mm in diameter, as shown in Figure 25.3. At the centre of each lobule lies the central vein, which empties into the hepatic vein. At the

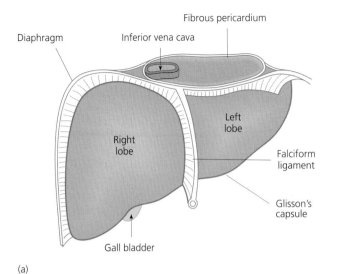

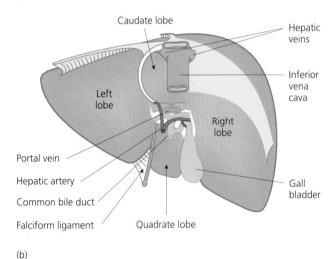

Figure 25.1 The gross anatomy of the liver. (a) The view from the front and (b) the view from the back. Note the major division by the falciform ligament and the position of the gall bladder, here shown in green.

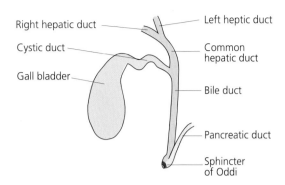

Figure 25.2 The anatomical relationships between the hepatic ducts, the cystic duct and the bile duct. The pancreatic duct joins with the bile duct as it reaches the ampulla. The smooth muscle of the ampulla forms the sphincter of Oddi, which controls the secretion of bile and pancreatic juice into the duodenum.

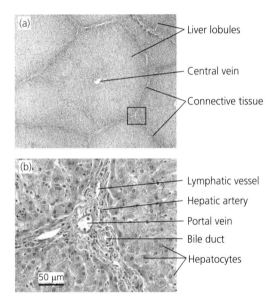

Figure 25.3 The histological structure of the liver. (a) The outlines of several liver lobules, each of which is bounded by connective tissue septa. The central vein of the principal lobule shown in the field of view is also clearly visible. (b) The anatomical relationships of the hepatic artery, portal vein and bile duct that constitute a portal triad.

boundaries with other lobules, there are the **portal triads**, so-called because three structures are always present there: a branch of the hepatic artery, a branch of the portal vein and a bile duct, as shown in Figure 25.4.

The **hepatocytes** (liver cells) radiate from the central vein and are arranged rather like the bricks in a wall but with many gaps between them so that they form a sponge-like structure, filled with blood from the portal

vein and hepatic artery. These irregularly shaped blood vessels are known as **sinusoids**. Macrophages, called **Kupffer cells**, are present on the inner (luminal) surface of the endothelium, and phagocytose old cell fragments and bacteria (see also Chapter 21). The hepatocytes are

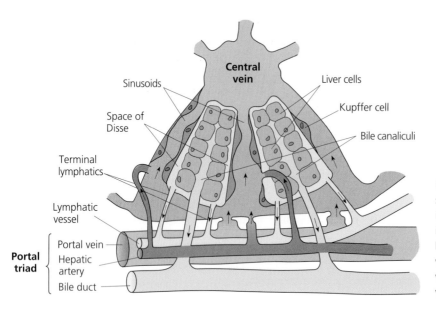

Figure 25.4 A diagrammatic representation of part of a liver lobule and its vascular supply. The small arrows indicate the direction of flow for the blood, bile and lymph. The central vein drains into the main hepatic vein, which empties into the inferior vena cava.

separated from the endothelial cells of blood vessels by a narrow space called the space of Disse, which contains fluid that is freely exchanged with the plasma via numerous gaps in the endothelial wall (fenestrations). Excess fluid drains into lymphatics. Between the columns of hepatocytes is a narrow space called a **bile canaliculus**, which communicates with a bile duct.

The hepatocytes themselves are large polyhedral cells, richly endowed with rough and smooth endoplasmic reticulum, mitochondria and lysosomes, emphasizing their importance in synthetic and metabolic processes. Microvilli are present on that part of the plasma membrane that faces the blood sinusoids. This adaptation provides a large surface area for the exchange of substances with the blood.

KEY POINTS:

- The liver is the largest internal organ. It is formed from units called lobules, which are made up of sheets of liver cells (hepatocytes).
- Macrophages (Kupffer cells) are present on the walls of the blood sinusoids. They phagocytose cell debris and bacteria.
- The liver performs a large variety of metabolic functions, including protein synthesis, glucose storage (as glycogen), gluconeogenesis and detoxification.
- The liver secretes several hormones and forms the bile, which is essential for the digestion of fat.

25.3   The hepatic circulation

At rest, the liver receives around a quarter of the cardiac output. It has a double blood supply – the hepatic artery, which carries around $400\,\mathrm{ml\,min^{-1}}$ of oxygenated blood, and the portal vein, which supplies about $1000\,\mathrm{ml\,min^{-1}}$ of blood from the gastrointestinal (GI) tract that is nutrient-rich but low in oxygen (see Chapter 24). The portal vein divides to form small portal venules that run in the septa between the lobules described earlier. From these venules, blood flows into the sinusoids that surround the hepatocytes. These sinusoids form a leaky capillary network from which blood flows into the central vein of each lobule. Deoxygenated blood from the central veins flows into the hepatic veins and thence into the inferior vena cava. The hepatic artery branches to form arterioles that run in the interlobular septa to drain directly into the sinusoids. Their oxygen-saturated blood is therefore mixed with deoxygenated blood from the portal venules. The gall bladder is supplied by the cystic artery, which arises from the right hepatic artery and is drained by cystic veins.

The blood flow to the liver is regulated by sympathetic nerves and by circulating hormones. Cholecystokinin and secretin enhance blood flow to the liver during digestion.

Conversely, the activation of the sympathetic nerves to the liver that occurs in exercise causes an increase in the vascular tone of the arterioles and veins. This decreases blood flow to the liver and mobilizes blood from the liver for use by the exercising muscle. A similar redistribution of blood occurs following haemorrhage.

> **KEY POINT:**
>
> - The liver has a dual blood supply. It receives blood from the gut via the portal vein and from the hepatic artery. Blood returns to the systemic circulation via the hepatic vein.

25.4 The role of bile in the digestion and absorption of fats

Bile production by the hepatocytes

The hepatocytes secrete a straw-coloured, isotonic fluid, known as hepatic bile, into the bile canaliculi. This fluid has a pH of between 7 and 8 and an ionic composition similar to that of plasma. It also contains bile salts, bile pigments, cholesterol, lecithin and mucus. As it passes along the bile ducts, ductal cells secrete a watery, bicarbonate-rich fluid, which modifies the composition of the primary secretion and increases its volume.

When it is not required for digestion, bile is stored in the gall bladder where it is concentrated. This occurs through the reabsorption of water and electrolytes by the gall bladder. This process results in gall bladder bile having greater concentrations of bile salts, bile pigments and cholesterol than hepatic bile.

The secretion of bile salts

Each day the liver synthesizes around half a gram of bile salts, which are important in the processing of fats within the small intestine. Bile salts are derived from bile acids, which in turn are synthesized from cholesterol. The primary bile acids, **cholic acid** and **chenodeoxycholic acid**, are synthesized by the hepatocytes, which subsequently link them to either glycine or taurine to form water-soluble

bile salts before secreting them into the bile. Bile salts have both hydrophobic and hydrophilic regions (i.e. they are amphipathic) and they aggregate to form micelles when they reach a certain concentration in the bile (the critical micellar concentration). The micelles are organized so that the hydrophilic groups of the bile salts face the aqueous medium and the hydrophobic groups form the core. The importance of micelles in the absorption of fats is discussed below.

Bile salts are recycled via the enterohepatic circulation

About 94% of all of the bile salts that enter the intestine in the bile are reabsorbed into the portal circulation by active transport from the lower ileum. Many of the bile salts return to the liver unchanged and are recycled. Others are deconjugated by intestinal bacteria before recycling. Only a small fraction of the material secreted into the bile is excreted in the faeces. This recycling of bile salts is known as the enterohepatic circulation.

The role of bile in the digestion of dietary fat

Because fats are so insoluble in water, their digestion and absorption pose a special problem for the GI tract. Bile salts are important in both processes. In the stomach, dietary fat forms large globules. As these globules enter the duodenum, they become coated with bile salts. The non-polar regions of the bile salts cling to the fat molecules while their negatively charged hydrophilic regions allow them to interact with water. As a result, the fat globules are pulled apart to form a stable emulsion (an aqueous suspension of fatty droplets, each about 1 μm in diameter). These small fat droplets have a greater surface area than the original fat globules and are therefore much more accessible to the pancreatic lipases. Consequently, the process of emulsification facilitates the breakdown of fats to monoglycerides and free fatty acids, as discussed on p. 519.

The role of the gall bladder

The gall bladder is a thin-walled, muscular sac that protrudes from the inferior margin of the liver (Figure 25.1). It appears green in colour due to the concentrated bile contained within it. Up to 60 ml of bile may be stored in the gall bladder during the period between meals, when it is not required for the digestion of fats. Hepatic bile is diverted into the gall bladder when the sphincter of Oddi is closed. During its stay in the gall bladder, the composition of the bile is modified and the bile salts are concentrated as the gall bladder absorbs sodium, chloride and bicarbonate. Water follows by osmosis. The concentration of bile by the gall bladder is so effective that, of the 800–1000 ml of bile secreted each day by the hepatocytes, only around 500 ml is directly secreted into the intestine.

The regulation of the secretion of bile

The major factor regulating the rate of hepatic bile secretion is the return of bile salts to the hepatocytes via the enterohepatic circulation. A further stimulus is thought to be the increase in hepatic blood flow that occurs in response to a meal. The production of hepatic bile itself is not under hormonal control, but secretin and, to a lesser extent, glucagon and gastrin enhance the secretion of fluid by the ductal cells.

Within a few minutes of starting a meal, particularly one that is rich in fats, the smooth muscle of the gall bladder contracts to force bile towards the duodenum. Although the initial response is mediated by the vagal (parasympathetic) innervation of the gall bladder, the principal stimulus for contraction is the hormone **cholecystokinin** (**CCK**), which is secreted by the duodenum in response to the arrival of chyme from the stomach. CCK also relaxes the sphincter of Oddi so that bile flows into the duodenum. The gall bladder normally empties completely within an hour of eating a fat-rich meal.

> **KEY POINTS:**
> - Bile is stored in the gall bladder between meals, where it is concentrated by the absorption of water.
> - The bile salts act as an emulsifier and are required for the complete digestion of fats.
> - The secretion of bile from the gall bladder is regulated by cholecystokinin secreted by cells in the duodenal epithelium.

25.5 The excretory role of the bile

The elimination of bile pigments

Bile pigments are the excretory products of the haem portion of haemoglobin. They are formed when old red blood cells are broken down in the spleen and they are responsible for the characteristic colour of faeces. The major bile pigment is bilirubin, which is relatively insoluble and is carried to the liver bound to the plasma protein albumin. In the hepatocytes, most of the bilirubin is conjugated with glucuronic acid to form bilirubin diglucuronide. This is water-soluble and enters the bile, giving it its characteristic yellow-green colour. Once it enters the intestine, particularly the colon, bilirubin diglucuronide is hydrolysed by bacteria to form three products. These are urobilinogen, which is highly water-soluble and colourless, and stercobilin and urobilin, which give the faeces their brown colour. Some of the urobilinogen is absorbed from the intestine back into the bloodstream. From there, it is either re-secreted back into the bile or excreted in the urine. The handling of bilirubin by the body is illustrated in Figure 25.5.

In addition to the bile pigments, many other substances, particularly drugs, are excreted via the bile. They then enter the small intestine where they may be recycled via the enterohepatic circulation (see Section 25.4 above). Examples are morphine and the antituberculosis drug rifampicin.

Jaundice

If the concentration of bilirubin in the blood exceeds a value of around $34\,\mu\text{mol}\,l^{-1}$, jaundice (**icterus**) develops. This is characterized by yellow discoloration of the skin, sclera of the eyes and the deep tissues. Plasma bilirubin may rise as a result of excessively high rates of red cell haemolysis, impaired uptake of bilirubin by liver cells, or blockage of the bile canaliculi or bile ducts.

Excessive rates of haemolysis may occur in the newborn, in certain hereditary conditions, or following a poorly matched blood transfusion. The jaundice that develops in these conditions is called **haemolytic jaundice**. Failure of the liver to take up or conjugate bilirubin results in **hepatic jaundice** and may be caused by cirrhosis or hepatitis. **Obstructive jaundice** occurs when bile is prevented from reaching the small intestine (cholestasis), either by gallstones, strictures or tumours of the bile duct or pancreas. This form of jaundice may be accompanied by itching (pruritis), caused by the accumulation of bile acids in the blood. In obstructive jaundice, the faeces will be pale in colour due to the absence of bile pigments and may contain fatty streaks as a result of the lowered absorption of dietary fat. Conversely, the urine is likely to be darker in colour than normal due to the increased excretion of bilirubin by the kidneys.

Excretion of phospholipids and cholesterol in the bile

Bile is the major route for the excretion of cholesterol from the body. Phospholipids, especially lecithin, are also excreted via the bile. Both are secreted into the bile as lipid vesicles, which are then emulsified by the bile salts. Any excess cholesterol that cannot be dispersed may form crystals in the bile. In some instances, calcium and phosphate deposits accumulate around these crystals to form gallstones which may block the common bile duct. Gallstones can be imaged by ultrasound, as in the example shown in Figure 25.6.

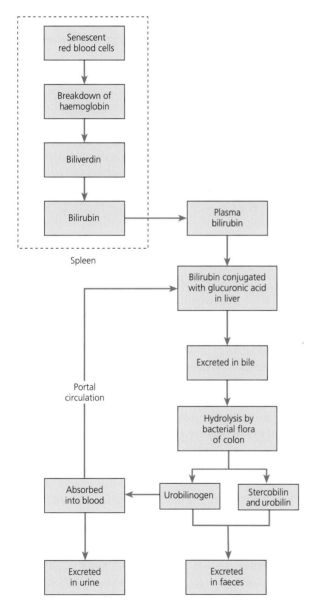

Figure 25.5 An outline of the formation and excretion of bilirubin.

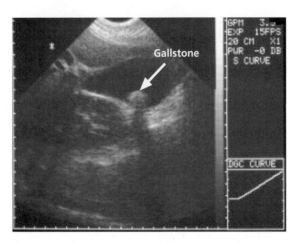

Figure 25.6 An ultrasound image of a gallstone in the gall bladder. Note the shadow cast by the gall stone. The outline of the gall bladder is clearly visible.

KEY POINTS:

- The colour of bile is due to the presence of bile pigments formed from the breakdown of the haem derived from the red cells.

- The bile pigments, cholesterol and other substances are excreted in the bile.

- If the bile duct is obstructed, the bile pigments build up in the blood and cause a yellowing of the skin and sclera of the eyes. This is known as obstructive jaundice.

- Other factors that cause bile pigments to accumulate in the blood give rise to other types of jaundice (haemolytic jaundice and hepatic jaundice).

25.6 Energy metabolism and the liver

Glycogen storage and breakdown

Following a meal, blood glucose is high and this stimulates the secretion of insulin by the β- cells of the pancreas (see Chapter 16). Insulin promotes the uptake of glucose by the liver, which stores it as glycogen. Glycogen is a large polysaccharide, and each glycogen molecule contains many glucose residues. The liver's store of glycogen is an important reserve of carbohydrate for the whole body, and is not solely for use by the liver itself. After a meal, plasma glucose levels begin to fall and the secretion of insulin declines while that of glucagon increases. The rise in the secretion of glucagon stimulates the breakdown of glycogen to glucose, which is released into the blood for use by the brain and other tissues.

During exercise, the secretion of adrenaline (epinephrine) from the adrenal medulla is increased. This hormone also increases the breakdown of glycogen to glucose. The main features of this cascade are summarized in Figure 25.7.

Gluconeogenesis

Proteins are the main structural components of cells and the principal use for the protein of the diet is the synthesis of new protein. However, those amino acids that are not required for protein synthesis are used for energy metabolism. Before this can happen, the amino acids must first have their amino group ($-NH_2$) replaced by a keto group ($>C=O$). This process occurs in the liver and is called **deamination**. It results in the liberation of ammonia, which is subsequently metabolized to urea (see below). After deamination, most amino acids can be used to synthesize glucose (a process known as **gluconeogenesis**) and those that can be so utilized are known as the glucogenic amino acids. The few that are not metabolized in this way are known as ketogenic amino acids. The glucogenic amino acids are ultimately metabolized to generate glucose for release into the blood to supply those tissues (especially the brain) that rely on glucose for ATP generation via the tricarboxylic acid cycle.

The ketogenic amino acids are oxidized by β-oxidation to form acetyl CoA, which, like the fatty acids, is metabolized via the tricarboxylic acid cycle to form ATP. Although the amino acids are generally considered to be either glucogenic or ketogenic, several, including the aromatic amino acids phenylalanine, tryptophan and tyrosine, can be either glucogenic or ketogenic depending on the metabolic pathway by which they are broken down.

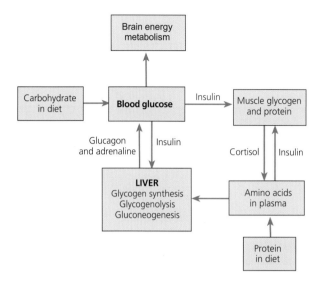

Figure 25.7 An outline of the central part played by the liver in the regulation of plasma glucose. During a meal, plasma glucose rises. This stimulates the secretion of insulin, which promotes the uptake of glucose by the liver (as well as other organs). Within the liver, glucose is stored as glycogen. As the plasma glucose begins to fall, the secretion of insulin falls and that of glucagon rises. The glucagon promotes the breakdown of stored glycogen (glycogenolysis) and the release of glucose into the blood. In starvation, muscle protein is broken down under the influence of cortisol and the amino acids are transported to the liver for conversion to glucose by gluconeogenesis.

Synthesis of plasma proteins

The liver secretes more protein than any other organ in the body. It secretes most of the plasma proteins including albumins, globulins and the clotting factors.

The **albumins** are the smallest and the most abundant, accounting for about 60% of the total plasma protein. They are transport proteins for lipids and steroid hormones. They are also important in body fluid balance as they provide most of the colloid osmotic pressure (the **oncotic pressure**) that determines the flow of water and solutes across the walls of the capillaries (see Chapters 19 and 23).

Globulins account for about 40% of total plasma protein. They are classified as **alpha (α), beta (β) and gamma (γ) globulins**. The α- and β-globulins are made in the liver and they transport lipids and fat-soluble vitamins in the blood while the γ-globulins are antibodies produced by lymphocytes (white blood cells) in response to exposure to antigens. Most of the clotting factors (e.g. fibrinogen and prothrombin) are synthesized by the liver, the main exceptions being Factor III (tissue factor, also called thromboplastin) and calcium ions.

> **KEY POINTS:**
>
> - The liver is able to store glucose as glycogen, and release it as required. The storage is promoted by insulin while the release of glucose is promoted by glucagon and adrenaline.
>
> - The liver is also able to form glucose from amino acids (gluconeogenesis).
>
> - The liver synthesizes and secretes the plasma proteins, including most of the clotting factors.

25.7 Endocrine functions of the liver

The liver produces a number of hormones. It synthesizes and secretes **angiotensinogen**, which is the precursor of angiotensins I and II (see Chapter 23, p. 480), **thrombopoietin**, which regulates the number of platelets circulating in the blood, the **insulin-like growth factors**, and **hepcidin**, which regulates iron metabolism.

Insulin-like growth factors

Although growth hormone is secreted by the anterior pituitary gland, it exerts many of its actions indirectly via peptide hormone intermediaries known as insulin-like growth factors (IGF-1 and IGF-2), which are synthesized and secreted by the liver and some other tissues. IGF-1 and IGF-2 are structurally similar to proinsulin (hence their names). The principal actions of IGF-1 are mediated via a specific receptor known as IGF-1R, although it is also known to bind to the insulin receptor with a much lower affinity. The blood level of IGF-1 is low in infancy, rises gradually until puberty, then increases more swiftly to reach a peak, which coincides with the peak height increase (see Figure 16.6, p. 292) after which it falls to its adult (and prepubertal) value. IGF-2 also binds to IGF-1R as well as to its own specific receptor, which acts as a binding site to regulate the quantity of IGF-2 available in the blood. The vast majority of IGF-1 (>97%) is carried in the blood bound to specific binding proteins, which are also synthesized by the liver.

Hepcidin and iron metabolism

Iron is an essential component of the body. The body of an adult man contains, in total, about 4.5 g of iron, of which about two-thirds is within the haemoglobin of red blood cells. A further 5% or so is contained within myoglobin and enzymes, while the remainder is stored in the form of **ferritin**, largely by the liver.

When red blood cells become senescent, they are removed from the blood by phagocytes in the liver and spleen. Much of the iron derived from haemoglobin is recycled by the body, as illustrated in Figure 25.8. The iron from the digested haemoglobin is either returned to the plasma where it binds to **transferrin** (an iron-carrying protein) or is stored in the liver as ferritin.

In iron-deficiency, or following haemorrhage, the capacity of the small intestine to absorb iron is increased. After severe blood loss, there is a time lag of 3 or 4 days before absorption is enhanced. This is the time needed for the enterocytes to migrate from their sites of origin in the mucosal glands to the tips of the villi, where they are best

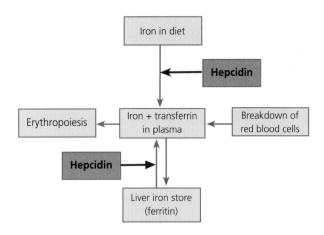

Figure 25.8 A summary of the role played by the liver in the metabolism of iron. Hepcidin is a hormone secreted by the liver that regulates the absorption of iron from the small intestine.

able to participate in iron absorption. The enterocytes of iron-deficient animals absorb iron from the intestinal lumen more rapidly than normal, a process controlled by a hormone called **hepcidin**, which is synthesized by the liver. Hepcidin reduces the capacity of the enterocytes to transport iron. When demand for iron is low, hepcidin levels are high and iron absorption decreases. However, when demand for iron is high, for example following haemorrhage, circulating levels of hepcidin fall and iron uptake from the small intestine is increased.

> **KEY POINT:**
>
> - The liver secretes the hormones hepcidin, IGF-1, IGF-2 and thrombopoietin as well as angiotensinogen, the precursor of angiotensin I and angiotensin II.

25.8 Detoxification mechanisms

Urea synthesis

The deamination of amino acids that occurs during gluconeogenesis and some other metabolic pathways results in the formation of ammonia, which is highly toxic. Any rise in the concentration of ammonia in the

plasma causes a disturbance of the acid–base balance, which may result in coma. To defend against this, the liver converts ammonia to urea, which is non-toxic and can be excreted by the kidneys. The liver is the only organ able to perform this function.

The ammonia is converted to urea by the urea cycle, shown in Figure 25.9. Ammonium ions and bicarbonate ions combine under the influence of an enzyme, carbamyl phosphate synthetase, to form carbamyl phosphate, which then combines with ornithine to form citrulline. In turn, citrulline is converted to arginosuccinate and arginine, which gives rise to urea and ornithine. The ornithine is then able to re-enter the cycle to participate in the synthesis of more urea.

Biotransformation of drugs and poisons

Those substances that are not normally present in the body are known as **xenobiotics**. In addition to drugs administered for therapeutic purposes, xenobiotics may enter the body as constituents of plants that are not part of the normal diet (many plants are highly toxic to man) or as a result of contamination of food by industrial chemicals such as dioxin and plasticizers. The liver is the main organ responsible for neutralizing xenobiotics. This it achieves by modifying their chemical constitution to render them inactive and more easily excreted. These changes are known as **biotransformation** or **detoxification**.

The biotransformation of most foreign substances occurs in two main stages known as phase I and phase II. In phase I, the foreign molecule is modified by oxidation, reduction or hydrolysis. In phase II, the modified molecule is joined to another molecule that inactivates it and makes it easier to excrete. This is known as **conjugation**. Commonly, the molecules formed in phase I are conjugated with glucuronic acid, sulphate or acetate, which increases their water solubility and thus facilitates their excretion via the kidneys. The well-known anti-inflammatory drug aspirin provides a simple example of biotransformation. The aspirin is first hydrolysed to salicylic acid, which is then conjugated with glucuronic acid to form salicylglucuronide (Figure 25.10). Other substances may be oxidized by one of the family of cytochrome P_{450} enzymes prior to their conjugation.

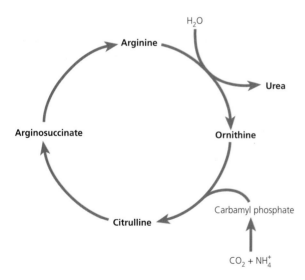

Figure 25.9 An outline of the urea cycle. The initial step is the formation of carbamyl phosphate from carbon dioxide and ammonium ions formed following the deamination of amino acids. The carbamyl phosphate enters the cycle by combining with ornithine to form citrulline. The splitting of urea from arginine results in the further formation of ornithine, which is then able to participate in another turn of the cycle.

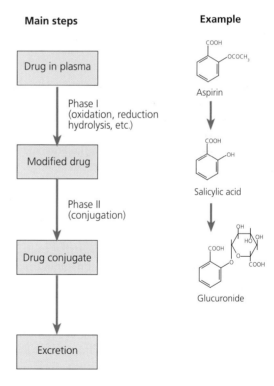

Figure 25.10 The main stages of drug metabolism. Although phase I and phase II are shown occurring sequentially, this is not always the case. In the case of aspirin shown here, phase I occurs by hydrolysis of the bond between the acetate residue and a hydroxyl group on the benzene ring.

In most cases, the phase I reactions result in a reduction or loss of biological activity, but in a few unfortunate cases, the phase I reactions sharply increase the toxicity of the xenobiotic material. A well-known example is the toxicity of paracetamol, which is due to the transformation of the parent compound by cytochrome P_{450} to a toxic metabolite N-acetyl-p-benzoquinine. Liver failure occurs a few days after an overdose of paracetamol. This disastrous outcome highlights the importance of following the correct guidelines for drug administration.

In addition to the inactivation and excretion of foreign molecules, the liver plays an important part in the inactivation and excretion of many hormones. In particular, many of the steroid hormones are oxidized by cytochrome P_{450} enzymes to make them more water-soluble and inactive. They are subsequently conjugated with glucuronic acid before being excreted by the kidneys.

KEY POINTS:

- The liver is the only organ in the body able to convert the ammonia liberated during amino acid metabolism to urea, which is non-toxic.
- The liver protects the body from toxins by modifying their chemical constitution to render them inactive and more easily excreted. This is known as biotransformation or detoxification.

25.9 Liver failure

Despite the great importance of the liver in metabolism, hepatic disease only rarely leads to serious illness. This fortunate circumstance arises because the liver has a great reserve capacity and a remarkable ability to regenerate. However, there are circumstances in which the damage to the liver is so extensive that its function is impaired. Liver failure is diagnosed when there is evidence of jaundice, fluid accumulation in the peritoneal cavity (ascites), failure of blood clotting and marked psychological changes called hepatic encephalopathy.

The many functions of the liver are summarized in Table 25.1 (p. 534), from which it should be evident that the consequences of liver failure or hepatic insufficiency will be widespread and extremely grave. The liver is subject to most of the disease processes that can affect other body structures including inflammation, vascular disorders, metabolic disease, toxic injury and neoplasms. Two of the most common hepatic diseases are hepatitis (characterized by inflammation of the liver) and cirrhosis (particularly, although not necessarily, due to alcohol abuse). Cirrhosis is characterized by fibrosis and a loss of normal structure and function.

The manifestations of liver failure are widespread and reflect its various functions:

- In liver failure, severe hypoglycaemia develops rapidly because of the central role played by the liver in the maintenance of plasma glucose.

- Bile production will slow or cease and this will impair the digestion and absorption of fats.

- As the ammonia formed by deamination of amino acids is only converted to urea in the liver, the level of ammonia in the plasma will rise while the concentration of urea will fall.

- The protein profile of the blood will change radically following liver failure. The production of plasma proteins by the liver will be impaired, leading to a loss of albumin and of certain globulins. If the total plasma protein level falls significantly, there will be a reduction in the plasma oncotic pressure, which may lead to generalized oedema, or fluid accumulation in the peritoneal cavity (ascites). At the same time, many of the enzymes contained within the hepatocytes will enter the blood after leaking out of the damaged cells. These enzymes include the amino-transferases (or transaminases), alkaline phosphatase and lactate dehydrogenase, which catalyses the conversion of pyruvic acid to lactic acid. The levels of these enzymes in the plasma may be used clinically as liver function tests to provide an index of the extent of liver damage.

- As the liver plays a vital role in the excretion of bilirubin derived from the breakdown of red blood cells, unconjugated bilirubin accumulates in the plasma and the sufferer will develop jaundice. In severe cases, damage to the brain may occur (hepatic encephalopathy).

- As the liver starts to fail, the production of clotting factors will fall, leading to an impairment of blood clotting, which may result in spontaneous bleeding from the skin and mucous membranes. The prothrombin clotting time will be increased.

- In chronic liver disease, the vascular resistance of the portal circulation increases resulting in **portal hypertension**. As the portal pressure rises above about 12 mmHg, the venous anastomoses linking the portal and systemic circulations dilate and form 'varices' (swollen veins) that may project into the lumen of the oesophagus and stomach. These may rupture and cause gastrointestinal bleeding.

- Finally, a failing liver is less able to carry out detoxification of poisonous chemicals. For this reason, drug actions are likely to be prolonged in patients in liver failure.

The treatment of liver failure

The treatment of liver failure is, of necessity, complex and must address the many physiological processes throughout the body that are disturbed. If the initial problem was alcoholic cirrhosis, the first step in treatment will be to eliminate alcohol intake. Sufficient carbohydrate and calories will be required to prevent protein breakdown and fluid and electrolyte imbalances must be corrected, while protein intake may be limited to inhibit the production of ammonia. For patients in the final stages of liver failure, liver transplantation is now a realistic form of treatment.

> **KEY POINTS:**
>
> - During liver failure, many physiological processes are disrupted. In acute hepatic failure there is hypoglycaemia, disordered lipid metabolism and decreased protein synthesis. The diminished synthesis of albumin leads to the formation of ascites and peripheral oedema. The loss of those clotting factors that are synthesized in the liver leads to disorders of blood clotting.
>
> - As the liver plays a central role in the recycling of iron from senescent red blood cells, in liver failure there is an increased level of bilirubin in the plasma, which results in jaundice.
>
> - Liver failure leads to failure of detoxification mechanisms, which prolongs the action of the steroid hormones and gives rise to the symptoms of endocrine disease.

 # Recommended reading

Histology

Junquieira, L.C. and Carneiro, J. (2005). *Basic Histology*, 11th edn. McGraw-Hill: New York. Chapter 16.

Physiology of the liver

Johnson, L.R. (2007). *Gastrointestinal Physiology*, 7th edn. Mosby: Philadelphia. Chapter 10, pp. 97–106. [A clear explanation of bile production and the control of its secretion.]

Medicine

Kumar, P. and Clarke, M. (2005). *Clinical Medicine*, 6th edn. Saunders: Edinburgh. Chapter 7. [A brief introduction to the functions of the liver and billiary tract with a detailed account of liver diseases and appropriate clinical investigations.]

Self-assessment questions

Which of the following statements are true and which are false? Answers are given at the end of the book (p. 755).

1 a) The liver has two lobes.
   b) The falciform ligament attaches the liver to the diaphragm.
   c) The gall bladder is part of the liver.
   d) The liver is made up of functional units called lobules.
   e) The principal liver cells are called hepatocytes.

2 a) At rest, the liver receives as much as a quarter of the cardiac output.
   b) The portal vein provides most of the blood flow to the liver.
   c) Blood flow to the liver falls during haemorrhage.
   d) Hepatic bile is an ultrafiltrate of plasma containing bile salts.
   e) The secretion of hepatic bile is under the control of cholecystokinin.

3 a) Bile pigments are the breakdown products of haemoglobin.
   b) The reabsorption of bile pigments in the intestine stimulates bile secretion.
   c) Most bile salts are absorbed in the terminal ileum.

   d) Loss of bile salts will lead to poor absorption of vitamin E.
   e) Bile salts are hydrophobic molecules.

4 a) The accumulation of bilirubin in the blood can lead to jaundice.
   b) Jaundice can be caused by gallstones.
   c) In obstructive jaundice, the faeces are normal in colour.
   d) In liver failure, there is jaundice and fluid accumulation in the abdominal cavity.
   e) Blood clotting is normal in liver failure.

5 a) The liver stores glucose as glycogen.
   b) The liver can synthesize glucose from fatty acids.
   c) The liver and kidneys convert ammonia to urea.
   d) The liver synthesizes almost all of the plasma proteins.
   e) The liver promotes the excretion of many toxic substances by conjugating them with glucuronic acid.

The pancreas

26

After reading this chapter you should have gained an understanding of:

- The anatomy of the pancreas
- The role of the exocrine pancreas in digestion
- The control of pancreatic secretion
- Common diseases affecting the pancreas
- The endocrine functions of the pancreas

26.1   Introduction

The pancreas performs two distinct functions in the body: it is an endocrine gland, secreting insulin and glucagon into the bloodstream, and it is an exocrine organ that secretes the enzymes necessary to digest food. Its endocrine role in the regulation of blood glucose is discussed in detail in Chapter 16, pp. 315–324.

26.2   Anatomy and histology of the pancreas

The pancreas lies posterior to the greater curvature of the stomach and extends across the abdomen for 15–20 cm. It consists of three main sections: a **head**, **body** and **tail**. The expanded head lies within the arc formed by the duodenum while the body and tail extend towards the spleen, as shown in Figure 26.1. The exocrine portion of the pancreas, which accounts for more than 80% of total pancreatic tissue, consists of secretory units called **acini** (singular acinus), which are similar in appearance to those of the salivary glands. The endocrine part of the pancreas takes the form of scattered clumps of cells known as the **islets of Langerhans**, which contain the

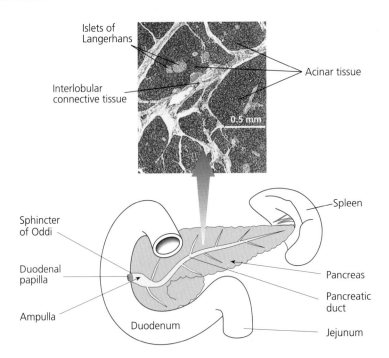

Figure 26.1 A diagrammatic representation showing the position of the pancreas in relation to the duodenum and spleen. The inset image shows a section from the middle of the pancreas. Note the pale staining of the islets of Langerhans and the connective tissue that divides the pancreas into lobules. The bulk of the pancreas is made up of densely staining acinar tissue.

cells that secrete glucagon – the α-cells – and those that secrete insulin – the β-cells.

The pancreatic tissue is loosely divided into lobules by sheets of connective tissue called the interlobular septa (Figure 26.2). Within each lobule lie the individual acini whose cells are arranged around a central space so that each acinus resembles a ball. The acinar cells themselves are filled with densely staining **zymogen granules** containing the digestive enzymes that are secreted by the pancreas. These digestive enzymes, together with a little fluid, pass from the acini into microscopic ducts known as intercalated ducts. These form the first part of the duct system that collects the exocrine secretions of the pancreas before they are discharged into the duodenum. The ducts are lined by epithelial cells, which secrete the fluid that forms the bulk of the pancreatic juice. The intercalated ducts drain into larger intralobular ducts, which, in turn, drain into interlobular ducts (ducts linking the pancreatic lobules). Finally, these ducts empty their contents

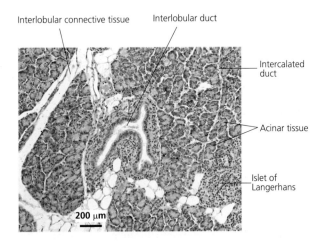

Figure 26.2 The microscopical structure of the pancreas. Note the distinctive appearance of the acinar tissue and the paler-staining islet tissue (on the lower right-hand side of the picture). Part of the branching interlobular duct system can be seen near the centre of the picture. Its epithelial lining is very clearly visible.

into the main pancreatic duct that runs along the axis of the pancreas, as shown in Figures 26.1 and 26.3. Before it empties into the duodenum at the duodenal papilla, the main pancreatic duct fuses with the common bile duct at the ampulla, as shown in Figure 25.2 (p. 535). Surrounding the ampulla is the **sphincter of Oddi**, which regulates the flow of pancreatic juice and bile into the duodenum.

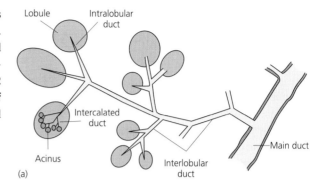

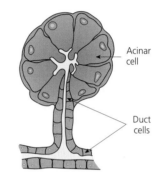

Figure 26.3 A simple diagrammatic representation of the arrangement of the duct system of the pancreas (a). The cells lining the ducts are responsible for most of the volume of fluid secreted by the pancreas. (b) The arrangement of the acinar cells around the central space and the intercalated duct, which conduct the acinar secretion towards the duodenum.

26.3 The blood supply and innervation of the pancreas

The pancreas has a rich blood supply; oxygenated blood is carried to the pancreas by branches of the coeliac and superior mesenteric arteries while its venous blood drains into the portal vein, from where it travels to the liver. The acini and the ducts are surrounded by capillaries. The pancreatic blood vessels are innervated by sympathetic vasoconstrictor fibres from the coeliac and superior mesenteric plexuses. The islet tissue receives sympathetic fibres, which act to increases the secretion of glucagon from the α-cells of the islet tissue (see below). Parasympathetic nerves from the vagus nerves innervate the pancreatic acinar cells and β-cells of the islets. Stimulation of these parasympathetic nerves increases the secretion of both enzymes and fluid by the exocrine part of the pancreas and the secretion of insulin by the β-cells.

> **KEY POINTS:**
>
> - The pancreas is both an endocrine and an exocrine gland.
> - It secretes all the enzymes necessary to digest the food, together with an alkaline fluid.
> - The principal hormones secreted by the islet tissue of the pancreas are insulin and glucagon, both of which are important in the regulation of blood glucose.

26.4　The pancreatic juice and its functions

The pancreas secretes between 1000 and 1500 ml of enzyme-rich fluid each day. As mentioned above, this juice consists of an aqueous component, secreted almost entirely by the cells lining the ducts, and enzymes, secreted by the acinar cells. As discussed in Chapter 24 (p. 519), the pancreas secretes most of the enzymes necessary for the digestion of fats, proteins and carbohydrates.

The aqueous component. The fluid secreted by the duct cells is slightly hypertonic. It is rich in bicarbonate and has a pH of about 8, which helps to neutralize the acidic chyme arriving in the duodenum from the stomach. As the fluid passes along the ducts, bicarbonate ions are reabsorbed from the fluid in exchange for chloride ions. As a result, the bicarbonate concentration of the fluid is lower at slower flow rates than at faster rates, when the fluid spends less time in the ducts and is therefore scarcely modified.

The pancreatic enzymes. A list of the principal enzymes present in the pancreatic juice is given in Table 26.1. Many of them, including the proteolytic enzymes and some of the lipolytic enzymes, are stored and secreted as zymogens, which are inactive. The zymogens are subsequently activated by trypsin within the lumen of the small intestine. In this way, the pancreas, like the stomach, avoids self-digestion.

Proteolytic enzymes of the pancreas

The principal proteolytic enzyme of pancreatic juice is **trypsin**, which, like many other pancreatic enzymes, is secreted in the form of an inactive precursor – trypsinogen. As mentioned above, this adaptation is necessary to prevent the pancreas digesting itself. The trypsinogen can be activated spontaneously in response to the alkaline environment of the duodenum. It is also activated by **enteropeptidase** (formerly known as enterokinase), one of the brush border enzymes of the duodenum. Once it has been activated, trypsin activates chymotrypsinogens and other pancreatic enzymes, as outlined in Figure 26.4. Trypsin and the chymotrypsins hydrolyse peptide bonds within protein molecules to liberate free amino acids and polypeptides of various sizes. Other enzymes, such as carboxypeptidase, aminopeptidase and elastase, digest the polypeptides further to release small peptides and free amino acids.

Pancreatic amylase

Pancreatic α-amylase is responsible for the continuation of the digestion of starch that was initiated by salivary α-amylase in the mouth. Like salivary amylase, pancreatic amylase is secreted in its active form and, within

Table 26.1 The major pancreatic enzymes

Enzyme	Zymogen	Activator	Action
Trypsin	Trypsinogen	Enteropeptidase of brush border	Cleaves internal peptide bonds
Chymotrypsin	Chymotrypsinogen	Trypsin	Cleaves internal peptide bonds
Amylase		Secreted in its active form	Digests starch to maltose and oligosaccharides
Lipase		Secreted in its active form	Cleaves glycerides liberating fatty acids and glycerol
Colipase	Procolipase	Trypsin	Binds to micelles to anchor lipase to lipid

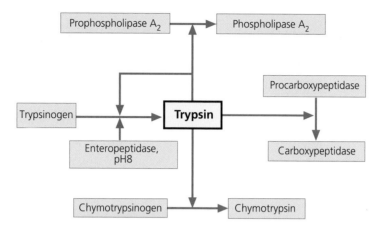

Figure 26.4 The sequence of activation of the main pancreatic enzymes. Note the key role played by trypsin. Trypsin itself is activated in the duodenum by the relatively high pH and by enteropeptidase located on the brush border of the cells lining the duodenum (the duodenal enterocytes).

about 10 minutes of entering the small intestine, starch is entirely converted to various disaccharides and oligosaccharides, chiefly maltose and maltriose. (An oligosaccharide consists of a few monosaccharides combined together; see Chapter 2.) Enzymes on the intestinal brush border then hydrolyse the disaccharides and oligosaccharides, completing the breakdown of starch to glucose. Unlike salivary amylase, pancreatic amylase can digest uncooked starch.

Pancreatic lipases

Although pancreatic **lipase** is secreted in its active form, the other principal lipolytic enzymes, **colipase**, **cholesterol esterase** and **phospholipase A₂**, are secreted as inactive zymogens. These are then activated by trypsin within the duodenum. Pancreatic lipase hydrolyses water-insoluble

triglycerides to release free fatty acids and glycerol (see Figure 24.23, p. 520). Bile is also required for the normal digestion of fats (see p. 537) and colipase works with bile to facilitate the actions of lipase.

> **KEY POINTS:**
>
> - The pancreas secretes all the enzymes necessary for the digestion of the dietary fats, proteins and starch.
> - Most of these enzymes are secreted as inactive precursors (zymogens), which are activated by trypsin in the duodenum.
> - Trypsin itself is secreted in its inactive form and is activated in the duodenum by enteropeptidase, a brush border enzyme.

26.5 The regulation of pancreatic secretion

Although the secretion of pancreatic juice is regulated by both nerves and hormones, the hormonal regulation is quantitatively the most important. As for gastric secretion (p. 511), pancreatic secretion occurs in three phases: cephalic, gastric and intestinal. The **cephalic phase** is

largely regulated by the autonomic nervous system, while the **gastric phase** and the **intestinal phase** are controlled both by nerves and by hormones, as shown in Figure 26.5, which summarizes the main factors that regulate the different phases of pancreatic secretion.

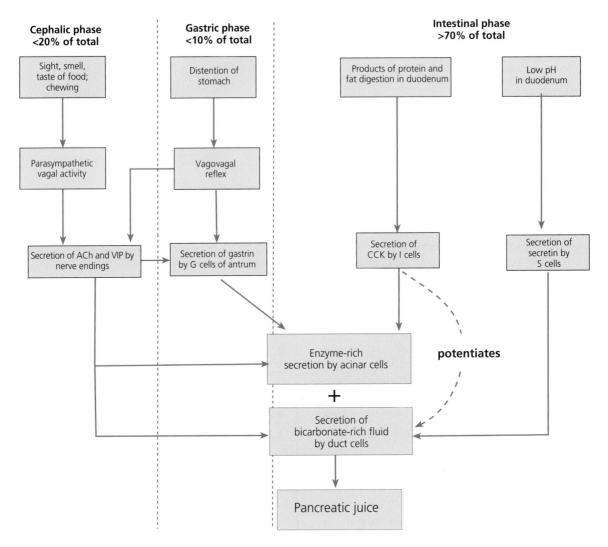

Figure 26.5 A schematic diagram showing the factors that regulate pancreatic secretion. Note the importance of hormonal regulation during the intestinal phase, which accounts for approximately 70% of the total secretion. ACh, acetylcholine; CCK, cholecystokinin; VIP, vascoactive intestinal polypeptide.

The cephalic phase occurs in response to the sight, smell and taste of food and accounts for about one-fifth of the total exocrine secretion of the pancreas. These stimuli increase parasympathetic activity, which increases both blood flow to the pancreatic acinar tissue and the rate of secretion of enzymes by the acinar cells. In contrast, sympathetic stimulation causes vasoconstriction and a reduction in the rate of pancreatic secretion.

The **gastric phase** represents only a small part of the total pancreatic secretion in response to a meal (less

than 10%). It occurs mainly in response to the gastrin secreted by the stomach as it distends and to the presence of amino acids and peptides in the pyloric region, particularly the antrum.

The **intestinal phase** of pancreatic secretion represents about 70% of the total exocrine pancreatic secretion that occurs following a meal. A number of hormones secreted by the upper intestinal mucosa stimulate the production of pancreatic enzymes and fluid. The most important of these are **cholecystokinin**

(**CCK**) and **secretin**. CCK is secreted when fatty acids, peptides and amino acids enter the duodenum, while secretin is secreted in response to acid chyme coming into contact with the wall of the duodenum. CCK enhances the secretion of enzymes by the exocrine pancreas, while secretin stimulates the production of bicarbonate-rich fluid by the ductal cells. The actions of secretin also appear to be facilitated by CCK.

> **KEY POINTS:**
> - The secretion of pancreatic juice is regulated by both hormones and nerves.
> - The hormones cholecystokinin and secretin are secreted in response to the arrival of acid chyme in the upper duodenum.

26.6 Disorders of pancreatic exocrine secretion and their treatment

Pancreatitis and cystic fibrosis are conditions in which the exocrine function of the pancreas is impaired. In **chronic pancreatitis**, there is defective secretion of alkaline fluid by the ducts, which results in a high protein concentration within the pancreatic juice. Over time, protein precipitates out of the pancreatic juice, leading to blockage of the ducts and back flow of juice. This causes intense pain in the epigastric region (often radiating to the back) and progressive damage to pancreatic tissue. Pancreatic insufficiency follows. Acute pancreatitis is often characterized by high circulating levels of pancreatic amylase, and measurements of the level of pancreatic amylase in the plasma can give important diagnostic information regarding the extent and progression of the disease.

Cystic fibrosis is an autosomal recessive condition caused by mutations in a gene on chromosome 7 (see Chapter 30). The mutation results in the production of a defective transmembrane protein that is involved in the transport of chloride ions in epithelial cells of the respiratory, gastrointestinal and reproductive tracts. The reduction in chloride transport, is accompanied by a fall in salt and water transport, and in the pancreas this results in the production of a dehydrated, viscous exocrine secretion, which often obstructs the pancreatic ducts and causes damage and destruction of pancreatic tissue. As with pancreatitis, pancreatic insufficiency will ensue.

As the pancreatic juice contains so many important digestive enzymes, individuals with pancreatic insufficiency are often treated with oral pancreatic supplements such as **pancreatin**, an extract of pancreas that contains protease, lipase and amylase. These enzymes are inactivated at the low pH encountered in the stomach, so the extracts are often given as enteric-coated preparations that can protect the enzymes from degradation and hence deliver a higher concentration of active product to the duodenum. Alternatively, gastric acid secretion can be inhibited by an H_2-receptor antagonist (e.g. cimetidine) or a proton pump inhibitor (e.g. omeprazole) before administering the pancreatin.

Carcinoma of the pancreas. Like other organs, the pancreas may develop cancer. Such cancers are more common in men than in women and their incidence increases with age. They are very aggressive and survival rates are low. Most are adenocarcinomas that arise in the epithelium of the pancreatic ducts. They frequently involve the head of the pancreas and invade other local tissues including the duodenum. When a carcinoma invades the head of the pancreas, the cancerous mass may obstruct the flow of pancreatic juice and bile, so giving rise to obstructive jaundice with its attendant symptoms (see p. 539).

26.7   Endocrine functions of the pancreas

As mentioned earlier, small clumps of endocrine tissue known as the islets of Langerhans are dispersed throughout the pancreas (Figure 26.1). The islets are between 100 and 200 µm across and occupy about 2% of the total mass of the pancreas. They are highly vascular and are innervated by both the sympathetic and the parasympathetic branches of the autonomic nervous system.

At least four different cell types have been identified within the islets: α-cells, β-cells, δ-cells and PP cells. The α-cells synthesize and secrete **glucagon** and the β-cells synthesize and secrete **insulin**. The δ-cells secrete **somatostatin** and PP cells synthesize **pancreatic polypeptide**. Islet tissue stained for glucagon and insulin is shown in Figure 26.6, from which it is clear that the most common cells in the islets are the β-cells (about 75% of the total) followed by the α-cells (up to 20% of the total). The δ-cells account for about 5% while the PP cells are relatively rare, accounting for only 1% or so of the islet cells.

Insulin and glucagon play an essential role in the regulation of plasma glucose levels. They act in a push–pull manner to provide minute-to-minute control of the level of glucose in the plasma. Sympathetic stimulation increases the secretion of glucagon, as does a fall in blood glucose. In contrast, the main stimulus to the secretion of insulin is a rise in blood glucose. Insulin is the only hormone able to lower blood glucose, which it does by promoting the uptake of glucose by the cells (except those of the brain, where glucose uptake is independent of insulin) and stimulating glycogen synthesis. It also plays an important role in the growth and maintenance of the tissues by promoting the uptake of amino acids from the plasma and the synthesis of new protein.

Lack of insulin results in a condition known as **diabetes mellitus** (see p. 321). In contrast, glucagon acts to mobilize glucose from the body's glycogen stores. It is the most potent hyperglycaemic hormone and acts to promote the release of glucose into the blood. Low blood glucose (hypoglycaemia) is the principal stimulus for the secretion of glucagon,

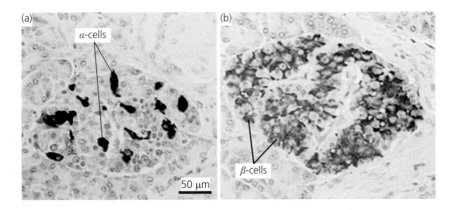

Figure 26.6 Pancreatic islet tissue stained for glucagon (a) and insulin (b). The cells containing glucagon and insulin stain black or brown in contrast to the other islet cells and surrounding acinar tissue, which is counterstained with a pale blue dye. Note that the islets contain more insulin-containing β-cells than α-cells (which contain glucagon).

which then stimulates glycogenolysis, lipolysis and gluconeogenesis. The long-term regulation of plasma glucose depends on many other factors, which are discussed in detail in Chapter 16 (p. 320). The somatostatin secreted by the δ-cells inhibits the secretion of both insulin and glucagon via a paracrine action and perhaps plays a role in regulating the secretion of both hormones in response to blood glucose. The function of pancreatic polypeptide remains unclear.

> **KEY POINTS:**
>
> - The endocrine tissue of the pancreas is formed by the islets of Langerhans.
> - The β-cells secrete insulin, which acts to lower blood glucose.
> - The α-cells secrete glucagon, which mobilizes glucose from the body's stores.

Recommended reading

Histology

Junqueira, L.C. and Carneiro, J. (2005). *Basic Histology*, 11th edn. McGraw-Hill: New York. Chapter 16, pp. 303–306.

Physiology of the pancreas

Johnson, L.R. (2007). *Gastrointestinal Physiology*, 7th edn. Mosby: Philadelphia. Chapter 9. [A clear account of the exocrine functions of the pancreas.]

Laycock, J. and Wise, P. (1996). *Essential Endocrinology*, 3rd edn. Oxford University Press: Oxford. Chapters 11 and 12. [A detailed account of the endocrine functions of the pancreas.]

Medicine

Kumar, P. and Clarke, M. (2005). *Clinical Medicine*, 6th edn. Saunders: Edinburgh. Chapter 7. [A brief introduction to modern diagnostic and imaging methods applied to the pancreas.]

Self-assessment questions

Which of the following statements are true and which are false? Answers are given at the end of the book (p. 755).

a) The exocrine pancreas consists entirely of acinar cells.
b) The islet tissue secretes some of the enzymes necessary to digest the food.
c) The introduction of acid into the duodenum stimulates pancreatic secretion.
d) Activation of the parasympathetic (vagal) nerves increases the secretion of pancreatic juice.
e) Cholecystokinin increases the flow of bile but inhibits secretion of pancreatic juice.

f) Loss of pancreatic enzymes will result in weight loss due to poor protein digestion.
g) Pancreatic acinar cells secrete trypsinogen.
h) Pancreatic amylase is activated by trypsin.
i) The β-cells of the islets secrete insulin.
j) Insulin promotes the release of glucose from the glycogen stores of the body.

27 Skin structure and function

After reading this chapter you should have gained an understanding of:

- The structure of human skin and the process of keratinization
- Different types of skin
- Histological features of skin
- The circulation of the skin (the cutaneous circulation)
- The role of the skin as an interface between the body and the external environment
- The role of the skin in thermoregulation
- The role of the skin in preventing infection
- The role of the skin as a sensory organ
- Pain perception
- Wound healing
- The pharmacology of the skin

27.1 Introduction

The skin (or cutis) is the largest organ in the body, covering its entire outer surface and shielding most of its organs. Together with the hair, nails and glands, the skin forms the **integumentary system**. In an adult of average size, the skin has a surface area of around 2 m² and accounts for around 15% of body weight. It varies in thickness from 0.5 to 5 mm depending upon the region of the body and the age of the person. The thickest skin is found in areas that are subject to abrasion, such as the soles of the feet, while the thinnest is found around the eyes.

As the interface between the internal structures of the body and the external environment, the skin serves a number of important functions including protection against environmental hazards and disease-producing agents (pathogens), insulation, thermoregulation, sensation and the synthesis of vitamin D. Integrity of the skin is also vital for normal fluid balance – indeed the severe fluid loss that occurs in patients who have suffered extensive burns constitutes a medical emergency. The structure and functions of the skin and its appendages will be discussed in this chapter but the reader is referred to Chapter 33 for a fuller treatment of the role of the skin in the regulation of body temperature.

Because the skin is on the outside of the body, and may therefore be seen and touched, a great deal of useful information regarding a person's age, nutritional status, hydration status, circulation and emotional state can be obtained by careful inspection. Furthermore, the characteristics and colour of a person's skin can sometimes yield information regarding certain disease states. Examples include the yellow hue of jaundiced skin or the blue skin colour (**cyanosis**) characteristic of a reduction in the amount of oxygen being carried by the blood. Furthermore, skin rashes and other changes may represent external manifestations of systemic disease.

The skin is highly susceptible to traumatic injury and wounds often become infected by micro-organisms. The complex processes that bring about the healing of wounds will be described in this chapter in addition to some of the drugs and clinical interventions used to treat skin disorders and to promote wound healing.

27.2 The main structural features of the skin

Structurally, the skin is made up of two layers of tissue that have different histological and functional characteristics and are of different embryological origin. These are illustrated in Figure 27.1. The thin outer layer is the **epidermis** and the thicker underlying layer is the **dermis**. Beneath the dermis lies a third layer, consisting largely of adipose tissue (fat). Although this is not part of the skin itself, this layer serves to bind the skin to underlying tissues and supplies it with nerves and blood vessels. This layer is called the **hypodermis** (subdermal or subcutaneous layer).

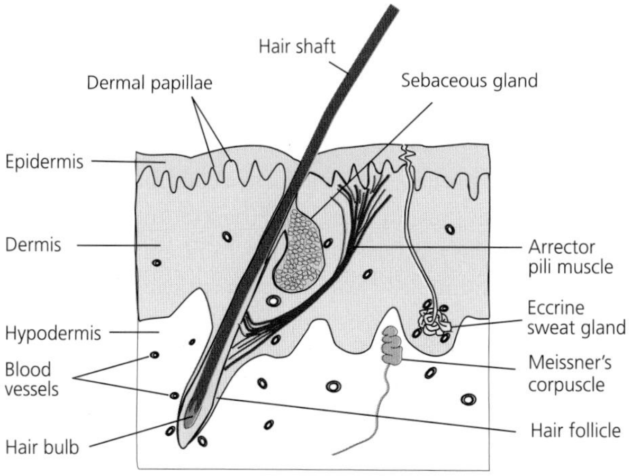

Hairy skin is found over most areas of the body, but in some locations, such as the soles of the feet, the palms of the hands and parts of the external genitalia, the skin is hairless. Smooth skin of this kind is known as **glabrous skin**. Hairy skin varies in different parts of the body with regard to both the number of hairs present and their physical characteristics – compare for example the hairy skin of the scalp with that of the arm.

Figure 27.1 A simple diagram of the principal structures of the skin.

> **KEY POINTS:**
>
> - The skin is the interface between the internal structures of the body and the outside world.
>
> - It is a barrier to infection, a major sense organ and a waterproof outer layer. It is also important in the regulation of body temperature.
>
> - It consists of two main layers: the epidermis and the dermis. The dermis is bound to the underlying tissues by the hypodermis.

27.3 The epidermis

The epidermis is a **keratinized**, **stratified squamous epithelium** (see Chapter 6 for a description of the different kinds of epithelium), which varies in thickness from 0.1 mm around the eyes to around 1.4 mm on the soles of the feet. It forms the surface layer of the skin and is derived from ectodermal tissue in the embryo. It contains no blood vessels and no nerve endings. Structurally, the epidermis is well adapted to its key functions of preventing fluid loss and providing an effective physical barrier against environmental hazards.

As Figure 27.2 shows, the epidermis consists of five distinct layers of cells. Starting from the deepest layer and working towards the surface, these are:

- Stratum basale (basal layer)
- Stratum spinosum
- Stratum granulosum
- Stratum lucidum
- Stratum corneum.

The **stratum basale** is formed by a single layer of cells that lie on a basement membrane (the basal lamina)

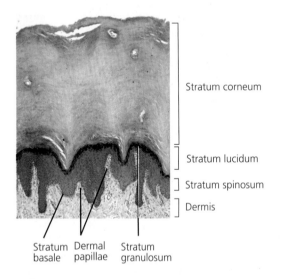

Stratum corneum

Stratum lucidum

Stratum spinosum

Dermis

Stratum basale   Dermal papillae   Stratum granulosum

Figure 27.2 Section through the outer layers of the skin showing the layers of the epidermis. This section is from skin taken from the sole of the foot.

that separates the epidermis and dermis. These cells may be cuboidal or columnar in character and they are the **germinal cells** of the epidermis, dividing rapidly to provide a continuous supply of new cells, known as **keratinocytes**, to the more superficial layers. As the cells migrate upwards towards the skin surface, they alter in shape and biochemical characteristics. In the **stratum spinosum**, the cells become irregular in shape and become separated by narrow gaps or clefts spanned by thin cytoplasmic processes or spines that give this layer its name. The cells of this layer are sometimes called **prickle cells**. Membrane-bound granules containing lipids (**lamellar granules**) begin to appear in their cytoplasm, together with accumulations (fine grains) of keratohyalin, a protein that will eventually form the keratin of the outer epidermis. As the cells enter the **stratum granulosum**, the contents of the lamellar granules begin to be released from the cells and to fill the interstitial space with lipid, which is important for the barrier function of the epidermis.

As cells reach the outer margins of the stratum granulosum, their nuclei begin to degenerate and the cells begin to die as they are separated from their blood supply. The **stratum lucidum** consists of a few layers of flattened, dead cells with some filaments of keratin present inside them. The cells are difficult to visualize microscopically and appear clear and poorly defined. Cells in the **stratum corneum** are known as 'horny cells'. They are completely filled with keratin filaments and the spaces between them are filled with lipids (released from the lamellar granules), which bind the cells together to form a continuous membrane. Towards the outer surface of the stratum corneum, the intercellular connections begin to loosen and cells slough off continuously from the skin surface (a process called desquamation), to be replaced by new cells migrating upwards from deeper layers. The process whereby cells gradually become filled with keratin is called **keratinization** and it normally takes around 30 days for cells to complete their journey from the stratum basale to the outer surface of the stratum corneum, from which they are shed.

Psoriasis is an inflammatory skin condition characterized by an increased rate of production of keratinized cells, which form silvery, scale-like plaques on the skin surface. **Dandruff** (also called scurf) is a relatively minor, though common, scalp condition in which dead epidermal cells are shed from the scalp skin in large clumps or flakes.

Other cells of the epidermis

Although the majority of cells within the epidermis are keratinocytes, three further cell types are also found. They are **melanocytes** (pigment cells), **Merkel** cells and **Langerhans cells** (which are part of the immune system).

Melanocytes

Melanocytes are derived from embryonic neural ectoderm. Typically, between 1000 and 2000 melanocytes are found per mm^2 of skin, predominantly in the stratum basale. They can be seen in the histological section shown in Figure 27.3. They possess fine processes, which make contact with keratinocytes. They synthesize a brown pigment called **melanin**, the primary determinant of skin colour. Depth of skin colour is determined principally by the concentration of melanin within the melanocytes rather than by the number of melanocytes present within

the skin. Melanin levels vary among individuals and ethnic groups. Melanin is also found in the hair, iris, parts of the adrenal gland and in certain regions of the brain, such as the substantia nigra.

The melanocytes synthesize melanin from the amino acid **tyrosine** within cytoplasmic organelles called melanosomes. Melanin is then transferred to the keratinocytes of the basal layer. The exact mechanism by which melanin is transferred to the epidermal cells is not clear but it is thought that the fine processes of a melanocyte may actually invade a keratinocyte and then bud off to 'deposit' the melanin inside the cell.

The regulation of melanin synthesis

The synthesis of melanin by melanocytes is controlled by certain hormones and by light. **Melanocyte-stimulating hormone** (**MSH**) is secreted by the anterior lobe of the pituitary gland and, as its name suggests, it increases the production of melanin by melanocytes. It is believed to work by stimulating the activity of tyrosinase. MSH is structurally related to adrenocorticotrophic hormone (ACTH), which also has a weak stimulatory action on melanocytes. In conditions such as Addison's disease, in which circulating levels of ACTH are raised (see Chapter 16), there is often increased pigmentation of the skin and mucous membranes.

Melanin absorbs ultraviolet (UV) light, regulating the amount that enters the deeper layers of the skin and protecting the underlying structures from its harmful effects. Melanin synthesis is stimulated when the skin is exposed to UV light, which increases the activity of tyrosinase. This explains why people living in parts of the world where there is a lot of sunshine tend to have darker skins than those who live at higher latitudes. It also accounts for the 'tanning' of the skin that occurs after sunbathing. It has recently been shown that α-MSH binds to a receptor (MC-1) on the human melanocyte membrane and this binding activates tyrosinase. Melanocytes taken from individuals who tan poorly (such as those with red hair) appear to have mutations in their MC-1 receptor.

Although over-exposure of the skin to UV light can be harmful, limited regular exposure to sunlight is necessary for the skin to synthesize vitamin D. UV light penetrates the epidermis and acts on 7-dehydrocholesterol, converting it to vitamin D, through a series of steps. By absorbing UV light, melanin reduces the ability of

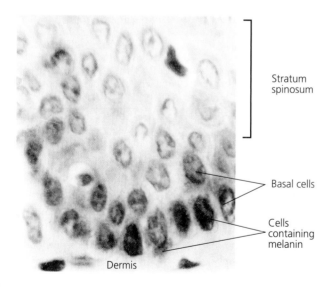

Figure 27.3 Section showing cells in the basal layer of the epidermis that contain large amounts of melanin. The pale-staining cells at the top of the image are prickle cells.

the skin to synthesize vitamin D. For this reason, dark-skinned people, particularly those living in relatively sunless climates, may be unable to produce their daily requirement of vitamin D. For such individuals, dietary intake of vitamin D is essential.

Disorders of melanin synthesis may occur. For example, in the inherited condition known as **albinism**, the ability to synthesize the enzyme tyrosinase is lacking, while the acquired condition **vitiligo** is characterized by progressive destruction of melanocytes. Here, large areas of depigmentation appear over the skin and the hair growing on these areas may also turn white.

Excessive exposure to UV light is known to increase the risk of developing cancers of the skin, and areas of the body most frequently exposed to the sun are the most vulnerable (such as the face, neck, hands and, in women, the lower legs). The three most common skin cancers are **basal cell carcinomas**, **squamous cell carcinomas** and **malignant melanomas**. Melanomas are aggressive tumours derived from melanocytes and are the most lethal of the skin cancers because, if left untreated, they spread rapidly to other tissues (**metastasize**). Initially, the tumour spreads to regional (nearby) lymph nodes, but later tends to disseminate more widely via the blood.

Merkel cells

Merkel cells are found in the deeper layers of the epidermis. They have small, dense granules in their cytoplasm and are associated with free nerve endings. For this reason, it is thought that they play a role as touch receptors. They can give rise to a rare but highly aggressive form of skin cancer variously known as Merkel cell carcinoma, apudoma of the skin, primary small-cell carcinoma or small-cell neuroepithelial tumour of the skin.

Langerhans cells

Langerhans cells are similar to melanocytes in appearance but behave rather like macrophages, participating in immune responses of the skin. Langerhans cells appear to bind to antigens and then migrate to a regional lymph node where they then 'present' the antigen to the lymphocytes of the immune system so that an appropriate immune response can be initiated (see Chapter 21).

KEY POINTS:

- The epidermis consists of five main cell layers. The outermost layer (stratum corneum) consists of dead cells filled with keratin and a fatty matrix. This provides a waterproof, abrasion-resistant barrier.

- In addition to the keratinocytes that make up most of the epidermis, the skin contains melanocytes, which are responsible for skin pigmentation, and Langerhans cells, which play an important role in combating infection.

- The epidermis has no nerve endings or blood vessels.

27.4 The dermis and hypodermis

The dermis lies directly beneath the epidermis. It ranges in thickness from 0.6 to 3 mm, being thickest on the soles of the feet and thinnest on the eyelids and prepuce. It contains blood vessels, lymphatic vessels, sensory receptors and their afferent nerve fibres, hair follicles, sebaceous glands and sweat glands. It is made up of dense connective tissue, which provides the skin with mechanical strength and supports the structures within it. At its deep border, the dermis merges with the hypodermis without a sharply defined boundary. The dermis itself is composed of two layers, the **papillary layer** and the **reticular layer**. The papillary layer lies adjacent to the stratum basale of the epidermis and forms a series of ridges and hollows (the dermal papillae) as shown in Figures 27.1 and 27.2. Although under most normal circumstances this arrangement helps to prevent the dermis and the epidermis from separating, excessive or repeated shearing forces (e.g. from the rubbing of a tight shoe) can cause the epidermal layer to be torn off, creating a blister. The reticular layer is thicker and lies beneath the papillary layer, although the boundary between the two is indistinct.

Like all structural connective tissues (see Chapter 6), the dermis consists of a non-cellular matrix (or ground substance) within which lie fibres and cells. Cells of the dermis include fibroblasts, which synthesize collagen, and elastin fibres, macrophages, mast cells and other cells of the immune system. The matrix is a gel-like substance that consists of hyaluronic acid, mucopolysaccharides and chondroitin sulphate.

Collagen is a fibrous protein with an extremely high tensile strength. It is arranged in bundles whose orientation lies predominantly parallel to the skin surface. Wherever possible, surgical incisions are made along these orientation lines (the line of cleavage), because the resulting wound is finer, heals more quickly and scars less noticeably than an incision made across the line of cleavage. Elastin fibres are branching protein fibres found in both the papillary and reticular dermal layers. They give the skin its flexibility (elasticity) but they can rupture if subjected to excessive stretch, as in rapid weight gain, fluid retention or pregnancy. Ruptured elastic fibres create silvery scars on the skin surface. These scars are known as **striae** or, colloquially, as stretch marks.

The hypodermis

Beneath the dermis, and attaching the skin to the underlying muscles, is a layer of tissue known as the hypodermis or subcutaneous layer. It consists largely of adipose tissue (fat) and provides the body with both mechanical and thermal insulation. The stored fat also forms a nutritional reserve which may be called upon to provide energy during periods of fasting.

> **KEY POINTS:**
> - The dermis is made up of dense connective tissue containing collagen and elastic fibres.
> - It also contains the skin blood vessels, sensory receptors, hair follicles, sebaceous glands and the coiled regions of the eccrine sweat glands.

27.5 The accessory structures of the skin – hairs, nails and glands

The hairs

Hairs (**pili**) are present over the entire body surface except for the plantar surfaces of the feet and toes, the palmar surfaces of the hands and fingers and certain areas of the external genitalia. In some areas (such as the scalp) the hair grows thickly, while in others it is so sparse as to be virtually invisible. Hairs grow from **follicles** located at the lower border of the dermis. Typically, around 5 million follicles are distributed over the skin surface, of which about 1 million are in the scalp, and all are present within the skin at the time of birth. Figure 27.4 shows the anatomical features of a hair follicle and the hair contained within it.

The follicle itself is made up of an outer connective tissue sheath lined by an epithelial membrane that is continuous with the stratum basale (the germinal layer) of the epidermis. The lower portion of each hair follicle is formed by the **hair bulb**, at the base of which is an indentation called the **papilla**. This contains connective tissue and blood vessels to nourish the follicle. It also contains a population of cells arising from the stratum basale and responsible for the growth of existing hairs and the development of new hairs (the **matrix cells**).

A network of nerve fibres surrounds each hair follicle (the hair root plexus). These fibres are stimulated by movements of the hair shaft and contribute to the touch sensitivity of hairy skin. A bundle of smooth muscle fibres (the arrector pili muscle) is also associated with the hair follicle and extends from the follicle to the upper border of the dermis (Figure 27.4). This smooth muscle contracts in response to activation of its nerve supply, which forms part of the sympathetic division of the autonomic nervous system. Contraction causes the hair to become erect, for example in response to cold or fright.

The hair itself is made up of a **root** and a **shaft**. The root forms the base of the hair and penetrates deep into the dermis, while the shaft is the more superficial part of

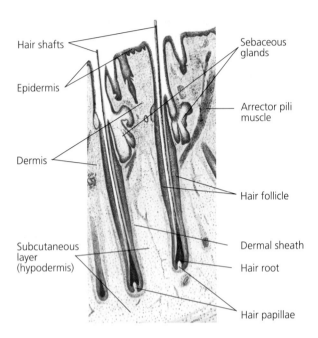

Hair shafts

Epidermis

Dermis

Subcutaneous
layer
(hypodermis)

Sebaceous
glands

Arrector pili
muscle

Hair follicle

Dermal sheath

Hair root

Hair papillae

Figure 27.4 A camera lucida drawing of the skin of the scalp. Two hair follicles are shown. The arrector pili muscle of the right-hand follicle is clearly visible as are the sebaceous glands of both follicles. Note that the hair root extends deep into the subcutaneous layer but is surrounded by a sheath of dermal tissue.

the hair, and penetrates the skin and projects above its surface. Both consist of three layers, the inner medulla, the middle cortex and the outer cuticle. The cells of the medulla and cortex contain melanin granules that determine the hair colour, while the outer cuticle is formed by a single layer of flattened, highly keratinized cells arranged rather like the scales of a fish.

Hair growth

Hair tends to grow in cycles and, at any one time, about 85% of the scalp hairs are likely to be growing actively. The rest will be in a state of rest or regression (prior to being lost). Between 25 and 100 hairs are shed from the scalp each day, pushed out by a new hair developing within its follicle. Hair growth and loss is affected by a number of factors including severe illness, nutritional status, childbirth and exposure to radiation. Certain hormones, including testosterone and thyroxine, also influence the growth and texture of hair.

The nails

The nails are a form of modified hair and consist of tough, translucent plates of keratin that protect the ends of the digits and permit fine manipulation of small objects. Careful inspection of the nails may reveal information concerning nutritional deficiencies or certain disease states, for example, the 'clubbing' of the fingernails that occurs in some cardiovascular and respiratory disorders.

Figure 27.5 shows the different regions of the dorsal surface of a nail. The visible part of the nail is called the nail body, the free edge of which may extend beyond the end of the digit. The nail root is the part of the nail that lies under the fold of skin (the cuticle) situated at the base of the nail. Beneath the nail root is a layer of epithelial cells (the nail matrix) that divides to bring about nail growth.

The glands of the skin

Three types of gland are found within the skin. These are:

- Sebaceous (oil) glands
- Sudoriferous (sweat) glands
- Ceruminous glands.

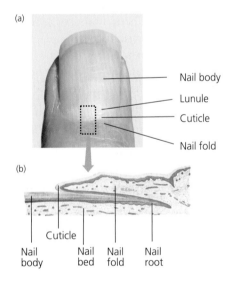

(a)

Nail body

Lunule

Cuticle

Nail fold

(b)

Cuticle

Nail
body

Nail
bed

Nail
fold

Nail
root

Figure 27.5 (a) Dorsal view of a human fingernail showing the main features. (b) A diagrammatic representation of a sectional view of the area outlined in (a).

Sebaceous glands are found over all parts of the skin except for the palms of the hands and soles of the feet. They are mostly found in association with hair follicles, as may be seen in Figure 27.4. These glands secrete an oily substance called sebum, either onto the hair shaft with which they are associated or directly onto the skin surface. Sebum helps to reduce evaporative water loss from hair and skin, preventing hair from becoming brittle and keeping skin soft and flexible. It also contributes to the slightly acidic pH of the skin surface (pH 5.5–7.0), which helps to inhibit the growth of certain micro-organisms.

Sudoriferous glands are further subdivided into two groups, according to their structure, location and the nature of their secretions: the **eccrine glands** and the **apocrine glands**.

Eccrine glands are found in the dermis of the skin over the entire body. There are around 2.5 million of these, about half of which are situated in the skin of the back and chest. As Figure 27.1 shows, eccrine sweat glands are simple tubular structures consisting of a coiled portion deep in the dermis and an unbranched duct that opens on to the body surface via a sweat pore. In cold conditions, around 500 ml of sweat are secreted each day but in warm conditions this may rise to between 1.5 and 6 litres per hour for short periods. Sweating is stimulated by cholinergic sympathetic fibres that innervate the sweat glands. Circulating adrenaline and noradrenaline from the adrenal medulla are also thought to stimulate the production of sweat. The concentration of salts in sweat depends upon the rate of sweat production. The faster the sweat rate, the higher the concentration of sodium and chloride in the sweat as there is less time for these salts to be reabsorbed as sweat flows along the duct towards the skin surface. Consequently, at high sweat rates there is a potentially dangerous loss of both water and sodium chloride from the body. Unless both are replaced, heavy sweating can quickly lead to dehydration (heat exhaustion). The role of sweating in thermoregulation is discussed further in Chapter 33.

Apocrine glands are confined to the skin at specific locations, notably the axillae, the feet and the genital area. Although their general structure is similar to the eccrine glands, the apocrine glands are larger (their secretory portion is 3–5 mm in diameter compared with

c. 0.4 mm diameter for the eccrine glands) and their coiled region lies within the subcutaneous tissue, rather than in the dermis. Their secretion differs from that of the eccrine glands in that it contains fatty materials. The characteristic odour of sweat is produced by the action of skin bacteria on the fatty components of apocrine secretions. Eccrine sweat, by contrast, is normally odourless.

Ceruminous glands are specialized sweat glands found in the skin lining the ear canal. They secrete a thick, waxy material known as **cerumen** (or colloquially as ear wax), which helps to prevent drying of the tissue and to maintain cleanliness of the ear canal by trapping dust, debris and dead cells. Occasionally, over-production and impaction of cerumen can occur, leading to impaired hearing. In some cases, in order to remove the cerumen, it is necessary to irrigate the ear canal using a syringe after first softening the wax by the regular application of ear drops. Preparations that may be used to soften ear wax include:

- Solutions of sodium bicarbonate
- Aqueous acetic acid
- Triethanolamine
- Cerumol (a combination of chlorobutanol, dichlorobenzene and arachis (peanut) oil
- Olive oil
- 5% urea hydrogen peroxide in glycerol.

The faecal softener docusate sodium is also used occasionally to soften cerumen in preparation for syringing.

> **KEY POINTS:**
> - The accessory structures of the skin include the hairs, nails and three types of gland: sebaceous (oil) glands, sudoriferous (sweat) glands and the ceruminous glands of the ear canal.
> - The sweat glands are subdivided into the eccrine glands, which play an important role in regulating body temperature, and the apocrine glands, which secrete a thick fluid that gives rise to body odour.

27.6 The cutaneous circulation

Figure 27.6 illustrates the principal features of the blood supply to the skin. The epidermis contains no blood vessels (it is avascular), but receives its nourishment from the rich blood supply of the dermis. Under normal (thermoneutral) conditions, blood flow to the skin is around $0.15 \, \mathrm{l \, min^{-1}}$ but this varies considerably in response to changes in the environmental temperature. Blood enters the dermis via arterioles, which branch to form capillary loops that carry blood into the dermal papillae. The venous limbs of these capillary loops then drain into venules that form a venous plexus (network) within the dermis. Subsequently, blood is carried away from the dermis by veins, as shown in Figure 27.6. In some areas of the skin, particularly the hands, feet, ears, nose and lips, additional vessels are found. These are called arteriovenous (AV) anastomoses and they provide a route for blood to pass directly between the arterioles and the venules, thus bypassing the capillary loops. AV anastomoses are shown in Figure 27.6. They are richly innervated by sympathetic nerve fibres, which can bring about powerful vasoconstriction.

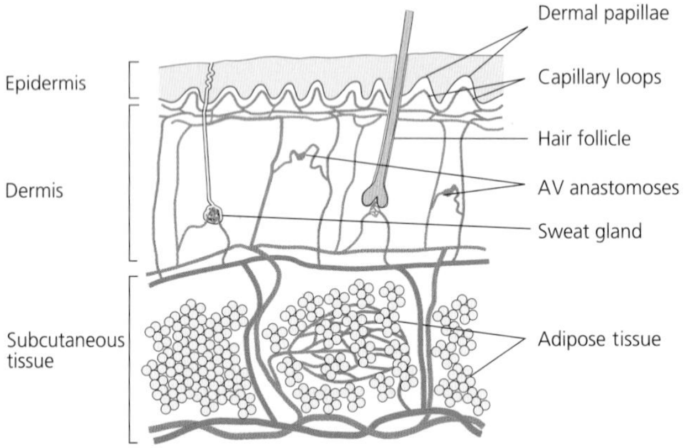

The most important regulator of blood flow to the skin is not the supply of oxygen and nutrients as it is in most tissues. Indeed, the nutritional requirements of the skin are small. Instead, blood flow to the skin is determined mainly by the level of sympathetic nervous activity, which is continuously adjusted so that body temperature is kept largely constant. If body temperature starts to fall, sympathetic activity increases, the arterioles constrict and the AV anastomoses close. This diverts blood away from the body surface and acts to conserve body heat. When body temperature starts to rise, sympathetic activity is reduced and the arterioles and AV anastomoses dilate, blood flow to the body surface is increased and this facilitates heat loss. In circumstances that require a greater degree of heat loss, sweating is stimulated. Sweat contains an enzyme whose action produces the powerful vasodilator **bradykinin**, which acts locally to bring about vasodilatation of the arterioles. The arterioles also dilate in response to increased local levels of metabolites, a situation likely to occur when body temperature is elevated and metabolic rate is increased (reactive hyperaemia; see also Chapter 33).

Figure 27.6 The arrangement of the vessels of the cutaneous circulation. There are three networks of blood vessels. One is very deep in the subcutaneous tissue, another lies below the dermis and the third lies just below the epidermis. The outermost capillaries do not pass into the epidermis but remain within the dermis. The blood flow through the dermis determines the rate of heat loss from the skin. The arteriovenous (A–V) anastomoses are found only in certain areas of the skin, such as the tip of the nose and the fingertips.

KEY POINTS:

- The cutaneous circulation consists of three main networks of vessels: one in the subcutaneous tissue, one deep in the dermis and a superficial network just below the epidermis.

- The epidermis has no blood supply of its own and the oxygen and nutrients it requires are supplied by diffusion from the outermost dermal capillaries.

- Cutaneous blood flow is mainly controlled by the sympathetic nervous system and can be varied by more than a 100-fold. This ability is vital for the regulation of body temperature.

27.7 The role of the skin as a sense organ

The general principles of sensory transduction and perception were introduced in Chapter 14. Receptors sensitive to a specific environmental stimulus (light, sound, temperature and so on) send information regarding that stimulus to the CNS in the form of action potentials carried by afferent nerve fibres. These electrical signals are subsequently processed by the CNS to provide a constant stream of information regarding many aspects of our external and internal environments.

The skin provides information regarding the tactile environment (touch, pressure, vibration, etc.), temperature, itch and pain. It is richly endowed with receptors sensitive to these stimuli, which relay their information by way of the afferent nerves of the somatosensory system to the brain and spinal cord. The various kinds of receptor present in the skin are illustrated in Figure 27.7. Both bare nerve endings and encapsulated receptors are present and each kind of receptor responds to a specific type of cutaneous sensation, as summarized in Table 27.1.

The largest receptors found in the skin are the **Pacinian corpuscles**, which are around 1 mm in length. They are located deep in the dermis or subcutaneous layer and consist of a bulb made up of many concentric layers (lamellae) of fibrous connective tissue in the centre of which is an afferent nerve fibre within a fluid-filled cavity (the inner core). When the lamellae are compressed by pressure on the skin, action potentials are triggered in the afferent nerve fibre and relayed to the spinal cord and onwards to the somatosensory cortex for processing. Pacinian corpuscles are called rapidly adapting receptors as they respond best to rapid changes in pressure and to vibration. They are relatively insensitive to sustained pressures. A section of human skin with two Pacinian corpuscles is shown in Figure 27.8. Other cutaneous receptors include Merkel disks, which respond to touch and pressure, Meissner's corpuscles, which respond to vibration, Ruffini corpuscle, which are sensitive to pressure and stretch, and numerous free nerve endings

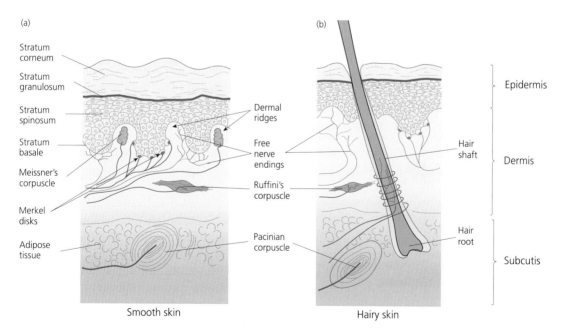

Figure 27.7 The arrangement of the various types of sensory receptor in smooth (glabrous) skin (a) and in hairy skin (b). Note the greater thickness of the epidermis in glabrous skin and the location of the touch-sensitive Meissner's corpuscles between the dermal ridges. Pacinian corpuscles are located in the upper part of the subcutaneous tissue (hypodermis or subcutis).

Table 27.1 Receptor types and modalities of the somatosensory system

Modality	Receptor type
Touch	Rapidly adapting mechanoreceptor, e.g. hair follicle receptors, bare nerve endings, Pacinian corpuscles
Touch and pressure	Slowly adapting mechanoreceptors, e.g. Merkel's cells, Ruffini end organs, Bare nerve endings
Vibration	Meissner's corpuscles
	Pacinian corpuscles
Temperature	Cold receptors (bare nerve endings)
	Warm receptors (bare nerve endings)
Pain	Fast 'pricking' pain – bare nerve endings (Aδ fibres)
	Slow burning pain, itch – bare nerve endings (C-fibres)

that respond to warm or cold (thermoreceptors), to pain (nociceptors) or to itch-provoking stimuli.

The skin is not uniformly sensitive to touch. For example, the skin of the fingertips is much more sensitive to touch than the skin of the back. This is explained by the relative density of cutaneous receptors in different areas of skin. Skin sensitivity may be determined indirectly by measuring the distance between two points on the skin that can just be detected as separate stimuli. This is known as the **two-point discrimination threshold**, and is discussed in Section 14.2, p. 229.

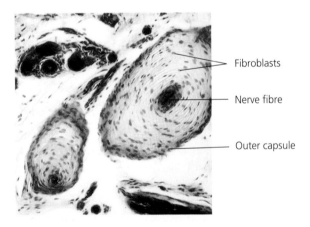

Figure 27.8 A section through the skin showing two Pacinian corpuscles within the subcutaneous tissue. The inner core consists of a fluid-filled capsule surrounding a single nerve ending. Note the concentric arrangement of the fibroblasts of the capsule surrounding the inner core.

Fibroblasts

Nerve fibre

Outer capsule

Pain detection by the skin

Pain is an unpleasant experience associated with acute tissue damage and is the main sensation experienced by people following injury or surgery. Pain also accompanies certain diseases such as rheumatoid arthritis and advanced cancer. The neural pathways involved in the processing of pain are discussed in Chapter 14 and only the part played by the skin will be considered here. The receptors responsible for pain sensation are bare nerve endings (also called **nociceptors**) in the skin and some internal organs. In the skin, these receptors are distributed in a punctate fashion with discrete areas of dense innervation separated by areas of skin in which there are no pain endings. The pain receptors are stimulated by pain-producing chemicals (**algogens**) that are released following injury to the skin. Chemicals that are known to cause the sensation of pain include ATP, bradykinin, histamine, serotonin and hydrogen ions. A number of inflammatory mediators such as prostaglandin E$_2$ increase the sensitivity of the skin to noxious stimuli.

KEY POINTS:

- The skin provides the CNS with information concerning touch, peripheral temperature and tissue damage (pain perception).

- Individual skin receptors respond to different kinds of stimuli. The encapsulated nerve endings such as Pacinian corpuscles and the nerves surrounding hair follicles respond to touch. Temperature and pain are sensed by specialized bare nerve endings.

- Pain is associated with actual or impending tissue damage triggered by strong thermal, mechanical or chemical stimuli.

27.8 The triple response

If a small area of skin is injured, for example by a burn, there is an inflammatory response, which starts with a local vasodilatation that elicits a reddening of the skin (the 'red reaction'), followed by a swelling (a weal or wheal) localized to the site of the injury and its immediate surroundings. The original site of injury is then surrounded by a much wider area of less intense vasodilatation known as the 'flare'. The red reaction, flare and weal formation comprise the triple response. The red reaction is due to the dilatation of arterioles in response to vasodilator substances released from the damaged skin. The weal is the result of local oedema caused by the accumulation of fluid in the damaged area and the flare is due to dilatation of arterioles in the area surrounding the site of injury.

Within the injured area and across the surrounding weal, the sensitivity to mildly painful stimuli such as a pinprick is much greater than before the injury. This is known as **primary hyperalgesia**, which may persist for many days. In the region covered by the flare, outside the area of tissue damage, there is also an increased sensitivity to pain, which may last for some hours. This is known as **secondary hyperalgesia**.

27.9 Wound healing

Injury to the body results in a wound, which must subsequently heal in order to restore normal structure and function to the area involved. The healing of skin wounds involves regeneration of the epithelium and the formation of a connective tissue scar. The exact course of the healing process depends upon the nature and extent of the wound. Clean, uninfected wounds, such as those made by surgical incisions, heal by a process called 'primary union' or 'healing by first intention' whereas larger wounds, such as ulcers, abscesses or other gaping wounds, in which there is more extensive loss of tissue, heal by a more complex process known as 'secondary union' or 'healing by second intention'. Each will be described briefly here but for a fuller discussion you should consult a textbook of pathology.

Figure 27.9 illustrates the healing of a simple cut. Within about 8 hours of making the incision, the space caused by the cut has filled with clotted blood. The tissue on either side of the incision becomes necrotic as the cells are deprived of their normal blood supply and die. This necrotic tissue forms a scab that covers and protects the injured site.

Immediately following the injury, there is an inflammatory response (see p. 418) and the damaged site becomes invaded by neutrophils, which begin to clear it of debris by phagocytosis. At about the same time, cells of the

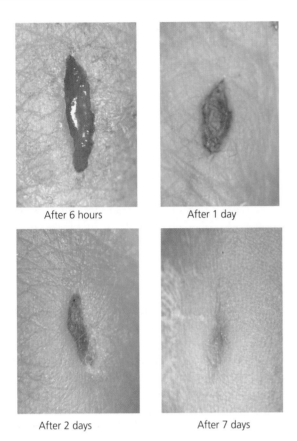

After 6 hours   After 1 day

After 2 days   After 7 days

Figure 27.9 The stages by which a simple incision wound heals. After 6 hours, the clot is fresh, soft and red but by the next day a scab has formed and healing has begun. On day 2, the wound has retracted and by day 7 it is almost completely healed. In this case, the initial wound was approximately 1 cm in length.

stratum basale at the cut edges of the epidermis begin to divide rapidly (the proliferative phase) and the newly formed epithelial cells begin to migrate over the wound and gradually unite the cut edges of the incision beneath the scab. Over the next few days, epithelial cell proliferation continues and the neutrophils are largely replaced by macrophages, which, in addition to their phagocytic actions, secrete growth factors that stimulate the activity of fibroblasts and promote the formation of new blood vessels (**angiogenesis**). Between 3 and 7 days after the injury, fibroblasts have begun to synthesize and secrete collagen at the edges of the cut and new capillaries grow into the area from surrounding blood vessels to create a region of soft, pink tissue called **granulation tissue** that

fills the wound. Later, the fibroblasts contract to close up the wound and form a firm fibrous scar. Roughly 4 weeks after the original incision was made, the scar consists of connective tissue covered by a virtually normal epidermis – the cut edges of the incision have been united and the wound is healed.

Larger, more extensive, open wounds such as burns and pressure sores require much more tissue replacement than simple cuts. The stages of healing follow the same pattern but wounds of this kind take much longer to heal as the proliferation of cells from the surrounding undamaged tissue is not sufficient by itself to restore the original architecture of the damaged area. Instead, the lost tissue must first be replaced by granulation tissue, which has to grow inwards from the margins of the wound. The epithelial cells then migrate across the surface of the granulation tissue to complete the healing process. As they do so, they secrete enzymes that help to dissolve any scab that covers the wound. Not surprisingly, the extent of the wound determines the extent of scarring. Compared with the healing of simple cuts, there is a much greater degree of wound contraction, which can lead to severe deformation of the skin. An example of healing after such a wound is shown in Figure 27.10.

Burn and scald injuries to the skin

Each year in the UK, around 100 000 people attend accident and emergency departments as a result of a burn or scald injury to the skin. These injuries are of various types and include thermal injuries (caused by flames, contact with a hot object, steam or hot liquids), electrical burns from high-voltage currents or lightning, cold injury due to frostbite or freezing materials, radiation burns (including sunburn) and burns caused by corrosive chemicals.

The number of skin layers affected by a burn (the burn thickness) determines its severity. Burns are normally categorized as:

- Superficial (first degree)
- Partial thickness (second degree)
- Full thickness (third degree).

Superficial burns are those in which only the epidermis is affected. They cause a typical local inflammatory reaction with pain, redness and swelling (oedema) but

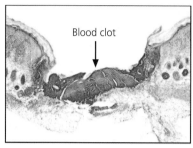

1 day post-wounding

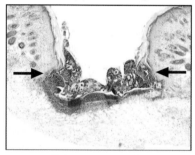

2 days post-wounding

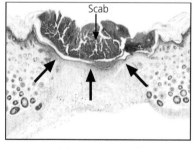

4 days post-wounding

Figure 27.10 The stages by which an area of damaged skin heals. At day 1, the base of the wound is covered by a blood clot and by day 2 there is growth of replacement epithelium at the sides of the wound (arrows). At day 4, the new epithelium has formed beneath the scab, which is in the process of being shed (arrows).

no blistering of the skin. The reddened area turns white when touched. Superficial burns normally heal within 2–7 days. A superficial burn might result, for example, from touching a hot pan or iron, or from moderate sunburn.

Partial-thickness burns involve part of the dermis as well as the epidermis of the skin. If damage to the dermis is relatively superficial, the burned area will appear red

or mottled with some swelling and blistering but hairs remain intact. Healing normally takes 1–2 weeks but if blisters rupture, the burned area may become infected and this will impair the healing process. In deeper partial-thickness burns, hairs are usually destroyed. If nerve endings and sweat glands have also been damaged, the area may be painless and dry.

Full-thickness burns are the most severe. All of the skin layers are affected, with damage extending into tissues below the dermis. Nerve endings, blood vessels, sweat glands and hair follicles are destroyed and underlying bone and muscle may also be damaged. Burns of this type are normally painless and show no touch sensitivity (because of the damage to nerve endings). They appear dry and either greyish-white or charred. Full thickness burns cannot repair because granulation tissue cannot form and epithelial cells cannot cover the wound. Skin grafts from undamaged areas of the patient's own skin are usually needed to cover severely burned skin and allow healing. Significant scarring is, however, inevitable. (Recently, tissue-engineered skin replacements have become available for the treatment of severe burns.)

Burns can damage large areas of skin and may affect its ability to carry out all of its normal functions including protection, fluid and electrolyte balance, touch sensation, thermoregulation and vitamin D synthesis. Burns affecting more than 20% of the skin surface (10% in children and the elderly) are considered critical. Two potentially life-threatening processes are seen following serious or extensive burns injuries. These are:

• Dehydration and electrolyte imbalance

• Infection.

Fluid is lost from the circulation into burned tissue because of a fluid shift from blood to interstitial fluid caused by increased capillary permeability to both fluid and proteins and a small increase in hydrostatic pressure within the small blood vessels supplying the area. To prevent dangerous fluid and electrolyte loss and the possibility of dehydration and circulatory shock, it is essential to replace lost fluids immediately. The amount of replacement fluid required is calculated indirectly from the percentage of total skin area affected by the burn and the weight of the patient.

Burns often lead to infection because the skin's protective barrier is disrupted and bacteria and fungi nourished by the dead skin cells colonize the injured site rapidly. Topical antibiotics (creams or ointments) are used to prevent and treat such infection. Examples include silver nitrate, sulphadiazine and mafenide acetate. However, if bacterial infection becomes widespread, serious systemic symptoms such as fever and hypotension may develop. This is known as septicaemia and it is a leading cause of death in burn injury, with a mortality rate of 30–50% depending upon the causative micro-organisms.

Pressure sores

Pressure sores, also known as **bed sores** or **decubitus ulcers**, are areas of skin that have become ischaemic as a result of maintained pressure – usually over a bony joint such as the heel, or the lower back around the sacrum. The skin is initially reddened but becomes blue and then quickly develops an ulcer in which necrotic skin sloughs off leaving a raw area, which may spread unless carefully treated. Most pressure sores occur in hospital as a consequence of lying in the same position for prolonged periods. They are a serious problem as around 80% of patients who develop deep skin ulcers die within 4 months. The most satisfactory treatment is prevention. Careful inspection of the skin for redness night and morning and scrupulous attention to its cleanliness will minimize the risk of pressure ulcers, as will regular changing of an invalid's position to avoid constant pressure over a particular bony prominence.

Factors influencing wound healing

Apart from the size of the original wound and its location, a number of factors may hinder or delay the process of wound healing. These include:

- Poor nutritional status
- Age
- Inadequate blood supply to the injured area
- The circulating concentration of cortisol
- Infection.

Energy and macronutrients are needed for all phases of wound healing. Amino acids are required for cellular regeneration and an adequate dietary intake of protein is therefore essential for normal wound healing. Vitamin C is needed for normal protein metabolism and for the synthesis of collagen, while vitamin A is essential for the growth and repair of epithelia. Deficiencies of certain minerals, including selenium, manganese, zinc and copper, have been linked with poor wound healing (particularly pressure ulcers) and a reduction in the tensile strength of fibrous scar tissue.

As a person ages, wound healing tends to occur more slowly than in earlier adulthood. Fibrous scars also tend to be weaker. Several factors contribute to this deterioration in the processes of tissue repair. Collagen formation is impaired due to a reduction in the activity of fibroblasts and cell division in the stratum basale is reduced. Furthermore, other pathologies that may influence wound healing (diabetes and peripheral vascular disease for example) are increasingly likely to co-exist in the elderly. A good blood supply is essential for normal healing and tissue regeneration.

Cortisol secreted by the adrenal cortex suppresses the immune system and reduces the inflammatory response that begins the process of wound healing. Consequently, healing is delayed and infection is more likely in individuals who have a high circulating level of cortisol. Cortisol secretion is elevated both in physiological and emotional stress, and in Cushing's disease (see Chapter 16). Steroidal medications, used to treat chronic inflammatory diseases such as rheumatoid arthritis, also lead to increased fragility of the skin and delayed wound healing.

Infection is an extremely important cause of delayed or inadequate wound healing. It causes increased inflammation and tissue necrosis at the wound site and prolongs all the stages of repair. Staphylococci and streptococci are the most commonly encountered pathogenic organisms in community-acquired injuries (although more unusual organisms may be present in bite wounds). Organisms infecting surgical wounds vary according to the anatomical site but antibiotic-resistant microbes such as methicillin-resistant staphylococcus aureus (MRSA) and other so-called 'superbugs' pose an increasing problem in hospital and community-acquired wounds. For this reason, infection control is of paramount importance in the hospital setting.

27.10 Pharmacological aspects of skin

Numerous skin conditions are recognized and many drugs are available to treat them. Table 27.2 provides a summary of some of the more common dermatological problems and their drug treatments.

Transdermal administration of systemic drugs

Most of the drugs that are applied to the skin are intended to act directly on the skin itself. It is also possible, however, to administer certain systemically acting drugs via patches placed on or just under the skin. This is called transdermal drug administration. The efficiency of drug administration by this route will be limited by the rate at which the drug can penetrate the skin and enter the bloodstream. This is determined by the chemical nature of the drug itself, the thickness of the skin and the blood supply to the area. Transdermal drug administration is particularly useful for drugs that are quickly eliminated

Table 27.2 Drugs used in the treatment of skin disorders

Condition	Type of medication	Typical drugs prescribed
Dry skin/pruritis	Emollient	Aqueous cream (in combination with treatment for underlying cause of pruritis)
Impetigo/other bacterial infections	Antibacterial	Mupirocin, fusidic acid (as ointment or cream)
Headlice/crab lice	Antiparasitic	Carbaryl (carylderm as topical preparation)
Scabies	Antiparasitic	Permethrin (e.g. Lyclear dermal cream)
Psoriasis	Keratolytic and antiseptic	Salicylic acid, 1% coal tar preparations
	Topical anti-inflammatory	Dithranol
	Topical anti-psoriatic (slows down the rate of new cell production in stratum basale)	Calcipotriene (a derivative of vitamin D)
	Anti-inflammatory	Topical corticosteroids
Ringworm, athlete's foot	Antifungal	Clotrimazole, imidazole (creams)
Eczema	Emollients	Aqueous cream, E45
	Topical steroid preparations	Hydrocortisone 0.5–1% (low potency)
		Betnovate, Eumovate, synalar (medium/high potency)
		Dermovate (high potency)
Acne	Antibacterial	Erythromycin, tetracycline (as liquid, gel or lotion)
	Keratolytics	Isotretinoin (a retinoid)
		Benzoyl peroxide

from the body as, through a patch, the drug can be delivered slowly and continuously for many hours, days or even longer, ensuring that blood levels of the drug are kept relatively constant. In addition to the familiar **nicotine patch** used by many people while attempting to give up smoking, examples of drugs administered by this route include **scopolamine**, to prevent motion sickness, **fentanyl**, used to alleviate cancer pain or pain caused by other chronic conditions, **nitroglycerine** for angina and **clonidine** for high blood pressure.

Recommended reading

Histology

Junquieira, L.C. and Carneiro, J. (2005). *Basic Histology*, 11th edn. McGraw-Hill: New York. Chapter 18.

Dermatology

Kumar, P. and Clarke, M. (2005). *Clinical Medicine*, 6th edn. Saunders: Edinburgh. pp.1315–1365. [A brief introduction to the physiology of the skin with a detailed account of common skin conditions and their treatment.]

Skin circulation

Levick, J.R. (2003). *An Introduction to Cardiovascular Physiology*, 4th edn. Hodder Arnold: London. pp. 259–265. [A good, clear account of the basic organization and properties of the cutaneous circulation.]

Self-assessment questions

Which of the following statements are true and which are false? Answers are given at the end of the book (p. 755).

1. a) The skin consists of three main layers.
   b) The epidermis is innervated by sensory nerve fibres.
   c) All cells in the stratum corneum originate from the stratum basale.
   d) All the cells of the epidermis are keratinocytes.
   e) The dermis contains collagen fibres that give skin its strength.

2. a) Melanocytes protect the skin from ultraviolet radiation.
   b) Dark-skinned people have more melanocytes per mm^2 than white people.
   c) People are able to synthesize sufficient vitamin D if they receive adequate amounts of sunlight.
   d) Melanin is synthesized from the amino acid tryptophan.
   e) The synthesis of melanin is regulated by melanocyte-stimulating hormone secreted by the anterior pituitary gland.

3. a) The hair follicles are lined by an epithelial membrane that is formed from cells of the dermis.
   b) Hairs are important in the sense of touch.
   c) Hairs follicles are formed from cells of the hypodermis.
   d) Poor nutrition results in hair loss.

4 a) The apocrine glands of the skin secrete an oily mixture on to the skin to keep the surface free of bacteria.

b) The sweat glands are innervated by sympathetic cholinergic nerve fibres.

c) The cutaneous circulation is controlled largely by circulating adrenaline.

d) The skin blood flow is controlled by the arteriovenous anastomoses.

e) The epidermis has no direct blood supply of its own.

5 a) The skin plays an important role in temperature sensations.

b) All pain sensations arise from pain receptors in the skin.

c) Wound healing and hair growth are impaired in people suffering from poor nutrition.

d) Increased levels of circulating cortisol speed up the progress of wound healing.

e) Wound healing becomes slower with age.

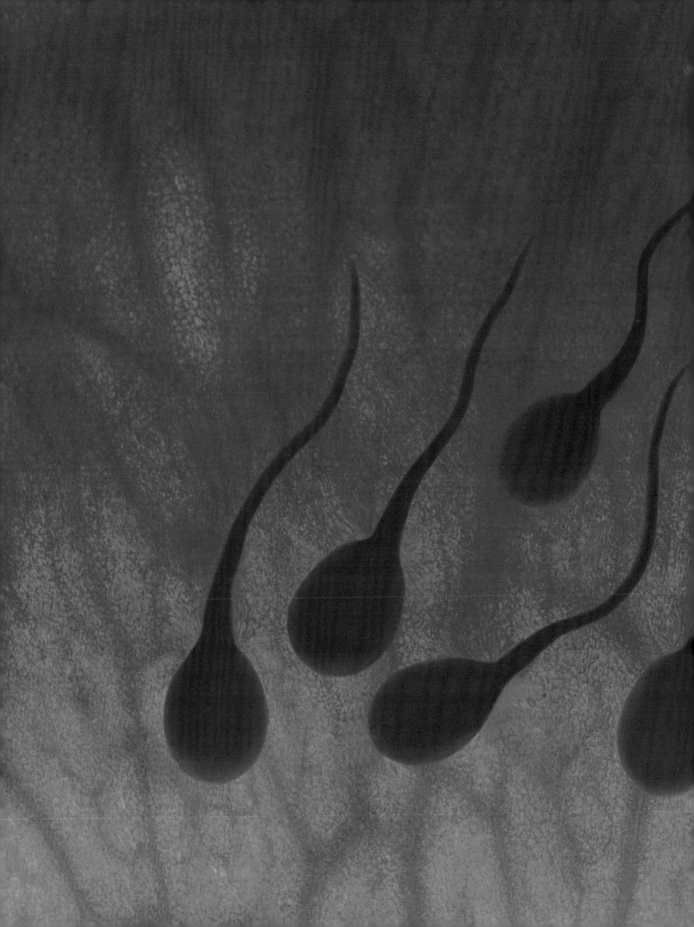

Section 5

Reproduction and inheritance

28 The physiology of reproduction **576**

29 Fertilization and pregnancy **607**

30 The genetic basis of inheritance **646**

28 The physiology of reproduction

After reading this chapter you should have gained an understanding of:

- The significance of sexual reproduction
- The anatomy of the male and female reproductive systems
- The formation of sperm (spermatogenesis) and its regulation by hormones
- The ovarian cycle and its regulation by hormones
- The peripheral actions of the testicular and ovarian steroid hormones
- Puberty and the menopause

28.1 Introduction

Reproduction is the ability of one generation to produce a new generation of individuals of the same species and is one of the fundamental characteristics of living organisms. In mammals, including man, genetic material is transmitted from parents to their offspring by sexual reproduction. This ensures that the children differ genetically from their parents. A strong and genetically varied population is maintained that is able to show resilience in the face of environmental challenges.

The essential feature of sexual reproduction is the mixing of chromosomes from two separate individuals to produce offspring that differ genetically from their parents. The creation and fusion of the male and female gametes are the first essential steps in the process. **Gametes** are specialized cells, produced by the male

and female gonads, which provide a link between one generation and the next. The male gametes are called **spermatozoa** (usually abbreviated to sperm) and are produced by the testes. The female gametes are called **ova** (eggs) and these are produced by the ovaries. As explained in Chapter 30, the nuclei of the gametes contain 23 unpaired chromosomes – they are **haploid**. Haploid cells are created when a cell divides by meiosis (see p. 583). At fertilization, when the nucleus of a sperm fuses with the nucleus of an egg, a new cell, called a **zygote**, is created. This cell possesses a full set of chromosomes – 23 pairs, the **diploid** number. One member of each pair of chromosomes is **maternal** in origin (i.e. it came from the egg) while the other is **paternal** in origin (it came from the sperm). During meiosis, there is extensive reshuffling of genetic material so that all gametes are genetically different. This explains why children born to the same parents have different biological characteristics.

This chapter will explain the processes by which male and female gametes are produced and the hormonal mechanisms that control their production. The nervous and endocrine control of reproductive activity, including puberty and the menopause, will also be discussed.

> **KEY POINTS:**
>
> - Gametes are specialized sex cells produced by the testes and ovaries.
> - The male gametes are the spermatozoa and the female gametes are the eggs or ova.
> - The nuclei of gametes are haploid, i.e. they contain a set of 23 unpaired chromosomes.

28.2 The anatomy of the male reproductive system

Figure 28.1 is a schematic illustration of the adult male reproductive tract showing the major organs that lie within it. The **testes**, or male gonads, lie outside the abdominal cavity, within the **scrotum**, which is a sac of thick skin with underlying connective tissue separated by a septum into a right and left compartment. Each adult testis is about 4.5 cm long by 2.5 cm wide and weighs around 15 g. The testes are encapsulated within a layer of fibrous connective tissue, the **tunica albuginea**. A coiled tube, the **epididymis**, leads away from the testis before straightening to become the **vas deferens** (or **deferent duct**). The vas deferens is a tube of around 35–40 cm in length, which enters the pelvic cavity and terminates in an enlarged region, the ampulla, before merging with the **ejaculatory duct**. The right and left vas deferens each join their ejaculatory ducts posterior to the urinary bladder, close to the prostate gland. The ejaculatory ducts are 2–3 cm in length and deliver sperm into the **urethra**. This is the terminal duct of the reproductive system and, in males, carries both urine and semen.

The male urethra is around 20 cm in length and is subdivided into three parts. These are:

- The prostatic urethra, which is around 2.5 cm long and is surrounded by the prostate gland
- The membranous urethra, which is about 1.5 cm in length and runs between the prostate and the penis
- The spongy or cavernous urethra, which runs the length of the penis (roughly 16 cm in most men). This enters the head of the penis and terminates at the urethral opening or orifice.

The **prostate** gland is normally about the size of a walnut. It secretes an alkaline fluid containing citrate and several enzymes. The alkalinity of the prostatic fluid (pH 7.2–7.6) helps to counteract the acidic environment of the vagina while the citrate is thought to be used as a nutrient by the sperm. One of the prostatic enzymes is called PSA (or prostate-specific antigen). This enzyme helps to liquefy clotted semen in the female reproductive tract to liberate sperm. Clinically, measurements of blood PSA levels are used as a marker in initial screening for prostate cancer. Typically, the larger the prostate gland, the more PSA it secretes. Although high levels of the enzyme are a sign of an enlarged prostate and indicate a need for further investigation, in many cases no cancer is present. To check for enlargement or other changes, the prostate gland can be palpated through the rectal wall.

The location of the prostate gland also has important clinical implications. As shown in Figure 28.1, the gland lies in close proximity to the urinary bladder and the start of the urethra. During later life, it is quite common for the prostate gland to enlarge. Although this enlargement may be the result of prostate cancer, it is more often a benign condition (benign prostatic hyperplasia or BPH). As the gland enlarges, it tends to press upon the bladder causing a sense of urgency to urinate. However, because an enlarged gland also tends to compress the urethra, the flow of urine during micturition may be impeded. This combination of urgency coupled with a reduced urine flow rate is seen in many middle aged and elderly men.

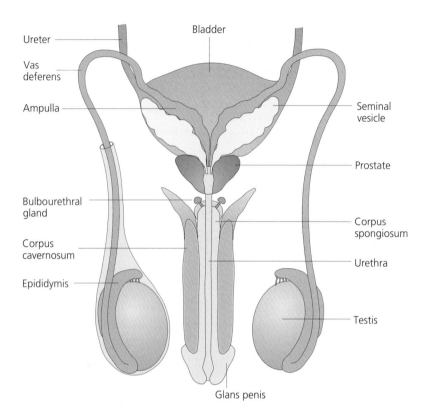

Figure 28.1 A diagrammatic representation of the adult human male reproductive system showing the principal structures viewed from the posterior aspect.

Two small round glands, the **bulbourethral glands** (or Cowper's glands) are situated beneath the prostate gland. These secrete a sticky, mucus-rich lubricating fluid that is released just before ejaculation. Two further accessory glands, the **seminal vesicles**, are located on either side of the prostate gland. These glands empty their secretions into the ejaculatory duct and, together with sperm from the testes and fluid from the prostate, they form semen.

As Figure 28.1 shows, the urethra traverses the length of the penis before opening at the surface at the glans penis. Erectile tissue, called the **corpus spongiosum**, surrounds the urethra and two larger masses of erectile tissue, the **corpora cavernosa**, run side by side and above the corpus spongiosum. The erectile tissues are essentially large venous sinusoids surrounded by a coat of strong fibrous tissue. Their physiology will be discussed in Chapter 29 (p. 608).

The location of the testes outside the abdominal cavity has physiological importance. Although the testes develop during fetal life within the abdomen, they descend through the inguinal canal at around the time of birth, to lie within the scrotal sac. Here, the ambient temperature is around 34°C, some 2–3°C lower than normal core temperature. Spermatogenesis proceeds normally at this lower temperature but is inhibited at the higher temperature typical of the body core. For this reason, failure of the testes to migrate to the scrotal sac, a condition called **cryptorchidism**, leads to infertility should it persist until puberty.

28.3 The adult testis makes sperm and male steroid hormones (androgens)

The testis performs two important functions in the adult male, both of which are vital to his fertility and sexual competency. They are:

- The production of sperm – the male gametes, which carry genetic material and can fertilize an egg

- The secretion of steroid hormones, particularly testosterone, which are needed for normal sperm production and to bring about full masculine development.

The two principal products of the testes, sperm and testosterone, are synthesized in different regions within the organ. Figure 28.2 shows the internal structure of the testis and its compartmental organization. The testis is made up of a large number of coiled tubes called **seminiferous tubules**. The testis is divided internally into 250–300 lobules or segments, each of which contains two or three seminiferous tubules. At the apex of each lobule, the tubules join and pass first into short straight tubes, the

tubuli recti, and subsequently into the coiled **epididymis**. Together, the tubuli recti form a region of the testis called the **rete testis** (Figure 28.2).

The walls of the seminiferous tubules have an outer layer of smooth muscle and an inner layer of epithelial cells. The latter are called **Sertoli cells** and they are responsible for controlling the formation and development of sperm. The lumen of the tubule is filled with fluid that serves both as a medium for the development of the sperm and for their transportation. The smooth muscle layer is able to contract to move sperm and fluid through the tubule. Between the seminiferous tubules lies connective tissue that supports the tubules and contains clusters of **Leydig cells** (also called interstitial cells of Leydig). These features are easily seen in the cross section of the seminiferous tubules shown in Figure 28.3. The Leydig cells are responsible for the synthesis and secretion of testosterone and other androgens.

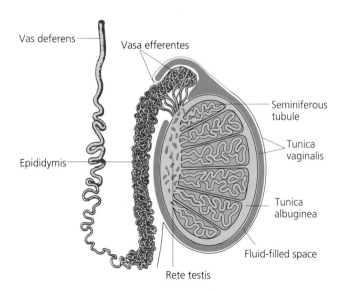

Figure 28.2 A diagrammatic representation of the adult testis, epididymis and vas deferens.

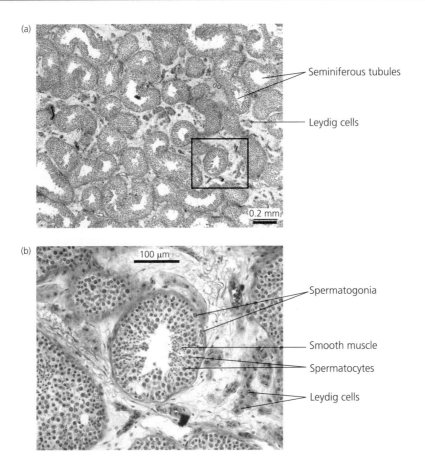

Figure 28.3 A section of the testis showing seminiferous tubules cut across at various angles: (a) a low-power view and (b) the area bounded by the square at a higher magnification. The dark clumps of cells scattered between the tubules are Leydig cells. The section was stained with iron haematoxylin and eosin.

The seminiferous tubules and the Leydig cells form two compartments within the testis, which are separated functionally and anatomically. Water-soluble materials are unable to pass freely between the two compartments because extremely tight junctional complexes exist between the basal regions of adjacent Sertoli cells. These complexes may be seen in the diagram shown in Figure 28.4. They form the so-called blood–testis barrier, which has two important functions. Firstly it protects the sperm from any harmful blood-borne substances that may interfere with their normal development. Secondly, it prevents any antigenic materials (e.g. proteins) arising from spermatogenesis, from passing into the circulation and triggering an auto-immune response to the sperm. Such a response could lead to infertility. Although the manufacture of sperm and steroid hormones takes place in separate compartments of the testis, their production is nevertheless closely related. Indeed, the production of mature sperm is only possible if androgen secretion is normal.

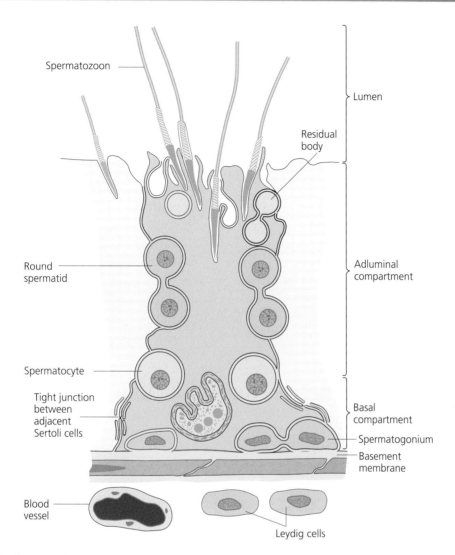

Figure 28.4 This figure is a diagrammatic representation of part of the wall of a seminiferous tubule to show the relationship between the Sertoli cells and the developing spermatozoa. The tight junctions between the basal regions of the Sertoli cells separate the basal compartment from the adluminal compartment.

28.4 Hormone secretion by the testis

Testosterone is the principal steroid hormone produced by the testis. It is synthesized and secreted by the Leydig cells and adult males secrete about 4–10 mg of testosterone each day. Most of the hormone passes into the bloodstream but a small amount enters the seminiferous tubules where it plays an important role in the development of sperm in the Sertoli cells (see below). As it is a steroid hormone and thus fat-soluble (see Chapter 7), testosterone is able to pass across the blood–testis barrier by passive diffusion. The secretion of testosterone is regulated by luteinizing hormone (LH), a gonadotrophic hormone secreted by the anterior pituitary gland (see Chapter 16 and p. 587 below).

The peripheral actions of testosterone

The mode of action of testosterone is broadly similar to that of other steroid hormones, as described in Chapter 7. Testosterone enters cells freely and once inside may be converted to dihydrotestosterone or to 5α-androstene-dione. All three androgens bind to specific intracellular receptors to form steroid–receptor complexes that interact with chromosomal DNA to regulate gene expression. Androgen receptors are most numerous in those tissues that depend upon androgens for their growth, maturation and/or function. Such tissues include the prostate gland, seminal vesicles and epididymis.

Dihydrotestosterone is important in the fetus for the differentiation of the external genitalia, and, at puberty, for the growth of the scrotum and prostate and for the development of sexual hair. Testosterone is important during fetal life for the development of the epididymus, vas deferens and seminal vesicles, and at puberty it is responsible for the development of the male secondary sexual characteristics (see p. 604).

> **KEY POINTS:**
>
> - The adult testis produces sperm and secretes steroid hormones known as androgens, which bring about full masculine development.
> - The principal androgen is testosterone.

28.5 Spermatogenesis – the production of sperm by the testes

Sperm are highly specialized cells, adapted to carry their genetic material a considerable distance within the female reproductive tract (roughly 40 cm or around 100 000 times the length of a sperm) in order to optimize the chances of fertilization. A sexually mature man produces around 200 million spermatozoa each day. Spermatogenesis is a complex process that involves the generation of huge numbers of germinal cells by mitosis and the halving of their chromosomal number by meiosis to form the mature sperm (male gametes).

The development of mature sperm cells is summarized in Figure 28.5. At puberty, germ cells (**spermatogonia**) within the epithelium of the seminiferous tubules start to undergo rapid mitotic cell division to produce a small population (or clone) of cells known as **primary spermatocytes**. These cells are diploid, that is they possess 46 chromosomes. As for all human diploid cells, there are 22 pairs of autosomes (chromosomes not concerned with the determination of gender) and one pair of sex chromosomes. The 22 pairs of autosomes are homologous,

that is they consist of one chromosome of maternal origin (the maternal homologue) and one of paternal origin (the paternal homologue) that carry genes for the same characteristics. The sex chromosomes of the primary spermatocytes consist of one X chromosome and one Y chromosome. These chromosomes are not homologous and carry different genetic information (see Chapter 30). The mitotic cell divisions take place within the basal compartment of the seminiferous tubule. The primary spermatocytes then migrate through the tight junctions at the base of the Sertoli cells and enter the adluminal tubular compartment (Figure 28.4). Following a period of growth, each primary spermatocyte then undergoes cell division by **meiosis** in which the chromosome number is halved (see Box 28.1).

The principal events of meiosis in the testis

The main events taking place during meiosis are illustrated in Figure 28.6. Essentially, meiosis consists of two

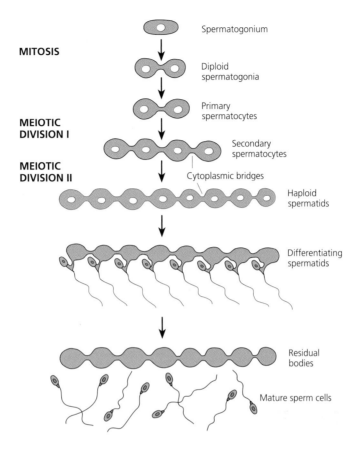

Figure 28.5 A simplified scheme to show the principal stages of spermatogenesis. The primordial cells divide to form spermatogonia, which undergo two more divisions to form primary spermatocytes. The primary spermatocytes undergo meiotic divisions to form secondary spermatocytes and spermatids. During meiosis, the chromosome number is halved. Note the cytoplasmic bridges between the differentiating spermatids, which permit the exchange of metabolites between adjacent cells.

Box 28.1 Meiosis

In man, somatic cells (i.e. those of the body) are *diploid* with 46 chromosomes (23 pairs). The gametes (eggs and sperm) have 23 (unpaired) chromosomes. They are *haploid* cells. In diploid cells, half the chromosomes originated from the father (the paternal chromosomes) and half from the mother (maternal chromosomes). The maternal and paternal forms of a specific chromosome are known as *homologues*. With the exception of the X and Y chromosomes (the chromosomes that determine sex), homologous chromosomes carry identical sets of genes arranged in precisely the same order along their length. This arrangement is crucial for the genetic recombination that takes place during meiosis.

During meiosis, the chromosome number is halved to form the gametes. This process consists of two successive cell divisions. As for mitosis, meiosis begins with

(Continued)

Box 28.1 *(Continued)*

DNA replication. The first stage of the first cell division is called **prophase I**.

During prophase, the chromosomes condense in a similar manner to that seen in mitosis. The homologous pairs of chromosomes then become aligned along their length so that the order of their genes exactly matches. Each pair of chromosomes is bound together by proteins to form a **bivalent** (i.e. a complex with four chromosomes; Figure 1). When all 23 pairs of chromosomes have become aligned, new combinations of maternal and paternal genes occur by the crossing over of chromosomal segments before the homologous pairs of chromosomes begin to separate. They remain joined by two DNA linkages. It is at this stage that the eggs (ova) arrest their meiotic division and start to accumulate material. The final stage of prophase I is similar to mitosis with spindle formation and repulsion of the chromosome pairs (diakinesis).

The dissolution of the nuclear envelope marks the start of **metaphase I**. The bivalents align themselves so that each pair of homologues faces opposite poles of the spindle. This process is random so the homologues that face any one pole of the spindle are a mixture of maternal and paternal chromosomes. At **anaphase I**, the chromosome pairs separate. As sister chromatids remain joined, they move towards the spindle poles as a unit so that the daughter cells each receive two copies of one of the two homologues.

Formation of the gametes now proceeds by a second cell division (cell division II), which resembles a normal mitotic cell division. It differs in that (a) it occurs without further DNA replication and (b) the daughter cells have half the number of chromosomes of the parent. Occasionally, some chromosomes do not separate properly and the daughter cells will either lack one or more chromosomes or have a greater number than normal. This is known as **non-disjunction**. Non-disjunction causes a number of genetic diseases, of which Down syndrome is perhaps the best known (see p. 649).

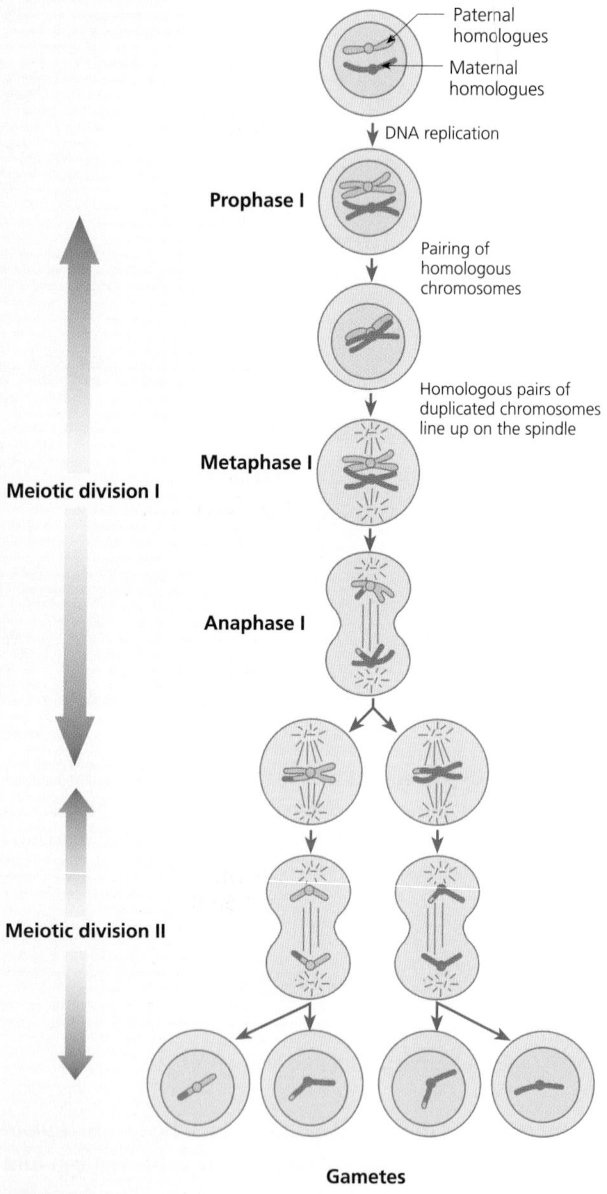

Figure 1 The main stages in the formation of the gametes by meiosis.

consecutive cell divisions (called meiosis I and meiosis II), which halves the chromosomal number (from 46 to 23) and recombines the genetic information so that all the gametes are genetically different. This is known as genetic recombination. Meiosis I begins with replication of the chromosomal DNA so that each chromosome then consists of two identical sister chromatids. Homologous pairs of chromosomes then become aligned along their

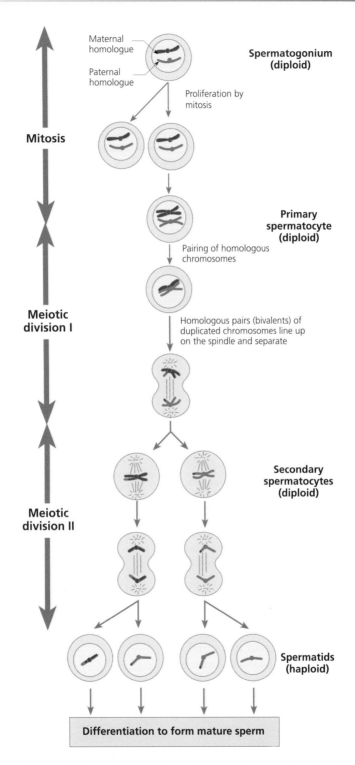

Figure 28.6 The chromosomal changes that occur in spermatogenesis. Intially, the spermatogonia proliferate by mitosis to form primary spermatocytes. The primary spermatocytes then undergo two meiotic divisions to form spermatids, which are haploid cells. The spermatids then undergo spermiogenesis to form the mature sperm.

length so that the order of their genes exactly matches. The pairs are bound together by proteins to form a complex consisting of four chromatids (called a bivalent). Once all 23 pairs of chromosomes have become aligned, a process known as 'crossing over' takes place in which small segments of chromosomal material are exchanged between the chromatids of homologous chromosomes at locations (chiasmata) where there is close contact between chromatids. This process allows new combinations of genes to be generated while preserving the integrity of individual genes. The nuclear envelope then dissolves and the bivalents align themselves so that each pair of homologues faces opposite poles of the cell. This process is random so the homologues facing any one pole are a mixture of maternal and paternal chromosomes. The two members of each bivalent then separate, one moving to each pole of the cell, and the cytoplasm divides. The cells formed by this division possess 23 chromosomes each of which consists of a pair of chromatids differing from one another only as a result of crossing over. These cells are called **secondary spermatocytes**.

A second cell division then occurs, which is similar to a normal mitotic division except that there is no further replication of DNA. Consequently, the daughter cells (gametes) resulting from this second meiotic division have only 23 chromosomes – they are haploid. As a result of crossing over and the random assortment of chromosomes that occurs during the first meiotic division, all the gametes are genetically different. In the male, the gametes are called spermatids and they remain joined together by a thin bridge of cytoplasm. The nucleus of each spermatid contains 22 autosomes and either an X or a Y sex chromosome. The genetic events of spermatogenesis are now complete but extensive remodelling is still required to convert the small, round spermatids into mature, motile sperm. This process is called **spermiogenesis**.

Figure 28.7 illustrates the main features of a mature spermatozoon – a very different cell from the round spermatid from which it arose. The process of spermiogenesis involves reorganization of both the nucleus and the cytoplasm of the cell, as well as the acquisition of a flagellum ('tail'). The whole process takes place in close association with the Sertoli cells.

The sperm consists of three morphologically and functionally distinct regions – the head, the midpiece and the tail. These regions are illustrated in Figure 28.7. The

head contains the haploid nucleus with its genetic material and a specialized secretory vesicle – the acrosomal vesicle. This contains enzymes that will help the sperm to penetrate the oocyte prior to fertilization. The **tail** region consists of a long flagellum, around 50 μm in length (the last few microns of the tail are sometimes called the endpiece, while the main length is called the principal piece). This is made up of a central **axoneme**, originating from a **basal body** that lies just behind the nucleus. The axoneme consists of two central microtubules surrounded by nine evenly spaced pairs of microtubules. Active bending of the flagellum is brought about by the sliding of adjacent pairs of microtubules past one another. Whiplash movements of the flagellum propel the sperm along. The **mid-piece** of the sperm contains mitochondria that

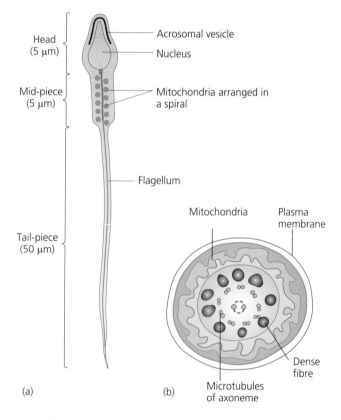

(a)   (b)

Figure 28.7 Diagram of a mature spermatozoon in longitudinal section (a) and transverse section through the mid-piece (b). Note the spiral arrangement of the mitochondria and the arrangement of the microtubules in the mid-piece. The main tail region has the same structure as the cilium shown in Figure 4.10 (p. 53).

generate ATP, to provide the energy needed to bring about these movements.

The process of differentiation of a spermatocyte to a motile sperm takes approximately 70 days to complete. About 300–600 sperm per gram of testis are produced each second, although not all of these survive. Newly formed sperm are released from the adluminal compartment of the Sertoli cells into the lumen of the seminiferous tubule and from there they progress towards the rete testis and into the epididymis. Here, sperm undergo further maturation and acquire the capacity for sustained motility. As a result of fluid absorption in the epididymis, sperm also become highly concentrated. Passage of sperm through the epididymis can take anything from 1 to 21 days, which means that the epididymis can act as a reservoir for sperm. Sperm and other testicular secretions are then transported along the vas deferens and into the ejaculatory ducts. The volume of the seminal fluid is greatly increased by the addition of fluid from the prostate gland (about 20% of total volume) and the seminal vesicles (about 60% of total volume). These fluids provide nutrients for the sperm and create an alkaline environment that helps to counteract the normal acidity of the vagina and thus increase motility and fertility of the sperm, both of which are optimal at a pH of around 6.5. Seminal fluid is released as **semen** from the penis at ejaculation during sexual intercourse. This process is discussed further in the next chapter.

KEY POINTS:

- Spermatogenesis takes place in the Sertoli cells of the seminiferous tubules while the androgens are secreted by the Leydig cells.

- Spermatogenesis involves the generation of huge numbers of cells by mitosis and the halving of their chromosomal complement by meiosis. It culminates in the formation of a highly specialized cell, the mature, motile sperm.

- The sperm are then mixed with secretions from the seminal vesicles and prostate to form seminal fluid. This is released as semen from the penis at ejaculation during sexual intercourse.

28.6 The control of spermatogenesis by pituitary and testicular hormones

In Chapter 16, some of the general mechanisms involved in the regulation of hormone secretion within the body were discussed, including the importance of the hypothalamic releasing hormones and the concept of negative feedback control. These key regulatory processes are known to operate in the hormonal control of male reproductive function and are summarized in Figure 28.8.

The anterior pituitary gland secretes two hormones that are crucial for the normal reproductive processes of both men and women. They are **follicle-stimulating hormone** (**FSH**) and **luteinizing hormone** (**LH**), which are known collectively as **gonadotrophins**. The secretion of FSH and LH is, in turn, controlled by the hypothalamus. **Gonadotrophin-releasing hormone** (**GnRH**) is synthesized by neurones in the hypothalamus and secreted into the blood of the hypophyseal portal vessels that run in the pituitary stalk. When GnRH arrives at the anterior pituitary, it stimulates the secretion of both FSH and LH into the systemic circulation. In the male, FSH acts mainly on the Sertoli cells of the testis while LH acts particularly on the Leydig cells.

FSH is required to initiate the process of spermatogenesis and to mediate the development of spermatids to spermatozoa. It also stimulates the Sertoli cells to secrete androgen-binding protein and a hormone called **inhibin**, which depresses the secretion of FSH by the anterior pituitary gland through a negative feedback loop (Figure 28.8). This internal control system helps to maintain a relatively constant blood concentration of FSH.

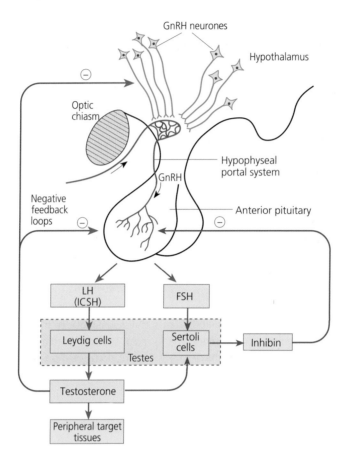

Figure 28.8 A schematic diagram to illustrate the relationship between the hormonal secretions of the hypothalamus, pituitary gland and testes.

LH stimulates the secretion of testosterone by the testicular Leydig cells. For this reason, LH was originally called interstitial cell-stimulating hormone or ICSH in males. Testosterone, in turn, inhibits the further secretion of pituitary LH by exerting a negative feedback inhibition at the level of both the hypothalamus (to inhibit GnRH secretion) and the anterior pituitary gland itself, as shown in Figure 28.8.

Testosterone exerts a variety of important effects throughout the body. It is responsible for the development of the male secondary sexual characteristics at puberty; it maintains the function of the accessory reproductive organs during adult life and acts on the brain to promote the sex drive. Furthermore, testosterone is essential for normal sperm production by the Sertoli cells. Some of the testosterone secreted by the Leydig cells enters the intratubular compartment where it is bound by androgen-binding protein secreted by the Sertoli cells. By mechanisms that are not yet fully understood, this bound testosterone helps to maintain sperm production.

KEY POINTS:

- Spermatogenesis is regulated by a variety of hormones including FSH, LH and testicular testosterone.

- The blood hormone levels are regulated by negative feedback loops in which testosterone and inhibin regulate the secretion of FSH and LH.

28.7 The anatomy of the female reproductive system

Figure 28.9 shows a simplified diagram of the adult female reproductive organs. The ovaries are the female gonads and, like the testes in the male, they are responsible for the production of gametes, in this case the eggs or ova. They also have an important endocrine function and the hormones secreted by the ovaries regulate gamete production and prepare the reproductive tract for gamete transport and fertilization. These functions are discussed in Section 28.9. The rest of the female reproductive organs are concerned with the processes of fertilization and nurture of an embryo.

The **ovaries** lie in the ovarian fossa of the pelvis and are attached to the posterior wall of the abdomen by a mesentery (a fold of the peritoneum) called the **mesovarium**. They are 3–4 cm long and each weighs around 15 g. The ovaries consist of a cortical region, where follicles at various stages of development are found within a stroma of connective tissue, and a medullary region of loose connective tissue, which has a rich blood supply.

The **Fallopian tubes** (or oviducts) are thin tubes roughly 12 cm in length that transport the ova released at ovulation towards the uterus. They also receive any sperm that have swum through the uterus so that fertilization normally takes place within one of the Fallopian tubes. The fertilized egg is then transported along the remainder of the tube to the uterus, where implantation takes place. The tube itself is muscular and covered with peritoneum. As Figure 28.10 shows, the Fallopian tubes have three main layers: an outer serosa derived from the visceral peritoneum, a layer of smooth muscle and a mucosa that is covered by ciliated, columnar secretory epithelial cells. Movements of the cilia, coupled with contractile activity of the smooth muscle of the wall, facilitate gamete transport by the Fallopian tube.

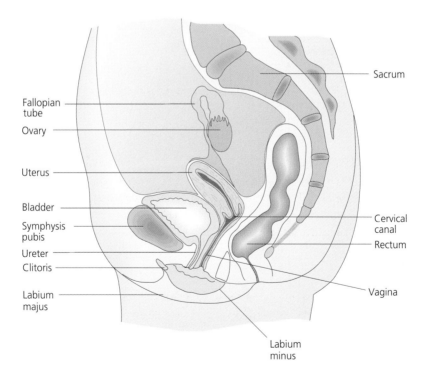

Figure 28.9 A diagrammatic view of a midline section showing the main female reproductive organs and their spatial relationship to other pelvic structures.

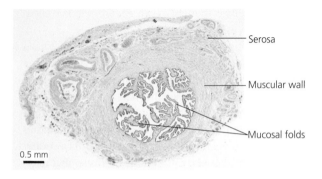

0.5 mm

Figure 28.10 A cross-section of a human Fallopian tube showing its three main layers: the outer serosa, the muscular wall and the highly folded mucosa. (Haematoxylin and OG erythrosin stain.)

The opening (or ostium) of each tube is expanded and split into fringes called fimbriae, which house numerous cilia. At ovulation, the fimbriae move closer to the ovary and the movements of the cilia create currents in the peritoneal cavity, which direct the ovulated ovum towards the ostium of the Fallopian tube. The egg is thus prevented from floating out into the peritoneal cavity but is instead 'captured' by the tube.

The **uterus** is adapted to receive the early embryo and permit implantation during the second half of the menstrual cycle. Subsequently, it permits the formation of a placenta and houses the developing fetus during **gestation** (pregnancy), which normally lasts for 38 weeks after fertilization (or 40 weeks from the first day of the last menstrual period). Although it has to contract powerfully to give birth to the baby at the end of the gestation period (labour), the uterus must remain relatively relaxed (quiescent) throughout pregnancy to allow full fetal development and to avoid premature delivery.

The wall of the uterus has three layers, as can be seen in Figure 28.11. These are:

- An outer serous coat, the **perimetrium**
- A thick middle layer of smooth muscle, the **myometrium**
- An inner endometrial layer or **endometrium**.

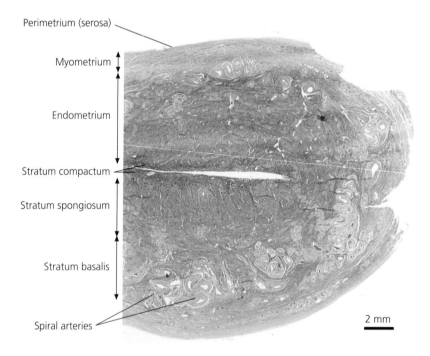

2 mm

Figure 28.11 A low power view of a section of the wall of the human uterus during the proliferative phase. The top half of the image is labelled to show the three main layers: the outer serosa, the myometrium and the endometrium. The bottom half of the image is labelled to show the layers of the endometrium: the inner stratum compactum, the stratum spongiosum and the outer stratum basalis, in which the spiral arteries are clearly visible. (haematoxylin and eosin stain.)

The endometrium is composed of epithelial cells, simple tubular glands and the spiral arteries that supply the cells. Its characteristics and thickness alter considerably during each menstrual cycle.

The neck of the uterus is formed by the **cervix**, a ring of smooth muscle containing many mucus-secreting cells. The cervix forms the start of the birth canal and is also traversed by sperm deposited in the vagina during sexual intercourse. The mucus-secreting cells alter their activity during each menstrual cycle in order to optimize conditions for fertilization (see Section 28.9, p. 601).

The **vagina** is a muscular tube that forms the final internal structure of the female reproductive tract. It receives the penis at intercourse and is the canal along which the baby must travel to be delivered. It is therefore subject to considerable frictional force and is lined by a stratified squamous epithelium, the cells of which show characteristic variations during each menstrual cycle. The vaginal epithelium is covered by a layer of fluid composed of secretions from the cervix, endometrium and Fallopian tubes. The vaginal fluid is relatively acidic (pH 4–4.5) due to its high content of organic acids such as lactic acid,

and this acidity is thought to protect the vagina against pathogens.

The remainder of the female sexual organs are known collectively as the **external genitalia** and consist of the vulva and the clitoris. The vulva protects the vaginal orifice, the urethral orifice and the clitoris. Within the walls of the vulva lie vestibular glands, which secrete mucus during sexual arousal and help to lubricate the movements of the penis within the vagina during sexual intercourse. The vulva consists of two sets of tissue folds. The larger outer folds are called the labia majora and the more delicate, inner leaflet-like folds are the labia minora. The **labia majora** consist largely of adipose tissue covered by a layer of skin. Their primary function is to protect and enclose the other external genitalia. The **labia minora** are usually thin and are hidden, at least partially, by the labia majora. They are rich in nerve endings and may swell and moisten during sexual arousal. The labia minora meet at the top of the vulva to form the clitoral hood beneath which lies the **clitoris** itself (Figure 28.12). This is a small erectile structure, which is homologous with the male penis and which in many women is the area of greatest sexual sensitivity.

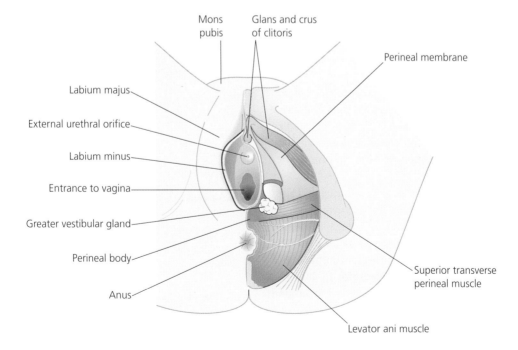

Figure 28.12 An illustration of the perineal region in the female.

The wedge of tissue that lies between the vaginal opening and the anus is called the **perineal body**. It is attached to the pelvic floor by the levator ani muscle and to the muscles and fascia of the perineum. During childbirth, this region may sustain damage. If so, careful repair must be carried out to avoid weakness of the pelvic floor and possible prolapse of the uterus, urethra or rectum. To prevent ragged tearing of the tissue during childbirth, the region is sometimes cut surgically (an episiotomy). When this is done, the incision is curved and angled to avoid the perineal body itself.

28.8 The physiology of the ovaries

The ovaries play a key role in the reproductive physiology of the female. During the fertile years of a woman's life, between the onset of menstruation (menarche) at around the age of 12 or 13 and the cessation of menstruation (menopause) at around the age of 50, the ovaries are responsible for the development and regular release of gametes (ova) for fertilization. They also secrete steroid hormones that not only regulate the ovaries themselves but also prepare the rest of the reproductive system for fertilization and pregnancy.

It is well known that the activity of a woman's ovaries occurs in a cyclic fashion. The orderly sequence of events that underlies this cyclical behaviour is called the ovarian cycle. At the same time, under the influence of the ovarian hormones, the uterine endometrium undergoes a pattern of change known as the uterine cycle. Together, the ovarian and uterine cycles form the familiar menstrual cycle, during which there is remarkable co-ordination between physical and endocrine changes within the woman's body. The onset of bleeding at menstruation is the visible sign that the ovaries and uterus have completed one cycle and that pregnancy has not occurred.

The events of the menstrual cycle are rather complicated and it will be helpful to consider first the ovarian events (**ovarian cycle**) and then the changes taking place within the uterus (the **uterine cycle**) and the rest of the reproductive tract under the influence of the ovarian hormones.

The ovarian cycle may be divided broadly into two halves, the events leading up to ovulation at around the mid-point of the cycle and the events occurring in the ovary following the release of an egg at ovulation. The first half of the cycle is known as the follicular phase while the second half is the luteal phase. In this section, the physical changes of the ovarian cycle will be described together with the hormonal mechanisms that regulate these changes.

The follicle is the fundamental unit of the ovary

As Figure 28.13 shows, the ovary possesses a large number of follicles at various stages of development. During fetal life, the germ cells of the ovary are laid down and continue to proliferate by mitotic division throughout gestation. This process is similar to that described earlier for the male in which the germ cells divide by mitosis to

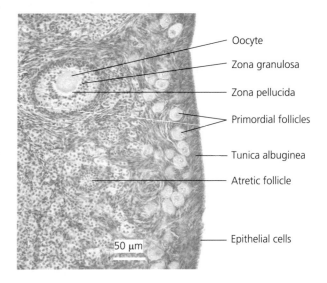

Oocyte
Zona granulosa
Zona pellucida
Primordial follicles
Tunica albuginea
Atretic follicle

50 μm

Epithelial cells

Figure 28.13 A section of human ovary showing the main structural features. Visible in this section are an antral follicle with a maturing oocyte, an atretic follicle and a number of primordial follicles. The outer connective tissue layer (tunica albuginea) and its covering epithelium are clearly shown.

generate a population of primary spermatocytes. In the male, new germ cells are produced throughout adult life. In the female, however, germ cell proliferation is complete by the time of birth so that when a baby girl is born, her ovaries will contain all the germ cells she will ever have. These primordial germ cells are called **oogonia**. Once mitosis is complete, the oogonia enter their first meiotic division (meiosis I) and become known as **primary oocytes**. At the same time, they become surrounded by several layers of cells to form primordial follicles. The inner layer of slightly flattened granulosa cells is located on a basement membrane called the membrana propria. Surrounding this layer are several layers of stromal cells. The first meiotic division is not completed at this stage but the oocytes arrest just before the chromosomal bivalents separate (see Box 28.1). They remain in this arrested state until signalled to resume further development. This might occur at any time during the reproductive life of the woman.

Throughout the reproductive years of a woman's life, between puberty and the menopause, the pool of primordial follicles that was established during fetal life is gradually depleted as each day around one to four follicles are recruited in a steady trickle to undergo further development. During each ovarian cycle, a follicle progresses through a series of developmental stages: growth and maturation, ovulation, corpus luteum formation and, in the absence of fertilization, degeneration. This is shown schematically in Figure 28.14. In most women, each cycle takes between 25 and 35 days (a typical menstrual cycle

lasting for 28 days) although wider variations sometimes occur.

Although several follicles begin to undergo this developmental cycle each day, usually only one, or occasionally two, follicles will reach the point of ovulation. The rest will degenerate and die, a process called **atresia**. This means that for every ovarian cycle, around 100 follicles are lost from the ovary. The following section will describe the physical changes undergone by a single follicle as it passes through the various stages of one ovarian cycle. The endocrine mechanisms that control these physical changes will also be considered. For simplicity, a cycle length of 28 days will be assumed. By convention, the first day of bleeding at the start of the menstrual period represents day 1 of the cycle.

> **KEY POINTS:**
>
> - The first half of the ovarian or menstrual cycle is known as the follicular phase and is the period during which a follicle undergoes the growth and development that culminates in ovulation – rupture of the follicle and release of the oocyte from the ovary.
>
> - A number of follicles start to develop each day but normally only one, the dominant follicle, matures to ovulation in each cycle. The remainder become atretic and die.

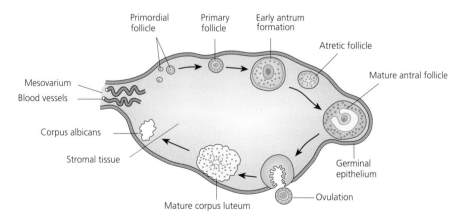

Figure 28.14 A diagram of the stages of follicular development, ovulation, the formation of the corpus luteum and its subsequent regression to a corpus albicans. In reality, not all stages would be seen at the same time.

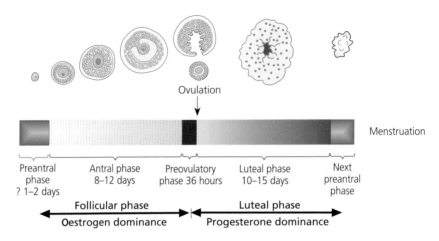

Figure 28.15 A summary of the principal phases of the ovarian cycle. The top panel shows the sequence of follicular development, ovulation, formation of the corpus luteum and formation of the corpus albicans. The lower part of the figure shows the relationship between the follicular and luteal phases of the cycle in relation to follicular development.

The hormonal control of the physical development of a follicle

During the first half of the cycle (the 13–14 days that lead up to ovulation), the follicle grows and matures. This is called the follicular phase of the cycle and it may be subdivided further into phases that have names reflecting the changing structure and function of the follicle throughout this time. These phases are the preantral, antral and preovulatory phases. The various phases are shown schematically in Figure 28.15 and are described briefly below.

The **preantral phase** normally lasts for about 2 days (although its exact duration is uncertain) and during this time a primordial follicle is recruited into the developing pool and converted to a preantral follicle. During the preantral phase, the follicle increases in diameter from around 20 μm to 200–400 μm, while the primary oocyte it contains grows to around 120 μm. As it grows, the oocyte carries out synthetic activity that loads its cytoplasm with the nutrients required for its subsequent maturation.

During the preantral phase of follicular development, there are also important changes to the cells surrounding the oocyte. The flattened granulosa cells adjacent to the oocyte divide to form several layers. These cells secrete a glycoprotein, which forms a cell-free region around the oocyte known as the **zona pellucida**; this can be seen clearly in Figure 28.13. The stromal cells around the outside of the follicle also multiply and differentiate to form concentric layers called the theca. The outermost thecal

cells are flattened and fibromuscular in character (the theca externa) while the inner layers are more cuboidal (the theca interna).

The factors that control the development of the preantral follicle are not well understood but the recruitment of primordial follicles into the preantral phase seems to be independent of hormonal regulation. However, if a preantral follicle is to progress into the next stage of development, its cells must acquire specific hormone receptors: the granulosa cells develop receptors for oestrogens and for anterior pituitary FSH (follicle stimulating hormone), while the thecal cells develop receptors for pituitary LH (luteinizing hormone). Subsequent stages of development are absolutely dependent on control by these hormones and any preantral follicles that have failed to acquire the appropriate receptors will undergo atresia (degenerate and die) at this stage.

Follicles that have acquired receptors for the pituitary gonadotrophins and for oestrogens will enter the next stage of development, the **antral phase**, which normally lasts for 8–10 days. Stimulated by FSH and LH, the granulosa and thecal cells continue to divide and the layers increase in thickness. The granulosa cells also secrete follicular fluid all around the oocyte. This fluid forms the antrum, which gives this phase its name. A fully developed antral follicle (also called a **Graafian follicle**) is illustrated in Figure 28.16. By now the follicle is around 5 mm in diameter and the oocyte within it is suspended in follicular fluid, remaining attached to the rim of granulosa cells by only a thin stalk.

Antral follicle

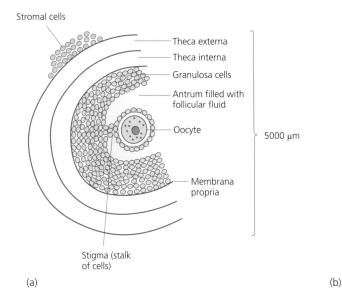

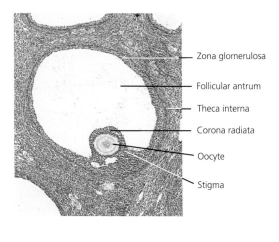

(a)

(b)

Figure 28.16 This figure shows a diagrammatic representation of a late antral (or Graafian) follicle (a) and a section through a mature antral follicle (b). Note the large fluid-filled space (the follicular antrum), the layer of granulosa cells and the theca interna. The oocyte itself is covered with a layer of granulosa cells (the corona radiata), which link it to the zona granulosa by a stalk or stigma.

During the antral phase, the follicular cells start to secrete large quantities of hormones, under the influence of the pituitary gonadotrophins. LH stimulates the cells of the theca interna to synthesize and secrete the androgens testosterone and androstenedione together with small amounts of oestrogens. The granulosa cells are stimulated by FSH to convert these androgens to oestrogens (in particular oestradiol 17β). The result of this secretory activity is a significant increase in the blood levels of androgens and oestrogens, especially the latter, during this part of the ovarian cycle.

These oestrogens have important actions within the follicle itself and also throughout the reproductive tract of the woman. Within the follicle, the granulosa cells possess receptors for oestrogens. Binding of oestrogens to these receptors appears to stimulate the proliferation of more granulosa cells. There is, therefore, an increase in the number of granulosa cells available for converting androgens to oestrogens. As a result of this internal potentiation mechanism, circulating oestrogen levels rise sharply towards the end of the antral phase (at around day 12 of the cycle) to give the so-called 'oestrogen surge'.

A simplified profile of oestrogen secretion during the ovarian cycle is shown in Figure 28.17. The effects of high circulating level of oestrogens on the reproductive tract will be discussed later (see pp. 599–601).

KEY POINTS:

- The physical changes occurring as the follicle develops are closely regulated by hormones, particularly FSH, LH and the oestrogens produced by the follicle itself.

- The follicular phase may be subdivided into preantral, antral and preovulatory stages. The preantral stage appears to be hormone-independent. The antral stage is dependent upon FSH and LH.

- Under the influence of FSH and LH, the follicle secretes large amounts of oestrogens and fluid. By the end of the antral stage, the oocyte is suspended in fluid and attached to the outer rim of follicular cells by a thin stalk.

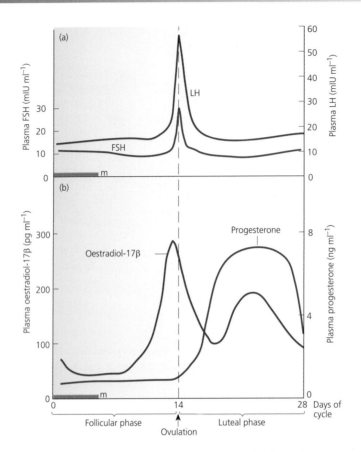

Figure 28.17 The changes in blood hormone levels during the menstrual cycle; (a) shows the pattern of secretion shown by the gonadotrophins (FSH and LH) while (b) shows the changes in the plasma levels of oestradiol-17β and progesterone. The solid red bar marked 'm' represents the period of menstruation.

The preovulatory follicle

Towards the end of the antral phase, at around the time of the oestrogen surge, two important events must coincide if the follicle is to progress into the next stage of development, the brief but dramatic preovulatory phase:

- The granulosa cells must now acquire receptors for pituitary LH.
- Circulating levels of LH must rise steeply.

LH receptors are synthesized by the granulosa cells in response to oestrogens and pituitary FSH. Oestrogens are also believed to be responsible for a sharp rise in LH secretion. Any follicles that do not acquire LH receptors will undergo atresia at this stage. Usually only one follicle (the dominant follicle) will proceed into the preovulatory phase and undergo ovulation in each cycle.

The preovulatory phase only lasts for around 36 hours, but during this time the follicle shows marked changes that culminate in the rupture of the follicle and the release of the oocyte – the process of **ovulation**. Soon after the start of the rise in LH secretion at the start of the preovulatory phase, the oocyte completes its first meiotic division. Remember that at the end of meiosis I, the two members of each bivalent separate, one moving to each pole of the cell and the cytoplasm divides (Table 28.1). In males, each of the resulting cells possesses a similar quantity

Table 28.1 Comparison between spermatogenesis and oogenesis

Meiosis	Spermatogenesis	Oogenesis
Begins in diploid cells (cells with 46 chromosomes) and is only seen in the production of gametes	A primordial cell called a spermatogonium undergoes mitosis and produces primary spermatocytes	A primordial cell called an oogonium produces a primary oocyte
In meiosis I, the original chromosome number of 46 is reduced to 23 chromosomes in each daughter cell	One primary spermatocyte with 46 chromosomes produces two secondary spermatocytes, each with 23 chromosomes	One primary oocyte with 46 chromosomes produces one secondary oocyte and one polar body (the first polar body), each with 23 chromosomes
In meiosis II, each daughter cell divides and four haploid cells are produced, each with 23 chromosomes (one chromosome from each original homologous pair)	Each of the two secondary spermatocytes divides so that a total of four spermatids are produced	The secondary oocyte divides and produces one ovum and one polar body (the second polar body), each with 23 chromosomes (i.e. haploid)
The final outcome of meiosis is the production of four haploid cells for each original diploid cell	The 1:4 ratio is observed	A 1:1 ratio results (polar bodies are not cells)

of cytoplasm. In the female, however, the cytoplasm divides unequally so that half of the chromosomes but almost all of the cytoplasm is contained within one cell, the **secondary oocyte**, while the remaining chromosomes are discarded, along with a very small amount of cytoplasm, in the form of the first polar body. Meiosis then arrests once more and the secondary oocyte undergoes ovulation at this stage of development. Completion of the second meiotic division will not occur unless the egg is penetrated by a sperm (see Chapter 29). The meiotic events taking place within a developing oocyte, alongside the relevant stages of follicular development, are summarized in Figure 28.18.

During the antral phase of follicular development, the granulosa cells were concerned mainly with converting androgens to oestrogens under the influence of FSH. During the preovulatory phase, the granulosa cells alter their secretory activity and, under the influence of LH, begin to secrete progesterone instead. Consequently, blood levels of oestrogens start to fall slightly while those of progesterone increase.

Rupture of the follicle – ovulation

By the end of the preovulatory phase, the oocyte is suspended within the antral fluid attached to the rim of granulosa cells by merely a thin stalk, as shown in Figure 28.16. At ovulation, under the influence of the high circulating levels of LH, the cells of the stalk dissociate (separate from one another) and the follicle ruptures. Follicular fluid flows out onto the surface of the ovary carrying with it the secondary oocyte together with a few surrounding cells. The egg mass is swept into the Fallopian tube by currents set up by the movements of cilia on the fimbriae at the mouth of the tube (the ostium) and the first half of the ovarian cycle is complete.

Formation of the corpus luteum

The second half of the ovarian cycle is known as the **luteal phase** and normally lasts for between 10 and 14 days. Following departure of the egg at ovulation, the remainder of the follicle left behind within the ovary

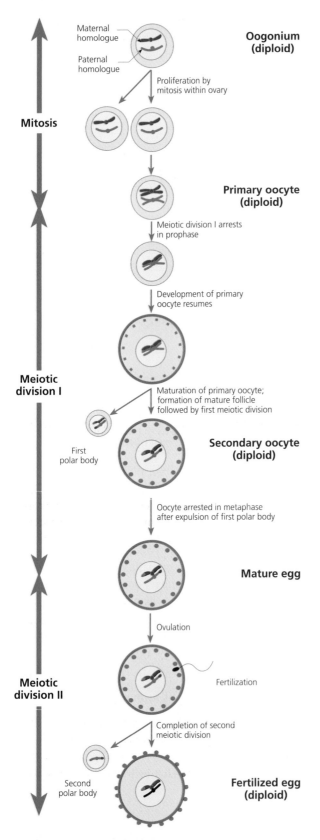

Maternal homologue

Paternal homologue

Oogonium (diploid)

Proliferation by mitosis within ovary

Mitosis

Primary oocyte (diploid)

Meiotic division I arrests in prophase

Development of primary oocyte resumes

Meiotic division I

Maturation of primary oocyte; formation of mature follicle followed by first meiotic division

First polar body

Secondary oocyte (diploid)

Oocyte arrested in metaphase after expulsion of first polar body

Mature egg

Ovulation

Meiotic division II

Fertilization

Completion of second meiotic division

Second polar body

Fertilized egg (diploid)

collapses into the space that was occupied by the oocyte and follicular fluid and a blood clot forms within the cavity. The collapsed follicle then undergoes transformation to become a **corpus luteum**, which, in the event of fertilization, will secrete the appropriate balance of steroid hormones to ensure successful implantation and maintenance of the embryo during the first weeks of pregnancy.

Formation of the corpus luteum appears to depend entirely on the high level of LH present during the preovulatory phase, but the exact hormonal mechanisms that control luteal function are not clear. In the hours immediately following ovulation, the follicular cells undergo a process called luteinization, in which they enlarge and begin to secrete large amounts of progesterone. The corpus luteum may be as large as 15–30 mm in diameter by the middle of the luteal phase (6–8 days after ovulation) and at this time shows its peak secretory activity. Although progesterone is the principal steroid hormone secreted by the corpus luteum throughout the luteal phase, significant quantities of oestrogens are also secreted, and a second oestrogen peak is seen around the middle of the luteal phase (as shown in Figure 28.17). The progesterone secreted by the corpus luteum plays a crucial role in preparing the uterus for pregnancy (see pp. 599–601).

Degeneration of the corpus luteum (luteolysis)

In the absence of fertilization, the corpus luteum has a limited lifespan and degenerates after 10–14 days. This

Figure 28.18 The meiotic events leading to the formation of a mature egg. Initially, the diploid oogonia proliferate by mitosis to form primary oocytes. The primary oocytes enter the first phase of meiosis but are arrested in prophase I. At some time during the fertile years, a primary oocyte resumes development and enlarges considerably in size. During this phase of development, the oocyte acquires an outer coat (shown in green) and cortical granules develop (represented by the small red circles). It then completes the first of two meiotic divisions in which the first polar body is extruded and the secondary oocyte is formed. The oocyte then arrests in metaphase as a mature egg, which is ovulated at mid-cycle. If the egg is fertilized, it undergoes a cortical reaction to prevent a second sperm entering and completes its second meiotic division before the head of the sperm penetrates the nucleus to restore the diploid chromosome number. The second polar body is extruded at this time.

process is called **luteolysis** and involves collapse of the luteinized cells, ischaemia and cell death. As the luteinized cells die, there is a fall in the output of progesterone and oestrogens. This rapid decline in steroid hormone secretion may be seen in Figure 28.17. The degenerated corpus luteum leaves a whitish scar called the corpus albicans (white body), which persists for several months within the ovarian tissue. The mechanisms that cause the corpus luteum to degenerate remain unclear but it has been suggested that the high levels of oestrogen secretion occurring 8–10 days after ovulation may trigger the decline in luteal function.

KEY POINTS:

- The first meiotic division of the oocyte is completed during the preovulatory stage under the influence of high circulating LH levels.
- Progesterone secretion begins to rise as the follicle ruptures to release the egg mass – ovulation.
- The second half of the ovarian cycle is known as the luteal phase, during which the postovulatory follicle is transformed into a corpus luteum under the influence of anterior pituitary LH.
- The luteal phase is characterized by the secretion of large amounts of progesterone and oestrogens.
- In the absence of fertilization, the corpus luteum degenerates after 10–14 days and the output of progesterone and oestrogens falls to very low levels. This is the process of luteolysis, which marks the end of one ovarian cycle.

28.9 Hormonal regulation of the female reproductive tract

The oestrogens and progesterone secreted by the ovaries during each ovarian cycle have a crucial role to play within the female reproductive tract, preparing it for fertilization of an ovum and subsequent pregnancy. All the reproductive tissues are sensitive to the ovarian steroids, but the most obvious effects are seen in the uterus. The uterus prepares first to receive and transport sperm from the cervix to the Fallopian tubes and, later, to receive and nourish the embryo. The changes in appearance and function occurring during each cycle reflect these roles and constitute the **uterine cycle**. Both the endometrium and the myometrium are highly sensitive to the steroid hormones secreted by the developing follicle and the corpus luteum during each ovarian cycle.

Changes to the uterine endometrium

During the follicular phase of the ovarian cycle, the developing follicle secretes increasing amounts of oestrogens so that, by the time of ovulation, plasma oestrogen levels are very high (Figure 28.17). Oestrogens stimulate the uterine endometrium to proliferate so that it increases in thickness from around 2 mm (its thickness just after menstruation) to as much as 10 mm by the time of ovulation, as shown in Figure 28.19. Furthermore, the surface epithelium increases in area and the oestrogen-primed epithelial cells secrete an alkaline watery fluid. At the same time, the coiled spiral arteries that supply the uterine wall start to enlarge. This phase of the uterine cycle, which corresponds to the oestrogen-dominated follicular phase of the ovarian cycle, is called the **proliferative phase** and during this time the uterine endometrium is being prepared to receive a fertilized egg. Oestrogens also stimulate the development of progesterone receptors on the endometrial cells so that, by the time of ovulation, the endometrium is primed to respond to progesterone.

As Figure 28.17 shows, progesterone, secreted by the corpus luteum, dominates the luteal phase of the ovarian

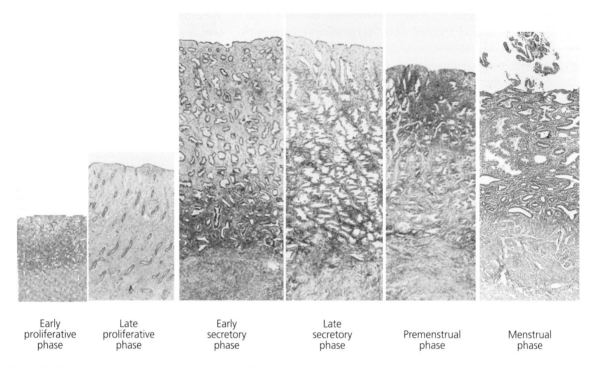

| Early proliferative phase | Late proliferative phase | Early secretory phase | Late secretory phase | Premenstrual phase | Menstrual phase |

Figure 28.19 The changes seen in the endometrium of the uterus during the menstrual cycle. Note the increased depth of tissue and the increased development of the endometrial glands during the late proliferative and secretory phases.

cycle. This hormone is essential for implantation of the newly fertilized egg and development of the placenta. Indeed, adequate levels of progesterone are required throughout the whole of gestation to ensure that the pregnancy has a successful outcome. As progesterone levels rise during the luteal phase of the cycle, the endometrium continues to proliferate, though more slowly, and the spiral arteries develop fully. In the event of pregnancy, these arteries will form the maternal blood supply to the placenta. The endometrial glands start to secrete a thick fluid, rich in sugars, amino acids and glycoprotein, which helps to create a favourable, nutrient-rich environment for implantation of an early embryo. The second half of the uterine cycle, which corresponds to the luteal phase of the ovarian cycle, is called the **secretory phase**.

In the absence of a fertilized egg, the corpus luteum degenerates after 10–14 days and the secretion of progesterone falls steeply. Once the uterine endometrium is deprived of its hormonal support, its elaborate secretory epithelium collapses. The endometrial layers are shed along with blood from the ruptured spiral arteries. This process is known as **menstruation** and marks the completion of one cycle of ovarian activity and the beginning of a new one. Contraction of the spiral arteries may be the cause of the pain experienced by some women at the start of menstruation (dysmenorrhoea). Bleeding continues for several days (usually 3–8) and the total blood loss is normally between 30 and 200 ml. After this time, the endometrial epithelium has been completely repaired and proliferation can begin again.

Changes to the uterine myometrium

The smooth muscle of the uterine wall is also sensitive to the ovarian steroid hormones. Under the influence of the oestrogens secreted by the developing follicle, the excitability of the myometrium is increased and the smooth muscle shows an increase in spontaneous contractility. This may help sperm to swim through the uterus towards the Fallopian tubes and thus improve the chances of successful fertilization. However, once an embryo has entered the uterus, too much excitability could result in spontaneous abortion of the fetus – a 'miscarriage'. Progesterone has the very important action of relaxing the myometrium and thus reducing the likelihood of spontaneous contractions.

Actions of ovarian steroids on other reproductive tissues

The oestrogens and progesterone secreted during the ovarian cycle have important actions on other tissues of the female reproductive system. These contribute to the preparations being made by the body for fertilization and pregnancy in a variety of ways.

Fallopian tubes. Under the influence of oestrogens, tubal ciliary and muscular activity appears to be enhanced in preparation for transporting the egg towards the uterus. Progesterone, by contrast, appears to relax the smooth muscle of the tubes and reduce their contractility.

Secretions of the cervix. The endocervical glands secrete mucus that varies in quantity and consistency throughout each ovarian cycle. These changes are controlled by the ovarian steroids, and have important consequences for fertility. When plasma levels of oestrogens are high, cervical mucus is secreted in large amounts and it is thin, clear and stretchy. Peak volume and elasticity coincide with the peak levels of oestrogens, seen just before ovulation. Mucus with these characteristics is readily penetrated by sperm and so the endocrine action of the ovary optimizes the conditions for successful fertilization. When progesterone levels are high, during the luteal phase, the cervical mucus is produced in much smaller amounts and becomes thick, sticky and difficult for sperm to penetrate. This action of progesterone forms part of the mechanism of action of the progesterone-only contraceptive pill (see Box 29.1).

Vaginal epithelium. Changes in the histological appearance of the epithelium lining the vagina are associated with the changing levels of oestrogens and progesterone occurring throughout the ovarian cycle. During the follicular phase, while oestrogens are the dominant hormones, there is proliferation and maturation of the vaginal mucosa. Maximal thickness is reached at the time of ovulation. During the luteal phase, when progesterone secretion is high, there is shedding of the superficial cell layers. Examination of the vaginal epithelium can be used as an indicator of the stage of the menstrual cycle that has been reached.

Some non-reproductive tissues are also influenced by the ovarian steroids

In addition to their specific actions on the reproductive tract, oestrogens and progesterone exert widespread and generalized effects throughout the whole body. Table 28.2 lists the principal actions of the ovarian steroid hormones on non-reproductive tissues. In addition to the physical effects listed above, both oestrogens and progesterone exert profound effects on mood. The mechanisms by which these psychological effects are brought about are not understood.

Table 28.2 A summary of the actions of the oestrogens and progesterone

Oestrogens	Progesterone
Mildly anabolic	Mildly catabolic
Tend to depress appetite	Tends to stimulate appetite
Reduce plasma levels of cholesterol	Causes a rise in core body temperature of 0.2–0.6°C when levels are elevated during the luteal phase of the menstrual cycle
Reduce capillary fragility	Promotes the development of the lobules and alveoli of the breast
Cause proliferation of the ductal tissue of the breast	Causes breasts to swell due to fluid retention
Contribute to the maintenance of the skeleton	

> **KEY POINTS:**
>
> - The follicular phase of the ovarian or menstrual cycle is dominated by oestrogens secreted by the developing follicle. These hormones prepare the reproductive tract for gamete transport, fertilization and the implantation of an embryo.
> - During the second half of the ovarian cycle, the luteal phase, progesterone is secreted by the corpus luteum. This maintains the endometrium in a favourable condition for implantation.
> - In the absence of an embryo, the corpus luteum degenerates after 10–14 days and the output of steroid hormones falls steeply. The elaborate endometrium, which was built up during the cycle, is sloughed off and shed together with blood from the spiral arteries. This process is called menstruation and its onset marks the beginning of a new ovarian cycle.

28.10 Why do the plasma concentrations of gonadotrophins and ovarian steroids vary during the ovarian cycle?

The previous sections have been concerned mainly with the structural and functional changes that occur throughout the 28-day menstrual cycle. Cyclical alterations in the plasma levels of FSH and LH (shown in Figure 28.17) play a crucial role in controlling follicular development and the growth and function of the corpus luteum. It is, therefore, important to understand how these fluctuations come about and how they regulate ovarian activity. A key point to remember is that not only do the gonadotrophins regulate ovarian function, but also that the ovarian steroids, in their turn, influence the secretion of the gonadotrophins. This complex interaction (called feedback) between the hypothalamus, the anterior pituitary gland and the ovaries is illustrated in Figure 28.20.

Ovarian steroids can exert either a negative or a positive feedback effect on the secretion of LH and FSH, depending upon their concentration in the plasma. Low to moderate concentrations of oestrogens inhibit the secretion of the gonadotrophins (negative feedback) while the high levels, seen during the days just prior to ovulation, stimulate gonadotrophin secretion (positive feedback). By contrast, high levels of progesterone tend to inhibit gonadotrophin secretion, while lower concentrations tend to enhance the positive feedback effects of oestrogens. The feedback effects

of the ovarian steroids are probably exerted mainly at the level of the anterior pituitary gland itself although they may also have a direct effect on the output of the hypothalamic gonadotrophin-releasing hormone (GnRH).

The following gives a brief description of the hormonal interactions taking place throughout one ovarian cycle and to make this easier to understand, it will be helpful to refer back to Figure 28.17. The ovarian cycle begins on day 1 of menstrual bleeding. At this time, plasma levels of progesterone and oestrogens are low, and, released from negative feedback inhibition, secretion of the anterior pituitary gonadotrophins FSH and LH starts to rise. Under the influence of FSH, a follicle grows and begins to secrete ever-increasing quantities of oestrogens. As the plasma concentration of oestrogens rise, their negative feedback effects on gonadotrophin secretion are replaced by a positive feedback that results in a sharp increase in the output of both FSH and LH, but especially LH (the 'gonadotrophin' or 'LH surge'). This surge is responsible for ovulation, after which time, oestrogen levels fall steeply as the luteal cells switch to progesterone secretion. The high circulating levels of progesterone that dominate the luteal phase of the ovarian cycle maintain a strong negative feedback effect on gonadotrophin

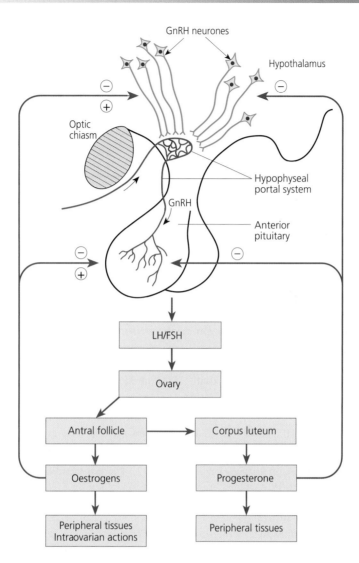

Figure 28.20 The positive (+) and negative (−) feedback loops that control the hormonal secretions of the hypothalamus, pituitary gland and ovaries.

output so that during the second half of the cycle plasma levels of FSH and LH remain low.

In the absence of fertilization, the corpus luteum regresses and steroid secretion falls dramatically. This is when the uterine endometrium, deprived of its hormonal support, sloughs off and menstrual bleeding begins. Released from negative feedback inhibition, FSH and LH levels then start to rise slowly once more to trigger the entry of new follicles into the antral phase and begin a new ovarian cycle.

KEY POINTS:

- Although the gonadotrophins regulate ovarian function, the ovarian steroids, in their turn, influence the secretion of the gonadotrophins.

- This complex interaction between the hypothalamus, anterior pituitary gland and the ovaries involves both positive and negative feedback.

28.11   Puberty and the menopause

In the female, the fertile years are defined by two events. These are the onset of menstruation at puberty (**menarche**) and the cessation of cyclical ovarian activity at around the age of 50 (the **menopause** or **climacteric**).

Puberty in the female

Puberty is a collective term that includes the many changes taking place within the body of an adolescent girl as her ovaries mature. Menarche is the outward sign that these changes have taken place and that the cyclical secretion of ovarian steroids has begun. The average age of menarche in the United Kingdom is around 12 years (with a range of 10–16 years) but it may be several more months before regular ovulatory menstrual cycles begin and fertility is established. Some of the other changes that take place during the 2–3 years prior to menarche are listed in Table 28.3, together with the hormones that

are thought to be responsible for them. These include the development of the female secondary sexual characteristics (pubic hair and breast development) and changes in body composition – adult females have about twice as much body fat as males of similar size, and a smaller mass of skeletal muscle.

The menopause

The menopause is the cessation of menstrual periods that marks the end of a woman's fertile years. It normally occurs between the ages of 45 and 55 years. However, for several years leading up to this point there is progressive failure of the reproductive system during which the responsiveness of the ovaries to gonadotrophins gradually declines. By now there are very few oocytes left in the ovaries and menstrual cycles often become irregular and anovulatory before ceasing altogether. After the

Table 28.3 Summary of the principal changes of puberty

Characteristic	Age range of first appearance (years)	Principal hormones responsible for development
Girls		
Breast bud	8–13	Oestrogens, progesterone, growth hormon (GH)
Pubic hair	8–14	Adrenal androgens
Menarche	10–16	Oestrogens, progesterone
Growth spurt	10–14	Oestrogens, GH
Boys		
Growth of testis	10–14	Testosterone, FSH, GH
Growth of penis	11–15	Testosterone
Pubic hair	10–15	Testosterone
Facial and axillary hair	12–17	Testosterone
Enlargement of larynx	11–16	Testosterone
Growth spurt and male pattern of musculo-skeletal development	12–16	Testosterone, GH

menopause, plasma levels of FSH and LH are relatively high because of the loss of negative feedback inhibition by oestrogens. LH surges, however, no longer occur.

The years leading up to and just after the menopause itself are often called the 'perimenopause' and are characterized by a variety of somatic and emotional changes, most of which are a direct consequence of the loss of ovarian steroid hormones. Among the more obvious changes are a loss of breast tissue, vaginal dryness, night sweats, hot flushes, depression and an increased susceptibility to myocardial infarction. There is often an increase in bone weakness due to the loss of bone mass that occurs following the dramatic fall in plasma concentrations of oestrogens. As a result, fractures, particularly of the wrist and neck of the femur, are more frequently seen in postmenopausal women. Hormone replacement therapy (HRT) may be used to treat most of the undesirable consequences of the menopause if they are serious enough to warrant medical intervention.

Puberty in the male

Testosterone is the key to reproductive function in the male. From early infancy until the start of puberty, testosterone secretion by the testes is low. Between the ages of 10 and 16 (on average), secretion rises so that adult testosterone levels are normally reached by the age of around 17 and full reproductive capacity is established. Testosterone is also responsible for the development of male secondary sexual characteristics such as deepening of the voice and the growth of facial and pubic hair, the growth spurt of puberty and the development of the adult musculature. These maturational events take place over a period of several years and Table 28.3 shows their average timing during puberty.

Is there a male 'menopause'?

While there is no obvious event marking the loss of reproductive capacity in the male comparable to the female menopause, sperm production does decline after the age of around 50. There is also a small reduction in plasma levels of testosterone in elderly men. This decline in sexual capacity, which may also be associated with emotional problems and a loss of libido, is sometimes called the **andropause**. Nevertheless, many elderly men maintain active sex lives and are able to father children.

Recommended reading

Anatomy

MacKinnon, P. and Morris, J. (2005). *Oxford Textbook of Functional Anatomy*, 2nd edn, Vol. 2. Oxford University Press: Oxford. pp. 179–211.

Cell biology of germ cells

Alberts, B., Johnson, A., Lewis, J., Raff, M., Roberts, K. and Walter, P. (2007). *Molecular Biology of the Cell*, 5th edn. Garland: New York. Chapter 21.

Histology

Junquieira, L.C. and Carneiro, J. (2005). *Basic Histology*, 11th edn. McGraw-Hill: New York. Chapters 22 and 23.

Young, B., Lowe, J., Stevens, A., Heath, J.W. and Deakin, P. (2006). *Wheater's Functional Histology*, 5th edn. Churchill-Livingstone: Edinburgh. Chapters 18 and 19.

Physiology

Griffin, N.E. and Ojeda, S.R. (2004). *Textbook of Endocrine Physiology*, 5th edn. Oxford University Press: Oxford.

Johnson, M.H. (2007). *Essential Reproduction*, 6th edn. Blackwell Scientific: Oxford.

Pocock, G. and Richards, C.D. (2006). *Human Physiology: the Basis of Medicine*, 3rd edn. Oxford University Press: Oxford. Chapter 20.

Self-assessment questions

Which of the following statements are true and which are false? Answers are given at the end of the book (p. 755).

1 The following statements apply to human reproduction:
a) The male gametes are known as sperm.
b) A mature sperm contains a full complement of chromosomes.
c) At birth, an ovary contains all the oocytes it will ever have.
d) Primary and secondary spermatocytes divide by mitosis to give rise to spermatids.

2 In the testis:
a) Leydig cells secrete testosterone.
b) spermatogenesis requires testosterone.
c) Sertoli cells prevent free diffusion of water-soluble substances between the seminiferous tubules and the blood.
d) Sertoli cells respond to testosterone by synthesizing androgen-binding protein.
e) Sertoli cells line the seminiferous tubules and directly give rise to the developing sperm.

3 During the ovarian cycle:
a) the initiation of the preantral phase of follicular development is under the control of LH.

b) the development of the antral follicle depends on the expression of receptors for FSH and LH.
c) ovulation occurs after about 14 days.
d) ovulation occurs in response to a sudden increase in plasma LH.
e) the period before ovulation is known as the luteal phase.
f) the luteal phase is associated with a large increase in plasma progesterone.
g) the myometrium proliferates under the influence of oestrogens.

4 The following statements relate to the hormonal control of the menstrual cycle:
a) Oestrogens are synthesized mainly by the cells of the theca interna.
b) In the absence of receptors for FSH and LH, preantral follicles undergo atresia.
c) After ovulation, the ruptured follicle is converted into a corpus albicans, which secretes progesterone.
d) Progesterone promotes full development of the endometrium.
e) If no egg is fertilized, progesterone levels fall and this is the trigger for menstruation.

Fertilization and pregnancy

29

 After reading this chapter you should have gained an understanding of:

- The sexual reflexes of males and females
- The processes of fertilization and implantation
- The formation and functions of the placenta
- The role of the placental hormones during pregnancy
- Important changes in maternal physiology during pregnancy
- The nutritional demands of pregnancy
- Some important aspects of fetal physiology
- Labour and delivery (parturition)
- Lactation
- The physiological adaptations in the newborn
- Methods of contraception

29.1 Introduction

In the previous chapter, the discussion of female reproductive physiology assumed a situation in which cyclical ovarian activity continues uninterrupted by pregnancy. In this case, each cycle ends with regression of the corpus luteum and shedding of the uterine endometrium at menstruation. If, however, an oocyte is fertilized, a completely different set of events must be initiated to prevent loss of the endometrium so that the embryo is able to implant and gestation can proceed. This chapter is concerned with the processes of fertilization, pregnancy and delivery of a baby and with the important changes that take place in the body of a woman during pregnancy and lactation. Some aspects of fetal and neonatal physiology will also be discussed.

29.2   The sexual reflexes

Fertilization of an egg requires that sperm are deposited in the vagina of a woman close to the time of ovulation. Except for the case of artificial insemination, this is achieved through sexual intercourse in which the penis of the male becomes erect and seminal fluid is ejaculated within the vagina. In men and women, sexual responses are controlled by autonomic and somatic nerves originating in the lumbar and sacral regions of the spinal cord (see Chapter 13).

The main stages of the male sexual act are:

- Erection (also called tumescence) of the penis
- Secretion of mucus by the bulbourethral glands
- Emission of fluid from the seminal vesicles, vas deferens and prostate gland
- Ejaculation of semen from the penis.

Penile **erection** results from either stimulation of the skin of the genital region (reflexogenic erection) or from nerve activity originating in higher centres of the brain (psychogenic erection). Erection itself is a simple hydraulic process resulting from engorgement of the erectile tissues of the penis with blood. The penis contains three columns of erectile tissue – two corpora cavernosa and the corpus spongiosum, which surrounds the urethra (see Chapter 28 for more details). The erectile tissues are essentially large venous sinusoids surrounded by a strong fibrous capsule. In response to appropriate stimulation, parasympathetic activity causes the internal pudendal artery to dilate, thereby increasing blood flow to the penis. The venous outflow is unchanged, so there is pooling of blood in the erectile tissues leading to engorgement. As the pressure of blood in the penis increases, the venous drainage is partially occluded so that the penis becomes rigid and erect. Dilatation of the pudendal artery is mediated by nitric oxide derived from its parasympathetic innervation.

Emission and **ejaculation** are the final stages of the male sexual act. When sexual stimulation becomes very intense, rhythmic contractions of the vas deferens begin to drive sperm towards the ejaculatory duct. The prostate gland secretes prostatic fluid and the seminal vesicles contract. This is known as emission. Ejaculation is the expulsion of the semen from the penis and it is accompanied by an intense sensation called orgasm. During ejaculation, the internal urethra closes to prevent sperm from entering the urinary bladder, prostate or seminal vesicles. The bulbo-cavernous and ischiocavernous muscles contract to drive seminal fluid along the urethra and out of the penis. Between 2 and 5 ml of semen are released in each ejaculation and this will normally contain around 200 million sperm.

In females, the neural pathways involved in the sexual response are the same as those of the male. Sexual reflexes are integrated in the sacral region of the spinal cord but can be modified by cerebral influences. During sexual excitement, parasympathetic stimulation causes the erectile tissue in the clitoris and around the vaginal opening to become engorged with blood. Mucus, secreted by the vaginal epithelium and by glands adjacent to the labia minora (Bartholin's glands), facilitates entry of the penis into the vagina and lubricates the movements of the penis during intercourse. The tactile stimulation of the female genitalia, coupled with psychological stimuli, will normally trigger an orgasm. Orgasm is associated with contraction of the perineal muscles, dilatation of the cervical canal and an increase in motility of the uterus, and possibly also the Fallopian tubes. These changes may facilitate the progression of sperm through the female reproductive tract, but female orgasm is not a requirement for fertilization.

Contraception

The prevention of unwanted pregnancy (contraception) is a major issue in most societies. The principal means of preventing conception rely either on preventing ovulation, or preventing the egg and sperm from meeting (barrier methods). They are summarized in Box 29.1.

> **KEY POINTS:**
>
> - Sexual intercourse involves penile erection, penetration of the vagina by the penis and ejaculation of semen containing around 200 million sperm.
> - Male and female sexual responses are mediated by sacral reflexes involving the autonomic nervous system.

Box 29.1 An outline of contraceptive methods

Method	Effectiveness (estimated as % of couples remaining childless after 1 year)	Comments
Oral contraceptives (the birth control pill)	99.5%	Contraceptive preparations are of two types: (a) Combination: consisting of oestrogens and progesterone in sufficiently high concentrations to exert powerful negative feedback on gonadotrophin output. They therefore mimic the luteal phase of the menstrual cycle and prevent ovulation. (b) The mini-pill: this contains only progesterone and acts by modifying the secretions and internal environment of the reproductive tract.
		Side-effects include minor symptoms of early pregnancy – nausea, breast tenderness, fluid retention, hypertension and, in rare cases, thromboses (mainly in smokers over 35 years of age). There is still debate over whether or not the risks of endometrial, ovarian and breast cancer are altered by these drugs.
Transdermal contraceptive patches	99.5% (lower in obese women)	Delivers oestrogens and progesterone into the blood stream for 3 weeks followed by a week without the patch.
Norplant (two silicone rods implanted just under the skin)	Better than 99.5%	Release progesterone over a 5-year period.
Depo-Provera (an injectible form of progesterone)	Better than 99.5%	Lasts for 3 months.
Postcoital contraception (the 'morning afterpill)	Around 75–95% (depending upon how soon after intercourse the pills are taken)	Taken within 3 days of unprotected sexual intercourse these contraceptive pills contain high concentrations of hormones, which prevent fertilization and/or implantation.
Intrauterine device (IUD)	98.5%	Probably works by creating an environment within the uterus that is hostile to fertilization or implantation. Newer forms of IUD deliver a sustained low dose of progesterone to the endometrium.
		May cause uterine bleeding.
		Increased risk of inflammatory disease in the pelvic region.
Condom (sheath)	96%	A barrier method that is more effective if used with a spermicidal gel, sponge or foam.
		Its chief advantage is that it protects against AIDS and other sexually transmitted diseases.
		It may also protect against cervical cancer.
		Its disadvantage is that it may split in use.

(Continued)

Box 29.1 *(Continued)*

Method	Effectiveness (estimated as % of couples remaining childless after 1 year)	Comments
Diaphragm	98% when used with spermicide	An alternative barrier method. Needs to be inserted before intercourse.
		It fits over the cervix and blocks the entrance to the uterus.
		Its use carries a small risk of infection.
Rhythm method	Highly variable, dependent on regularity of cycle and accuracy with which mid-cycle is calculated	Couple must refrain from intercourse during the fertile period of the cycle (i.e. the 2 or 3 days on either side of ovulation).
		The time of ovulation must be calculated from the previous cycle assuming that ovulation occurred on day 14 before menstruation.
		Other indicators, such as the change in body temperature at mid-cycle and the constitution of cervical mucus, may also be used.
Sterilization (vasectomy in men and ligation of the Fallopian tubes in women)	c. 100%	Requires surgery and is not always reversible.
		Risks are similar to those of other minor surgical procedures.

29.3 Fertilization

Sperm deposited in the vagina during sexual intercourse swim through the cervix and uterus to reach a Fallopian tube, which is the site of fertilization. Both the male and female gametes have a limited period of viability within the female reproductive tract. Sperm are thought to remain viable for up to 72 hours but eggs probably only remain fertile for about 24 hours after ovulation. This means that there is only a relatively short time during each menstrual cycle in which pregnancy can be achieved. Furthermore, a very small number (possibly as few as 200) of the ejaculated sperm complete the journey to a Fallopian tube.

Before a sperm can fertilize an oocyte, it must undergo a process called **capacitation**. This involves a series of modifications to the plasma membrane of the sperm head that enable it to penetrate the membrane of the egg. Capacitation normally takes place in the female reproductive tract, within about 6 hours of ejaculation. The mechanisms of capacitation are not fully understood but the process depends upon the high concentration of bicarbonate ions found in the fluid of the uterus and Fallopian tubes.

Should an activated sperm meet a viable egg in the Fallopian tube, the process of fertilization may occur. Figure 29.1 illustrates the structure of the secondary oocyte and shows that it is surrounded by several layers of follicular (granulosa) cells. Between these cells and the plasma membrane of the oocyte is a region called the **zona pellucida**, which consists of glycoproteins produced by the oocyte. To achieve union with the oocyte, a sperm must first swim between the granulosa cells and then undergo a reaction called the **acrosome reaction** as it binds to the zona pellucida. The acrosome reaction is a calcium-dependent process in which the acrosomal vesicle fuses with the head of the sperm, leading to the secretion of enzymes that aid the passage of the sperm through the zona pellucida and into the oocyte itself.

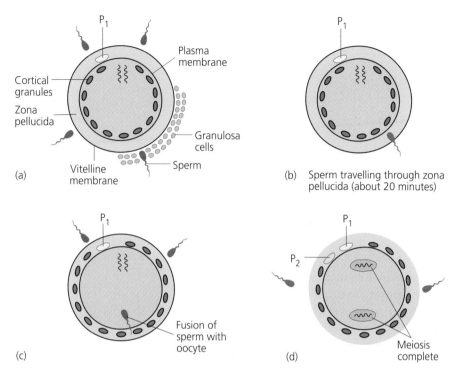

Figure 29.1 Diagrammatic representation of the changes that occur during fertilization. For a sperm to fuse with an egg it must first penetrate the layer of granulosa cells and the zona pellucida as shown in (a) and (b). Fusion with the egg is followed by the cortical reaction (c) in which the structure of the zona pellucida is altered to prevent the entry of any further sperm. This is followed by the completion of meiosis and extrusion of the second polar body (d). P1 and P2 are the first and second polar bodies.

Although fusion of the sperm with the plasma membrane of the oocyte completes the first stage of fertilization, a number of further important tasks must be completed by the fertilized egg to ensure its continued successful development:

1. *It must complete its second meiotic division* (remember that the secondary oocyte arrests after extrusion of the first polar body at the time of ovulation; see Chapter 28). Meiosis is normally completed within 2 or 3 hours of fertilization and the second polar body is extruded (Figure 29.1). The egg is now haploid. If meiosis fails to complete, the fertilized egg will possess too much chromosomal material and will not develop further.

2. *It must avoid fusing with any other sperm (polyspermy).* Although many sperm may undergo acrosome reactions as they arrive in the vicinity of an oocyte, it is essential that only one sperm fuses with the plasma membrane of the oocyte. Polyspermy is prevented by the oocyte itself by the **cortical reaction** (also called

the **zonal reaction**). Entry of the sperm head into the oocyte triggers the release of calcium ions from internal stores within the oocyte. Lying close to the plasma membrane of the oocyte are a number of granules containing a variety of enzymes. These are known as cortical granules, which secrete their contents into the extracellular space beneath the zona pellucida (Figure 29.1c) in response to the increase in intracellular calcium. The enzymes released appear to alter the structure of the zona pellucida so that sperm can no longer bind to it – the so-called block to polyspermy.

3. *Regression of the corpus luteum must be prevented.* For successful implantation of the early embryo, it is essential that the thickened, secretory endometrial lining of the uterus remains intact. To prevent endometrial shedding, the corpus luteum must go on secreting progesterone. This means that instead of regressing 10–14 days after ovulation, as it would in the absence of fertilization, its life and secretory

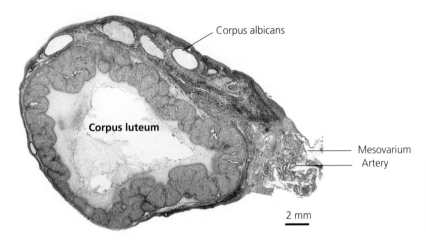

Corpus albicans

Corpus luteum

Mesovarium

Artery

2 mm

Figure 29.2 Section through a human ovary during pregnancy (stained with haematoxylin and eosin) showing the corpus luteum occupying almost the entire space within the organ.

capacity must be prolonged. Within a few days of fertilization, the zygote (early embryo) begins to secrete a hormone called **human chorionic gonadotrophin (hCG)**, which is similar to LH but very much more potent. It is a powerful luteotrophic hormone that prevents the corpus luteum from degenerating. Instead, it continues to secrete the progesterone needed for the continuation of pregnancy. Figure 29.2 shows a histological section through an ovary in early pregnancy. The corpus luteum occupies a large fraction of the ovarian tissue.

hCG appears in the maternal circulation within a few days of fertilization and can be detected in the maternal urine by 10–14 days after fertilization. Levels of hCG then continue to rise steadily up until 8–10 weeks of gestation before falling quite steeply over the next few weeks. This pattern of secretion is illustrated in Figure 29.3. By the time that hCG levels start to fall, the placenta has become established and secretes the large amounts of progesterone that are required to maintain the rest of the pregnancy. The progesterone contributed by the corpus luteum becomes insignificant once the placenta is fully functional.

Clinically, hCG is a very important hormone: its presence in the maternal urine within 2 weeks of ovulation is used as a reliable test for pregnancy, which is so simple that it can be carried out, using a kit, by a woman herself at home. More sophisticated measurements can be performed by clinicians to gain information about the pregnancy. Elevated levels of hCG, for example, suggest the presence of twins.

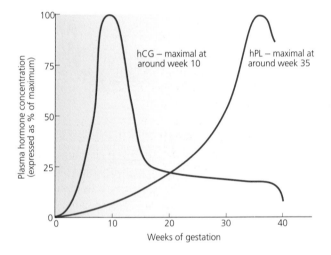

hCG – maximal at around week 10

hPL – maximal at around week 35

Plasma hormone concentration (expressed as % of maximum)

Weeks of gestation

Figure 29.3 The changes in the plasma concentrations of human chorionic gonadotrophin (hCG) and human placental lactogen (hPL) that occur during gestation.

KEY POINTS:

- Before a sperm can fuse with an egg to bring about fertilization, it must first undergo capacitation followed by the acrosome reaction.

- Fusion of the egg and sperm takes place in the Fallopian tube. After fusion, the egg completes its second meiotic division and undergoes the cortical reaction, which serves to prevent any other sperm from fusing with it.

- The newly fertilized egg secretes hCG, which prolongs the secretory life of the corpus luteum and ensures the continued production of progesterone.

29.4   Implantation and the development of the placenta

At fertilization, the nuclei of a sperm and an oocyte fuse to produce a diploid cell called a **zygote**. Almost immediately, this cell starts to divide by mitosis and by day 4 after fertilization the zygote has become a cluster of 16–32 cells forming a solid ball called a **morula**. During the next 24 hours or so, the morula hollows out so that the embryonic cells surround a fluid-filled space. At this stage, the organism is called a **blastocyst**. The cells of the blastocyst form a layer (the trophoblast) enclosing the fluid-filled space, with one end of the blastocyst possessing a thicker accumulation of cells, the embryonic disc or inner cell mass (Figure 29.4a). The embryo will form from the cells of this region. The outer cell layer will form the placenta and other extra-embryonic structures. At around day 6–7 after fertilization, the blastocyst will have reached the uterus and will begin to implant within the endometrium.

Implantation occurs in three stages. Firstly the blastocyst makes contact with the receptive uterine endometrium at its embryonic disc region and becomes pressed up against it (**apposition** or adplantation). Microvilli on the surface of the blastocyst cells then interact via cell-surface glycoproteins with the endometrial cells to bring about **adhesion** of the blastocyst to the uterine lining. Invasion of the uterine endometrium by the trophoblast then begins. Shortly before the blastocyst makes contact with the endometrium, the trophoblast differentiates into two regions of cells, the inner **cytotrophoblast**, a layer of cells undergoing rapid mitotic activity, and the outer **syncytiotrophoblast**, a multinucleated layer without cell boundaries, formed by the fusion of trophoblast cells. The cells in this region secrete enzymes that break down the endometrial cells and the walls of capillaries. This invasive behaviour allows the embryo to penetrate the endometrium – it has now become implanted. The syncytiotrophoblast develops further as implantation proceeds so that it surrounds the embryo completely by 9–10 days after fertilization (Figure 29.4c).

The uterine mucosa reacts to implantation by undergoing the so-called **decidual reaction**. The stromal cells enlarge and become filled with glycogen, which provides nourishment for the embryo until the blood vessels of the placenta develop.

All organisms need a large and continuous supply of nutrients throughout their embryonic development. They must also be able to respire and to dispose of the waste products of their metabolism. Most mammalian species, including humans, accomplish this by the process of placentation in which a specialized organ, **the placenta**, is developed. This provides an association between the uterine endometrium (now called the **decidua**) and the embryonic membranes (derived from the trophoblast), which allows the blood supply of the fetus to be brought into close proximity with that of its mother, thereby permitting the exchange of substances between the two circulations. The placenta performs those functions carried out by the lungs, kidneys and gastrointestinal organs in the adult – indeed, it is the only source of nourishment, gas exchange and waste disposal available to the fetus. For this reason, normal development of the placenta is crucial to the success of a pregnancy.

Implantation is complete by around day 12 after fertilization and, over the next few weeks, the placental structures are established. The placenta is fully formed (the so-called 'definitive placenta') by the end of the third month of gestation (the first **trimester**).

The following brief account will explain how the placenta is adapted structurally for carrying out its role as an organ of exchange. Figure 29.5 shows an enlarged view of a section of human placenta soon after the start of implantation. This is known as the stem villus stage of development because the fetal tissue grows up into the maternal decidual tissue in the form of finger-like projections or **villi** (Figure 29.4f). Blood vessels form within these villi, giving rise to the fetal component of the placental circulation, which will later become the umbilical vessels of the umbilical cord. The enzymes secreted by the invading syncytiotrophoblast erode the spiral arteries of the uterine lining and allow blood to spill out and create blood-filled spaces between the villi. These are called **intervillous blood spaces** and form part of the maternal circulation to the placenta. These

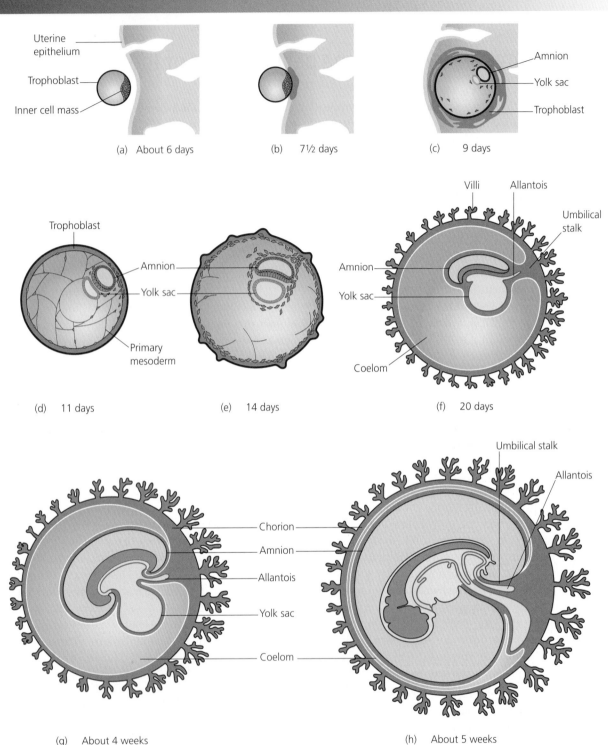

Figure 29.4 A seriews of diagrams to illustrate the events that take place in the first 5 weeks after fertilization. By 6 days after fertilization (a), the embryo (blastocyst) consists of a hollowed-out ball of cells with a thickening at one end (the inner cell mass). Over the next few days (b and c), the blastocyst implants into the uterine endometrium so that by 11 days after fertilization (d) the blastocyst is completely surrounded by maternal tissue. Subsequently (e and f), the fetal membranes (chorion and amnion) develop and villi of embryonic tissue invade the uterine tissue to begin the formation of the placenta. Further invasion of the uterus by embryonic tissue and the development of the umbilical stalk are shown in (f–h). In this way, the crucial interface between the maternal and fetal circulations is established.

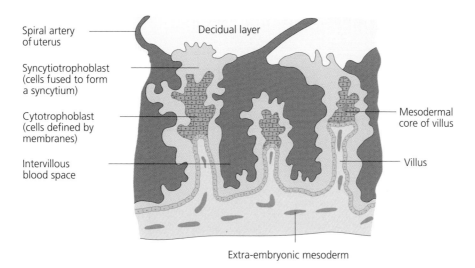

Spiral artery of uterus

Decidual layer

Syncytiotrophoblast (cells fused to form a syncytium)

Cytotrophoblast (cells defined by membranes)

Intervillous blood space

Mesodermal core of villus

Villus

Extra-embryonic mesoderm

Figure 29.5 The stem villus stage of placental development at around 3–4 weeks of gestation. The finger-like projections of embryonic tissue can be seen invading the decidual layer of the uterus to establish a close association between the fetal and maternal blood supplies.

changes establish the essential interface between the maternal and fetal circulations.

Invasion of the endometrium and villus formation take place during the first month after conception to establish proximity between the maternal and fetal blood (Figure 29.4h). In the next 2 months, the richly vascularized villi become much more highly branched, thereby increasing the surface area of the fetal capillaries available for transplacental exchange of nutrients and waste products. Although the placenta is now fully formed and will show few further changes to its basic structure – it is 'definitive' – it will continue to increase in size until the last weeks of gestation to keep pace with the increasing metabolic demands of the growing fetus.

A simplified diagram of the placenta at 3 months' gestation is shown in Figure 29.6. It shows that the arrangement of the placental blood supply is such that the fetal capillaries essentially dip into the maternal blood spaces so that they are virtually surrounded by maternal blood. This arrangement of the two circulations is termed a **dialysis pattern** and it allows movement of substances

in either direction, according to the concentration gradient, over the entire surface of the fetal placental capillaries.

KEY POINTS:

- The newly fertilized egg (zygote) divides to form a blastocyst consisting of a layer of cells surrounding a fluid-filled space – the blastocyst.

- The blastocyst implants within the prepared endometrium of the uterus to establish the interface between maternal and fetal circulations – the placenta.

- The placenta is fully developed by the end of the first trimester (3 months) of pregnancy. The maternal and fetal blood flows are arranged in a dialysis pattern in which the fetal capillaries dip into spaces filled with maternal blood (intervillous blood spaces).

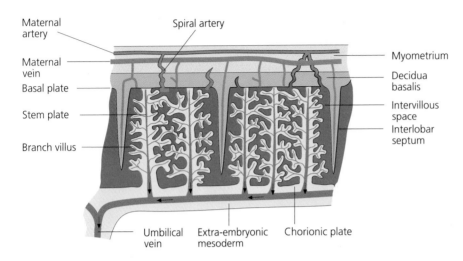

Maternal artery
Spiral artery
Maternal vein
Basal plate
Stem plate
Branch villus
Myometrium
Decidua basalis
Intervillous space
Interlobar septum
Umbilical vein
Extra-embryonic mesoderm
Chorionic plate

Figure 29.6 A diagrammatic representation of the organization of the placenta at the end of the first trimester – the 'definitive' placenta.

29.5 Exchange of substances across the placenta

Figures 29.5 and 29.6 show that for a substance to diffuse across the placenta it must cross the syncytiotrophoblast and the fetal capillary endothelial layer. These layers constitute the so-called 'placental barrier' between the maternal and fetal blood. Although the fetal capillary endothelium is only one cell thick, the syncytiotrophoblast is a relatively thick layer with no paracellular pathways to act as shortcuts for diffusion. The placental barrier is therefore rather impermeable. However, this low solute permeability is largely offset by the enormous surface area available for placental exchange created by the extensive branches within the villi and the dialysis arrangement of the two circulations.

Another important factor that determines the rate and extent of diffusion across a cellular barrier is the **concentration gradient** that exists for the substance. Within the placenta, the concentration gradient of any substance between the fetal and maternal blood will be influenced by the relative blood flow rates to either side. Although the rate of entry of blood to the maternal intervillous spaces is difficult to measure accurately, it is thought to be of the order of 300–500 ml min^{-1}. The maternal blood space has a total volume of about 250 ml, so the blood on the maternal side of the placenta is exchanged roughly twice each minute. Furthermore, blood enters the intervillous

spaces under relatively high pressure (about 13 kPa or 100 mmHg) as it is discharged by the eroded spiral arteries, thus ensuring a fair degree of turbulence and good mixing.

In a full-term fetus, blood flow in the fetal capillaries of the placental villi is estimated at around 360 ml min^{-1} (roughly half of the fetal cardiac output). The total volume of blood in the capillaries is about 45 ml so the blood in the fetal placental compartment is exchanged about eight times each minute. In summary, the maternal blood spaces represent a large volume of well-mixed blood, with a moderate turnover, while the fetal capillaries have a much smaller volume but a greater turnover.

This pattern of blood flow emphasizes the dialysis arrangement of placental blood flow, which, by maximizing concentration gradients, optimizes the conditions for passive exchange of solutes. By way of illustration, consider the diffusion of a solute from maternal to fetal blood. As the fetal blood flow is very high, solute diffusing into the fetal blood from the maternal side will be removed from the placenta rapidly, keeping its concentration low in the fetal capillaries. The maternal blood, however, has a much larger volume, so, despite its lower flow, will not become depleted of solute readily. By this arrangement, the concentration gradient for the solute between the maternal and fetal blood is maintained to ensure efficient diffusion. The transport of some specific solutes across the placenta will be considered briefly in the following sections.

Gas exchange across the placenta occurs by diffusion

The placenta performs the vital function of providing the fetus with the oxygen it requires for aerobic metabolism. It also removes the carbon dioxide generated by metabolizing cells. Measurements of the gas tensions on either side of the placenta show that, although oxygen diffuses passively from maternal to fetal blood, full equilibration is not achieved (Figure 29.7). There are two important reasons for this: firstly, not all of the maternal blood is in direct contact with the villi (which is the area of gas exchange) so that 'shunts' exist, which are analogous with a ventilation/perfusion mismatch in the lungs (see p. 449). Secondly, the placental tissue itself, which is highly metabolically active, uses around 20% of the oxygen in the maternal blood before it reaches the fetal capillaries. The placenta is therefore a less efficient organ of gas exchange than the lung. It is, however, able to satisfy the oxygen demand of the fetus because of a variety of specific adaptations that ensure that the transfer of oxygen to the fetal tissues is maximized (see below).

In addition to supplying oxygen to the fetus, the placenta is also responsible for the removal of fetal carbon dioxide. As for oxygen, the passive diffusion of carbon dioxide depends on blood flow and diffusion gradients, but the placental barrier is more permeable to carbon dioxide than it is to oxygen, and exchange is virtually complete.

Placental exchange of glucose and amino acids is carrier mediated

The placenta is the sole source of nutrients essential for the growth of the fetus. The most important of these is **glucose**. Glucose is rather insoluble in lipids and so cannot simply diffuse across the placenta. Instead, it travels from the maternal bloodstream to the fetal bloodstream by a process called facilitated diffusion (see Chapter 5 for further discussion). This mechanism is mediated by a carrier protein located in the membrane of the cells of the syncytiotrophoblast. Unless the carrier is saturated, fetal plasma glucose levels will be in equilibrium with those of the maternal blood. While this is normally desirable, if the mother is diabetic and has hyperglycaemia, the fetal plasma glucose concentration will also be elevated. This can result in over-nourishment of the fetus with the result that babies of diabetic mothers are often big for their gestational age (see below).

Amino acids are needed by the fetus to support the high rate of protein synthesis that occurs during development. Amino acids are transported actively from mother to fetus by specific transporters located in the placenta. The fetus receives most of its lipid in the form of **free fatty acids**, which pass readily across the placental barrier.

Fetal waste products are excreted via the placenta

The waste products generated by fetal metabolism cross the placenta to enter the maternal circulation, before being excreted along with those of the mother herself. **Urea** is the waste product of protein metabolism and is an important fetal waste material. Of the amino acids that

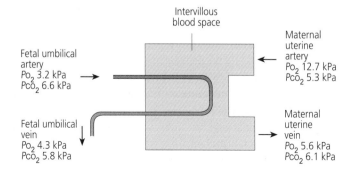

Figure 29.7 The dialysis organization of the placenta with typical blood gas values. Note that the maternal and fetal circulations are separate and that the blood leaving the placenta via the umbilical vein is not fully equilibrated with the maternal blood.

enter the fetal circulation from the mother, around 40% are deaminated and form urea, which must be excreted. Urea diffuses passively down its concentration gradient, from fetal to maternal blood.

Another important fetal waste product is **bilirubin**, which is formed by the breakdown of haemoglobin. In the adult, bilirubin is normally conjugated by enzymes in the liver to bilirubin diglucuronide, which is water soluble and easily excreted. However, the fetal liver is immature and does not possess sufficient quantities of hepatic enzymes to process the bilirubin produced by red cell breakdown. Instead, the bilirubin diffuses across the placenta into the mother's bloodstream from where it is carried to the liver for conjugation and excretion via the bile.

> **KEY POINTS:**
>
> - During fetal life, the placenta carries out the functions normally performed in the adult by the lungs, kidneys and gastrointestinal tract. The dialysis pattern of fetal and maternal blood flow ensures efficient exchange of substances across the placenta.
>
> - Oxygen, carbon dioxide and fatty acids cross the placenta by means of passive diffusion. Fetal waste products such as urea and bilirubin diffuse from fetal to maternal blood down their concentration gradients.
>
> - Glucose and amino acids cross the placenta by carrier-mediated transport.

29.6   The placenta is an important endocrine organ

The placenta secretes a number of peptide and steroid hormones that are crucial for the maintenance of pregnancy and for the preparation of the mother's body for labour and lactation. The major peptide hormones are:

- Human chorionic gonadotrophin (hCG)
- Human placental lactogen (hPL).

The major placental steroids are:

- Progesterone
- Oestrogens.

The physiological role of each of these hormones will be considered briefly.

Human chorionic gonadotrophin (hCG)

This hormone is secreted by the trophoblast of the embryo from only a few days after conception. The pattern of hCG secretion is illustrated in Figure 29.3. As discussed earlier, hCG is powerfully luteotrophic hormone and prevents regression of the corpus luteum so that it continues to secrete the progesterone needed to prevent endometrial shedding and inhibit myometrial contractions. By about 6 weeks after fertilization, the placenta is established and takes over as the main source of progesterone. The pregnancy is now said to have become autonomous and progesterone from the corpus luteum is no longer needed. hCG output declines sharply at this time although it continues to be secreted in small amounts throughout the entire pregnancy.

Human placental lactogen (hPL)

As Figure 29.3 shows, hPL first appears in the maternal circulation at around the time that hCG levels are starting to fall (around 8 weeks of gestation). hPL concentrations then rise progressively to reach a peak at around week 35. This hormone is secreted by cells of the syncytiotrophoblast and is secreted preferentially into the maternal blood. Its principal action is to encourage proliferation of the breast tissue in preparation for lactation following delivery. In addition to this **mammotrophic** action, hPL also exerts some important metabolic effects whose net effect is to adjust maternal levels of metabolites so that placental exchange is favoured without serious depletion of the maternal blood. Monitoring of hPL levels during pregnancy provides a valuable indication of placental sufficiency.

Progesterone

Adequate amounts of progesterone are essential if a pregnancy is to proceed successfully. In the first weeks of gestation, this progesterone is supplied by the corpus luteum, under the luteotrophic influence of hCG. By about 8 weeks, however, the placenta has become well established and has begun to secrete large amounts of progesterone. The pattern of progesterone secretion by the placenta is illustrated in Figure 29.8. By late gestation, progesterone is produced at a rate of 250–350 mg day^{-1}, which is about 15 times higher than the rate of secretion seen during the luteal phase of the menstrual cycle.

Exactly why progesterone is so important during pregnancy is not clear, but one of its more important actions appears to be the maintenance of a quiescent myometrium and the prevention of endometrial shedding prior to placentation. It also plays a role in the stimulation of breast development in preparation for lactation.

Placental oestrogens

The placenta secretes large amounts of a number of oestrogenic hormones including oestradiol-17β, oestriol and oestrone. Their secretory patterns are shown in Figure 29.8. During pregnancy, the plasma concentrations of these hormones are in excess of 100 times those seen at the peak of follicular development during the menstrual cycle. Oestrogens seem to have a role in preparing the body for giving birth and for lactation. They bring about relaxation of the symphysis pubis and act with hPL to stimulate proliferation of the breast tissue. Oestrogens may also play a part in the initiation of parturition (see below).

> **KEY POINTS:**
>
> - The placenta secretes a wide variety of peptide and steroid hormones.
>
> - The major peptide hormones are human chorionic gonadotrophin (hCG) and human placental lactogen (hPL). hCG prevents luteal regression and hPL exerts important metabolic effects in the mother. It also contributes to preparation of the breast for lactation.
>
> - Steroid hormones are secreted in large amounts by the placenta. Progesterone is essential for the maintenance of pregnancy.

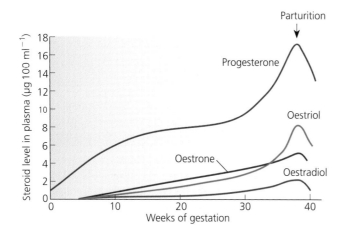

Figure 29.8 The plasma levels of various steroid hormones during pregnancy. Note that progesterone secretion by the placenta dominates the period of gestation, falling only after parturition.

29.7   Delivery of the infant – parturition

Under normal conditions, the duration of the gestation period is remarkably constant, with parturition occurring 38 weeks after conception (or 40 weeks after the start of the last menstrual period). This suggests that the onset of labour is a well-co-ordinated event. Much research effort has been directed towards discovering the nature of the signals that bring pregnancy to an end and it now seems likely that there is no single trigger for the initiation of parturition but rather a combination of different factors both physical and endocrine. Intriguingly, the fetus itself appears to play a part in the initiation of its own birth.

Two important **physical factors** are thought to contribute to the onset of parturition:

- Stretching of the myometrium. As the fetus grows during gestation, the uterine muscle is stretched more and more. As it stretches, it thins and its excitability increases until eventually spontaneous contractions occur, which contribute to the onset of labour. Preliminary contractions may be experienced for several weeks prior to the onset of true labour. These are called **Braxton–Hicks contractions**.

- Placental insufficiency. As pregnancy progresses, the demands of the fetus eventually outstrip the ability of the placenta to meet those demands. The fetal capillaries tend to become blocked with debris and placental efficiency falls. This may also contribute to the onset of labour.

In addition to these physical factors, a variety of endocrine factors are believed to play a part in triggering parturition. These involve both the maternal and fetal endocrine systems. Throughout most of gestation, placental progesterone secretion exceeds that of oestrogens and maintains the myometrium in the relaxed state that is essential for a successful pregnancy. Oestrogens, however, enhance the excitability of the myometrium and also increase its sensitivity to other substances, such as histamine, oxytocin and prostaglandins, all of which are known to stimulate contractions of the uterus. Close to full term, there is a change in the pattern of steroid secretion by the placenta in which levels of progesterone fall and those of the oestrogens rise. Oestrogenic hormones thus begin to dominate the hormonal profile and the 'progesterone block' to myometrial contraction is removed. Oestrogens and other agents, such as prostaglandin $F_{2\alpha}$ and oxytocin (from the posterior pituitary), may then increase the contractility of the myometrium still further, allowing contractions to become more and more powerful until they bring about delivery of the infant.

What causes this shift from progesterone to oestrogen secretion?

Observations made in animals and, more recently, humans strongly suggest that cortisol, secreted by the fetal adrenal gland, may play a key role in the initiation of parturition. It is known that cortisol stimulates the conversion of progesterone to oestrogens in the placenta, and just prior to term there is an increase in the secretion of adrenocorticotrophic hormone (ACTH) from the fetal pituitary gland, which stimulates the secretion of fetal cortisol. This hormone is thought to switch placental steroid secretion in favour of oestrogens rather than progesterone. Anencephalic fetuses, in whom the brain and pituitary gland are severely malformed, are often born post-mature, that is after the normal gestation time, presumably because the switch to oestrogen production fails to occur or is delayed. It must be emphasized that these ideas are highly speculative and that the precise neural and hormonal interactions that occur as pregnancy nears term are far from clear, especially in humans.

> **KEY POINTS:**
>
> - Parturition is a complex process involving both the maternal and fetal nervous and endocrine systems.
>
> - Fetal cortisol appears to initiate a switch in the placenta away from progesterone production, to allow oestrogens to dominate in the last few days of pregnancy.
>
> - Oestrogens and other agents, such as oxytocin, then increase the contractility of the myometrium to bring about delivery of the baby.

29.8 Important changes to maternal physiology in pregnancy

Pregnancy (or gestation) normally lasts for 38 weeks from conception. It is conventional, however, to begin counting the days of pregnancy from the first day of the last menstrual period. Calculated in this way, pregnancy lasts for 40 weeks (280 days). The gestation period is generally divided into three terms or **trimesters**, each lasting for about 3 months. Full-term is achieved 40 weeks after the first day of the last menstrual period.

During the months of pregnancy, the anatomy, physiology and metabolism of a pregnant woman's body undergo a number of significant changes. These are designed to meet the considerable metabolic demands of the fetus, to create favourable conditions in which the fetus can grow and to prepare the mother's body for delivery and breastfeeding. While a detailed account of maternal physiology is beyond the scope of this book, the following will provide a brief description of some of the more important anatomical and physiological changes that take place within the body of a pregnant woman.

Anatomical changes associated with the growing fetus

The uterus shows an astonishing degree of enlargement as the fetus grows during development. A non-pregnant uterus is about the size of a pear, but by 16 weeks of gestation it has grown to occupy most of the pelvic cavity. As pregnancy progress, the uterus continues to enlarge so that by full term, it occupies almost the entire abdomen. Other abdominal organs are displaced and compressed against the diaphragm, which is itself pushed up into the thoracic cavity causing the ribs to move outwards and the thorax to widen.

Figure 29.9 illustrates the growth of the fetus and the space occupied by the uterus within the abdomen during gestation. The changing body shape and the extra mass associated with pregnancy shift the trunk's centre of mass anterior to the hips and thus alter the mechanical demands placed upon the structures of the musculoskeletal system. Many women show a compensatory increase in the curvature of the spine (lordosis) and this can be a cause of lower back pain, especially in late pregnancy.

Cardiovascular changes in pregnancy

During pregnancy, there are some characteristic changes to a number of cardiovascular parameters including:

- Cardiac output
- Blood volume and composition
- Arterial blood pressure.

Between weeks 6 and 32 of pregnancy, maternal **cardiac output** increases by 30–50%. Typically, this represents a rise from about 4.5 to 6.0 l min^{-1}. Increases in both heart rate (from 70 b.p.m. to 80 or 90 b.p.m.) and stroke volume (around 10%) contribute to this rise. From mid-pregnancy, the enlarged uterus starts to compress the inferior vena cava and the lower aorta (so-called aortocaval compression), particularly when the woman is lying supine. Obstruction of the inferior vena cava reduces venous return to the heart with the result that cardiac output may be reduced in late pregnancy (see Chapter 19 for further explanation of the relationship between venous return and cardiac output). The increased cardiac output occurring during gestation supplies the increasing demands of the uterus and the placenta for nutrients and ensures the efficient removal of waste products across the placenta.

Changes in **plasma volume** and **red cell mass** occurring during pregnancy are illustrated in Figure 29.10. Plasma volume increases by 40–50% so that total circulating blood volume is raised. The red cell mass increases by only 20–30%, so that there is dilution of the blood (haemodilution) and a fall in the haemoglobin content. Published figures vary but a typical drop in haemoglobin content might be from 13.3 to 12.1 g dl^{-1} of blood. If haemoglobin content should fall below this value, supplemental iron and folic acid may be required.

Leukocyte counts tend to be somewhat variable during pregnancy, but usually remain within the upper limits of normal (see Table 20.1). Platelet numbers vary little but

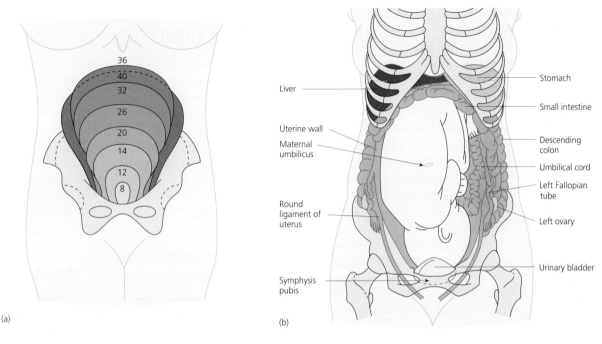

(a)

(b)

Figure 29.9 (a) The height of the fundus of the uterus throughout gestation. Heights vary considerably between women but a general rule is that by 20 weeks the fundus is usually around the height of the umbilicus. (b) The position and space occupied by the fetus at full term. Note the displacement of the gastrointestinal structures and the pressure on the bladder and diaphragm.

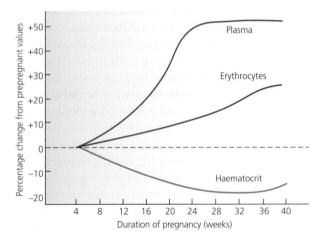

Figure 29.10 Changes in plasma volume, red blood cell (erythrocyte) volume and haematocrit during pregnancy.

may fall slightly. Pregnancy is sometimes regarded as a 'hypercoaguable state' because levels of fibrinogen and clotting factors VII, X and XII increase markedly. This may help to prevent excessive blood loss at delivery but increases the risk of deep vein thrombosis.

Systemic arterial blood pressure is never significantly increased during a normal pregnancy. In fact, despite the increased cardiac output described above, arterial blood pressure often falls from the prepregnant value during the second trimester. This is because the placental circulation is expanding and peripheral vascular resistance falls (see Chapter 19). A typical fall in midpregnancy would be 5–10 mmHg (0.66–1.32 kPa) for systolic pressure and 10–15 mmHg (1.32–1.98 kPa) for diastolic pressure.

Blood pressure is usually monitored very carefully during the third trimester as any significant increase, particularly if seen in conjunction with oedema or the urinary excretion of protein (proteinuria), may indicate a risk of **pre-eclampsia**, a potentially life-threatening condition for both mother and fetus.

Changes to the respiratory system in pregnancy

Most of the changes in respiratory function that occur during pregnancy can be attributed to the positional changes associated with increasing uterine size, or to the effects of placental progesterone. Arterial blood gas levels alter little, if at all, throughout pregnancy. During late pregnancy, the uterus causes an elevation of the diaphragm of about 4 cm. At the same time, the ribs flare and the chest widens. The resulting increase in thoracic circumference largely compensates for the displacement of the diaphragm and, consequently, total lung capacity falls only very slightly (by about 5%) towards the end of gestation. Functional residual capacity, expiratory reserve volume and residual volume all fall by about 20% from around the middle of the second trimester, probably because of the obstruction caused by the uterus as it pushes up under the diaphragm.

Hormonal effects on the vasculature of the respiratory mucosa lead to capillary engorgement and swelling of the epithelial lining of the nose, oropharynx, larynx and trachea. Symptoms of nasal congestion, voice change and increased susceptibility to nosebleeds are often reported by pregnant women.

Changes to maternal renal function and body fluid balance

Renal plasma flow and glomerular filtration rate (GFR) increase progressively during the first trimester in parallel with the increases in cardiac output and blood volume described above. Peak GFR values occur around weeks 16–24 of gestation. As renal function increases, the concentrations of both urea and creatinine in the maternal plasma fall from their prepregnancy values of 4–7 mmol l^{-1} and 100 mmol l^{-1} to around 3.6 mmol l^{-1} and 60 mmol l^{-1}, respectively.

In response to oestrogens secreted by the placenta, there is an increase in the secretion of renin from granulosa cells of the afferent arteriole (see Chapter 23, p. 480). Renin elicits a rise in angiotensin levels and angiotensin, in turn, stimulates the secretion of aldosterone from the adrenal cortex. By the end of pregnancy, aldosterone secretion may be as much as six to eight times higher than in the non-pregnant state. Aldosterone stimulates the reabsorption of salt and water from the distal tubular fluid, more than balancing out the increase in filtered water and salt resulting from the increased GFR. Consequently, there is a small overall retention of salt and water. Indeed, total body water increases by 6–8 litres in a normal pregnancy, contributing a significant fraction of the overall weight gain.

From week 12 of gestation, placental progesterone relaxes the smooth muscle of the ureters and causes the ureters and renal calyces to dilate. As the uterus enlarges progressively, it may also compress the ureters and cause further dilatation by obstructing the flow of urine. These changes increase the risk of urinary tract infections during pregnancy.

Many pregnant women experience an increase in the frequency of micturition. There are several reasons for this. In the first and second trimesters of pregnancy, the increased urinary flow rate probably reflects increased renal function. In later weeks, as the fetus increases in both size and activity, there is likely to be increased pressure on the bladder causing urgency and discomfort.

Changes to maternal gastrointestinal function

The changes to gastrointestinal function that take place throughout pregnancy are largely of minor significance. Most aspects of gastrointestinal function are reduced or slowed, principally as a result of relaxation of the smooth muscle of the gut wall in response to the elevated plasma levels of progesterone. There is an overall decrease in gastrointestinal motility, for example, which may cause constipation, a common complaint in pregnancy. Relaxation of the lower oesophageal sphincter may result in gastro-oesophageal reflux and heartburn, particularly at night when the weight of the fetus presses down on the diaphragm and intra-abdominal organs.

Most women experience nausea and vomiting during pregnancy. Typically, this is most severe in the first 12–14 weeks, after which symptoms often become less severe and may disappear completely. Although the cause of sickness is not clear, its onset and severity does seem to parallel the rising levels of hCG in the bloodstream.

In rare cases, the nausea and vomiting are extremely severe and persist throughout pregnancy. This condition is called **hyperemesis gravidarum** and can lead to significant changes in body fluid balance and plasma pH. The most severe cases will require hospital admission.

Changes to endocrine function in pregnancy

Pregnancy alters the function of many endocrine glands. In addition to the principal steroid and peptide hormones discussed earlier (pp. 618–619), the placenta also secretes a number of other peptide hormones, some of which have actions similar to those of the anterior pituitary hormones. Such hormones include:

- **Human chorionic thyrotropin**: this may stimulate the secretion of thyroxine from the thyroid gland and can occasionally lead to symptoms of hyperthyroidism – tachycardia, anxiety and excessive sweating.
- A hormone with **ACTH-like activity** (see Chapter 16): this stimulates the secretion of adrenal cortical hormones and melanocyte-stimulating hormone (MSH). The latter causes increased pigmentation of the skin and probably explains why some pregnant women develop blotchy, brown pigmentation on their face – the so-called 'mask of pregnancy' (or melasma).

Secretion of growth hormone and the anterior pituitary gonadotrophins FSH and LH are decreased during pregnancy, as a consequence of the negative feedback actions of placental hPL (on growth hormone) and the placental sex steroids (on FSH and LH). By contrast, prolactin secretion shows an eightfold increase by the end of gestation. This hormone contributes to the preparatory changes that take place in the mammary glands in readiness for milk production.

Relaxin is a large polypeptide hormone secreted by the corpus luteum and the decidual tissue of the uterus during pregnancy. It appears to ripen the cervix and to relax the pelvic ligaments and the symphysis pubis to accommodate the expanding uterus and to prepare for childbirth.

The fetus represents a significant drain on maternal stores of calcium, largely because of the demands of its developing skeleton. To prevent any fall in maternal plasma calcium, there is normally an increase in the rate of secretion of **parathyroid hormone** (**PTH**) during gestation. As described in Chapter 16, PTH reduces the urinary loss of calcium. It also augments plasma levels of calcitriol and this, in turn, stimulates the intestinal absorption of calcium.

Changes to insulin sensitivity during pregnancy

In early pregnancy, maternal tissues typically show an increased sensitivity to insulin. Consequently, plasma glucose may fall slightly. Later in gestation, insulin sensitivity falls and plasma glucose may rise. These changes reflect the initial increase in the requirement for glucose of the maternal tissues and the later needs of the developing fetal tissues. Mild glycosuria is often seen in normal pregnancy and may be explained by the combination of raised plasma glucose, increased GFR (see above) and a reduction in the tubular reabsorption of glucose.

Gestational diabetes occurs in between 1 and 3% of pregnancies. In this condition, carbohydrate intolerance of varying severity develops during the pregnancy and often (although not invariably) resolves once the baby has been delivered. Women who develop this form of diabetes are unable to respond effectively to the metabolic stress of pregnancy. Furthermore, hPL may increase the risk of developing gestational diabetes in susceptible women. The fetus of a diabetic mother with hyperglycaemia will also be hyperglycaemic as glucose is in equilibrium between the fetal and maternal sides of the placental circulation. Not surprisingly, therefore, in many cases, the fetus has a higher-than-normal weight for dates (which can pose difficulties during labour and delivery). Also associated with maternal gestational diabetes is an increased risk of fetal respiratory distress due to retarded lung maturation and surfactant production. The incidence of fetal abnormalities is also increased. Early recognition of gestational diabetes and good maternal plasma glucose control are therefore vital if such problems are to be minimized. Glucose-tolerance tests are routine in many antenatal clinics.

Alterations in maternal metabolism

During early pregnancy, the mother's body prepares for the metabolic demands to come. About 3 kg of fat is laid down during the first 20 weeks or so, which will provide a store of energy for the final trimester during which fetal growth is particularly rapid. At the same time, there is an increase in the rate of maternal protein synthesis, the net effect of which is to stimulate growth of the uterus, breasts and essential musculature of the pregnant woman. From about week 6 of gestation, there is an increase in the sensitivity of maternal tissues to insulin. Because of this, carbohydrate loads are more readily assimilated. During the second half of gestation, the fetal requirements for oxygen and nutrients increase steeply. From now on, the mother's metabolism enters a state sometimes called 'accelerated starvation' in order to meet these increased needs.

Glucose is the chief metabolic fuel for the developing fetus, which, at full-term, uses up to 25 g of glucose each day. In late pregnancy, maternal tissues show a fall in insulin sensitivity. This ensures that the maternal plasma glucose concentration is adequate to maintain a constant supply of glucose both for transfer across the placenta and for the needs of the mother's CNS. At the same time, mobilization of lipids is facilitated, possibly by hPL, to ensure an alternative source of energy for the mother. During late gestation, the mother's body requires a small quantity of additional protein (no more than 6–10 g day $^{-1}$), mainly to provide the substrate for the synthesis of fetal protein. By the time of its birth, the fetus will normally have accumulated 400–500 g of protein. In most cases, a normal diet is adequate to supply the additional protein requirements of pregnancy.

> **KEY POINTS:**
>
> - All the maternal body systems show certain changes throughout pregnancy. Some create favourable conditions for fetal development while others prepare the mother's body for delivery and lactation.
> - Maternal cardiac output increases to satisfy the increasing demands of the uterus and placenta for nutrients and waste disposal.
> - Changes in maternal respiratory function are largely due to the positional changes associated with increasing uterine size and with compression of the diaphragm.
> - Glomerular filtration rate increases during pregnancy. Total body water increases by 6–8 litres.
> - Pregnancy alters the function of most endocrine glands.

29.9 Nutritional requirements of pregnancy

Adequate nutrition, both prior to conception and during pregnancy itself, is essential in order to ensure that optimal intrauterine conditions are maintained for fetal development and that the health of the mother is not compromised by the additional metabolic and nutritional demands of the growing fetus. Poor maternal nutrition is associated with an increased risk of low birth weight, fetal abnormalities and neonatal mortality.

Most women carrying a single fetus will gain between 7 and 14 kg in weight during their pregnancy. Table 29.1 shows how this weight gain is distributed for a typical pregnancy with a weight increase of 12 kg.

In order to supply the demands of both mother and fetus, an increased daily energy intake of about 1050–1250 kJ (250–300 kcal) is required. This will normally represent an increase of about 15% in energy requirement compared with the prepregnant state. Many women report an increase in appetite while they are pregnant, which may be due to the high circulating level of progesterone.

Maternal requirement for micronutrients during pregnancy

Table 31.2 gives figures for the typical adult daily requirements for vitamins and minerals (so-called micronutrients). In pregnancy, there is a small increase in the body's requirement for certain of these micronutrients. Average daily recommendations for a pregnant woman are given in Table 29.2.

Folic acid is now recognized as a very important dietary requirement of pregnancy. It is needed for normal cell proliferation, and maternal deficiency has been linked to an increased risk of neural tube defects in the fetus. These are abnormalities in the closure of the neural tube, which normally occurs between weeks 3 and 6 of gestation. Neural tube defects include **anencephaly** (in which the cerebral hemispheres fail to develop normally) and **spina bifida**, in which there is defective fusion of the vertebral arches, most often in the lumbar region. Women are encouraged to supplement their folic acid intake both in the weeks prior to conception and throughout

Table 29.1 The distribution of maternal weight gain at 38 weeks of gestation

	Weight (kg)
Fetus	3.3–3.5
Placenta	0.65
Additional blood volume	1.3
Amniotic fluid	0.8
Weight gain of uterus and breasts	1.3
Additional fluid retention and fat deposits	4.2–6.0
Total	11.5–13.5

pregnancy, by eating leafy green vegetables, wheat grains and legumes.

Vitamin A is essential for normal fetal development but there is some concern that a high daily intake of this vitamin may be teratogenic (i.e. it may cause congenital malformations of the fetus). A vitamin intake of around 500 µg per day is recommended for adult females and this requirement remains unchanged in pregnancy.

Calcium is a very important mineral in pregnancy, particularly for women under 25 years of age in whom the skeletal mass is still increasing. During the third trimester, the fetus needs about 0.3 g of calcium each day to permit calcification of the skeleton. All of this must

Table 29.2 Recommended daily protein and micronutrient intake in pregnancy

Nutrient	USA	Canada	UK	Australia
Protein (g)	60	75	51	51
Vitamin A (mg)	800	800	700	750
Vitamin D (mg)	10	5	10	Not set
Vitamin E (mg)	10	8	Not set	7
Vitamin C (mg)	70	40	50	60
Vitamin B_1 (thiamin) (mg)	1.5	0.9	0.9	1.0
Vitamin B_2 (riboflavin) (mg)	1.6	1.3	1.4	1.5
Vitamin B_3 (niacin or nicotinic acid) (mg)	17	16	13	15
Folate (mg)	400	385	300	400
Vitamin B_{12} (mg)	2.2	1.2	1.5	3.0
Calcium (g)	1.2	1.2	0.7	1.1
Magnesium (g)	0.32	0.25	0.27	0.30
Iron (mg)	30	13	14.8	22–36
Zinc (mg)	15	15	7	16
Iodine (mg)	175	185	140	150

come from maternal reserves. Although increased secretion of PTH helps to maintain plasma calcium levels (see above), a pregnant woman must still increase her intake of calcium by about 70%, to around 1200 mg per day. If she does not, the fetus will draw on calcium stored in its mother's skeleton. This may increase her susceptibility to osteoporosis in later life, particularly if repeated in several pregnancies.

Iron is of great importance during pregnancy. The fetus and placenta use about 300 mg of iron during a typical pregnancy, while the increased red blood cell mass of the mother requires a further 500 mg. Provided that the mother's diet contains sufficient amounts of meat and vitamin C (which enhances the absorption of iron in the gut), this extra demand for iron can normally be met by maternal stores. If, however, absorption of dietary iron falls short of the demands of pregnancy, maternal iron stores will become depleted and iron-deficiency anaemia may develop. This can be avoided by regular monitoring of the maternal haemoglobin content so that iron supplements can be given if necessary.

Zinc plays an important part in a number of metabolic processes, including the synthesis of proteins and nucleic acids. It is also essential for the synthesis and activity of insulin. Not surprisingly, therefore, there is a small increase in the requirement for this trace element during pregnancy. In most women, a normal mixed diet containing meat and fish will meet this demand but strict vegans may be unable to absorb enough zinc from their food. Women taking iron supplements may also become deficient in zinc because iron:zinc ratios greater than 3:1 have been shown to interfere with the intestinal absorption of zinc.

> **KEY POINTS:**
>
> - To meet the demands of pregnancy, an increased daily energy intake of about 1050–1250 kJ (250–300 kcal) is required.
> - During pregnancy, there is a small increase in the mother's requirement for a variety of micronutrients, in particular calcium, folic acid and iron.

29.10 Some important aspects of fetal physiology

The fetus is totally dependent upon the placenta for gas exchange, nutrition and waste disposal and is adapted in a number of important ways to life within a fluid-filled bag. The following sections are intended to provide a brief introduction to fetal physiology. Some of the important physiological adaptations that take place at or soon after birth, to enable the baby to make the transition from its uterine existence to a semi-independent, air-breathing life, will also be considered. The special thermoregulatory problems of the newborn are considered in Chapter 33.

This section will focus on the following aspects of fetal and neonatal physiology:

- How and why the fetal cardiovascular system differs from that of the adult
- The first breath and the cardiovascular changes that follow the initiation of breathing

- Respiratory characteristics of the neonate
- Differences between the fetal, neonatal and adult gut, kidneys and adrenal glands
- Differentiation of the male and female reproductive tracts.

Organization of the fetal circulation

The fetal circulation is arranged to make the best use of a relatively poor oxygen supply and to bypass organs such as the lungs and liver, which are virtually non-functional. There are, therefore, a number of differences between the fetal circulation and that of the adult.

The fetal heartbeat is detectable by week 4–5 of gestation and the fetal cardiovascular system is fully developed (in miniature) by week 11. Figure 29.11 shows the organization of the fetal cardiovascular system and

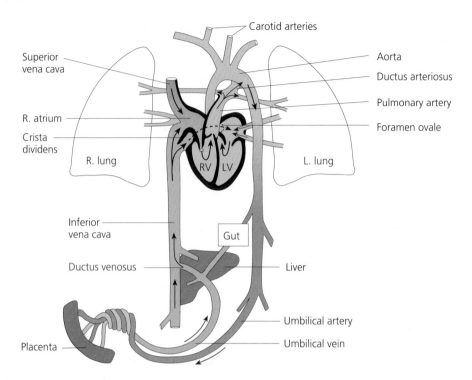

Figure 29.11 The organization of the fetal cardiovascular system. Note that the structure of the heart has been distorted to show the operation of the crista dividens. The broken arrow indicates the direct flow of blood from the inferior vena cava through the foramen ovale to the left atrium.

illustrates the three important **shunts** (or 'shortcuts') that differentiate it from the adult circulation. These are:

- The **foramen ovale** – a gap between the right and left atria formed by incomplete fusion of the septum. In the adult, there is no mixing of blood between the two sides of the heart, but in the fetus, blood can pass directly between the inferior vena cava and the left atrium. The upper portion of the foramen ovale forms the crista dividens, a flap that functionally separates the entrance to the right atrium into an upper and a lower portion. This can be seen in Figure 29.12.

- The **ductus arteriosus** – a direct link between the pulmonary artery and the aorta.

- The **ductus venosus** – a link between the umbilical vein and the inferior vena cava, which allows blood to bypass the fetal liver.

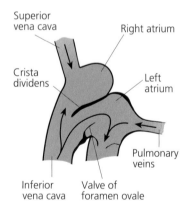

Figure 29.12 The pattern of blood flow in the right atrium of the fetal heart. Note how the crista dividens effectively splits the stream of blood in the inferior vena cava to divert the majority of the oxygenated blood through the foramen ovale into the left atrium. Only a small proportion of the oxygenated blood in the inferior vena cava enters the right atrium.

These shunts allow the two sides of the fetal heart to work in parallel, with mixing of the right and left ventricular inputs. This is in contrast to the adult pattern in which the pulmonary and systemic circulations are perfused entirely separately (Figure 29.13). What follows is a description of the route taken by blood as it flows around the fetal circulation. It will be helpful to refer to Figure 29.11 at the same time.

The fetus receives oxygenated blood from the placenta via the umbilical veins. A significant fraction of this blood passes directly into the inferior vena cava via the ductus venosus, thus bypassing the immature fetal liver. Only a small amount of blood enters the liver, via the portal vein, but this is sufficient to support hepatic growth and development. This arrangement means that the oxygenated blood arriving from the placenta is mixed almost immediately with deoxygenated blood returning from the lower parts of the fetal body in the inferior vena cava.

In the adult, blood in the inferior vena cava mixes with blood in the superior vena cava as it enters the right atrium. If this also happened in the fetus, the oxygenated blood from the umbilical veins, already mixed with deoxygenated blood in the inferior vena cava, would be further diluted by deoxygenated blood from the superior vena cava. This is prevented by the action of the crista dividens, which essentially splits the bloodstream entering the right atrium and allows most of the blood in the inferior vena cava to pass directly to the left atrium via the foramen ovale. Only a small amount of the blood from the inferior vena cava enters the right atrium, but all of the blood from the superior vena cava does so.

Blood from the left ventricle (a mixture of oxygenated and deoxygenated blood) is pumped to the ascending aorta. Deoxygenated blood in the right ventricle is pumped into the pulmonary artery. However, soon after the start of the pulmonary artery, the ductus arteriosus branches off to provide a direct link between the pulmonary artery and the aorta. At this point in the circulation, therefore, blood may take one of two routes. It may either travel to the fetal lungs via the pulmonary arterial vessels or it may flow directly into the aorta via the ductus arteriosus. Because the fetal lungs are collapsed and fluid filled and they are not carrying out gas exchange, there is a high resistance to blood flow in the pulmonary capillaries.

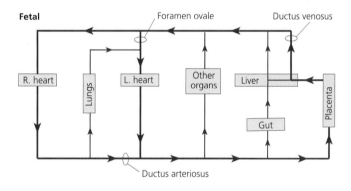

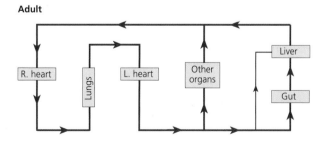

Figure 29.13 A comparison of the pattern of circulation in the fetus and adult. The three fetal shunts allow blood to bypass those organs that have little or no function.

Resistance in the ductus arteriosus and aorta is lower, so most of the blood from the right ventricle takes this route. In fact, about 80% passes directly to the aorta via the ductus arteriosus while only 20% goes to the lungs.

This arrangement means that blood in the descending aorta has a relatively low oxygen content. However, from Figure 29.11 it may be seen that the carotid arteries, which carry blood to the brain, branch off from the aorta before the point of entry of the ductus arteriosus. This is very important as it ensures that blood reaching the fetal brain has a higher oxygen content than that which perfuses other fetal tissues.

> **KEY POINT:**
> - The circulation of the fetus is adapted to make the best of a poor oxygen supply and to ensure that the fetal brain is comparatively well oxygenated.

Control of the fetal circulation

By the 11th week of gestation, the fetal heart is beating at around 160 beats per minute, which is much higher than the resting adult heart rate of around 70 beats per minute. At this stage, there is no autonomic innervation of the heart but, in the final trimester, parasympathetic innervation becomes established and the fetal heart rate falls to about 140 beats per minute. This gradual increase in autonomic control of the cardiovascular system is also evident in the changes in blood pressure that occur during fetal life.

Pressure is comparatively low – around 9/6 kPa (c. 70/45 mmHg) in the early months of gestation when there is very little peripheral vascular tone and therefore a low total peripheral resistance. Blood pressure gradually rises as autonomic activity becomes established and vascular tone is increased. At the same time, aortic and carotid baroreceptors begin to function. This increase in blood pressure and drop in heart rate continues after delivery, until by the age of about 7 years, values similar to those of the adult are achieved.

The fetal respiratory system

As far as gas exchange is concerned, the fetal lungs are non-functional and the alveoli are almost collapsed and filled with fluid. This fluid is secreted by type I alveolar cells (epithelial cells that overlie the pulmonary capillaries) and its composition differs from that of the amniotic fluid. It first appears around mid-gestation, and by full term the lungs contain a total of about 40 ml of this fluid. Because the alveoli are collapsed, their capillaries are tortuous and offer a high resistance to blood flow. Consequently, the lungs are relatively poorly perfused.

Breathing movements develop before birth

Although the fetal lungs do not participate in gas exchange, breathing movements do occur during gestation. Ultrasound scans have revealed that these breathing movements begin at around the 10th week of gestation. They remain shallow and irregular up until around week 34 of gestation, after which time they start to display a more rhythmical pattern, with periods of activity interspersed with periods when movements are absent. Occasionally, gasping movements are seen, especially if the fetus experiences hypercapnia – for example as a result of placental insufficiency or compression of the umbilical cord. This response suggests that chemoreceptors (see Chapter 22) are functional during the latter part of gestation. It is now believed that fetal breathing movements are important in the preparation of the respiratory system for its postnatal function of gas exchange.

Fetal hiccups are also common, particularly during the later stages of gestation. Many pregnant women notice distinct episodic movements of their unborn babies, which may go on for several minutes at a time. Non-invasive measurements of fetal diaphragmatic activity have demonstrated that these are due to spasmodic contractions of the diaphragm, similar to those that cause hiccups in children and adults.

> **KEY POINTS:**
> - The fetal circulation is organized so that the two sides of the heart work in parallel. Three shunts permit blood to bypass organs with little or no function.
> - The fetal heart rate is high and the blood pressure relatively low.
> - The fetal lungs are collapsed and fluid filled. Pulmonary vascular resistance is high so pulmonary blood flow is only 20% of the right ventricular output. The remaining 80% passes directly to the aorta through the ductus arteriosus.

29.11 Respiratory and cardiovascular changes at birth

When a baby is born, it is separated from the placenta, which has acted as the site of gas exchange throughout gestation. It is vital therefore that the baby starts to breathe independently. The establishment of air breathing is accompanied by closure of the fetal cardiovascular shunts so that the circulation adapts to pulmonary gas exchange.

The first breath

The mechanisms responsible for triggering a baby's first breath are not known with certainty but a number of possible factors may contribute. These include:

- The drop in ambient temperature experienced by the baby following delivery.
- Hypercapnia experienced during delivery, particularly if labour is difficult or prolonged. Measurements of the P_{CO_2} of scalp blood sampled during and just after delivery have revealed a marked increase, which may provide the neonate with an important stimulus to gasp for air.
- Increased sensory input, including tactile, auditory and visual stimuli, which may stimulate breathing movements.

In order to inflate its lungs for the first time, the newborn baby must overcome the enormous surface tension forces at the gas–liquid interface of the alveoli. This can only be achieved if sufficient **surfactant** is present in the lungs. At around week 20 of gestation, **type 2 epithelial cells** start to appear within the walls of the developing alveoli. Some weeks later (between weeks 26 and 30), these cells begin to secrete phospholipid surfactants. Surfactant molecules reduce the surface tension forces that oppose lung inflation (see p. 445) thus making it possible for the baby to take its first inspiration. Nevertheless, a considerable mechanical effort is still required by the newborn to inflate its lungs. Figure 29.14 illustrates the pressure–volume relationships for the neonatal lung during the first and subsequent breaths. A large negative pressure must be generated within the chest and this is brought about by strong contraction of the diaphragm. The ribs and sternum, which are very flexible at this stage, become slightly concave during the first few breaths. An equally large positive pressure must be generated to bring about expiration because lung compliance is relatively low (see Chapter 22).

After the first breath, the lungs do not empty completely but a little air remains in them after expiration

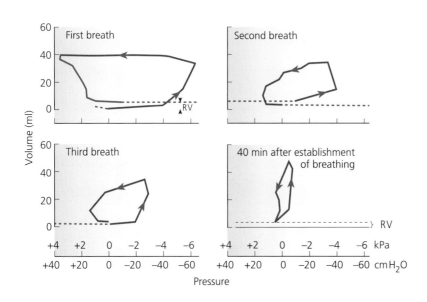

Figure 29.14 The lung pressure–volume relationships during the first, second and third breaths, together with that for breathing about 40 minutes after the first breath. Note that the residual volume (RV) is established with the first breath and that the compliance increases with subsequent breaths (i.e. the pressure change required for a given change in volume falls after the first breath).

to begin to establish the residual volume, which persists throughout life (see Chapter 22). As Figure 29.14 shows, subsequent breaths are achieved with less mechanical effort as the compliance of the newborn lungs increases.

Respiratory problems in the newborn

To breathe independently, a baby must have sufficient levels of surfactant in its alveoli. Babies born before about week 28 of gestation have few functioning type 2 cells and therefore have difficulty in overcoming the surface tension forces that oppose lung inflation. They will suffer respiratory distress and will need to be ventilated artificially until their lungs are sufficiently mature to permit independent breathing. Babies born to diabetic mothers often show delayed type 2 alveolar cell development and even those born after 30 weeks of gestation may suffer from respiratory distress. In some cases, administration of cortisol to the mother can accelerate the maturation of the fetal type 2 cells and prevent the occurrence of respiratory distress. Aerosol surfactants can also be administered to some newborn babies at risk of respiratory distress.

Characteristics of neonatal respiration and its control

Neonatal respiration differs in certain respects from that of the older child or adult. The principal differences are listed in Table 29.3, from which it is evident that the respiratory rate of the neonate is higher but more erratic than that of the adult, airway resistance is higher and lung compliance is lower. These differences mean that the overall work of breathing is greater in the newborn than in adults or older children.

The ventilatory response to hypercapnia is well developed in both the fetus and the neonate, with central medullary chemoreceptor activity present from about mid-gestation. In the fetus, the peripheral chemoreceptors have a very low level of activity but they start to respond to reductions in oxygen tension following delivery.

Table 29.3 Comparison of respiratory variables between the neonate and the adult

Variable		Neonate	Adult
Body weight (kg)		3.3	70
Ventilation rate (breaths min^{-1})		20–50	12–15
Minute volume (ml)		c. 500	c. 6500
Tidal volume (ml)		18	500
Vital capacity (ml)		120	4500
Surface area for gas exchange in lungs (m^2)		3	60
Compliance	($l\, kPa^{-1}$)	0.051	1.7
	($ml\, cmH_2O^{-1}$)	5	165
Bronchiole diameter (mm)		0.1	0.2
Oxygen diffusion capacity ($ml\, s^{-1} kPa^{-1}$)		0.6	6
	($ml\, min^{-1} mmHg^{-1}$)	2.5	25
Energy expended in breathing as % of total O_2 consumption		6	2

29.12 Adaptations of the neonatal circulation to pulmonary gas exchange – closure of the fetal shunts

The fetal circulation is adapted to placental gas exchange. The three shunts described earlier allow the two sides of the circulation to work in parallel and permit blood to bypass, to a large extent, those organs with little or no function. As the newborn baby takes its first breaths of air, the fetal circulation must start to adapt to the adult pattern so that blood no longer bypasses the lungs. To achieve this, the fetal shunts must close.

Closure of the shunts depends upon ventilation itself. As the baby starts to breathe, pulmonary blood flow increases dramatically. This happens for two reasons:

- Firstly, as the alveoli inflate, the tortuosity of the pulmonary capillaries is reduced and the resistance of the pulmonary vessels falls.

- Secondly, as a result of breathing air, there is a significant increase in the partial pressure of oxygen in the blood perfusing the lungs. In response to this rise, there is vasodilatation of the pulmonary vessels and a corresponding fall in resistance.

At the same time, the baby is separated from its placental blood supply and the umbilical vessels shut down (either spontaneously or as a result of clamping by the obstetrician). These changes in the blood flow pattern following delivery contribute to the closure of the fetal shunts. Each shunt will be considered in turn.

The foramen ovale

During fetal life, right atrial pressure is similar to, or just exceeds, left atrial pressure because the pulmonary resistance is relatively high and the systemic resistance is relatively low. After birth, as described above, pulmonary perfusion increases substantially and there is an increase in venous return to the left atrium. At the same time, loss of the umbilical blood supply reduces the venous return to the right atrium (via the inferior vena cava). Consequently, left atrial pressure rises above right atrial pressure. The foramen ovale consists of two unfused septa. All the time that right atrial pressure exceeds left atrial pressure, the septa are kept apart and the shunt remains open. Once the atrial pressures are reversed, the septa are forced against one another and the shunt is functionally closed. Within a few days, the septa fuse permanently and anatomical closure of the foramen ovale is complete. Figure 29.15

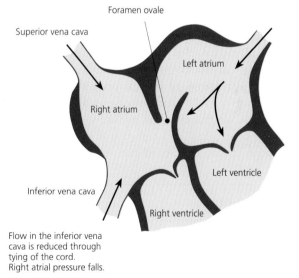

1. Flow in the pulmonary veins increases.
2. Left atrial pressure then rises above right atrial pressure.
3. The septum closes.

Foramen ovale

Superior vena cava

Left atrium

Right atrium

Left ventricle

Inferior vena cava

Right ventricle

Flow in the inferior vena cava is reduced through tying of the cord. Right atrial pressure falls.

Figure 29.15 The changes in the pressures of the right and left atria that lead to closure of the foramen ovale. While the lungs are non-functional with respect to gas exchange, the pressure in the right atrium exceeds that in the left atrium and blood passes through the foramen ovale. Following the first breath, the lungs expand and pulmonary vascular resistance falls. Pressure in the right atrium falls below that in the left atrium and this leads to closure of the foramen ovale.

illustrates how the pressure changes bring about closure of the foramen ovale.

The ductus arteriosus

This is an extremely wide channel, almost as large as the aorta itself, and the mechanisms that bring about its closure are not completely understood. It is thought that the smooth muscle of the ductus arteriosus constricts in response to the substantial rise in P_{O_2} seen after the first breaths. Permanent closure occurs after 10 days or so as a result of fibrosis within the lumen of the vessel.

The ductus venosus

During fetal life, this shunt carries a significant fraction of the blood in the umbilical veins directly to the inferior vena cava (thus bypassing the liver). Closure is thought to occur as an extension of constriction of the umbilical vessels following delivery.

Occasionally the fetal shunts fail to close

Although there have been no reports of the ductus venosus failing to close, persistent fetal connections associated with the foramen ovale and ductus arteriosus are seen. Each accounts for about 15–20% of congenital heart defects. Although intermittent flow through the fetal shunts is not uncommon during early neonatal life, if the shunts remain open for a prolonged period, circulatory function is impaired, and surgical intervention will be required to correct the defect.

KEY POINTS:

- After delivery, the circulation of the infant must adapt to pulmonary gas exchange.

- The three fetal shunts close and the circulation adopts the adult pattern.

- Shunt closure depends upon ventilation. The first breaths require the generation of large negative intrathoracic pressures. Surfactant is vital to reduce the surface tension forces tending to oppose lung inflation.

The fetal adrenal glands

The adrenal glands are vital endocrine organs in the adult. They consist of two distinct regions, the cortex, which synthesizes and secretes a variety of steroids (see Chapter 16), and the medulla, which produces adrenaline and noradrenaline (epinephrine and norepinephrine). During fetal life, the adrenal glands appear to be, if anything, even more important as they play a key role in the development of many of the fetal organ systems and are important in the initiation of parturition (see above p.620).

In relation to the overall body size of the fetus, the fetal adrenal gland is much bigger than that of the adult. Furthermore, it is organized in a different way. Unlike the adult gland, which consists of a medulla and a zoned cortex, the fetal adrenal is divided into three areas with differing characteristics. These are: a small region of medullary tissue; a small zoned cortex – the so-called definitive cortex – which resembles that of the adult; and a third, very large region, the **fetal zone**. The relative sizes of these areas are shown in Figure 29.16.

The functions of the different regions of the fetal adrenal gland are not fully established but the medulla is known to secrete adrenaline and noradrenaline in response to stress such as hypoxia. The fetal zone seems to be very important in producing the precursors required for the placental synthesis of oestrogens. The 'definitive cortex' seems to carry out little in the way of steroid synthesis itself during fetal life but it does perform one very important task – it converts progesterone to cortisol, especially during the last 3 months of pregnancy.

Fetal cortisol has a number of crucial functions:

- It is linked with the production of surfactant by type 2 alveolar cells.

- It accelerates maturation of the fetal liver.

- It plays an important role in the triggering of parturition.

After delivery, the fetal zone of the adrenal gland regresses while the definitive zoned cortex grows rapidly to establish the organizational pattern of the adult.

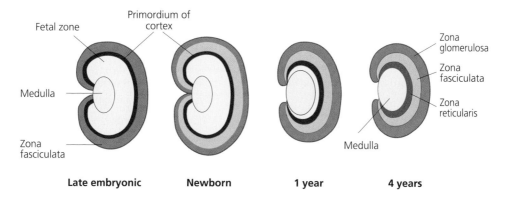

Figure 29.16 The changes in the adrenal gland during early postnatal life. Note that the large fetal zone present at birth regresses during the first years of life as the adrenal gland acquires the adult organization.

29.13 Renal function and fluid balance in the fetus and neonate

Although the placenta is the major organ of homeostasis and excretion of metabolic waste products during gestation, the fetal kidneys do play a role in the regulation of fluid balance and the control of fetal arterial blood pressure.

The human fetus begins to produce urine at about week 8 of gestation. Its volume increases progressively throughout gestation and is roughly equivalent to the volume of amniotic fluid swallowed by the fetus (around $28\,\text{ml}\,\text{h}^{-1}$ in late gestation). Fetal urine is usually hypotonic with respect to the plasma. Indeed, the ability of the kidney to concentrate the urine is not fully developed until after birth when the organ matures, the loops of Henle increase in length and sensitivity of the tubules to antidiuretic hormone (ADH) increases. In the adult, virtually all the filtered sodium is reabsorbed by the renal tubules. In the fetus, sodium reabsorption is lower (85–95% of the filtered load), probably because the renal tubules have a relatively low sensitivity to aldosterone. Fetal glucose reabsorption is thought to occur by sodium-dependent transport – as in the adult.

After delivery, the kidneys of the newborn infant become solely responsible for maintaining fluid balance and disposing of waste products. GFR and urine output increase over the first few weeks of life – although adult levels (relative to body surface area) are not reached for 2–3 years. Young babies cannot concentrate their urine as effectively as adults and this means that they can quickly become dehydrated, particularly during episodes of diarrhoea and vomiting. It is therefore essential that the lost fluids are replenished by mouth and, if this is not possible, intravenous fluid replacement may be needed.

29.14 The gastrointestinal tract of the fetus

The fetus obtains glucose, amino acids and fatty acids from its mother via the placenta (see Section 29.5). Towards the end of gestation, glycogen is stored in the muscles and liver of the fetus, while deposits of both brown and white fat are laid down. These stores will be crucial to the survival of the infant immediately after its birth.

The gut of the fetus is relatively immature, with limited movements and secretion of digestive enzymes. Some salivary and pancreatic secretion commences during the second half of gestation. Gastric glands appear at around the same time, although they do not appear to be secretory as the gastric contents are neutral at birth. Most of the major gastrointestinal hormones are secreted during fetal life although at a low level.

The fetus passes little, if any, faeces while it remains in the uterus. The contents of the large intestine accumulate as **meconium**, a sticky greenish-black substance. Meconium does not normally enter the amniotic fluid, although if the fetus becomes distressed, gut motility increases and meconium is passed. This may occur before birth, for example if there is placental insufficiency, maternal hypertension or pre-eclampsia. It may also occur during a prolonged or difficult delivery in which the fetus becomes hypoxic. Meconium-stained amniotic fluid is recognized as a sign of fetal distress and can cause damage to the lungs if it is inhaled (meconium aspiration syndrome or MAS).

> **KEY POINTS:**
>
> - The fetal adrenal gland consists of a cortex, medulla and a large fetal zone. Fetal cortisol has an important role in surfactant production and plays a role in triggering parturition.
>
> - Although the placenta is responsible for the excretion of fetal metabolic waste products, the fetal kidneys excrete dilute urine and participate in body fluid balance.
>
> - The fetal gut is immature. It passes little by way of faeces and the contents of the large intestine accumulate as meconium.

29.15 Development of the fetal reproductive organs

Humans have 46 chromosomes, one pair of which are the sex chromosomes. As explained in Chapter 28, females possess two X chromosomes (46, XX) and males possess one X and one Y chromosome (46, XY). As far as gonadal development in the fetus is concerned, the presence of a Y chromosome is the critical determinant of 'maleness'. A small region of the Y chromosome (called the sex-determining region or SRY gene) is required for the normal differentiation of male reproductive organs. In its absence, female reproductive organs will develop.

The male and female reproductive organs arise from embryonic mesodermal tissue and during weeks 3 and 4 of gestation two ridges of tissue (primitive sex chords) develop on either side of the aorta. Within these ridges, the gonads appear as bulges of tissue, which, by week 6 of gestation, begin to protrude into the ventral cavity of the body. At about the same time, primordial germ cells migrate to lie within and between the sex chords. This population of germ cells expands progressively by mitotic proliferation.

Adjacent to the primitive gonads are two pairs of ducts, the **Wolffian ducts** and the **Müllerian ducts**, which will eventually develop into the male or female reproductive tracts, respectively. Both sets of ducts empty into a region called the urogenital sinus. This arrangement can be seen in Figure 29.17. Up until around 6 weeks of gestation, the gonads of males and females develop identically – they are indistinguishable and are described at this stage as 'indifferent'. Subsequent development of the reproductive organs depends on the genetic sex of the fetus.

Development of male gonads and genital ducts

If a Y chromosome is present, the indifferent gonad is converted to a testis. The sex chords proliferate considerably and, together with tissue from the mesonephros (the early embryonic excretory system), form a structured organ surrounded by a fibrous layer. The cells of the sex chords secrete a basement membrane and become known as **seminiferous chords**. These incorporate the primordial germ cells and will give rise to the seminiferous

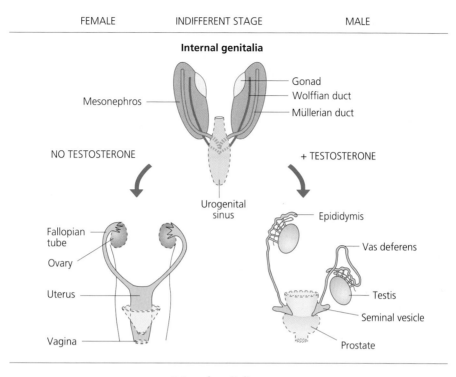

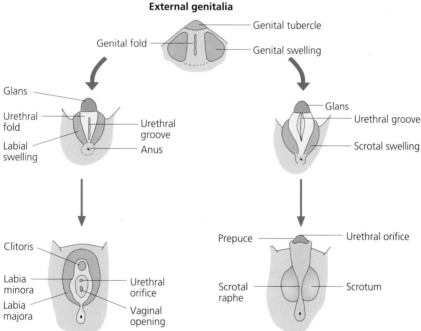

Figure 29.17 A simplified diagrammatic representation of the stages of development of the internal and external genitalia of males and females. Testosterone plays a crucial part in the development of the male genitalia. In the absence of testosterone, female structures develop. In the upper panel, one of the testes is shown in the process of descent.

tubules of the testis. The primordial germ cells will give rise to spermatozoa and the mesenchymal chord cells will form the Sertoli cells. The endocrine cells of the testis (the Leydig cells) form as clusters between the chords.

Further differentiation of the male genital structures depends upon the endocrine activity of the testes themselves. The fetal testes secrete two important hormones – testosterone (from the developing Leydig cells) and a substance called Müllerian inhibiting hormone (MIH), from the Sertoli cells. The testosterone appears to be secreted in response to hCG and in turn stimulates further development of the Wolffian ducts to form the epididymis, seminal vesicles and vas deferens. At the same time, MIH stimulates regression of the Müllerian ducts, which would otherwise develop into female reproductive structures.

Development of female gonads and genital ducts

The Y chromosome is responsible for initiating the conversion of an indifferent gonad into a testis. In the absence of a Y chromosome, however, the changes described above do not occur. The developing female gonad appears to remain indifferent. The primordial germ cells continue to proliferate mitotically and the primitive sex cords disappear. Clusters of cells in the cortical (outer) region of the gonad differentiate to form granulosa cells that surround the germ cells. In this way, the primitive follicles that characterize the ovary are laid down. At around weeks 10–12 of gestation, the primary oocytes contained within the primitive follicles begin their first meiotic division. As explained in Chapter 28, this division arrests at the diplotene stage and is only completed at ovulation.

In the female fetus, the Wolffian ducts degenerate in the absence of testosterone but the Müllerian ducts remain and eventually develop into the paired Fallopian tubes, midline uterus, cervix and upper vagina.

Development of the male and female external genitalia

For the first 8 weeks of gestation, the external genitalia of the male and female fetus are indistinguishable. They consist of a genital (also called phallic) tubercle and lateral swellings – the genital or labioscrotal swellings, behind which lies the urogenital sinus. The Wolffian and Müllerian ducts of the developing reproductive system and the ureters of the embryonic urinary tract open into the urogenital sinus. Between the genital swellings lie the urethral folds, containing the urethral groove (Figure 29.17).

In male embryos, androgens from the testis bring about fusion of the urethral folds to enclose the urethral tube, and fusion of the genital swellings to form the scrotum. The genital tubercle enlarges to form the penis. In female embryos, the urethral folds and genital swellings remain separate to form the labia, while the genital tubercle forms the small clitoris. These stages of development are illustrated in Figure 29.17.

KEY POINT:

- During development, the male pattern of differentiation must be actively induced. In the absence of intervention (by the Y chromosome and later by male hormones), the female pattern develops inherently.

29.16 Lactation – the synthesis and secretion of milk after delivery

While the fetus is developing within its mother's uterus, it receives all the nutrients it requires via the placenta. Once it has been delivered, however, the baby needs a regular and plentiful supply of milk. Human infants have a long gestation period and are born at a relatively advanced stage of development. Consequently, the baby makes considerable nutritional demands upon its mother. To ensure that sufficient milk of adequate calorific value is produced from the very start of lactation, preparatory changes must occur within the mammary glands during pregnancy. These changes are regulated by hormones from the placenta, pituitary and adrenal glands and will be discussed briefly in the following paragraphs. First, however, it will be helpful to consider the growth and development of the breasts prior to pregnancy.

Anatomy and physiology of the non-pregnant mammary gland

Until puberty, the immature breast consists almost entirely of ducts known as **lactiferous ducts**. At puberty, the ovaries start to increase their production of oestrogenic hormones, which initiate further breast development – in particular, the ducts begin to sprout and to become more highly branched. Once menstruation has commenced, progesterone, secreted during the luteal phase of each cycle, stimulates the formation of small, spherical masses of granular cells at the end of each duct. These are known as immature **alveoli** and are the cells that, in the event of a successful pregnancy, will develop into the milk-secreting alveoli of the lactating gland. During this phase of development, fat and connective tissue are deposited in the breast, causing a significant increase in size. In each gland, there are 15–20 lobes, separated by fat. Each lobe consists of clusters of granular cells at the ends of lactiferous ducts. The ducts dilate near to the areola (the area of brownish pigment surrounding the nipple) to form lactiferous sinuses. Each sinus runs up into the nipple and opens onto its surface. Dotted around the nipple are small sebaceous glands, called **Montgomery glands**.

Development of the mammary gland during pregnancy

Most of the growth and structural changes that are essential for successful lactation take place during the first 4 months or so of pregnancy. By midterm, the mammary gland is fully developed for milk secretion. The ducts proliferate further and the alveoli mature – the balls of granular cells become hollowed out so that alveolar cells surround a central lumen, which is drained by a branch of one of the lactiferous ducts. Figure 29.18 shows the organization of the mammary gland during pregnancy and Figure 29.19 illustrates the histological appearance of

the breast tissue. The hormones thought to be responsible for these changes are the placental steroids progesterone and oestrogens and the placental peptide hormone hPL (human placental lactogen). Progesterone in particular seems to be required for the alveolar changes that take place during early pregnancy. Additional adipose tissue is also deposited between the lobules of the gland during early pregnancy, adding further to the size and weight of the breast.

The alveoli are the primary sites of milk production. The structure of a mature alveolus is shown in Figure 29.20. The alveolar wall is formed by a single layer of epithelial cells, which are responsible for the synthesis and secretion of the constituents of milk. Adjacent alveolar cells are connected by junctional complexes near to the luminal surfaces, and between the basement membrane and the secretory alveolar cells are specialized cells called **myoepithelial cells**. These cells are, as their name suggests, contractile, and they are important for moving milk into the lactiferous ducts prior to ejection from the nipple when the baby suckles.

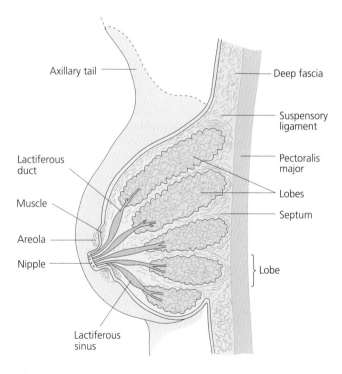

Figure 29.18 Sectional view of the mammary gland during pregnancy. Note the development of the alveoli.

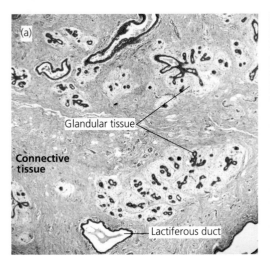
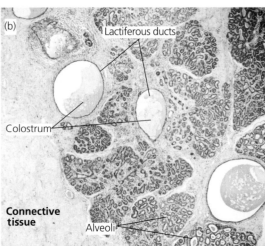

Figure 29.19 Thin tissue sections illustrating the differences between resting breast tissue (a) and its appearance during pregnancy (b). In (a), the alveoli are immature and there is a large amount of connective tissue between the lobules and around the glandular cells. In (b), note how the alveoli have proliferated and developed during pregnancy. Mature alveoli can be seen filled with colostrum in panel (b). During lactation, the alveoli develop further and swell to occupy most of the breast tissue.

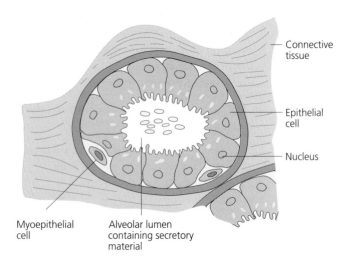

Figure 29.20 A diagrammatic representation of a cross-section of a mature (lactiferous) alveolus showing the glandular epithelial cells. The contraction of the myoepithelial cells helps to squeeze milk into the lactiferous sinuses during suckling.

Lactation is triggered by the fall in steroid secretion that follows delivery

Although the breast is fully developed for lactation by the middle of pregnancy, no significant milk production (lactogenesis) takes place until the baby has been born. After delivery, a series of endocrine changes activates the prepared mammary gland to synthesize and secrete milk.

Prolactin, secreted by the anterior pituitary gland, is the principal lactogenic hormone and high levels of this hormone are needed to sustain milk secretion. Prolactin is secreted in significant amounts throughout pregnancy but, during this time, levels of placental steroids (oestrogens and progesterone) are also very high and these hormones appear to inhibit the secretory activity of the mammary tissue. After birth, however, the placental steroid levels fall dramatically and prolactin is able to initiate milk production by the fully prepared breast.

The composition of human breast milk

The composition of human milk changes gradually over the first few weeks after birth. In the first few days, a fluid called **colostrum** is secreted, at a rate of about 40 ml a day. This fluid is rich in protein, minerals and vitamins A, D, E and K. It also contains significant quantities of immunoglobulins, which provide the infant with some resistance to infection for the first few months of its life. During the second and third weeks after birth, the composition of the milk changes. This so-called 'transitional milk' is much richer in fats and sugars than colostrum but contains fewer immunoglobulins and other proteins.

Table 29.4 The composition of human breast milk

	Colostrum	Transitional milk	Mature milk
Total fats (g l^{-1})	30	35	45
Total protein (g l^{-1})	23	16	11
Lactose (g l^{-1})	57	64	71
Total solids (g l^{-1})	128	133	130
Calorific value (MJ l^{-1})	2.81	3.08	3.13

Note that these values are approximate as the composition changes both during a single feed and during the course of the day. In general, the fat content rises from the beginning to the end of a feed.

By about the fourth week after delivery, the milk has attained its 'mature' composition. It is rich in fats, sugars and essential amino acids and has a calorific value of around 3.1 MJ l^{-1} (75 kcal per 100 ml). Table 29.4 shows the principal constituents of colostrum, transitional milk and mature milk.

Most of the fat in human milk consists of **medium-chain (10–12 carbons) fatty acids**, which are synthesized within the alveoli themselves under the control of prolactin and insulin. Prolactin also stimulates the secretion of the fatty acids into the central lumen of the alveolus.

The major milk proteins are **casein**, **α-lactalbumin** and **lactogobulin**. They have both nutritional and immunological significance. Furthermore, α-lactalbumin has a specific role in the synthesis of lactose, the principal sugar of milk. Amino acids, the precursors of protein synthesis, are supplied to the mammary tissue by the maternal circulation and pass from the blood into the alveolar cells via specific carrier systems. The proteins are synthesized in the usual way by the endoplasmic reticulum (see Chapter 4) and are packaged into vesicles that bud off from the Golgi apparatus and pass into the cytoplasm of the alveolar cell. Prolactin stimulates the release of the contents of these vesicles into the alveolar lumen.

The most abundant milk sugar is **lactose**, which is synthesized in the Golgi apparatus from α-lactalbumin. This protein forms a complex with galactosyltransferase, an enzyme within the Golgi membranes. The enzyme complex then metabolizes blood glucose to lactose, which is packaged, together with the proteins, into vesicles that are secreted by exocytosis into the alveolar lumen. Human breast milk contains more than 50 different oligosaccharides, most of which are synthesized from lactose. In addition to providing the source of many of the other milk sugars, lactose also promotes the growth of intestinal flora, which is very important to the newborn infant. Furthermore, galactose, one of the digestion products of lactose, is an essential component of the myelin that surrounds many nerve axons (see Chapter 8).

Nutritional requirements of lactating women

A baby weighing 5–6 kg will typically drink around 750 ml of breast milk each day. This volume of mature milk has an energy equivalent of about 2.3 MJ (560 kcal). The mother needs to provide sufficient nutrients to sustain this level of milk production as well as provide for her own metabolic needs. Although a proportion of the additional energy requirement will come from the mobilization of maternal fat, a lactating woman will normally need to increase her daily calorific intake by around 2 MJ (400–500 kcal).

Table 29.5 illustrates some other important nutritional requirements of lactation. Of particular importance is the need for an adequate intake of calcium and phosphate. The normal requirement for a woman of childbearing age is about 800 mg a day for both calcium and phosphate. During lactation, an extra 400 mg of each mineral is needed to match the quantities secreted in the milk.

Table 29.5 Recommended daily intake of protein and micronutrients during lactation

Nutrient	USA	UK
Protein (g)	65	56
Vitamin A (mg)	1.3	0.95
Vitamin D (mg)	10	10
Vitamin E (mg)	12	10
Vitamin C (mg)	95	70
Vitamin B$_1$ (thiamin) (mg)	1.6	1.0
Vitamin B$_2$ (riboflavin) (mg)	1.8	1.6
Vitamin B$_3$ (niacin or nicotinamide) (mg)	20	15
Vitamin B$_{12}$	2.6	2.0
Folate (mg)	280	260
Calcium (g)	1.2	1.25
Magnesium (mg)	355	320
Iron (mg)	15	15
Zinc (mg)	19	13
Iodine (mg)	200	140

KEY POINTS:

- Until puberty the mammary gland is composed largely of lactiferous ducts.

- After puberty, progesterone stimulates the development of alveoli at the ends of the ducts. These contain the cells that will secrete milk.

- The alveoli mature during pregnancy, under the influence of oestrogens, progesterone and hPL from the placenta, and acquire the potential to secrete milk.

- After birth of the baby, lactation commences. Colostrum is secreted in the first few days, followed by transitional and then mature milk after 2–3 weeks.

Lactation is maintained through suckling

The fall in steroid hormone secretion that occurs after delivery allows prolactin to exert its lactogenic effect on the breast and triggers the start of milk production. Babies may, however, be breastfed for several months or even years after delivery.

Sustained milk production is only possible if circulating levels of prolactin remain high. During breastfeeding, this is made possible by suckling itself. When the baby feeds, stimulation of the nipple induces the release of prolactin from the anterior pituitary gland via a **neuroendocrine reflex**. The components of this reflex are shown in Figure 29.21. Nerve impulses set up by the mechanical stimulation of the nipple pass via the spinal cord and brainstem to the hypothalamus. As a result, it appears that there is a fall in the

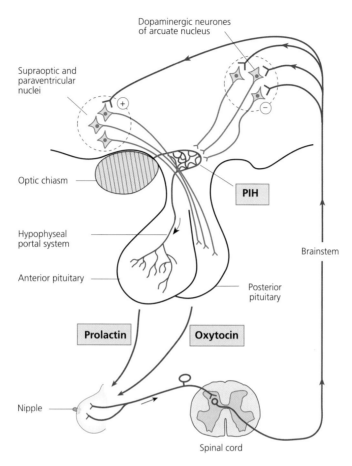

Figure 29.21 Summary of the principal neuroendocrine pathways responsible for the reflex release of prolactin and oxytocin during suckling.

secretion of the hypothalamic prolactin inhibitory hormone (PIH or dopamine; see Chapter 16, p. 288). Subsequently, prolactin secretion is increased and this hormone stimulates the further synthesis and secretion of milk. In a sense, therefore, the baby ensures its own continued nutrition through regular feeding. Mothers with twin babies are usually able to produce enough milk to feed them both as the quantity of milk produced is related to the intensity and duration of the suckling stimulus.

Sick or premature babies may be unable to suckle. Alternatively, a mother may have to spend time away from her baby. In such situations, an electric or manual breast pump can be used to provide expressed milk for

the baby and to maintain prolactin secretion (and thus lactogenesis) through nipple stimulation.

Milk let-down is stimulated by suckling

Under the influence of prolactin, milk is synthesized by the alveolar epithelial cells and secreted into the alveolar lumen. For the baby to feed, however, the milk must be moved to the nipple. This process is called milk 'let-down' and is another example of a neuroendocrine reflex. This reflex is also illustrated in Figure 29.21. Here, suckling triggers the release of **oxytocin** from the posterior pituitary gland. This hormone stimulates contraction of the myoepithelial cells that lie within the alveolar basement membrane (Figure 29.20) and milk is squeezed from the alveolar lumen into the lactiferous ducts. As the ducts fill with milk, the intramammary pressure rises and milk is ejected from the nipple to the suckling baby. This reflex is readily conditioned and the cry of a hungry baby may induce the secretion of oxytocin by the mother.

As well as stimulating milk let-down, oxytocin also enhances uterine contractions during labour. Indeed, some women experience quite painful uterine contractions when they breastfeed their baby for the first time after birth ('after-contractions') as oxytocin secreted in response to the baby suckling stimulates renewed contraction of the uterus, which is still highly excitable following labour.

Cessation of milk production at weaning

Lactation normally ceases within 2 or 3 weeks of weaning the baby onto a bottle or solid foods. This is entirely due to the loss of the suckling stimulus. In the absence of mechanical stimulation of the nipple, prolactin secretion declines and lactogenesis gradually slows down. Although milk production itself stops relatively quickly, it is several months before the mammary glands return to their prepregnant structure. At first, milk accumulates in the alveoli and small lactiferous ducts causing distension of the epithelial structures. The alveolar cells are ruptured and hollow spaces form within the mammary tissue. The distension also causes compression of the capillary network supplying the alveoli and, as a result of the reduced blood flow, the alveolar cells become hypoxic and lack nutrients. Milk production is depressed and the alveolar cells begin to disappear. The ductal system starts to dominate once more and the alveolar epithelial cells revert to the granular, non-secretory-type characteristic of the non-pregnant state. All these changes occur quite naturally as a direct result of removing the suckling stimulus at the time of weaning.

The human mammary gland is fully prepared for lactation by the fourth month of pregnancy. This means that, in the event of a miscarriage or abortion after this time, milk production will commence because of the decline in steroid secretion following removal of the placenta. Under these circumstances, it is clearly desirable to inhibit lactation as rapidly as possible. This is achieved through the administration of drugs such as **bromocriptine**, which acts as a dopamine agonist to inhibit the secretion of prolactin.

> **KEY POINTS:**
>
> - Milk production is maintained by regular suckling. Prolactin is secreted in response to nipple stimulation.
>
> - Milk let-down occurs in response to oxytocin secreted by the posterior pituitary gland during suckling.
>
> - Once the baby is weaned and the suckling stimulus is lost, prolactin secretion falls and lactation slows and then stops.

 Recommended reading

Griffin, N.E. and Ojeda, S.R. (2004). *Textbook of Endocrine Physiology*, 5th edn. Oxford University Press: Oxford. Chapter 11.

Johnson, M.H. and Everitt, B.J. (1999). *Essential Reproduction*, 5th edn. Blackwell Scientific: Oxford.

Pocock, G. and Richards, C.D. (2006). *Human Physiology – the Basis of Medicine*, 3rd edn. Oxford University Press: Oxford. Chapters 21 and 22.

Self-assessment questions

Which of the following statements are true and which are false? Answers are given at the end of the book (p. 755).

1 a) Penile erection is a parasympathetic reflex.
   b) Sexual reflexes are integrated in the thoracic region of the spinal cord.
   c) Sperm remain viable for up to a week.
   d) Capacitation of a sperm normally takes place in the vas deferens.
   e) Fertilization takes place in the Fallopian tube.

2 a) The egg completes its first meiotic division after fertilization.
   b) hCG is secreted by the anterior pituitary gland.
   c) Luteal regression is prevented by hCG.
   d) The zygote is a diploid cell.
   e) The blastocyst implants into the myometrium of the uterus.

3 a) Oxygen crosses the placenta by passive diffusion.
   b) Intervillous blood spaces of the placenta are filled with fetal blood.
   c) The placenta is 'definitive' at around 3 months of gestation.
   d) Placental progesterone is essential for the maintenance of pregnancy.
   e) hPL is a steroid hormone.

4 a) Arterial blood pressure often falls in the second trimester of pregnancy.
   b) Maternal heart rate increases in pregnancy.
   c) There is a reduced requirement for calcium during pregnancy.
   d) Residual volume shows an increase in pregnancy.
   e) Thyroid function is increased during pregnancy.

5 a) The ductus arteriosus is a fetal shunt between the right and left atria.
   b) The umbilical artery carries oxygenated blood.
   c) The ductus venosus allows blood to bypass the fetal liver.
   d) Surfactant-secreting cells start to develop at around 20 weeks of gestation.
   e) Pulmonary vascular resistance is higher in a fetus than in an adult.

6 a) Suckling increases the output of prolactin.
   b) Oxytocin is secreted by the anterior pituitary gland.
   c) Following a miscarriage at 5 months of gestation, lactation will commence.
   d) Colostrum has a higher calorific content than mature milk.
   e) During pregnancy, progesterone stimulates development of the alveoli.

30 The genetic basis of inheritance

After reading this chapter you should have gained an understanding of:

- Basic properties of chromosomes, genes and DNA
- How genetic information is coded by the nucleotide sequences within DNA
- How the genetic code is used by cells to manufacture proteins
- The terms phenotype, genotype and allele
- Dominant and recessive alleles
- The laws that govern simple Mendelian inheritance
- Co-dominance, polygene inheritance and sex-linked inheritance
- Some specific examples of inheritance patterns
- Mitochondrial inheritance
- The importance of genetic counselling and gene therapy

30.1 Introduction

In Chapter 4, the principal characteristics of human cells were discussed. Located within the cytoplasm of most human cells is the nucleus, which is separated from the cytoplasm by the nuclear membrane (also called the nuclear envelope). The nuclear membrane contains pores, which permit communication between the nucleus and the cytoplasm. Contained within the nucleus are **chromosomes**. These contain **genes**, the basic units of inheritance. Somatic cells (those cells that are not gametes, i.e. not eggs or sperm) possess 46 chromosomes organized into 23 pairs – the **human diploid number**. These pairs are known as **homologous pairs** because each member of the pair is of similar size and carries genetic information about the same biological characteristics. One member of each pair is of maternal origin, i.e. it is derived from the egg of the mother at fertilization. The other is of paternal origin and derived from the sperm of the father at fertilization. At certain stages in the cell cycle, and with appropriate chemical treatment, it is possible to view chromosomes using a light microscope. Figure 30.1 illustrates a set of human chromosomes from a somatic cell displayed in homologous pairs. Such an image is called a **karyotype.**

(a)

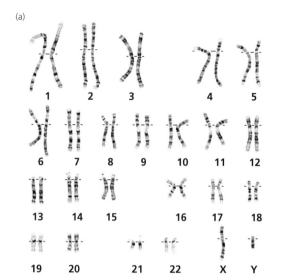

(b)

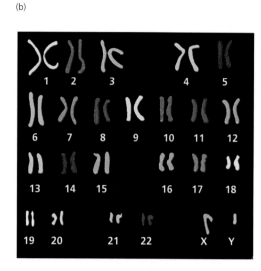

Figure 30.1 (a) A normal male karyotype (46, XY) stained with Giemsa dye to show the G banding pattern. The dotted lines indicate the positions of the centromeres. The chromosomes are arranged in order: the longest autosome pair are placed first and the shortest last. The sex chromosomes (X and Y) are placed at the end. Note the marked difference in size between the X and Y chromosomes. (b) A spectral karyotype – a technique that permits the visualization of each pair of chromosomes with a different fluorescent colour.

Of the 23 pairs of chromosomes that make up the human karyotype, 22 pairs are called **autosomes**. These are not involved in the determination of gender but carry information about other biological characteristics. The 23rd pair carries information that determines a person's sex and these are called the **sex chromosomes**. There are two types of sex chromosome, called X and Y. Females possess two X chromosomes whereas males have one X chromosome and one Y chromosome. As can be seen in Figure 30.1, unlike the autosomal pairs, the X and Y chromosomes are very different in size and carry different genetic information. The X chromosome is much larger than the Y chromosome and carries many more genes. Those characteristics whose genes are carried by the X chromosome but not the Y are known as sex-linked characteristics and will be discussed in Section 30.8. The Y chromosome carries the SRY gene, which governs the anatomical and functional development of the male sexual organs as well as other specific male characteristics. Thus, the presence of a Y chromosome in a person's karyotype determines 'maleness'.

Throughout the life of a person, tissues grow (during childhood) and cells are replaced when necessary. This is accomplished by a process called **mitosis**, a form of cell division that results in the production of **genetically identical** diploid daughter cells (see Chapter 4, p. 45). The chromosomes within the nucleus of a dividing cell replicate so that the genetic information contained within it is passed to the two daughter cells.

Humans reproduce sexually, which means that genetic information is passed from one generation to its offspring through the process of the fusion of male and female gametes that takes place at fertilization. The events leading to the formation of the male and female gametes are discussed in Chapter 28. Ova (eggs) and spermatozoa (sperm) are produced by a process of cell division called **meiosis** (described in Chapter 28, p. 583). This results in the formation of cells whose nuclei contain 23 unpaired chromosomes. This is the human **haploid number** – for this reason, the gametes are known as haploid cells. At fertilization, the haploid nucleus of a sperm fuses with the haploid nucleus of an egg to form a new cell (the zygote), which is diploid, that is it possesses 46 chromosomes, 23 of maternal origin and 23 of paternal origin. Because all gametes are genetically different, all individuals, with the exception of identical twins, are genetically unique (for a fuller explanation, see p. 584). The sex of the new individual is determined by whether the fertilizing sperm is carrying an X or a Y chromosome.

> **KEY POINTS:**
>
> - Most human cells have 23 pairs of chromosomes making 46 chromosomes in all (the diploid number).
> - There are 22 pairs of autosomes and one pair of sex chromosomes.
> - The gametes (i.e. eggs and sperm) are haploid cells and have 23 unpaired chromosomes.

30.2   Chromosomes, genes and the genetic code

Chromosomes are composed of DNA (deoxyribonucleic acid) and an array of structural proteins called **histones**, which help to pack extremely long, coiled lengths of DNA into the relatively small space of the nucleus, as shown in Figure 30.2. This packaging allows the DNA strands to be correctly distributed between the daughter cells during cell division.

To visualize the chromosomes under a microscope, dividing cells are treated with a drug, colchicine, which inhibits the formation of the mitotic spindle and arrests

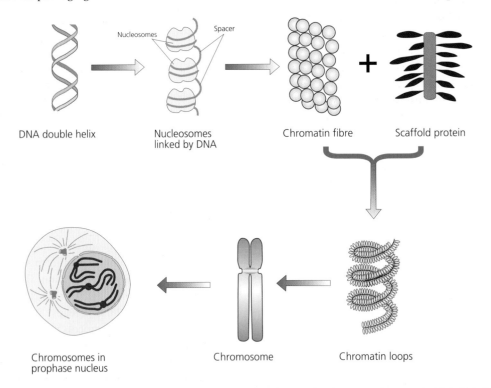

Figure 30.2 The stages by which long strands of DNA are condensed to form the chromosomes. The DNA double helix is first wrapped 1¾ turns around protein molecules known as histones to form nucleosomes, which are arranged rather like beads on a string. The nucleosomes then form chromatin fibres, which are supported by a protein scaffold to form the loops of chromatin that make up the individual chromatids of a chromosome. The chromosomes are only clearly seen in the cell nucleus during cell division. In this case, the nucleus is shown in prophase, just after the nuclear chromatin has become condensed into well-defined chromosomes.

cell division in metaphase (for an explanation of mitosis, see Chapter 4, p. 45). (The chromosomes are prepared by the cell's internal machinery for mitosis but, in the absence of the spindle, cell division cannot proceed.) The cells are then treated with a hypotonic solution to cause them to swell before chemically denaturing their proteins to preserve their internal structures, including that of the chromosomes (a process called fixation). This stage of preparation is followed by gentle treatment with trypsin and staining with various dyes, most commonly Giemsa. The staining reveals a series of bands, which can be used to identify particular chromosomes reliably. The images of chromosomes are then taken and the individual chromosomes paired and displayed as in Figure 30.1 to identify the specific karyotype (e.g. 46, XX for a normal female; 46, XY for a normal male).

Each chromosome consists of two arms, a short arm (called the p arm) and a long arm (the q arm) separated by a constricted region, called the **centromere**, at the junction between the two sister chromatids (Figure 30.3). Within the centromere is a structure called the **kinetochore**, which attaches the chromosome to the spindle during cell division. At the ends of the arms of each chromosome are specialized regions called telomeres, formed from repeated nucleotide sequences. The telomeres act to preserve the integrity of each chromosome and prevent chromosomes fusing in a random manner.

Depending on the position of the centromere, chromosomes are classified as metacentric (centromere near the middle, e.g. chromosomes 1 and 3), submetacentric (centromere towards one end, e.g. chromosomes 9 and 10) and acrocentric (chromosomes with a very short p arm, e.g. chromosomes 13 and 14). By convention, and for the purposes of accurate localization of a band within a chromosome, the arms are further divided into regions numbered outwards from the centromere, as shown in Figure 30.3 for chromosome 17. The chromosomes can then be represented diagrammatically as a series of ideograms, such as that shown in Figure 30.10 for chromosome X (p. 662).

Chromosomal abnormalities. Occasionally, during meiosis, some chromosomes fail to separate properly and the daughter cells (gametes) will either lack one or more chromosomes or have a greater number than normal.

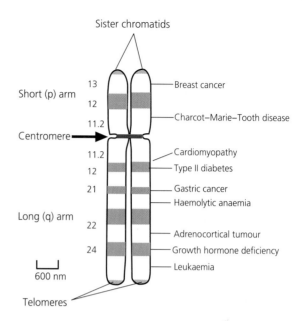

Figure 30.3 A schematic diagram of a typical chromosome (chromosome 17); the chromosome consists of two sister chromatids linked at their centromeres. The centromere divides the chromosome into two unequal segments known as the p arm and the q arm. The individual arms of each chromatid terminate in DNA sequences known as telomeres (shown in green). A characteristic banding pattern is seen after staining the chromosome spread with Giemsa dye (shown here by the blue bands). The standardized numbering system used to locate specific segments on the chromosome is shown on the left side. In common with all other chromosomes, a large number of genetic disorders have been traced to chromosome 17. The mapping of a few examples is shown in the figure.

KEY POINTS:

- Each chromosome consists of a pair of sister chromatids joined at a region called the centromere.

- The centromere divides each chromatid into two segments: a short p arm and a longer q arm.

- Individual chromosomes can be identified by the position of the centromere and the banding pattern revealed by staining with specific dyes.

This is known as **non-disjunction**. Non-disjunction causes a number of genetic diseases, of which Down syndrome is perhaps the best known. In this disease, the sufferer

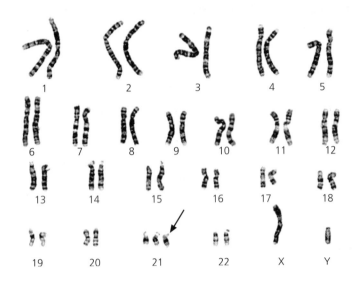

Figure 30.4 A male karyotype from an individual with Down syndrome (47, XY, +21). Note the extra copy of chromosome 21 indicated by the arrow.

has inherited an extra copy of chromosome 21 and for this reason Down syndrome is also known as trisomy 21. An example of a male Down syndrome karyotype (designated 47, XY, +21) is shown in Figure 30.4.

Genes are segments of DNA within a chromosome that determine specific inherited characters. Each gene carries information (genetic code) that instructs the cell to manufacture a specific protein or a group of proteins.

The totality of all the genes present within an animal or human is known as its **genome**. Genes may thus be considered to be the physical units of inheritance. They occupy a specific position or locus on a chromosome, which can be described using the standard convention described above. Thus, a gene might occupy position 13.2 on the p arm of chromosome 7. This is its specific **locus**. The work of the Human Genome Project (see Box 30.1) has

Box 30.1 The Human Genome Project

During the closing years of the last century, a huge effort was put into determining the sequence of all the DNA in human cells. This was known as the **Human Genome Project**, which succeeded in determining the entire DNA sequence that is required to make a human being. There are some 3.2 billion (3 200 000 000) base pairs in human DNA, and the DNA sequence of different individuals differs by only 1 base in every 1000 or so. From this we can say that 99.9% of the human DNA sequence is shared between all members of the human population. The 0.1% difference between individuals corresponds to about 3 million variations in sequence. As each of these variations is independent and arises from random changes, each person is genetically unique (except identical twins who share the same DNA sequence). Most of the DNA

(around 97%) has no known function and is sometimes referred to as 'junk' DNA. The remaining 3% or so codes for all the genes (around 22 000 in all) that are required to make a human being.

Each amino acid is coded by a sequence of three bases (a sequence triplet or **codon** – see main text). If one base is substituted for another in the DNA sequence, the amino acid sequence of any protein encoded by that region of DNA is likely to be changed. Depending on the position of the base substitution, the effect of such a mutation on the function of the protein may either be trivial or severe. Examples of genetic disorders that arise from the substitution of one base for another (**single nucleotide polymorphism** – a DNA base is equivalent to a nucleotide) are sickle-cell disease and haemophilia B.

revealed that humans have a total of around 22 000 genes, carrying the information needed to manufacture around 100 000 different proteins. Although it was once believed that each gene coded for a single protein, some genes are able to code for more than one protein sequence. For example, the synthesis of antibodies by the immune system occurs by combining different gene segments to synthesize the four polypeptide chains of each antibody. As a result, there are many different DNA sequences from which the polypeptide chains of the antibodies can ultimately be synthesized. Most of the DNA (around 97%) has no obvious coding function and is sometimes called 'junk' DNA; the remaining 3% codes for all the genes that are required to make a human being.

> **KEY POINTS:**
> - Genes are the physical units of inheritance.
> - Each gene is a region of a DNA strand that instructs a cell to manufacture a specific protein or a group of proteins.
> - The totality of all the genes present within an animal or human is known as its genome.

30.3 DNA and the genetic code

As explained in Chapter 3, proteins are assembled from a set of 20 α-amino acids. Protein synthesis is carried out by intracellular organelles called **ribosomes**, located in the cytoplasm of the cell or attached to endoplasmic reticulum (rough endoplasmic reticulum; see Chapter 4). Amino acids can be linked together in any number and in any sequence to make an infinite number of possible protein structures. Thus, there is endless scope for variation between individuals. As a person's biological characteristics are determined by which proteins are synthesized by that person's cells, then the number of possible characteristics or combinations of characteristics must also be essentially infinite.

Key questions to be addressed are:

- In what form do genes carry the instructions for making a particular protein or set of proteins?
- How are these instructions used by the ribosomes to assemble the protein correctly?

The structure of DNA

The basic structure of DNA was described in Chapter 3. In essence, a molecule of DNA consists of a pair of nucleotide chains coiled into a double helix. Each nucleotide consists of a sugar (deoxyribose), a phosphate group and one of four organic bases – adenine, thymine, cytosine or guanine (abbreviated to A, T, C and G). This organization may be seen in Figure 3.13. Adjacent bases are linked by hydrogen bonds to hold the two nucleotide chains of the DNA molecule together. As explained in Chapter 3, adenine always pairs with thymine and cytosine always pairs with guanine. A–T and C–G are referred to as complementary base pairs. The two DNA strands are referred to as the primary strand and the complementary strand.

The instructions for manufacturing proteins are contained within the sequence of organic bases along the primary strand (sometimes called the template strand) of the DNA molecule. Each amino acid is coded by a sequence of three organic bases, known as a **triplet**. For example, the triplet TAC represents the code for the amino acid methionine, while the triplet ACC is the code for tryptophan. Many of the amino acids have more than one triplet code. For example, the DNA sequences TTT and TTC both represent lysine. Thus, a linear sequence of information within the DNA molecule (the order of organic bases on the primary strand) is used to specify a chain of amino acids.

Reading the code – gene transcription and translation

As we have seen, the coded information that instructs the cell to manufacture proteins having the correct number of amino acids, in the correct order, is lodged within the DNA molecule itself. However, the ribosomes, which perform the task of putting together the protein according to these instructions, are in the cytoplasm of the cell. DNA

Table 30.1 The genetic code

First position ↓	← Second position →				Third position ↓
	U	**C**	**A**	**G**	
U	Phenylalanine	Serine	Tyrosine	Cysteine	U
	Phenylalanine	Serine	Tyrosine	Cysteine	C
	Leucine	Serine	**Stop**	**Stop**	A
	Leucine	Serine	**Stop**	Tryptophan	G
C	Leucine	Proline	Histamine	Arginine	U
	Leucine	Proline	Histamine	Arginine	C
	Leucine	Proline	Glutamine	Arginine	A
	Leucine	Proline	Glutamine	Arginine	G
A	Isoleucine	Threonine	Asparagine	Serine	U
	Isoleucine	Threonine	Asparagine	Serine	C
	Isoleucine	Threonine	Lysine	Arginine	A
	Methionine	Threonine	Lysine	Arginine	G
G	Valine	Alanine	Aspartate	Glycine	U
	Valine	Alanine	Aspartate	Glycine	C
	Valine	Alanine	Glutamate	Glycine	A
	Valine	Alanine	Glutamate	Glycine	G

This table summarizes the genetic code for mRNA. The first base of the triplet that codes for an amino acid (a codon) is given by the column on the left, the second is given by the row across the top of the table and the third is given by the column on the right. To take methionine as an example, the first base is A, the second is U and the third is G, so the mRNA code for methionine is AUG. Note that most amino acids are coded by more than one triplet. For example, lysine is coded by AAA and by AAG. Stop codons tell the transcription process when a peptide chain is completed.

is far too large a molecule to leave the nucleus, so to overcome this, the genetic information encoded by the DNA triplets is first copied (by a process called **transcription**) on to a strand of **messenger RNA** (**mRNA**). RNA (ribonucleic acid) is a single-stranded nucleic acid similar to DNA except that its sugar is ribose and the organic base thymine is replaced by **uracil**.

Transcription takes place in the nucleus and the mRNA then migrates to the cytoplasm where it becomes attached to a ribosome and provides a template for the synthesis of a specific protein. The position of each amino acid is coded by the sequence of three bases (called a **codon**) on the mRNA strand. The base sequence in mRNA that specifies the sequence of amino acids in proteins constitutes the **genetic code**. It is shown in Table 30.1. To form a protein, the amino acids must be assembled in the correct order as dictated by the code. This process is called **translation** and it involves another type of RNA molecule, called **transfer RNA** (**tRNA**). This type of RNA differs from mRNA in several important aspects. Firstly, it is relatively small, possessing only around 70 nucleotides. Secondly, each molecule contains a binding site for a specific amino

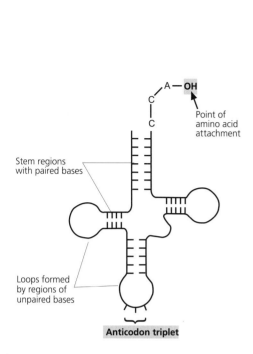

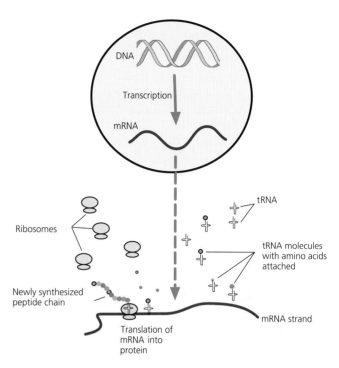

Figure 30.5 The clover-leaf structure of transfer RNA (tRNA). Each amino acid binds only to tRNA molecules with the correct genetic code for that amino acid as specified by the anticodon triplet.

Figure 30.6 The principal steps in the conversion of the genetic information encoded by DNA to the synthesis of specific proteins. The DNA molecules are too large to leave the nucleus so the genetic information is transcribed onto messenger RNA (mRNA), which passes into the cytoplasm and acts as a template for protein synthesis.

acid and a specific triplet of bases (called an **anticodon**) that matches its complementary sequence on the mRNA. A simplified diagram of tRNA is shown in Figure 30.5.

To assemble the protein, the ribosome matches the anticodon on the tRNA with the codon on the mRNA. In this way, amino acids are brought into line along the ribosome according to the base sequence originally specified by the DNA of the gene responsible for that protein. Peptide bonds are then formed between adjacent amino acids and the protein is progressively synthesized. The main steps of protein synthesis are summarized in Figure 30.6.

Errors of coding

It is perhaps not surprising, given the complexity of the processes described above, that from time to time errors occur in the coding of genetic information or in the processes of transcription and translation. In particular, errors

may occur during the replication of DNA that accompanies cell division. These errors are called **mutations**. For example, the substitution of one base for another (a point mutation) in one of the triplets within a gene may result in the substitution of one amino acid for another in the amino acid sequence of the protein encoded by that gene. This is called a **substitution** mutation and, depending upon the position of the base substitution, its effect on the protein may be either trivial or disastrous. Point mutations most commonly arise during spermatogenesis. Examples of genetic disorders arising from base substitutions include sickle-cell anaemia, haemophilia B and familial hypercholesterolaemia.

Another type of mutation is a **deletion** mutation, in which certain base pairs are omitted (deleted) from the DNA of a gene. The most common cystic fibrosis mutation, for example, involves the deletion of three base pairs within one of the genes that regulates the function of a particular chloride channel (see Chapter 22).

KEY POINTS:

- Point mutations most commonly arise during spermatogenesis.
- Most abnormalities of chromosome number occur during the halving of chromosome number that occurs during oogenesis.

30.4 Principles of inheritance

The discussion above provides a brief explanation of how cells use the information coded within their genes to manufacture the proteins that determine a person's biological characteristics. Further details of the control of gene expression are beyond the scope of this book but can be found in several of the texts listed at the end of this chapter. Each individual has their own unique set of instructions within their DNA. Although the vast majority of our DNA is the same (all normal healthy people manufacture insulin, haemoglobin, thyroxine and so on), a small fraction of each person's DNA is unique to them. This explains why we all look, sound and behave differently from one another. It also explains why it is now possible to identify individuals, with an extremely high degree of certainty, from their DNA – so-called 'DNA fingerprinting'. Nevertheless, certain characteristics are passed from one generation to the next – it is common knowledge that members of the same family often exhibit similar physical characteristics. The following will consider the important principles that govern this process – the mechanisms of inheritance or **heredity**.

As explained earlier, chromosome numbers are halved at meiosis so that the gametes (ova and spermatozoa) are haploid cells with 23 chromosomes. Furthermore, during meiosis there is random assortment of genes and crossing over of chromosomal segments (see p. 584) that ensures that all gametes have a different genetic constitution. At fertilization, the diploid state is restored, as 23 maternal chromosomes from the egg and 23 paternal chromosomes from the sperm come together to form 22 homologous pairs of autosomes and one pair of sex chromosomes. Everyone, therefore, receives half of their genetic information from one of their parents and half from the other. Members of the homologous pairs of autosomes carry genes that code for particular proteins at the same specific loci. This means that the nucleus of a newly fertilized egg possesses two genes for each protein, one paternal in origin and one maternal. The X and Y chromosomes are an exception to this rule as they possess different genetic information.

It is important to realize that even those genes that occupy the same specific locus on a chromosome can exist in different forms. These are called **alleles**. For example, the gene that determines a person's ability to roll their tongue exists as two alleles (tongue-rolling or non-tongue-rolling) while the gene that determines a person's ABO blood group exists as three different alleles, the A allele, the B allele and the O allele. When an individual inherits two copies of an allele they are said to be **homozygous**. If they inherit two different alleles, they are **heterozygous** for that gene (see Box 30.2).

The genetic constitution of a person is called their **genotype** and describes all the genes within their chromosomes. Not all genes are expressed in the person's physical characteristics. When two different alleles are present, one form of the gene can be expressed in preference to the other. The expressed gene is said to be **dominant** and the non-expressed gene is **recessive**. Those genes that are expressed determine a person's **phenotype**, that is whether they are straight haired or curly haired, have fixed or free earlobes, have blood group A, B, AB or O, and so on. These characteristics represent the physical expression of the genotype.

Box 30.2 The maintenance of genetic variation within a population

When there are two forms of a gene (alleles) in a population and one is dominant while another is recessive, why does the recessive allele not die out with successive generations? The answer is to be found in the fact that each allele is a distinct unit of heredity. The frequency of the different forms can be shown mathematically to remain constant within a given population provided that the population is large, mating within the population is random and there is no biological selection in favour of one form of the allele over the other. When these conditions are met, the frequency of the alleles within the population will be stable and the situation is called the **Hardy–Weinberg equilibrium**.

As an example, consider the inheritance of the D antigen (the Rhesus factor gene). Individuals inherit one of two forms of the RHD gene. Those who are rhesus positive have the dominant D gene and are able to synthesize the RhD polypeptide found on the surface of the red cells. Those who possess the non-functional form of the gene (d) are unable to synthesize the RhD polypeptide. It is obvious that the sum of the frequency probabilities ($p + q$) of the two alleles must add up to 1 (as people are either rhesus positive or rhesus negative). There are three possible genotypes: DD, Dd and dd with relative frequencies of p^2, $2pq$ and q^2. The numerical values for the European Caucasian population are given in the table below:

frequency of the two alleles. If this condition is not met, the frequency of the individual alleles will change as the incomers mate with the original population.

As mutation is always occurring, and as the chance that a mutation will lead to a non-functional gene is higher than the chance that a non-functional gene will give rise to a functional gene, it is possible for the frequency of alleles to change. However, the relatively low mutation rate of human populations helps to maintain stability in the gene pool.

As the Hardy–Weinberg equilibrium applies to genetically stable populations, it can be used to estimate the frequency of carriers of a recessive gene provided that its frequency of expression is known (i.e. the frequency with which both chromosomes carry the defective gene). If the frequency of expression is f, the frequency of the recessive allele is given by the relationship $f = q^2$ (remember that, in this case, both chromosomes must carry the recessive gene). The frequency of carriers for a Hardy–Weinberg equilibrium is $2pq$ (see above for the Rhesus antigen). To work out the occurrence of a specific defective gene, consider cystic fibrosis. This disorder has a frequency of occurrence of approximately 1 in 2000. The probability of finding a person with the disease in the population at large is therefore 0.0005. The gene frequency q is given by the relationship $f = q^2$, in this case

Genotype	Phenotype	Frequency	Probability of occurrence
DD	Rhesus positive	p^2	0.36
Dd	Rhesus positive	$2pq = 0.48$	0.48
dd	Rhesus negative	q^2	0.16

The frequencies for other populations are different; for example, European Basques have a much higher chance of being rhesus negative (35% rather than 16%) while the chance of a black African being rhesus negative is much lower (~1%). This highlights a key aspect of Hardy–Weinberg equilibria: there should not be significant interbreeding with a population that has a different

$f = 0.0224$. The frequency of carriers is ~2q (as p ≈ 1.0) or 0.0448. To put it another way, the chance of someone carrying the defective gene is about 1 in 22, a surprisingly high frequency. Similar calculations can be made for the frequency of carriers for other inherited recessive disorders if the approximate frequency of the disease within the population is known.

KEY POINTS:

- Two copies of each autosomal gene are inherited, one from each parent.
- The genotype of a person is the totality of the genes they possess.
- Their phenotype is the physical expression of those genes.
- When a gene exists in more than one form, the different types are called alleles.

30.5   Simple (Mendelian) inheritance

The laws that explain how many characteristics are inherited were first elucidated by Mendel in the mid-nineteenth century and can be summarized as follows:

- Characteristics are determined by genes.
- Many characteristics are determined by a pair of genes, one of paternal origin and one of maternal origin.
- Genes at the same specific locus on a chromosome can exist in different forms (alleles).
- Alleles may be dominant or recessive.
- If a dominant allele is present in the genotype, it will be expressed in the phenotype.
- Recessive alleles can only be expressed in the phenotype if there is no dominant allele in the genotype.

These laws provide a powerful framework on which to base an understanding of simple inheritance but they do not explain all aspects of inheritance. Many characteristics are determined by more complex mechanisms involving groups of genes (polygenic inheritance), incomplete dominance and so on. Some of these will be considered briefly here but for a full explanation you will need to consult a textbook of genetics.

Tongue-rolling is a good example of how the Mendelian laws may be used to explain inheritance patterns. Most people are able to roll their tongues into a U-shape. These people are said to be tongue-rollers (this is their phenotype). Some, however, cannot achieve this – they are, phenotypically, non-tongue-rollers. Thus, for the purposes of this discussion, we shall assume that the gene responsible for this aspect of muscular control

exists as two alleles. The tongue-rolling allele is dominant and the non-tongue-rolling allele is recessive. It is conventional in genetics to denote the dominant allele using a capital letter (here T) and the recessive allele using the same letter in its lower case (here t). What are the genotypes of tongue-rollers and non-tongue-rollers? As recessive alleles can only be expressed in the absence of the dominant allele, a person who cannot roll their tongue must have inherited recessive alleles from each of their parents. In other words, their genotype for this trait must be tt. Such a person is said to be homozygous for this trait. By contrast, a tongue-roller may have inherited either a dominant allele from each parent (genotype TT) or one dominant and one recessive allele (genotype Tt). In the latter instance, the recessive allele from one parent, although present in the genotype, is not expressed in the phenotype. The person is described as being heterozygous for the tongue-rolling trait.

Using a Punnett square to show inheritance patterns

We can now make some predictions regarding the inheritance patterns for this trait.

If a man and a woman who are both non-tongue-rollers (genotype tt) have a child, it is possible to predict with certainty that the child will also be a non-tongue-roller as he or she will inherit a recessive (t) allele from each parent. However, the situation is not always this straightforward. Suppose that a man and a woman who are both tongue-rollers have a child. Both may be genotype TT, one may be Tt and the other TT, or both may be

Tt. The possible outcomes can be displayed in a **Punnett square**, as shown in Figure 30.7. Here, the genes carried within the eggs and sperm of the parents are specified as four squares that represent the possible combinations of genes that might arise in the offspring following fertilization.

In Figure 30.7a, both parents are tongue-rollers with genotype TT, so every egg from the mother and every sperm from the father will have the T gene (or more precisely, the T allele). Inevitably, all children will therefore have genotype TT and will be tongue-rollers. If both parents are non-tongue-rollers (genotype tt), all their eggs and sperm will have the t gene and their children will be non-tongue-rollers (genotype tt) as shown in Figure 30.7f. The situation gets a little more complicated if one parent is a tongue-roller with genotype TT and the other is

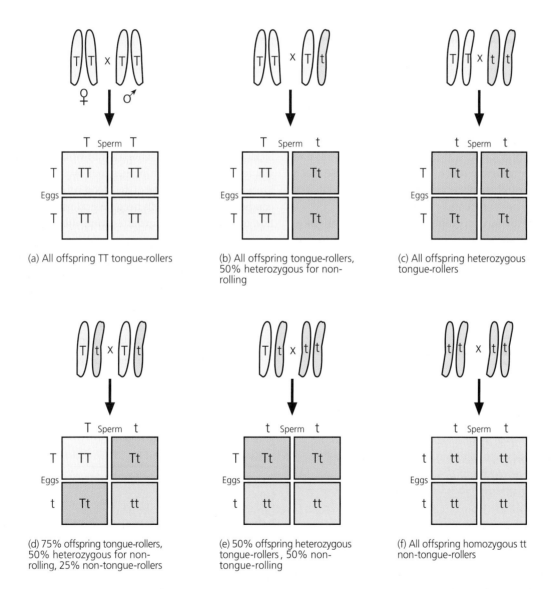

(a) All offspring TT tongue-rollers

(b) All offspring tongue-rollers, 50% heterozygous for non-rolling

(c) All offspring heterozygous tongue-rollers

(d) 75% offspring tongue-rollers, 50% heterozygous for non-rolling, 25% non-tongue-rollers

(e) 50% offspring heterozygous tongue-rollers, 50% non-tongue-rolling

(f) All offspring homozygous tt non-tongue-rollers

Figure 30.7 The inheritance pattern of tongue rolling plotted as Punnett squares. Capital T indicates the ability to form a U-shape with the tongue while lower-case t indicates the inability to do so. The tongue-rolling allele is dominant, so both TT and Tt genotypes are tongue-rollers. The inheritance of other dominant autosomal genes shows a similar pattern.

a non-tongue-roller (tt). In this case, their offspring will inherit one dominant allele (T) and one recessive allele (t). The recessive allele is masked and is not expressed in the phenotype. Consequently, they will all be tongue-rollers with the genotype Tt (Figure 30.7c).

Now consider the case where one parent has the genotype Tt and the other has genotype tt. There is an equal probability that a child will have genotype Tt (tongue-roller) or tt (non-tongue-roller) because half the gametes of the parent of genotype Tt will carry the T gene and half will carry the t gene (Figure 30.7e). If both parents are tongue-rollers with the Tt genotype, half the gametes of each parent will carry the t gene and half the T gene. So the probability that a child will be a tongue-roller with genotype TT is 0.25 (1 in 4), the probability of genotype Tt is 0.5 (1 in 2 or 50%) and the probability of genotype tt is 0.25 (Figure 30.7d). So there is a 1 in 4 chance that a child will inherit recessive alleles from both parents and will be a non-tongue-roller (tt). Thus, it is possible that a child may exhibit a phenotype that is different to that of both of his or her parents.

Tongue-rolling is an example of **autosomal inheritance**. The specific genes that determine the trait are located on a pair of autosomes. For this reason, the sex of the individual has no influence on the pattern of inheritance. Some other traits that show similar inheritance patterns are earlobe structure (free or fixed), cheek dimples (presence or absence) and foot arches (flat or normal).

> **KEY POINTS:**
>
> - When a person inherits different alleles from each of their parents, they are heterozygous for that gene.
> - If one allele is expressed in preference to the other, it is said to be dominant and the non-expressed allele is recessive.

Diseases caused by single gene mutations

None of the traits mentioned above is of particular significance to a person's health. However, of much greater clinical importance are those diseases whose inheritance is governed by the same laws. Indeed, the incidence of serious single gene disorders is estimated at around 1 in 200 live births. A few of these disorders are associated with the

mitochondrial DNA (see below), some are sex-linked and the rest are either autosomal dominant or autosomal recessive.

An **autosomal dominant disorder** is one in which the mutant allele is dominant. Examples include Huntington's disease, a serious degenerative condition affecting the nervous system, and familial hypercholesterolaemia, in which plasma levels of cholesterol are abnormally high. For this type of autosomal dominant genetic disorder, each child of an affected parent has a 50% chance of inheriting the disease. The inheritance pattern for Huntington's disease is shown in the pedigree chart of Figure 30.8.

Examples of **autosomal recessive** genetic disorders include phenylketonuria (PKU), sickle-cell anaemia and cystic fibrosis. PKU is the congenital absence of an enzyme, phenylalanine hydroxylase, that normally converts phenylalanine to tyrosine. In its absence, phenylalanine accumulates in the blood and impairs neuronal development. Sickle-cell anaemia is a condition in which the haemoglobin has abnormal properties that cause it to crystallize when the oxygen level of the blood is low. Under these conditions, red blood cells become sickle-shaped and the flow of blood through capillaries is impaired. Cystic fibrosis is characterized by alterations to chloride channel function and results in serious abnormalities of exocrine gland function. Its clinical aspects are discussed in Chapter 22.

In autosomal recessive inheritance, the parents of an affected child may not themselves express the disease. Both, however, must carry the mutated recessive allele in their genotype. If both parents are carriers, there will be a 1 in 4 chance that any child they have will express the disease. There is a 1 in 2 chance that the child will be a carrier and a 1 in 4 chance that they will be neither a carrier nor a sufferer of the disease. The inheritance pattern for cystic fibrosis is similar to that shown in the Punnett square for tongue-rolling (Figure 30.7).

> **KEY POINTS:**
>
> - A defective autosomal dominant gene will be expressed in a person who has at least one copy of that gene (i.e. a heterozygote).
> - Both men and women can have such a disorder and pass it on to their children.
> - People who have two copies of the defective gene (homozygotes) are generally more severely affected.

Generation

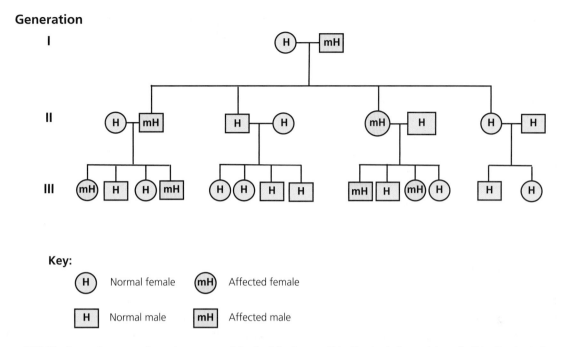

Key:

(H) Normal female (mH) Affected female

[H] Normal male [mH] Affected male

Figure 30.8 The figure shows a pedigree (or genogram) for the inheritance of Huntingdon's disease (also called Huntingdon's chorea). H designates the normal gene (HTT), which is found on the short arm of chromosome 4 and codes for a protein called huntingtin. mH indicates a defective gene. As the gene is autosomal and dominant, the children of an affected individual have a 50% chance of inheriting the condition.

30.6 Co-dominance and the inheritance of ABO blood groups

Many traits are inherited through mechanisms that are more complex than those that simply involve a pair of alleles, one dominant and one recessive. Some other inheritance patterns will now be considered briefly.

For some genes, more than two alleles exist. The gene that determines a person's ABO blood group codes for an enzyme called glycosyltransferase, which modifies an antigen on the surface of the red blood cells (the H antigen). The A and B forms of the gene modify the H antigen in different ways but the O form is inactive. This situation results in three alleles: the A, B and o alleles, sometimes represented as I^A, I^B and i. The A and B alleles (denoted A and B) are both dominant, that is they exhibit **co-dominance** while the O allele (denoted o) is recessive. This means that, even though only two alleles are inherited (one from each parent), four different blood groups exist: A, B, AB (in which both the dominant alleles are expressed in the phenotype) and O. The Punnett squares in Figure 30.9 illustrate some examples of blood group inheritance patterns. Notice in particular the example in which one parent has blood group A, with genotype Ao, and the other parent has blood group B, with genotype Bo (Figure 30.9d). Children of these parents have an equal chance of inheriting any of the four ABO groups and a 50:50 chance of inheriting a blood group that is different to that of either of its parents. The expression of the ABO genotypes is summarized in Table 30.2.

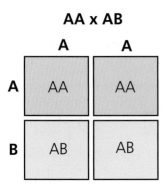

AA x AB

(a) ½ AA, ½ AB.

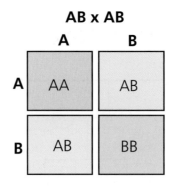

AB x AB

(b) ¼ AA, ½AB, ¼BB.

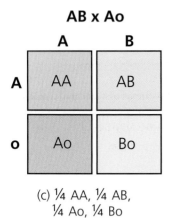

AB x Ao

(c) ¼ AA, ¼ AB,
¼ Ao, ¼ Bo

Ao x Bo

(d) ¼ AB, ¼ Ao,
¼ Bo, ¼ oo

Figure 30.9 The inheritance of ABO blood groups. There are three alleles designated I^A, I^B and i (A, B or o). The possible genotypes are AA, BB, AB, Ao, Bo and oo (more formally expressed as I^AI^A, I^BI^B, I^AI^B, I^Ai, I^Bi, ii). As the genes show codominance, there are four possible blood types: A, B, AB and O. The inheritance patterns of four different parental combinations are shown. (See also Table 30.2.)

Table 30.2 The genotypes and phenotypes of ABO blood groups

Genotype	Phenotype	Red cell antigen(s)	Serum red cell antibodies
AA	A	A	Anti-B
Ao	A	A	Anti-B
BB	B	B	Anti-A
Bo	B	B	Anti-A
AB	AB	A and B	None
oo	O	Neither A nor B	Anti-A and anti-B

Note that while there are six possible genotypes, there are only four different phenotypes (i.e. four ABO blood groups). A and B are co-dominant and both are dominant over O.

30.7 Incomplete dominance and polygenic inheritance

Some traits show a property known as **incomplete dominance** in which neither one of a pair of alleles shows complete dominance over the other. Instead, both are expressed when present together (i.e. in a heterozygous individual), giving rise to a characteristic that is intermediate between the homozygous forms. This is often seen in plants, where, for example, pink carnations result from fertilization between red- and white-flowered plants.

An example of incomplete dominance in humans is hair type. Alleles for this trait determine either curly hair or straight hair but when present together, that is when one of each allele is inherited, the individual has wavy hair. Another example is thought to be the pitch of the male voice. Homozygous men have either the lowest (bass) or highest (tenor) pitched voices while heterozygous men have voices in the intermediate (baritone) range.

Some diseases also exhibit incomplete dominance. For example, individuals who are homozygous for the sickle form of haemoglobin (haemoglobin-S) suffer from sickle-cell anaemia. Individuals who are heterozygous (one normal allele and one haemoglobin-S allele) do not have the disease but their erythrocytes do contain both normal haemoglobin and haemoglobin-S. Both alleles are coded for, and the individual is said to have the sickle trait. The fatal Tay–Sachs disease (a condition affecting one of the enzymes found in lysosomes, see p. 44) is also characterized by incomplete dominance. Heterozygous people produce around 50% of the normal amount of enzyme – enough to maintain health.

Many normal genetic traits as well as some diseases are controlled by several genes at different loci. Such traits display what is known as **polygenic inheritance**. Familiar examples of normal polygenic characteristics include eye colour, hair colour, skin colour and body shape. Many polygenic characteristics are also heavily influenced by environmental factors. A large number of diseases, including coronary heart disease, hypertension and certain forms of cancer, involve the interactions of several genes and environmental factors. Such disorders are said to be multifactorial in origin.

30.8 Sex-linked inheritance

As explained earlier, the 23rd pair of chromosomes within the nucleus of a diploid human cell determines the person's genetic sex. The sex chromosomes take two forms, the X chromosome and the much smaller Y chromosome. Females (XX) inherit an X chromosome from each of their parents while males (XY) inherit an X chromosome from their mother and a Y chromosome from their father. As roughly 50% of sperm carry an X chromosome and 50% a Y chromosome, the ratio of boys to girls at birth is around 1:1.

The male karyotype illustrated in Figure 30.1 shows the difference in size between the X and Y chromosomes. Because of its large size, the X chromosome can accommodate more genes than the smaller Y chromosome. However, the Y chromosome carries the gene specifically responsible for controlling the development of the male.

This gene occupies the **sex-determining region** (**SRY**) of the Y chromosome and is not found on the X chromosome. It follows, therefore, that the X and Y chromosomes are not homologous. Their genetic information is very different and a number of characteristics are determined by genes that are present only on the X chromosomes. They are called X-linked characteristics.

Sex-linked genetic disorders

Disorders arising from mutations of the genes confined to the X chromosome are called X-linked (or simply sex-linked) disorders. The best-known of these are probably colour blindness (see Chapter 14) and haemophilia A (which is discussed in Chapter 20) but a range of other conditions are also associated with the X chromosome.

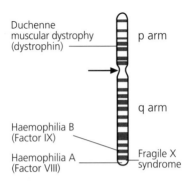

Human X chromosome

Figure 30.10 An ideogram of the human X chromosome showing the location of the genes associated with some well-known genetic diseases. The arrow indicates the position of the centromere.

Figure 30.10 shows a diagrammatic representation of an X chromosome illustrating some of the diseases associated with its genes.

To illustrate the mechanism of sex-linked inheritance, consider haemophilia A. The disease is caused by a lack of Factor VIII, which is important in the blood clotting cascade (see p. 405). Those lacking Factor VIII have a significantly increased tendency to bleed following injury (haemophiliacs). They may even bleed spontaneously. The mutated gene responsible for the condition is a recessive allele located on the q arm of the X chromosome, as shown in Figure 30.10.

The incidence of haemophilia A is much higher in males (1 in 10 000) than in females (1 in 100 000 000), because a male inherits only one X chromosome. If this single X chromosome carries a mutated gene for haemophilia, there will be no normal dominant allele to mask it and the male will be a haemophiliac. A female who inherits a mutated gene for haemophilia on one of her X chromosomes is likely to possess a normal dominant allele on her second X chromosome. The mutated allele will be masked and, although the female will be a 'carrier' for haemophilia, she herself will have blood that clots normally. The Punnett squares in Figure 30.11 show how haemophilia A is inherited through the X chromosome. Figure 30.11 also illustrates how it is possible for a female haemophiliac to arise. Although comparatively rare, if the mother is a carrier and the father is a haemophiliac,

their female children have a 1 in 2 chance of being a haemophiliac. The sperm will carry a mutated gene on its X chromosome and there is a 50% chance that it will fertilize an egg carrying a mutated gene.

A famous example of inherited haemophilia A is provided by the British royal family (see Chapter 20). It arose for the first time in the children of Queen Victoria and later spread to the royal families of Europe through Victoria's descendents. Figure 30.12 illustrates the inheritance of this sex-linked condition between successive generations of some of Queen Victoria's descendants and emphasizes the role of the female carriers in passing the disease to their male offspring.

KEY POINTS:

- An X-linked disorder is caused by a defective gene on the X chromosome, which affects men at a much higher frequency than women.

- For an X-linked disease, there is a 50:50 chance that a son of a carrier woman will be affected and that her daughters will be carriers.

- Among the offspring of a man affected by an X-linked condition and an unaffected woman, all of the daughters will be carriers and all the sons will be normal.

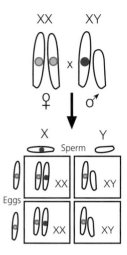

(a) All daughters are carriers. All sons are normal.

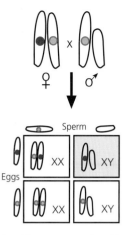

(b) Half daughters are normal and half are carriers.
Half sons are haemophiliacs and half are normal.

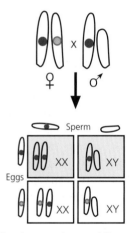

(c) Half daughters are haemophiliacs and half are carriers.
Half sons are haemophiliacs and half are normal.

⬤ Normal Factor VIII gene

● Defective Factor VIII gene

Figure 30.11 The inheritance of haemophilia A. The disease is associated with a defective gene for blood clotting Factor VIII. The gene is located on the X chromosome (see Figure 30.10) and the disease is an example of X-linked inheritance. The Punnett square in (a) shows that if the father is a haemophiliac, all his daughters will be carriers but his sons will be normal. The Punnett square in (b) shows that if the mother carries the defective gene on one of her X chromosomes, half her sons will be haemophiliacs and half her daughters will be carriers. The Punnett square in (c) shows that if the mother is a carrier and the father is a haemophiliac, their daughters have an equal chance of being a carrier or a haemophiliac. Their sons have a 50% chance of being a haemophiliac.

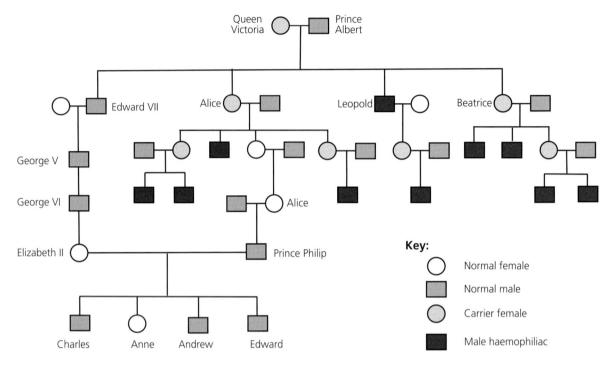

Figure 30.12 The transmission of haemophilia in some of the descendants of Queen Victoria. In this scheme, the carriers (all female) and the sufferers (all male) are both shown. Note that there are no carriers or haemophiliacs in the present generation of the British Royal Family.

30.9   Mitochondrial inheritance

Although the majority of human DNA is present in the chromosomes of the nucleus, a small amount is also found associated with the mitochondria and is known as **mitochondrial DNA**. (**mtDNA**). It takes the form of a small, round chromosome housing 37 genes. These genes contain the genetic instructions for making transfer RNA (see above), ribosomal RNA and a number of the enzymes involved in oxidative phosphorylation (see Chapter 4). Consequently, healthy mitochondrial genes are necessary for normal mitochondrial function.

In humans, mtDNA is inherited from the mother as ova contain mitochondria while the head of a sperm has none. This means that mutations of mtDNA can be passed from the mother to a child of either sex through a process known as **cytoplasmic** or **mitochondrial inheritance**. However,

men with mitochondrial disorders do not to transmit their condition to their offspring. Consequently, mitochondrial inheritance is transmitted exclusively via the maternal line, as shown in Figure 30.13. Diseases caused by mutations of mtDNA disrupt the production of energy. Consequently, they most often affect the growth and development of organs and tissues that have high energy requirements, for example heart, brain and muscle. The most widely observed features of mitochondrial disease include muscle weakness and wasting, heart disease, the development of certain cancers, kidney failure, dementia and abnormalities of the eyes and vision. A specific mitochondrial disorder affecting the eyes is Leber's hereditary optic neuropathy (LHON), which leads to progressive and irreversible loss of sight.

Generation

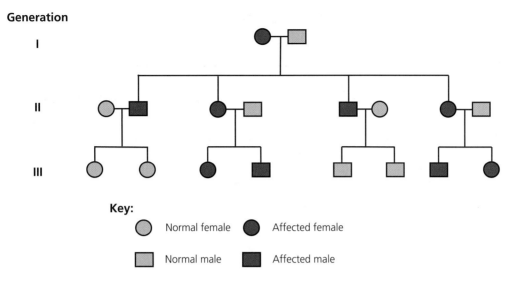

Key:

⬤ Normal female   ⬤ Affected female

⬛ Normal male   ⬛ Affected male

Figure 30.13 A pedigree to show how mitochondrial disorders can be transmitted by a woman to her descendants. These generally affect muscle and nervous tissues. Examples are Lebers hereditary optical neuropathy (LHON) and myoclonic epilepsy with ragged red fibres (MERRF). As mitochondria are passed between the generations via the egg, the diseases are inherited by all of an affected woman's children. However, only the children of her daughters will be affected, not those of her sons.

Not all mitochondrial disorders occur as a consequence of mitochondrial inheritance, some are caused by a defect in a nuclear gene. The type of genetic defect responsible can be deduced from the pattern of inheritance.

> **KEY POINT:**
>
> - As mitochondria are inherited from the mother via the egg, a defective mitochondrial gene can be transmitted to her children. A man cannot transmit such a gene to his children.

30.10 Genetic testing, genetic counselling and gene therapy

These topics all come within the branch of medicine known as clinical genetics and concern inherited disorders in families. Genetic testing includes carrier detection, predictive testing and prenatal diagnosis.

Carrier detection may be appropriate for people who are likely to have an increased risk of carrying a chromosomal anomaly (rearrangement) or an autosomal recessive condition. A variety of methods for detecting carriers of such disorders is available, including the identification of specific mutations, detailed blood tests and biochemical measurements of enzyme activity.

Predictive testing is most often used to provide information about conditions that develop in later life, particularly those, like Huntington's disease or adult polycystic kidney disease, that show autosomal dominance. Such testing is preclinical, that is it is carried out before the disease has started to produce symptoms. If preventative treatment or regular surveillance is available, this kind of testing may have considerable health benefits. If no treatment is available, however, test results could potentially cause great distress. For this reason, it is essential that anybody choosing to undergo predictive genetic testing is fully informed of the consequences of a positive diagnosis.

Prenatal diagnosis refers to the detection of a genetic disorder before birth. Testing may be carried out by means of maternal blood tests, ultrasound scans, amniocentesis (analysis of amniotic fluid samples, usually at around 16–18 weeks of gestation) or chorionic villus sampling (CVS) in which a small portion of tissue is removed from the placenta (usually at around weeks 11–14 of gestation) to provide fetal DNA for chromosomal analysis. Although prenatal testing may give prospective parents the opportunity to take informed decisions regarding the pregnancy, it also has the potential to cause them great distress. Furthermore, both amniocentesis and CVS carry a small risk of miscarriage. Examples of genetic conditions

that can be detected prenatally include Down syndrome, cystic fibrosis and Duchenne muscular dystrophy.

With increased knowledge of genetics and the inheritance of genetic diseases come a number of dilemmas. For example, who should have access to the information? How should it be acted upon? **Genetic counselling** is the process whereby individuals or couples can receive information and education about the nature of a genetic condition and the risks that the condition may pose to their own health and that of their children. Should it be necessary, they will also be offered appropriate psychological and emotional support to help them in making informed decisions regarding possible interventions.

Gene therapy refers to the insertion of a normal copy of a gene into the cells of a person suffering from a genetic disorder. Although extensive research has been and continues to be carried out in this area, efforts have, as yet, met with very little success. The inserted genes have shown poor levels of expression and many of the patients who have been treated experimentally in this way have suffered a variety of serious side-effects caused by the viruses used to insert the genes or by new mutations resulting from insertion.

Recommended reading

Alberts, B., Johnson, A., Lewis, J., Raff, M., Roberts, K. and Walter, P. (2008). *Molecular Biology of the Cell*, 5th edn. Garland: New York. Chapters 1 and 4.

Elliott, W.H. and Elliott, D.C. (2008). *Biochemistry and Molecular Biology*, 4th edn. Oxford University Press: Oxford. Chapters 22–25.

Lewis, R. (2006). *Human Genetics*, 7th edn. McGraw-Hill.

Read, A. and Donnai, D. (2006). *New Clinical Genetics*. Scion Publishing: Oxford.

Young, I.D. (2005). *Medical Genetics*. Oxford University Press: Oxford. Chapters 1–8, 13 and 14.

Self-assessment questions

Which of the following statements are true and which are false? Answers are given at the end of the book (p. 755).

1
a) Chromosomes consist of a single large strand of DNA.
b) Human cells have 23 pairs of chromosomes.
c) A karyotype is the description of a person's chromosomes.
d) Transcription involves the synthesis of proteins on a mRNA template.
e) A gene consists of a single triplet of bases.

2 a) Each autosomal chromosome consists of a pair of homologous sister chromatids.

b) An egg or sperm has half the number of chromosomes of an autosomal cell.

c) During mitosis, the chromosome number is halved.

d) When a cell undergoes cell division, the chromosomes become attached to the spindle via the kinetochore.

e) Some genetic disorders are the result of eggs having more chromosomes than they should.

3 a) Errors in genetic coding are called mutations.

b) Sickle-cell anaemia is caused by a mutation in the gene that codes for haemoglobin.

c) The gene is the basic unit of inheritance.

d) A phenotype is a description of an individual's genetic makeup.

e) An allele is one of two or more forms of a gene.

4 a) Mendelian laws of inheritance precisely specify the inheritance of a characteristic coded by a single gene.

b) Down syndrome is an example of a disease caused by mutation of a single gene.

c) A dominant gene is one that is always expressed as a physical characteristic.

d) As haemophilia A is always passed to the male line, it is caused by a defective gene on the Y chromosome.

e) Mitochondrial defects are always the result of gene transmission from the mother to her child.

Section 6

Integrative aspects of physiology and pharmacology

31 The nutritional needs of the body **670**

32 Energy balance and exercise **682**

33 The regulation of body temperature **695**

34 Body fluid and acid–base balance **707**

35 The uptake, distribution and elimination of drugs **726**

31 The nutritional needs of the body

After reading this chapter you should have gained an understanding of:

- The importance of a mixed, balanced diet

- Dietary sources of nutrients

- The role of dietary fats, proteins and carbohydrates (the macronutrients)

- The importance of vitamins, minerals and trace elements (the micronutrients)

- The factors that regulate hunger, appetite and satiety

- Some aspects of malnutrition

- Eating disorders such as anorexia nervosa, bulimia and obesity

- The assessment of nutritional status

31.1 Introduction

Nutrition is fundamental to healthy body function. It involves the intake of food, its digestion and the subsequent utilization of its constituents by the body for energy production and the building or repair of the tissues. The particular selection of foods eaten by an individual is called the **diet**.

Nutrients are substances that are absorbed and utilized to promote cellular function. They are often classified as **macronutrients** and **micronutrients**. Carbohydrates, fats and proteins are required in large amounts by the body and are thus macronutrients, whereas vitamins, minerals and trace elements are needed in much smaller quantities and are therefore known as micronutrients. It must also be remembered that water is an essential nutrient and aspects of fluid balance are considered on pp. 714–715.

A balanced diet contains all the essential nutrients in the appropriate amounts and proportions. By convention, essential nutrient requirements are assessed by determining the amount per day needed to prevent clinical deficiency. An additional 30% is added to this amount to give the RDA (recommended daily amount). Deficiencies often develop progressively and it is therefore difficult to establish an RDA with precision. For this reason, published values often differ significantly (particularly between countries). It is also important to remember that a number of factors may influence the nutritional requirements of an individual. These include age, gender, activity level, pregnancy, lactation and general state of health. Furthermore, social, psychological and environ- mental factors may influence appetite and thus nutri- tion, particularly during a stay in hospital.

31.2 Macronutrients

Carbohydrates

Most people consume between 50 and 70% of their calorific intake in the form of carbohydrates. Cereals, pasta, fruits and vegetables are the principal sources of carbohydrates but they are also found in dairy products. Except for liver, very little carbohydrate is found in meat. Carbohydrates are digested in the gut to monosaccha- rides (see Chapter 24), which are then utilized by cells to generate energy. However, the carbohydrate com- ponents of plant cell walls, the celluloses and lignins, are not digested in the human gut but pass through unchanged. This indigestible carbohydrate is known as fibre or roughage and helps to facilitate the movement of material through the gut. Glucose is stored by the body as glycogen (chiefly in the liver and skeletal mus- cle). Once the glycogen stores are full, the excess glucose is converted to fat, which is stored in adipose tissue, as indicated in Figure 31.1.

Each gram of pure carbohydrate yields 17.2 kJ (4.1 kcal) of energy when metabolized and the amount of carbo- hydrate required each day depends on the level of physi- cal activity of the individual. For a person in a relatively sedentary job, the daily energy requirement is met with a carbohydrate intake of 400–500 g (see Chapter 32 for further details).

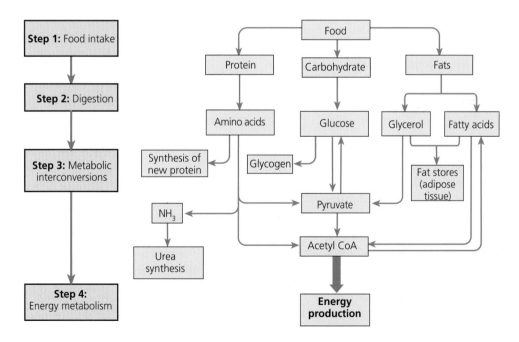

Figure 31.1 The major pathways by which dietary proteins, carbohydrates and fats are broken down and converted for energy produc- tion and storage.

Proteins

Proteins are the main structural components of the tissues, so the principal use for the protein of the diet is the synthesis of new protein. Consequently, the body's requirement for protein is determined by its need for amino acids, which are the building blocks for the structural protein necessary for growth and the maintenance and repair of the tissues. Amino acids are also required for the synthesis of functional proteins such as enzymes and hormones.

Table 31.1 lists the daily protein requirements of males and females at different ages. During illness, both protein synthesis and degradation occur at an increased rate but in most cases there is a net loss of body protein with muscle wasting. During recovery, this must be restored and the protein requirement during this period is increased.

Many people, particularly those in Western societies, eat far more protein each day than they need for the building or repair of the tissues. Those amino acids that are not required for protein synthesis are deaminated in the liver and kidneys prior to their subsequent metabolism. Deamination results in the liberation of ammonia, which is metabolized to urea by the liver (Figure 31.1) before being excreted in the urine. The carbon skeleton of most of the deaminated amino acids can be used to synthesize glucose (a process known as **gluconeogenesis**) and those that can be utilized in this way are known as the glucogenic amino acids. Of the 20 amino acids found in proteins, only leucine and lysine cannot be used for

gluconeogenesis. These amino acids are oxidized via the same pathway as fats and are classified as ketogenic amino acids. On average, each gram of pure protein will yield 17.2 kJ (4.1 kcal) of energy.

The nutritional value of protein depends upon the amino acids that it contains. Of the 20 α-amino acids that make up the proteins of the body, eight cannot be synthesized by cells and must be included in the diet. These are the **essential amino acids**. The remaining 12 are non-essential as they can be synthesized by the liver from other amino acids and carbohydrate precursors. Therefore they need not be included in the diet. Protein foods that contain all the essential amino acids in the proportions required to maintain health are known as **complete proteins**. Examples are meat, fish, soy beans, milk and eggs. Most plant proteins are incomplete and lack one or more of the essential amino acids. However, by eating a wide variety of incomplete protein foods, it is possible to avoid amino acid deficiencies – an important consideration for vegetarians and vegans.

A diet that is deficient in both protein and energy may result in a range of clinical syndromes known collectively as **protein–energy malnutrition** (**PEM**). In Western societies, it generally only occurs in patients who have, or who are recovering from, severe illness such as cancers of the gastrointestinel tract, malabsorption diarrhoea, or chronic conditions such as AIDS and chronic obstructive pulmonary disease. It may also be caused by inadequate food intake due to physical, psychological or social factors. In developing countries, PEM is more often caused by a poor diet and may be exacerbated by infections that cause diarrhoea. Chronic deprivation may result in stunting of height as well as inadequate weight gain. **Marasmus** and **kwashiorkor** are clinical syndromes resulting from a diet lacking in protein. In the case of marasmus, the energy content of the diet is also inadequate.

Table 31.1 Recommended daily intake of dietary protein (in grams) for different age groups

Age (years)	Males	Females
1–3	15	15
4–6	20	20
7–10	28	28
11–14	42	41
15–18	55	45
19–50	56	45
Over 50	53	47

Fats

Fats are divided into two groups, saturated and unsaturated. Vegetable oils contain mainly unsaturated fat while saturated fat is found in milk, eggs, meat and oily fish. Linoleic, linolenic and arachidonic acid are polyunsaturated fatty acids that cannot be synthesized by the body. They are therefore **essential fatty acids** and must be included in the diet. Plant oils are good sources of linoleic and linolenic acids while animal tissues contain small amounts of arachidonic acid. Cholesterol, an important constituent of the plasma membranes of cells, is found in fatty meat, egg yolk and dairy products but it is also synthesized by the body. Fatty acid deficiencies are rare, because most people consume more fat than they need. In a typical Western diet, fats contribute around 30% to the total calorific intake. Of this, around one-third will be in the form of saturated fat. Complete oxidation of 1 g of fat will yield 38.9 kJ (9.3 kcal) of energy. Fats are thus a very rich source of energy.

In addition to their role as a source of energy, fats have a number of important uses in the body. Fats in the form of phospholipids are an important constituent of cell membranes and the myelin of nerve sheaths, while cholesterol is the starting point for the synthesis of the steroid hormones (see also Chapter 2). Fatty tissue supports and protects organs such as the kidneys and eyes. It also provides a layer of thermal insulation beneath the skin. Finally, fat depots store the fat-soluble vitamins A, D, E and K.

31.3 Micronutrients

Vitamins

Vitamins are essential for normal metabolism and health. They are subdivided into the fat-soluble vitamins (A, D, E and K) and water-soluble vitamins (C and B complex). Table 31.2 lists the recommended daily amounts (RDA), major sources, functions and deficiency disorders of each vitamin.

Fat-soluble vitamins (A, D, E and K)

Vitamin A (retinol)

Vitamin A is needed for the formation of the visual pigments, for the maintenance of the epithelia lining the gut, respiratory tract and urogenital system, for efficient operation of the immune system and for normal limb development in the fetus. Vitamin A deficiency is associated with a number of abnormalities. It is a major cause of blindness in tropical countries, and causes diarrhoea and respiratory disease and an increased susceptibility to infection. Vitamin A deficiency during pregnancy may result in abnormalities of the fetus.

Vitamin A toxicity is potentially serious. Acute symptoms of excessive intake include vomiting, vertigo, headache, blurred vision and raised CSF pressure. Longer-term consequences are more variable but include hyperlipidaemia, bone and muscle pain, and skin disorders. Excess vitamin A is now known to be teratogenic during the first 3 months of pregnancy (the first trimester). It may also cause miscarriage (spontaneous abortion). For this reason, pregnant women are advised to avoid eating foods such as liver, which are rich in vitamin A.

Vitamin D (cholecalciferol)

The metabolism of vitamin D and its importance in the normal mineralization of the skeleton are discussed in Chapter 16 (p. 310). Vitamin D deficiency has different consequences at different stages of life. In toddlers, a lack of vitamin D leads to a condition called **rickets** in which the bone fails to mineralize normally because of poor calcium absorption in the gut. As the infant starts to stand and bear weight on his or her legs, the weakened bones become deformed and bowed, as shown in Figure 31.2. In adulthood, a lack of vitamin D causes a related condition known as **osteomalacia**, which is characterized by demineralization and weakening of the skeleton. The elderly and women who have had several children are most at risk of developing osteomalacia, especially if they have little exposure to sunlight.

Vitamin D is toxic if consumed in amounts far in excess of the RDA. Symptoms are caused by the high plasma calcium (hypercalcaemia) that results from enhanced intestinal calcium absorption. There may be hypertension, and calcification of soft tissues such as the kidneys, heart, lungs and blood vessel walls.

Table 31.2 Actions and daily requirements of vitamins

Vitamin	Daily requirement (adults)	Major dietary sources	Deficiency disorders
A (retinol)	700 μg	Dairy products, oily fish, eggs, liver	Night blindness, epithelial atrophy, susceptibility to infection
D (cholecalciferol)	10 μg	Fish oils, dairy products; synthesized in skin	Rickets (children) Osteomalacia (adults)
E (α-tocopherol)	8 μg	Nuts, egg yolk, wheatgerm, milk, cabbage	Haemolytic anaemias
K (coagulation vitamin)	100 μg	Green vegetables, pig liver; also synthesized by intestinal flora	Bruising, bleeding
B_1 (thiamine)	1.6 mg	Lean meat, fish, eggs, legumes, green vegetables	Beriberi with many neurological symptoms, disturbances of metabolism
B_2 (riboflavin)	1.8 mg	Milk, liver, kidneys, heart, meat, green vegetables	Dermatitis, hypersensitivity to light
B_3 (niacin or nicotinamide)	15 mg	Most foods; can be synthesized from tryptophan	Pellagra, listlessness, nausea, dermatitis, neurological disorders
B_6 (pyridoxine)	2 mg	Meat, fish	Irritability, convulsions, anaemia, vomiting, skin lesions
Pantothenic acid	5–10 mg	Most foods	Neuropathy, abdominal pain
Biotin (vitamin H)	100 μg	Liver, egg yolk, nuts, legumes	Muscle pain, scaly skin, elevated blood cholesterol
B_{12} (cobalamin)	1.2 mg	Liver, meat, fish (NOT plants)	Pernicious anaemia
Folic acid or folate (vitamin B_9)	250 μg	Liver, dark green vegetables; also synthesized by intestinal bacteria	Macrocytic anaemia, gastrointestinal disturbances, diarrhoea
C (ascorbic acid)	50 mg	Fresh fruits (especially citrus fruits), vegetables	Scurvy, liability to infection, poor wound healing, anaemia

Vitamin E (α-tocopherol)

Vitamin E is a powerful antioxidant that protects cell membranes from free radical damage. Deficiency is characterized by damage to the membranes of nerve and muscle cells, which leads to a loss of reflexes, abnormal gait and ataxia. In infants with very fragile red blood cell membranes, deficiency can lead to the development of haemolytic anaemia.

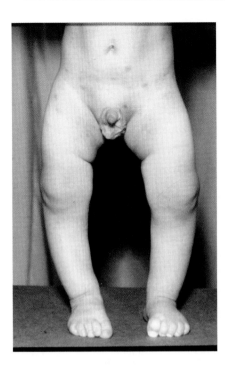

Figure 31.2 Typical curvature of the legs in a young child suffering from rickets.

Vitamin K

The principal dietary sources of vitamin K are green leafy vegetables but it is also synthesized by commensal bacteria of the large bowel (see p. 526). Consequently, primary vitamin K deficiency is uncommon in healthy adults. Vitamin K is needed for the normal coagulation of blood (see Chapter 20) and deficiency of vitamin K causes bleeding disorders (haemorrhagic disease). The widely used anticoagulant drug warfarin is a vitamin K antagonist. Vitamin K is also known to be required for the synthesis of osteocalcin, a protein that enhances the activity of osteoblasts (bone-forming cells) and is an important calcium-binding protein of the bone matrix.

Many elderly patients admitted to hospital with a fractured neck of the femur have been found to be low in vitamin K. This suggests that vitamin K deficiency may predispose to osteoporosis. Newborn infants, especially those who are born prematurely, are often lacking in vitamin K as they have no gut bacteria and the vitamin does not cross the placenta. As a result, babies are at risk of a disorder known as **haemorrhagic disease of the newborn**

(HDN), which may lead to bleeding from fragile capillaries into the subcutaneous tissues, gastrointestinal tract, lungs or, most damaging of all, the brain. For this reason, newborn babies are often given an injection of vitamin K as a prophylactic measure.

Water-soluble vitamins (B group and vitamin C)

Vitamin B_1 (thiamine)

This vitamin has a key role in the metabolism of carbohydrates. Deficiency results in a range of metabolic disturbances including acidosis as a result of the accumulation of lactic acid. **Beriberi** is the name given to describe the clinical responses to thiamine deficiency and the condition is further subdivided into wet and dry beriberi. In wet beriberi, there is vasodilatation that leads to right-sided heart failure and generalized oedema. Dry beriberi is normally associated with chronic thiamine deficiency and is characterized by peripheral neuropathies and altered consciousness, headaches, vomiting and confusion. In prolonged, untreated vitamin B_1 deficiency (often associated with chronic alcohol abuse), a condition called **Korsakoff's psychosis** may develop in which there is short-term memory loss and, eventually, signs of physical damage to the nervous system (**Wernicke's encephalopathy**).

Vitamin B_2 (riboflavin)

This vitamin is one of a group of coenzymes that participate in oxidation and reduction reactions of metabolism. Its main dietary sources are meats, eggs and milk products. Deficiency is rare and not fatal although symptoms include cracking at the corners of the mouth (angular stomatitis), loss of epithelium over the tongue (glossitis) and skin lesions.

Vitamin B_3 (niacin)

Niacin can be synthesized by the body from the amino acid tryptophan found in most proteins. Deficiencies are normally confined to populations whose principal food is maize, a cereal that does not contain tryptophan. The deficiency syndrome is called **pellagra** and may be fatal if untreated. Its symptoms include diarrhoea, dementia and dermatitis, especially around the neck region (Casal's collar), as illustrated in Figure 31.3.

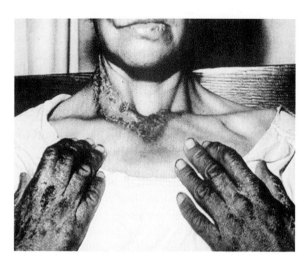

Figure 31.3 A patient with pellagra showing Casal's collar and severe dermatitis on the back of each hand.

Vitamin B₆ (pyridoxine)

This vitamin is involved in amino acid metabolism and the release of glucose from glycogen stores, particularly those of muscle. A lack of vitamin B_6 gives rise to an increased sensitivity of target tissues to steroid hormones such as oestrogens, androgens, cortisol and vitamin D. It is also needed for the synthesis of GABA, an important inhibitory neurotransmitter of the CNS (see Chapter 7). Vitamin B_6 deficiency may result in sleeplessness, convulsions and other neurological changes. Vitamin B_6 overdose may lead to peripheral nerve damage.

Folic acid

Folic acid (also known by the name of its anion, folate, and occasionally as vitamin B_9) is a water-soluble vitamin found principally in liver and green leafy vegetables. It is involved in a number of metabolic reactions and is needed, along with vitamin B_{12}, for normal red blood cell maturation. Folic acid deficiency can result in a condition called megaloblastic anaemia in which immature red cells are released prematurely into the circulation. Folic acid is also known to be essential for normal closure of the neural tube during development of the nervous system. Women who have folate deficiency and who become pregnant are more likely to have small babies, which are at an increased risk of having neural tube defects. Because of this, folic acid supplements are now recommended for women prior to conception and during early pregnancy. The use of such supplements has resulted in a significant reduction in the incidence of neural tube defects such as anencephaly and **spina bifida**, and other neurological deficits.

Vitamin B₁₂ (cobalamin)

The term vitamin B_{12} refers to a group of cobalt-containing compounds known as the cobalamins. Vitamin B_{12} is found only in foods of animal origin so vegans are at risk of developing vitamin B_{12} deficiency unless it is added to their diet as a supplement.

The absorption of vitamin B_{12} depends on its binding to intrinsic factor secreted by the stomach (see Chapter 24). Consequently, individuals lacking intrinsic factor may become deficient in vitamin B_{12}. As vitamin B_{12} is required for the production of red cells, deficiency results in anaemia, specifically **macrocytic anaemia** (or pernicious anaemia). Vitamin B_{12} is also required for normal neurological development and deficiency can result in degeneration of nerve fibres and irreversible neurological damage.

Vitamin C (ascorbic acid)

Vitamin C is needed for the synthesis of collagen, which is the principal connective tissue of tendons, arteries, bone, skin and muscle. It is also required for the manufacture of noradrenaline (norepinephrine) and aids the absorption of dietary iron. **Scurvy**, the disorder caused by vitamin C deficiency, is characterized by a reduced ability to make new connective tissue and its symptoms include bleeding from gums, nose and hair follicles. Bleeding may also occur in the joints, the bladder or gut. Gum disease, loose teeth and listlessness are also typical features of scurvy (Figure 31.4). There is no evidence that excessive ingestion of vitamin C is harmful but reports that high doses of vitamin C may protect against cancers or viral infections have proved unfounded.

Minerals

A wide variety of inorganic ions is needed for normal cellular function. Important examples are sodium (the major extracellular cation), potassium (the major intracellular cation), calcium, phosphorus, iron and iodine. In addition, smaller amounts of a number of other elements

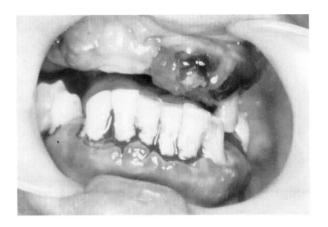

Figure 31.4 The typical symptoms of scurvy. Note the bleeding and swollen gums and the loss of teeth from the upper jaw.

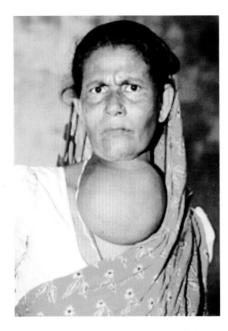

Figure 31.5 A Bangladeshi woman with a very pronounced goitre.

are also needed. These are referred to as trace elements and include zinc, copper and manganese. In general, mineral deficiencies are seen only in regions where the food is grown on soil lacking a particular mineral. Most **iodine deficiency** is seen in places where iodine is scarce, such as parts of Africa, Asia and South America. In these regions, the thyroid gland may become enlarged and give rise to **goitre**. A particularly clear example is shown in Figure 31.5. Most other important minerals are present in a wide range of foods so those eating a balanced diet are unlikely to be at risk of deficiencies.

Iron is an essential component of haemoglobin and myoglobin, as well as of certain pigments and enzymes. As the recycling of iron is very efficient, the need for dietary iron in adults arises mainly from loss by bleeding and the death of intestinal cells. It therefore follows that the dietary requirement for iron is greater in menstruating women than in men, being about 1 mg per day in men and 2 mg per day in women of child-bearing age. Furthermore, children and pregnant women need relatively more iron because of their expanding circulatory volume. Dietary sources of iron include meat (specifically the myoglobin of the muscle), vegetables and fruits. The normal Western diet contains adequate quantities of iron (around 15 mg day^{-1}) but strict vegetarians and vegans risk iron deficiency as much of their dietary iron is unavailable for absorption.

31.4 The regulation of food intake – appetite, hunger and satiety

Although related, hunger and appetite are different sensations. Hunger refers to the physiological sensation of emptiness while appetite refers to the feelings associated with the anticipation of food. Appetite may be altered by nervousness, fear or other emotional or social factors. Satiety refers to the sensation of fullness that accompanies the satisfaction of hunger.

The physiological basis for the regulation of food intake is not completely understood but it is clear that the hypothalamus plays a key role. The lateral hypothalamus contains a region called the feeding centre. Lesions here cause aphagia (cessation of feeding) in rats. The ventromedial hypothalamus contains an area known as the satiety centre as damage to this region of the brain can induce excessive eating (hyperphagia).

Overweight and obesity

People who have excessive fat stores in relation to their height, gender and race are classified as being overweight or obese (see below). Obesity is a growing problem, especially in affluent societies such as parts of Europe and the USA and has a number of physical, physiological and psychological consequences. The increasing number of obese children and young adults is of particular concern. Obese individuals are at increased risk of diabetes, gallstones, fatty liver, infertility, osteoarthritis, spinal problems, obstructive sleep apnoea and a range of cardiovascular problems such as coronary heart disease, varicose veins, peripheral oedema and hypertension. They are also more likely to die while undergoing surgery and to suffer postoperative complications. Many obese people, particularly those with mobility problems, also suffer from feelings of isolation and low self-esteem.

Weight gain reflects an imbalance between energy intake and expenditure so the most important contributory factor is the type and amount of food eaten and the amount of physical activity carried out. There is some evidence to suggest that the tendency to be overweight is at least partly determined by genetic factors. In the early 1990s, it was shown that a gene (the *ob* gene) is responsible for the production of a satiety factor in genetically engineered obese (*ob/ob*) mice. Its protein product, leptin, has been shown to reduce the weight of obese mice, possibly by speeding up the rate at which fat is metabolized. Interactions between leptin, the satiety centre of the hypothalamus and the thyroid gland are the focus of much study in this important area of nutrition.

Apart from advising reduced food intake, weight loss can be facilitated by the use of the drug **Orlistat**, which is a lipase inhibitor. It is used in combination with a calorie-restricted diet in individuals who have diabetes, high blood pressure, high cholesterol or heart disease. Orlistat prevents the digestion and absorption of part of the dietary fat. The undigested fat is eliminated in the faeces.

Anorexia nervosa and bulimia nervosa

Anorexia nervosa is an eating disorder characterized by the ingestion of very small amounts of food, typically 2.5–3.37 MJ day^{-1} (600–900 kcal day^{-1}), often accompanied by a vigorous exercise regime. The disorder is mainly confined to young women aged between 10 and 30 years but males account for around 10% of all cases and older people may also be affected. Patients have extreme concerns about weight gain and have a distorted body image. Feelings of worthlessness and depression are common. Once the body mass index (BMI, see below) falls below about 17, a number of physical and physiological symptoms are seen. These include amenorrhoea in females of reproductive age, increased susceptibility to cold, osteoporosis, and gastrointestinal and cardiovascular problems. If left untreated, death may result as a consequence of heart failure caused by atrophy of the cardiac muscle.

Bulimia nervosa is a disorder in which periods of starvation are interspersed with binges that may involve the consumption of enormous quantities of food (up to 200 MJ or 50 000 kcal at a time). After a binge, bulimics commonly purge by vomiting or by using laxatives. Although there are fewer obvious physical consequences of bulimia (indeed BMI is usually in the normal range), purging may lead to electrolyte imbalances and persistent vomiting can cause erosion of the enamel of the teeth.

Enteral and parenteral nutritional support

Nutritional support may be required for people who are unable to eat normally for some reason. Dysphagia (difficulty swallowing), impaired intestinal absorption, a reduced level of consciousness or the consequences of treatments may all interfere with normal nutrition. Enteral feeding (via the gastrointestinal tract) may be possible if the gut is healthy, but parenteral nutrition (bypassing the gastrointestinal tract) may be necessary if it is not.

Enteral feeding may be achieved either by giving liquid formula feeds orally, or via a nasogastric tube. If the upper gastrointestinal or respiratory tract is obstructed, a percutaneous endoscopically guided gastrostomy tube (PEG tube) may be used. Such feeds will contain appropriate amounts of macro- and micronutrients.

In parenteral feeding, the gut is bypassed and nutrients are infused directly into a large central vein (usually either the subclavian or the jugular vein). Nutrients are in a form that can be used directly by cells, that is amino acids, glucose and an emulsion of triglycerides. As parenteral feeds are hypertonic, it is essential that appropriate fluids are also given and that the individual is monitored regularly for fluid and electrolyte status. Regular weight checks are also important.

31.5 Assessment of nutritional status

Individuals suffering from any form of malnutrition, including obesity, must be monitored regularly and accurately to assess their progress in terms of weight gain or loss. A number of assessment methods are available.

The simplest way to assess nutritional status is to weigh the person. Weighing alone, however, is of little use without knowledge of the person's height. A person weighing 70 kg will be normal if he or she is 1.8 m tall but severely overweight if he or she is 1.3 m tall. A more accurate assessment can be made from determination of the BMI (body mass index) in which weight is viewed in the context of height. BMI is calculated as follows:

$$BMI = \frac{\text{Weight (kg)}}{(\text{Height in m})^2}$$

Thus, for a person 1.7 m tall weighing 70 kg:

$$BMI = \frac{70}{1.7 \times 1.7} = 24.2 \text{ kg m}^{-2}$$

Values between 18.5 and 25 kg m^{-2} are in the desirable range and are associated with optimum life expectancy while values less than 18.5 are considered underweight and may be associated with an increased risk of health problems in some individuals. Values above 25 are considered to represent varying degrees of overweight and obesity, although most health-care practitioners nowadays regard vales of BMI between 25 and 28 as acceptable in terms of health risks. When BMI exceeds 30, the individual is classified as obese and values above 40 represent severe obesity.

BMI is not always a reliable indicator of a person's true nutritional status. In heavily muscled individuals, such as trained athletes, BMI often exceeds normal values because muscle tissue weighs more than adipose tissue. Furthermore, in children, BMI measurements are of little value because of the changes in build that take place at various times during development (e.g. the growth spurt at puberty). In children, longer-term monitoring of height and weight as well as of cognitive function is of greater value.

A number of other physical measurements may also yield useful information regarding nutritional status. Skinfold thickness reflects the size of the subcutaneous fat depot and may be measured using callipers. Skinfold thickness can be measured at a variety of body sites, and a typical determination might involve four sites, the biceps, triceps, subscapular (below the shoulder blade) and suprailiac (waist) regions. The measurements from these four sites are added together and, by using tables constructed from the results obtained from a large population of men and women, it is possible to relate skinfold thickness to total body fat content (expressed as a percentage of body mass). Some values are given in Table 31.3.

Other physical measurements include mid-upper-arm circumference and mid-arm circumference, which are useful indicators of the body's protein reserves and are particularly helpful in monitoring weight gain in severely undernourished children.

Often, physical measurements of the kind described above are carried out in combination with biochemical measurements. The appearance of ketones in the urine, for example, indicates that fats are being broken down for use in metabolism, while a negative nitrogen balance indicates that proteins are being broken down faster than new proteins are being synthesized.

Finally, devices are available that can measure the relative proportions of fat, lean tissue and water in the body by measuring its conductivity. Such devices are often used in fitness clubs and gyms. It is important to remember that all methods used to assess nutritional status are subject to errors and that these may be considerable in people suffering from oedema.

Table 31.3 Percentage of body fat estimated from skinfold thickness

Skin fold thickness (mm)	Percentage body fat		
	Age 17–19 years	Age 30–39 years	Age 50+ years
Male subjects:			
20	8.3	12	12.5
30	13	16	18.5
40	16.5	19	23
50	19	21.5	26
60	21	23.5	29
70	23	25	32
Female subjects:			
20	14	17	21
30	19	22	26.5
40	23	25.5	30
50	26	28	33
60	28	31	35.5
70	30.5	32.5	37.5

Skinfold thickness is measured at four locations: the biceps, triceps, below the shoulder blade (subscapular skinfold) and at the waist (suprailiac skinfold). The values are added together to determine the final skinfold thickness. For males, the desirable percentage of body fat lies between 15 and 21% depending on age. For females, the desirable range is 17–25%.

KEY POINTS:

- The diet consists of all the food eaten by an individual. Broadly, it may be divided into macronutrients and micronutrients.
- The macronutrients either provide energy for daily activities (carbohydrates and fats) or provide the necessary materials for the maintenance and growth of the body tissues (proteins).
- The micronutrients are accessory food factors that are required in small quantities and include the vitamins and certain minerals. The vitamins are divided into the fat-soluble vitamins (vitamins A, D, E and K) and the water-soluble vitamins (B group vitamins and vitamin C).
- The nutritional status of a person can be assessed by their body weight in relation to their height and age. More useful measures for adults are the body mass index and skinfold thickness.

Recommended reading

Bender, D.A. (2007). *Introduction to Nutrition and Metabolism*, 4th edn. Taylor & Francis: London.

Mann, J.A. and Truswell, S. (eds) (2007). *Essentials of Human Nutrition*, 3rd edn. Oxford Medical Publications: Oxford.

Self-assessment questions

Which of the following statements are true and which are false? Answers are given at the end of the book (p.755).

1. a) Within the body, carbohydrates are stored as glycogen.
   b) All essential amino acids are found in meat.
   c) Plant oils mainly consist of unsaturated fats.
   d) All fatty acids can be synthesized by the body from glucose.
   e) Cholesterol is an essential constituent of the plasma membrane of cells.

2. a) Excess intake of vitamin A can cause severe toxic reactions.
   b) Vitamin K deficiency may increase the risk of thrombosis.
   c) Vitamin B_3 deficiency may cause severe neurological disorders.
   d) Vitamin B_{12} deficiency results in aplastic anaemia.
   e) Rickets is caused by vitamin C deficiency.

3. a) A diet that is deficient in iodine may cause goitre.
   b) The hypothalamus plays an important role in the regulation of food intake.
   c) A person suffering from bulimia nervosa is likely to be severely underweight.
   d) An individual with a BMI of $25\,kg\,m^{-2}$ is overweight.
   e) A growing child may have a BMI significantly below $20\,kg\,m^{-2}$ and yet be perfectly healthy.

32 Energy balance and exercise

After reading this chapter you should have gained an understanding of:

- The general concepts of energy balance, metabolism, metabolic rate and basal metabolic rate
- The concept of the energy equivalent of oxygen for carbohydrates, fats and proteins
- The relationship between oxygen consumption and carbon dioxide production – the respiratory quotient
- The physiological factors that influence metabolic rate
- The energy requirements of exercise
- The cardiovascular adjustments that permit the delivery of adequate quantities of oxygen, fatty acids and glucose to the muscles during exercise

32.1 Introduction

A human being at rest uses about 360 litres of oxygen a day to 'burn' several hundred grams of carbohydrates and fats, liberating about 7500 kJ (1792 kcal) of heat. This is roughly equivalent to the heat output of an 80-watt electric bulb burning all day. During physical activity, the amount of oxygen consumed and the associated heat output may increase substantially. The biochemical processes that are responsible for generating this heat constitute the **metabolism** of the body. Metabolism may be divided into two categories: anabolic and catabolic. **Anabolic metabolism** involves the synthesis of complex molecules from simpler ones. A well-known example is the synthesis of new protein from amino acids that occurs during growth. **Catabolic** **metabolism** involves the breaking down of large, complex molecules to smaller, simpler ones. This process is accompanied by the liberation of energy (remember that heat is a form of energy). The oxidative metabolism of glucose to carbon dioxide and water is a typical example.

In considering the energy balance of the body, this chapter will first discuss how much energy can be obtained from the carbohydrates, fats and proteins of the diet. It will then discuss the factors that influence the rate of energy production and the energy requirements of various daily tasks. The final section will discuss the adjustments to the circulatory system that occurs in exercise.

32.2   The chemical processes of the body produce heat

In the body, anabolic and catabolic reactions take place side by side. Glucose and fatty acids are constantly being taken up by the cells for energy production. The energy released by the oxidative metabolism of these substances is harnessed to drive anabolic processes such as protein synthesis and other important cellular activities such as the maintenance of ionic gradients, muscle contraction and secretion. Adenosine triphosphate (ATP) plays a crucial role in all these processes. It is manufactured in large amounts in the mitochondria during the oxidative metabolism of glucose and fats, as described in Chapter 3. ATP enables the cells to utilize the energy liberated by catabolism to

provide energy for their activities, as shown schematically in Figure 32.1. Virtually all the work performed by the body eventually ends up as heat, which is lost from the body. Only the work done outside the body, such as that involved in lifting a heavy object, is not directly converted to heat. Even the energy consumed in anabolic reactions represents a store of chemical energy that is eventually lost as heat when, for example, a cellular protein is eventually broken down and its constituent amino acids are utilized for energy production. The heat produced by catabolism is essential for the maintenance of normal body temperature (see Chapter 33).

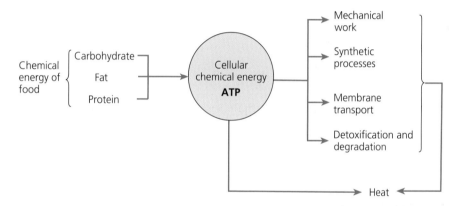

Figure 32.1 An overview of the biological transformation of various forms of energy.

KEY POINTS:

- The chemical processes of the body constitute its metabolism.
- Metabolism gives rise to heat and provides energy for cellular activities in the form of ATP.

32.3   Energy balance

The unit of energy used in the study of energy metabolism is the kilojoule (kJ). The **joule** is defined as the amount of work done when a force of 1 newton moves through

a distance of 1 metre in the direction of the force. The rate at which work is done is known as the power of the system and is measured in watts, where 1 watt is equivalent to 1 joule of work per second. In the earlier literature, and in the popular press, the unit of energy is often given as **kilocalories** (or Calories with a capital C). One kilocalorie is equivalent to 1000 calories (with a small c). In terms of heat, 1 kilocalorie is the amount of heat needed to raise the temperature of 1 kg of water by 1°C and is equivalent to 4187 joules (4.187 kJ). The energy value of prepackaged foodstuffs is usually printed on the side of the packet in both kJ and kcal. A serving of 50 g of breakfast cereal would have an energy content of around 950 kJ or 227 kcal.

The First Law of Thermodynamics states that energy can neither be created nor destroyed. Applied to the human body, this means that the total amount of energy taken in by the body must be accounted for by the energy put out by the body. Thus:

ENERGY INPUT = ENERGY OUTPUT

This may be represented by the following equation:

Chemical energy = Heat energy + Work energy
of food   + Stored chemical energy   (1)

The rate at which chemical energy is expended by the body is known as the **metabolic rate**. When the amount of energy ingested as food is sufficient to balance the amount of energy put out in the forms of heat and work, the chemical energy (and mass) of the body remains constant. In reality, this is rarely the case. More often there is a small imbalance such that energy is either stored within the body or depleted by the catabolism of carbohydrates, fats and, in more prolonged fasting, proteins. Even during the course of one day, there are small fluctuations in the energy reserves of a normal individual. At night, the glycogen stored in the liver is slowly consumed but is restored following the first meal of the next day (see Chapter 25). During growth, there is substantial gain in body mass due to the synthesis of new proteins but in later life, as activity declines, there is a tendency for excess food intake to result in the deposition of fat, with consequential weight gain. The principal stages of energy metabolism are illustrated in Figure 32.2.

To study the energy balance of a person under the conditions normally encountered in life, three of the

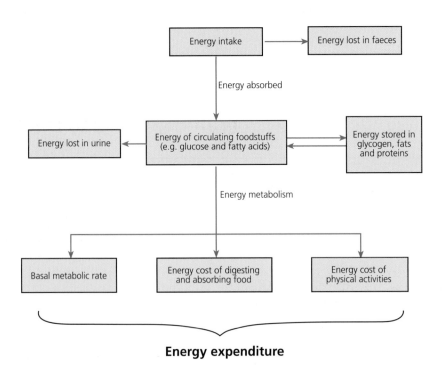

Energy expenditure

Figure 32.2 A schematic diagram of the factors that must be considered in determining energy balance.

variables in Equation 1 above would have to be measured so that the fourth may be calculated. This can be done in a whole body calorimeter, which measures the total heat output of a subject. This technique is impractical for the routine clinical assessment of patients. Instead, the measurement of energy balance may be simplified by eliminating some of the variables in Equation 1. By carrying out measurements on an individual 8–12 hours after a meal (the postabsorptive state of metabolism), the input of chemical energy from food may be taken as zero. If the person is at complete rest, energy output in the form of external work can also be disregarded. Under these conditions, all the chemical energy derived from the metabolic pool is used to maintain vital bodily functions and is ultimately completely converted to heat. Equation 1 now simplifies to:

Loss of stored chemical energy = Heat energy   (2)

Consequently, under these conditions the heat production is a direct measure of metabolic rate, which is said to represent the **basal metabolic rate** (**BMR**). *The BMR is defined as the energy requirement of the fasting body during complete rest in a thermoneutral environment (c. 20°C).* It is usually measured after a night's sleep but before eating, and measures the energy requirements of the basic body functions such as breathing, beating of the heart, maintenance of body temperature, etc. It is an index of metabolism under standardized conditions and does not represent the minimum metabolic rate of an individual, which is normally only achieved during deep sleep. As the BMR varies with body size, it is usual to relate it either to body surface area or to lean body mass (see Section 32.5 below).

KEY POINTS:

- The rate at which chemical energy is expended by the body is called the metabolic rate and is measured in watts (joules per second).

- During the postabsorptive period, under conditions of complete rest, the metabolic rate is referred to as the basal metabolic rate or BMR.

- The BMR varies with the age, sex and size of the individual.

32.4 How much heat is liberated by metabolism?

The answer to this question depends upon the nature and proportions of the foods being metabolized. One way of looking at the energy value of different foods is to consider how much energy is liberated by the complete oxidation (burning) of a standard amount. Thus, the complete oxidation of 1 g of carbohydrate or protein will liberate 17.16 kJ (4.1 kcal), while 1 g of fat will liberate 38.94 kJ (9.3 kcal). Considered from this point of view, fat represents a very efficient way of storing energy.

The energy equivalent of oxygen

If known amounts of pure fat, carbohydrate or protein are burned in a calorimeter together with 1 litre of oxygen, then different amounts of heat energy are liberated from each foodstuff. Pure carbohydrate, for example, will liberate 20.93 kJ (5 kcal) of energy per litre of oxygen consumed, whereas pure fat will liberate 19.68 kJ (4.7 kcal) and protein will liberate about 18.84 kJ (4.5 kcal). These values represent the **energy equivalent of oxygen** for the complete oxidation of carbohydrate, fat and protein.

The energy equivalent of oxygen will vary according to the relative amounts of protein, fat and carbohydrate being utilized by the body at a particular time. As these amounts cannot be determined directly, the energy equivalent of oxygen is calculated indirectly from the

respiratory quotient (RQ) (also known as the respiratory exchange ratio), which is the ratio of the volume of carbon dioxide breathed out to the volume of oxygen absorbed from the lungs in 1 min:

$$RQ = \frac{\text{volume of } CO_2 \text{ produced per minute}}{\text{volume of } O_2 \text{ consumed per minute}} \quad (3)$$

The RQ depends upon the nature of the foodstuffs undergoing metabolism. If pure carbohydrate is oxidized, an RQ of 1.0 is obtained while the specific oxidation of fat gives an RQ of 0.7 and that of protein an RQ of about 0.8. In reality, the body uses a variable mixture of all three metabolic fuels (although protein is usually a minor energy source, except in starvation). The RQ will then be a mean value weighted towards the principal fuel. The RQ of a subject on an ordinary mixed diet will be around 0.85 while that of a fasting individual is about 0.82, as relatively more fat (or even protein) is metabolized. Once the RQ value for the subject under investigation has been obtained, it is possible to determine the energy equivalent of oxygen from standard tables.

Calculation of metabolic rate

To determine the metabolic rate of an individual by indirect calorimetry, we must know:

1. The rate at which they are using up oxygen – his or her **oxygen consumption** (normally expressed in litres per minute).
2. The energy equivalent of oxygen for the food being metabolized (expressed in kJ per litre of oxygen).

Modern methods of measuring oxygen consumption rely on measuring the flow of air in and out of the lungs while determining the difference in the oxygen and carbon dioxide content between the inspired and expired air. As these can be measured for each breath, it is possible to determine the RQ and the energy equivalent for oxygen. Thus, if a person has an oxygen consumption of $300\,ml\,min^{-1}$ with an energy equivalent of $20.93\,kJ\,l^{-1}$, their metabolic rate is $0.3 \times 20.93 = 6.28\,kJ\,min^{-1}$. If the values are divided by 60, the number of seconds in a minute, their energy consumption can be expressed in kW (i.e. $kJ\,s^{-1}$). In this case, it would be $0.105\,kW$ (105 watts).

32.5 Basal metabolic rate

In adults, the BMR amounts to an average daily expenditure of 84–105 kJ (20–25 kcal) per kg of body weight. This requires the consumption of some 200–250 ml of oxygen each minute. About 20% of the BMR is accounted for by the activity of the CNS, 25% by the liver, 20–30% by the skeletal muscle mass and about 16% by the heart and kidneys.

The BMR depends on many different factors. It is determined genetically, at least in part, and is affected by a number of physiological variables including:

- Body weight
- Body surface area
- Lean body mass

- Age
- The sex of the individual.

As there is no simple, direct relationship between BMR and body size, it is customary to relate metabolism to the surface area (kJ per square metre per hour) as heat is produced in proportion to surface area. A standard nomogram is available for computing the body's surface area from an individual's height and weight. Tables are available that give the expected normal range of BMR values for people of different body size, age and sex. This information is valuable in clinical situations where it is necessary to determine whether a person's BMR lies outside the normal range.

Figure 32.3 illustrates the influence of age on the normal metabolic rates of both males and females. The BMR

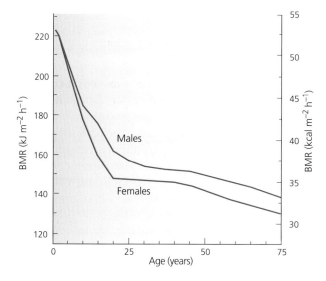

Figure 32.3 The average age-related changes in BMR for males and females corrected for body size.

of women is 6–10% lower than that of men of comparable size and age. For adult males between 20 and 60 years of age, the normal BMR ranges from 142 to 168 kJ m⁻² h⁻¹ (34–40 kcal m⁻²h⁻¹). For non-pregnant women, the range is 134–146 kJ m⁻² h⁻¹ (32–35 kcal m⁻²h⁻¹). BMR shows a significant increase during pregnancy because of the additional metabolic activity of the fetus.

In relation to its size, the metabolic rate of a young child is almost twice that of an elderly person. This high BMR reflects the relatively high rates of cellular reactions, rapid synthesis of cellular materials and growth, all of which require moderate quantities of energy. Part of the decline in BMR seen after the age of 40 years is due to a progressive rise in the proportion of adipose tissue, which has a lower metabolic rate than lean body tissue.

32.6 Physiological factors that affect overall metabolic rate

It is important to distinguish between the metabolic rate and the BMR (which refers to the metabolic activity of a subject who is awake but at complete rest, see above). The BMR provides a standard way of comparing the metabolic rate of different subjects, which is of particular value in the assessment of endocrine disorders. However, the metabolic rate varies from minute to minute depending on the nature of the activities being undertaken by an individual, and during normal waking activity the metabolic rate is greater than the BMR. Any increase in the physical activity of the body will increase the metabolic rate in proportion to the energy expended. This section is concerned with the influence of physiological factors other than exercise on metabolic rate. The effect of exercise on metabolic rate and energy balance is considered in Section 32.7 below.

Effects of hormones

A variety of hormones are able to alter the way in which fats, carbohydrates and proteins are utilized by the cells and tissues of the body. Changes in the pattern of secretion of these hormones alter the metabolic rate. The principal hormones that may affect BMR are:

- The thyroid hormones
- Growth hormone
- The male and female sex steroids (particularly the androgens)
- The catecholamines, adrenaline and noradrenaline (epinephrine and norepinephrine).

The thyroid hormones exert a powerful effect on the BMR. Depression of thyroid activity gives rise to a condition known as myxoedema (see Chapter 16) in which BMR may fall to half its normal value. In hyperthyroid individuals, BMR is markedly enhanced. Thyroid hormones raise metabolic rate because they stimulate many of the chemical reactions taking place within cells, including oxygen consumption by the mitochondria (see Chapter 16, p. 299). Growth hormone and the male and female sex steroids influence BMR to a minor degree, all having a mild stimulatory effect. Male sex steroids (e.g. testosterone) are more potent than oestrogens in this respect and probably account for the higher BMR seen in males (Figure 32.3). The adrenaline and noradrenaline (epinephrine and norepinephrine) secreted from the adrenal medulla and the noradrenaline secreted by the sympathetic postganglionic nerve fibres, stimulate the metabolic rates of virtually all the tissues of the body.

Ingestion of food

After eating a protein-rich meal, there is an increase in the rate of heat production by the body, that is an increase in metabolic rate. Although part of this rise may be due to the cellular processes involved in the digestion and storage of foods, it is believed to result mainly from specific effects exerted by certain amino acids derived from the protein in the meal. This is known as *the specific dynamic action of protein*. Heat production also increases following the ingestion of a meal rich in carbohydrates or fats, but to a lesser extent. It is thought that the liver is the major site of the extra heat production because of its importance in the intermediary metabolism of absorbed foods.

Fever

Whatever its cause, fever elicits an increase in metabolic rate. This is due simply to the fact that all chemical reactions, including those taking place within cells, proceed more rapidly as temperature increases. For every 0.5°C rise in body temperature, the BMR increases by about 7%. Thus, the metabolic rate of a person with a fever of 39°C will be nearly a third higher than normal.

Fasting and starvation

The basal energy requirement of a man with a surface area of 1.8 m² is about 7–7.5 MJ a day (1700–1800 kcal a day). This will normally be supplied by the daily ingestion of foods. When little or no food is being eaten, the glycogen stores in the liver are consumed first, followed by the mobilization of stored fats and the conversion of amino acids to glucose (gluconeogenesis, see Chapter 4, p. 52). If starvation is prolonged, muscle protein is broken down to provide amino acids for glucose production. Prolonged starvation also decreases the BMR, often by as much as 20–30%, as the availability of necessary food substances in the cells is reduced. This decrease serves to limit the drain on available resources. Such people are said to be in a state of catabolism, with stores of fat, carbohydrate and protein all diminishing.

Sleep

There is a fall in metabolic rate of between 10 and 15% during sleep as a result of a reduction in the level of muscular tone and a fall in activity of the sympathetic nervous system.

> **KEY POINTS:**
> - The metabolic rate may be calculated from oxygen consumption and the energy equivalent of oxygen for the foods being metabolized.
> - Metabolic rate is increased by any form of exercise, the ingestion of food and by fever. It is reduced in malnutrition and during sleep.
> - A variety of hormones can modify the metabolic rate. Catecholamines and thyroid hormones are potent stimulators of metabolism while growth hormone and the sex steroids exert a mild stimulatory effect.

32.7 Energy expenditure during exercise

Any degree of exercise, from that required simply to carry out the normal daily activities of life such as walking, eating, dressing, etc. to the severe exercise involved in manual labour, produces an increase in metabolic rate. Walking slowly (4 km h⁻¹) uses about three times as many kilojoules per hour as lying in bed asleep, while maximal muscular exercise can, in short bursts, increase the metabolic rate by around 20-fold. The energy requirements of different tasks and different occupations have been investigated very thoroughly. People such as office workers who are involved in sedentary work have a modest daily energy requirement – about 8.25 MJ (*c.* 2000 kcal), while manual workers in heavy industries such as mining or agriculture may have an energy requirement three times greater than this. Table 32.1 shows some examples of the energy requirements of particular activities. The daily energy requirement for any individual will depend on the amount of time spent on each of the activities undertaken in the day and their energy cost. A typical calculation is given in Box 32.1. The minute-by-minute energy cost of any activity expressed as a ratio of the BMR is called the physical activity ratio (PAR). The 24-hour energy expenditure expressed as a ratio of the BMR is called the physical activity level (PAL).

Delivery of oxygen and glucose to tissues in exercise

When exercise is undertaken by a healthy person, the circulatory and respiratory systems are adjusted to meet the increased metabolic demands. Cardiac output rises as a result of an increase in both heart rate and stroke volume and the proportion of the cardiac output distributed to the active muscles increases. There is a rise in oxygen uptake that is related to the amount of work done. This is achieved both by an increase in ventilation and by a greater extraction of oxygen from the circulating blood. In this section, the detailed mechanisms by which these changes are accomplished are considered.

Table 32.1 Approximate energy consumption of an adult male engaged in various physical activities

Activity	Energy consumed	
	watts	kcal min^{-1}
Sleeping	77	1.1
Sitting/office work	105	1.5
Standing	120	1.7
Slow walking (3 km h^{-1})	195	2.8
Washing, dressing etc	230	3.2
Household chores (bed-making, cleaning, ironing, etc.)	250	3.6
Brisk walking (6 km h^{-1})	363	5.2
Gardening	400	5.7
Cycling (16 km h^{-1})	433	6.2
Coal mining/loading	500	7.2
Dancing	533	7.6
Jogging (10 km h^{-1})	712	10.2
Jogging (14 km h^{-1})	963	13.8
Swimming (fast crawl)	872	12.5
Cross-country skiing	1033	14.8

1 watt = 1 joule s^{-1}. This is a measure of the power required for a given task.

Box 32.1 Daily energy balance calculation

A man spends 8 hours a day working with an average energy consumption of 700 kJ h^{-1}. He spends another 8 hours sleeping or lying down (276 kJ h^{-1}). He walks to and from work at a brisk pace (1250 kJ h^{-1}), which takes 30 minutes a day. In his remaining time, he spends 7 hours sitting and attending to personal needs (dressing, eating, etc.) at an average energy consumption of 376 kJ h^{-1} and exercises by jogging for 30 min each evening at an average energy consumption of 3000 kJ h^{-1}. His total energy requirement each day can be calculated as shown in the table:

or protein yields approximately 17 kJ of energy and each gram of fat yields approximately 39 kJ, his recommended dietary intake should be:

Carbohydrates:	$(12\,565 \times 0.6) \div 17 = 443\,g$
Fats:	$(12\,565 \times 0.3) \div 39 = 97\,g$
Protein:	$(12\,565 \times 0.1) \div 17 = 74\,g$

If the BMR is taken as 270 kJ h^{-1}, the basal energy requirement for a day is 6480 kJ. The physical activity

Activity	Time spent (hours)	Rate of energy consumption (kJ h^{-1})	Total energy requirement (kJ)
Sleeping and lying	8	276	2208
Sitting, eating, etc.	7	376	2632
Walking	0.5	1250	625
Jogging	0.5	3000	1500
Working	8	700	5600
Total	24	–	12 565

His total energy consumption is 12 565 kJ (12.56 MJ or 3000 kcal) a day, which must be met by the diet. For optimal nutrition, about 60% of the energy value of food should be in the form of carbohydrate, 30% as fats and the remainder as protein. As each gram of carbohydrate

level or PAL for this subject would be 12 565 ÷ 6480 = 1.93. Calculations of this kind are used for estimating the quantities of food required to feed populations, particularly in the developing countries of the world.

Oxygen consumption rises in proportion to the work done

As soon as exercise begins, the muscles start to expend energy in proportion to the work done. Oxygen consumption, however, does not rise immediately to match the energy requirements. Instead, it rises progressively over several minutes until it matches the needs of the exercising muscles. As the work continues, the oxygen uptake remains at a level appropriate to the severity of the exercise (the steady state). Thus, at the commence-

ment of exercise the body builds up an oxygen deficit (or 'oxygen debt') (Figure 32.4).

In the steady state, the oxygen consumption is proportional to the work done until the work rate approaches the maximum capacity. In the case of the example shown in Figure 32.5, the maximal oxygen uptake is 3.5 l min^{-1}, which provides the **maximal aerobic power**. Blood lactate levels begin to rise above 1–2 mmol l^{-1} (the normal plasma level) when the oxygen consumption rises above about 2 l min^{-1}. The increase in work from 250 to 300 watts is not

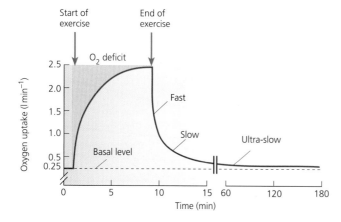

Start of exercise | End of exercise

Figure 32.4 Oxygen consumption during and after a period of exercise. (Note the progressive oxygen uptake after the start of exercise and the slow return to the resting level after exercise ceases.)

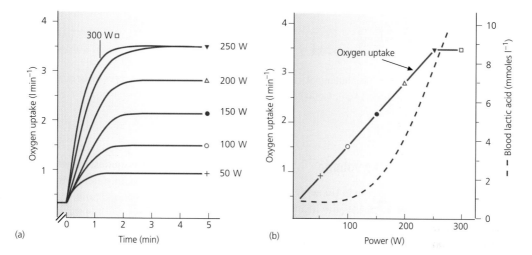

(a)

(b)

Figure 32.5 The relationship between oxygen uptake and energy expenditure expressed in watts (W). The highest levels of exercise are associated with large increases in blood lactate concentrations.

accompanied by additional oxygen uptake and depends on anaerobic metabolism. As this requires glycolysis, blood lactate rises steeply (see Chapter 4, p. 48).

At the end of a period of exercise, the oxygen consumption declines rapidly but does not reach normal resting levels for up to 60 minutes. The first phase of the decline in oxygen consumption is very fast. This is followed by a slower decline that takes tens of minutes to reach normal resting levels. Excess lactate is resynthesized into glucose and glycogen during this period. After severe and sustained exercise, oxygen consumption remains elevated for several hours, perhaps due to stimulation of metabolism as a result of the heat generated during the period of exercise.

Cardiovascular changes in exercise

At rest, cardiac output is about $5 \, l \, min^{-1}$. Of this, only 15–20% is distributed to the skeletal muscles. In heavy exercise, cardiac output may rise fivefold to more than $25 \, l \, min^{-1}$, of which approximately three-quarters is distributed to the exercising muscles (Figure 32.6). During this change, the blood flow to the brain remains essentially constant while that to the gut and kidneys declines substantially.

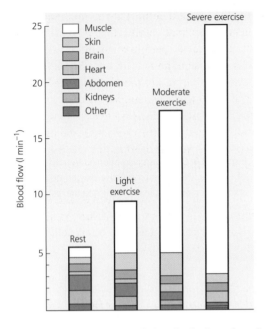

Figure 32.6 Estimated values of the distribution of cardiac output to various tissues at rest and during light, moderate and severe exercise in man.

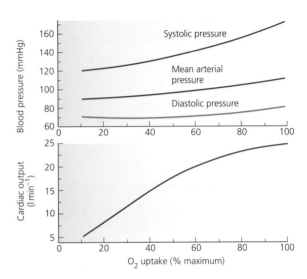

Figure 32.7 The changes in the cardiac output and arterial blood pressure with exercise of increasing intensity expressed as oxygen uptake.

Effects on heart rate and stroke volume

As exercise begins, there is an increase in the activity of the sympathetic nerves, which increases the heart rate and mobilizes blood from the large veins. In addition, the increased sympathetic activity leads to an increase in stroke volume. Consequently, cardiac output increases, as shown in Figure 32.7. At moderate levels of exercise (up to about 40% of maximal oxygen uptake), both heart rate and stroke volume increase in proportion to the work done. Above this level, stroke volume does not increase further and any additional increase in cardiac output is due to an increase in heart rate.

Blood pressure during exercise

During exercise, the increased force of ventricular contraction causes an increase in systolic pressure, which becomes more marked as exercise intensifies. The diastolic pressure remains relatively stable and may even decline as the peripheral resistance falls due to the dilatation of the arterioles in the skeletal muscles. Consequently, the rise in

mean arterial pressure is modest in dynamic exercise such as running (Figure 32.7). In static exercise such as weight lifting, however, arterial blood pressure rises substantially and systolic pressure may exceed 200 mmHg (27 kPa).

Blood gases in exercise

At rest, the haemoglobin is about 97% saturated and the oxygen content of the arterial blood is around $19.8\,\mathrm{ml\,dl^{-1}}$ (i.e. 19.8 ml of oxygen are carried in each 100 ml of blood). The venous blood is approximately 75% saturated and has an oxygen content of around $15.2\,\mathrm{ml\,dl^{-1}}$ so that approximately 4.6 ml of oxygen is extracted from each dl of blood as it passes through the tissues. At work loads below the anaerobic threshold, the partial pressures of oxygen and carbon dioxide of the arterial blood remain relatively constant while the oxygen content of the venous blood draining the active muscles falls progressively as the intensity of the exercise increases. At the same time, the $P\mathrm{co_2}$ of the venous blood rises from its normal value of around 46 mmHg. The rise in $P\mathrm{co_2}$ and the associated fall in pH favour the delivery of oxygen to the active tissues (the Bohr effect; see Chapter 20, p. 398). At workloads above the anaerobic threshold, there is a gradual fall in the P_{O_2} and pH of the arterial blood. Overall, the amount of oxygen extracted from the blood increases with the intensity of exercise, as shown in Figure 32.8.

Effects of training

The performance of exercise can be improved by training. This requires the regular undertaking of physical exercise that is of an appropriate intensity, duration and frequency. To achieve optimal results, the intensity of training must increase as performance improves. Frequent, regular exercise of an appropriate kind is important if the improvement in performance is to be maintained. Regular training with strenuous exercise will lower the resting heart rate and increase the size of the heart and the thickness of the ventricular wall. As the stroke volume increases, the resting cardiac output is maintained, despite the fall in resting heart rate. Maximal cardiac output increases from $20-25 \, l \, min^{-1}$ in untrained young adults to values that may exceed $35 \, l \, min^{-1}$ in highly trained athletes. The cardiovascular changes increase the maximal oxygen uptake and the capacity for physical work is thereby increased.

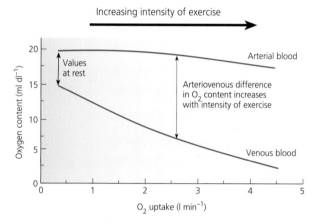

Figure 32.8 The changes in the oxygen content of arterial blood and mixed venous blood with exercise of increasing intensity. Note the progressive increase in the arterio-venous difference in oxygen content with the severity of exercise and that the oxygen content of arterial blood falls very little.

KEY POINTS:

- In mild and moderate exercise, rate and depth of breathing increase in direct proportion to the work done.

- The cardiac output increases in proportion to the metabolic demand.

- The oxygen required by the exercising muscles is met by an increased blood flow and by an increased extraction of oxygen from the circulating blood.

Recommended reading

McArdle, W.D., Katch, F.I. and Katch, V.L. (2006). *Exercise Physiology: Energy, Nutrition and Human Performance*, 6th edn. Lippincott, Williams & Wilkins. Chapters 4–10.

Self-assessment questions

Which of the following statements are true and which are false? Answers are given at the end of the book (p. 755).

1 a) Basal metabolic rate (BMR) is defined as the lowest metabolic rate that occurs in any 24-hour period.
 b) BMR decreases with age.

 c) BMR is increased in people suffering from hypothyroidism.
 d) The oxygen consumption of a normal adult at rest is around $250 \, ml \, min^{-1}$.

2 a) The oxygen consumption of a person is a good measure of their metabolic rate.
   b) The respiratory quotient (RQ) provides information about the foodstuffs being metabolized.
   c) Metabolic rate is decreased following a protein-rich meal.
   d) Carbohydrates provide the main source of energy in exercise.
   e) Plasma glucose falls significantly during prolonged exercise at half the maximal rate.

3 a) In fit individuals, the cardiac output increases more than fivefold in severe exercise.

   b) In severe exercise, the mean arterial pressure is unchanged because diastolic pressure falls.
   c) The increase in cardiac output during exercise is mainly due to an increase in heart rate.
   d) The arterial P_{CO_2} rises significantly during exercise.
   e) The activity of the joint receptors is an important factor driving the increase in ventilation.

The regulation of body temperature

33

After reading this chapter you should have gained an understanding of:

- The importance of regulating body temperature

- The concept of core and shell temperatures and the ways in which heat may be lost from the body surface to the environment

- The importance of the skin for regulating heat exchange between the core and the surface of the body

- The responses of the body to cold, including behavioural changes, shivering, vasoconstriction and non-shivering thermogenesis

- The responses of the body to heat, including behavioural changes, vasodilatation and sweating

- The consequences of hypothermia, hyperthermia and fever

33.1 Introduction

Despite wide fluctuations in the ambient (environmental) temperature, it is possible for humans to maintain their body temperature within very narrow limits. The brain and organs within the thoracic and abdominal cavities have the highest temperature. This is known as the **core temperature**, which is defined as the temperature of the blood flowing to vital internal organs. The core temperature is normally regulated within the range 35.5–37.7 °C. Under most conditions, the skin has the lowest temperature, the **shell temperature**. The skin can tolerate a much wider range of temperature than the internal organs and its surface temperature varies according to the temperature of the surroundings, as shown in Figure 33.1.

The maintenance of a constant core temperature is vital for normal body function. The rates of all metabolic processes are highly temperature dependent. An increase in temperature of 1 °C increases the rate of enzymatic chemical reactions by 10% and vice versa. Furthermore, most enzymes have an optimum temperature for activity above which their activity declines as they start to degrade (denature). Many enzymes and cellular proteins begin to become irreversibly damaged at temperatures above about 42 °C and for this reason a body temperature of 43 °C is close to the upper limit for human life. Similarly, although cells can withstand marked reductions in temperature, they do function more slowly as the

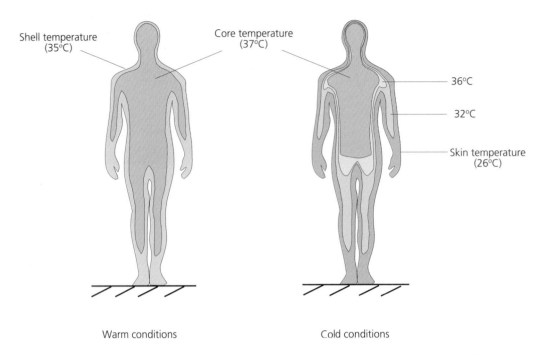

Figure 33.1 The body shell and core temperatures when a subject is in a warm environment (left) and in a cold environment (right). The shaded areas indicate areas of equal temperature. The innermost area is the core body temperature and includes most of the body under warm conditions. When the body is exposed to cold, the boundary of the core temperature shrinks to preserve heat in the brain, thorax and abdomen. The temperature of the limbs falls, with the greatest drop seen in the most distal regions.

temperature falls. A low core temperature affects the function of the brain and consciousness is normally lost when the core temperature falls below about 33°C. The process by which the body regulates its core temperature within the range that is compatible with normal function is known as **thermoregulation**.

In clinical practice, core temperature is measured routinely for a number of reasons including:

- To establish a baseline temperature on admission to a hospital or clinic
- To monitor the progress of patients recovering from hypothermia
- To monitor the temperature of people undergoing treatment for infections
- To monitor fluctuations in temperature as an indicator of developing infection or the presence of a deep vein thrombosis (this is especially important during the postoperative period)

- To monitor signs of incompatibility during a blood transfusion.

There are several locations in the body at which a value close to core temperature may be measured. These include the oral cavity (in one of the sublingual pockets), the axilla (armpit), the ear canal (tympanic temperature) and the rectum. The most commonly used sites are the ear canal and the oral cavity. A number of different types of thermometer are also available and it is important to remember that the reading obtained will depend upon both the device used and the site of measurement.

Natural variations in core temperature arise as a result of:

- Genetic factors
- Exercise
- Circadian variation
- The menstrual cycle and pregnancy.

A healthy person's core temperature varies throughout the day, with the lowest values occurring during the night (particularly between 3 a.m. and 5 a.m.) and the highest in the late afternoon or early evening. Differences of up to 1°C are not unusual. This 24-hour pattern is illustrated in Figure 33.2.

In women of reproductive age, core temperature fluctuates during each menstrual cycle. Between ovulation and the onset of the next menstrual bleed (the luteal phase of the cycle), core temperature rises by 0.2–0.8°C in response to the high levels of progesterone secretion occurring during this time (see Chapter 28). Core temperature falls just prior to menstruation as steroid levels fall following degeneration of the corpus luteum, as shown in Figure 33.3. In the event of pregnancy, however, the elevated core temperature is maintained until the baby has been delivered reflecting the large amounts of progesterone secreted by the plasma.

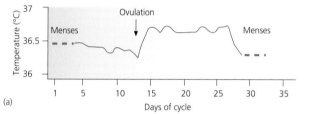

(a)

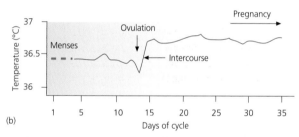

(b)

Figure 33.3 (a) The body temperature of a woman recorded first thing in the morning throughout a complete menstrual cycle. (b) A similar record for the same woman showing the persistence of the temperature rise once she became pregnant.

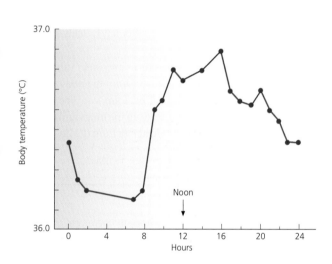

Figure 33.2 The change in core body temperature with time of day. The core temperature is at its lowest in the early morning and at its highest in mid- to late afternoon.

33.2 Mechanisms of heat exchange between the body and the environment

In order to maintain a normal core temperature, heat lost from the body must be balanced by heat gained or generated by the body. Heat is generated by all the active metabolic processes carried out by cells. The more active a tissue is, the more heat it generates, so that large, active organs such as the liver, skeletal muscle and brain generate the most heat. It is also possible for the body to gain heat from the environment if the ambient temperature is higher than that of the body: when lying in a hot bath, heat is gained by conduction; when sunbathing, heat is gained by radiation; when sitting by a fire, heat is mainly gained by radiation but some may reach the body by convection of warm air. When the environmental temperature is lower than that of the body, heat is lost from the body surface to its surroundings, by radiation, conduction and convection, and by the evaporation of sweat, as shown in Figure 33.4.

Radiation is the loss of heat in the form of infrared rays (thermal energy) and accounts for around 60% of the heat lost by the body to its surroundings. Conduction is the transfer of heat between objects that are in direct contact with one another, such as the warming of a chair by sitting on it or the cooling of oral tissues by cold drinks. Heat is also conducted to the air immediately in contact with the body and this air is removed by convection currents and replaced by cooler air so that further heat loss occurs. This explains why it feels colder when a breeze is blowing than when the air is still (the 'wind chill' effect). By restricting the flow of air, clothing can significantly reduce the heat lost by this route.

As water can absorb much more heat than air, the body loses heat more rapidly when it is immersed in water than it does when in air of the same temperature. Indeed, a person can become hypothermic extremely quickly after falling into cold water.

Evaporation occurs when water turns into water vapour. For each litre of water that evaporates from the body surface, around 2.26 MJ (540 kcal) of heat are lost. For this reason, sweating and evaporation from the lungs and oral mucosa are important routes of heat loss from the body.

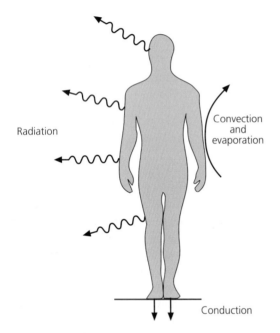

Figure 33.4 A simple diagram to show the routes by which heat is lost from the body. Heat can also be gained by convection, conduction and radiation if the surroundings are at a higher temperature than the body.

KEY POINTS:

• Human beings maintain core body temperature between 36 and 38°C, which requires heat loss to be balanced by heat generation.

• Heat is generated within the body by metabolic reactions and is lost from the body surface mainly by radiation, but also by conduction and convection, and by evaporation of water.

33.3   The role of the skin in thermoregulation

The skin forms the interface between the internal structures of the body and the outside world. It is the surface from which the rate of heat loss from the body core can be regulated. Between the outermost layers of the skin (the epidermis and dermis) and the muscles and internal organs lies the subcutaneous tissue (also known as the hypodermis), which is rich in fat cells (the subcutaneous adipose tissue). This fatty tissue acts as an insulator, minimizing heat loss from the deeper tissues of the body.

Figure 33.5 shows the organization of the circulation of the skin (the cutaneous circulation). There are

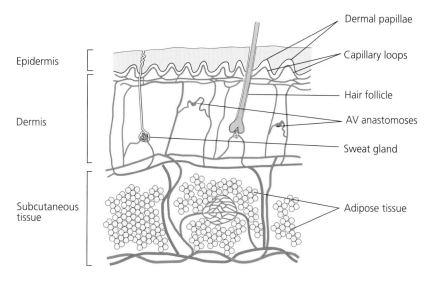

Figure 33.5 The arrangement of the blood vessels supplying the skin. The rate of heat loss is determined by how much blood is flowing through the outer arteriovenous (AV) plexuses and their associated capillaries.

three networks of blood vessels. One is very deep in the subcutaneous tissue, another lies below the dermis and the third lies just below the epidermis. The outermost capillaries do not pass into the epidermis but remain within the dermis. The oxygen and nutrients required by the epidermal cells reach them by diffusion from the outermost capillaries.

By altering the amount of blood flowing through the deeper and superficial layers of the skin, the heat lost by the body to its surroundings can be regulated. This is achieved by the activity of the sympathetic nerves that supply the arterioles. In this and the following section, the mechanisms responsible for regulating this heat loss are discussed. The blood flow through the skin varies more than 100-fold between extremely cold and very warm conditions. The heat lost from the skin in very cold conditions is less than 5% of that lost under warm conditions.

In cold conditions, the arterioles supplying the superficial capillaries constrict and blood flow to the skin surface decreases. This adaptation conserves heat. In warm conditions, the opposite happens: the arterioles dilate, and more blood is diverted to the superficial layers of the skin to increase heat loss.

In certain areas of the body, such as the finger tips, the ear lobes and the tip of the nose, blood flow to the skin may virtually cease altogether for brief periods. Here there are vessels (arteriovenous anastomoses) that link the arterioles and the venules directly. When these constrict, blood flow to the superficial parts of the skin is considerably reduced so heat is conserved and these parts of the body will feel numb. The skin is able to tolerate a severely reduced blood supply for fairly long periods but eventually, if circulation is not restored, frostbite will ensue in which tissue freezes and dies.

KEY POINTS:

- The circulation of blood through the skin plays an important role in the regulation of body temperature.

- Heat loss is promoted by dilatation of the skin blood vessels while constriction of the skin vessels reduces heat loss.

33.4   Body temperature is controlled around a 'set-point'

The normal core temperature of around 37°C is known as the set-point. A region deep within the brain called the hypothalamus behaves as a kind of thermostat to regulate temperature around this set-point. The set-point can be altered in response to fever-producing agents known as **pyrogens**. The hypothalamus receives information from temperature receptors (thermoreceptors) throughout the body and, when body temperature starts to alter, the hypothalamus initiates physiological responses that act to restore core temperature to normal through mechanisms of heat loss, heat conservation or heat generation (Figure 33.6).

Both core and shell temperatures are monitored by specific temperature-sensitive neurones. Those that respond to changes in shell temperature receive information from thermoreceptors located in the skin (**peripheral thermoreceptors**) while core temperature is monitored by thermosensitive neurones located in the spinal cord and in the hypothalamus itself (**central thermoreceptors**). Thermoreceptors can respond to rising temperature (the warm receptors) or to falling temperature (the cold receptors).

The thermoreceptors of the skin are probably bare nerve endings. They relay their afferent information to the spinal cord via unmyelinated C-fibres (warm receptors) or Aδ fibres (cold receptors). The ascending fibres from the spinal cord relay this temperature information to the hypothalamus, which exerts its thermoregulatory actions on the vasculature of the skin and the sweat glands, mainly via the sympathetic nervous system.

β-Blockers used in the treatment of hypertension and other cardiovascular disorders antagonize the normal sympathetic vasodilator action and reduce the blood flow to the skin. As a result, cold extremities are a common side-effect of their action.

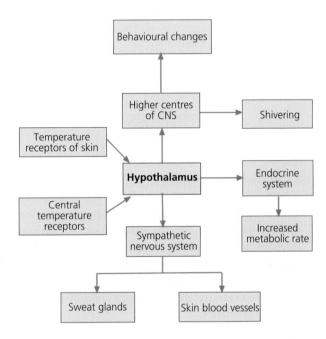

Figure 33.6 A flow chart showing the factors that determine the heat loss and heat production of the body. Note the central role of the hypothalamus.

33.5   Thermoregulatory responses to cold

As body temperature starts to fall, both physiological and behavioural homeostatic responses are initiated that either generate or conserve heat. Behavioural responses include seeking a warmer environment, adding clothing, taking a warm drink, increasing physical activity and so on. Physiological responses to a falling temperature include cutaneous vasoconstriction, shivering and heat

generation that does not require contraction of individual muscles (non-shivering thermogenesis).

Cutaneous vasoconstriction occurs in response to an increase in the activity of the noradrenergic sympathetic fibres that innervate the skin vessels, particularly the arteriovenous anastomoses. This diverts blood away from the skin surface and, as a result, heat loss from the body surface is reduced. **Shivering** begins in response to cold when vasoconstriction alone is insufficient to restore core temperature to normal. Shivering is a specialized form of muscular activity in which the muscles perform no external work and virtually all of the energy of contraction is converted directly to heat. It is predominantly an involuntary activity in which small groups of antagonistic muscles contract and relax rapidly. Shivering is normally preceded by a generalized increase in skeletal muscle tone.

Although shivering generates large amounts of heat, it cannot sustain a normal core temperature for long periods of time if the ambient temperature is very low.

Furthermore, it is very exhausting for the individual. Other mechanisms of heat production, known as non-shivering thermogenesis, take over. **Non-shivering thermogenesis** is the production of heat through processes other than muscle contraction. A number of hormones have calorigenic (heat-producing) actions. For example, adrenaline and noradrenaline (epinephrine and norepinephrine) stimulate brown fat metabolism in which ATP is hydrolysed to liberate heat. During prolonged exposure to cold, the secretion of thyroid hormone is increased and this increases the metabolic activity of most cells. The heat generated contributes to non-shivering thermogenesis and the maintenance of body temperature.

33.6   Thermoregulatory responses to heat

When core temperature starts to rise, behavioural and physiological responses are initiated that serve to enhance the rate of heat loss from the body. Behavioural changes include a reduction in physical activity, the shedding of clothes, taking cool drinks, etc. The principal physiological responses are cutaneous vasodilatation and sweating. The **cutaneous vasodilatation** is mediated by the sympathetic nervous system and allows blood flow to the surface of the skin to increase considerably. This facilitates heat loss to the surroundings.

Sweating allows heat to be lost by evaporation from the surface of the skin. It is particularly efficient when the air is dry and there is a breeze but, in humid conditions, evaporation occurs more slowly and sweating is a less effective means of heat loss. Heat loss by evaporation of water from the lungs and skin occurs even when the ambient temperature is comfortable. This is known as **insensible water loss** and amounts to about 500 ml a day. In warm conditions, however, sweat rates of 2–6 litres an hour may occur for short periods.

Sweat is secreted by the **eccrine sweat glands**. These are coiled tubular structures situated in the dermis of the skin and open on to the skin surface at sweat pores (Figure 33.5). The sweat glands are innervated by sympathetic cholinergic fibres and activation of these fibres leads to an increase in the rate of sweat production. The composition of sweat depends upon the sweat rate, but

sweat is normally hypotonic to plasma, with a sodium concentration of 30–90 mmol l^{-1}. Although sweating is an efficient mechanism of heat loss, it results in the loss of considerable amounts of water and electrolytes from the body. Consequently, it is important that the lost fluid and salts are replaced quickly to prevent dehydration and heat exhaustion.

> **KEY POINTS:**
>
> - The regulation of body temperature is governed by the hypothalamus, which is located at the base of the brain.
>
> - The hypothalamus receives input from thermoreceptors in the skin and the body core and initiates the appropriate mechanisms to conserve, generate or lose heat.
>
> - Mechanisms that help to maintain body temperature during exposure to cold include cutaneous vasoconstriction, shivering and non-shivering thermogenesis.
>
> - Heat loss is promoted by cutaneous vasodilatation and sweating, which brings about heat loss through evaporation.

33.7   Disorders of thermoregulation

Although the body employs a range of strategies to prevent significant changes in core temperature, in extreme conditions these compensatory mechanisms may prove insufficient to achieve thermoregulation. Hypo- or hyperthermia may then ensue.

Hypothermia

This is defined as a core temperature that is below 35°C. Figure 33.7 summarizes the effects of extreme core temperatures and shows that as the body starts to become hypothermic, the normal heat-generating mechanisms are lost. The muscles weaken and both voluntary activity and shivering are reduced. The core temperature will then start to fall more rapidly. Below 34°C, mental confusion is seen, often with hallucinations and a feeling of well-being. Soon afterwards, consciousness is lost. As the core

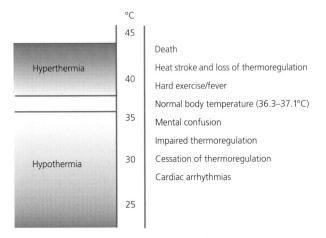

Figure 33.7 A summary of the effects of changing the core temperature. Both extreme heat and extreme cold will lead to death.

temperature falls below 28°C or so, heart rate falls and a fatal ventricular fibrillation may develop.

A hypothermic person should be warmed slowly and from the 'inside out'. Vigorous rubbing of the skin or the administration of alcohol is inadvisable as either may cause peripheral vasodilatation, which will lead to further heat loss from the skin surface and may compromise blood flow to vital organs. Instead, if possible, it is better to warm the person by peritoneal lavage with warmed fluids or dialysis.

Although comatose hypothermic individuals often appear to be dead, it is important that attempts at resuscitation are not discontinued too early as the reduced metabolic needs of the cooled tissues mean that recovery is often possible even when core temperature is as low as 18°C. Indeed, hypothermia has important clinical uses. In open heart surgery, for example, it is now routine to reduce the body temperature of the patient in order to minimize the metabolic requirements of the tissues and allow more time for the surgical procedures to be completed without the risk of hypoxic tissue damage.

Hyperthermia

Prolonged exposure to temperatures above about 40°C can result in the serious condition of **heat stroke** in which compensatory mechanisms such as sweating are lost and core temperature soars. The skin feels hot and dry, there is a fall in blood pressure and respiration becomes weak. As neural damage begins to occur, there is mental confusion, delerium and coma. Death will follow unless cooling is rapidly achieved. Heat stroke should not be confused with heat exhaustion. The latter is due to dehydration as a result of heavy sweating and is characterized by fatigue, dizziness and muscle cramps. It can be treated by oral rehydration with both salt and water.

33.8   Special thermoregulatory problems

Certain members of the general population experience particular problems of thermoregulation. These include newborn babies, the elderly and people with mobility problems. Furthermore, those who have just undergone surgery may also be significantly hypothermic and require special care.

The newborn

Newborn babies are at greater risk of both hypo- and hyperthermia than older children or adults. Hypothermia is the more commonly encountered problem because infants lose heat readily and their heat conservation mechanisms are poorly developed. Premature babies are especially susceptible to heat loss and often need to be kept in a temperature-controlled incubator until they are sufficiently mature to regulate their core temperature.

The risk of hypothermia arises because very young babies have a relatively thin skin and little or no subcutaneous fat to provide insulation. Consequently, vasoconstriction is less effective at minimizing heat loss than it is in adults. Moreover, their shivering mechanisms are poorly developed and their surface area to volume ratio is high, which means that heat loss occurs readily. Furthermore, they are unable to increase their voluntary muscular activity significantly and have no control over their clothing and bedding. Fortunately, because they have a very high proportion of brown adipose tissue relative to body weight, the capacity of a full-term neonate to generate heat by non-shivering pathways is four to five times greater (per unit body weight) than that of an adult. Figure 33.8 illustrates the location of the principal brown fat deposits in the infant. Brown fat metabolism generates large amounts of heat in response to activation of the sympathetic nervous system and to circulating adrenaline and noradrenaline (epinephrine and norepinephrine).

When a baby is exposed to a hot environment, its body temperature may rise above 37.5°C and lead to the development of hyperthermia. It should not be confused with

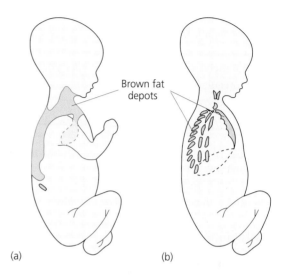

Figure 33.8 The distribution of brown fat in a neonate showing the interscapular pad (a) and the thoracic depots (b).

fever, which is an increase in body temperature caused by a viral or bacterial infection. Although it is less common than hypothermia, hyperthermia is equally dangerous. It increases the metabolic rate and water loss by evaporation. The loss of water can cause dehydration, which is serious and should be treated by getting the baby to drink, or if necessary, by providing fluid via a nasogastric tube or an intravenous infusion. The commonest causes are wrapping the baby in too many layers of clothes for the climate, leaving the baby in direct strong sunlight, for example in a parked car, and putting the baby too near to a source of heat. Severe hyperthermia may cause circulatory shock, convulsions and coma. Unless promptly treated, it may result in neurological damage.

The elderly

The ability to respond to thermal stress declines with age. There is both a reduction in awareness of temperature changes and an impairment of thermoregulatory responses. Shivering in response to moderate cooling is reduced and the change in metabolic rate seen in response to cold is small in comparison with that shown by younger adults. In addition to these age-related changes, the elderly are often relatively immobile and may suffer from disorders that make it difficult for them to increase their voluntary muscular activity. Social and financial considerations may also contribute to thermoregulatory problems.

Although hypothermia is more common in the elderly, they are also vulnerable to hyperthermia. Older people are less sensitive to thirst and have a decreased ability to sweat. Moreover, drugs such as anticholinergics (e.g. atropine) inhibit sweating. Each year, many people die during prolonged spells of very hot weather and the majority of these are 50 years of age or older.

Postoperative hypothermia

Many patients returning to the ward following a surgical procedure have a core temperature below 35.5°C. Not only is hypothermia unpleasant for the individual concerned, but the cardiovascular changes and the shivering caused by the lowered core temperature may also compromise recovery. A number of factors contribute to postoperative hypothermia including:

- Exposure of internal tissues to the air – this enhances loss of heat by evaporation
- Replacement of warm blood with fluids that have not been warmed to 37°C

- Cool operating theatre temperature (usually around 17°C)
- Inhibition of thermoregulatory responses by the anaesthetics and neuromuscular blockers used during surgery
- Hospital gowns and linen that are often inadequate to maintain core temperature perioperatively
- Prolonged immobility of the individual.

Recognition of this problem has led to more strenuous efforts being made by health-care professionals to monitor core temperature and keep patients warm during the postoperative period.

> **KEY POINTS:**
>
> - Hypothermia occurs when core temperature falls below 35°C. As the body temperature drops, heat-conserving mechanisms start to fail; there is mental confusion and increased risk of cardiovascular complications, followed by a loss of consciousness.
> - Newborn infants and the elderly are particularly at risk of hypothermia.
> - Hyperthermia (a core temperature in excess of 40°C that is not due to fever) may have grave consequences. As the body's heat loss mechanisms fail, there may be irreversible neurological damage.

33.9 Pyrexia (fever)

This is defined as a rise in temperature above the normal range that is not associated with exercise or a high ambient temperature. It is usually associated with infectious disease and is caused by chemicals called **pyrogens**, which are sometimes derived from bacteria (bacterial endotoxins) but more often are secreted by cells of the immune system in response to infection. An example of a pyrogen is interleukin-6 (IL-6). Pyrogens appear to induce fever by resetting the hypothalamic thermostat so

that heat conservation mechanisms are initiated to raise the core temperature to the new, higher value. Hence the person with a rising temperature will shiver, show cutaneous vasoconstriction and will feel cold. Once the cause of the infection has been removed from the body, the hypothalamic set-point returns to normal and heat loss mechanisms will be initiated. The person sweats, shows vasodilatation and feels hot (previously called the 'flush' or 'crisis') until his or her core temperature returns

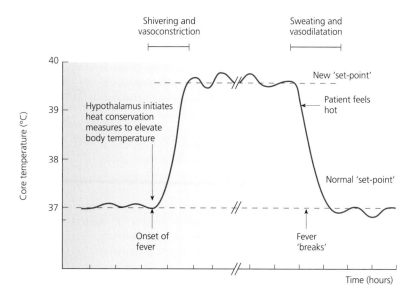

Figure 33.9 A diagrammatic representation of the time course of a typical febrile episode.

to normal. The time course of a typical febrile episode is shown in Figure 33.9.

Drugs used to lower temperature during an infection are called **antipyretics**. These include **aspirin**, **paracetamol** and other non-steroidal anti-inflammatory drugs (NSAIDs), which also have analgesic actions. The resetting of the hypothalamic thermostat that occurs in response to pyrogens is associated with an increased rate of synthesis of prostaglandins, especially PGE_1, by the hypothalamus. Aspirin and paracetamol inhibit cydo-oxygenase and thus reduce the synthesis of prostaglandins, thereby exerting an antipyretic effect.

Although pyrexia may be regarded as part of a natural response to infection as it reduces the rate at which viruses in particular replicate, antipyretic drugs are often used to reduce the temperature so that the patient feels better. In children, particularly the under-fives, it is important to reduce a fever as in this age group there is an increased risk of febrile convulsions, or even neurological damage.

> **KEY POINTS:**
>
> - Fever is an elevation of body temperature usually associated with the presence of infectious agents. In response to pyrogens, the body conserves heat and body temperature rises.
>
> - Antipyretic drugs include aspirin, paracetamol and other non-steroidal anti-inflammatory drugs (NSAIDs).

 # Recommended reading

Astrand, P.-O., Rodahl, K., Dahl, H.A. and Stromme, S. (2003). *Textbook of Work Physiology, Physiological Bases of Exercise*, 4th edn. Human Kinetics: Champaign, IL. Chapter 13.

Case, R.M. and Waterhouse, J.M. (eds) (1994). *Human Physiology: Age, Stress, and the Environment*. Oxford Science Publications: Oxford. Chapter 6.

Hardy, J.D. (1980). Chapter 59, Body temperature regulation. In: *Medical Physiology*, 14th edn. Ed Mountcastle, V.B. Mosby: St Louis.

Self-assessment questions

Which of the following statements are true and which are false? Answers are given at the end of the book (p. 755).

1 a) In a temperate climate under normal conditions, the greatest loss of body heat occurs through radiation.
   b) Heat is lost less readily from the body surface during water immersion than in air.
   c) The skin temperature is regulated more closely than the core temperature.
   d) The hypothalamus is the centre for integrating thermal information.
   e) Vasodilatation of skin's blood vessels promotes heat loss from the body.

2 a) Non-shivering thermogenesis is more efficient in neonates than in adults.
   b) Heat loss through sweating is controlled by the parasympathetic nervous system.
   c) The skin temperature is generally higher than the temperature within the abdominal cavity.

   d) Profuse sweating can lead to heat exhaustion.
   e) The efficacy of sweating as a mechanism of heat loss is increased in humid conditions.
   f) If environmental cold is prolonged, the thyroid gland secretes more thyroxine.

3 A man moves from a temperate to a tropical area to live:
   a) During the first few weeks in the tropical area, his sweat rate will rise.
   b) The salt content of his sweat will fall.
   c) His rate of urine production will fall.
   d) The rate at which aldosterone is secreted from his adrenal cortex will fall.

Body fluid and acid–base balance

34

After reading this chapter you should have gained an understanding of:

- How body water is distributed between the various body fluid compartments
- How the volumes of the body compartments can be measured
- The mechanisms involved in maintaining the normal fluid balance between compartments
- The importance of sodium in the determination of body fluid volume
- How total body fluid is sensed and regulated
- Common disorders of fluid balance: dehydration and oedema
- The principal physiological mechanisms that regulate the pH of the body fluids
- The common disorders of acid–base metabolism
- The compensatory mechanisms that minimize the effects of acid–base disorders

34.1 Introduction

Water is the principal constituent of the human body and is essential for life. Although the proportion of total body weight contributed by water varies with the age and sex of an individual, in both men and women, the water content of the lean body mass (i.e. the non-adipose tissues) is about 73%. Adipose tissue, however, contains relatively little water and, as it is laid down, the proportion of body weight contributed by water decreases. Newborn infants possess little body fat and water accounts for nearly 75% of their body weight. By the end of the first year, body water accounts for only about 65% of body weight. For a fit adult male in his twenties, water makes up about 60% of the total body weight. As women have more adipose tissue than men, their body water accounts for only about half of their body weight.

Broadly speaking, body water is distributed between the intracellular fluid (ICF) and the extracellular fluid (ECF), as discussed in Chapter 3 (p. 27). The space between the cells (the interstitial space) consists of connective tissue together with an ultrafiltrate of plasma. The water of the interstitial fluid hydrates the connective tissue so that it becomes like a thin jelly in consistency. Normal, healthy tissues have very little free liquid. Without this important adaptation, the extracellular fluid would flow to the lower regions of the body under the influence of gravity. Nevertheless, there is some bulk flow of isotonic fluid between the capillaries and the interstitium, which is normally returned to the blood via the lymphatics. When the lymphatic drainage is obstructed, the tissues swell and free fluid is found within the interstitial space.

34.2   The distribution of body water between compartments

The amount of water in the main fluid compartments can be determined by the dilution of specific markers. For a marker to permit the accurate measurement of the volume of a particular compartment, it must be evenly distributed throughout that compartment and it should be physiologically inert (i.e. it should not be metabolized or alter any physiological variable). In practice, it is necessary to correct for the loss of the markers in the urine. Fortunately, it is not difficult to make the appropriate corrections.

The **plasma volume** can be estimated from the dilution of the dye Evans Blue, which binds to plasma albumin and so does not readily pass from the capillaries into the interstitial space. As the amount of marker injected is known, it is a simple matter to calculate the volume in which it has been diluted (the principle is explained in Box 34.1).

To determine the total body water, a known amount of weakly radioactive water (^3H_2O) is injected and sufficient time allowed for it to distribute throughout the body. A sample of blood is then taken and its radioactivity measured. Measurement of the extracellular fluid volume requires a substance that passes freely between the circulation and the interstitial fluid but does not

---

Box 34.1   The use of dilution methods to estimate the volume of fluid compartments

Evans Blue does not enter the red cells and is largely retained within the circulation. This dye is therefore useful for estimating the plasma volume. As an example, assume that a patient with a body weight of 70 kg was injected with 10 ml of a 1% (w/v) solution of the dye. Further assume that a sample of blood was taken after 10 min and the plasma was found to contain 0.037 mg ml^{-1} of dye. What is the plasma volume?

As:

$$\text{Concentration} = \frac{\text{Amount of dye}}{\text{Volume}}$$

$$\text{Volume} = \frac{\text{Amount of dye}}{\text{Concentration}}$$

The total amount of dye injected was 0.1 g (or 100 mg) and the concentration in the plasma 10 min after injection was 0.037 mg ml^{-1}. Therefore:

$$\text{Plasma volume} = \frac{100}{0.037} = 2702 \, \text{ml}$$

Note that this calculation assumes: (i) that the dye is evenly distributed, and (ii) that all of the dye remains in the circulation. In practice, some dye is lost from the circulation and corrections for the lost dye need to be applied to improve the accuracy of the estimate. Similar limitations apply to estimates of the ECF using inulin (and other markers) and to estimates of total body water.

enter the cells. These requirements are met by the polysaccharide inulin and by mannitol – although other markers have been used. The volume of the intracellular fluid is simply the difference between the total body water and the volume of the extracellular fluid. Thus:

Total body water = Extracellular fluid + Intracellular fluid

and

Extracellular fluid = Plasma + Interstitial water

> **KEY POINTS:**
>
> - Body water is distributed between the intracellular fluid and the extracellular fluid. The extracellular fluid is further subdivided into the plasma and the interstitial fluid.
> - The distribution of body water can be estimated by measuring the dilution of markers that stay within particular body compartments.

34.3 Body fluid osmolality and volume are regulated independently

If a person drinks a litre of water, their urine output will increase rapidly. This is known as a **diuresis**. Urine output peaks within an hour of drinking and returns to normal about an hour later, by which time the excess water has been eliminated (Figure 34.1). If the same person were to drink a litre of isotonic saline, there would be no diuresis and only a very small rise in urine output. In this case, it takes many hours for the body to eliminate the excess fluid. In both situations, there is an initial increase in total body water but when there is a pure water load, the osmolality of the body fluids falls. To restore the normal osmolality of the plasma, only the excess water needs to be excreted. In contrast, when a litre of isotonic saline is drunk, the osmolality of the tissue fluids is unchanged but the total volume of the ECF is increased. To restore the normal ECF volume, the body must eliminate both the excess salt and the excess water. This simple experiment illustrates an important principle: the osmolality and volume of the body fluids are regulated by separate mechanisms.

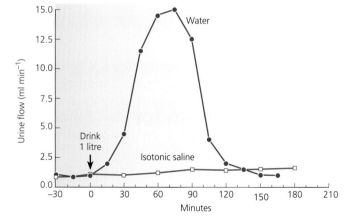

Figure 34.1 The effect of drinking 1 litre of water or 1 litre of isotonic saline on the urine flow rate of a normal subject in water balance. The subject drank 1 litre of water or saline at the time indicated by the arrow. After drinking the water, the urine flow rate increased from about 1 ml min^{-1} to nearly 15 ml min^{-1} within an hour and the excess water was all excreted in about 2 hours, by which time the urine flow rate had returned to normal levels. In contrast, drinking 1 litre of isotonic saline had little effect on urine flow rate.

34.4   Body water balance

Water requirements depend on body surface area as this determines the extent of water loss via the skin and lungs. This loss, together with the water loss in the urine and faeces, is replaced by water in the diet and by that generated during metabolism. For a normal adult male with a body weight of 70 kg, a typical balance sheet could be as shown in Table 34.1.

The amount of water lost from the lungs and skin obviously depends on prevailing conditions. In temperate climates, water is lost from the lungs and from the skin without sweating. This is called **insensible water loss** and, as it cannot be reduced, any restriction of fluid intake must be balanced by a decline in urine output. As the urine osmolality cannot be greater than about 1250 mOsmol kg^{-1} and as the quantity of solids excreted in the urine each day is between 50 and 70 g (chiefly as sodium chloride and urea), the minimum volume of urine required for excretion is about 700 ml per day. To balance this and the other losses, a minimum fluid intake of about 1.75 l each day is required just to maintain water balance. As noted above, this figure is for a 70-kg adult in a temperate environment. In hot environments, there is additional water loss via sweating, which may amount to several litres a day and which must be matched by an appropriate intake of water.

In addition to the obvious water exchanges that occur between the body and the environment, the gastrointestinal tract secretes and reabsorbs about 7 litres of fluid each day and the kidneys form about 180 litres of filtrate a day, of which about 99% is normally reabsorbed by the nephrons. Consequently, any reduction in fluid reabsorption arising from a disturbance of gastrointestinal or renal function will have dramatic consequences for water balance.

Table 34.1 A typical body water balance sheet for a normal adult male with a body weight of 70 kg

Source	Water gains (ml 24 h^{-1})	Source	Water losses (ml 24 h^{-1})
Fluid intake (drinking)	2000	Urine	1900
Water content of food	500	Skin	500
Water generated by metabolism	500	Lungs	500
		Faeces	100
Total	3000	Total	3000

34.5   Thirst is the physiological mechanism for replacement of lost water

Excessive loss of water from the body is known as **dehydration** and occurs when water intake is not sufficient to balance water loss. Two sources provide water: oxidative metabolism of fats and carbohydrates, and water intake via the diet. In humans, the generation of water during metabolism is not sufficient to meet the needs of the body and drinking is essential for the maintenance of water balance. The stimulus for water intake is thirst, which may be defined as the appetite for water.

What factors stimulate drinking?

The state of water balance is monitored by the osmoreceptors of the anterior hypothalamus (see pp. 289–291). These receptors regulate the amount of antidiuretic hormone (ADH; vasopressin) secreted by the posterior pituitary, increasing ADH secretion to conserve water in dehydration and decreasing ADH secretion during a water load. Current evidence suggests that the osmoreceptors that stimulate ADH secretion also play an important part in the regulation of water intake. When isotonic fluid is lost through diarrhoea or haemorrhage, there is a reduction in the circulating volume and this also stimulates thirst – probably as a result of increased plasma levels of angiotensin II (see below). Thus, an increase in the osmolality of the ECF or a decrease in the circulating volume will lead to an increased thirst and to an increase in water intake by drinking.

During dehydration, water is initially lost from the extracellular compartment but, as there is an osmotic equilibrium between the extracellular and intracellular compartments, the loss of water will ultimately result in cellular dehydration. The increase in plasma osmolality during dehydration is detected by the osmoreceptors of the hypothalamus, which, in turn, stimulate ADH secretion from the posterior pituitary. ADH then acts on the distal nephron to increase water reabsorption, as described in Chapter 23 (p. 483). There is a reduction in urine flow rate and an increase in urine osmolality. As a result, body water is conserved. However, restoration of water balance requires an increase in water intake and the high plasma osmolality stimulates thirst. The increased water intake restores plasma osmolality to normal. These processes are summarized on the right-hand side of Figure 34.2.

A water load causes a fall in plasma osmolality. This is detected by the osmoreceptors, which inhibit the secretion of ADH from the posterior pituitary. As a result, water reabsorption in the distal nephron declines and the urine flow rate increases while urine osmolality falls. The net effect is an increase in solute-free water excretion and the restoration of plasma osmolality to its normal value, as illustrated in Figure 34.2.

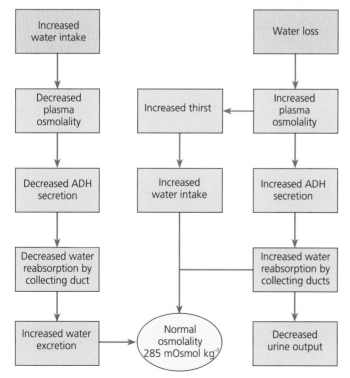

Figure 34.2 Summary of the principal steps by which the osmolality of the body fluids is restored following a water load (left-hand side) or pure water loss (right-hand side).

34.6 Alterations to the effective circulating volume regulate sodium balance

As it is the most abundant ion in the extracellular fluid, sodium is the principal determinant of plasma osmolality and, as the osmolality of the plasma is closely regulated, total body sodium is also the principal determinant of body fluid volume. Moreover, as the equilibrium between the extracellular fluid and plasma volume is determined by Starling forces, any change in the total body sodium will affect both the volume of the extracellular fluid and that of the plasma.

In healthy people, the **effective circulating volume** or **ECV** (the degree of 'fullness' of the circulatory system) is essentially constant and the sodium chloride and water losses are balanced by dietary intake. When the ECV is changed, the whole-body sodium content is adjusted to restore the normal situation.

The ECV itself is sensed by **low-pressure receptors** (also known as volume receptors) in the great veins and right atrium. In addition, the secretion of ADH by the posterior pituitary gland is also modulated by the central volume receptors so that a decrease in extracellular volume (**hypovolaemia**) leads to an increase in the secretion of ADH while an increase in volume (**hypervolaemia**) leads to a fall in the secretion of vasopressin.

When the ECV falls, there is an increase in the activity of the renal sympathetic nerves, which acts to promote sodium reabsorption and thus restore ECV. The principal mechanisms involved are illustrated in Figure 34.3.

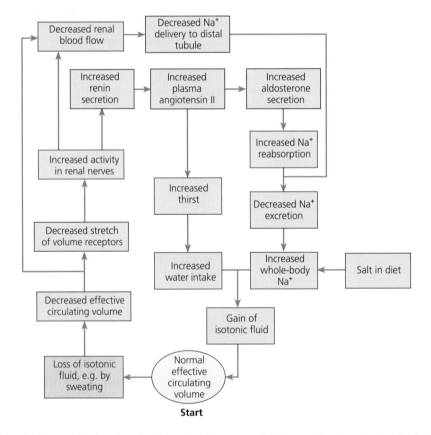

Figure 34.3 A schematic diagram showing the principal mechanisms responsible for restoring the effective circulating volume (ECV) following loss of isotonic fluid (e.g. as a result of heavy sweating or diarrhoea).

- Firstly, in response to the reduced filling pressures of the right side of the heart, there is vasoconstriction of the arterioles of the muscle and splanchnic vascular beds. This is a reflex response that is triggered by the central volume receptors and occurs before there is any significant change in arterial blood pressure. The afferent and efferent arterioles of the nephrons become constricted and this reduces the glomerular filtration rate. Consequently, the filtered load of sodium is reduced and a higher proportion of the filtered sodium is reabsorbed.

- Secondly, the increased sympathetic activity leads to an increase in renin secretion by the granular cells of the renal arterioles. This leads to an increase in the amount of aldosterone in the circulation, which acts to increase the absorption of sodium by the distal nephron. In addition, the secretion of ADH is increased, promoting the reabsorption of water in the distal nephron.

- Finally, the increased plasma levels of angiotensin II will stimulate thirst and drinking. The increased water

intake together with the increased retention of dietary sodium leads to an increase in ECV and the restoration of body fluid volume.

When the ECV rises above normal, full correction requires elimination of both the excess sodium and the excess water. This is achieved by the following processes:

- An increase in ECV activates the arterial baroreceptors. This triggers the baroreceptor reflex and the activity of the renal sympathetic nerves is decreased so that both renal blood flow and glomerular filtration rate increase. This increases the filtered load of sodium and the delivery of sodium to the distal tubule. Renin secretion is inhibited and the concentration of aldosterone in the plasma falls. As a result, less of the filtered sodium is reabsorbed. In addition, the secretion of ADH is decreased. These changes promote the excretion of salt and water by the kidneys, as shown in Figure 34.4.

- The increase in end diastolic volume triggers the secretion of a hormone called atrial natriuretic

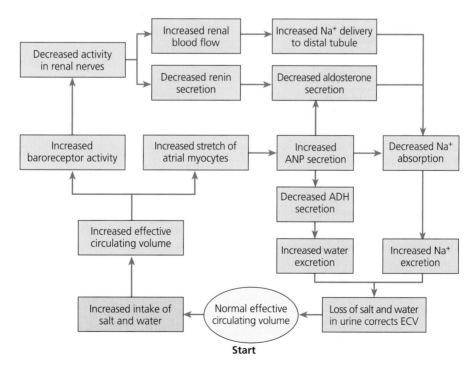

Figure 34.4 A schematic diagram showing the principal steps responsible for restoring the ECV after its expansion by the intake of salt and water. ANP, atrial natriuretic peptide.

peptide (ANP), which increases sodium excretion. ANP has actions that antagonize those of the renin–angiotensin–aldosterone system. It inhibits aldosterone formation and directly inhibits sodium chloride uptake by the distal tubule and collecting duct. It also inhibits both ADH secretion and its action on the distal nephron, so promoting the loss of water. Overall, these effects lead to a loss of sodium chloride and water and re-establishment of the normal ECV and body fluid volumes.

> **KEY POINTS:**
>
> - The osmolality of the body fluids is regulated by the activity of the osmoreceptors in the hypothalamus, which control the secretion of ADH from the posterior pituitary.
>
> - Extracellular fluid volume is sensed via the effective circulating volume or ECV and it is regulated by adjustment of sodium intake and excretion. Thus, osmolality and fluid volume are regulated independently of each other.

34.7   Dehydration and disorders of water balance

Dehydration has a number of causes:

- Excessive water loss from the lungs and sweat glands
- Excessive urine production
- Excessive loss of fluids from the gastrointestinal tract either by persistent vomiting or as a result of chronic diarrhoea
- Inadequate intake of fluid and electrolytes.

In most of these situations, *both water and electrolytes are lost.* For example, after heavy exercise, significant quantities of water and salt are lost in the sweat and both need to be replaced. (Sweat contains 30–90 mmol sodium chloride per litre.) If the lost water is replaced by drinking but the lost salt is not replaced, there is a deficiency of sodium chloride and the osmolality of the tissue fluids falls. The fall in tissue fluid osmolality results in muscle cramps (known as heat cramp), which may be relieved by drinking a weak saline solution (0.5 % sodium chloride) or by drinking water and taking salt tablets.

Pure water loss can occur when drinking water is unavailable or in situations where the kidneys are unable to reabsorb water from the distal tubule and collecting ducts. The classical example of this is **diabetes insipidus** where ADH secretion by the posterior pituitary is insufficient (see Chapter 16, p. 290). Despite their deficiency, people with diabetes insipidus remain in good health provided that enough water is available for them to drink. If they are rendered unconscious, however, their situation rapidly becomes perilous. In the normal course of events, pure water loss causes a rise in plasma osmolality and an increased thirst.

When water intake is persistently less than water loss, there is a progressive dehydration of the tissues. When more than 6–10% of body water has been lost, the plasma volume falls and circulatory failure commences. The poor circulation causes a failure of urine production and a metabolic acidosis (see below) develops. In addition, during severe dehydration the lack of water leads to a reduction in evaporative heat loss and fever may occur. The fever may be associated with drowsiness and delirium and, unless the lost fluid is replaced, these signs will be followed by coma and eventually death.

Over-hydration of the tissues (**water intoxication**) is much less common. Nevertheless, when it occurs, a new osmotic equilibrium between the plasma and the tissues is established and the cells swell. When the brain cells swell, the intracranial pressure increases and brain function becomes impaired. The clinical signs include nausea, headache, fits and coma. The commonest cause is acute renal failure where the ingested water cannot be excreted.

Inappropriate ADH secretion either from the posterior pituitary (perhaps as a result of a head injury) or from tumours can also lead to water retention. Failure of the anterior pituitary gland to secrete adrenocorticotrophic hormone also leads to water intoxication as the excretion of water depends on normal circulating levels of glucocorticoids. The exact mechanism of this effect is not known but may reflect a role of glucocorticoids in regulating the secretion of ADH.

34.8 Oral rehydration therapy

In vomiting and diarrhoea, there is a large loss of water and electrolytes as the fluids of the gastrointestinal tract are isotonic with the blood. Consequently, both dehydration and changes in acid–base balance are associated with these conditions (see below for discussion of acid–base balance). As the average person secretes about 7 litres of fluid into the gastrointestinal tract each day, persistent diarrhoea can lead to severe dehydration, even in mild gastroenteritis. In cholera, the fluid loss is greater than with other causes of diarrhoea as the causative organism, *Vibrio cholerae*, stimulates intestinal secretions. Consequently, fluid loss in the stools during an attack of cholera may be 10 litres or more each day. Clearly, unless the effects of the diarrhoea are rapidly countered, death will inevitably occur. Indeed, in many poor countries, dehydration caused by fluid loss in the stools is a common cause of death, particularly in children.

Drinking water will not be sufficient by itself to combat the loss of fluid, as the lost fluid will have been isotonic with the plasma, and both sodium chloride and water will be required to rehydrate the tissues. While intravenous infusions of isotonic sodium chloride (0.9% NaCl) could be used to restore body fluid volume, this approach needs suitable resources. Oral rehydration has been found to provide a very effective alternative therapy. It relies on the fact that the intestinal absorption of glucose is unimpaired despite the high fluid loss in the stools during bouts of diarrhoea. Rehydration is achieved by giving the patients a solution of salt and sugar to drink. The sugar (as glucose) is absorbed across the intestinal wall by cotransport with sodium, and water follows osmotically. The solution must not be significantly hypertonic, otherwise there is a risk that water loss will be enhanced. The use of sucrose or starch has the advantage that these sugars are readily available. Moreover, as they are broken down to glucose in the intestine before being absorbed, the amount of glucose available for absorption (together with sodium and water) can be increased without making the rehydration fluid hypertonic.

34.9 Oedema

Oedema is the abnormal accumulation of fluid in the interstitial space. It arises when alterations to the Starling forces occur as a result of various pathologies. As described in Chapter 19, fluid moves from the plasma to the interstitium when the capillary hydrostatic pressure exceeds the sum of the plasma oncotic pressure and the hydrostatic pressure within the tissues. Fluid moves from the tissues to the plasma when the sum of the plasma oncotic pressure and the tissue hydrostatic pressure exceeds the hydrostatic pressure in the capillaries. Under normal conditions, estimates suggest that about 8 litres of fluid per day pass from the circulation into the

interstitium. Of this, about half is reabsorbed by the circulation either in the tissues or in the lymph nodes. The remainder is returned to the circulation as lymph via the thoracic duct into the subclavian vein on the left side.

The hydrostatic pressure in the capillaries is normally closely regulated by the tone of the afferent arterioles. However, the average capillary pressure also depends on the venous pressure. When venous pressure is elevated as a result of a venous thrombosis, or as a result of chronic right-sided heart failure, the average capillary pressure is increased and more fluid passes from the plasma to the tissues. The resulting fall in plasma volume is reflected in a fall in the ECV and this in turn leads to the retention of sodium and water by the mechanisms discussed above (p. 712). A situation thus exists in which fluid can progressively accumulate in the tissues.

As about half of the fluid passing from the capillaries to the interstitial space is returned to the circulation via the lymphatic drainage, any obstruction of the flow of lymph will lead to fluid accumulation in the affected region. In the industrialized countries of Western Europe and North America, lymphatic insufficiency is relatively rare but is seen when the lymph nodes have been damaged during radical surgery (as in the example shown in Figure 34.5) or where cancerous growths have invaded the lymph glands (**lymphomas**). In Third World countries, oedema is frequently the result of the invasion of the lymph nodes by parasitic nematode worms (**filariasis**). This results in obstruction of lymph flow from the limbs and is manifest by a gross oedema (**lymphoedema**) known as elephantiasis.

Oedema also occurs when the plasma oncotic pressure is low. In this situation, the net filtration pressure rises and fluid accumulates in the tissues. This can occur during nephritis when significant quantities of protein are lost in the urine. It may also arise when the liver is unable to synthesize adequate quantities of the plasma proteins.

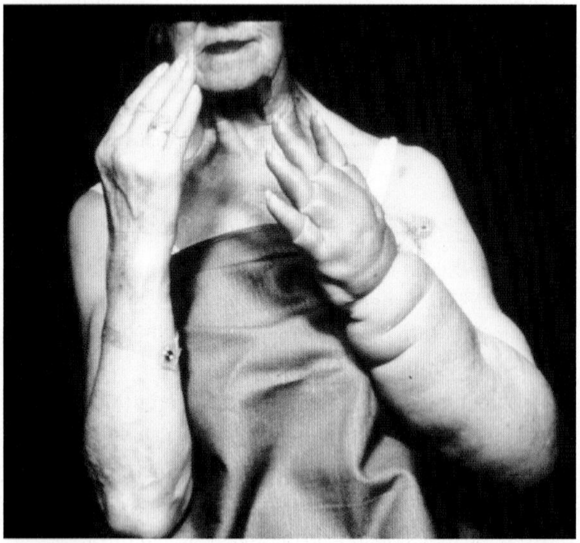

A similar situation arises during severe malnutrition when the diet may be rich in carbohydrate but contains little or no protein. This gives rise to a disease common in children in the poorer parts of the world known as **kwashiorkor**. A typical example is shown in Figure 34.6.

Systemic oedema first appears in the lower parts of the body (the **dependent regions** of the body), particularly in the ankles as the venous pressure in the legs is elevated during standing. It is easy to distinguish oedema in the ankles from local tissue fat by applying firm pressure to the affected area with a finger or thumb for a short period. If oedema is present, the pressure will have forced fluid from the area and a depression in the skin will remain for some time after the pressure has been removed

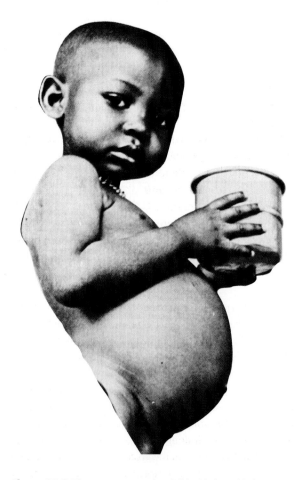

Figure 34.5 Severe lymphoedema in the left arm following radical surgery to treat breast cancer.

Figure 34.6 The appearance of a child with kwashiorkor. Note the widespread oedema, particularly the swollen abdomen.

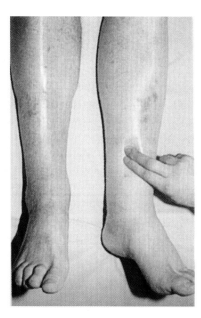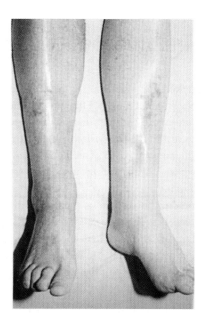

Figure 34.7 An example of pitting oedema.

(**pitting oedema**; Figure 34.7). If the swelling is due simply to tissue fat, the skin springs back as soon as the pressure is removed.

In normal people, hydrostatic oedema can arise when the leg muscles are relatively inactive and the muscle pumps contribute little to the venous return from the lower body. There is venous pooling and swelling of the ankles. This situation is exacerbated in persons with varicose veins where the walls of the veins have become stretched, rendering their valves incompetent. The accumulation of excess fluid is readily reversed by a short period of rest in a horizontal position or by mild exercise.

The fluids of the serosal spaces are separated from the extracellular fluid by an epithelial layer. These include the fluids of the pericardial, pleural and peritoneal spaces. These fluids are essentially ultrafiltrates of plasma and their formation is governed by Starling forces. Normally, the amount of fluid in these spaces is relatively small as the plasma oncotic pressure exceeds the hydrostatic pressure in the capillaries, but, in certain disease states, abnormal accumulations of fluid occur. For example, the volume of fluid between the visceral and parietal pleural membranes of the chest is normally only about 10 ml but when fluid formation exceeds reabsorption, fluid accu-

mulates between the pleural membranes in a process known as **pleural effusion**.

An accumulation of excessive amounts of fluid in the peritoneal cavity known as **ascites** can arise when there is a rise in pressure within the hepatic venous circulation, when there is obstruction of hepatic lymph flow or when plasma albumin is abnormally low (as in kwashiorkor; see above). It also occurs during right-sided heart failure when the pressure within the systemic veins rises.

> **KEY POINTS:**
>
> - Oedema occurs when there is an abnormal accumulation of fluid in the tissues.
>
> - A pleural effusion is fluid accumulation between the pleura of the lungs. Fluid accumulation in the abdominal cavity is called ascites.
>
> - Oedema and other kinds of fluid accumulation have a number of possible causes, but all result from a change in the balance of Starling forces. Most commonly, there is an increase in capillary pressure, a reduction in plasma oncotic pressure or lymphatic obstruction.

34.10   Treatment of oedema with diuretics

From the previous section, it is clear that oedema can arise as a result of various pathologies. Effective treatment requires identification and elimination of the underlying cause. Nevertheless, it may be desirable to eliminate the oedema and, in many cases, this can be achieved by treating the patient with drugs that promote the loss of *both sodium and water* in the urine. As this results in an increase in urine output known as a diuresis, these drugs are called **diuretics**. They are classified according to their modes of action. Diuretics may act indirectly by exerting an osmotic pressure that is sufficient to inhibit the reabsorption of water and sodium chloride from the renal tubules or they may act directly by inhibiting active transport in various parts of the nephron. Note that lymphoedema cannot be treated in this way.

Osmotic diuretics, such as the sugar mannitol, are filtered at the glomerulus but they are not transported by the cells of the proximal tubule. In consequence, as other substances are transported and the proportion of the original filtered volume falls, these substances accumulate and exert sufficient osmotic pressure to inhibit tubular reabsorption of water. As absorption by the proximal tubule is iso-osmotic, a decrease in fluid reabsorption allows more sodium to reach the distal nephron so that there is an increase in sodium excretion. There is, therefore, a loss of both sodium and water. Nevertheless, the osmotic diuretics are more effective in increasing water excretion than they are in increasing sodium excretion.

Diuretics that act by inhibiting active transport are exemplified by **loop diuretics**, such as **furosemide** (frusemide). These compounds inhibit the cotransport of sodium, potassium and chloride by the ascending thick limb of the loop of Henle (see p. 479). Inhibition of this transport decreases the ability of the nephron to concentrate urine. The effects of the loop diuretics are thus twofold, an increase in sodium excretion by inhibiting sodium chloride transport and an increase in water loss through impairment of the countercurrent mechanism. They are the most potent diuretics in current clinical use and produce a pronounced increase in sodium excretion (known as a **natriuresis**).

One of the consequences of inhibiting the cotransport of sodium, potassium and chloride in the ascending thick limb is an increase in potassium excretion. Unless this is carefully monitored, potassium balance will be disturbed and cardiac arrhythmias may result. To avoid this, a group of **potassium-sparing diuretics** has been developed. These drugs, of which **amiloride** is an example, act on the distal tubule, connecting tubules and collecting ducts to inhibit both sodium absorption and potassium excretion. The diuretic **spironolactone** exerts its effect on the distal tubule by antagonizing the sodium-retaining action of aldosterone.

> **KEY POINTS:**
>
> - Several forms of oedema can be treated by administration of drugs known as diuretics that promote the excretion of both water and sodium.
>
> - Diuretics can act either directly by inhibiting sodium transport by the nephron (e.g. loop diuretics such as furosemide) or indirectly by modifying the filtrate (osmotic diuretics such as mannitol). Lymphoedema cannot be treated in this way.

34.11   Acid–base balance

Although the body continually produces carbon dioxide and non-volatile acids as a result of metabolic activity, the blood hydrogen ion concentration $[H^+]$ is normally maintained within the relatively narrow range of 40–45 nmol (40–45×10^{-9} mol) of free hydrogen ions per litre. This corresponds to a blood pH between 7.35 and 7.4. The extreme limits that are generally held to be compatible with life range from pH 6.8 to pH 7.7. This regulation is achieved in two ways: hydrogen ions are absorbed by other molecules in a process known as **buffering** and acid products are subsequently eliminated from the body via the lungs and kidneys. *The concept of acid–base balance refers to the processes that maintain the hydrogen ion concentration of the body fluids within its normal limits.*

An average, healthy person, eating a typical Western diet, produces between 12 and 15 mol of carbon dioxide a day and excretes about 50 mmol of acid in the urine. The carbon dioxide dissolved in the body fluids leads to the formation of hydrogen ions in solution. However, as it is excreted via the lungs as a gas, carbon dioxide is often referred to as **volatile acid**. The acid that is excreted in the urine is chiefly sulphate, derived from the metabolism of sulphur-containing amino acids (cysteine and methionine). This is known as **non-volatile acid** and must be excreted in the urine. Under normal conditions, no organic acids appear in the urine but, in severe uncontrolled diabetic ketosis, large quantities of β-hydroxybutyric acid and acetoacetic acid are produced each day and a significant portion of these acids may be eliminated in the urine as 'ketone bodies'.

As mentioned above, the kidneys must excrete around 50 mmol of hydrogen ions each day if they are to maintain the pH of the plasma within normal limits. As the average volume of urine produced each day is 1–1.5 litres and as the pH of the urine is usually between 5 and 6, less than 0.05% of this acid is excreted as free hydrogen ions. Most hydrogen ions are eliminated either as ammonium ions or in combination with the urinary buffers of which the most important is phosphate. A normal person, eating a typical Western diet, excretes about 30 mmol of phosphate a day and this enables them to excrete about 24 mmol of hydrogen ions. This accounts for about half of the hydrogen ions derived from non-volatile acids. The remainder is excreted as ammonium ions (NH_4^+), which are derived from the metabolism of the amino acid glutamine, in a process known as **ammoniagenesis**.

The total amount of non-volatile acid that is excreted is the sum of that buffered by the urinary buffers and the amount of ammonium in the urine. It can be measured by titrating the urine with an alkali until the urine pH is the same as that of the plasma (the **titratable acid**) and by separately measuring the total amount of ammonium excreted. The total urinary acid excretion is the sum of the two figures.

34.12 Primary disturbances in acid–base balance

When the pH of the arterial blood is less than 7.35, it is regarded as being acid with respect to normal. Patients with such blood are said to have **acidaemia** (literally acid in the blood). The processes responsible for this increase in plasma hydrogen ion concentration can be attributed to increased amounts of non-volatile acid (**metabolic acidosis**) or to a failure to remove carbon dioxide from the blood (**respiratory acidosis**). Conversely, when the pH of the arterial blood is greater than 7.45, it is regarded as being alkaline with respect to normal. Patients with such a blood pH are said to have **alkalaemia**. The underlying causes define the kind of **alkalosis** responsible for

the condition (**respiratory** or **metabolic alkalosis**). Some of the common causes of acid–base imbalance are given in Table 34.2.

Respiratory acidosis is the result of an increase in the P_{CO_2} of the plasma and is usually caused by inadequate alveolar ventilation. The rise in plasma P_{CO_2} results in an increase in the formation of carbonic acid, which dissociates giving rise to H^+ and HCO_3^-. The increase in hydrogen ion concentration is directly related to the P_{CO_2} and the plasma bicarbonate increases in proportion to the fall in plasma pH. These relationships are indicated in Figure 34.8 by the line linking point A

Table 34.2 Some causes of acid–base disturbance

Respiratory acidosis (alveolar hypoventilation)	Impaired ventilation due to airway obstruction
	Impaired alveolar gas exchange
	Decreased respiratory drive
	Inhalation of carbon dioxide
Respiratory alkalosis (alveolar hyperventilation)	Hypoxia (e.g. while living at high altitude)
	Increased respiratory drive due to cerebrovascular disease
	Hepatic failure
	Effects of drugs and poisons
Metabolic acidosis	Endogenous acid loading (e.g. diabetic ketoacidosis)
	Loss of base from the gut (e.g. diarrhoea)
	Impaired acid secretion by the renal tubules (renal tubular acidosis)
	Exogenous acid loading (e.g. methanol ingestion)
Metabolic alkalosis	Loss of gastric juice (e.g. by vomiting)
	Excessive base ingestion
	Aldosterone excess

to the normal value ($P_{a}CO_{2}$ 40 mmHg, pH 7.4 and HCO_{3}^{-} 24 mmol l^{-1}).

Respiratory alkalosis is the result of a fall in plasma $P_{CO_{2}}$ due to an increase in alveolar ventilation. The fall in $P_{CO_{2}}$ shifts the $[CO_{2}]$–$[HCO_{3}^{-}]$ equilibrium and leads to a decreased concentration of carbonic acid and so to a rise in plasma pH and a fall in plasma bicarbonate. The magnitude of the pH change is directly related to the increase in ventilation. This condition is a common feature of life at high altitude where the fall in atmospheric $P_{O_{2}}$ increases ventilation. The production of carbon dioxide does not increase so the $P_{a}CO_{2}$ falls. The net result is an increase in plasma pH and a decrease in bicarbonate. These changes are illustrated in Figure 34.8 by the line linking the normal value to point B.

In **metabolic acidosis**, the fall in plasma pH is accompanied by a fall in plasma bicarbonate. There are many causes (Table 34.2), including an increase in metabolically derived acids, a loss of base ($NaHCO_{3}$) from the gut during diarrhoea and a failure of the renal tubules to excrete hydrogen ions (renal tubular acidosis). The pH and HCO_{3}^{-} changes in metabolic acidosis are indicated in Figure 34.8 by the line linking the normal value to point C.

When diabetes mellitus is inadequately controlled, energy metabolism shifts from carbohydrates to fats and the amounts of β-hydroxybutyric acid and acetoacetic acid in the plasma increase. As a result, there is a fall in plasma pH. These changes are known as **ketoacidosis**. The hydrogen ions react with plasma bicarbonate to form carbonic acid and the carbon dioxide liberated is excreted via the lungs.

The bones of the skeleton contain a vast number of mineral crystallites bound together by cells, collagen and ground substance rich in mucopolysaccharides. The mineral crystallites consist of calcium phosphate

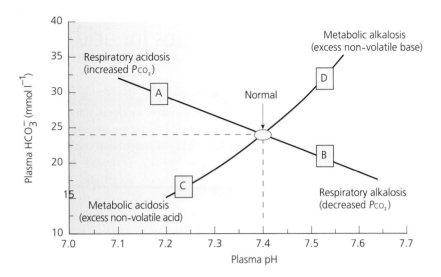

Figure 34.8 The pH–[HCO_3^-] diagram for blood plasma. The line A–B shows the change in plasma pH as P_{CO_2} is varied. Values below that line reflect increased metabolic acid while values above it reflect excess base. If alveolar ventilation is less than normal, pH falls and [HCO_3^-] rises because CO_2 accumulates (from normal to point A). This is respiratory acidosis. If alveolar ventilation is higher than needed to remove CO_2 from the body, both plasma P_{CO_2} and [HCO_3^-] fall while pH increases (from normal to point B). This is respiratory alkalosis. The line C–D shows the change in pH that occurs as non-volatile acid is gained in metabolic acidosis (from normal to point C) or lost in metabolic alkalosis (from normal to point D). Note that in this case, the partial pressure of CO_2 is maintained at a constant value of 40 mmHg.

and calcium carbonate and their surface is negatively charged. Normally these charges are neutralized largely by sodium and potassium ions, but when plasma pH falls, these ions are displaced by protons so that the mineral phase of the skeleton provides an additional source of extracellular buffering. This additional buffering is bought at a price – the ions displaced (sodium, potassium and calcium) are excreted in the urine and there can be a significant loss of calcium during chronic metabolic acidosis leading to a slow dissolution of the bone.

Metabolic alkalosis is caused by an excess of non-volatile base in the plasma, which may arise from a number of factors (Table 34.2). Very commonly, metabolic alkalosis arises as a result of vomiting and can be attributed to the loss of HCl from the stomach. As the P_{CO_2} is unchanged, the fall in hydrogen ion concentration that results from this loss is accompanied by an increase in plasma bicarbonate. In metabolic alkalosis, the rise in pH is associated with a rise in plasma bicarbonate. The

pH and bicarbonate changes in metabolic alkalosis are summarized by the line linking the normal value to point D in Figure 34.8.

KEY POINTS:

- Normal arterial blood pH is generally taken as 7.4.

- If plasma pH falls because of inadequate alveolar ventilation, there is a respiratory acidosis. If it rises because of hyperventilation, the accompanying disorder is a respiratory alkalosis.

- All other disorders of acid–base balance are classified as metabolic irrespective of their underlying cause. Thus, a metabolic acidosis develops if the production of non-volatile acids exceeds their rate of excretion via the kidneys, or if there is a loss of non-volatile base from the gut. Conversely, loss of acid from the stomach or ingestion of non-volatile base gives rise to a metabolic alkalosis.

34.13 How the body compensates for acid–base balance disorders

When acid–base balance has become disturbed, various mechanisms operate to bring plasma pH closer to the normal range in a process called **compensation**. The mechanisms that act to restore plasma pH can be grouped under two headings:

1. **Respiratory compensation**, which is fast but not very sensitive (pH adjustment occurs in minutes but the changes in P_{CO_2} offset the original stimulus).
2. **Renal compensation**, which is sensitive but slow (pH adjustment takes hours to days).

While compensatory mechanisms operate to minimize the change in plasma pH, complete restoration of acid–base balance (i.e. correction) requires treatment or elimination of the underlying cause.

Compensation of chronic respiratory acidosis and alkalosis can only occur by renal means, as the primary deficit is due to a change in alveolar ventilation. While

respiratory acidosis and alkalosis can be produced voluntarily by breath holding or hyperventilation, the effects of short-term voluntary changes to ventilation are readily reversed by resumption of normal patterns of breathing. In chronic (i.e. long-term) respiratory acidosis, the plasma P_{CO_2} is elevated as the alveolar ventilation is insufficient to eliminate all of the carbon dioxide generated during metabolism. This leads to a fall in plasma pH and an increased hydrogen ion secretion into the proximal tubule and collecting ducts. Overall, the higher the P_{CO_2} the greater the secretion of hydrogen ions and the larger the quantity of bicarbonate generated and reabsorbed. Thus, in chronic respiratory acidosis, increased renal hydrogen ion secretion leads to an increase in plasma bicarbonate (this is shown by the line joining points A and B in Figure 34.9).

In chronic respiratory alkalosis, the situation is reversed. If the kidneys are to restore plasma pH to normal they must

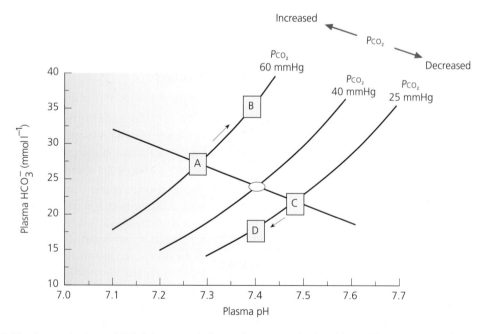

Figure 34.9 The changes in plasma [HCO_3^-] that occur during renal compensation in subjects with respiratory acidosis (points A to B) and respiratory alkalosis (points C to D). The white ellipse represents the range of normal values. See text for further information.

excrete bicarbonate and conserve hydrogen ions. This is accomplished by a reduction in hydrogen ion secretion in the proximal tubule so that less of the filtered load of bicarbonate is reabsorbed and plasma bicarbonate falls. This is shown by the line joining points C and D in Figure 34.9. Consequently, during the early stages of compensation, the kidneys excrete bicarbonate and the urine is relatively alkaline. Renal compensation for respiratory hyperventilation is sufficiently powerful to enable healthy people living at high altitude to have a normal plasma pH.

In metabolic acidosis and metabolic alkalosis, the changes in plasma pH are first minimized by respiratory compensation. Fine adjustment occurs over a longer period by altering the amount of H^+ or HCO_3^- excreted by the kidneys.

As **metabolic acidosis** is usually due to an increase in the production of metabolic acid, a loss of base from the lower gut or a reduced ability to excrete acid (renal tubular acidosis), the compensatory mechanisms employed will depend on the underlying cause. In the initial stages, the decrease in blood pH stimulates respiration and this increases the

loss of carbon dioxide from the lungs. The resulting fall in P_{CO_2} causes the pH of the plasma to rise towards normal. This restoration of pH is relatively rapid (it occurs within minutes) but as the pH approaches 7.3, the stimulus to the chemoreceptors becomes less and the hyperventilation declines. Consequently, respiratory compensation can only bring plasma pH within about 0.1 of a pH unit of its normal range. Moreover, this is achieved at the cost of a fall in the plasma bicarbonate concentration, as shown by the change from point A to point B in Figure 34.10.

Secondly, the fall in plasma pH results in the excretion of urine that is more acid than usual. Low plasma pH also stimulates the production and excretion of ammonium ions by the kidneys. These mechanisms may eventually bring plasma pH back to normal but, as it takes time to filter plasma and excrete the excess acid, it takes hours or days for full compensation to occur. However, if metabolic acid production exceeds the ability of the kidneys to form and excrete ammonium ions, full compensation will not occur until the underlying cause is treated.

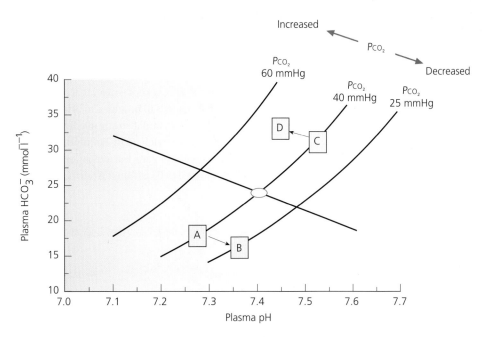

Figure 34.10 The changes in plasma [HCO_3^-] during metabolic acidosis and metabolic alkalosis and the effects of respiratory compensation. In metabolic acidosis (point A), the P_{CO_2} is initially normal (40 mmHg) but the low plasma pH stimulates respiration and the P_{CO_2} falls, thereby shifting the plasma pH closer to normal (point B). In metabolic alkalosis, the loss of fixed acid initially shifts the pH in the alkaline direction even though P_{CO_2} remains normal (point C). The increase in pH tends to depress respiration and leads to CO_2 retention and a fall in plasma pH (point D). As in the previous figure, the white ellipse represents the range of normal values.

In acidosis caused by loss of base from the gut, there is a fall in the filtered load of Na^+, which stimulates aldosterone production. This leads to an increase in Na^+ reabsorption by the mechanism discussed in Chapter 23 (p. 480).

Metabolic alkalosis is commonly caused by vomiting gastric juice or by the ingestion of alkali. Initially, respiratory compensation occurs as the high plasma pH depresses respiration. In consequence, the P_aCO_2 rises and the plasma pH tends to fall towards normal as shown by the shift from point C to point D in Figure 34.10. Nevertheless, as plasma pH approaches the normal range, the fall in pH offsets the depression of ventilation so that respiratory compensation is only partial. Final correction of metabolic alkalosis due to loss of gastric juice requires excretion of HCO_3^- and retention of Cl^- by the proximal tubule. Alkalosis due to ingestion of base is corrected by renal excretion of the excess base.

KEY POINTS:

- Following a disturbance of acid–base balance, compensatory mechanisms come into play to bring plasma pH within the normal range.

- Respiratory disorders are compensated by renal adjustments of plasma bicarbonate, which may take days to complete.

- Metabolic disorders are initially compensated by alterations to the rate of alveolar ventilation (respiratory compensation) but this is always insufficient to restore plasma pH to the normal range.

- Complete compensation occurs via renal mechanisms. Full correction requires treatment or elimination of the underlying cause.

34.14   The clinical evaluation of acid–base status

The state of acid–base balance in any person can be deduced from the pH–HCO_3^- diagram discussed in the previous two sections. By measuring the arterial pH, bicarbonate and P_aCO_2, an unambiguous interpretation can be made. The various conditions are set out in Table 34.3.

Another common way of determining the extent of metabolic disturbance of acid–base balance is to measure the **base excess** or **base deficit**. This is measured by titrating the blood or plasma to a pH of 7.4 with a strong acid or base while the P_{CO_2} is kept constant at 40 mmHg. If strong acid (e.g. HCl) needs to be added to bring the pH to 7.4, there is a base excess, and if strong alkali (e.g. NaOH) is required, there is a base deficit. By definition, in normal people the base excess is zero but deviations of $\pm 2.5\,mmol\,l^{-1}$ are considered to lie within the normal range.

Table 34.3 The changes in plasma pH, P_{CO_2} and bicarbonate that characterize the primary metabolic and respiratory disturbances of acid–base balance

Condition	Plasma pH	Plasma P_{CO_2}	Plasma HCO_3^-
Respiratory acidosis	Decreased	Increased	Increased
Metabolic acidosis	Decreased	Normal	Decreased
Respiratory alkalosis	Increased	Decreased	Decreased
Metabolic alkalosis	Increased	Normal	Increased

Recommended reading

Holmes, O. (1993). *Human Acid–Base Physiology: a Student Text.* Chapman & Hall: London.

Pocock, G. and Richards, C.D. (2006). *Human Physiology: the Basis of Medicine*, 3rd edn. Oxford University Press: Oxford. Chapter 29.

Self-assessment questions

Which of the following statements are true and which are false? Answers are given at the end of the book (p. 755).

1 a) The water content (as a percentage of total body mass) of a healthy young adult human male is mainly due to the volume of the blood.

b) Water accounts for a greater percentage of total body mass in a healthy adult male than it does in a healthy adult female.

c) The extracellular fluid is hypotonic with respect to the intracellular fluid.

d) The extracellular fluid has the same ionic composition as the plasma.

e) The intracellular fluid contains around 140 mmol l^{-1} of sodium ions

2 a) The volume of the plasma can be determined using the dye Evans Blue.

b) The extracellular fluid volume (plasma, interstitial fluid and transcellular fluid) accounts for about half of total body water.

c) The principal determinant of effective circulating volume is total body sodium.

d) The volume of the fluid in the circulation is detected by the arterial baroreceptors.

e) A lack of antidiuretic hormone (ADH) can result in severe dehydration.

Interpretive questions regarding acid–base balance.

3 In a man undergoing surgery, it was necessary to aspirate the contents of the upper gastrointestinal tract. After surgery, the following values were obtained from an arterial blood sample: pH 7.55, P_{CO_2} 52 mmHg and HCO$_3^-$ 40 mmol l^{-1}. What is the underlying disorder? (a) Metabolic acidosis; (b) respiratory alkalosis; (c) metabolic alkalosis; or (d) respiratory acidosis.

4 A person was admitted to hospital in a coma. Analysis of the arterial blood gave the following values: P_{CO_2} 16 mmHg, HCO$_3^-$ 5 mmol l^{-1} and pH 7.1. What is the underlying acid–base disorder? (a) Metabolic acidosis; (b) respiratory alkalosis; (c) metabolic alkalosis; or (d) respiratory acidosis.

5 A climber attempts an assault on a high mountain in the Andes and reaches an altitude of 5000 metres (16 400 ft) above sea level. What will happen to his arterial P_{CO_2} and pH? (a) Both will be lower than normal; (b) the pH will rise and P_{CO_2} will fall; (c) both will be higher than normal due to the physical exertion; or (d) the pH will fall and P_{CO_2} will rise.

35 The uptake, distribution and elimination of drugs

After reading this chapter you should have gained an understanding of:

- How drugs can be administered
- The factors influencing their absorption and distribution
- The role of metabolism and excretion in the elimination of a drug
- Effects of age on the potency of drugs
- Adverse effects of drugs: side-effects and toxicity

35.1 Introduction

Therapeutically useful drugs are given to treat specific disorders. Following their administration, they must reach their target and remain there in sufficient quantity for them to have the desired effect. This chapter is concerned with this aspect of pharmacology. It begins with an explanation of the various routes of drug administration and will then discuss how drugs are taken up and distributed between the various tissues. The fundamental processes of drug metabolism and excretion will be then be briefly surveyed before a short discussion of the effects of age on drug distribution, metabolism and excretion. The chapter ends with a discussion of the adverse effects of drugs.

The study of uptake, distribution, metabolism and excretion of drugs within the body is known as **pharmacokinetics** while **pharmacodynamics** is the study of the actions of drugs on the physiological systems of the body (see Chapter 7, p. 95). In their textbook, *Pharmacology*, Rang and his colleagues succinctly summarize the difference as follows: '*Pharmacokinetics is what the body does to a drug while pharmacodynamics is concerned with what a drug does to the body*' (see recommended reading).

35.2 Routes of drug administration

Drugs can be administered via the gastrointestinal tract, via the skin (either by direct application or by injection) or via the respiratory system. Most drugs are taken by mouth in liquid form, as capsules or as tablets. Common examples are analgesics such as aspirin and paracetamol, antacids such as magnesium hydroxide, and inhibitors of acid secretion such as cimetidine, ranitidine and omeprazole. Some drugs are applied to the mucous membrane of the oral cavity from where they are directly absorbed into the blood. For example, during an attack of angina, glyceryl trinitrate is usually sprayed into the mouth from where it is absorbed. Alternatively, a tablet of glyceryl trinitrate can be placed under the tongue (sublingual administration). The antiemetic prochlorperazine (Buccastem) can be taken by placing a tablet between the upper lip and the gum.

Less commonly, drugs are administered via the rectum. This is how anticonvulsants and analgesics are given to children when intravenous injection is difficult or impossible. Rectal administration of the gastrointestinal stimulant bisacodyl is sometimes required to promote evacuation of the bowel prior to radiological examination or colonic surgery. It is given as a suppository. Inflammatory bowel disease can be treated by rectal administration of the corticosteroid **prednisolone**.

Application of drugs directly to the skin is often used to treat or prevent local infections or to reduce local inflammation such as that caused by an insect bite. A **eutectic mixture** of local anaesthetics is sometimes applied as a cream to make the skin numb prior to an injection or the insertion of an intravenous drip. (A eutectic mixture is one in which the constituents have their lowest melting point.) The loss of local sensation reduces the pain that can be associated with such procedures and is especially helpful in children. Transdermal administration is used when a slow and prolonged delivery of a drug is needed: well-known examples are the use of nicotine patches to help people give up smoking, and the use of patches to administer oestrogens in hormone replacement therapy. Patches of opiate analgesics (e.g. fentanyl) are also used extensively to give pain relief in terminal care. Subcutaneous injection (i.e. injection into the deeper parts of

the skin) is used to provide prolonged delivery of insulin via insulin–zinc suspensions. Intramuscular injections are also used to prolong the action of some antibiotics for the treatment of severe infections, for example benzyl penicillin and procaine penicillin. In dentistry, local anaesthetics (e.g. **lidocaine**, also known as lignocaine) are administered by injection at specific points to block nerve conduction. Some formulations contain adrenaline (epinephrine) to cause a local vasoconstriction, which slows the uptake of the anaesthetic by the blood and prolongs its action.

The respiratory system offers a large surface area that is well perfused with blood for the uptake and excretion of drugs. These advantages are exploited for the administration of volatile and gaseous anaesthetics by inhalation. Salbutamol and other drugs used to treat the bronchoconstriction of asthma are given by inhalation as fine powders or as aerosols. Some peptide hormones such as antidiuretic hormone (ADH) and gonadotrophin releasing hormone (GnRH) are given as nasal sprays. They are absorbed across the nasal mucosa and thus avoid rapid degradation by the enzymes of the gastrointestinal tract.

The mucosal membranes of the nose, mouth, pharynx, larynx and trachea can be anaesthetized by a localized spray of lidocaine. This is known as **topical anaesthesia** and is used to allow simple surgical procedures to be carried out on the affected areas or to allow the painless passage of an endotracheal tube or bronchoscope. Topical application of local anaesthetics to the conjunctiva can be used to permit minor surgical procedures to be carried out on the surface of the eye. Eye drops containing **dorzolamide** (an inhibitor of carbonic anhydrase) are used in the treatment of glaucoma. The dorzolamide is absorbed across the corneal epithelium and inhibits the production of the aqueous humour. All the above methods suffer from a degree of uncertainty with respect to the rate of uptake of a drug into the systemic circulation, which may result in unwanted complications.

The direct intravenous injection or infusion of drugs produces a rapid and relatively predictable rise in the concentration of a drug and is the preferred route of administration in many cases. Short-acting, general anaesthetics, such as **thiopentone** (thiopental), **propofol**, **etomidate** and **ketamine**, are given by intravenous injection. These anaesthetics are generally given as induction agents prior to administration of an inhalational anaesthetic such as isoflurane. Propofol is also used as a short-acting anaesthetic for day surgery. The treatment of some severe infections may require the administration of an antibiotic by a slow intravenous infusion.

Local anaesthetics may be injected into the subarachnoid space to produce spinal anaesthesia (**intrathecal injection**). This is almost always done in the lower lumbar region to avoid damage to the spinal cord. (The spinal cord extends only as far as the first or second lumbar vertebra; below this level, the spinal canal acts as a conduit for the spinal nerves – the cauda equina.) In **epidural anaesthesia** (also called extradural anaesthesia), a local anaesthetic is injected into the space between the dura covering the spinal cord and the bones of the spinal column – the epidural space. The spinal nerves pass through this space, which is filled with fat, and it is possible to insert a fine catheter (tube) into the epidural space through which a local anaesthetic with or without opiates can be infused as required. This technique can provide prolonged pain relief, which is useful during labour, surgical anaesthesia and for the management of chronic pain. Unlike spinal anaesthesia, epidural anaesthesia can be carried out at any level. For example, it can be used to block sensation in the dermatomes of T4–T12 to provide postoperative pain relief for an upper abdominal incision with only minor interference with bladder function or loss of motor control of the legs.

> **KEY POINT:**
> - Common methods of drug administration are:
>   - Oral (as liquid, capsule or tablets)
>   - Sublingual (as tablets)
>   - Rectal (as enemas or suppositories)
>   - Transcutaneous (as creams, ointments or patches)
>   - Injection (subcutaneous, intramuscular or intravenous)
>   - Inhalation (as a nasal spray, aerosol, or as a gas or vapour for uptake via the lungs)
>   - Topical – direct application to the skin or other exposed mucosal surfaces such as the cornea and vagina.

35.3 Absorption and distribution of drugs

For any drug to exert its effect upon the target tissue, it must first reach the site of action. This occurs by physical transport in the blood and by diffusion. Unless a drug is directly injected into the blood stream, it will first have to cross an epithelial layer, such as the wall of the intestine, before being taken up by the blood. It will then be distributed throughout the body via the circulation before diffusing out of the capillaries to reach the cells where it will exert its effect.

Whatever the route of administration, the uptake of a drug is determined by its physico-chemical characteristics. Its solubility in water will determine how quickly the drug can be distributed around the body to exert its effect and its lipid solubility will determine how readily it can cross an epithelial layer. A few drugs are absorbed by transporters present in epithelial cell membranes. For example, the antihypertensive agent captopril is absorbed in part by the carrier responsible for the absorption of small peptides from the small intestine (see Chapter 24, p. 522). Once in the blood, a drug becomes distributed through the various body systems outlined in Figure 35.1. From the plasma, a drug will first enter the interstitial space (the fluid that surrounds the cells) before coming into contact with those cells on which it will exert its effects. In addition, those drugs that are very lipophilic (fat soluble) will progressively accumulate in the adipose tissues. A well-known example of this tendency is the barbiturate thiopentone (see below). The degree to which a drug will accumulate in adipose tissues is determined by its lipid–water partition coefficient, which is the ratio of its solubility in fat compared with its solubility in water.

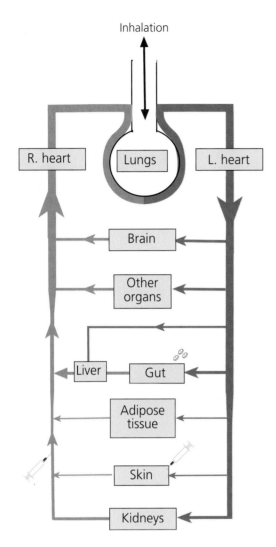

Figure 35.1 A schematic drawing of the circulation illustrating the main routes of drug administration. The thickness of the arrows indicates the approximate blood flow to the different organ systems. Adipose tissue has a relatively low blood flow and accumulates and releases lipid-soluble drugs relatively slowly. Although many drugs are applied directly to the skin, others are injected subcutaneously or intramuscularly. Intravenous injection is indicated separately.

The effect of ionization

Charged molecules do not cross cell membranes at all readily. Fortunately, most clinically useful drugs are either uncharged or are weak acids or weak bases. Consequently, they can be absorbed across epithelial membranes in their uncharged form. The ionization of a drug that is a weak acid can be represented by the following chemical reaction:

$$\mathrm{AH} \underset{}{\overset{K_a}{\rightleftharpoons}} \mathrm{H^+ + A^-}$$

where AH represents the unionized and A⁻ the ionized form of the drug. K_a is the **acid dissociation constant** (often given as its pK_a value by analogy with the pH scale described in Chapter 2). For a drug that is a weak base (B), the ionization reaction is:

$$\mathrm{BH^+} \underset{}{\overset{K_a}{\rightleftharpoons}} \mathrm{H^+ + B}$$

These chemical equations show that weak acids are proton donors (i.e. they release hydrogen ions from their neutral form) while weak bases are proton acceptors (they absorb hydrogen ions). As hydrogen ions are always present in aqueous solutions, the uncharged form of an ionizable drug is always present and can be absorbed across the epithelial layers of the body, as illustrated in Figure 35.2. However, the fraction of the drug in this form

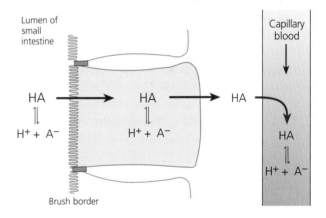

Figure 35.2 The absorption of an ionizable drug from the small intestine. Unless a drug has a permanent charge, both the ionized and unionized forms of the drug will be present. The position of the equilibrium will depend on the pK_a of the drug; the pH of the small intestine is around 7.5–8.0. The unionized form has no charge and can pass across the brush border into the enterocytes and, once within the cell, it can pass across the basolateral surface and into the blood stream. As the blood is constantly flushing away the drug from its site of absorption, the process continues until virtually all of the drug has been absorbed.

Table 35.1 pK_a values for some common drugs

Weak acids		Weak bases	
Drug	**pK_a**	**Drug**	**pK_a**
Aspirin	3.5	Allopurinol	9.4, 12.3†
Furosemide	3.9	Atropine	9.7
Ibuprofen	4.4, 5.2†	Bupivicaine	8.1
Levodopa	2.3	Ergotamine	6.3*
Phenytoin	8.3*	Lidocaine	7.9
Tolbutamide	5.3	Morphine	7.9
Warfarin	5.0	Propranolol	9.4

*Weak acids release hydrogen ions from their neutral form (they are proton donors) while weak bases absorb hydrogen ions (they are proton acceptors).

† These drugs have two ionizable groups.

depends on its pK_a and the pH of the solution it is in: acids are more ionized when the pH is high (i.e. alkaline; see p. 18) and bases are more ionized when the pH is low (i.e. acid). When the pH is equal to the pK_a of the drug, half of the drug will be in the unionized form.

For a weak acid, if the pH is greater than the pK_a by 1 log unit, the concentration of the ionized form will be ten times that of the uncharged form. If the pH is less than the pK_a by 1 log unit, the concentration of the uncharged form will be ten times that of the ionized form. Thus, aspirin, which is a weak acid with a pK_a of 3.5, is mainly present in its unionized form when it is in the acidic environment of the stomach, which has a pH of around 2.5, but is mainly in its ionized form in the small intestine, where the pH is around 7.5. Table 35.1 lists the pK_a values of a number of drugs of pharmacological interest.

Figure 35.3 shows the essential features of drug uptake and elimination following oral administration. As the drug reaches the small intestine, the concentration in the blood progressively rises to a maximum. It then declines as the drug becomes distributed to the various tissues of the body, metabolized and excreted (usually via the urine). For all drugs, there is a minimum effective concentration but excessively high concentrations increase the

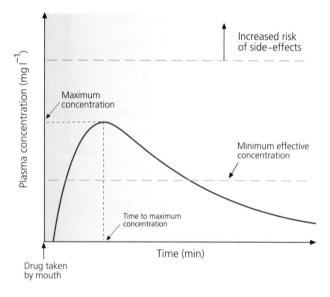

Figure 35.3 A simple diagram to show the change in the plasma level of a drug after it has been taken by mouth. The initial rate of uptake depends on the physico-chemical properties of the drug and its pharmaceutical formulation (see also Figure 35.7). The subsequent decline depends on its redistribution, metabolism and excretion.

risk of side-effects, some of which will be highly undesirable toxic reactions (see below). The ratio of the toxicity of a drug to the clinically effective concentration is called the **therapeutic index**. Drugs are often administered for many days in succession and, when a drug is given repeatedly, the dose regimen is chosen to maintain the drug concentration within the therapeutically useful range as far as possible (Figure 35.4). As an aid to this, the first dose is often doubled. This is called a **loading dose** and it is particularly helpful with some antibiotic treatments.

The exact time course of uptake and elimination of a drug depends on the nature of the drug and on its route of administration. Figure 35.5 shows the distribution of the barbiturate thiopentone after a single intravenous injection. After it has been injected, the blood level rises very rapidly to a peak. The blood concentration then rapidly falls as the thiopentone is taken up by the brain, which has a very high blood flow relative to its mass ($0.551\,kg^{-1}$ – about 15% of the resting cardiac output). Thiopentone is very lipid soluble and passes through the blood–brain barrier with ease. This rapid accumulation by the brain results in loss of consciousness. As the thiopentone

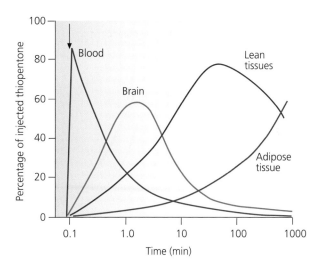

Figure 35.5 The uptake and distribution of thiopentone by various tissues following an intravenous injection. As the drug is directly injected into the blood, the initial blood level rises quickly (arrow) and then falls as it is distributed around the body. The brain rapidly accumulates the drug before it starts to redistribute to the larger mass of the lean tissues (such as skeletal muscle). The accumulation of the drug in the adipose tissue is a major factor in the short-term recovery from the drug's effect. Note that time is plotted on a logarithmic scale.

becomes redistributed to the lean tissues, the concentration in the brain starts to fall. At the same time, the concentration of the drug in the adipose tissue slowly rises. Although the blood flow to adipose tissue is relatively low, the high lipid solubility of thiopentone permits it to be accumulated in body fat and it is this redistribution that is responsible for the fall in its concentration in the blood, brain and lean tissues, rather than its elimination via the kidneys, which proceeds much more slowly.

Volatile anaesthetics are administered by inhalation. Like all gases, before they can reach the brain they must first dissolve in the fluid lining the lungs and diffuse across the alveolar membranes into the blood. Figure 35.6 shows the changes in the blood concentration of three inhalational anaesthetics over a period of 45 minutes. The blood concentration of each agent progressively rises as it equilibrates with the partial pressure of anaesthetic in the lungs. The differences in the rates of equilibration reflect the differences in solubility of the three anaesthetics. Elimination from the blood is also very rapid as anaesthetics are excreted via the lungs. (However, during prolonged application, a significant

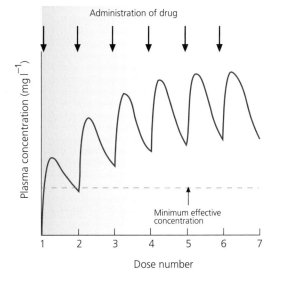

Figure 35.4 The effect of repeated drug administration on plasma drug concentration. The plasma concentration of the drug progressively rises as it accumulates in the body until a balance is reached between its rate of absorption and its elimination from the body. After that, although the plasma concentration fluctuates with each dose, the drug concentration remains within the therapeutically useful range.

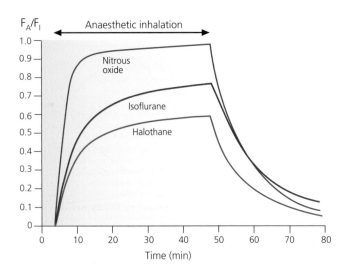

Figure 35.6 The uptake and elimination of the general anaesthetics halothane, isoflurane and nitrous oxide. This figure shows the alveolar concentration of anaesthetic (F_A) as a fraction of the inhaled partial pressure (F_I) plotted against time. The alveolar concentration is in equilibrium with that of the blood. Note how the speed of equilibration of each agent varies. These differences reflect the solubility of each agent in the blood. Halothane is more soluble in the blood than isoflurane, which is more soluble than nitrous oxide.

fraction of halothane (*c.* 30%) is metabolized by the liver and its metabolites excreted in the urine.)

When drugs are taken by mouth, their pattern of absorption is complicated by the need to dissolve in the fluid within the lumen of the gut before being absorbed across the intestinal wall and entering the blood stream. Apart from alcohol, few drugs are absorbed in the stomach. However, the low gastric pH converts aspirin into its neutral form and a small amount is absorbed through the gastric mucosa. Most drugs taken by mouth are absorbed in the small intestine, including aspirin. The rate of absorption is influenced by the motility of the gastrointestinal tract; for example, the stasis of the gut that occurs in migraine impairs drug absorption. Taking drugs immediately after a meal generally tends to slow their absorption, although the absorption of some drugs, such as propranolol is increased, probably due to the increase in the blood flow of the gastrointestinal tract that occurs after a meal. Conversely, when splanchnic blood flow is reduced following haemorrhage, there is a reduction in drug absorption from the gut.

Most drugs are absorbed across the intestinal epithelium by simple diffusion. (Recall that the surface area available for absorption in the small intestine is very large; see Chapter 24). Those drugs that are moderately lipid soluble pass through the brush border relatively easily but those that are very strong acids or bases are often poorly absorbed due to their ionization. However, the epithelium of the small intestine has a large number of transport proteins, which serve to convey specific molecules from the gut lumen to the blood and a number of drugs are absorbed by these transporters. For example, levodopa, which is used in the treatment of Parkinson's disease (see Chapter 12), is transported by the carrier normally used by phenylalanine. Captopril is absorbed by the carrier responsible for the absorption of small peptides (see above).

The rate at which a drug is absorbed from the gastrointestinal tract depends on its pharmaceutical formulation (i.e. whether it is in a slow-release or rapid-release preparation). Figure 35.7 illustrates the way in which different formulations of the same drug influence its pharmacokinetics. Rapid-release preparations produce an early peak in the

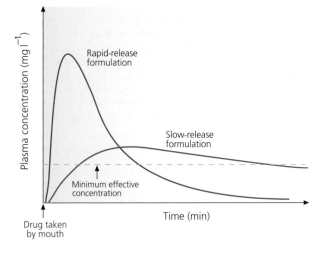

Figure 35.7 Different oral formulations of the same drug change its pharmacokinetics. The red line shows a pattern of rapid uptake and elimination. Here, the rapid kinetics may mean that for a short period the high plasma concentration gives an increased risk of adverse effects. The dark blue line shows the uptake and elimination for the same dose of the drug given as a slow-release preparation. The therapeutically useful level is reached more slowly but the drug's action is greatly prolonged.

plasma concentration followed by a sharp fall as the drug is eliminated from the circulation. In contrast, a slow release preparation reaches its effective concentration much more slowly but its action is significantly prolonged.

Any drug absorbed from the small intestine will first pass to the liver via the portal vein before entering the systemic circulation. The liver is able to metabolize certain drugs rapidly and, in these cases, the amount of such a drug reaching the general circulation is substantially reduced. This is known as **first-pass metabolism** and explains why some drugs need a much smaller dose when given intravenously compared with that required when the same drug is given orally.

When a drug enters the plasma, some of it may bind to plasma albumin (the most abundant of the plasma proteins). This reduces its effective concentration and slows its elimination from the blood. Drugs that bind to plasma albumin include tolbutamide, which is used in the treatment of type II diabetes, and the anticoagulant warfarin,

which is almost all bound (*c.* 98% of the total). As such a high proportion of warfarin is bound, taking another drug that also binds to plasma albumin will displace some warfarin and this has the potential to cause serious bleeding.

> **KEY POINTS:**
>
> - To reach their targets, drugs must cross various epithelial layers.
> - This requires them to traverse the lipid bilayer of cells.
> - They do so chiefly by passive diffusion or by carrier-mediated transport.
> - Those drugs that are weak acids or weak bases cross the membranes as the uncharged species.
> - Some drugs bind to plasma proteins and this slows their rate of elimination.

35.4 Drug metabolism and excretion

Most therapeutically active drugs are chemical species that are not normally present in the body (i.e. they are **xenobiotics**) and they must be eliminated from the body. The process of elimination begins as soon as the drug enters the bloodstream. The liver is the main organ responsible for metabolism and it acts by modifying a drug's chemical constitution to render it less active and more easily excreted, as described in Chapter 25 (p. 542). In brief, drugs are metabolized in two stages known as phase I and phase II. In phase I, the foreign molecule is modified by oxidation, reduction or hydrolysis. In phase II, the modified molecule is joined to another molecule that inactivates it and makes it easier to excrete via the bile or urine. Commonly, drug metabolites formed in phase I are conjugated with glucuronic acid to increase their water solubility and facilitate their excretion via the kidneys. Most drugs have a relatively small molecular mass and are freely filtered from the renal glomeruli into Bowman's capsule. Unless it is completely reabsorbed by the renal tubules, some, or all, of the filtered drug will appear in the urine unchanged. However, if a drug is bound to plasma albumin, only that fraction that is not bound will be freely filtered.

The clearance of a drug (C_{drug}) can be calculated in the usual way (see Box 23.1, p. 476):

$$C_{drug} = \frac{U_{drug} \times \dot{V}}{P_{drug}} \text{ ml min}^{-1} \tag{1}$$

where U_{drug} and P_{drug} are the drug concentration in the urine and plasma, respectively, and $\dot{V}$ is the urine flow rate. The clearance represents the volume of plasma from which a drug has been completely removed in 1 min. In other words, it is the minimum volume of plasma from which the excreted drug could have been derived. The clearance of some drugs is higher than the glomerular filtration rate (see Chapter 23, p. 476), so the renal tubules must secrete these drugs. Examples are penicillin, quinine and salicylates (metabolites of aspirin), all of which are excreted by a combination of filtration and tubular secretion.

Studies have shown that two transport systems are involved in the tubular secretion of drugs: one for anions such as penicillin and the conjugates of glucuronic acid and one for organic cations such as the H_2 antagonist cimetidine. In common with the transport of amino acids and glucose from the lumen into the tubular cells,

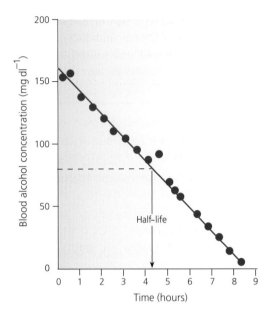

Figure 35.8 The elimination of alcohol from the blood after a large intravenous infusion. The blood concentration falls steadily with time. This is an example of zero-order kinetics, which indicates that the enzymes responsible for metabolizing the alcohol are saturated (see text). The half-life is the time it takes for the plasma concentration to fall by half.

these carriers are proteins and their transport capacity can be saturated. As infusion of penicillin can depress the secretion of *p*-aminohippuric acid and other organic anions (and vice versa), these molecules appear to be secreted by the same transport system. Active secretion provides the kidney with a very efficient means of elimination of protein-bound drugs that could otherwise be eliminated only very slowly by filtration. Some drugs are excreted in the bile (e.g. the muscle relaxant **vecuronium**), as are the metabolites of others (e.g. morphine).

A plot of the plasma concentration of a drug against time can be used to examine how a drug is handled by the body. Figure 35.8 shows that, after a large infusion of alcohol (ethyl alcohol), blood alcohol concentration falls linearly with time. The time it takes to fall by half its initial value is called its **half-life** ($t_{1/2}$). Other drugs show a more complex pattern: Figure 35.9a shows the change in plasma concentration of warfarin after a single dose. Unlike the decline in alcohol concentration seen in Figure 35.8, the rate of decline is initially very fast but slows progressively as the plasma concentration falls. This pattern is very common and is known as **exponential decay**. Here, the rate of decline is proportional to the plasma concentration – the higher the concentration, the faster the rate of decline.

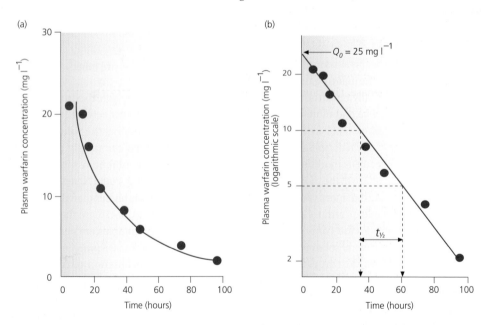

Figure 35.9 The time course of warfarin elimination following a single dose of 200 mg. (a) The rate at which the warfarin is eliminated depends on the concentration in the plasma (first-order kinetics). This pattern of elimination is exhibited by many drugs. (b) The same data plotted as a semilogarithmic plot (the concentration is plotted on a logarithmic rather than a linear axis). This kind of plot makes it much easier to determine the half-life of the drug in the plasma and provides an estimate of the initial plasma concentration.

The mathematical equation for exponential decay is:

$$Q_t = Q_0\, e^{-kt} \qquad (2)$$

where Q_t is the amount in the plasma at time t, Q_0 is the initial concentration and k is the time constant of decay; the larger the value of k, the faster the concentration of the drug declines. Expressing Equation 2 in logarithms:

$$\log_e Q_t = \log_e Q_0 - kt \qquad (3)$$

Plotting $\log_e Q_t$ as a function of time gives a straight line with a slope of k, the time constant for the elimination of the drug. Such a plot is shown for warfarin in Figure 35.9b, where $k = 0.027\,\mathrm{h}^{-1}$. The half-life of the drug can be calculated as $0.693/k$ and the value of the initial concentration of the drug (Q_0) can be determined by extrapolating the line of decay to time zero. From the data of Figure 35.9, the half-time for elimination of warfarin can be calculated as 25.5 hours and the initial warfarin concentration in the plasma was $25\,\mathrm{mg\,l}^{-1}$.

The time course of drug elimination can be classified according to the complexity of the underlying processes. A process that proceeds at a constant rate (such as the elimination of alcohol shown in Figure 35.8) is said to show **zero-order kinetics**; one that proceeds exponentially exhibits **first-order kinetics**. Zero-order kinetics are seen when the concentration of a drug overwhelms the enzymes responsible for its metabolism (i.e. the enzymes are fully saturated). This is the situation for the elimination of alcohol, which is metabolized in the liver by the enzyme alcohol dehydrogenase. The capacity of the liver to metabolize the alcohol becomes saturated at low concentrations ($c.\ 20\,\mathrm{mg\,dl}^{-1}$) so that at higher concentrations alcohol is metabolized at a constant rate. Most other drugs are present in the blood at much lower concentrations and their elimination proceeds by first order-kinetics.

Some drugs are eliminated with a time course that is more complicated than simple zero-order or first-order kinetics. An example is the time course for the elimination of aspirin, which shows an initial phase that is linear (i.e. it follows zero-order kinetics) followed by a period of exponential loss (Figure 35.10). The initial phase of zero-order kinetics occurs when the concentration of aspirin saturates the enzymes responsible for its metabolism but, as the concentration falls, the enzymes become less saturated and the rate of elimination increasingly reflects the prevailing concentration so that the kinetics progressively become first order. Frequently, the time course of

elimination of a drug is best explained by the sum of two or more exponentials; this indicates that more than one process is involved in that drug's elimination.

KEY POINTS:

- Drugs distribute throughout the main body compartments. These are the plasma, the interstitial fluid, the intracellular fluid and adipose tissue.

- Those drugs that are not very lipid soluble are chiefly confined to the plasma and interstitial fluid (the extracellular compartment).

- Lipid-soluble drugs are able to penetrate all body compartments. The higher their lipid solubility, the more they tend to accumulate in adipose tissue.

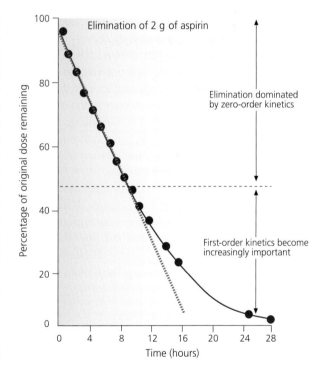

Figure 35.10 The time course of elimination of 2 g of aspirin taken by mouth. The ordinate is the plasma concentration of aspirin expressed as the percentage of the original amount of aspirin remaining in the body. The first 10–12 hours are dominated by zero-order kinetics (dotted green line) indicating that the aspirin has saturated the enzymes responsible for its metabolism. After that time, first-order kinetics become increasingly important.

35.5 Effects of age on drug potency

Human beings are not equally affected by individual drugs. There is individual variation in the response to a particular drug that is due to differences in genetic makeup. This is the province of **pharmacogenetics** and will not be considered further here. Moreover, for any individual, the pharmacokinetics of a given drug will vary according to their age. This has important implications when determining the dose of drug to given to an elderly individual.

Newborn babies have a different body composition to older children and adults. Furthermore, their liver and kidneys are immature and the protein binding of drugs is lower in neonates than it is in adults, partly because neonates have a lower concentration of plasma albumin. These differences are reflected in the pharmacokinetics of many drugs. For example, the half-life of diazepam in premature babies is 30–120 hours, that of normal neonates is 22–46 hours, while that of a 1-month-old infant is around 12 hours – less in fact than that of a young adult (which is around 24 hours). These differences must be taken into account when prescribing diazepam for young babies. It is not sufficient to determine the dose merely by body weight or body surface area. The glomerular filtration rate (GFR) of neonates is less than half that of adults when corrected for body surface area. This changes the pharmacokinetics of drugs such as penicillin that are excreted unchanged in the urine.

Older people form a much more heterogeneous group than neonates. Many suffer from chronic illness, often with multiple disorders, and there is the normal age-related decline in physiological function. Both liver function and renal function decline, slowing the elimination of drugs and prolonging their actions. The GFR and tubular

function progressively decline with age. The loss of tubular function is evident in the elimination of the loop diuretic furosemide (frusemide), which is significantly reduced in old men compared with that seen in young men. The half-life of diazepam progressively increases with age as its metabolism by the liver declines (Figure 35.11). Changes in the distribution of drugs with age also reflect the changes in body composition: older people tend to have a higher proportion of body fat and a lower extracellular volume.

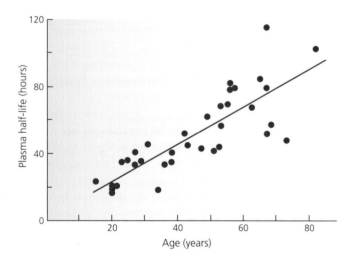

Figure 35.11 The change in the plasma half-life of diazepam with age. From the age of around 15 to 80 years, the half-life of diazepam increases progressively from about 20 hours to around 90 hours. Such a large difference in the half-life will require adjustment of the dose regimen to prevent undue accumulation of the drug in the body.

35.6 Adverse effects of drugs

Any unwanted effect of a drug can be considered an adverse reaction. There may be unwanted pharmacological effects – side-effects – which can be defined as any effect of the drug other than its principal action, or the unwanted reaction may reflect a toxic reaction due to the drug itself or to one of its metabolites (e.g. the toxic

effect of an excessive dose of paracetamol discussed in Chapter 25, p. 544).

Adverse reactions may be unrelated to the dose (e.g. fevers and rashes) or dose related. Many are due to changes in pharmacokinetic behaviour as a result of liver disease or renal insufficiency. There may be long-term adaptive changes, which result in **drug tolerance**. Morphine tolerance is perhaps the best-known example. Withdrawal effects are well known for opiate drugs (including morphine) and hypnotics such as the benzodiazepines. Less well known is the rebound hypertension seen after withdrawal of antihypertensive medication. The withdrawal of corticosteroids poses a special problem as these drugs suppress the normal regulation of the adrenal cortex and may cause it to atrophy. Some drugs are the cause of delayed reactions, such as an increased propensity to certain types of cancer or impaired fertility.

Finally, a number of drugs in common use cause fetal abnormalities. These are generally grouped under the heading of **teratogenic drugs**. Such drugs cause gross fetal malformations. Although the mechanisms responsible for abnormal development are largely unknown, it is important to avoid taking a known or suspected teratogen during the first trimester of pregnancy. This includes anticonvulsants, such as phenytoin and valproate, tetracycline, alcohol and warfarin. The most notorious example is **thalidomide**, which was prescribed in the 1950s as an antiemetic and sedative in early pregnancy with disastrous consequences. Over 10 000 children were born with major limb abnormalities after their mothers had taken thalidomide in early pregnancy. Nevertheless, thalidomide is currently licensed for the treatment of painful skin lesions experienced by those suffering from leprosy. To avoid any repetition of the earlier disaster, prescription of the drug is strictly regulated.

Recommended reading

Gibson, G.G. and Skett, P. (2001). *Introduction to Drug Metabolism*, 3rd edn. Nelson Thornes: Cheltenham.

Grahame-Smith, D.G. and Aronson, J.K. (2002). *Oxford Textbook of Clinical Pharmacology and Drug Therapy*, 3rd edn. Oxford University Press: Oxford. Chapters 1–3, 9 and 11.

Rang, H.P., Dale, M.M., Ritter, J.M. and Flower, R. (2007). *Pharmacology*, 6th edn. Churchill-Livingstone: Edinburgh. Chapters 7, 8 and 52.

Self-assessment questions

Which of the following statements are true and which are false? Answers are given at the end of the book (p. 755).

1
a) The direct application of a drug to the surface to be treated is known as topical application.
b) Drugs can be directly administered to the airways to treat respiratory diseases.
c) Drug application to the skin is only of use in treating disorders of the skin.
d) Intramuscular injection is used when a rapid effect of a drug is required.
e) Intrathecal injections are used for spinal anaesthesia.

2
a) Some water-soluble drugs cross the intestinal epithelium by specific transporters.
b) Molecules that carry an ionic charge can be successfully administered by mouth.
c) Alcohol and aspirin can be absorbed in the stomach.
d) Fat-soluble drugs tend to accumulate in the adipose tissues of the body.
e) The main effect of a drug only wears off after it has been completely excreted by the body.

3
a) Drugs that are weak bases cannot be absorbed by the small intestine.
b) Volatile anaesthetics readily cross cell membranes because they are very lipid soluble.
c) Protein binding of certain drugs prolongs their pharmacological action.
d) Alcohol is eliminated from the blood by a first-order kinetic process.
e) Non-metabolized drugs can be excreted from the kidneys by filtration.

Appendix: SI units

A system of units based on the metre, kilogram and second has now been adopted internationally. This system is known as the 'Système International des Unités' or SI system of units. There are seven basic units:

Physical quantity	Name of unit	Standard symbol
mass	kilogram	kg
length	metre	m
time	second	s
electric current	ampere	A
temperature	degree Kelvin	K
light intensity	candela	cd
amount of substance	mole	mol

All other units are derived from these base units. The principal derived SI units are:

Physical quantity	Unit	Standard symbol	Definition
electrical potential	volt	V	$J\,A^{-1}\,s^{-1}$
energy	joule	J	$kg\,m^2\,s^{-2}$
force	newton	N	$J\,m^{-1}$
frequency	hertz	Hz	s^{-1}
power	watt	W	$J\,s^{-1}$
pressure	pascal	Pa	$N\,m^{-2}$
volume	litre	l (or dm³)	$10^{-3}\,m^3$

The size and quantities of items of biological interest range over many orders of magnitude. To avoid writing very large or very small numbers, most measurements are expressed as powers of ten. The superscript (or exponent) indicates the number of zeros following the principal number when the number is greater than one. Thus, 10 is 10^1, 100 is 10^2 (10×10), 500 is 5×10^2, 1000 is 10^3 ($10 \times 10 \times 10$) and so on. As $10^0 = 1$, the superscript number becomes negative for numbers less than one. Thus, 0.5 can be written as 5×10^{-1}, 0.05 can be written as 5×10^{-2} and 0.0015 can be written as 1.5×10^{-3}. Negative superscript numbers thus indicate the position of the principal number to the right of the decimal point. The individual powers of 10 are given names, the most important in biology being given by the following table:

Multiple		Name	Symbol
1 000 000	10^6	Mega	M
1000	10^3	kilo	k
0.1	10^{-1}	deci	d
0.001	10^{-3}	milli	m
0.000001	10^{-6}	micro	μ
0.000000001	10^{-9}	nano	n
0.000000000001	10^{-12}	pico	p
0.000000000000001	10^{-15}	femto	f

Under the SI system, the standard volume for expressing concentrations is the litre. Thus, a plasma protein concentration of 7 g per 100 ml should be expressed as 70 g per litre ($70\,g\,l^{-1}$) although $7\,g\,dl^{-1}$ or 70 g/dl (7 grams per decilitre) is equally correct. Where the molecular weight of a constituent of one of the body fluids is known, its concentration should be expressed as its molar concentration (moles per litre). Thus, the plasma sodium concentration should be expressed as $0.14\,mol\,l^{-1}$ or $140\,mmol\,l^{-1}$. The same rule applies for expressing cell counts in blood so that a red cell count of 4.5×10^6 cells per microlitre on the old system is now expressed as 4.5×10^{12} cells per litre.

The unit of pressure in the SI system is the pascal, which is one newton per square metre ($N\,m^{-2}$) as pressure is force per unit area. The conventional unit of pressure is millimetres of mercury (mmHg), which is still widely used. To convert from mmHg to pascals, multiply by 133.325. For kilopascals, multiply by 0.133325. A pressure of 7.5 mmHg is equivalent to

$$7.5 \times 0.133325 \text{ kPa} = 0.9999375 \text{ kPa}$$

Thus, to a good approximation:

$$7.5 \text{ mmHg} = 1 \text{ kPa}$$
$$15 \text{ mmHg} = 2 \text{ kPa}$$
$$40 \text{ mmHg} = 5.3 \text{ kPa}$$
$$60 \text{ mmHg} = 8 \text{ kPa}$$
$$75 \text{ mmHg} = 10 \text{ kPa}$$
$$100 \text{ mmHg} = 13.3 \text{ kPa}$$
$$150 \text{ mmHg} = 20 \text{ kPa}$$
$$760 \text{ mmHg} = 101 \text{ kPa}$$

The unit of temperature is K (degrees Kelvin) but °C (degrees Celsius) is still commonly used. To convert from degrees Celsius to degrees Kelvin, add 273.15. Thus:

$$37°C = 37 + 273.15$$
$$= 310.15 \text{ K}$$

The calorie is not an SI unit as the joule is used as the unit for energy. Heat is merely one form of energy. To convert from calories to joules, multiply by 4.185. For example: the energy equivalent of 100 g of bread is 240 kilocalories or 240×4.185 joules (= 1004 kjoules).

Glossary of technical terms

Abduction: movement of a limb or other structure away from the midline or away from its original position. An abductor is a muscle that performs this action.

Absorption: to take up a substance from a compartment, e.g. to take up material from the central space of a renal tubule.

Accessory structures: structures associated with, but not directly part of, a body system.

Accommodation: (i) the adjustment of the thickness of the lens during focusing of the eyes; (ii) a change in excitability of a nerve during a prolonged depolarization.

Achlorhydria: lack of acid secretion by the stomach.

Acid: any molecule that dissociates to release a hydrogen ion (proton); a proton donor.

Acid dissociation constant: the equilibrium constant for the binding of protons (hydrogen ions) to a molecule. Written as K_a but frequently expressed as the pK_a value by analogy with the pH scale (i.e. as the logarithm of $1/K_a$).

Acid–base balance: refers to the processes responsible for maintaining the blood pH value within its normal range.

Acidaemia: a blood pH below 7.35 (the normal range is pH 7.35–7.45).

Acinus: of a secretory gland: a terminal sac-like group of cells that surrounds a central secretory space; the plural is **acini**.

Active transport: the transport of a substance against its concentration gradient.

Adaptation: the progressive change in the rate of discharge of a sensory receptor during a maintained stimulus.

Adduction: to move a limb or muscle towards the central axis of the body; an **adductor** is a muscle that performs this action.

Adenohypophysis: the anterior lobe of the pituitary gland.

Aerobic metabolism: metabolism that requires the participation of free oxygen (cf. anaerobic metabolism).

Afferent arteriole: an arteriole that carries blood towards the renal glomerulus.

Afferent lymph: the lymph of the lymphatic vessels flowing towards the lymph nodes.

Afferent nerves: nerves that carry information towards the brain or spinal cord; sensory nerves.

Afterload: the pressure in the aorta against which the ventricles must eject blood during systole.

Agglutination: the clumping together of red blood cells seen when incompatible blood samples are mixed (see agglutinin).

Agglutinin: an antibody that reacts with red blood cells to cause them to clump together.

Agglutinogen: an antigen on the surface of red blood cells that reacts with an agglutinin.

Agnosia: failure to recognize familiar objects or people as a consequence of brain damage.

Agonist: a molecule that binds to and activates a receptor.

Algogen: a substance that activates pain receptors; a pain-producing substance.

Alkalaemia: a blood pH greater (i.e. more alkaline) than pH 7.45 (the normal range is pH 7.35–7.45).

Allele: one of two or more forms of a gene that occupy the same region on a chromosome.

Allograft: a tissue graft from an individual unrelated to the recipient; the prefix *allo-* is from the Greek for other or different.

Ammoniagenesis: the liberation of ammonia from glutamine to buffer hydrogen ions in the urine.

Amnesia: loss of memory usually following damage to the brain following traumatic injury; anterograde amnesia refers to loss of memory of events that occur after the injury; retrograde amnesia refers to loss of memory of events occurring before the moment of injury.

Amniocentesis: sampling the amniotic fluid by inserting a hollow needle through the mother's abdominal wall into the amniotic cavity and withdrawing a small quantity for examination.

Anabolic: a metabolic process in which complex molecules are synthesized from simpler ones. Anabolic steroid: a steroid that stimulates muscle development; an androgen.

Anaemia: a reduction in the haemoglobin content of the blood either as a result of reduced numbers of red cells or a reduction in their haemoglobin content.

Anaerobic metabolism: metabolism that does not depend on free oxygen.

Analgesia: absence of pain; an analgesic is a substance that mitigates pain.

Anastomosis: a natural connection between two tubular structures; an arteriovenous anastomosis is a direct connection between an arteriole and a vein, so bypassing a network of capillaries.

Androgen: any steroid hormone that promotes the male pattern of development (e.g. testosterone).

Anencephaly: the absence of a properly developed brain in a newborn baby.

Angiogenesis: the formation of blood vessels.

Anion: an atom or molecule possessing a negative charge.

Anoxia: lack of oxygen; low oxygen in the blood or tissues is hypoxia.

Antagonist: (i) a drug that prevents the activation of a receptor or physiological system by another drug; (ii) a muscle that acts in opposition to another.

Antidromic propagation: the propagation of an action potential in a nerve in the direction opposite to its normal path. It is usually a consequence of the artificial stimulation of a nerve.

Antigen: any molecule that is able to elicit a response from the immune system such as the formation and secretion of an antibody.

Antiport: a transport process in which the movement of an ion or molecule in one direction across a membrane is coupled to the movement of another molecule or ion in the opposite direction.

Antipyretic: a drug that acts to reduce the increase in body temperature seen during a fever.

Aortic stenosis: an abnormal narrowing of the aorta.

Aphasia: loss of the ability to communicate either by speech or by writing.

Apical surface of a cell or epithelium: the surface that is exposed to the central cavity of a hollow organ.

Apnoea: Cessation of breathing; voluntary apnoea is breath holding.

Apoptosis: programmed cell death, e.g. during development.

Apraxia: a disorder of the nervous system in which the ability to perform purposeful movements is impaired.

Aquaporins: channels in cell membranes that permit the passage of water.

Arrhythmia: irregular beating of the heart; also called dysrhythmia.

Arthrology: the study of joints; arthroses are joints, hence the term arthritis for the painful conditions that affect the joints.

Ascites: fluid accumulation in the abdominal cavity.

Athetosis: abnormal writhing movements of the fingers, arms, head and tongue caused by damage to motor control systems.

Atomic mass: the average mass of an atom of an element relative to one-twelfth of the mass of carbon 12.

Atomic number: the number of protons in the nucleus of an atom of an element.

Atresia: (i) the absence of a natural opening or the abnormal closure of a channel by a membrane; (ii) the degeneration and subsequent resorption of immature ovarian follicles.

Auscultation: the determination of the condition of the internal organs by listening, especially to the chest. Auscultation is also used to determine the arterial blood pressure by means of a sphygmomanometer.

Autocoid: a locally secreted chemical signal; a paracrine signalling molecule.

Autograft: a graft of tissue from one part of the body to another as in grafting skin.

Autologous blood transfusion: transfusion of an individual's own blood after it has been stored in a blood bank.

Autonomic nervous system: that part of the nervous system concerned with the regulation of the internal organs.

Autoregulation: the local regulation of blood flow to an organ; regulation that does not depend on hormonal or nervous control systems.

Autosomes: those chromosomes not concerned with the determination of the sex of an individual.

Avogadro's number: a chemical constant that is the number of molecules in one mole of a substance (6.023×10^{23}).

Baroreceptors: sensory receptors that respond to the pressure within the blood vessels. They are of two kinds: high-pressure receptors, which are present in the carotid sinus and aortic arch, and low-pressure receptors, which are found in the walls of the right atrium and great veins. They are mechanoreceptors that respond to the degree of stretch in the vessel wall.

Basal metabolic rate: the metabolic rate of an individual in the postabsorptive state, in a thermoneutral environment and at complete rest. It is usually measured on awakening first thing in the morning.

Base: any molecule that will accept a proton when it is in an aqueous solution.

Basolateral surface of a cell: the surface that does not face the central cavity of a hollow organ but is exposed to the interstitial space and oriented towards the blood (cf. apical surface).

Basophil: a white blood cell (leukocyte) that contains granules that stain strongly with basic dyes such as methylene blue. Basophils are considered to be circulating mast cells and account for around 0.5% of the white cell population.

Biotransformation: the metabolic transformation of any chemical entity, usually applied to the metabolism of exogenous substances (xenobiotics) such as drugs and toxins.

Blastocyst: (in early development) a ball of cells surrounding a central fluid-filled cavity.

Body mass index (BMI): a measure of the weight of an individual relative to their height. It is used to assess whether a person is underweight, overweight or obese. The BMI is calculated as the body weight in kilograms divided by the square of the height in metres.

Bradycardia: a heart rate at rest that is below 60 beats per minute.

Buffering: the ability of a solution to resist a change in its hydrogen ion concentration.

Buffer: in aqueous solutions, buffers are ionic compounds that resist changes to the solutions' acidity or alkalinity, so stabilizing the pH. Buffers are usually salts of weak acids or bases such as bicarbonate and phosphate.

Catabolism: metabolism in which complex molecules are broken down to simpler ones with the release of energy (adjective: catabolic).

Catecholamine: one of a number of compounds that are chemically related to catechol. They are synthesized in the body from the amino acid tyrosine. Examples are adrenaline (epinephrine), noradrenaline (norepinephrine) and dopamine.

Cation: an ion or molecule carrying a positive charge. Cations of importance in physiology are sodium, potassium, magnesium and calcium.

Central nervous system or CNS: that part of the nervous system comprising the brain and spinal cord.

Central venous pressure: the hydrostatic pressure of the blood of the right atrium during diastole (when the heart is relaxed between beats).

Chelators: molecules that have a ring structure that enables them to bind cations (e.g. Ca^{2+}) tightly. Examples are EDTA and citrate.

Chemical elements: the fundamental units of chemical reactions. An element cannot be broken down into simpler materials by chemical means. Each element can be represented by a specific abbreviation or **chemical symbol** of one or two letters. Examples are carbon (C), oxygen (O), calcium (Ca), chlorine (Cl) and sodium (Na).

Chemical senses: taste (gustation) and smell (olfaction).

Chemical synapse: a synapse that operates by the release of a small amount of a signalling molecule (a neurotransmitter) from a nerve ending to influence the activity of another cell. (cf. an electrical synapse, which operates by transmitting an electrical disturbance from one cell to another).

Chorea: a motor disorder characterized by involuntary jerky movements.

Chromosomes: rod-like structures seen in the cell nucleus during cell division. Chromosomes are assemblies of scaffolding proteins and DNA that carry the units of inheritance – the genes.

Chronopharmacology: the study of the changes in drug potency during biological cycles.

Chronotropic effects: these are changes to the heart rate induced by hormones, nerves or drugs. A positive chronotropic effect is an increase in heart rate; a negative chronotropic effect is a fall in heart rate.

Chyme: the mixture of food and juices formed in the stomach and passed to the small intestine for further digestion.

Circadian rhythm: a biological rhythm having a period of around a day. The sleep–wakefulness cycle is the best known but many physiological processes show circadian rhythms.

Climacteric: an obsolete term sometimes applied to the onset of the menopause.

Clone: a group of cells that have arisen from a common precursor. All cells within a clone are genetically identical.

Clonus: a succession of involuntary muscular contractions that is usually caused by a sustained stretching of one of the limb muscles. Short spasmodic movements are referred to as clonic contractions.

Codon: a sequence of three nucleotides that represents an amino acid or that signifies the end of a peptide chain (a stop signal).

Commensal organisms: organisms that cause no harm but live in the gastrointestinal and respiratory tracts or on the skin.

Commissures: arrays of nerve fibres that connect one side of the brain to the other.

Condyle: a rounded protuberance at the end of a bone. The outer ends of the femur where they articulate with the tibia to form the knee joint are good examples.

Connective tissues: provide structural support to the various organs of the body and so are sometimes called structural tissues. They are very variable in appearance but all connective tissues are characterized by having an extensive extracellular matrix that is secreted by the cells that are dispersed within it. The main cells of the soft connective tissues are **fibroblasts**. Blood is sometimes considered to be a connective tissue but differs from most connective tissues in that it does not offer structural support and the extracellular matrix is not secreted by the blood cells.

Coronal: an anatomical term for a vertical plane that divides the body or internal structure into a front (anterior) and back (posterior) portion; also the name of the suture that divides the frontal and parietal bones of the skull.

Coronary: a term used to describe structures that encircle an organ rather like a crown. The term is most frequently applied to the arterial blood supply of the heart (the coronary arteries) and the associated circulation (the coronary circulation).

Corpus striatum: the nuclei of the basal ganglia comprising the caudate nucleus, globus pallidus and putamen.

Corticosteroids: the steroid hormones secreted by the adrenal cortex.

Cotransport: the linked transport of two or more molecules or ions.

Cranial nerves: the 12 pairs of nerves that originate directly from the brain.

Cretinism: a deficiency of thyroid hormone in childhood that leads to impaired physical and mental development.

Cyanosis: a bluish appearance to the skin or mucous membranes that is a sign of hypoxia.

Cytoskeleton: the microfilaments that give shape to a cell.

Cytosol: the fluid component within a cell.

Deamination: the removal of an amino group from an amino acid.

Decibel: a unit for measuring the intensity of a sound with respect to a reference sound. In auditory physiology, the reference sound has an intensity that is close to the threshold of hearing.

Deglutition: the act of swallowing.

Dementia: severe impairment of intellectual function accompanied by short-term memory loss.

Dendrite: a fine branching structure of a nerve cell that receives signals from other nerve cells; any structure that is highly branched, e.g. the dendritic cells of the immune system.

Depolarization: a decline in the membrane potential to a less negative value.

Dermatome: the area of the skin supplied by the nerves from a single spinal segment.

Diabetes insipidus: a disease caused by a lack of anti-diuretic hormone (ADH or vasopressin) or by a failure of the kidney to respond to circulating ADH. It is characterized by the production of large volumes of very dilute urine.

Diabetes mellitus: a disease caused by either a lack of insulin or a failure of insulin receptors to respond to it. It is characterized by a high blood sugar, loss of weight

and a high urine output. Two forms of the disease exist: type I, which is caused by destruction of the β-cells of the pancreatic islets, and type II, which reflects progressive loss of sensitivity to insulin.

Diapedesis: the process whereby white blood cells pass from the circulation into the tissues.

Diaphysis: the main shaft of a long bone.

Diastole: that phase of the cardiac cycle in which the ventricles are relaxed.

Diastolic pressure: the lowest value of the arterial blood pressure recorded during the cardiac cycle. It occurs when the ventricles are relaxed and filling with blood.

Diet: (i) the food and drink normally taken each day for nourishment; (ii) a specific regimen of food and drink intended to treat a medical condition or induce loss of weight.

Diploid cell: a cell having the full complement of chromosomes; in humans, this number is 46 arranged as 23 pairs (cf. haploid cell).

Diplopia: double vision.

Disaccharide: a molecule consisting of two simple sugars linked together. Examples are lactose and sucrose.

Diuresis: a marked increase in the production of urine.

Diuretic: any substance that increases the output of urine. The term is particularly applied to drugs that are administered to relieve systemic oedema by promoting the excretion of sodium.

Diverticulum: a sac or pouch in the wall of the gastro-intestinal tract. A diverticulum may arise during development or during later life. Inflammation of such a malformation is known as diverticulitis.

Drug: any substance that alters some aspect of biological function. Normally applied to those given therapeutically to ameliorate symptoms, relieve pain or treat a disease. The term is also used to denote substances of abuse, which are sometimes called recreational drugs.

Dyskinesia: difficulty in controlling movement.

Dysphagia: difficulty in swallowing.

Dyspnoea: any distress during breathing.

Dysrhythmia: an alternative name for arrhythmia.

Ectopic beats: ventricular contractions that occur prematurely and disturb the normal cardiac rhythm.

Effective circulating volume: the volume of blood in the circulation that contributes to the maintenance of blood pressure.

Effector: a muscle or secretory gland controlled by a nerve.

Efferent arterioles: an arteriole that carries blood away from the glomerular capillaries of the renal capsule.

Efferent lymph: the fluid in the lymphatic ducts as it leaves a lymph node.

Efferent nerves: motor nerves; nerve trunks that leave the central nervous system to supply the muscles or secretory glands.

Eicosanoid: any of the unsaturated organic molecules containing 20 carbon atoms that are derived from arachidonic acid.

Electrochemical gradient: the combined influence of ionic charge, concentration gradient and electrical potential that determines the direction in which ions can move across the cell membrane by passive diffusion.

Electrolyte: any ion in solution. Frequently refers to the inorganic ions of the plasma and extracellular fluid, that is bicarbonate, calcium, chloride, magnesium, potassium and sodium.

Emesis: vomiting; an **emetic** is a drug that induces vomiting.

Endocardium: the thin layer that lines the chambers of the heart comprising an endothelium and a layer of supporting connective tissue. Its endothelium is continuous with that of the blood vessels (the vascular endothelium).

Endocrine glands: the ductless hormone-secreting glands of the body.

Endocytosis: the process by which cells take up part of the plasma membrane to keep its surface area largely constant. During endocytosis some of the extracellular fluid becomes trapped in the vesicles that are formed (the **endosomes**) and imported into the cell. This is sometimes called **pinocytosis** (q.v.).

Endolymph: one of the two fluids of the inner ear (the other is perilymph). It is characterized by having a high concentration of potassium and is found in the scala media and fills the membranous labyrinth, which houses the sensory epithelia of the vestibular system.

Endometrium: the innermost layer of the uterus; the mucous membrane that lines the uterus.

Energy metabolism: the breakdown of complex foodstuffs (mainly glucose and fatty acids) to simpler components (chiefly carbon dioxide and water) to synthesize ATP for fuelling the energy-consuming processes within the cells.

Enteric nervous system: the complex of nerve cells and nerve fibres that lies wholly within the wall of the gastro-intestinal system. Two sets of ganglia are present: the submucosal (or Meissner's) plexus and the myenteric (or Auerbach's) plexus. The submucosal plexus is found within the submucosa while the myenteric plexus lies between the circular and longitudinal layers of smooth muscle of the gut wall.

Eosinophil: a white blood cell with granules that are stained by acidic dyes such as eosin. Eosinophils account for only about 1.5% of the total white cell population.

Epiphysis: the regions at the ends of the long bones. An epiphysis is mainly composed of spongy bone and is the region where one bone forms a joint with its neighbour. In young people, the epiphysis is separated from the main shaft of the long bone (the diaphysis) by a region of bone growth called the epiphyseal plate. In adults, the epiphyseal plate becomes mineralized to form the epiphyseal line.

Epithelium: a continuous sheet of cells that are bound tightly to one another. The cells lie on a matrix of connective tissue fibres called the **basement membrane**. Epithelia line the hollow organs of the body and separate one compartment of the body from another. However, the cell layer that lines the heart and blood vessels is called an **endothelium**.

Equilibrium potential: the potential at which the tendency of an ion to move down its concentration gradient is exactly balanced by the membrane potential. It can be calculated from the Nernst Equation.

Erythropoiesis: the formation of red blood cells.

Eupnoea: normal breathing.

Excretion: the elimination of waste from the body via the faeces, urine or, in the case of carbon dioxide, the lungs.

Exocrine glands: glands that have a duct that discharges the secretory product onto an epithelial surface. Examples are the salivary glands and the gastric glands of the stomach.

Exocytosis: the process by which the contents of a cell's secretory vesicles are released into the extracellular space.

Exponential decay: a process in which the rate of decline of a variable is proportional to its magnitude.

Extension: a backward or posterior movement of the trunk or movement of the distal part of a limb away from the body.

Extensor muscles: any of the muscles that will cause the extension of a limb.

Extracellular fluid: the fluid that lies outside the cells.

Facilitated diffusion: the passive transport of solutes across the plasma membrane of cells that is mediated by specific membrane transport proteins.

Fascia: a sheet of fibrous connective tissue that separates muscles or groups of muscles from one another.

Fascicle: a small bundle of nerve or muscle fibres (also known as a fasciculus).

Feedback: a process in which the output is used to regulate the process itself. In **negative feedback**, the output inhibits the process and so acts to maintain the level of some variable such as body temperature within a given range. In **positive feedback**, the initial change becomes amplified until the process reaches its limit. An example is the triggering of an action potential in which a small change in membrane potential results in a full action potential.

Filtration: the separation of suspended particles from a fluid by means of a porous barrier.

Flaccid: floppy, limp.

Flexion: a forward or anterior movement of the trunk or a bending motion of a limb.

Gastric: pertaining to the stomach (e.g. gastric glands, gastric ulcer).

Gene: a unit of inheritance. The physical entity is a sequence of DNA occupying a specific location on one of the chromosomes.

Genetic code: the sequence of nucleotides that specifies the positions of individual amino acids at specific points within a protein. Each amino acid is specified by a sequence of three nucleotides called a codon.

Genetics: the study of the mechanisms of inheritance.

Genome: the complete sequence of DNA for a specific organism.

Genotype: the genetic constitution of an organism (cf. phenotype).

Gestation: the period during which a fertilized egg develops before birth – around 266 days or 38 weeks in humans.

Glucocorticoid: any of the steroid hormones secreted by the adrenal cortex that acts to maintain the glucose level in the plasma. The major glucocorticoid is cortisol.

Gluconeogenesis: the synthesis of glucose from non-carbohydrate precursors – specifically any of the amino acids except for leucine and lysine (which cannot be used for gluconeogenesis).

Glycogen: a large polysaccharide that acts as a store of glucose within the body, particularly in the liver.

Glycogenesis: the formation of glycogen from glucose.

Glycogenolysis: the breakdown of glycogen to individual glucose molecules.

Glycolysis: the process by which glucose is broken down before being oxidized by the tricarboxylic acid cycle. Glycolysis can provide a limited supply of ATP in the absence of oxygen. This anaerobic metabolism is important in muscle during heavy exercise.

Glycosuria: the presence of glucose in the urine.

Goitre: an enlargement of the thyroid gland caused by iodine deficiency.

Gonads: the ovaries and testes.

Gustation: the sense of taste.

Gyrus: the surface of the human cerebral cortex that lies between two folds or sulci.

Haematocrit ratio: the ratio of the packed red cell volume to the total volume of the blood; normally around 0.47 for men and 0.42 for women.

Haematopoiesis: the production of the blood cells (cf. erythropoiesis).

Haematuria: the presence of blood in the urine.

Haemodilution: a reduction in the haemoglobin concentration of the blood.

Haemolysis: the disintegration of red blood cells with the release of haemoglobin into the plasma.

Haemorrhage: a significant loss of blood due to bleeding from a damaged vessel.

Haemostasis: the process by which loss of blood from a damaged blood vessel is limited; blood clotting.

Half-life: the time taken for half the quantity of a substance to be eliminated from the body.

Haploid cell: a cell having half the full number of chromosomes; the gametes (i.e. the mature eggs and spermatozoa).

Haptic touch: the determination of the shape and texture of an object by exploring it with the fingers.

Heart failure: a failure of the heart to pump sufficient blood to meet the demands of the body.

Heredity: the genetic transmission of characteristics from one generation to another.

Heterozygous: having different forms of the same gene (alleles) on the homologous chromosomes.

Homozygous: having the same allele for a given gene on the homologous chromosomes.

Hormone: a chemical signal released into the blood, usually by an endocrine gland, to regulate the activity of other organs.

Hormone replacement therapy: therapeutic replacement of one of the hormones normally supplied by one of the body's endocrine glands; commonly abbreviated to HRT, particularly for oestrogen replacement after the menopause.

Hydrocephalus: accumulation of cerebrospinal fluid within the brain due to obstruction of its normal outflow. In very young children, it results in an enlargement of the head.

Hypercapnia: a partial pressure of carbon dioxide in the blood that is higher than normal (*c.* 40–46 mmHg or 5.33–6.12 kPa); often a sign of respiratory obstruction.

Hyperphagia: excessive eating.

Hyperpolarization: an increase in the negativity of the membrane potential.

Hypertension: a blood pressure at rest that is higher than normal.

Hyperthermia: an elevated body temperature.

Hypertonic: (i) solutions that have a greater osmolality than the cells; cells placed in such a solution will shrink; (ii) hypertonic muscles are those that have an abnormally high tension.

Hypervolaemia: a blood volume that is abnormally large.

Hypocapnia: a low partial pressure of carbon dioxide in the blood, usually as the result of over-breathing (hyperventilation).

Hypotonic: (i) solutions that have a lower osmolality than the cells; cells placed in such a solution will swell; (ii) hypotonic muscles lack the normal resting tone and are flaccid.

Hypovolaemia: a blood volume that is abnormally low.

Hypoxaemia: a low oxygen partial pressure in the blood.

Hypoxia: a deficient delivery of oxygen to the tissues.

Hysteresis: a lag in the magnitude of a variable with respect to the physical force controlling it; in the lung, the difference in lung volume during inspiration and expiration at a given intrathoracic pressure.

Infarct: a localized area of dead tissue that is caused by a failure of the local blood supply.

Infradian rhythms: biological rhythms with a period that is longer than one day.

Inotropic effect: an effect on the strength of contraction of the heart during systole. A positive inotropic effect denotes an increase in the force of contraction while a negative inotropic effect denotes a weakening of the force of contraction.

International Normalized Ratio (INR): the INR is an individual's prothrombin time (q.v.) compared with a standard international reference preparation. Normal values for the INR lie between 0.9 and 1.2.

Interstital fluid: that part of the extracellular fluid that lies outside the circulation and bathes the cells. In the body, there is little free water so the interstitial fluid mainly consists of a gel formed by hydration of proteoglycan filaments.

Intracellular fluid: the fluid inside the cells.

Intrathecal injection: an injection into the subdural space (as in spinal anaesthesia).

Involution: degeneration of a structure during normal development.

Ion channel: a membrane protein through which ions can cross the plasma membrane of cells.

Ionization: in an aqueous solution, this is the loss or gain of a hydrogen ion by a molecule to give rise to a molecular entity carrying an electrical charge.

Ion: an electrically charged atom or molecule.

Ischaemia (or ischemia): an inadequate supply of blood to a tissue or organ.

Isometric contraction: a contraction in which the length of the muscle is kept constant.

Isotonic contraction: a contraction in which the load on a muscle is kept constant.

Isotopes: atoms of the same element having different atomic masses.

Karyotype: an image of the complete set of chromosomes of an individual in which the homologous chromosomes are laid out in pairs.

Ketoacidosis: an acidosis caused by the excessive production of ketone bodies; commonly seen in untreated diabetes mellitus.

Kinaesthesia: knowledge of the position of the limbs in space and of the loads placed on them obtained from the proprioceptors.

Kyphosis: abnormal backwards curvature of the thoracic spine; hunchback.

Labyrinth: the system of interconnecting tubes that form the vestibular system (that part of the inner ear concerned with balance).

Leukaemia (leukemia): a disease characterized by excessive proliferation of the white blood cells (leukocytes).

Lipid: any one of the naturally occurring organic compounds that are insoluble in water. They are all derivatives of fatty acids or of cholesterol.

Lipolysis: the breakdown (hydrolysis) of body fat to free fatty acids and glycerol, both of which can be utilized by the tissues of the body for the synthesis of ATP.

Liver failure: disease of the liver characterized by jaundice, fluid accumulation in the abdominal cavity (ascites), failure of blood clotting and marked psychological changes.

Lordosis: a forward curvature of the lumbar region of the spine.

Lower motoneurones: the motoneurones of the spinal cord.

Lumbar puncture: the sampling of cerebrospinal fluid by means of a needle inserted between two of the lower lumbar vertebrae.

Lymph: the clear fluid that is found in the vessels of the lymphatic system. It has the same ionic composition as plasma but less protein.

Lymph nodes: the swellings that lie along the course of the lymphatic vessels, which contain large numbers of cells concerned with combating infections. They are also sometimes called lymph glands.

Lymphatic system: the system of vessels and lymph nodes that drains excess fluid from the tissues into the left subclavian vein.

Macronutrients: those foodstuffs utilized by the body for energy production and growth; the carbohydrates, fats and proteins.

Macrophages: large phagocytic cells that remove cellular debris or foreign materials from the blood and tissues.

Mechanoreceptor: any sensory receptor that responds to mechanical stimuli, e.g. pressure and stretching.

Meiosis: the process of cell division by which the number of chromosomes is halved during the formation of the eggs and sperm (the gametes).

Membrane potential: the electrical potential that exists across the plasma membrane of a cell.

Menarche: the onset of menstruation at puberty.

Meninges: the membranes that cover the brain and spinal cord; the dura, arachnoid and pia.

Menopause: the cessation of menstruation, marking the end of the reproductive phase of a woman's life.

Mesentery: a double sheet of peritoneal membrane attached to the back wall of the abdominal cavity that supports the small intestine and the associated blood vessels and nerves.

Metabolism: the chemical processes that occur in living organisms.

Microcirculation: the circulation in the smaller blood vessels, namely the arterioles, capillaries and venules.

Micronutrients: those components of the diet that are required in small amounts: the vitamins, minerals and trace elements.

Micturition: the act of passing urine.

Mineralocorticoid: any of the steroids secreted by the adrenal cortex that regulate sodium balance in the body. Aldosterone is the major mineralocorticoid.

Mitosis: the process of cell division in which both daughter cells have the full complement of chromosomes.

Mixed nerves: a peripheral nerve having both motor and sensory nerve fibres.

Molarity: the concentration of a substance in solution expressed as moles per litre.

Mole: the basic SI unit of the amount of a substance. One mole of a substance is equal to the relative molecular mass of that substance in grams.

Molecule: any stable arrangement of two or more atoms.

Monoglycerides: an ester of glycerol in which one hydroxyl group is linked to a fatty acid.

Monosaccharide: a simple sugar such as glucose (cf. disaccharide, polysaccharide).

Morula: the solid ball of cells that arises from a fertilized egg as a result of cell division.

Motoneurones: neurones that directly control the activity of muscles or secretory glands.

Motor unit: a motoneurone and all the muscle fibres to which it is connected.

Mucosa: a mucous membrane.

Murmurs: abnormal soft heart sounds, often a sign of heart disease.

Mutation: any change in the genetic material of an organism.

Myocardium: the muscle that makes up the wall of the heart.

Myocyte: a muscle cell, as in cardiac myocyte.

Myoepithelium: an epithelium having contractile properties.

Myofibril: a bundle of protein filaments that form the contractile apparatus of a muscle fibre.

Myofilament: one of the protein strands that make up the contractile apparatus of muscle.

Myogenic contraction (muscle): a contraction that is initiated by pacemaker activity within the muscle itself.

Myometrium: the muscular part of the wall of the uterus.

Myxoedema: a disease caused by underactivity of the thyroid gland.

Narcolepsy: a condition characterized by uncontrollable episodes of deep sleep.

Natriuresis: an increase in sodium excretion by the kidneys.

Near point: the closest point an object can be brought into focus by the eye.

Negative feedback: see feedback.

Nephron: the basic functional unit of the kidney consisting of a renal corpuscle and its associated tubules.

Neuroendocrine reflex: a reflex that results in the secretion of a hormone. Examples are the secretion of oxytocin in response to suckling by an infant and the secretion of vasopressin (ADH) in response to an increase in the osmolality of the plasma.

Neurogenic contraction (muscle): a contraction elicited by activity in a motor nerve.

Neurohypophysis: the posterior lobe of the pituitary gland, also known as the neural lobe.

Neuromuscular junction: the region of contact between a motor nerve and the muscle it innervates.

Neurone: a nerve cell, also called a neuron.

Neurotransmitter: a chemical released from a nerve ending to regulate the activity of a target cell.

Neutropenia: a relative deficiency of neutrophils within the blood.

Neutrophil: the most numerous of the white blood cells, accounting for 40–75% of the total white cell population. They are phagocytes with granules that do not stain with basic or acidic dyes.

Nociceptor: a sensory receptor specifically sensitive to stimuli that give rise to the sensation of pain.

Non-disjunction: a failure of a dividing cell to distribute its chromosomes evenly between the daughter cells.

Nucleolus: a small rounded structure within the nucleus of a cell that is concerned with synthesizing ribosomes.

Nucleotides: molecules that consist of a purine or pyrimidine base linked to a pentose sugar, to which a phosphate group is attached. Nucleotides are the building blocks of the nucleic acids. A **codon** is a sequence of three nucleotides that codes for a specific amino acid. Some nucleotides are coenzymes and play an important role in cellular metabolism, e.g. ATP, cyclic AMP and GTP.

Nucleus: a large round or oval structure within a cell that contains the genetic material. The nucleus is bounded by a membrane and contains almost all of the DNA (a small amount of DNA is found in the mitochondria). The nucleus stains strongly with basic dyes.

Nystagmus: an involuntary eye movement in which the eye moves smoothly in one direction before flicking back to its original position.

Oedema: abnormal retention of fluid in the tissues (i.e. in the interstitial compartment), leading to swelling.

Oestrogens: steroid hormones produced by the ovaries.

Olfaction: the sense of smell.

Oncotic pressure: the osmotic pressure exerted by the plasma proteins. Sometimes called the colloid osmotic pressure.

Opsonization: coating of a foreign particle such as a bacterium with antibody to facilitate its ingestion by the phagocytes.

Oral cavity: the mouth.

Organ: a fully differentiated, functional unit of the body, e.g. the heart, the liver.

Organelle: a functional unit within a cell, e.g. the Golgi apparatus, mitochondria and the cell nucleus.

Orthodromic propagation: the passage of an action potential in its normal direction, i.e. from cell body to the target tissue (cf. antidromic).

Osmolality: the number of moles of osmotically active substances in 1 kg of water.

Osmolarity: the number of moles of osmotically active substances in 1 litre of a solution.

Osmoreceptors: receptors in the hypothalamus that respond to changes in the osmolality of the blood.

Osmosis: the passage of water through a semipermeable membrane driven by a difference in the concentration of solutes on either side of the membrane.

Osmotic pressure: the pressure that is just sufficient to prevent the movement of water from one compartment to another by means of osmosis.

Ossification: the mineralization of cartilage to form true bone.

Pacemaker potential: a regular change in membrane potential that generates an action potential as exemplified by the pacemaker potential of the cells of the sinoatrial node that establishes the rhythmic beating of the heart.

Paracrine signalling: the regulation of the activity of a group of cells by means of a locally released chemical signal.

Paraesthesia: an abnormal or inappropriate sensation in the body.

Paralysis: loss of voluntary movement of a muscle caused by a lesion to the nerves supplying that area of the body. Sensory paralysis is loss of sensation in the affected area.

Paraplegia: paralysis of the lower part of the body due to damage or disease of the thoracic or upper lumbar segments of the spinal cord.

Parasympathetic nervous system: the division of the autonomic nervous system that is concerned with the maintenance of bodily functions during rest.

Partial pressure: the pressure exerted by a gas in a mixture that it would exert if it occupied the same volume on its own.

Parturition: the physiological process of giving birth.

Peptide: a molecule formed by the joining of two or more amino acids by means of peptide bonds.

Pericardium: the membrane that surrounds the heart.

Perilymph: the fluid of the inner ear that fills the scala tympani and scala vestibuli and surrounds the membranous labyrinth (cf. endolymph).

Peripheral nervous system: that part of the nervous system that lies outside the skull and spinal canal. It comprises the autonomic nerves and ganglia, the enteric nervous system and the peripheral nerves.

Peristalsis: a movement of a localized region of the intestine that moves the contents towards the anus.

Peritoneum: the membrane that lines the abdominal cavity and covers the viscera. The part that lines the abdominal wall is called the parietal peritoneum while the part that covers the internal organs is the visceral peritoneum.

Pharmacodynamics: the study of the action of drugs on the body.

Pharmacogenetics: the study of the influence of genetics on the actions of drugs.

Pharmacokinetics: the study of the uptake, distribution, metabolism and elimination of drugs by the body.

Pharynx: the part of the digestive system between the mouth and the oesophagus; the throat.

Phenotype: the physical form of an organism that is determined by its genetic constitution.

Phonocardiogram: the recording of the heart sounds by means of a sensitive microphone placed over the chest.

Photopic: the process of vision in normal daylight.

Pinocytosis: the process by which a cell takes up fluid from the extracellular space; endocytosis.

Placentation: the formation of a placenta.

Plasma: the liquid part of the blood in which the cells are suspended. Its appearance is variable, especially after a meal, but in between meals it is normally a clear pale yellow fluid. After a fatty meal, it often has a milky appearance.

Plasma membrane: the outer bounding membrane of a cell.

Plasma proteins: the proteins normally present in the plasma of the blood: the albumins, globulins and clotting factors.

Polysaccharide: a molecule in which simple sugars are combined to form long branching chains.

Polyspermy: the penetration of an egg by more than one sperm.

Positive feedback: see feedback.

Preload: the pressure in the left ventricle at the end of diastole. The preload determines the length of the cardiac fibres prior to contraction and thus influences the force of contraction.

Pressure: force exerted by a fluid on the wall of a vessel in which it is contained. The SI unit is the pascal (Pa), which is defined as a force of one newton per square metre. Blood pressure is still measured in millimetres of mercury (mmHg) and a blood pressure of 120/80 mmHg corresponds to 15.9/10.6 kPa.

Pronation: a movement of the arm that allows the palm to face backwards (posteriorly) or downwards when the hand is resting on a surface.

Prone: lying flat on the chest, face down.

Proprioceptors: the sensory receptors of the muscles and joints.

Prostaglandins: organic molecules derived from arachidonic acid (an unsaturated fatty acid with a chain of 20 carbon atoms) that have potent effects on smooth muscle.

Protein kinases: enzymes that add a phosphate group to proteins to change their function.

Proteins: large molecules formed by linking many amino acids together by peptide bonds. They perform many different roles in the body including structural support and acting as enzymes and as hormones.

Proteinuria: the presence of protein in the urine.

Prothrombin time: the time taken for plasma to clot after the addition of tissue factor. This standardized test is a measure of the ability of the blood to clot and is of value in managing patients who require anticoagulant therapy.

Pulmonary: pertaining to the lung, as in pulmonary circulation.

Pyrexia: an elevated body temperature that is not caused by exercise or an extreme environmental temperature. It is usually due to disease; a fever.

Quadriplegia: paralysis of all four limbs due to disease of or damage to the upper (cervical) region of the spinal cord; also called tetraplegia.

Receptive field: the region of space over which a sensory receptor can be activated.

Receptor: (i) a cellular protein that responds to a chemical signal by activating a chain of physiological actions, e.g. by causing a depolarization of the cell or changing the production of a second messenger such as cyclic AMP. Most receptors of this kind are found in the plasma membrane (the notable exceptions being the nuclear receptors that respond to steroid and thyroid hormones); (ii) a sensory nerve ending that responds to a specific kind of stimulus within its receptive field, e.g. the nerve endings that respond to light touch or the photoreceptors of the eye.

Reflex: an automatic and stereotyped response to a given stimulus. Examples are the knee jerk in response to tapping the patellar tendon and the withdrawal reflex in response to touching a hot object.

Reflex arc: the neural pathways that are the basis of a reflex action. They have an afferent arm that conducts the information to the nervous system and an efferent arm that connects the nervous system to the appropriate effector. The simplest reflex arc has only one synapse (a monosynaptic reflex, e.g. the knee jerk) but frequently other neurones called interneurones are interposed between the afferent and efferent arms of a reflex arc. When interneurones are involved, more than one synapse is present in the pathway. According to the number of synapses involved, such pathways are known as di-, tri- or polysynaptic reflexes.

Refractory period: the period of reduced excitability in a nerve or muscle fibre that follows the passage of an action potential. The interval during which it is impossible to elicit a second action potential is known as the absolute refractory period. The subsequent period of reduced excitability is called the relative refractory period.

Relative molecular mass: the mass of a molecule relative to one-twelfth the mass of an atom of carbon 12; formerly called the molecular weight.

Renal failure: the inability of the kidneys to maintain the homeostasis of the body.

Respiratory exchange ratio (respiratory quotient or RQ): the ratio of the volume of carbon dioxide exhaled to the volume of oxygen taken up by the lungs in a given time. The RQ for carbohydrate is 1.0 while that for fats is 0.7 so measurement of the RQ provides a means of determining the relative proportions of fats and carbohydrates that are being metabolized at any given time.

Respiratory failure: inability of the respiratory system to maintain normal values of arterial P_{O_2} and P_{CO_2}.

Saccade: a rapid movement of the eye from one point of fixation to another.

Sagittal: an anatomical section that divides the body or internal structure into a right and left portion (also known as a median section). Parasagittal sections are those longitudinal sections that are parallel to the midline. The sagittal suture courses along the midline of the skull.

Saltatory conduction: the process by which myelinated nerve fibres conduct action potentials.

Sarcomere: the fundamental contractile unit of striated muscle. Each sarcomere is separated from its neighbours by Z lines and consists of two half I bands, one at each end, separated by a central A band. Actin is the principal protein of the I bands while myosin is the principal protein of the A bands.

Satiety: the sense of fullness, particularly with respect to the intake of food.

Scoliosis: an abnormal lateral curvature of the spine.

Scotopic: vision in the dark-adapted eye.

Second messenger: a chemical that is synthesized by a cell in response to a hormone or other external chemical signal (the first messenger). Second messengers regulate processes within the cell. Examples are cyclic AMP and inositol 1,4,5-trisphosphate (IP_3).

Secondary active transport: the process by which molecules or ions are moved against their electrochemical gradient by exploiting the sodium gradient that is established by the sodium pump (q.v.). Examples are glucose uptake by the small intestine and the regulation of intracellular calcium by sodium–calcium exchange.

Segmentation: a movement of the small intestine in which the circular muscle contracts to promote the mixing of food residues and digestive enzymes.

Sex chromosomes: the X and Y chromosomes.

Sex steroids: the steroid hormones secreted by the ovaries and testes to regulate the reproductive organs. The term usually applies to the oestrogens, progesterone and testosterone.

Shunt: the passage of blood through a channel that enables it to bypass its normal path. In many cases, shunts are a developmental abnormality.

Signalling cascade: the sequence of events that occurs within a cell after it has been activated by a hormone or neurotransmitter.

Sodium pump: the metabolic process that maintains a low sodium concentration within the cells; the classic example of primary active transport. It is an energy-requiring process that is driven by an enzyme known as the Na^+, K^+ ATPase.

Somatotopic map: a representation of the body within the CNS in which the different regions are represented in the same order as on the body itself.

Spatial summation: the combining of different inputs to the same cell or group of cells.

Sphincter: a ring of smooth muscle that controls the flow of a fluid along a hollow organ, notably the flow of chyme along the gastrointestinal tract.

Spinal shock: the loss of reflex activity that occurs after injury to the spinal cord, also known as neurogenic shock.

Splanchnic circulation: the circulation to the upper abdominal organs, including the stomach, small intestine, liver and spleen.

Starling forces: the hydrostatic forces acting on the plasma that determine its rate of filtration into the tissues or renal corpuscle.

Steatorrhoea: the presence of large amounts of fat in the faeces.

Stem cell: an undifferentiated cell that can give rise to a population of specialized cells.

Stereognosis: the perception of the three-dimensional shape of an object, e.g. by touch.

Steroids: organic molecules that have a structure with three six-membered rings and one five-membered ring linked together to form an arrangement of atoms known as the steroid nucleus. In the body, these compounds are synthesized from cholesterol.

Stroke: loss of blood supply to part of the brain leading to impaired brain function. Strokes may be haemorrhagic (caused by the rupture of a blood vessel) or ischaemic (caused by the blockage of a blood vessel by a blood clot).

Submucosa: the connective tissue beneath a mucous membrane.

Sulcus: an infolding of the surface of the brain separating one gyrus from another.

Supination: a movement of the arm that allows the palm to face forwards (anteriorly) or upwards.

Supine: lying on the back, face up.

Sutures: the fixed (fibrous) joints between the bones of the skull.

Sympathetic nervous system: the division of the auto-nomic nervous system concerned with preparing the body for activity, for example by increasing heart rate and increasing the breakdown of glycogen to glucose.

Symport: a protein that transports two or more ions or molecules in the same direction across the plasma membrane.

Synapse: the junction between a nerve terminal and its target cell.

Synaptic transmission: the process of transmitting infor-mation from a nerve cell to its target. In mammals, most synapses operate by secreting a small quantity of neuro-transmitter onto a specialized region of the target cell (the postsynaptic membrane).

Synovial joint: a joint in which the articulating surface is lubricated by synovial fluid.

Systemic circulation: the circulation of blood around the whole body except for the lungs, which are served by the pulmonary circulation.

Systole: the phase of contraction of the heart muscle.

Tachycardia: a resting heart rate in excess of 100 beats per minute.

Teratogenic: capable of inducing developmental abnormalities.

Testosterone: a steroid that is the predominant male sex hormone.

Tetanus: (i) the sustained contraction of a muscle in response to repeated stimulation; (ii) a disease caused by infection with the tetanus bacillus *Clostridium tetani* in which there is a painful and prolonged spasm of the voluntary muscles with difficulty in opening the mouth (lockjaw).

Tetany: hyperexcitability of the muscles caused by a fall in plasma calcium.

Tetraplegia: paralysis of all four limbs due to injury to the upper region of the spinal cord (see also quadriplegia).

Thermoreceptors: sensory receptors that respond to tem-perature. Those that are found in the skin are called periph-eral thermoreceptors while those in the hypothalamus and spinal cord are central thermoreceptors. Thermoreceptors respond either to low temperatures (cold receptors) or to non-painful heat (warm receptors).

Thermoregulation: the processes involved in maintaining and regulating core body temperature.

Threshold: the minimum depolarization required to elicit an action potential in a nerve or muscle.

Thrombocytes: another name for platelets.

Thrombosis: formation of a blood clot within a blood vessel or heart (as in coronary thrombosis).

Thrombus: a clot of blood, usually within a blood vessel.

Tissue: an aggregate of cells that perform a specific function.

Tonic activity: the steady activity of a nerve; the main-tenance of a steady tension in a muscle, e.g. a postural muscle.

Tonicity: the osmolality of a solution with respect to the cells.

Transverse section: a horizontal anatomical section that is at right angles to both the coronal and sagittal planes. It divides the body or internal structure into an upper (superior) and lower (inferior) portion. It is also known as a cross-section.

Triglyceride: an ester of glycerol in which all three hydroxyl groups are attached to fatty acids.

Trimester: a period of 3 months, usually applied to the gestation period.

Tuberosity: a protuberance on a bone, especially a point of attachment for a muscle or ligament.

Ulcer: a lesion in which the disintegration of a surface layer exposes the underlying tissues. Ulcers occur on the skin and in the gastrointestinal tract and are frequently difficult to heal.

Ultradian rhythm: any rhythm with a period of less than a day.

Ultrafiltration: the separation of small solutes from the plasma proteins as fluid passes through the walls of capil-laries as in the formation of the glomerular filtrate, for example.

Uniport: a membrane transport protein that transports one molecule or ion.

Upper motoneurone: one of the neurones of the primary motor cortex that give rise to the corticospinal tract.

Urinalysis (uranalysis): the chemical analysis of urine for diagnostic purposes.

Vaccinate: to inoculate an individual with a vaccine to produce immunity against a specific disease.

Vagus nerve: the tenth cranial nerve, which supplies the heart, lungs and viscera; part of the parasympathetic nerv-ous system.

Vascular endothelium: the layer of cells that lines the blood vessels.

Vasoconstriction: the narrowing of a blood vessel caused by contraction of its layer of circular smooth muscle.

Vasodilatation: an increase in the diameter of blood vessels brought about by the relaxation of the smooth muscle of their walls.

Vasopressin: the approved name for antidiuretic hormone (ADH). The hormone acts to maintain the circulating volume by increasing the absorption of water by the collecting ducts, so decreasing urine flow. It acts to maintain blood pressure, especially after haemorrhage, by constricting the blood vessels.

Venous admixture: the mixing of venous blood with fully oxygenated blood from the lungs.

Villus: a finger-like projection from the mucous membrane of the small intestine; any similar membranous process (the plural is villi). The villi greatly increase the surface area for absorption compared with a smooth membranous layer.

Vitamin: an organic substance present in the diet that is required in small amounts for the maintenance of normal healthy metabolism.

Work: the product of force applied to an object multiplied by the distance the object has moved in the direction of the force.

Xenobiotic: any foreign chemical entity in the body.

Zygote: a fertilized egg.

Zymogen: an inactive precursor of an enzyme stored in granules within a membrane-bound vesicle.

Answers to self-assessment questions

Chapter 1

1 a) True b) False c) False d) True e) False

The endocrine system uses chemicals (hormones) for communication. Organelles are intracellular structures that perform specific functions within cells. The lungs are part of the respiratory system.

2 a) False b) True c) True d) False e) True

Most physiological parameters are regulated by negative feedback mechanisms. Receptors detect changes in physiological variables; effectors are responsible for returning the variable to its steady-state value.

3 a) True b) True c) False d) True e) False

The stomach is located within the abdominal (abdominopelvic) cavity. The epigastric cavity lies above the umbilicus and is superior to the hypogastric cavity, which is situated below the umbilicus.

4 a) Lateral b) Medial c) Superior d) Distal or inferior e) Anterior or ventral

Chapter 2

1 a) 60 b) 180 c) 111.1 d) 260 e) 300

2

a) 0.5 molar (0.5 mol l^{-1})

b) 0.15 molar or 150 mmol l^{-1} (150 mmolar)

c) 0.3 molar or 300 mmol l^{-1} (300 mmolar)

d) 0.0055 molar or 5.55 mmol l^{-1} (5.55 mmolar)

e) 0.00065 molar or 650 micromol l^{-1} (0.65 mmolar)

3

a) 1 osmol l^{-1} (1 osmolar) (NaCl dissociates into its constituent ions so the osmolarity of a solution of NaCl is twice the molar concentration)

b) 300 osmol l^{-1} (300 mOsmoles per liter or 300 mOsmolar)

c) 300 osmol l^{-1} (urea does not dissociate so mol l^{-1} = osmol l^{-1})

d) 0.0055 osmol l^{-1} or 5.55 mOsmol l^{-1}

e) 0.00065 osmol l^{-1} or 0.65 mOsmol l^{-1}

Note the low osmolarity of the albumin solution for each gram dissolved compared with that calculated for NaCl. This quantity of albumin corresponds to the normal level in plasma.

4 a) 1.0 b) 2.3 c) 5.0 d) 7.4

5

a) The osmolarity of the solution is 308 mOsmol l^{-1}, approximately isotonic with the blood so the cells will neither swell nor shrink.

b) The osmolarity of the solution is 256 mOsmol l^{-1}, significantly hypotonic to the blood so the cells will swell and may burst (lyse).

c) The osmolarity of the solution is 360 mOsmol l^{-1}, hypertonic to the blood so the cells will shrink.

Chapter 3

1 a) True b) False c) False d) True e) True f) False

The intracellular and extracellular fluids are isotonic with each other and thus have the same osmolality (they are iso-osmotic with each other). Proteins account for only a small part of the osmotic pressure of plasma. Most of the body water is within the cells (i.e. in the intracellular fluid).

2 a) True b) False c) True d) False e) True

Glycogen is the body's main store of carbohydrate. Sucrose is a disaccharide.

3 a) True b) False c) True d) False e) True

Lipids account for around half of the mass of the cell membranes. Palmitic acid is a saturated fatty acid that is an important component of triglycerides. Essential fatty acids such as linolenic acid are unsaturated.

4 a) True b) True c) True d) False e) True

Some structural proteins are insoluble in water, e.g. those of the cell membranes.

5 a) False b) True c) False d) True e) False

Nucleotides consist of a purine or pyrimidine base linked to a pentose sugar (either ribose or deoxyribose). The phosphate group is linked to the sugar. DNA and RNA differ in the pentose sugars from which they are built, and in the bases they contain. In DNA, the sugar is deoxyribose and the bases are adenine, guanine, cytosine and thymine (A, G, C and T). In RNA, the sugar is ribose and the bases are adenine, guanine, cytosine and uracil (A, G, C and U).

Chapter 4

1 a) True b) False c) True d) True

The plasma membrane forms a barrier to polar molecules and ions but is relatively permeable to non-polar molecules and to water (which crosses via specific water channel proteins called aquaporins). The lipid bilayer is a relatively stable structure in aqueous environments.

2 a) False b) False c) True d) True

The most prominent feature of most cells is the nucleus. The red blood cells are an exception to this rule as they have no nucleus (in mammals). They also lack mitochondria. When the nucleus is large and pale, the DNA is dispersed (euchromatin) and is involved in wholesale mRNA production for protein synthesis.

3 a) False b) True c) False d) True

Halving of the chromosome number occurs only during meiosis. The nuclear membrane disperses at the beginning of prometaphase and the mitotic spindle disperses during telophase. The daughter cells separate during cytokinesis.

4 a) True b) True c) True d) False

Certain amino acids (tryptophan, tyrosine, leucine and isoleucine) are not glucogenic and are broken down via the same metabolic routes as the fatty acids.

5 a) True b) False c) False d) True

Fatty acids cannot be broken down by anaerobic pathways but are broken down to acetyl CoA before entering the TCA cycle to generate ATP.

Chapter 5

1 a) False b) True c) False

Gases are very lipid soluble and readily pass through the lipid bilayer. They diffuse down their concentration gradient by *passive transport*.

2 a) False b) True c) True

Unlike gases, ions and other polar molecules cannot diffuse through the lipid bilayer but cross the membrane via channel proteins or are transported from one side of the membrane to the other by carrier proteins.

3 a) True b) False c) False d) True

For any substance to be accumulated against its electrochemical gradient, energy must be expended. In active transport, this is provided either by the hydrolysis of ATP (e.g. the sodium pump) or by coupling the movement of one substance against its electrochemical gradient to the movement of another down its electrochemical gradient. This is secondary active transport. The Na^+-dependent uptake of glucose by the enterocytes is one example.

4 a) True b) True c) True d) False

The sodium pump is a carrier protein, not an ion channel, and is an example of an antiporter.

5 a) False b) True c) False d) True

The resting membrane potential is determined by the K^+ gradient because there are many more open K^+ channels than Na^+ channels. Consequently, the resting membrane potential is close to the K^+ equilibrium potential.

6 a) False b) True c) True

Lipid-soluble molecules such as the steroids and prostaglandins are secreted as they are synthesized and pass across the lipid bilayer. Vesicle-mediated secretion is called exocytosis.

7 a) False b) True c) True

Cells take up large proteins from the extracellular fluid by receptor-mediated endocytosis. Bacteria and cell debris are taken up by phagocytosis.

Chapter 6

1 a) False b) True c) True d) False e) False

Organs invariably consist of several tissue types. Although the blood does allow for communication between different tissues (via the circulation), it is not a supporting tissue. Unlike the peripheral nerves, the central nervous system has very little connective tissue.

2 a) False b) True c) False d) True e) True

Connective tissues frequently contain both collagen and elastic fibres. The supporting fibres of the liver cells are collagen, not elastin.

3 a) True b) False c) True d) True e) False

The alveoli are lined with a simple squamous epithelium. The synovial cells do not form a continuous layer; they do not lie on a basement membrane and do not separate one compartment from another. Therefore they do not form an epithelium.

4 a) False b) True c) True d) True e) True

Epithelial cells are linked together via a junctional complex of tight junctions, adherens junctions and desmosomes. Hemidesmosomes link the cells to the basal lamina.

Chapter 7

1 a) True b) True c) False d) True

Hormones are secreted by endocrine glands into the bloodstream to act on cells at a distance. Paracrine signals act locally.

2 a) True b) True c) True d) True e) False
    f) False g) False

Nitric oxide and prostaglandins are paracrine signalling molecules. Insulin and adrenaline are hormones. Ca^{2+} and cyclic AMP are both intracellular mediators but cyclic AMP can spread from cell to cell via gap junctions. Glucose is a substrate for metabolism.

3 a) True b) False c) True d) True

Many receptors are membrane proteins but some are intracellular proteins (e.g. steroid hormone receptors). Receptors may directly activate ion channels but some receptors are protein kinases that are activated when they bind their ligand. Many receptors activate G proteins and thereby modulate the levels of second messengers.

4 a) False b) True c) False d) True

Prostaglandins are metabolites of arachidonic acid, which is derived from membrane phospholipids by the action of phospholipases. They are lipid-soluble molecules that are secreted as they are formed. They are paracrine and autocrine signalling molecules that activate G protein-linked receptors in the plasma membrane.

5 a) True b) False c) True d) True

Steroid hormones are lipid signalling molecules that act mainly by altering gene transcription. As they are lipids, they cannot be stored in membrane-bound vesicles and are therefore secreted as they are synthesized.

Chapter 8

1 a) True b) True c) False d) False

All peripheral axons are associated with Schwann cells but only the larger fibres are covered by layers of myelin. In the intact nervous system, axons normally conduct action potentials in one direction (orthodromic conduction) but, if they are artificially stimulated, they can conduct an action potential in the opposite direction (antidromic conduction).

2 a) True b) True c) False d) False

Once the stimulus intensity has reached the threshold, an action potential is elicited but the amplitude of the action potential does not increase further with increasing stimulus strength. The action potential is all-or-none. When one action potential has passed along an axon, the nerve passes through its refractory period and a second action potential cannot occur until the membrane regains its excitability. For this reason, action potentials cannot summate.

3 10 ms

4 a) True b) True c) False d) True e) False f) True

The actions of acetylcholine are terminated by enzymatic destruction; other neurotransmitters are inactivated by diffusion away from the synaptic region or uptake by the cells surrounding the synapse. IPSPs hyperpolarize the membrane potential and this makes the post-synaptic cell less excitable. The release of a synaptic transmitter in response to a nerve impulse occurs by exocytosis, which is triggered by the entry of calcium into the nerve terminal.

Chapter 9

1 a) True b) False c) True d) True e) False

Skeletal muscle has a highly ordered array of protein filaments that gives it its striated appearance. Smooth muscle has a more diffuse arrangement of actin and myosin filaments. Skeletal muscle fibres are large multinucleated cells with a well-developed T-system; smooth muscle cells have a small diameter and no T-system. Unlike smooth muscle, individual skeletal muscle fibres are not coupled via gap junctions.

2 a) True b) True c) False d) False

Muscles relax when Ca^{2+} is pumped out of the sarcoplasm (i.e. when intracellular Ca^{2+} falls). A rise in intracellular Ca^{2+} permits actin to interact with myosin.

3 a) True b) True c) True d) False e) False

Skeletal muscle fibres act independently of each other unlike cardiac and smooth muscle cells, which are linked by gap junctions. Acetylcholine is a neurotransmitter; the energy for contraction is provided by the hydrolysis of ATP.

4 a) True b) False c) False d) True e) True

A motor nerve together with the muscle fibres it innervates is called a motor unit. The end-plate potential is the result of an increase in the permeability of the junctional membrane to sodium and potassium leading to depolarization. The junctional receptors are nicotinic cholinergic receptors and these receptors can be blocked by curare. If acetylcholinesterase is inhibited, the acetylcholine released by the motor nerve endings is not

destroyed and can continue to stimulate the nicotinic receptors. Very prolonged action of acetylcholine ultimately leads to depolarization of the muscle and block of neuromuscular transmission.

5 a) False b) True c) False d) True e) True

Autonomic nerves have varicosities in the terminal regions, which usually secrete neurotransmitter in a diffuse manner close to several fibres, but in the piloerector muscles and in those of the iris, each varicosity is closely associated with a single fibre. Smooth muscles can be broadly considered to act either as a single unit (e.g. the bladder) or as a multiunit muscle (e.g. the muscle fibres of the iris).

Chapter 10

1 a) True b) False c) True d) False e) True

The appendicular skeleton consists of the pectoral and pelvic girdles and the bones of the limbs. The sternum is a flat bone. The pituitary gland is situated in a depression of the sphenoid bone called the sella turcica. The auditory tube is located in the petrous part of the temporal bone. Metatarsals are further from the trunk than tarsals.

2 a) False b) False c) True d) True e) True

The acetabulum is a depression in the hip bone that houses the head of the femur. The tibia is the weight-bearing bone of the lower leg.

3 a) True b) False c) True d) True e) False

Bone consists of a non-cellular matrix in which cells are situated. It is therefore a connective tissue. Osteoid is secreted by osteoblasts. Haematopoietic tissue is located predominantly in the marrow cavities of short, flat and irregular bones. The shaft of a long bone is called the diaphysis.

4 a) False b) True c) True d) False e) False

Osteoporosis is more common in women, especially after the menopause. Trabecular (spongy) bone contains large spaces filled with bone marrow. Calcium and phosphate are the principal inorganic components of bone. Bone mass reaches a peak at about the age of 25 years and then remains stable for about 10 years. Bone mass starts to fall progressively after this time.

5 a) False b) True c) True d) False e) False

Some joints are fixed or only slightly moveable. Synovial fluid is found only in synovial joints. The rotator cuff (composed of muscles and tendons) stabilizes the shoulder joint.

6 a) True b) False c) True d) True e) False

Intervertebral discs are made of cartilage. Gout is caused by elevated levels of plasma uric acid.

7 a) True b) True c) True d) False e) True

The biceps brachii is the flexor muscle of the elbow, i.e. it bends (flexes) the elbow.

8 a) False b) True c) False d) True e) True

The pectoralis muscle covers the upper chest. The quadriceps muscle forms the fleshy part of the anterior upper leg (the front of the thigh).

Chapter 11

1 a) False b) True c) True d) True

The white matter contains large numbers of myelinated nerve fibres. The grey matter contains large numbers of nerve cell bodies. The end feet of the astrocytes form part of the blood–brain barrier, which isolates the extracellular fluid around the neurones from changes in the composition of the blood.

2 a) True b) True c) False d) False

Peripheral axons are protected by the epineurium, the perineurium and the endoneurium. All peripheral axons are associated with Schwann cells but only the larger fibres are covered by layers of myelin. Efferent (motor) fibres convey action potentials from the CNS to target organs while afferent (sensory) fibres convey action potentials from sense organs and other receptors to the CNS.

3 a) False b) False c) True d) True

Glial cells do not generate action potentials. The end feet of astrocyte processes that end on capillaries do form the blood–brain barrier, but this barrier is freely permeable to oxygen and other gases. Oligodendrocytes form the myelin sheaths of axons within the CNS. They thus perform the same function as the Schwann cells of the myelinated axons found in peripheral nerves.

4 a) False b) True c) False d) True

Each spinal segment gives rise to four spinal roots (two dorsal and two ventral roots) but the dorsal and ventral roots fuse to form the spinal nerves. Therefore each segment gives rise to a pair of spinal nerves. In the cerebral hemispheres, the grey matter lies on the outside with the white matter beneath. In the spinal cord, the grey matter forms the central butterfly-shaped region and is surrounded by white matter.

5 a) False b) True c) True d) False

Each cerebral hemisphere is divided into four lobes: the frontal, parietal, occipital and temporal lobes. The CSF is secreted by the choroid plexus and has a different ionic composition to that of the plasma (see Table 11.2).

Chapter 12

1 a) False b) True c) True d) True e) False

White matter is made up of nerve axons and oligodendrocytes. The anterior horn cells are motoneurones, the loss of which will lead to motor, not sensory, impairment.

2 a) True b) False c) True d) False e) True

Muscle proprioceptors play an important role in both reflex and goal-directed movements. Muscle spindles monitor the length of muscle fibres while the tendon organs monitor tension. Spindle afferents relay information regarding stretch to higher centres. The efferent (fusimotor) fibres stimulate contraction of the intrafusal fibres, thereby altering their sensitivity to stretch throughout the duration of a muscle contraction. There is no equivalent in Golgi tendon organs.

3 a) False b) False c) True d) False e) True

The sensory neurone of a monosynaptic reflex arc makes direct contact with the motoneurone. The withdrawal reflex is organized within the spinal cord. Flaccid paralysis is a characteristic of spinal shock, although recovery is often significant. The tendon tap reflex (knee-jerk) is a monosynaptic reflex.

4 a) False b) True c) True d) True e) False

Most corticospinal neurones make contact with spinal interneurones. The precentral gyrus of the right cerebral hemisphere contains the motor areas that control the muscles of the left (contralateral) side of the body. The patient would therefore experience a loss of movement on the left side.

5 a) True b) False c) False d) True e) False

In addition to the corticospinal tract, descending extrapyramidal pathways connect the motor cortex with the spinal cord. The cerebellum has no direct projection to the cerebral cortex . It influences the execution of motor commands via the thalamus and red nucleus. Hemiballismus is a feature of damage to the basal ganglia.

Chapter 13

1 a) True b) True c) False d) True e) False

Although the sympathetic preganglionic neurones are found only in the thoracic and upper lumbar spinal segments, paired sympathetic ganglia are found on both sides of the spinal cord from the cervical region to the sacral region. Sympathetic preganglionic fibres secrete acetylcholine at their nerve endings. Although the majority of blood vessels are innervated by sympathetic vasoconstrictor fibres, some, such as those in the salivary glands and exocrine pancreas, also have a vasodilator parasympathetic innervation.

2 a) True b) False c) False d) False e) True

Parasympathetic preganglionic fibres are found in cranial nerves III, VII, IX and X. Stimulation of the vagus nerve slows the heart rate. The salivary glands are innervated by parasympathetic vasodilator fibres (parasympathetic stimulation promotes salivary secretion). Both parasympathetic preganglionic and parasympathetic postganglionic fibres secrete acetylcholine.

3 a) False b) True c) True d) True e) False

Stimulation of the sympathetic nerves to the eyes results in dilatation of the pupils. Although activation of the sympathetic nerves results in generalized vasoconstriction, adrenaline secreted by the adrenal medulla acts on the β-adrenoceptors of the blood vessels of skeletal muscle to cause vasodilatation. The vagus nerve exerts a tonic inhibitory effect on heart rate. Cutting the nerve releases this tonic inhibition and heart rate increases.

4 a) False b) True c) False d) True e) True

The postsynaptic receptors in the autonomic ganglia are nicotinic not muscarinic cholinergic receptors. Mecamylamine is an antagonist at cholinergic receptors.

5 a) False b) True c) True d) False

Noradrenaline acts preferentially on α-adrenoceptors, while adrenaline acts preferentially on β-adrenoceptors. It is, however, important to realize that noradrenaline and adrenaline will activate both types of adrenoceptor. Adrenaline and noradrenaline are secreted by the adrenal medulla in response to stimulation of sympathetic preganglionic fibres travelling in the splanchnic nerves.

The chromaffin cells of the adrenal medulla are homologous with sympathetic postganglionic neurones. As atenolol is a β-blocker, it antagonizes the action of adrenaline and slows the heart.

Chapter 14

1 a) False b) True c) True d) False e) True

Afferent fibres carry information towards the CNS. The stimulus to which a receptor is most sensitive is called its adequate stimulus. The initial step in sensory transduction varies with receptor type. There is then a change in the number of open ion channels, which results in the generation of the receptor potential.

2 a) True b) False c) True d) True e) True

Nociceptors are bare nerve endings.

3 a) True b) True c) False d) False e) True

Signals from the somatosensory system reach consciousness in the postcentral gyrus. Fast (pricking) pain rarely elicits autonomic responses.

4 a) False b) False c) True d) True e) True

The outer layer of the eye is the sclera. The iris is the coloured part of the eye. Extraocular muscles are responsible for eye movements. Pupil diameter is controlled by intraocular muscles.

5 a) False b) False c) True d) False

There are no receptors at the optic disk (blind spot). Density is greatest at the fovea centralis. Full colour vision is not possible in dim light as the cones require a high light intensity for stimulation. Intraocular pressure is normally around 2 kPa (15 mmHg).

6 a) True b) True c) True d) False e) False

The wavelength of a sound determines its frequency and therefore its pitch. The Organ of Corti is located on the basilar membrane.

7 a) False b) True c) False d) False e) True

Sensorineural hearing loss is associated with defects of the cochlea or auditory fibres. A middle ear problem would cause conduction deafness. Hair cells sensitive to low-pitched tones are located near to the apex of the cochlea. The endolymph has a high potassium content and a low sodium content, unlike plasma, which has a high sodium and a low potassium content.

8 a) True b) True c) False d) False e) False

The utricle and saccule detect linear accelerations of the head such as gravity. Although the tongue is the principal organ of taste, a few taste receptors are found scattered around the oral cavity and soft palate. The olfactory epithelium lies high in the nasal cavity, above the turbinate bones.

Chapter 15

1 a) True b) True c) False d) True e) False

The neural regions concerned with the control of speech are located in the left hemisphere in most people including those who are left-handed. In only a small minority of left-handed people are the regions concerned with speech located in the right hemisphere. Broca's aphasia results from damage to the frontal speech area (Broca's area).

2 a) False b) False c) True d) True e) False

The EEG of a normal awake adult is dominated by beta waves. During deep sleep, the EEG is dominated by delta-wave activity but this is interrupted by beta-like activity during episodes of REM sleep. Delta waves are larger in amplitude than beta waves.

3 a) False b) True c) True d) False e) True

Melatonin is secreted by the pineal gland. Its secretion is stimulated by darkness and inhibited by light. The menstrual cycle lasts for 28 days and is therefore an example of an infradian rhythm. Ultradian rhythms last for less than 1 day.

Chapter 16

1 a) True b) False c) False d) True e) False

Blood flows from the hypothalamus to the anterior pituitary gland via the hypothalamo-hypophyseal portal system. Although the releasing hormones are secreted into the portal system, negligible quantities reach the general circulation. Dopamine acts as an inhibitory hormone on the pituitary lactotrophs that secrete prolactin. GH secretion is under the control of both somatostatin and GHRH. The latter hormone increases GH secretion.

2 a) False b) True c) False d) True e) True

Acromegaly is a disorder caused by the hypersecretion of GH in adults. Deficient secretion of GH in childhood results in pituitary dwarfism in which body proportions are normal. Lack of ADH from the *posterior* pituitary increases the risk of dehydration. Lack of pituitary gonadotrophins will lead to delay or failure of sexual maturation due to lack of stimulation of the gonads. Deficient anterior pituitary function will lead to low TSH levels and so thyroid hormone secretion will be depressed leading to a reduction in BMR.

3 a) True b) True c) False d) True e) False

Adrenaline and noradrenaline raise arterial blood pressure through their actions on both the heart and vasculature. The catecholamines increase heart rate. Catecholamines activate lipases, which mobilize fats to increase the production of free fatty acids and glycerol. Adrenaline and noradrenaline are secreted by the adrenal medulla.

4 a) True b) True c) False d) False e) True

When plasma calcium levels are normal, PTH secretion stimulates bone formation through its action on osteoblasts (cells that secrete bone matrix). While PTH stimulates calcium absorption by the renal tubules, it has no direct action on the gut. It does, however, stimulate the production of 1,25-dihydroxycholecalciferol, which stimulates calcium absorption from the gut. Low plasma calcium levels stimulate PTH secretion; high plasma levels stimulate the secretion of calcitonin. When PTH levels are elevated, osteoclast activity is stimulated, which leads to the demineralization of bone and release of calcium into the plasma.

5 a) False b) True c) True d) False e) True

Oxytocin stimulates the ejection of milk from the lactiferous ducts. Synthesis of milk is controlled by prolactin. In the absence of ADH, very little water is reabsorbed by the collecting ducts of the kidney leading to the production of a large volume of dilute urine. Oxytocin is released in response to the mechanical stimulus of suckling while ADH is released in response to changes in the volume and osmolality of the plasma, which are detected by specific osmoreceptors. The releasing hormones regulate the secretion of hormones from the anterior pituitary. ADH exerts its actions by binding to G protein-coupled V2 receptors on the P cells of the collecting ducts.

6 a) False b) True c) False d) True e) True

All the hormones secreted by the adrenal cortex are steroids. Pituitary ACTH acts to maintain the structural integrity and function of the adrenal cortex. Aldosterone plays an important role in the regulation of plasma sodium and thereby in the regulation of total body water. Cortisol secretion shows a circadian rhythm that is controlled by ACTH released from the anterior pituitary. Secretion is at its lowest around midnight and rises to a peak around dawn. It then declines during the course of the day. Cortisol stimulates gluconeogenesis and maintains reserves of glycogen for use in periods of fasting.

7 a) True b) False c) True d) False e) True

Lack of thyroid hormone in fetal and early neonatal life leads to a condition called cretinism in which there is mental retardation. Thyroid hormones *inhibit* the secretion of TSH by negative feedback. Thyroid hormones stimulate metabolism and when TH levels are low, BMR is reduced. Thyrotoxicosis is caused by an overactive thyroid gland. High levels of thyroid hormone will increase BMR and elevate the heart rate. The value given is normal. The thyroid gland avidly takes up iodine and incorporates it into thyroglobulin.

8 a) False b) True c) False d) True e) False

When plasma glucose levels are high and cellular ATP reserves are adequate, glucose is stored in the form of glycogen and triglyceride. When plasma glucose levels fall, glycogen is broken down (glycogenolysis) to liberate glucose and glucose is synthesized from glucogenic amino acids (gluconeogenesis) but not from stored fat. The liver carries out both these processes. When the demand for ATP is high (as in exercise), glycogenolysis is stimulated and glucose is released into the circulation.

9 a) True b) True c) False d) True e) False

In the absorptive state (after a meal), plasma glucose is high, insulin secretion is stimulated and glucose entry and utilization by cells is enhanced. In the postabsorptive state, plasma glucose falls, output of the hyperglycaemic hormones (including GH and glucagon) is stimulated and reserves of fat and glycogen are mobilized.

10 a) True b) True c) False d) False e) True

During fasting, plasma glucose is likely to be low. Levels of all the hyperglycaemic hormones (including the catecholamines) will be increased. Glycogen synthesis will be inhibited but there will be an increase in lipolysis leading to the liberation of fatty acids from adipose tissue. In starvation, the mobilization of fats for cellular energy metabolism will lead to the production of ketones in the liver. Although some of these are metabolized by the tissues (e.g. the heart), excess ketones will appear in the plasma and urine.

11 a) True b) False c) False d) True e) True f) True

After eating, plasma glucose is elevated so glycogen synthesis is favoured. The secretion of hyperglycaemic hormones such as cortisol

is depressed. Following an overnight fast, hyperglycaemic hormones, particularly GH, are secreted. There is enhanced synthesis of glucose from non-carbohydrate sources (gluconeogenesis) and an increase in the utilization of fats.

12 a) True b) False c) True d) False e) True

Diabetes is the result of either a lack of pancreatic insulin or a reduction in the sensitivity to insulin. It is characterized by raised plasma glucose, which leads to glycosuria and an increased urine output (an osmotic diuresis). In the absence of insulin, glucose cannot be utilized by many cells, so fats are broken down to create alternative sources of energy. Acromegaly is caused by the excessive secretion of GH in adulthood. Because of the anti-insulin action of GH, acromegaly can result in diabetes.

Chapter 17

1 a) True b) False c) False d) True e) True

The action potential of skeletal muscle is about 2 ms in duration while the action potentials of the heart cells range from about 150 ms in the cells of the SA node to about 300 ms in a Purkinje fibre. The membrane potential of the SA node cells is low and constantly changing. It has a maximum value of about −60 mV, while that of ventricular cells is about −90 mV. Although the conduction of the cardiac impulse through the atria occurs preferentially via certain fibre bundles, these cells are, nevertheless, normal atrial myocytes. In the ventricles, the specialized conducting cells are the bundle cells and the Purkinje fibres, which transmit their action potentials to the ventricular myocytes via gap junctions.

2 a) True b) True c) False d) True e) True

The volume of blood in the ventricles at the end of diastole is about 120 ml. Of this, about 70 ml is ejected (the stroke volume). The ratio of the stroke volume to the end-diastolic volume is called the ejection fraction and is usually about 60% at rest.

3 a) True b) False c) True d) False

As the first heart sound is caused by closure of the atrioventricular valves, it must occur after the onset of ventricular contraction, which follows the R-wave. The second heart sound is caused by the closure of the aortic and pulmonary valves. This occurs at the end of systole when the ventricles have repolarized. The second heart sound therefore occurs just after the T-wave.

4 a) True b) True c) False d) True e) False f) False

Both circulating catecholamines and activation of the cardiac sympathetic nerves increase cardiac output. Both cause an increase in heart rate and an increase in the force of contraction (i.e. they have a positive inotropic action). The arterial blood pressure is determined by the cardiac output and the peripheral resistance, not the other way around. Sympathetic stimulation has a positive inotropic effect (see above). Stimulation of the vagus nerve slows the heart.

Chapter 18

1 a) False b) True c) True d) False e) True

The P-wave reflects atrial depolarization, which precedes atrial contraction. The T-wave reflects the repolarization of the ventricles and has the same polarity as the R-wave because the outermost myocytes begin to repolarize before the innermost ones.

2 a) False b) True c) False d) True e) False

In atrial fibrillation, the ECG shows an irregular fluctuating baseline. In second-degree heart block, the conduction between the atria and ventricles is impaired so that not every wave of atrial depolarization is conducted to the ventricles. Following a heart attack (a cardiac infarct), the S–T segment becomes elevated above the isoelectric line. It lies below the isoelectric line in cardiac ischaemia.

Chapter 19

1 a) True b) False c) False d) False e) True

Arterial blood pressure depends on cardiac output and the total peripheral resistance. Mean blood pressure is given by the sum of the diastolic pressure plus one-third of the pulse pressure (the difference between the diastolic and systolic pressures). The main site of vascular resistance is the arterioles. The pressure changes in the right atrium give rise to distinct pressure waves in the large veins (see Figure 19.6).

2 a) False b) False c) True d) True e) True

The arterioles are under nervous, hormonal and local control. Autoregulation occurs in denervated blood vessels and is an intrinsic property. Two mechanisms appear to be involved – local metabolites and the myogenic response of the smooth muscle to stretch.

3 a) True b) True c) False d) False e) True

The main resistance vessels are the arterioles. A fraction of the fluid that leaves the capillaries at the arteriolar end is reabsorbed by the capillaries during constriction of the arterioles. The fraction that is not absorbed is returned to the circulation via the lymph.

4 a) True b) False c) True d) True

When arterial blood pressure falls, the baroreceptors are less active and this increases sympathetic activity in the cardiac and vasoconstrictor nerves. Heart rate and peripheral vascular resistance increase, restoring the blood pressure to normal.

Chapter 20

1 a) True b) False c) True d) True e) False

While blood volume accounts for 7–8% of body weight, the plasma accounts for just over half of this, i.e. about 4%. Normal plasma osmolality is around $290\,mOsmol\,kg^{-1}$. Albumins account for 60% of total plasma protein or about 5% of plasma by weight. Plasma is 95% water.

2 a) True b) False c) True d) False e) False

Blood pH varies between 7.35 (mixed venous blood) and 7.4 (arterial blood) under normal circumstances. There are about $5–7 \times 10^9$ leukocytes l^{-1} and the red cells account for about 65% of total body iron.

3 a) True b) True c) True d) False e) False

Red cells are released into the circulation as reticulocytes. Healthy red cells survive in the circulation for about 120 days before being destroyed by macrophages in the spleen and liver.

4 a) True b) False c) True d) False e) False

Mixed venous blood contains around $53\,ml\,CO_2\,dl^{-1}$. The affinity of haemoglobin for carbon monoxide is greater than its affinity for oxygen. As P_{CO_2} increases, the affinity for oxygen decreases. This is the Bohr Effect.

5 a) True b) False c) True d) False

The platelet count is normally about 250×10^9 (i.e. about one platelet for every 200 red cells). While platelets play an essential role in clotting, the main clotting factors are plasma proteins secreted by the liver.

6 a) False b) True c) False d) True e) False

Both extrinsic and intrinsic pathways lead to activation of Factor X, which leads in turn to the activation of prothrombin. Haemostasis is inhibited in healthy blood vessels by heparin, prostacyclin and thrombomodulin secreted by the vessel wall. Warfarin is given to patients at risk of developing deep vein thrombosis. It inhibits the synthesis of clotting factors by the liver. While a deficiency of Factor VIII will impair the clotting process, deficiencies of other factors or a reduction in platelet count will also result in a failure of clotting.

Chapter 21

1 a) True b) False c) False d) True e) True

Neutrophils pass from the circulation into the tissues in response to chemotactic stimuli. Once in the tissues, they phagocytose and kill invading micro-organisms. Macrophages use nitric oxide (and reactive oxygen intermediates) to kill organisms they have ingested.

2 a) False b) True c) True d) False e) True

The first stage of the inflammatory response is vasodilatation (the red reaction of the triple response). Mast cells (not macrophages) secrete vasoactive substances (prostaglandins and leukotrienes) during the inflammatory response.

3 a) False b) False c) True d) True e) False

The neutrophils are the most abundant of the white cells (about 70%). Lymphocytes account for about 25% of the white cells of the blood. Lymphocytes leave the circulation to migrate through the tissues, finally re-entering the blood via the thoracic duct. Although some B cells become transformed into plasma cells and secrete antibody, others remain in the lymphoid tissue as memory cells.

4 a) True b) True c) False d) False e) True

T lymphocytes are derived from stem cells in the bone marrow and mature in the thymus gland. T cells are of two main kinds, helper cells and cytotoxic cells. Only the cytotoxic cells secrete cytotoxic materials to kill target cells. Helper cells secrete cytokines. T cells respond to MHC molecules that incorporate a foreign peptide.

5 a) True b) True c) True d) False e) True

IgM is the largest of the plasma proteins and cannot cross the placenta. IgG, however, can cross the placenta and it is this immunoglobulin that helps to protect the fetus and neonate from infection.

Chapter 22

1 a) True   b) False   c) True   d) True

All of the output of the right ventricle passes through the pulmonary circulation and the output of both right and left heart will be the same over any significant period. The systolic and diastolic pressures in the pulmonary arteries are about 25 and 8 mmHg, respectively, compared with 120/80 mmHg for the systemic arteries. As the pressure in the pulmonary arteries is low, the blood flow in the upright lung is strongly influenced by gravity. At the top of the lung, blood flow is low, particularly during diastole.

2 a) False   b) True   c) False   d) False   e) True

The trachea and bronchi have cartilage rings or plates in their walls that allow them to stay open despite changes in intrapleural pressure, but the bronchioles have no cartilage and may collapse when the intrathoracic pressure exceeds the pressure in the airways. The principal sites of airways resistance are the upper airways (mainly the nose and airway generations 1 to 6 – the larger bronchi). The resistance falls as the airways branch as there is a very large increase in the total cross-sectional area of the airways. While the alveoli are the principal site of gas exchange, the respiratory bronchioles and alveolar ducts also contribute to gas exchange.

3 a) True   b) False   c) False   d) True   e) True

The vital capacity is equal to the total lung volume *minus* the residual volume and the expiratory reserve volume is equal to the functional residual capacity *minus* the residual volume, which is about 1.2 litres. So vital capacity would be about 4.8 litres and functional residual capacity would be about 2 litres. The FEV_1 should be greater than 70% of vital capacity, i.e. greater than $0.7 \times 4.8 = 3.3$ litres.

4 a) True   b) False   c) False   d) True   e) True

The compliance of the chest is determined by that of the chest wall and the lungs. Lung recoil assists expiration not inspiration. The elasticity of the lung parenchyma tends to collapse the lungs, as does the surface tension in the alveoli. The intrathoracic pressure must therefore overcome both forces if the lungs are to expand during inspiration. The smaller the surface tension and the lower the elasticity of the parenchyma, the easier it will be to expand the lungs. Pulmonary surfactant reduces the work of breathing by lowering the surface tension of the air–liquid interface in the alveoli. This is particularly important during the first few breaths following birth.

5 a) False   b) True   c) True   d) True

The anatomical dead space is the volume of air that does not enter the alveoli (i.e. it is the volume of the conducting airways). The physiological dead space is always greater than the anatomical dead space as it includes the volume of air taken up by alveoli that are not perfused by blood. These alveoli cannot take part in gas exchange. In healthy subjects, the difference is not great. Alveolar ventilation is greatest at the base of the lungs for lung volumes at or above FRC.

6 a) True   b) True   c) True   d) False   e) False

The central chemoreceptors respond to the partial pressure of carbon dioxide in the arterial blood and provide most of the respiratory drive. The receptors in the aortic arch and carotid sinus are baroreceptors that monitor the blood pressure. The peripheral chemoreceptors are the carotid bodies, which sense plasma pH as well as the partial pressures of oxygen and carbon dioxide. They are the only chemoreceptors that respond to the partial pressure of oxygen in the arterial blood.

Answers to quantitative problems

7   10.8 litres

8   Respiratory exchange ratio $= \dfrac{\text{volume of } CO_2 \text{ exhaled}}{\text{volume of } O_2 \text{ consumed}}$

Volume of CO_2 exhaled $= 25 \times 0.044 = 1.1$ litres
and volume of O_2 consumed $= 25 \times (0.209 - 0.158)$
$= 1.275$ litres

So the respiratory exchange ratio $= 1.1/1.275 = 0.86$

9   Using the equation for calculating the FRC:
Residual volume $= ((4.5/3.8) - 1) \times 6.0 = 1.1$ litres

10 $6.6 \, \text{l min}^{-1}$

11   Using the Bohr equation:
$V_D = 450 \times (1 - (3.3/5.05))$
$= 450 \times (1 - 0.65)$
$= 157 \, \text{ml}$

12 $5.45 \, \text{l min}^{-1}$

Chapter 23

1 a) True   b) False   c) False   d) False   e) True

Urine pH is normally much lower than that of plasma and contains no bicarbonate. If plasma pH is elevated (alkalaemia), then the kidneys will excrete bicarbonate so that urine pH rises.

2 a) True   b) True   c) False   d) True   e) False

As cardiac output rises in exercise, renal blood flow does not increase. It usually falls due to vasoconstrictor activity in the sympathetic nerves supplying the smooth muscle of the afferent arterioles.

3 a) False   b) True   c) False   d) True   e) True

The concentration of small molecules in the glomerular filtrate is the same as plasma but the filtrate contains very little protein (unlike plasma). A clearance less than that for creatinine can be due either to a barrier to filtration (e.g. plasma proteins) or to reabsorption by the renal tubules (e.g. glucose).

4 a) True   b) False   c) False   d) True

Glucose is absorbed by sodium-dependent secondary active transport. The transport maximum for glucose is about $360 \, \text{mg min}^{-1}$.

5 a) False   b) True   c) False   d) False   e) True

Sodium balance is regulated by the activity of the cells of the distal tubules. The sodium uptake by the proximal tubules is obligatory and linked to the absorption of small organic molecules, chloride and bicarbonate. Renin is an enzyme that is secreted by the cells of the macula densa in response to low sodium in the fluid reaching the distal tubule. It cleaves angiotensin I from a precursor molecule called angiotensinogen, which is always present in the plasma. Angiotensin II acts on the zona glomerulosa of the adrenal cortex, which responds by secreting aldosterone.

Chapter 24

1 a) False b) False c) True d) False e) True

Swallowing is a complex reflex, the first part of which is under voluntary control. The bolus is moved along the oesophagus by a wave of peristaltic contraction. The saliva contains an α-amylase, which is able to digest starch. The activation of the parasympathetic nerves increases salivary secretion.

2 a) True b) True c) False d) True e) False

The pepsinogens are secreted by the chief cells; the parietal cells secrete hydrochloric acid and intrinsic factor. The venous blood draining the stomach has a higher pH (it is more alkaline) as it contains a greater concentration of bicarbonate than normal.

3 a) False b) True c) True d) False e) False

Gastric secretion begins in response to the sight and smell of food – this is the cephalic phase of secretion. The stomach plays little role in absorption, only absorbing lipophilic substances such as alcohol and those that become lipophilic because of the low gastric pH, e.g. aspirin. The stomach secretes intrinsic factor, which is required for the absorption of vitamin B_{12}, but the absorption itself takes place in the terminal region of the ileum.

4 a) False b) True c) True d) True e) False

Following prolonged vomiting, the plasma bicarbonate will be higher and give rise to a metabolic alkalosis (vomiting results in loss of gastric acid). Bile salts, not bile pigments, are required for the digestion of fats.

5 a) True b) False c) True d) True e) False

The stomach has an additional oblique layer of smooth muscle in addition to the circular and longitudinal layers. The taeniae coli of the colon are formed by local thickening of the longitudinal muscle layer.

6 a) False b) False c) True d) True e) True

All of the digested carbohydrate is absorbed in the small intestine. The digestive enzymes of the small intestine are mostly derived from the pancreas but some are brush border enzymes.

7 a) False b) True c) True d) True e) False

Increased activity in the parasympathetic nerves increases intestinal motility. It is decreased by stimulation of the sympathetic nerves. The mucosa of the anal canal is lined by a stratified squamous epithelium, which is able to resist the abrasion caused by the passage of stools.

Chapter 25

1 a) False b) True c) False d) True e) True

The liver has four lobes: the large right and left lobes and the smaller caudate and quadrate lobes. The gall bladder is associated with the liver but is anatomically distinct.

2 a) True b) True c) True d) False e) False

Bile is actively secreted by the hepatocytes into the bile canaliculi. Its composition differs significantly from plasma. The secretion of bile

from the gall bladder is stimulated by cholecystokinin but the main stimulus for the secretion of hepatic bile is the recirculation of bile salts via the enterohepatic circulation.

3 a) True b) False c) True d) True e) False

As mentioned in the previous answer, bile salts (not bile pigments) stimulate the secretion of hepatic bile. Bile salts are amphipathic molecules (part hydrophilic and part hydrophobic). This gives them a detergent-like action so they are able to break up globules of fat to form a stable emulsion. The fats (mainly triglycerides) can then be broken down by pancreatic lipases.

4 a) True b) True c) False d) True e) False

In obstructive jaundice, the normal flow of bile is prevented so bile pigments do not enter the intestine but accumulate in the plasma. The secretion of bile salts is also prevented, with the result that fats are not processed properly by the pancreatic enzymes. As a result, the faeces are pale in colour (lack of bile pigments) and have fatty streaks (due to the presence of undigested fats). The liver secretes almost all of the clotting factors, so blood clotting is abnormally slow in liver failure.

5 a) True b) False c) False d) True e) True

Animals cannot synthesize glucose from fatty acids. Only the liver is able to convert the ammonia derived from the deamination of amino acids to urea, which is non-toxic.

Chapter 26

a) **False**: the exocrine tissue consists of acinar tissue divided into lobules by connective tissue. The secretions of the acini flow to the pancreas by means of ducts lined by epithelial cells that secrete an alkaline fluid.

b) **False**: the islet tissue is endocrine in function and secretes the hormones insulin and glucagon.

c) **True**: acid stimulates the release of secretin from the duodenal mucosa.

d) **True**

e) **False**: cholecystokinin increases the secretion of enzyme-rich juice from the pancreas.

f) **True**

g) **True**: the trypsinogen is activated by enteropeptidase in the duodenum.

h) **False**: amylase is secreted as the active enzyme.

i) **True**

j) **False**: insulin promotes the uptake of glucose by the cells and so acts to lower blood glucose.

Chapter 27

1 a) False b) False c) True d) False e) True

The skin consists of two main layers: the epidermis and the dermis. The hypodermis is a separate structure linking the skin with the deeper tissues. The epidermis does not contain nerve fibres, which are located in the dermal ridges. In addition to the keratinocytes, the deepest part of the epidermis contains melanocytes, Merkel cells and Langerhans cells.

2 a) True b) False c) True d) False e) True

Dark-skinned people have a similar number of melanocytes to white people but the melanocytes contain a larger amount of melanin. Melanin is synthesized from the amino acid tyrosine.

3 a) False b) True c) False d) True

The hair follicles are lined by an epithelial membrane that is continuous with the germinal layer of the epidermis. Although the deepest parts of the hair follicles lie in the hypodermis, the cells are derived from the dermis, and each follicle is surrounded by a sheath of dermal tissue.

4 a) False b) True c) False d) False e) True

The apocrine glands are located in the axillae and the anal regions and secrete a milky fluid that gives rise to body odour. It is the sebaceous glands that secrete an oily mixture to maintain the flexibility of the epidermis. The skin circulation is controlled by the sympathetic nervous system, mainly to maintain body temperature within normal limits. The arteriovenous anastomoses are only found in a few regions such as the fingertips and nose; the blood flow to the bulk of the skin is controlled by the calibre of the skin arterioles, which is mainly determined by the activity in the sympathetic nerves.

5 a) True b) False c) True d) False e) True

Some pain sensations arise from within the body – such as the pain of angina, which arises from the heart but is often felt as pain radiating to the left arm. Increased cortisol exerts an anti-inflammatory effect and impairs wound healing.

Chapter 28

1 a) True b) False c) True d) False

During spermatogenesis, the number of chromosomes is halved by meiosis. The primary spermatocytes undergo two meiotic divisions to give rise to spermatids. As a result, mature sperm are haploid. This is also true of a mature ovum, which undergoes meiosis prior to ovulation. Following fertilization, the full complement of chromosomes is restored.

2 a) True b) False c) True d) False e) False

Sertoli cells synthesize androgen-binding protein in response to FSH. Testosterone binds to the androgen-binding protein and the hormone–protein complex enables the Sertoli cells to maintain the production of spermatozoa. While the spermatozoa are nurtured by the Sertoli cells until they develop into mature sperm, they are derived from germ cells known as spermatogonia, which lie within the basal compartment of the seminiferous tubules.

3 a) False b) True c) True d) True e) False f) True
  g) True

The initial recruitment of primordial follicles into the preantral phase is hormone independent. During the preantral phase, the follicular cells acquire receptors for various hormones. The granulosa cells develop receptors for FSH and oestrogen while the thecal cells acquire receptors for LH. The period before ovulation is known as the follicular phase while the period after ovulation is called the luteal phase.

4 a) False b) True c) False d) True e) True

The cells of the theca interna secrete mainly androgens, which are converted to oestrogens by the granulosa cells in response to FSH.

After ovulation, the follicle collapses around a blood clot to form a corpus luteum. If no pregnancy occurs, the corpus luteum regresses to form a corpus albicans, which has no secretory role.

Chapter 29

1 a) True b) False c) False d) False e) True

Sexual reflexes in both men and women are integrated in the sacral region of the spinal cord. Sperm remain viable for up to 72 hours. Capacitation of a sperm takes place as it moves through the female reproductive tract.

2 a) False b) False c) True d) True e) False

The egg completes its second meiotic division after fusion with a sperm. hCG is secreted by the early embryo. The zygote is diploid (i.e. it possesses 23 pairs of chromosomes). Each gamete contributes one member of each pair. The blastocyst implants into the endometrial lining of the uterus.

3 a) True b) False c) True d) True e) False

Intervillous blood spaces are filled with maternal blood derived from the eroded spiral arteries. hPL is a peptide hormone.

4 a) True b) True c) False d) True e) True

The maternal requirement for calcium increases during pregnancy to meet the demands of the growing fetal skeleton.

5 a) False b) False c) True d) True e) True

The ductus arteriosus is a shunt that allows blood to pass directly from the pulmonary artery to the aorta. The umbilical artery carries deoxygenated blood from the fetus to the placenta. Surfactant-secreting cells start to develop at around 20 weeks of gestation but are not fully functional for several more weeks.

6 a) True b) False c) True d) False e) True

Oxytocin is secreted by the posterior pituitary gland. The breast is prepared for lactation by about 4 months' gestation. Colostrum is relatively low in calories but is rich in protein, including immunoglobulins.

Chapter 30

1 a) False b) True c) True d) False e) False

In the formation of a chromosome, a strand of DNA is wrapped around a series of protein molecules known as histones. This forms a string of nucleosomes, rather like beads on a string. The nucleosomes form a chromatin fibre, which coils around a protein scaffold to form a single chromatid. The transfer of genetic information from DNA to mRNA is transcription. The mRNA is then translated into the sequence of amino acids that makes a specific protein. Each triplet of bases codes for a single amino acid.

2 a) True b) True c) False d) True e) True

Chromosome number is halved in meiosis, which occurs in the formation of the eggs and sperm.

3 a) True b) True c) True d) False e) True

A genotype specifies the genetic makeup of an individual; the phenotype is the physical expression of the genotype.

4 a) True b) False c) True d) False e) False

Down syndrome is caused by an extra copy of chromosome 21 (trisomy 21). Haemophilia A is caused by a defective gene on the X chromosome and is transmitted from a woman to her sons who have only one X chromosome. If a son carries the defective gene on his X chromosome, he will be a haemophiliac. As each daughter inherits one X chromosome from her mother and one from her father, her daughters are unlikely to be affected but they may be carriers. Mitochondrial diseases may arise from mutations either in nuclear DNA or in mitochondrial DNA. However, a mutation in mitochondrial DNA is always transmitted via the maternal line. Note the distinction between mitochondrial disease and mitochondrial inheritance.

Chapter 31

1 a) True b) True c) True d) False e) True

Animal protein is complete protein and contains all 20 of the amino acids used by the body to synthesize new proteins. Plant oils are rich in unsaturated fats and contain the essential fatty acids (i.e. those that cannot be synthesized by the body). Ketones are by-products of fat metabolism and appear in the urine of individuals who are eating little or no carbohydrate.

2 a) True b) False c) True d) False e) False

Excess vitamin A is highly toxic and can cause a variety of neurological problems. It also causes abnormalities of development. Vitamin K is required for normal blood clotting. Bleeding disorders are associated with its deficiency. Vitamin B_3 deficiency results in pellagra. This disease is associated with a variety of neurological symptoms and skin disorders. Vitamin B_{12} deficiency leads to pernicious anaemia. Rickets is caused by a deficiency of vitamin D. Scurvy is caused by a deficiency of vitamin C.

3 a) True b) True c) False d) False e) True

Iodine deficiency may lead to the development of goitre as the thyroid gland enlarges to trap any available iodine from the plasma. Individuals with bulimia are usually of normal weight. BMI is a measurement that relates weight to height. As a measure of bodily health, it is of most value in adults. Normal values range from 18.5 to 25 $kg\,m^{-2}$. Values in excess of 30 represent increasing levels of obesity. During periods of accelerated growth in childhood, the body is in positive nitrogen balance, which means that synthesis of new protein exceeds protein breakdown or fat deposition.

Chapter 32

1 a) False b) True c) False d) True

The basal metabolic rate is the energy requirement of the fasting body during complete rest in a thermoneutral environment. The metabolic rate may be lower during deep sleep. It is decreased in people suffering from hypothyroidism.

2 a) True b) True c) False d) True e) False

Metabolic rate is increased after a meal, especially if it is rich in protein. Although carbohydrates (especially glucose) are the main source of energy during exercise, plasma glucose changes very little unless the exercise is exceptionally severe and prolonged.

3 a) True b) False c) True d) False e) True

The mean arterial pressure rises in exercise, but less than the systolic pressure as the rise in diastolic pressure is relatively small. During exercise, there is little change in arterial P_{O_2} or P_{CO_2} but the arteriovenous difference in oxygen content increases with the severity of the exercise.

Chapter 33

1 a) True b) False c) False d) True e) True

In a temperate climate, more than half of the heat loss from the body occurs by radiation. Water has a greater thermal conductivity than air allowing heat to be lost more readily from a body immersed in it. Core temperature (surrounding the internal organs) is closely regulated. The skin temperature fluctuates and may be as much as 20°C lower than the normal core temperature, which is around 37°C. The hypothalamus acts as the 'thermostat' of the body. It receives input from thermoreceptors and initiates appropriate heat-conserving or heat-loss mechanisms. Increased blood flow to the skin brings warm blood closer to the surface allowing more heat to be lost by radiation, conduction and convection.

2 a) True b) False c) False d) True e) False f) True

Neonates can raise their metabolic rates more effectively than adults, largely through the metabolism of brown fat. Sweat rates are controlled by the sympathetic nervous system. Skin temperature is usually lower than core temperature. During profuse sweating, fluid and salts are lost from the body in large amounts. This can lead to a variety of symptoms collectively known as heat exhaustion. In humid conditions (when the atmosphere is moist), sweat evaporates more slowly and is therefore a less effective means of heat loss. Thyroid hormone stimulates oxidative metabolism and generates heat. Its secretion is enhanced during prolonged exposure to cold.

3 a) True b) True c) True d) False

Sweat rates rise progressively during the initial weeks of acclimatization to high temperatures. Aldosterone secretion is increased. This leads to increased reabsorption of salt from the sweat and a fall in its salt content. As the sweat rate increases, the body will lose water. ADH secretion will be increased in response to this loss and urine output will fall.

Chapter 34

1 a) False b) True c) False d) True e) False

Most of the body water is in the intracellular compartment (around 28 litres or two-thirds of body water) while the blood volume accounts for around 5 litres. The intracellular and extracellular fluids are isotonic with each other. The extracellular fluid contains around 140 $mmol\,l^{-1}$ of sodium ions, while the intracellular fluid is rich in potassium ions (about 150 $mmol\,l^{-1}$ water).

2 a) True b) False c) True d) False e) True

The extracellular fluid volume accounts for about one-third of total body water. The volume of fluid in the circulation (the effective circulating volume) is sensed by the volume receptors (low-pressure baroreceptors), which are found in the venae cavae and right atrium. The arterial baroreceptors regulate the arterial blood pressure; they are high-pressure receptors.

Answers to interpretive questions

3   The pH is higher than normal, so there is an alkalosis. In addition the P_{CO_2} and bicarbonate concentration are higher than normal, so the patient is suffering from a metabolic alkalosis (option c).

4   The pH is below normal so there is an acidosis. The P_{CO_2} and HCO_3^- are both below normal, so the underlying condition is a metabolic acidosis (option a).

5   The decline in the P_{CO_2} with altitude will stimulate breathing to offset the hypoxia. Carbon dioxide is driven from the blood faster than it is produced in the tissues so P_{CO_2} falls and pH rises. The climber will suffer from a respiratory alkalosis (option b).

Chapter 35

1   a) True   b) True   c) False   d) False   e) True

Some drugs are applied to the skin as patches to have a systemic (whole-body) effect, e.g. oestrogens in hormone replacement therapy. Intramuscular injection is used to provide a depot for slow release of a drug – in insulin–zinc injections for example. For rapid drug onset, intravenous injection would be used.

2   a) True   b) False   c) True   d) True   e) False

Molecules that carry a permanent charge cannot be successfully given by mouth as their rate of absorption is slow and uncertain. Such drugs (e.g. neuromuscular blockers) are given intravenously. The main effect of some drugs is terminated by redistribution in the body tissues. The short duration of thiopentone anaesthesia is due to the drug's redistribution from the brain first into the lean tissues and then into body fat deposits.

3   a) False   b) True   c) True   d) False   e) True

Weak bases are absorbed from the small intestine in their neutral form. Alcohol is eliminated from the body by zero-order kinetics.

A

A band 120–21, 331
ABO blood group
    409–410, 411
  inheritance of 659–60
abdomen 507
abdominal cavity 11–12
abduction 10, 741
absorption 741
  disorders of 524
  of electrolytes 523–524
  of amino acids 522
  of fats 523
  of glucose 521–522
  of vitamins 523–524
  water 521, 523–524, 525
absorptive state 320
accommodation 741
  of eye 243
  reflex 243
  sensory adaptation 228
ACE inhibitors 352, 386, 480
acetabulum 152, 165
acetoacetate/acetoacetic
    acid 51, 719
acetazolamide 491
acetyl CoA 48–51, 540
acetylcholine (ACh) 85,
    95, 113
  autonomic nervous
    system 221–22
  gastric acid secretion 512
  and heart rate 341, 342
  pancreatic secretion 552
  salivary secretion 503
  neuromuscular
    junction 86, 129
  smooth muscle 136
acetylcholine esterase 116,
    129–31, 530
acetylcholine receptors 86, 95
achilles tendon 177
achlorhydria 513, 741
achondroplasic dwarfism 161
acid (definition) 21, 719, 741
acid dissociation
    constant 729, 741

acid-base balance 481–482,
    718–24, 741
  disorders of 543, 719–21
  evaluation of 724
  factors influencing 719
acidaemia 346, 719, 741
acid excretion 474, 477–478,
    481–482, 720
acidosis 489, 719–720
  compensation 722–724
acinar cells 503, 548, 549
acinus/acini 76, 547, 548,
    549, 741
acne, treatment of 571
acromegaly 295–296
acromion 149–50
acrosome reaction (of
    sperm) 610
ACTH (adrenocortocotrophic
    hormone; corticotrophin)
    286, 287, 288, 289,
    306–307
  circadian rhythm 306
  corticotrophin releasing
    hormone 284, 289, 307
  corticotrophs 286–287, 288
  cortisol regulation 306–307
  fetal 620
  glucocorticoid
    secretion 306–307
  pituitary gland 286, 287
  regulation of 286–287, 307
  secretion 306–307, 624
actin,
  filaments 44, 52, 120–21
  in muscle
    contraction 122–23
action potential
  of cardiac muscle, 334,
    335–336
  compound 110–11
  of nerve 106–12
  of skeletal muscle 125, 130
  propagation of 109
  properties of 106–7
  refractory period 107–8
  saltatory conduction 109
  of smooth muscle 134

active tension (of muscle)
    127–28
active transport 57, 59–61, 741
  calcium pump 60
  primary 59
  proton pump 509, 510, 513
  secondary 59–61
  sodium pump 59
acute phase proteins 418
adaptation of sensory
    receptors 228, 741
adaptive immune
    system 419–426
adderall 275
Addison's disease 309
adduction 10, 741
adenohypophysis *see pituitary
    gland: anterior lobe*
adenosine 35, 351, 363
adenosine triphosphate
    see ATP
adenylyl cyclase 89, 224, 484
ADH *see antidiuretic
    hormone, vasopressin*
adipocytes 70
adipose tissue 70
  brown fat 70, 703
  as energy reserve 561,
    684, 685
  subcutaneous 561
  in thermoregulation
    698, 703
  white adipose tissue 70
ADP (adenosine
    diphosphate) 49,
    123, 124
adrenal cortex 304–9
  ACTH regulation 306–307
  fetal zone 634–35
  histology 304
  structure of 303–4
adrenal cortical
    hormones 304–9
  aldosterone 304, 305
  cortisol 304, 305–6
  deficiency 309
  excess secretion 308
  sex steroids 304, 306

adrenal glands 303–310
  blood supply 303
  of fetus 634
  histology 304
  innervation 309
  structure of 303
adrenal medulla 309–10
  autonomic nervous
    system 223, 309
  chromaffin cells 87–88
  disorders 309–10
  hormones of 223, 309–10
  phaeochromocytoma
    310, 385
  secretion by 309
adrenaline (epinephrine) 85
  cardiovascular
    effects 309–10, 343, 345,
    346, 380–381, 472
  exercise 346, 540
  and gastrointestinal
    tract 309–310
  glycogenolysis 309–10,
    320, 540, 541
  lipolysis 309–10, 320
  metabolic effect 688
  respiratory system
    309–10, 440
  secretion of 309
adrenoceptors
  autonomic nervous
    system 223
  classification of 223
  effects of (table) 310
  and metabolism 688
adrenoceptor antagonists
  α-blockers 223, 309
  β-blockers 223, 350, 352,
    363, 386
adrenocorticotrophic
    hormone, *see ACTH*
aerobic metabolism 50, 741
afferent arterioles 467, 469,
    471–472, 741
afferent lymph 383, 384, 741
afferent nerves 190, 226, 229,
    231, 499, 741
  visceral 224

afterload 344, 347
age/ageing
 and drug potency 736
 see also growth
agglutination 409, 410, 411, 741
agglutinins 410, 741
agglutinogens 410, 411, 741
agnosia 265, 741
agonist 62, 86, 96
 definition 741
 partial 96
agranulocytes 392, 402–403
AIDS (acquired immunodeficiency syndrome) 428
airway(s) 434–438, 444, 446, 448–449, 458, 459, 460
 dead space 446
 generations of 434–435
 innervation 440
 resistance of 444, 446, 458, 459, 460
 structure of 435–438
albinism 560
albumin 389–390, 538, 541
 drug binding 733
 and fluid exchange 541
 hormone binding 541
 osmotic pressure 390, 541
 plasma concentration 390
 Starling forces 383
aldosterone 304, 305, 381, 386, 480–481
 actions 305, 480–481
 antagonists 490, 491
 haemorrhage 381
 hypertension 386
 in pregnancy 623
 renal effect 381, 480–481
 volume regulation 712
alendronate 162
algogens 235, 566, 741
alimentary canal, *see digestive system*
alkalaemia 719, 741
alkaline tide 509, 510
alkalosis 719–21
 compensation 722–24
allantois 614
alleles 654–61, 741
allergic reactions 427
allograft 741
allopurinol 167, 491
alpha (*α*-) cells (pancreas) 317, 548, 554
alpha waves (EEG) 270, 271

alveolar air 438
 composition 434
alveolar ducts 435
alveolar epithelial cells 437–438
alveolar ventilation 446–448, 449–451
alveoli
 of lung 435, 437–438
 of mammary gland 639
Alzheimer's disease 278
amacrine cells 240
amenorrhoea 678
amilodipine 350, 351, 386
amiloride 308, 490, 491
amino acids
 absorption of (intestine) 522
 carrier proteins 522
 cortisol and 305
 deamination 52
 energy metabolism 52, 540, 672, 684, 688
 essential 672
 glucogenic 52, 540, 672
 insulin and 319, 322
 ketogenic 52, 541, 672
 metabolism 52, 540, 684, 688
 protein synthesis 33, 651–53
 secondary active transport 477, 478, 522
 structure of 33–4
 tubular transport 477, 478
aminoglutethimide 308
aminopeptidase 519, 550
amiodarone 351, 363
amitriptyline 237
ammonia 542–543, 544
ammoniagenesis 719, 741
amnesia 281, 741
amniocentesis 741
amnion 614
amniotic fluid 626
amoxicillin 463
amphetamines 275
ampicillin 463
ampulla
 hepatopancreatic 515, 516, 548, 549
 vestibular system 257
amygdala 280
amyl nitrate 92
α-amylase 503, 519, 550–551
β-amyloid 278
anabolism/anabolic metabolism 682, 741

anaemia 400–401, 741
 aplastic 401, 408
 megaloblastic (macrocytic) 401, 676
 microcytic 401
 pernicious 509
 sickle cell 401
 thalassaemia 401
anaerobic metabolism 48, 741
anaesthetics,
 general 727, 728
 local 111–12, 727, 728
 uptake and distribution 731–32
anal canal, *see anus*
anakinra 167
analgesia 235, 741
analgesics 237
anastomoses, arterio-venous 371, 699, 742
anatomy 4
anatomical position 9
anatomical planes 9–10
anatomical descriptive terms 8–12
androgen 742
 adrenal cortex 306, 308
 binding protein 587
 excess 308
 testosterone 306, 579, 581–82, 588, 595, 637–38
andropause 605
anencephaly 625, 676, 742
angina pectoris 92, 350, 351
angiogenesis 568, 742
angioplasty, coronary 350
angiotensin 480–481, 541
 see also renin
angiotensin converting enzyme (ACE) 386, 480, 481
angiotensin I 381, 480, 481
angiotensin II
 pressor effect 379, 381
 renal blood flow 472
 sodium balance 480–481
 thirst 711
 volume regulation 712
angiotensinogen 381, 480, 481
anion 18, 742
ankle 153
anorexia nervosa 678
anoxia, definition 455, 742
antacids 505, 513, 528, 530
antagonist 742
 competitive 96–99
 muscle 125, 170, 171

non-competitive 96–98
 receptor 86
 see also specific drugs
anterior lobe (of pituitary) 288–289, 291–296
 see also pituitary gland
anterograde amnesia 281
anti-arrhythmic drugs 352, 363
 classification 363
antibiotics 570
antibodies 410, 411, 422–424, 426, 541
 classification 422
 diversity of 422–423
 function of 422
 opsonisation 418
 see also adaptive immune system
antibody secretion 423–424, 426
 memory cells 420, 421, 424, 426
 primary response 423, 426
 secondary response 424, 426
anticholinesterase drugs 130, 131, 180, 278
anti-coagulants 351, 352
anticodon 653
anticonvulsant drugs 271–272, 737
anti-D immunoglobulin 411
antidepressant drugs 275
anti-diabetic drugs 323
anti-diuretic hormone (ADH, vasopressin) 287, 289–291, 379, 381, 472
 blood pressure 290, 379, 381
 deficiency 290–291, 484, 714
 dehydration 290, 484, 711
 diabetes insipidus 290–291, 484, 714
 haemorrhage 290, 379, 381
 hypothalamus 286, 287, 289–290, 711
 nasal spray 727
 posterior pituitary 289–290
 secretion 483
 thirst 711
 volume regulation 290, 712–13
 water retention 290, 711, 713, 714
 water reabsorption 290

anti-emetic drugs 352, 515
antigens 410, 411, 423, 742
antigen presentation 403, 423
anti-hypertensive drugs 386
anti-inflammatory drugs 429
anti-platelet drugs 350–352
antiport 58–59, 742
antipyretic drugs 705, 742
antral follicle 594–95
antrum, pyloric 506, 507
anus/anal canal 494, 525,
    526, 527, 528, 529
  haemorrhoids 526
  muscles 527
  nerve supply 526, 528, 529
  reflexes 528, 529
  sphincters 526, 528, 529
  venous plexuses 526
aorta 328, 329, 330, 366, 367,
    368, 371, 516
aortic baroreceptors
    379–380, 381
aortic bodies 454
aortic stenosis 347, 742
aortic valve 329–330,
    338–339, 372
aphagia 678
aphasia 742
  expressive (Broca's) 268
  receptive (Wernicke's) 268
apical surface (definition)
    74, 742
aplastic anaemia 401, 408
Aplysia 277
apnoea 456, 457, 460
  definition 456, 742
  deglutition 456, 504
  obstructive 460
  sleep 460
apocrine glands 563
apoferritin 395
apoptosis 47, 417, 423,
    425, 742
appendicitis 527
appendicular skeleton 141,
    149–53
appendix 494, 526, 527
appetite 625, 677
apraxia 265, 742
aquaporins (water
    channels) 57, 484, 742
aqueous humour 241
arachidonic acid 30–31,
    429, 672
arachnoid membrane 185,
    192–93
arachnoid villi 192, 193

arginine 92, 543
arginosuccinate 543
Aricept (donepizil) 278
arm
  bones of 149–50
  muscles of 171–73, 176, 177
arrhythmias
    (dysrhythmias) 360,
    361–362, 742
  see also antiarrhythmic
    drugs, heart
arterial baroreceptors
    379–380, 381
arterial blood pressure
    371–373
  in exercise 692
  measurement of 367, 372
  postural effects 374,
    379–381
  see also hypertension
arterial pulse 372
arterioles 370, 371, 382
  afferent 467, 469, 471–472
  efferent 467, 469, 471–472
  histology 370
  hormonal regulation
    380–381
  innervation of 378–379
  local control 377–378
  microcirculation 382–384
  as resistance vessels
    371, 373
  structure of 371
arteriosclerosis 372
arterio-venous
    anastomoses 371,
    564, 699
arterio-venous shunts 371
  fetal 628–30
artery/arteries 366, 370–371
  anterior cerebral 194
  anterior choroidal 194
  anterior
    communicating 194
  anterior spinal 194
  aorta 328, 329, 330, 366,
    367, 368, 369
  arcuate 471
  axillary 366
  basilar 194
  brachiocephalic 330, 366,
    367, 368
  brachial 366, 367, 371
  bronchial 366
  cerebral 371
  carotid 297, 330, 366, 367,
    435, 454

  coeliac 366, 367, 497, 498,
    507, 516
  colic 497
  coronary 331–334, 350,
    352, 366
  cystic 537
  elastic 371
  femoral 366, 368
  gastric 498
  hepatic 498, 535, 536, 537
  iliac 366, 367, 368
  ileocolic 497
  inferior mesenteric 367,
    497, 498, 527
  intercostals 366, 367
  internal carotid 194
  marginal 497
  middle cerebral 194
  muscular 371
  phrenic 366
  popliteal 366, 368, 371
  posterior cerebral 194
  posterior
    communicating 194
  posterior inferior
    cerebellar 194
  pulmonary 329, 330, 368,
    369, 371
  radial 366
  renal 366, 367, 368, 467, 468
  spiral 599
  splenic 498
  structure of 370–371
  subclavian 330, 366, 367,
    368, 435
  superior cerebellar 194
  superior mesenteric 367,
    497, 498, 516, 527
  tibial 366, 368
  ulnar 366
  umbilical 628
  vertebral 194
arthrology 163, 742
arthroses, *see joints*
ascending colon 494, 525, 526
ascites 544
ascorbate/ascorbic acid
    (vitamin C)
  absorption of 523
  deficiency 676
  nutritional
    requirements 674
aspartate/aspartic acid 34
aspirin
  absorption 514
  anti-inflammatory
    action 429

  anti-platelet action 350,
    351, 407
  antipyretic action 705
  chemical structure 543
  elimination kinetics 735
  metabolism of 543
  see also NSAIDs
association cortex 264–265
associative learning 277
asthma 427 440, 458, 459,
    460, 462
astigmatism 243–44
astrocyte 184
ataxia 212–13
atenolol 223, 309, 351,
    363, 386
atherosclerosis 350, 406
athetosis 214, 742
athlete's foot, treatment
    of 571
atlas 147–48
atom 15
atomic mass 15, 742
atomic number 15, 742
ATP (adenosine triphosphate)
  active transport 59–61
  calcium pump 60
  energy metabolism 47–51,
    316, 683
  in muscle
    contraction 122–24
  in pain 235, 565
  proton pump 509,
    510, 513
  sodium pump 59–60
  structure 35
  synthesis 43, 47–52, 316
atresia (in ovary) 593, 742
atria
  action potentials 335–336
  contraction 337, 338, 339
  pressure in 338, 339
  structure 329
atrial fibrillation 361, 362
atrial flutter 361, 362
atrial natriuretic peptide
    (ANP) 324, 713, 714
atrial tachycardia 362
atrio-ventricular node 335,
    336, 354
atropine 95, 222, 342, 503
audiogram 250
auditory system 251–56
  cochlea 252
  deafness 266–56
  ear 251–52
  neural pathways 255

Auerbach's (myenteric)
plexus 221, 495, 496
auscultation 367, 372, 742
autocoid 742
autocrine signalling 81, 285
autograft 428, 742
autoimmune diseases
(autoimmunity) 427
autologous blood
transfusion 411–412, 742
automatic bladder 486
autonomic ganglia 216–19
synaptic
transmission 221–22
autonomic nervous
system 183, 216–224, 742
central control of 224
defence reaction 224
dual innervation 220
effects of (table) 220
neurotransmitters 221–24
organization 217–19
reflexes 224
autoregulation 377, 742
mechanisms of
377–378, 472
renal blood flow 472
autorhythmicity (heart) 334
autosomal inheritance 658
autosomes 647, 742
AV (atrio-ventricular)
node 335, 336, 354
Avogadro's number 17, 742
axial skeleton 141, 143–48
axis 147–48
axon 104–112
action potential 106–8
conduction along 108–9
demyelinating
diseases 111
myelinated 105–6
terminal 104–5, 112–13, 129
unmyelinated 105–6
axoneme 53
azathioprine 180, 429

B

B cells (B lymphocytes) 393,
403, 420, 421–422, 424, 426
Babinski sign 206, 208
bacteria
commensal 415–416
in GI tract 415–416,
527–528, 539
bacterial infection
fever 704

immune response to
423, 425
of wounds 570
balance 256–58
see also vesibular system
ball and socket joints 165
barbiturate 346, 728
bare nerve endings 227, 229,
230, 235, 238, 565–66
baroreceptor reflex
379–380, 381
baroreceptors 379–380, 381
arterial 379–380, 381
in fetus 630
low pressure 379
baroreflex *see baroreceptor
reflex*
Bartholin's glands 608
basal body 586
basal ganglia 186, 213–14
basal lamina 78
basal metabolic rate.
(BMR) 685, 686–7, 742
base
definition 21, 742
deficit/excess 724
see also acid-base balance
basement membrane 74, 78
basilar membrane 253–4
basolateral membrane/surface
(definition) 74, 742
basophils 392, 402, 742
beclomethazone 462
bed sores (pressure sores) 570
bendroflumethiazide 386
benzocaine 112
benzodiazepines 275, 737
benzoyl peroxide 571
benzylpenicillin 463
beriberi 675
beta (β-) cells (pancreas) 317,
548, 554
beta oxidation of fatty
acids 50–51
beta waves (EEG) 270, 271
betahistine 515
bicarbonate
acid secretion 481–482
and ammonia 543
in bile 537
carbonic anhydrase 399,
477, 478
carbon dioxide
carriage 399
pancreatic juice 550,
552, 553
plasma 389 390

and renal system 477, 478,
481, 482
in small intestine 521
*see also acid, acid-base
balance, buffer*
biceps brachii 171, 177
biceps femoris 175, 179
bicuspid (mitral) valve 329
biguanides 323
bile
acids 537
canaliculi 536
duct 534–535
excretion of drugs 538
gall bladder 537, 538
hepatic 537–538
pigments 537, 538–539
production of 537
role in digestion 523
salts 523, 537–538
secretion of 537, 538
see also bilirubin
biliary colic 236
biliary obstruction 539
bilirubin
blood 538
diglucuronide 538, 539
in fetus 618
formation 538, 539
and jaundice 544
liver failure 544
renal excretion 488
biliverdin 539
binding proteins,
hormone 285
bioassay *see hormone
measurement*
biological clock *see circadian
rhythms*
biotransformation 543–544,
545, 743
bipolar cells (retina) 240
birth (parturition) 620
fetal cardiovascular
changes 633
first breaths 631
bisacodyl 530
bismuth chelate 513
2,3 bisphosphoglycerate 398
bisphosphonates 162
bitter (taste modality) 260
bivalent (chromosome) 584
bladder 467, 485–486, 697
control of 485–486
detrusor muscle 485, 486
guarding reflexes 485
micturition reflex 485–486

muscle of 485–486
neck (posterior
urethra) 485
nerve supply 485, 486
sphincters 485
trigone 485
voiding 485–486
blastocyst 613, 743
blind spot 240–244
blink reflex 242
Bliss, T. 277
blood 388–413
cells of 391–393
changes in
pregnancy 621–22
coagulation (clotting)
404–409, 545
composition 388–393
cross-matching 411–412
distribution 373–374, 389
disorders 400–401, 403,
407–409, 544
flow 367, 373, 375–376,
448–451, 699
gases, carriage of 396–400
glucose 315, 320–321
groups 409–412, 659–60
haematocrit 391, 400
haematopoiesis 391–393
haemostasis 404–409
iron metabolism 394–395
oxygen content 397, 399
pH 389, 390, 719–24
pressure 371–375,
379–381, 384–386,
448–449
transfusion 395, 408–412
vessels 328, 329,330,
376–381, 404
viscosity 400
volume 379, 381, 389
blood brain barrier 185
blood clotting 404–409
hereditary disorders
of 408–409, 662–64
blood gases 399
in exercise 692–93
blood glucose
regulation 315–321
blood groups 409–412
cross-matching 411–412
inheritance 659–60
blood pressure *see arterial
blood pressure*
blood-testis barrier 580
blood transfusion 395,
409–412

blood vessels 366–371
autonomic
innervation 378–379
endothelium 370, 371
hormonal control 378,
380–381
local control 377–378
nervous control 378–380
structure of 370–371
types of 366
B lymphocytes 420, 421,
422–423, 424
BMI *see body mass index*
BMR *see basal metabolic rate*
body cavities 10–11
body fluid volume 708–10
dehydration 710–11, 715
disorders of water
balance 714
effective circulating volume
(ECV) 389, 712–14
haemorrhage 379, 381
hypovolaemia 712
oedema 383, 715–717
osmolality and 483–484, 709
regulation of 709, 712–14
body mass index (BMI)
679, 743
body temperature
control 700
core temperature 695
disorders 702
diurnal variation 697
fever 704–5
heat exchange and 697–98
measurement of 696
in menstrual cycle 601, 697
pyrogens and 700
regulation 695–705
role of skin 698–99
skin temperature 696
shell temperature 695
shivering 701
thermogenesis 697
thermoreceptors 231, 700
body water
distribution of 27, 708–9
in neonates 707
measurement of 708–9
sex differences 707
Bohr effect/shift 398
Bohr equation 447
bolus 504, 505
bone 154–163
blood supply 155
calcification 157–58
cancellous/spongy 155–56

cells of 156–57, 293
compact 155–56
diaphysis 154, 293
diseases of 161–62
epiphyseal plate 155, 158
epiphysis 154, 293
growth 157–60, 293
healing after
fracture 159–60
marrow 155, 391, 393,
414, 415
osteoid 154, 293
remodelling 158–60
structure of 154–156
trabecular 155
see also calcium balance,
parathyroid gland,
skeleton
bony labyrinth (of ear) 252
Bowman's capsule 467,
468, 469
Boyle's law 432, 433
bradycardia 341
bradykinin 136, 235, 564, 566
brain 183
anatomy 185–86
blood supply 194–95
cerebral hemispheres 185,
210, 233, 263–72
cerebral ventricles 186,
192–93
hypothalamus 186, 224, 700
meninges 185
thalamus 186, 210, 232–33,
273, 274
brainstem 273, 274
Braxton-Hicks
contractions 620
breast
lactation 640–44
structure of 639
breathing
chest movements 442
Cheyne-Stokes 460–461
control of 452–457
disorders of 459–463
effects of gas mixtures
454–455
fetal 630
lung volumes 441, 442, 443
mechanics of 441–442
periodic 460–461
pressure changes
in 443–445
reflexes 453, 456–457
respiratory muscles
439, 442

ventilation 446–448,
449–451
voluntary control 452–453
bridging veins 192
Broca's aphasia 268
Broca's area 268
bromocriptine 644
bronchi 434–437
innervation of 440
structure of 435–437
bronchial circulation 448–449
bronchioles 435–438
bronchoconstriction 440, 442
bronchodilatation 440
bronchospasm 440, 460
brown adipose tissue (brown
fat) 71, 703
"brown tumours" 313
Brunner's glands 516, 517, 521
brush border 61
enzymes 521
of proximal tubule 470,
477, 478
of small intestine 516, 517
buccastem 727
budesonide 462
buffering 719, 743
buffers 21, 743
buffy coat 388–389
bulbo-urethral glands 578
bulimia nervosa 678
bumetanide 490, 490, 491
bundle block 360
bundle of His 335, 336, 354
bungarotoxin 222
bupivicaine 112
burns
classification of 568
fluid loss 569
healing of 569–70
infection of 570
bursa 164
bursitis 164
bypass (coronary artery) 350

C

C cells (of thyroid
gland) 297–8
C fibres (unmyelinated nerve
fibres) 109, 111, 230
structure of 105
C-reactive protein 418
c.s.f. *see cerebrospinal fluid*
cadherins 77
caecum 497, 526, 527
caffeine 511

calcaneus 142, 152–53
calcification of bone 155,
157, 158
calcipotriene 571
calcitonin 298, 314–315
actions of 314–315
calcium balance 314–315
clinical uses 314–315
regulation of 314
calcitriol 312, 466
calcium
absorption 311
ATPase 60
balance 311
blood clotting 405, 406
bone mineral 155
cell signalling 87–88, 90, 92
channels 63–5, 87–8,
113, 129
channel antagonists 350,
351, 363, 386
and distal tubule 482
exocytosis 64–5
and fertilization 611
hormonal regulation
310–315
intracellular
regulation 59–60
muscle
contraction 121–22,
133–34
neuromuscular
transmission 129
nutritional
requirement 626, 676
plasma concentration
389, 390
pump 59–60
secretion 64–5
synaptic transmission 113
in urine 487
calcium chelators 406
calcium pump 59–60
sarcoplasmic
reticulum 121–22
calorific requirement,
daily 690
calorigenesis (heat
production) 685
calorimetry 685–86
canaliculi 509, 510
canal of Schlemm 241
capacitance vessels 374
capacitation of sperm 610
capillary/capillaries
cluster 382
endothelium 371, 382

capillary/capillaries (*Contd.*)
  as exchange vessels 371,
    382–384
  fenestrations 383
  filtration 383
  glomerular 467, 469
  lymphatic 383, 384
  module 382
  peritubular 471
  permeability 383
  pressure 373, 382, 383
  pulmonary 438, 448
  structure of 371, 382–383
  see also microcirculation,
    Starling forces
capitulum 150
capsular ligaments 165
capsule (joint) 165
captopril 386, 480, 522
carbamazepine 271
carbamino compounds 399
carbamyl phosphate 543
carbaryl 571
carbimazole 301
carbohydrate(s)
  absorption 521–22
  active transport 61
  digestion 504, 519–521
  disaccharides 520
  and energy balance
    540–541, 671, 684, 690
  fructose 28, 316
  glucose 28, 315–316, 519,
    520, 540–541
  gluconeogenesis 52, 316,
    540, 541
  glycogen 28–9, 315, 316,
    540, 541
  hormonal regulation
    316–321, 541
  intolerance 524
  lactose 28–9, 641
  metabolism 47–50, 52,
    315–316
  nutritional
    requirements 671
  polysaccharides 28–29
  storage 315, 316, 540, 541
  structure of 27–30, 520
  sucrose 28
  see also energy metabolism,
    glucose, glycogen
carbon dioxide
  acid-base balance 719–22
  blood carriage of 398–400
  dissociation curve 400
  in exercise 692

in fetus 617
Haldane effect 400
hydrogen ion secretion 482
production 433
in respiration 433–434,
  453–455
respiratory quotient 433, 686
solubility in blood 398–399
carbonic acid 399
carbonic anhydrase 399, 481,
  491, 510
carbon monoxide
  haemoglobin affinity 398
  poisoning 398
carboxypeptidases 519,
  550, 551
cardiac action potential 334,
  335–336
cardia (stomach) 506, 507
cardiac cycle 337–341
  action of valves 338–339
  heart sounds in 338–339
  pressure changes 338,
    339–341
  venous pulse 339, 340
  ventricular volume
    changes 338, 339–340
  see also heart
cardiac ischaemia 362
cardiac muscle
  action potentials 334,
    335–336
  chronotropic effects
    341–343
  gap junctions of 330–331
  histology 330–331, 336
  inotropic effects 345–346
  length-tension
    relationship 343–345
  pacemaker activity
    334–336
  properties 330–331
  structure of 330–331
  see also heart
cardiac output 341–347, 376
  measurement of 342
cardiac sphincter 504,
  506, 507
cardiovascular changes
  at birth 633–34
  in exercise 346, 691–92
  in pregnancy 621–22
cardiovascular system 327
  anatomy of 327–328
  baroreceptors 379–380, 381
  baroreflex 379–80
  blood distribution 373, 374

blood pressure 371–376,
  384–386
blood vessels 370–371
chemoreceptors 452,
  453–54
fetal 627–30
heart 327–353
hormonal effects 343, 345,
  346, 378, 380–381
low pressure receptors
  379, 712
lymphatic vessels 383–384
microcirculation 382–384
regulation of 376–381
see also circulation
carotid bodies 452, 453–54
carotid sinus 454
  baroreceptors 379–380, 454
  nerve 379, 454
  role in haemorrhage 379
carpal bones/carpus 141,
  150–51
carrier
  proteins (transporters)
    56, 57–9
cartilage 435–436
Casal's collar 675
catabolism/catabolic
  metabolism 682, 743
catalytic receptors 87–8
catecholamine secretion
  adrenal medulla 223, 309
  disorders of 309–10
  sympathetic nervous
    system 221, 223
  see also adrenaline,
    adrenoceptors,
    noradrenaline
catechol-O-methyl
  transferase 309
cation 17, 743
cation channels 113, 130
cauda equina 728
caudate lobe (liver) 534, 535
caudate nucleus 213–14
causalgia 236
CCK *see cholecystokinin*
celecoxib 237
cell mediated immunity
  423, 425
cell/cells 40–53
  active transport 59–61
  cytoskeleton 49, 44–5
  division 45–7
  endoplasmic
    reticulum 43–4
  epithelia 74–6

gap junctions 77
Golgi apparatus 44
membrane bound
  vesicles 44
membrane potential 56–7
metabolism 47–52
mitochondria 43
motility 52–3
nucleus 43, 652
organelles 41–5
plasma membrane 40–43,
  55–66
ribosomes 44
tight junctions 77
cell signalling 80–94
  cell-cell recognition 76–7
  gap junctions 77–8
  G proteins and 88, 89–90
  hormones 82–3
  paracrine 81–2
  receptors 86–8
  second messengers 88,
    89–90
  steroid and thyroid
    hormones 93
  synaptic 83–4, 112–16,
    129–30
cellular respiration (internal
  respiration) 431
cellulose 527, 528, 671
central chemoreceptors
  454–455
central deafness 256
central nervous system 71,
  183, 743
  anatomy 185–90
  brain 185–86
  cells of 71, 184
  cerebral circulation
    194–95
  spinal cord 188–90
  spinal nerves 189
  see also specific structures,
    motor systems, sensory
    systems
central venous pressure
  344, 345
centrioles 45
centromere 649
centrosome 44
cerebellar ataxia 212–13
cerebellar peduncles 212
cerebellum 185, 187
  anatomy 212
  and motor control 211–13
  nuclei of 212
cerebral circulation 194–95

cerebral cortex
  association areas 264–265
  corpus callosum 266
  dominance 264
  and handedness 263–264,
    266–267
  lobes of 185
  motor control 209–10
  post-central gyrus 233–34
  precentral gyrus 209
  somatic sensation 233–34
  specialization of 263–67
  speech 268–69
  split brain operation
    266, 267
cerebral hemispheres
    263–272
  see also cerebral cortex
cerebral haemorrhage
    (stroke) 195, 208
cerebral ventricles 185,
    192–93
cerebrospinal fluid 185
  circulation of 192–93
  composition 192
  formation and
    functions 192
cerumen/ceruminous
    glands 563
cerumol 563
cervical mucus 601
cervical vertebrae 147, 148
cervix 591, 601
cetirizine 238
Charles' law 433
chemical constitution of
    body 26–36
  amino acids and
    proteins 33–35
  carbohydrates
    (sugars) 27–30
  lipids 30–32
  minerals 26
  nucleic acids 36
  nucleotides 36
  water 27
chemical elements 14–15, 743
chemical formulae 16–17
chemical senses 259–60, 743
chemical signalling 81–94
  autocrine 81
  endocrine 82
  paracrine 81
  synaptic 83, 112–116
  see also cell signalling,
    chemical synapses,
    specific hormones

chemical symbols
    14–15, 743
chemical synapses 112–116
  autonomic nervous
    system 221–223
  excitatory 113
  inhibitory 114
  neuromuscular
    junction 129–30
chemiluminescence
    assay 285
chemoreceptors
  carotid bodies 452,
    453–454
  central 454–455
  peripheral 453–455
chemoreceptor trigger zone
    (CTZ) 515
chenodeoxycholic acid 537
chest
  anatomy 148–9, 435
  leads (ECG) 355, 357
  muscles of 174, 439
  nerve supply 440
  wall 439–440
  see also respiration
chewing 501, 502
Cheyne-Stokes
    breathing 460–461
chief cells
  of gastric glands 508,
    509, 512
  of parathyroid glands 312
chlorambucil 429
chloride/chloride ions
  -bicarbonate
    exchange 477–478,
    481–482, 510
  channels 114
  cystic fibrosis 460
  extracellular 56
  gastric acid
    secretion 509–10
  intracellular 56
  and ipsps 114
  metabolic alkalosis
    720, 721
  plasma 391, 392
  shift, (Hamburger
    effect) 399
  tubular transport
    479–480, 481–482
cholecalciferol
    (Vitamin D) 312
cholecystokinin (CCK) 318,
    324, 500, 512, 514, 537,
    538, 552–553

cholera
  fluid loss in 529, 715
  toxin 529
cholesterol
  in bile 523, 537, 539–540
  blood plasma 390
  chemical structure 33
  familial
    hypercholesterolaemia
    653
  and gallstones 539–540
  nutrition 673
  plasma membrane 42
  steroid hormone
    synthesis 31
cholic acid 537
cholinergic receptors 86
  in autonomic nervous
    system 221–22
  at neuromuscular
    junction 129–30
cholinergic synapses
  of autonomic nervous
    system 221–223
  neuromuscular junction
    129–30
chondroblasts 293
chondrocytes 293
chordae tendinae 329
chorea 214, 743
chorion 614
choroid (eye) 238
choroid plexus 192–93
Christmas disease
    (haemophilia B) 408
chromaffin cell 87–8
chromatids 584, 649
chromatin 648
chromosomes
  abnormalities of 43,
    646, 743
  bivalent 584
  in cell division 45–7, 648
  classification 649
  homologous pairs 583, 646
  in meiosis 582–85,
    596–98, 654
  in mitosis 45–7
  structure of 649
chronic obstructive
    pulmonary disease
    (COPD) 462–463
chronopharmacology
    276, 743
chronotropic effects 341,
    343, 743
chylomicrons 523

chyme 496, 513, 514, 525, 743
chymotrypsin 519, 550, 551
chymotrypsinogens 550, 551
cilia 45, 53, 436–37
ciliary body 239, 241
ciliary muscle (eye) 239, 243
cimetidine 513, 553, 727
  excretion of 733
cinnarizine 515
circadian rhythm
    275–277, 743
  body temperature 696–97
  and hormone
    secretion 286,
    291–292, 306
  sleep wakefulness
    cycle 273–274, 276
circle of Willis 194
circular folds (intestine)
    516, 517
circulation 327–328, 365–386
  cerebral 194
  coronary 331–334
  cutaneous 698–99, 700–1
  enterohepatic 537
  and exercise 691–92
  fetal 627–30
  GI tract 497 498
  pulmonary 328, 369, 374
  renal 471–72
  splanchnic 497, 498
  systemic 328, 365–369,
    371–381
  see also blood vessels, heart
cirrhosis (liver) 491, 544–545
citrulline 543
clarithromycin 511
claudins 77
claustrum 213–14
clavicle 141, 149
clearance
  of drugs 733
  renal 476
climacteric 604, 743
clitoris 591, 608
clonazepam 271
clone 420, 421, 743
clonidine 242, 572
clonus 743
clopidogrel 350, 351
Clostridium difficile 416
Clot/clotting 404–409
  anticoagulants 352,
    406, 407
  cascade 405–406
  disorders 407–409, 545
  dissolution 406

Clot/clotting (*Contd.*)
  factors 405, 406, 407, 408, 412, 541
  fibrin 405, 406
  prevention 404, 406–407
  retraction 406
  role of calcium 405, 406
  role of platelets 404–406
  thrombosis 352
clotrimazole 571
club foot 153
coagulation pathway 405–406
cobalamin (vitamin B$_{12}$) 523–4, 676
coccyx 141, 147–48, 151
cochlea 252
codeine 237, 463, 528, 530
co-dominance
    (genetics) 659–60
codon 652, 653, 743
coeliac disease 524
celiac ganglion 500
coenzyme A (CoA) 48–51
coenzymes 36
collateral fibres 105, 199
colchicine 167
cold
  receptors 231
  stress 298
  thermoregulatory
    response 700–1
colic 236
colipase 519, 550, 551
collagen 70, 561
collecting ducts 468, 470, 471, 473–474, 481–482, 483–485
  acid excretion 481–482
  structure 470
  water reabsorption 483–84
colloid osmotic pressure
    (oncotic pressure) 23, 382, 383, 390
colon 494, 525–526, 527–528, 530
  absorption in 525, 527
  anatomy 526, 527
  autonomic innervation 526
  bacterial flora 527–528
  blood supply 527
  histology 526
  haustra 526, 528
  haustration 528
  movements of 528, 530
  structure of 526, 527
  vitamin synthesis 528
colostrum 522, 640–41

colour blindness 247–48
colour vision 246–47
commensal organisms 415–416
commisures 266, 743
compact bone 155–56
competitive
    antagonism 97–99
complement 417–418, 419, 422, 423
complete proteins 672
complex learning 277
compliance (pulmonary) 443–444, 448
compound action
    potential 110–11
conditioned reflex *see associative learning*
conducting airways 446
conduction velocity
    (nerve) 109, 111
conductive hearing loss 255
condyle(s) 141, 743
cones 240, 247
congenital heart disease 634
conjugate eye
    movements 248
conjugation 538–539, 543, 544
conjunctiva 239, 241
connective tissues 70–1, 388, 743
connexons 94
Conn's syndrome 308, 385, 491
constipation 529–530, 623
constitutive secretion 64
contraception, methods
    of 609–10
contre-coup injury 143
convergent eye
    movement 248
co-operativity 396
coracoid process 149–50
core temperature 695–96
cornea 239, 429
corneal reflex 242
coronal section 9, 744
coronary circulation 331–334, 744
  anatomy 331–333
  blood flow 332, 333
  disease of 352
  regulation of 334
  thrombosis in 352
coronary sinus 332
coronary sulcus 332

coronoid process 150
corpora cavernosa 578
corpus albicans 599
corpus callosum 185, 187
corpus luteum 593, 598, 611
corpus spongiosum 578
corpus striatum 186, 213, 744
cortex
  adrenal 304–9
  cerebral 186, 263–272
  lymph node 421, 422
  renal 467, 468, 470
Corti, organ of 251, 252, 253
cortical nephron 468
cortical reaction 611
cortico-spinal tract 206–8, 210
corticosteroids 304, 744
corticotrophin, *see ACTH*
corticotrophs 286, 287, 288
cortisol
  anti-inflammatory
    action 305, 306
  deficiency 309
  excess 308
  diurnal variation 306
  fetal 306, 620
  and glucose
    metabolism 319
  metabolic actions
    of 305–306
  regulation of 307
  secretion 306–307
  in stress 306
  synthesis 304
cortisone 304
corticosterone 304
corticotrophin releasing
    hormone (CRH) 289, 306, 307
costal cartilage 148–49
cotransport (coupled
    transport) 522, 744
covalent bond 17
cough reflex 456, 463
Cowper's glands *see bulbo-urethral glands*
cranial nerves 744
  anatomy 187–89
  functions of 189
cranium/cranial
    bones 143–45
creatine kinase 124
creatine phosphate 122–24
creatinine clearance 475
cretinism 300, 302–303, 744

CRH *see corticotrophin
    releasing hormone*
cribriform plate 145, 259
crista ampullaris 257
crista dividens 628
cristae (of mitochondria) 41, 43
Crohn's disease 524, 530
crossed-extensor reflex 203
cryptorchidism 578
crypts of Lieberkuhn 516, 517, 521
c.s.f. *see cerebrospinal fluid*
curare 95
Cushing's disease/syndrome 308, 385, 408
Cushing's reflex 193, 195
cutaneous circulation 564, 698–99
cuticle 562
cutis *see skin*
cyanosis 396, 557, 744
cyclic AMP 89, 484
cyclic GMP 92, 245–46
cyclo-oxygenase (COX)
    enzymes 237
cyclizine 352
cyclopentolate 223
cyclophosphamide 95, 429
cyclosporine 180, 429
cystic duct 534, 535
cystic fibrosis 460, 553, 655, 658
cytochrome P450
    enzymes 543–544
cytokines 420, 423
cytoplasm/cytosol 40, 43–45, 744
cytoskeleton 40, 41, 44–45, 744
cytotoxic T cells 420, 423, 425
cytotrophoblast 613, 615

D

D antigen (Rh factor) 410, 411
DAG (diacylglycerol) 90
daily energy requirement 690
daily protein
    requirement 672
Dalton's law of partial
    pressures 432
dandruff 559
dark adaptation 244
dazzle reflex 242
dead space 446–447
  anatomical 446
  Bohr equation 447

measurement of 447
physiological 446, 447
deafness 255–56
deamination 52, 540, 542, 744
decibel scale 250, 744
decubitus ulcers, *see pressure sores*
deep vein thrombosis 375, 400, 407
defecation 528–529
defence reaction 224
definitive placenta 615–16
deglutition (swallowing) 504–506, 744
deglutition apnoea 504
dehydration 290, 484, 529, 569, 710, 714–15
delta waves (EEG) 270, 271
dementia 278, 744
demyelinating diseases 111
dendrite 71, 744
deoxyhaemoglobin 396
deoxyribose 35
depolarization 62, 744
dermal papilla 557, 558, 560
dermatome(s) 189, 232, 742
dermis
  blood supply 564
  hair follicles 561–62
  sensory receptors 565–66
  structure of 560
  sweat glands 562–63
descending colon 494, 525, 526
desensitization 98
desmosomes 77
desoxyn 275
desquamation 558
detoxification, *see biotransformation*
detrusor muscle 485, 486
dexadrine 275
dexamethasone 429
dextropropoxyphene 237
diabetes insipidus 290–291, 484, 489, 714, 744
  see also antidiuretic hormone
diabetes mellitus 321–324, 488, 744
  complications of 321–322, 323–324, 530
  gestational 624
  insulin dependent 321, 323
  ketosis 321, 322, 488
  management of 323

symptoms of 321–322, 488, 530
treatment 323
type 1 (insulin dependent) 321, 323
type 2 (non-insulin dependent) 321
  see also blood glucose regulation, insulin
diacylglycerol (DAG) 90
diamorphine 237, 352
diapedesis 418, 420, 744
diaphragm 439, 440, 442, 535
diaphysis 154, 744
  see also bone
diarrhoea 527, 528, 529–530, 714, 715
diarthroses, *see synovial joints*
diastole 337, 338, 340, 744
diastolic pressure 367, 371, 372
  see also arterial blood pressure, cardiac cycle
diazepam 352, 736
diclofenac 237, 429
dicrotic notch (incisura) 338, 340, 372
diet 670, 744
  essential fatty acids 672–73
  macronutrients 671–73
  malnutrition (PEM) 672
  micronutrients 673–77
  minerals 676–77
  protein requirement 672
  vitamins 673–76
dietary fibre 671
differential white cell count 392
diffusing capacity 451
diffusion 24
digestion
  of carbohydrates 504, 519–521, 550–551
  of fats 519, 520, 537–538,
  of proteins 509, 519, 550, 551
  see also bile, pancreas
digestive enzymes
  pancreatic 519, 520
  salivary 503, 504
  small intestine 519, 521
  stomach 509
digestive system (alimentary canal) 493–532
diglycerides 31, 520
digoxin 345, 351, 352

dihydrocodeine 237
dihydrotestosterone 582
dilator pupillae *see iris*
diltiazem 350, 351, 363
di-octyl sodium sulphosuccinate (DSS) 530
diphenhydramine 275
dipeptide 33–34
diploid cell 583, 646, 744
diplopia 744
direct pupillary response 242
disaccharide 28, 744
disopyramide 363
distal tubule 468, 469, 470, 473, 474, 480–482, 483–484
  intercalated cells 470, 481–482
  principal cells 470, 480–481, 484
  sodium absorption by 480–481
  structure 469, 470
disynaptic reflex 202
dithranol 571
diuresis 484, 709, 744
diuretics 386, 490–491, 718, 744
divalent metal transporter 395
diverticulae 527, 744
diving response 657
DNA
  and chromosomes 648
  fingerprinting 654
  mitochondrial 664
  genetic code 651–52
  protein synthesis 651, 653
  structure of 36, 651
dobutamine 352
docusate 563
dominance,
  cerebral 264
  genetic 656–59
domperidone 515, 530
donepizil (Aricept) 278
l-dopa (levodopa) 214, 730, 732
dopamine 214, 515, 643, 644
dorsal column pathway 232
dorsal horn 189–90
dorsal root 189–90, 232
dorsal root ganglion 189
dorsal venous arch 368, 369
dorzolamide 727
dose-response curve 96

dihydrocodeine 237
Down syndrome 278, 584, 649, 650
doxylamine 275
drinking (thirst) 710, 711
drug/drugs 95, 744
  absorption of 728
  adminstration 727–733
  adverse effects 736–37
  distribution 731–32
  efficacy 96
  and lipid solubility 112, 731
  elimination
    kinetics 733–35
  ionization 729
  metabolism 542, 733
  patches 727
  potency 96
  protein binding of 733
  receptor activation 96–98
  resistance 98
  targets 95
  teratogenic effects of 737
  tolerance 98, 737
  transdermal administration 571–72
  uptake 728–33
  see also specific drugs
Duchenne muscular dystrophy 180
ductus arteriosus 628, 629, 634
ductus venosus 628, 634
dumping 514
duodenal ulcer 521
duodenum 494, 514, 515, 516, 548
dura mater 185, 193
dwarfism
  achondroplastic 161
  Laron 295
  pituitary 295
dynorphin 235
dyskinesia 214, 744
  see also movement disorders
dysmenorrhoea 600
dysphagia 505–506
dyspnoea definition 459
dysrhythmia *see* arrhythmia 744

E

ear
  anatomy 251–52
  balance 256–58
  deafness 255–56

ear (*Contd.*)
 drum (tympanic membrane) 251
 fluids of 252
 ossicles 252
 semicircular canals 252, 256–57
 wax (cerumen) 563
 see also auditory system, vestibular system
Ebixa (memantine) 278
eccrine (sweat) glands 563, 701
ECF *see extracellular fluid*
ECG (electrocardiogram) 336, 338, 339, 354–362
eclampsia/pre-eclampsia 385
ectopic beats 361, 744
eczema, treatment of 571
EDTA 406
EEG (electroencephalogram) 269–273
 in epilepsy 271–272
 in sleep 270–271, 272–273
 waves of 269–271
effective circulating volume (ECV) 389, 744
 regulation of 712–714
effector 190, 744
efferent arterioles 467, 469, 471–472, 744
efferent lymph 383, 384, 421, 422, 745
efferent nerves 190, 745
egg (ovum) 576, 597
 development 592–95
 fertilization 610–11
eicosanoids 745
ejaculation 608
ejaculatory duct 577–78, 587
elastic arteries 371
elastic fibres
 of arteries 370, 371
 of lungs 445
 of skin 561
elastin 70
elbow 150, 169, 170
electrocardiogram (ECG) 336, 338, 339, 354–363
 arrhythmias 361–362, 363
 augmented limb leads 355
 bipolar leads 355, 356
 characteristics of 358–359
 chest leads 355, 357
 origin of 354
 unipolar leads 355, 357

electrochemical gradient 56, 745
electroencephalogram (EEG) 269–273
electrolytes
 absorption of 523–24, 525, 527
 definition 20, 745
 in plasma 389, 390
 tubular transport of 477–82
 in urine 487
electron 15
elctron transport chain 43
embolism 407
emesis 745
emission 608
emphysema 451, 459, 460
emulsification (of fats) 519, 538
encainide 363
end-diastolic volume 337, 344, 345
 see also cardiac cycle
endochondral ossification 157
endocardium 745
endocrine system 283–326
 anatomy 284
 glands 75, 81, 284, 745
 growth 291–96
 hormone classification 285
 signalling 82–3, 84, 85, 93
 see also specific glands and hormones
endocrinology 284
endocytosis 65, 745
endolymph 252, 515, 745
endometrium 590–91, 599–600, 745
endomysium 120
endoneurium 191
endopeptidase 509, 519
endoplasmic reticulum 43–4, 523
endorphin 235
endosomes 745
endosteum 155
endothelin 136
endothelium 74, 370, 371, 404, 745
 see also blood vessels, structure of
endplate potential (epp) 113
end-systolic volume 337
 see also cardiac cycle
energy
 balance 683–85, 690

body stores 684
 equivalent of oxygen 685
 heat production 685–86
 physical activity ratio 689
 requirements in exercise 688–91
energy metabolism 47–52, 671, 683, 745
 aerobic 50
 anaerobic 47–49
 basal metabolic rate 685, 686–87
 fatty acids 50–51
 glycolysis 47–49
 tricarboxylic acid cycle 49–50
enkephalin 235
enteral nutrition 678
enteric nervous system 183, 221, 494, 499, 518, 526, 745
enterocytes 516, 517
entero-endocrine cells 508, 511, 512, 516, 517
entero-gastric reflex 514
enterohepatic circulation 537
enteropeptidase 521, 550, 551
enzyme(s) 35
 gastric 509
 pancreatic 550–551
 salivary 504
 zymogens 519–548, 550
eosinophils 392, 402, 417, 745
ependyma/ependymal cells 185
epidermis
 germinal cells 558
 melanocytes 559
 structure of 557–59
 see also skin
epididymis 577, 579, 587
epidural anaesthesia 728
epidural space 185
epiglottis 504, 505
epilepsy 266, 271–272
 causes 271, 665
 drug treatments 271–272
 EEG changes 272
 grand mal 271, 272
 iodiopathic 271
 Jacksonian 209
 petit mal 271, 272
epimysium 120
epinephrine, s*ee adrenaline*
epineurium 191

epiphyseal growth plates 154, 157–58, 293
epiphyseal line 155
epiphysis 154, 293, 745
 see also bone
episiotomy 592
epithelia/epithelium 74–76, 745
 basal lamina 78
 basement membrane 74, 78
 cell junctions 77–78
 classification 74
 desmosomes 77–78
 gap junctions 77
 gastric 511
 glandular 75–76
 hemidesmosomes 77–8
 intestinal 517–518
 lamina propria 74
 of nephron 470
 respiratory 435–438, 457
 see also gastrointestinal system, renal system, respiratory system, skin
epithelial transport 60–1
epsp (excitatory post-synaptic potential) 113–14
equilibrium potential 56, 745
erectile tissues 578, 591, 608
erection 608
erythroblasts 393
erythrocytes (red blood cells) 391–401, 402, 409–412
erythromycin 463, 571
erythropoiesis 393, 542, 745
erythropoietin 324, 393, 466
escape rhythms 360
eserine 95, 130
essential amino acids 672
essential fatty acids 31, 672–73
essential hypertension, *see primary hypertension*
etanercept 167
ethanol, elimination kinetics 734–35
ethmoid bone 143–144
ethmoid sinus 146
etidromate 162
etomidate 728
eupnoea 452 745
eustachian tube 251
eutectic mixture 727
excitation-contraction coupling

skeletal muscle 121–22
smooth muscle 133–134
excitatory post-synaptic
   potential (epsp) 113–14
excretion 745
Exelon (rivastigmine) 278
exercise 224, 688–93
   cardiovascular changes
      in 691–92
   effects of training 693
   energy requirements
      of 540, 688–89
exocrine glands 75–6, 745
exocytosis 64–65, 745
exophthalmos 301
expiratory reserve
   volume 441, 442
expired air, composition
   of 433–434
exponential decay
   734–35, 745
extension 10, 745
extensor muscles 199, 745
external intercostal
   muscles 439, 442
external respiration 431–463
   definition 431
external sphincter
   anal 528, 529
   bladder 485
extracapsular
   ligament 164–65
extracellular fluid (ECF) 27,
   707, 745
   composition 56
   measurement of 708–9
   volume of 27
extracellular space/
   compartment 55
extradural anaesthesia *see*
   epidural anaesthesia
extrafusal muscle fibres 201,
   204–205
extraocular muscles 248–49
extrapyramidal tracts 207–8,
   210–11
extrasystole 361
eye and vision 238–249
   accommodation 243
   anatomy 238–40
   aqueous humour 241
   autonomic innervation 239
   blind spot 240–41
   choroid 238
   colour vision 247–48
   cornea 239
   dark adaptation 244

eyelids 241–2
extra-ocular
   muscles 248–49
eye movements 248
focussing 243
glaucoma 242
image formation 243
intraocular
   pressure 241–42
iris 239
lachrymal/lacrimal
   gland 241
lens 239
near point 243
neural pathways 245
perimetry 244
phototransduction 245–46
pupil 239
reflexes 242
refractive errors 243–44
retina 239
sclera 238
visual field 244
vitreous body/humour 239

F

facial bones 143–46
facial muscles 170–71
facial nerve 187–89
facilitated diffusion 56,
   522, 745
faeces 525, 528
falciform ligament 534, 535
Fallopian tubes
   anatomy 589
   effect of ovarian
      steroids 601
   fertilization 610
   histology 590
familial hyperchole-
   sterolaemia 658
famotidine 513
fascia 745
fascicle 120, 745
fasting
   glucose regulation 320–21
   metabolic effects 688
fat(s)
   absorption 523
   brown adipose tissue 71
   composition of 30–2
   dietary 672–73
   digestion of 519, 520,
      537–538
   diglycerides 31, 520
   emulsification of 519, 538

as energy reserve 673
fatty acids 519, 523
metabolism 50–51
monoglycerides 31, 519,
   520, 523
myelin 105–6, 109
subcutaneous 561
triglycerides 31, 520
white adipose tissue 71
fatigue 124
fatty acids
   absorption 523
   essential 31, 672
   metabolism of 50–1
   of milk 641
   nutritional
      requirements 673
   phospholipids 31
   saturated 30, 672
   unsaturated 31, 672
feedback control 7–8,
   204, 745
female reproductive
   system 589–605
   anatomy of 589–92
   contraception 608–10
   hormonal regulation
      599–601
   menarche and
      menopause 604–5
   menstrual cycle 592–93
   see also ovary, ovarian cycle,
      pregnancy
femur 141, 152–53
fenbufen 237
fenestrations (capillary) 383
fentanyl 237, 572, 727
ferritin 394, 395, 541
ferroportin 395
   see also iron metabolism
fertilization 589, 610–12
   implantation of
      embryo 613–15
   sexual reflexes 608–10
fetal circulation 627–630
   changes after birth
      633–634
   control of 630
fetal reproductive system
   development of 636–38
fetal shunts 628–30
   closure of 633–34
fetal zone (adrenal
   gland) 634
fetus
   adrenal cortex 634–35
   bilirubin 618

birth, timing of 620
changes at birth 631–36
and cortisol 634
development 614
excretion 617–18
gas exchange 617
gastrointestinal
   tract 635–36
growth of 621
renal function 617–18
role of placenta 616–18
respiratory system 630
sex determination 636
FEV_1 458
fever
   antipyretics 705
   metabolic rate and 688
   pyrogens 704–5
fexofenadine 238
fibrillation
   atrial 361–62
   ventricular 352, 361–62
fibrin 405–6
fibrinogen 390, 405, 406
fibroblasts 70.561
fibrosis
   cystic 460
   pulmonary 459–460
fibula 141, 152–53
Fick principle 342
filariasis 716
filtered load (kidneys) 475,
   476, 477
filtration 383, 473,
   474, 746
filtration fraction 476
fimbriae 590
fingers 151
first breath 631
first order kinetics (drug
   elimination) 735
first pass metabolism
   (drugs) 733
flaccid 746
flagella 45, 586
flare 418
flat bones 141
flat feet 153
flatus 527–528
flavine coenzymes 50
flecainide 351, 363
flexion 10, 746
flexion reflex 203
flexor muscles 199
flurazepam 275
folate/folic acid 393,
   625, 676

follicle stimulating
hormone (FSH)
female reproductive
system 594–97, 602–3
gonadotrophs 287, 288
male reproductive
system 587–88
ovarian cycle 603–3
in pregnancy 624
follicles, ovarian
atresia of 593
development of 592–95
Graafian 594
follicular phase 592, 594
fontanelle 145–46
food/foodstuffs *see nutrition*
foot 153,177
foramen magnum 143, 145
foramen ovale 628, 633
force-velocity curve 127
forced expiratory volume
(FEV) 458
forced vital capacity
(FVC) 458
formoteral 462
fossae 143, 145
fovea centralis 240
fracture, healing of 159–60
Frank-Starling
relationship 343–345
frenulum 502
fresh frozen plasma 412
frontal bone 143–45
frontal lobe 185, 187
frontal cortex 264–265,
268–269
association areas 264–265
Broca's area 268
functions of 264–65
premotor area 268
frontal leucotomy 265
frostbite 699
fructose 28, 522
frusemide *see furosemide*
FSH *see follicle stimulating
hormone*
functional hyperaemia 378
functional residual
capacity 441, 442, 443
fundus
of bladder 485
of stomach 506, 507
of uterus 622
furosemide 95, 255, 351, 352,
490, 491, 718, 736
fused tetanus 125
fusdic acid 571

G

GABA (gamma amino butyric
acid) 113, 114
G cells 508, 511, 512
G protein coupled
receptor 88
G proteins
and adenylyl cyclase 89
autonomic nervous
system 222–223
cell signalling 88–90
cyclic AMP 89
cyclic GMP 92, 245
olfactory receptors 259
phospholipase C 90
second messengers 88
in vision 245–46
Gage, Phineas 264–265
galantamine (Reminyl) 278
gall bladder 494, 534, 535,
538, 540
gallstones 539, 540
gametes 576, 579, 583, 589
gap junctions 77, 94, 131
gas/gases
alveolar 434
arterial blood
concentrations 399
carbon dioxide 398–400
carriage 396–400
exchange 438, 451
laws 432–433
mixtures, effect of
breathing 454–55
nitric oxide 85, 91–2, 115
nitrogen 433, 434
oxygen 396–398
partial pressures in
blood 431
and pH 719–21
placenta 617
solubility 432
transport 396–400
venous blood
concentrations 399
gastrectomy 508, 509, 514
gastric acid 509, 510–513
antacids 505, 513
inhibitors of secretion 513
regulation 511–512
secretion of 509, 510
gastric arteries 498, 507
gastric emptying 513–514
gastric glands 508–511
gastric inhibitory peptide
(GIP) 512

gastric juice 509–511
gastric lipase 509
gastric mucosa 507–508
gastric pits 508
gastric secretion 509–513
cephalic phase 511, 512
disorders 513
gastric phase 511, 512
intestinal phase 512
hormonal regulation 512
and vagus nerve 511, 512
gastric ulcers 511, 513
gastrin 324, 500, 511, 512,
513, 538
acid secretion 511, 512, 513
action on pancreas 552
G cells 508, 511, 512, 552
secretion of 511, 512, 513
gastritis 511
gastro-ileal reflex 518
gastrointestinal (GI)
tract 493–531
absorption 514, 521–525,
527
accessory structures
494, 507
anatomy 494, 507, 515–516,
525, 526, 527, 528
blood supply 497–498, 507,
526, 527
components of 494, 507
digestion 504, 519–521
enteric nervous
system 495, 499, 518, 526
enzymes of 504, 509,
519–521
functions 506
of fetus 635–36
histology 516, 517, 526
hormones of 500, 511, 512
innervation of 499–500,
526, 528–529
parasympathetic
innervation 499, 500
sphincters 496, 503, 504,
505, 507, 526, 528, 529
sympathetic
innervation 499, 500
wall of 495–496, 507–508,
516–518, 526
*see also specific structures,
liver, pancreas*
gastro-oesophageal
reflux 505
gender
body water
distribution 707

genetic determination
of 636–38
metabolic rate 686–87
nutritional
requirements 671
gene(s) 646, 650, 746
alleles 645, 741
co-dominance 659
dominant 654
genotype 654, 746
locus 650
mutation 653
phenotype 654, 750
recessive 654
sex determining 636, 647
therapy 666
transcription 93, 652
general anaesthetic(s) 731
metabolism of 732
uptake and
distribution 731–32
genetic code 651–53, 746
genetic counselling 666
genetic diseases/
disorders 653, 658,
661–64
genetic dominance 654
genetic testing 665–66
genome 650, 746
genotype 654, 746
gentamicin 463
germinal centre (of lymph
node) 421, 422
gestation 590, 746
gestational diabetes 624
GFR *see glomerular filtration
rate*
GH *see growth hormone*
GHIH, s*ee growth hormone
inhibiting hormone*
GHRH, *see growth hormone
releasing hormone*
GIP s*ee gastric inhibitory
peptide*
GI tract *see gastrointestinal
tract*
Giemsa stain 649
glands
endocrine 75, 284
exocrine 75–76
glaucoma 242, 491, 727
glenoid cavity 149–50
glial cells (neuroglia) 71
glibenclamide 323
glicazide 323
glipizide 323
Glisson's capsule 534, 535

glitazones 323
globulins 390, 541
globus pallidus 213–14
glomerular filtration rate
    (GFR) 474–475, 476
    measurement of
        474–475, 476
    in pregnancy 623
glomerulonephritis 427
glossopharyngeal
    nerve 187–89
glomerulus 467, 469,
    473–475
glossitis 675
glottis 504, 505
glucagon 317, 318–319,
    554, 555
    actions of 318–319, 538, 555
    alpha cells (pancreas) 317,
        548, 554
    glycogenolysis 94, 318, 315,
        540, 541, 555
    plasma glucose
        regulation 318–319,
        554–555
    secretion of 318, 554
glucagon like peptide-1
    (GLP-1) 514
glucocorticoid(s) 304, 305,
    306, 319, 429, 746
glucogenic hormones 316
gluconeogenesis 51–2, 316,
    672, 746
glucose
    absorption of 60–61,
        521–522
    cellular uptake 522
    diabetes mellitus 321–24,
        475, 488, 744
    energy production 47–50,
        315, 540–541
    gluconeogenesis 51–2, 316,
        540, 541, 555
    glycolysis 47–9
    glycogenesis 540, 541
    hormonal
        regulation 316–20
    insulin and 316–18
    metabolism 47–52, 315–16
    placental exchange 617
    plasma 315, 320–321
    tolerance test 321
    transporters 477–478, 522
    tubular transport 475, 476,
        477, 478
    in urine (glycosuria) 475,
        476, 487, 488

glucuronic acid 538, 539,
    543, 544
glue ear 255
GLUT2 522
glutamine 719
glutamate 85, 113
glyceryl trinitrate (GTN) 92,
    572, 727
glycine 33, 85
glycogen
    breakdown 315, 540, 541
    as energy store 28, 540,
        541, 746
    structure 28–9
    synthesis 540, 541
glycogenesis 315, 540, 541,
    746
    hormonal regulation
        540, 541
glycogenolysis 315, 540, 541,
    555, 746
    adrenaline 89, 541
    hormonal regulation 541
glycolipids 31
glycolysis 315, 392
glycosuria 321, 746
glycosylated haemoglobin
    (HbA₁c) 315–316, 323
glycosylation 315–316
goal-directed
    movements 208–10
goblet cells 436–437,
    517, 525
goitre 301, 677, 746
Golgi tendon organs 201, 203
Golgi tendon reflex 203–4
gomphosis 163
gonadotrophin
    releasing hormone
        (GnRH) 288–89, 587–88,
        602–3
gonadotrophins (FSH, LH)
    gonadotrophs 287, 288
    ovarian cycle 594–98,
        602–3
    spermatogenesis 587–88
    see also follicle stimulating
        hormone, luteinizing
        hormone
gonads 746
    ovary 592–95
    testis 579–81
    see also reproductive system
gout 166, 167
Graafian follicle 594
grand mal 271, 272
granisetron 515

granular cells see
    juxtaglomerular cells
granulation tissue 568
granulocytes 392, 401, 402
granuloma 418
granulosa cells
    (ovary) 594–97
Graves' disease 301
greater curvature
    (stomach) 506, 507
greater omentum 506,
    507, 527
grey matter 186, 188
growth factors
    IGF-1 293, 294
    IGF-2 293
    in wound healing 293
growth hormone (GH) 85,
    288–289, 291–296
    actions of 292–296
    in childhood 292–293
    deficiency 295
    diabetogenic action
        of 294
    direct actions 292, 293, 294
    excess 295–296
    in fasting 294, 320
    glucose metabolism
        293–294, 320
    growth, control of
        292–293
    and IGFs 293, 294
    indirect actions of
        293, 294
    metabolic effects
        293–294, 687
    secretion 291–292, 294
    in sleep 291–292
    somatotrophs 287, 288
    somatostatin 288–289
growth hormone inhibiting
    hormone (GHIH,
    somatostatin) 288–289
growth hormone releasing
    hormone (GHRH) 288,
    289, 727
growth patterns 292–293
growth plate 293
growth spurt 292, 293
GTP binding protein
    see G protein
guanylyl cyclase 91–2
gustation (taste) 259–60
gut see GI tract
gut associated lymphoid tissue
    (GALT) 420, 501
gyrus 185, 746

H

H antigen 659
H⁺, K⁺-ATPase 509, 510, 513
habituation 277
haematocrit ratio 391, 400, 746
haematopoiesis 391–393, 746
haematuria 489, 746
haemochromatosis 395
haemodilution 746
haemodynamics (blood flow
    and pressure) 371–76
    arterial blood
        pressure 371–3
    arterioles and vascular
        resistance 373
    blood viscosity 400
    capillary pressure 373, 383
    skeletal muscle pump 375
    vascular resistance 373
    venous pressure 373–74
haemoglobin 394, 396–398,
    399, 400, 401, 538
    blood content 396
    Bohr effect 398
    buffering 399
    carbon dioxide dissociation
        curve 400
    carbon monoxide
        binding 398
    glycosylation 315–316
    Haldane effect 400
    oxygen dissociation
        curve 396–398, 455
    percentage saturation
        with oxygen (SpO₂)
        396–397, 398
    red cells 392, 394
    see also anaemias
haemolysis 539, 746
haemolytic disease of the
    newborn 411, 675
haemolytic jaundice 539
haemophilia A 408, 409,
    662, 663
haemophilia B 408, 662
haemorrhage 379, 381,
    537, 746
haemostasis 404–409, 746
    see also blood clotting
hair 561–2
hair cells
    cochelar 253–54
    vestibular 257–58
hair follicle 561
Haldane effect 400
    see also gas transport

half-life of drugs 734, 746
halothane 346, 727
Hamburger shift (chloride shift) 399
hamstrings 169, 175, 179
hand 150–51, 173–75
handedness 263–264, 266–267
haploid cells 746
  meiosis 583–85, 647
  oogenesis 596–98
  spermatogenesis 585
haptic touch 231, 746
hard callus 160
hard palate 501, 504, 505
Hardy-Weinberg equilibrium 655
Hashimoto's thyroiditis 301
haustra/haustration 525, 528
Haversian system 155
hCG (human chorionic gonadotrophin) 612, 618
headlice, treatment of 571
hearing loss 255–56
heart 327–353
  action potentials 334–336
  anatomy 329–339
  arrhythmia 359, 360, 361, 362
  atrial fibrillation 361, 362
  atrio-ventricular (AV) valves 329, 338–339
  attack (myocardial infarction) 352, 362, 406
  block 360, 361
  bundle block 360
  cardiac cycle 337–341
  cells of 330–331
  chronotropic effects 341, 343
  coronary circulation 331–334
  diastole 337, 338
  effect of drugs 350–352, 363
  endocardium 330
  escape rhythms 360
  extrasystole 361
  failure 347–349
  Frank-Starling relationship 343–345
  gap junctions 94
  histology 330–331, 336
  and hormones 343, 345, 346
  inotropic effects 345, 346, 347
  interventricular septum 329
  myocardium 329

pacemaker activity 334, 335, 336
pacemaker potential 334, 335, 336
parasympathetic innervation 218, 220, 343
pericardium 329
pressure changes in 338, 339–340
rate, control of 341–343
semilunar valves 329–330, 338–339
sick sinus syndrome 360
sinus rhythm 360
sounds 338–339
stroke volume 337, 341, 343–347
supraventricular rhythms 361
sympathetic innervation 218, 220, 343
systole 337, 338
ventricular fibrillation 352, 361, 362
ventricular tachycardia 361, 362, 363
  see also circulation, ECG
heartburn 236, 505, 623
heart failure 347–352, 386
  acute 348
  chronic 348, 349, 489
  oedema in 348–349, 489
  treatment of 350–352
heat
  body temperature 695–97
  exchange 697–98
  exhaustion 702
  metabolic production 683–87
  stroke 702
  see also thermoregulation
Helicobacter pylori 511, 513
helper T cells 423, 424, 425
hemiballismus 214
hemidesmosome 77
Henry's law 432
heparin 350, 351, 352, 406, 407
hepatic bile 537–538
hepatic circulation 536–537
hepatic duct 534, 535
hepatic encephalopathy 544
hepatic jaundice 539
hepatitis 544
hepatocytes 535, 536, 537
hepcidin 324, 395, 541, 542
  see also iron metabolism

heredity 654, 746
Hering-Breuer reflex 453, 456
herpesvirus varicella-zoster virus 232
Hess, R. 273
heterotrimeric G protein, see G protein
heterozygous individuals 654, 746
hexose 28
hiatus hernia 505, 506
high blood pressure, see hypertension
hilus (kidney) 467, 468
hinge joint 165
hip
  bone 141, 151–52
  joint 151–52, 165, 175
hippocampus 277, 278, 279
Hirschsprung's disease 499, 530
histamine 235, 285
  antagonists 238, 513
  gastric secretion 511
  H_2 receptors 511, 513
  mast cell 81, 511
histology 4
histones 648
HIV (human immunodeficiency virus) 428
HLA antigens/HLA complex 416, 429
homeostasis 7
homozygous individuals 654, 746
horizontal cells 240
hormonal signalling 81–3
hormone/hormones 81–3, 283–284, 285–286, 746
  binding proteins 285
  chemical nature of 83–4, 85, 285
  and gene transcription 93
  measurement of 285
  of GI tract 324, 500
  measurement of 285
  protein binding 285, 389
  replacement therapy (HRT) 301–302, 303, 315, 605, 727, 746
  see also endocrine system, pituitary gland and specific hormones
hormone regulation
  negative feedback 286, 307, 312, 587–8, 602–3

positive feedback 286, 602–3
  see also under specific hormones
host cells (immunology) 416, 427
housekeeper contractions 518
hPL see human placental lactogen
human chorionic gonadotrophin (hCG) 612, 618
  pregnancy test 612
human chorionic thyrotropin 624
human genome project 650
human milk
  composition 641
  secretion of 640
  suckling 642–44
human placental lactogen (hPL) 612, 618
humerus 141, 149–50
hunger 677
Huntingdon's chorea/disease 214, 658, 659
hyaline cartilage 163, 164
hydrocephalus 193, 746
hydrochloric acid, secretion of 509–510
hydrochlorothiazide 490
hydrocortisone 429, 571
hydrogen bond 19–20
hydrogen ion 21, 235, 566
β-hydroxybutyric acid 51, 719
5-hydroxytryptamine (5-HT, serotonin) 85, 235, 515
hyoid bone 501
hyoscine 222, 503, 515
hyperalgesia 567
hypercalcaemia 313
hypercapnia
  acid-base balance 719–20
  definition 454–455, 746
  effect of 454–455
  fetal 631–32
  obstructive lung disease 460
hyperemesis gravidarum 623
hyperglycaemia 321–322
  see also diabetes mellitus
hypermetropia/hyperopia 243–44
hyperparathyroidism 313
hyperphagia 678, 746

hyperpnoea 452, 720
hyperpolarization 343
hypersensitivity 427
hypertension 372, 384–386, 746
hyperthermia 702, 746
hyperthyroidism 301
hypertonic fluid 23, 747
hyperventilation 452
hypervolemia 712, 747
hypnotic drugs 275
hypocalcaemia 313
hypocapnia 747
  acid-base balance 719–720
  definition 455, 747
hypodermis 557, 561
hypogastric nerve 485
hypogastric plexus 499
hypoglycaemia 323, 544, 554
  effects of 323–324
  growth hormone
    secretion 320
  insulin 317
hypokalaemia 509
hypoparathyroidism 313
hypotension, postural 375, 380
hypothalamic-hypophyseal
    portal system 287, 288
hypothalamus
  anatomy 186, 187, 287
  appetite regulation 503, 677
  autonomic nervous
    system 224
  body fluid regulation 290, 483–84, 710–2, 714
  defence reaction 224
  paraventricular
    nucleus 287, 290
  and pituitary gland 286–289
  osmoregulation 290–291
  releasing hormones 287, 288–289
  role in
    thermoregulation 700
  supraoptic nucleus 287, 290
  thirst 710–11
hypothermia 702, 704
hypothyroidism 301–303
hypotonic fluid 23, 747
hypovolemia 712, 747
hypoxemia, definition
    of 455, 747
hypoxia 455, 461–462, 747
  oxygen therapy 461–462
hysteresis 444, 445, 747

I

I band 331
ibuprofen 237, 429, 511, 730
ICSH *see luteinizing hormone*
icterus (jaundice) 539
ideal gas law 433
IGF-1/IGF-2 324, 541–542
ileo-caecal valve 518, 525, 527
ileo-gastric reflex 518
ileum 494, 515, 516, 517, 518, 521, 523, 527
ilium 142, 151–52
immune system 414–429
  acute phase proteins 418
  adaptive 419–426
  antibody secretion 422, 423–424, 426
  cells of 417, 419–425
  complement 417–418, 419, 422, 423, 427
  disorders of 427–428
  hypersensitivity 427
  innate 416–419
  natural 416–419
  proteins of 417–419, 422–423, 425
  response to infection 417–419
immunization 424, 426
immunodeficiency 428
immunoglobulins 411, 412, 522
  classification 422
  diversity 422–423
  secretion 422, 423–424, 426
immunosuppressive
    drugs 429
impetigo, treatment of 571
implantation of
    embryo 613–15
incisura 338, 340, 372
incomplete dominance 661
incontinence 486
incus 252
indomethacin 237
infarct 347, 362, 747
infection
  adaptive immune
    system 419–426
  antibody response
    to 423–424, 426
  bacterial 423, 570
  complement 417–418, 419, 422, 423
  fever 704

natural immune system
    416–419
  passive barriers to 415–416
  resistance to 419
  viral 417, 418, 423, 426, 428
  of wounds 570
inferior mesenteric
    ganglion 500
inferior vena cava 328, 329, 330, 367, 369, 516, 535
inflammatory response 418, 419, 429, 567
infradian rhythms 275
inheritance, principles
    of 654–64
inheritance
  autosomal 658
  mitochondrial 664–65
  polygenetic 661, 656
  sex-linked 661–662
inhibin 587
inhibitory post-synaptic
    potential (ipsp) 114
inhibitory synaptic
    transmission 114
injection (drug
    administration) 727
innate immune system *see
    natural immune system*
inner ear 252
inositol 1,4,5-trisphosphate
    (IP$_3$) 90
inotropic effects 345–46, 747
INR (International
    Normalized Ratio)
    407, 747
insensible water loss 710
insomnia 274–275
inspiratory reserve
    volume 441, 442
inspired air
    (composition) 433, 434
insulin 85, 316–317, 321–324, 554
  amino acid uptake 317
  beta (β-) cells 317, 548, 554
  diabetes mellitus
    321–24, 554
  effect of 317–18
  glucose uptake 317–18, 540–41
  glycogenesis 317, 540–41
  islets of Langerhans 317, 547–548, 554
  preparations 323
  secretion of 317–318
  sliding scale 323

insulin-like growth factors,
    see IGF-1, IGF-2
integrins 77
integumentary system 556
  see also skin
intention tremor 213
intercalated cells (I cells)
    470, 481–482
intercalated disks 331–32
intercostal arteries 366–7
intercostal muscles 439, 440, 442
intercostal nerves 440
interferons 418, 425
interleukins 425, 704
intermediate filaments 44, 132
intermediate lobe 286
internal capsule 186, 188, 208
internal intercostal muscles
    148, 174, 439, 442
internal respiration 431
internal anal sphincter
    528, 529
International Normalised
    Ratio (INR) 407, 747
interneurons 105, 202
interstitial cell stimulating
    hormone (ICSH), *see
    luteinising hormone*
interstitial fluid 382, 383, 747
interventricular septum 329
intervillous blood spaces 613
intervertebral disk 147
intestinal flora 415–416, 525–526, 527–528
intestinal fluid 516, 521, 525
intestinal reflexes 518
intestine
  large 525–530
  small 515–525
  *see also gastrointestinal (GI)
    tract*
intracapsular ligament 165
intracellular fluid 27, 55–6, 707, 747
  composition of 56
intracranial pressure
    192–93, 514
intrafusal muscle fibres 201, 204–5
intramembranous
    ossification 157
intraocular pressure 241–42
intrapleural fluid 439
intrapleural pressure
    443–444
intrathecal injection 728, 747

intrathoracic pressure 345,
 443–444
intravenous injection
 drug administration
  727–28, 729
 thiopentone
  distribution 731
intravesical pressure 486
intrinsic factor 509, 523, 524
intrinsic membrane
 proteins 43
inulin clearance 474, 476
involution 747
iodide trapping 298
iodine 298, 301
 deficiency 301
 nutritional
  requirement 298, 677
 thyroid hormones 298
ion(s) 17–8, 389–390, 747
ion channel(s) 56–7
 ligand gated 62–3
 naming of 57
 regulation of 62–3
 transduction 254
 voltage gated 62–3
ionization 20, 747
 of drugs 514, 729
 of weak acids and bases 21
IP$_3$ (inositol 1,4,5,
 trisphosphate) 90
Ipratropium 462
ipsp (inhibitory post-synaptic
 potential) 114
iris (eye) 239, 242
 pupillary reflexes 242
iron
 absorption 395
 anaemia 401
 dietary 394–395, 677
 ferritin 394, 395
 haemoglobin 394
 metabolism 394–395, 542
 nutritional
  requirements 394–395,
  627, 677
 regulation of 395
 red cells 394
 storage 394, 395
 toxicity 395
 transferrin 394, 395
 transport 395
irregular bones 141
ischaemia 236, 378, 747
ischium 142, 151–52
islets of Langerhans 317,
 547, 554

autonomic regulation
 317, 318
cells of 317, 547–548, 554
histology 317, 548, 554
innervation 549, 554
isoelectric line (ECG) 358
isoflurane 732
isometric contraction
 127, 747
isoniazid 463
isosorbide mono- and
 di-nitrates 350
isotonic fluid/solution 23
isotonic contraction 127, 747
isotopes 15, 747
isoretinoin 571
isovolumetric contraction
 (ventricle) 339
isovolumetric relaxation
 (ventricle) 340
itch/itching 238, 539

J

Jacksonian epilepsy 209
jargon 268
jaundice 539, 544, 557
jaw 145, 146, 172
jejunum 494, 515, 516,
 521, 523
joint(s) 163–67
 ball-and-socket 165
 capsule 164
 cartilagenous 163
 classification 163
 condyloid 166
 disorders 166–67
 fibrous 145, 163
 fixed 163
 gliding 165
 hinge 165
 pivot 165–66
 saddle 166
 synovial 163, 165–66
 see also specific joints
joule 683, 739
jugular pulse 339, 340
jugular vein 367–368, 369
junction, cellular 77
junctional rhythms 361
juxtaglomerular
 apparatus 469, 470,
 480–481
juxtaglomerular cells 469,
 470, 480–481
juxtamedullary
 nephrons 468, 471

K

kaolin 530
Kartagener's syndrome 53
karyotype 647, 649,
 650, 747
keratinisation 558
keratinocytes 558
ketamine 728
ketoacidosis 322, 488,
 720, 747
ketoconazole 308
α-ketoglutarate 50
ketone bodies 51
 formation 321, 322
 utilization 321
ketonuria 322, 488
kidney 466–492
 acid-base balance
  481–482, 720, 722–23
 aldosterone 480–481
 ANP 324
 anatomy 467, 468, 516
 autonomic innervation 220
 antidiuretic hormone
  (ADH) 483–84
 blood flow
  autoregulation 472
 blood supply 471–472
 Bowman's capsule 467,
  468, 469,
 calcitriol 466
 collecting ducts 468, 470,
  471, 473–74, 481–82,
  483–85
 cortex 467, 468, 470
 disorders 347, 385, 716,
  720, 751
 distal tubule 468, 469,
  470, 473, 474, 480–482,
  483–484
 diuretics 386, 490–491, 718
 erythropoietin 393, 466
 excretion 466
 failure 347, 751
 glomerular filtration
  473–475, 476
 gluconeogenesis 466
 hilus 467, 468
 histology 470
 loop of Henle 468, 470, 471,
  473–474, 479–480, 490
 medulla 467, 468, 470
 nephron 467, 468, 470, 471,
  473–485
 osmoregulation 483–484,
  709, 711

proximal tubule 467, 468,
 469, 470, 471, 473–474,
 477–479
regulation of internal
 environment 466, 712–14
tubular transport 475–485
water reabsorption
 473–474, 483–484, 711
urine 487–489
kilocalories 684
kilojoule 684
kilopascal 740
kinaesthesia 231, 747
kinetochore 649
knee-jerk reflex 202
knee 164, 175
Korsakoff's psychosis 675
Krebs cycle, *see tricarboxylic
 acid cycle*
Kupffer cells 417, 535, 536
kwashiorkor 672, 716
kyphosis 148, 747

L

labia majora/minora 591
labyrinth 252, 257, 747
lachrymal/lacrimal
 glands 241
lactalbumin 641
lactase deficiency 524
lactate 48, 124, 544, 690–91
lactation 638–44
 breast milk
  composition 641
 colostrum 640
 hormonal regulation 640,
  642–44
 mammary gland
  development 639
 milk ejection reflex 643–44
 milk production 640
 nutritional requirements
  of 641–42
 suckling 643–44
lacteals 516, 518, 523
lactic acid *see lactate*
lactiferous ducts 639
lactogenesis 640
lactose 529
 intolerance 524
 structure 29
 synthesis of 641
lactulose 530
lamellae
 bone 155, 156
 myelin 106

lamina propria 495, 496
laminar flow 446, 457
Langerhans cells 559, 560
Lansoprazole 513
Large intestine 525–530
  absorption from 525, 527
  anatomy 494, 525, 526, 527
  bacterial flora 526, 527–528
  blood supply 497–498,
    526, 527
  haustra 526, 528
  histology 526
  innervation of 526, 528, 529
  motility of 528–530
Laron dwarfism 295
larynx 434
latent learning 277
lateral cerebral fissure 187
lateral geniculate body *see*
  visual pathways
law of Mass Action 96
laxatives 530
learning and memory 277–281
  cellular mechanisms of 277
  long term
    potentiation 277, 279
Leber hereditary optic
  neuropathy 664
leg
  blood vessels 366, 367,
    368, 369
  bones 151–153
  muscles 175–79
length-tension relationship
  cardiac muscle 343–45
  skeletal muscle 123,
    127, 128
lens (eye) 239, 241, 243
leprosy 678
leptin 324
lesser curvature
  (stomach) 506, 507
lesser omentum 506, 534
leukaemia 403
leukocytes 69, 391, 392, 393,
  401–403, 488, 489
  differential count 392
  disorders 403
leukopenia 403
leukopoiesis (white cell
  production) 391, 393
leukotrienes 91
levodopa (L-DOPA) 214,
  730, 732
Leydig cells 579–81, 587–88
LH *see luteinizing hormone*
LH surge 596, 602

lidocaine (lignocaine) 112,
  351, 363, 727
ligaments 164–65
ligand binding 96
ligand gated ion
  channel 62–3, 86–7
  see also agonist, antagonist
lignocaine, *see lidocaine*
limb leads, *see ECG*
linoleic acid 31, 672
linolenic acid 31, 672
lipase 519, 520, 550, 551
lipid(s) 30–2, 747
  absorption 523
  bilayer 41–2
  cholesterol 31, 33
  digestion 519, 520
  eicosanoids 85, 91, 745
  fats 672–673
  fatty acids 30, 31
  glycerides 31, 523
  glycolipids 31
  phospholipids 31, 32
  steroids 31, 33
  structural 31, 32
  structure of 30–33
lipolysis 50, 747
  cortisol 305, 319
  catecholamines 320
  glucagon 318–19
  growth hormone 320
liver 494, 498, 533–545
  anatomy 533–536
  bile production 537
  blood supply 497, 498,
    536–537
  cirrhosis of 544–545
  drug metabolism
    543–544, 545
  enterohepatic
    circulation 537
  enzymes 544
  failure 544–545
  functions (table) 532, 534
  function tests 544
  gap juctions in 536
  glucose regulation
    540–541, 544
  hormones 541–542
  histology 534–536
  jaundice 539, 544
  lobules 534, 535, 536
  metabolism 540–541, 542
  plasma protein
    synthesis 408, 541, 544
  see also bile, gall bladder,
    gastrointestinal tract

local anaesthetics
  administration of 727
  epidural anaesthesia 728
  mechanism of action 111
  spinal anaesthesia 728
local hormones 81, 285, 378
locus coeruleus 274
Lomo, T. 277
long bones 141
longitudinal cerebral
  fissure 185
long-term potentiation 277
loop diuretics 490, 491, 718
loop of Henle 468, 470, 471,
  473–474, 479–480, 490
  loop diuretics 490, 491, 718
  osmotic gradient 479,
    480, 483
  structure of 470
  tubular transport 479–480
  and water balance 483
loperamide 530
loratadine 238
lordosis 148, 747
losartan 386
lower oesophageal (cardiac)
  sphincter 504, 506, 507
low pressure receptors 712
lower motoneuron
  lesion 205–6
lumbar puncture 192, 747
lung(s)
  anatomy 434–435
  airways 434–438
  blood supply 369, 438
  compliance 443–444, 448
  diseases 459–462
  elasticity 445–446
  fetal 630
  fibrosis 451, 459, 460
  histology 435–438
  inflation at birth 631–32
  intrathoracic pressure 345,
    443–444
  pulmonary circulation
    369, 438
  reflexes 453, 456–457
  surfactant 438, 445, 631
  volumes 441, 442, 443
  see also breathing,
    respiratory system
lupus erythematosus 427
luteal phase (menstrual
  cycle) 594, 597–98
luteinising hormone (LH,
  ICSH)
  LH surge 596, 602

negative feedback
  regulation 588, 602–3
  ovarian cycle 594–98, 602–3
  spermatogenesis 587–88
  see also gonadotrophins,
    reproductive system
luteinization 598
luteolysis 598–99
lymph 382–384, 523, 747
  afferent 382, 383, 384, 420,
    421, 422, 715
  cells in 383
  circulation 383–384
  composition 383
  efferent 383, 384, 421, 422
  nodes (glands) 383, 384,
    420, 421–422, 747
lymphatic system 382–384,
  420, 421–422, 748
lymphatic vessels/
  lymphatics 383–384, 420,
  421–422
lymphoedema 383, 716
lymphocytes 69, 383, 392,
  402, 403, 419–420,
  421–422, 423–426
  circulation of 420
lymphoid cells 391–393
lymphoid tissue 419–420
lymphoma 716
lysis (blood clot) 404, 406, 407
lysosomes 44
lysozyme 241, 415, 503

M

M cells 426
macrocytic anaemia 401
macronutrients 671–73
macrophages 402–403, 417,
  423, 425, 457, 561, 568, 747
mafenide acetate 570
magnesium hydroxide 95,
  513, 727
major histocompatability
  complex (MHC) 416,
  423, 424
malabsorptive states 524
malaria 401
male reproductive
  system 577–588
  anatomy of 577–80
  hormonal
    regulation 581–82
  puberty 605
  spermatogenesis 582–88
malleus (hammer) 252

malnutrition (PEM) 672
maltose 520
mammary gland
  anatomy 639, 640
  development of 639
  milk production 640–42
  milk ejection 643
  in pregnancy 640
  involution 644
mandible 142, 144, 146
mannitol 490, 491, 709, 718
marasmus 672
Marfan syndrome 161
mass movements 528
mass spectrometry 285
mast cells 81, 402, 417,
  427, 561
mastication (chewing)
  501, 502
mastoid process 144
maternal physiology 621–625
  body fluid balance 623
  cardiovascular
    changes 621–22
  endocrine changes 623–24
  gastrointestinal
    changes 623
  gestational diabetes 624
  lactation 638–44
  nutrition 625–27
  in pregnacy 621–25
  renal function 623
  respiratory changes 622–23
maxilla 144–45
maximal aerobic power 690
maximal (peak) expiratory
  flow rate 458–459
maximal stimulus 110
maximal ventilatory
  volume 459
maximum breathing
  capacity 459
MC-1 receptor 559
mean arterial blood pressure
  (MAP) 372–373, 376
mebendazole 530
mecamylamine 222
mechanoreceptor 227, 748
meconium 636
median eminence 288
mediastinum 439
medulla oblongata 186, 187,
  452, 515
megakaryocytes 391–393, 404
megaloblastic anaemia 401
meiosis 45, 576, 582–86,
  647, 748

Meissner's plexus
  (submucosal plexus) 221
melanin 559, 562
melanocytes 559
melanocyte-stimulating
  hormone (MSH) 559, 624
melanoma 560
melasma 624
melatonin 276
memantine (Ebixa) 278
membrane potential 56–7,
  62, 133, 748
  see also action potential,
  synaptic transmission
memory 278–281
  Alzheimer's disease 278
  declarative 280
  episodic 280
  immediate 279
  Korsakoff's syndrome 675
  long-term 280
  neural basis 280
  procedural 280
  retrieval 281
  semantic 280
  short-term 279–280
  temporal lobe 280–81
memory cells (immune
  system) 420, 421, 424
menarche 604, 748
Mendelian
  inheritance 656–58
Menières disease 258, 514, 515
meninges 185, 192–93, 748
meningitis 193, 249
menopause 604–5, 748
menstrual cycle 593–603
mepyramine 136
Merkel cells 559, 560
Merkel disc 229, 565
mesangeal cells 467, 469
mesentery 495, 527, 748
mesovarium 589
messenger RNA (mRNA)
  652, 653
metabolic acidosis 720
  compensation 723
metabolic alkalosis 721, 724
  causes of 515
metabolic hyperaemia 334,
  377–378
metabolic hypothesis
  (autoregulation) 472–473
metabolic rate 684
  basal 685, 686–87
  calculation of 686
  factors affecting 686–88

metabolism 18, 682, 748
  anabolic 682
  catabolic 682
  energy 683
metacarpal bones 141, 150
metaclopramide 352, 515, 530
metatarsal bones 141, 153
metformin 323
methadone 237
methotrexate 167, 180, 518
metolazone 490
MHC complex 416, 417, 423,
  424, 425, 426, 427
micelles 523, 537
microcirculation 382–384, 748
microcytic anaemia 401
microglia 184–85, 417
micronutrients 673–77
  in pregnancy 625–27
  in lactation 641–42
microtubules 44
microvilli (brush
  border) 516, 517
micturition 467, 485–486,
  623, 748
midbrain 187
middle ear 251–52
milk (human)
  composition of 641
  ejection (let-down) 643–44
  production 640
mineralocorticoid 304, 748
minerals, dietary 676–77
minute volume 447
mitochondria 43
mitochondrial
  inheritance 664–65
mitosis 45–47, 647, 748
mitral (bicuspid) valve 329
mixed nerves 191, 748
mixed venous blood 399
molarity 18–9, 748
mole (chemistry) 17, 748
molecule 16, 748
monoamine oxidase 116, 309
  inhibitors 275
monocytes 392, 402, 417
monoglycerides 31, 519,
  520, 748
  absorption of 523
monosaccharides 27, 28, 520
  absorption of 521–522
monosynaptic reflex 202
Montgomery glands 639
morphine 237, 352, 463, 528,
  530, 539, 734, 737
morula 613, 748

motion sickness 258, 514
motoneuron 105, 124, 199, 748
  alpha-gamma
    co-activation 204
  gamma–
    amotoneurons 201
motor cortex
  connections of 210
  effect of lesions 206, 208
  functions of 209
  somatotopic
    organization 209
  stroke 210
motor nerves 124, 130
motor nuclei
  cranial nerves 208
  spinal 199
motor pathways 198, 206–8
  corticospinal tract 208
  extrapyramidal tracts 207
motor system 198
motor unit 124, 198, 748
mouth 501
movement
  control of 198
  disorders
    (dyskinaesias) 214
  goal-directed 208–11
MSRA (methicillin resistant
  staphlococcus
  aureus) 570
MSH (melanocyte
  stimulating
  hormone) 559, 624
muco-ciliary escalator 415,
  437, 457
mucosa 495, 748
  caecum 527
  colon 528
  gastric 507–8,
  small intestine 516, 521
mucosa associated lymphoid
  tissue (MALT) 420, 426
mucosal strengtheners 513
mucous membranes 238,
  396, 408, 461, 545
  of mouth 415, 501, 502, 503
  colon 526
  and MSH 559
mucous neck cells 508,
  510–511
mucus 415, 457, 507, 508,
  510–511, 525
Mullerian ducts 636
Mullerian Inhibiting
  Hormone (MIH) 638
murmurs 339, 748

muscarine 222
muscarinic receptors 222, 503
muscle 72–3, 118–35
  blood supply 73
  insertion of 170
  naming rules 170
  paralysis of 130
  origin of 170
  types of 72–3, 119
muscle contraction
  ATP and 122–24
  calcium 121–22
  efficiency of 127–28
  isometric 127
  isotonic 127
  skeletal muscle 121–24
  sliding filament
    theory 122–23
  smooth muscle 133–34
  tetanus 125–26
  twitch tension 125
muscle fatigue 124
muscle fibres 120
muscle spindles 200–201
muscle twitch 125
muscles
  aconeus 176
  adductor brevis 178
  adductor longus 178
  adductor magnus 169,
    178, 179
  alaeque nasi 171
  arrector pili 557, 561–62
  brachialis 171, 177
  brachioradialis 176, 177
  biceps brachii 168, 170,
    171, 177
  biceps femoris 169, 175, 179
  buccinator 171, 172
  coracobrachialis 171, 177
  deltoid 168, 169, 171, 173,
    175, 177
  depressor anguli oris 171
  depressor labii
    inferioris 171
  diaphragm 439, 440, 442
  erector spinae 175
  extensor carpi radialis
    brevis 176
  extensor carpi radialis
    longus 176
  extensor carpi ulnaris 176
  extensor digitorum 176
  extensor digitorum
    brevis 178
  extensor digitorum
    longus 176, 178

extensor digiti minimi 176
extensor halucis longus 178
external oblique 168,
  174, 439
extrafusal fibres 201
extraocular 248–49
flexor carpi radialis 177
flexor carpi ulnaris 177
flexor digitorum longus 179
flexor digitorum
  superficialis 177
frontalis 168, 171, 172
gracilis 169, 176
gastrocnemius 168, 169,
  177, 178, 179
gluteus maximus 169,
  175, 179
gluteus medius 175, 179
gracilis 179
hamstrings 169, 175, 179
hypothenars 177
iliacus 175, 178
iliopsoas 175
infraspinatus 165, 169, 173
intercostals 174, 439, 442
internal oblique 174
intrafusal fibre 201, 204–5
latissimus dorsi 169, 171,
  173, 174
levator anguli oris 171
levator ani 527, 528, 592
levator labii superioris 171
levator scapulae 173
masseter 168, 171, 172
nasalis 171
occipitalis 169, 172
orbicularis oculi 168,
  171, 172
orbicularis oris 168,
  171, 172
palmaris longus 177
pectineus 178
pectoralis major 168, 171,
  173, 174
peroneus brevis 169, 176,
  177, 178, 179
peroneus longus 168, 169,
  176, 177, 178, 179
plantaris 177, 179
platysma 171, 172
pronator teres 177
psoas major 175, 178
puborectalis 527, 528
quadriceps femoris 176
quadratus lumbricum 175
rectus abdominis 168,
  174, 439

rectus femoris 168,
  176, 178
rhomboid major 169, 173
risoris 171
sartorius 168, 175, 176,
  178, 179
scalene 439
semimembranosus 169,
  175, 179
semitendinosus 169,
  175, 179
serratus anterior 168, 173
soleus 168 169, 177,
  178, 179
sternocleidomastoid 168,
  169, 172, 439
subcapularis 165
supraspinatus 165, 173
temporalis 168, 172
tensor fascia latae 178
teres major 169, 171
teres minor 165, 169
tibialis anterior 168, 176,
  177, 178
trapezius 168, 169, 173, 174
traversus abdominus 174
triceps 169, 171, 173, 176
vastus lateralis 168, 176,
  178, 179
vastus medialis 168,
  176, 178
zygomaticus major 171, 172
zygomaticus minor 171
muscular arteries 371
muscular dystrophy 180
muscularis externa 495
  of stomach 507–508
  of large intestine 526
muscularis mucosa 495, 496
musculoskeletal
  system 140–80
  anatomy 141–153, 167–179
  bone disorders 162
  bone growth 157–58
  bone fractures 159–60
  bone histology 155–157
  bone physiology 158–60
  joint classification 163, 165
  principal muscle
    groups 169–77
  mutations 653, 658, 748
myasthenia gravis 131, 180
myelin 105–6
myeloid cells 391–393
myenteric plexus 221, 495,
  496, 499
myocardial contractility 345

myocardial infarction
  347, 406
  ECG changes 362
  oxygen therapy 352
myocardial ischaemia
  348, 362
myocardium 329,
  330–31, 748
myocyte 72, 119,
  131–32, 748
  cardiac 329
myoepithelial cells 132,
  639, 640
myoepithelium 131, 748
myofibril 120, 748
myofilament 121, 748
myogenic contraction
  (autoregulation) 119,
  132, 472, 748
myoglobin 126
myometrium 590, 748
  excitability of 600
myopia 243–44
myosin 52, 120–23
myotactic reflex *see stretch
  reflex*
myxoedema 301–302, 748

N

N-CAM 77
nails 562
naloxone 237
naproxen 429
narcolepsy 274, 275
nasal bone 144–45
natriuresis 718, 748
natural (innate) immune
  system 416–419
  cells of 417
  proteins of 417–419
natural killer (NK) cells 417
nausea 514
near point 243, 748
negative feedback 7, 286,
  307, 312, 587–88,
  602–3, 745
negative inotropic
  agents 346
neglect syndrome 265
neologisms 268
neomycin 255
neonatal physiology
  632–36, 703
neostigmine 131, 180
nephron *see kidney*
Nernst equation 57, 62

nerve
  action potential 106–11
  afferent 105
  efferent 105
  refractory period 107, 111
  terminal 104, 105,
    112–13, 129
  threshold 106
nerve cell, *see neurone*
nerves
  abducens 188, 189
  facial 187–89
  glossopharyngeal
    187–89, 454
  hypogastric 485
  hypoglossal 188–89
  intercostal 440
  oculomotor 187–89,
    219, 249
  olfactory 188–89, 259
  optic 188–89, 239–40, 245
  pelvic 485
  phrenic 440
  pudendal 485
  spinal 189, 217
  spinal accessory 188–89
  splanchnic 217–18
  trigeminal 188–89, 502
  trochlear 188–89
  vagus 187–89, 218–19, 454,
    499–500
  vestibulo-cochlear 188–89,
    251, 255, 256
nervous system
  action potentials 106–111
  anatomy 183–95
  autonomic division 183
  cells of 71–2, 104–5, 184–85
  central 184–90
  divisions of 183
  enteric 183, 221, 499
  glial cells 184–85
  motor 197–214
  organization 183
  parasympathetic 217–19
  peripheral 183
  sensory 226–60
  spinal cord 185, 188–90,
    199, 205–7, 217–19,
    232–33
  spinal nerves 189–90
  sympathetic 217–19
  synapses 112–116
nervous tissue 71
net filtration pressure 383
neuralgia 236
neurocranium 143

neuroendocrine reflex
  643, 748
neurogenic contraction
  119, 748
neurogenic shock (spinal
  shock) 205
neuroglia 104, 184–5
neurohypophysis *see pituitary
  gland: neural lobe*
neuroma 236
neuromuscular block 130
neuromuscular
    junction 129–131
  drugs affecting 130
neurone 71, 104–5, 184, 748
  action potential 106–11
  axon 71, 104–6
  dendrite 71, 104
  membrane potential 107
  structure 71–2, 104–5
  *see also action potential,
    synaptic transmission*
neuropil 71, 104
neurotensin 136
neurotransmitter(s) 83,
    113, 748
  actions of 113–14, 129–30
neutropenia 403, 748
neutrophils 392, 402, 417, 748
niacin (vitamin B₃) 674, 675
nicorandil 350, 351
nicotine 221, 511
nicotine patch 572, 727
nicotinic receptors 86, 221–22
nifedipine 350, 351, 386
Nissl substance 71
nitric oxide 85, 91–2, 115,
    378, 440
  in autoregulation 378
  role in erection 608
nitrites 488, 489
nitroglycerine (GTN) 92,
    572, 727
nitrous oxide 732
nizaditine 513
nociceptors 203, 566, 749
node of Ranvier 105
non-competitive
    antagonism 97–8
non-disjunction 584, 649, 749
non-shivering
    thermogenesis 701, 703
non-steroidal anti-
    inflammatory drugs
    (NSAIDs) 162, 167, 237,
    429, 511, 705
  aspirin 237, 429, 511, 705

ibuprofen 237, 429, 511
  paracetamol 167, 429, 705
non-tropical sprue 524
noradrenaline
  adrenal medulla 223,
    309, 310
  autonomic nervous
    system 221, 223
  blood vessel control
    380–381, 472
  energy balance and
    metabolism 320, 688
  fetal and neonatal
    physiology 634, 703
  heart and circulation
    342, 343
  in sleep 274
  kidneys and 472
  as neurotransmitter 115,
    116, 221, 223
  in respiratory system 440
norepinephrine, *see
    noradrenaline*
nose 145, 434, 436
nucleic acids 35–6
nucleolus 43, 749
nucleotide coenzymes 36
nucleotides 35–36, 749
nucleus (of cell) 43, 646,749
nucleus ambiguus 208
nutrients 671–77
nutritional
    requirements 625–27,
    672, 674
nutritional status, assessment
    of 679–80
nutritional support 678
nystagmus 248, 749

O

O blood group 410, 411, 659
obesity 678
obligatory reabsorption
    (nephron) 474, 479
observational learning 277
obstructive jaundice 539, 553
occipital bone 142–44
occipital lobe (cerebral
    hemisphere) 185, 187
oculomotor nerve 187–89
oculomotor nuclei 245
oedema 749
  causes of 489 715–17
  diuretic treatment 490, 718
oesophagus 494, 504, 505, 506
oestradiol 17-β 33, 595, 596

oestrogen(s) 31, 85, 595, 749
  feedback effects 602–3
  hormone replacement
    therapy 727
  metabolic effects 688
  non-reproductive
    effects 601
  receptor 93
  regulation of plasma
    calcium 315
  secretion by placenta 619
  surge 595–96, 602–3
olecranon fossa
    (humerus) 150
olecranon process (ulna) 150
olfaction 259, 749
oligodendrocyte 105, 184–85
omentum 506, 507, 527, 534
omeprazole 505, 510, 513,
    553, 727
oncotic pressure 22–23, 382,
    383, 390, 716, 749
ondansetron 515
oocyte
  maturation 594–95
  primary 593
  secondary 597
oogonia 593
opiod drugs 237
opsonisation 418, 749
optic chiasm 245
optic disk (papilla) 240
optic nerve 188–89, 239, 245
optic radiation 245
oral cavity 494, 501, 749
oral contraceptives 384, 609
oral drug administration 727,
    730–31
oral rehydration therapy
    530, 715
orbit 238
organ 5, 68, 749
  systems 6
  transplantation 428–429
organelle 40, 43–45, 749
orgasm 608
orlistat 678
ornithine 543
orthodromic
    propagation 109, 749
osmolality
  definition of 22, 749
  regulation of 483–485
  and tonicity 23
osmolarity 22, 749
osmoreceptors 290, 483, 749
osmosis 22, 749

osmotic diuresis 475
osmotic diuretics 490, 491, 718
osmotic gradient (renal medulla) 479–480
formation of (loop of Henle) 479–480
and water balance 483–84
osmotic laxatives 530
osmotic pressure 22, 749
of biological fluids 22
of proteins 22–23, 716
ossa coxae (hip bones) 151
ossicles 252
ossification 157, 293, 312, 749
osteoarthritis 166, 167
osteoblasts 154, 156–57, 293
osteoclasts 154, 157, 293
osteocytes 154, 157, 293
osteoid 154
osteomalacia 312, 673
osteomyelitis 161–62
osteons (Haversian system) 155–56
osteoporosis 161, 315
otitis media 255
otoliths 257
otosclerosis 255
ouabain 351, 352
outer ear 251
ova, *see eggs*
oval window 251
ovarian cycle 594–99
ovarian follicle, maturation of 592–97
ovarian hormones/steroids 595, 599–600, 602–3
ovary/ovaries 589, 592–93
development of 636–38
oviducts, (Fallopian tubes) 590
ovulation 596–97
oxidative phosphorylation 48
oxitropium 462
oxygen carriage 396–398
oxygen consumption
in exercise 690–91
at rest 686
oxygen content of blood 397
oxygen dissociation curve 396, 397, 398, 455
oxygen solubility in blood 396
oxyhaemoglobin 396, 397, 398, 455
oxyntic (parietal) cells 508, 509, 510, 512
oxyphil cells (of parathyroid) 312

oxytocin 289–291
in suckling 291, 643–44
role in labour 291, 620
role in male 291

P

P wave 338, 339, 358, 359, 361
p-aminohippurate (PAH) 475, 476, 477
pacemaker cells 334–335
pacemaker potential 334–335, 343
Pacinian corpuscle 227, 229, 230, 565–66
Paget's disease 162, 314–5
pain 234–37
analgesics 237
central pathways 235
classification of 234
phantom limb 236
producing-substances 235, 566
projected 236
referred 235–36
triple response 567
TENS 235
visceral 235–36
palate 501, 504, 505
pancreas 494, 516, 547–555
autonomic innervation 549
blood supply 497, 498, 549
carcinoma of 553
enzymes of 519, 520, 550–551
structure of 547–549
pancreatic amylase 519, 520, 550, 551
in pancreatitis 553
pancreatic duct 516, 534, 535, 548, 549
pancreatic enzymes 519, 550–551, 552
pancreatic hormones 554–555
pancreatic juice 550–551, 552
pancreatic lipase 519, 520, 550, 551
pancreatic polypeptide 554, 555
pancreatic secretion 550–551
regulation of 551–553
pancreatin 553
pancreatitis 553
pancytopenia 401
Paneth cells 517
papilla (optic disk) 240

papillae (of tongue) 259–60
papillary muscles 329
paracetamol 237, 705, 707
metabolism of 544, 736
paracortex (lymph node) 421, 422
paracrine signalling 81, 285, 423, 749
paradoxical sleep (REM sleep) 270, 273, 274
paraesthesia 236, 313
paralysis 130, 205, 749
paranasal sinuses 143–146
paraplegia 205, 749
parasympathetic ganglia 219
parasympathetic nervous system 219, 749
cranial outflow 187, 219
digestive system 499–500, 503, 518, 521, 526, 528, 529, 538, 549, 552
heart and circulation 341–343
penile erection 608
pupil diameter 242
respiratory system 440
sacral outflow 219, 500, 526
synaptic transmission 222
parathyroid gland 312–313
chief cells 312
oxyphil cells 312
parathyroid hormone 312–313, 482
actions of 312–313, 482
disorders of 313
effect of hypercalcaemia 312, 313
in pregnancy 624
paraventricular nuclei 287, 290
parenchyma (lung) 438, 444
parenteral nutrition 678
parietal bone1 42–44
parietal (oxyntic) cells 508, 509, 510, 512
parietal lobe 185, 187, 264, 265
parietal peritoneum 495, 527
parietal pleura 439, 440
Parkinson's disease 214
parotid gland 494, 502, 503
partial pressure(s) 749
blood gases 399
Dalton's Law of 432
of respiratory gases 434
parturition 620, 749
pascal (unit of pressure) 740

passive tension 127–28
passive transport 56
patella 141, 152–53, 164, 168, 178
patella bursa 164
patellar reflex (knee jerk) 202
patellar ligament 168
PCO$_2$ (partial pressure of carbon dioxide)
alveolar 434
arterial blood 399
expired air 434
mixed venous blood 399
peak expiratory flow rate (PEFR) 458–459
pectoral girdle 149
pellagra 675–76
pelvic girdle 151–52
Penfield W. 209, 233
penicillin, excretion of 733
penis 577–78
erection 608
pepsinogen 509
pepsins 509
peptic ulcer *see gastric ulcer*
peptidases 509, 519
peptides 33, 749
absorption 522
digestion 509, 519
hormones 285
peptide bond 33–4
perfusion pressure 375
pericardium 329, 435, 535, 749
pericyte 382
perilymph 252, 749
perimetrium 590
perimetry 244
perimysium 120
perineal body 592
perineurium 191
periodontal ligament 501
periosteum 155
peripheral nerve
compound action potential 110–11
local anaesthesia 111–12
structure of 191
peripheral arterial chemoreceptors 452, 453–454
peripheral nervous system 183, 749
peristalsis 485, 496, 497, 504, 513, 518
peritoneum 495, 506, 507, 525, 527

peritonitis 527

peritubular capillaries 471

pernicious (macrocytic) anaemia 509

pethidine 237, 528

petit mal 271, 272

Peyer's patches 415, 420, 426, 516, 517

pH
  blood 389, 390, 719–24
  definition of 21
  gastric juice 509
  pancreatic juice 550
  scale 21

phaeochromocytoma 310, 385

phagocytosis 417, 425

phalanges 141, 150–51, 153

phantom limb pain 236

pharmacodynamics 95, 726, 749

pharmacogenetics 736, 749

pharmacokinetics 95, 726, 750
  effects of age 736

pharynx 434, 494, 501, 504, 505, 750

phenazocine 237

phenelzine 275

phenothiazines 489

phenotype 654, 750

phenoxybenzamine 310

phenylbutazone 237

phenylketonuria 658

phenytoin 271, 290

phonocardiogram 339

phosphate
  balance 311
  in bone 155, 157
  hormonal regulation 311, 482
  in plasma 390

phosphocreatine, *see creatine phosphate*

phospholipase A_2 551

phospholipase C 90, 91

phospholipids 31, 42, 539

photopic vision 244, 750

phototransduction 245–46

phrenic nerve 440

pia mater 185, 193

pilocarpine 223, 242, 504

piloerector muscles 557, 561–62

pineal gland 276

pinna 251

pinocytosis 65, 750

pioglitazone 323

piroxicam 237

pitch discrimination 253–55

pitting oedema 717

pituitary dwarfism 295

pituitary gigantism 295

pituitary gland 286–296
  anatomy 286, 288
  anterior lobe 288–289, 291–296
  feedback regulation 286
  histology 288
  hormones of 286–287
  hypothalamic control 287, 288–289
  intermediate lobe 286
  posterior (neural) lobe 289–291
  see also specific hormones

pKa (acid dissociation constant)
  of drugs 730

placenta
  blood supply 615–16
  definitive 615
  endocrine functions 618–19
  exchange 616–18
  formation of (placentation) 613–15
  oestrogenic hormones 619
  progesterone secretion 619

plasma 69, 389–391, 750
  composition of 192, 389–390
  pH 389, 390, 719–24
  proteins 383, 389–390, 477, 479, 541, 544

plasma calcium 389, 390
  regulation of 310–315

plasma glucose 315, 320–321
  hormonal regulation 294, 315–320
  hyperglycaemia 321–323
  hypoglycaemia 323–324

plasma membrane 55–66, 750
  endocytosis 65
  exocytosis 64
  ion channels 57, 62–64
  permeability 56
  structure of 41–3
  transporters 57–9

plasma osmolality 389
  renal regulation 483–484

plasma proteins 383, 750
  oncotic pressure 716–17
  synthesis of 541

plasma volume 389
  changes in pregnancy 621
  measurement of 708

plasmin 406

plasminogen 406

Plasmodium 401, 416

plateau phase (cardiac action potential) 335–336

platelet plug 405

platelets (thrombocytes) 389, 391, 392, 393, 404–409
  deficiency of (thrombocytopenia) 407–408
  von Willebrand's disease 408

pleura (pleural membranes) 439, 440

pleural effusion 451–452, 717

pneumonia 460

pneumotachograph 458

pneumothorax 443

PO_2 (partial pressure of oxygen) 432, 434
  alveolar 434
  arterial blood 399
  in atmosphere 434
  in exercise 692–93
  expired air 434
  mixed venous blood 399

podocytes 467, 469

polar molecules 18–9

polio
  vaccine 426
  virus 426

polycythaemia 400

polydipsia 322

polygenic inheritance 656, 661

polymorphonuclear leukocytes (granulocytes) 401, 402

polymyositis 180

polyol pathway 316

polypeptides 33

polysaccharide 28, 750

polyspermy, prevention of 611, 750

polysynaptic reflex 202

polyuria 321

pons 186–87, 188, 452, 453

portal hypertension 545

portal triad 534, 535, 536

portal vein 498, 516, 522, 535, 536

positive feedback 8, 286, 745
  of ovarian steroids 602–3

positive inotropic effect 345–346

post-absorptive state 320–321

post-capillary (pericytic) venules 371, 382

post-central gyrus (cerebral hemispheres) 187

post-ganglionic fibres
  in gut 499–500
  parasympathetic 219
  sympathetic 217

post-operative hypothermia 704

postural hypotension 375

posture, maintenance of 197, 208, 203, 258

potassium/potassium ion
  in body fluids 56
  equilibrium potential 62
  kidneys and internal environment 479–480, 481
  loss (hypokalaemia) 509
  Nernst equation 57
  plasma concentration 390
  potassium channels 57, 108
  potassium channel antagonists 351
  resting membrane potential 57
  sodium pump 59

potassium-sparing diuretics 490, 491

prazosin 223

prednisolone 167, 180, 429, 727

pre-eclampsia 385, 489, 622

preganglionic fibres
  in gut 499–500
  parasympathetic 219
  sympathetic 217

pregnancy
  blood pressure 385, 622
  blood volume 621
  cardiac output 621
  clotting factors 622
  endocrine function 623–24
  gastrointestinal function 623
  lower back pain 621
  metabolic changes 624–25
  nausea 514, 623
  nutritional requirements 625
  renal function 623
  respiratory changes 622
  test 285
  weight gain 623, 625, 626

preload 344, 347, 750
premotor cortex 210, 268
pre-ovulatory follicle 596–97
presbyacusis 255
pressure 750
pressure changes
   in heart 338, 339–341
   in lungs 442–44
pressure-volume loop
   of lungs 444–445
pressure sores 568, 570
presynaptic neuron 113–114
primary afferents *see nerve,*
   afferent
primary genetic
   haemochromatosis 395
primary hypertension
   384–385
primary lymphoid
   tissue 419–420
primary motor cortex 209
primary sensory cortex,
   (somatosensory
   cortex) 234
primary visual cortex 245–46
primitive gonads 636–38
primordial follicles 593
principal cells (P cells) 470,
   480–481, 484
PR interval (ECG) 358, 359
probenecid 167, 491
procainamide 351, 363
prochlorperazine 515, 727
progesterone 31, 33, 594,
   596–98
   effect on uterus 599–600
   fertilization and
      pregnancy 612, 619
   non-reproductive
      tissues 601
   in ovarian cycle 594,
      596–98
   placental secretion 619
projected pain 236
prolactin 286, 288, 289
   milk production 640
   pituitary secretion
      288–89, 643
   regulation of 642
proliferative phase
   menstrual cycle 599–600
   wound repair 568
pronation 10, 750
prone 750
propofol 728
propranolol 95, 223, 310,
   346, 351, 363, 386

proprioceptors 199–201,
   456, 750
propyluracil 301
prostacyclin 404, 406
prostaglandin(s) 511, 750
   effects of 91, 136
   in fever 705
   in pain 235, 237
   PGE_1 91, 705
   PGE_2 85, 91, 136
   $PGF_{2\alpha}$ 136, 620
   PGI_2 (prostacyclin) 91,
      136, 404
prostate gland 577–78
prostate-specific antigen 577
prostatic hyperplasia 577
protein C 406
protein glycosylation 315–316
protein kinases 87–8, 750
protein-energy malnutrition
   (PEM) 672
protein(s) 51, 750
   clotting factors 405,
      406, 407
   digestion of 509, 550–551
   functions of 33
   gluconeogenesis 51–2, 540,
      541, 672
   of immune system
      417–419, 422–423, 425
   liver synthesis 541
   of milk 640–41
   nutritional
      requirements 672
   plasma 389–390
   structure of 33–35
   synthesis 651–53
proteinuria 479, 489, 750
prothrombin 405, 406,
   407, 545
prothrombin time 750
proton pump
   of parietal cell 509, 510, 513
proximal tubule 467, 468, 469,
   470, 471, 473–474, 477–479
   absorption in 473–474,
      477–479
   structure of 467, 470
pruritis (itch) 238, 539
   treatment of 571
psoriasis 558
   treatment of 571
PTH, *see parathyroid hormone*
puberty 604–5
   precocious 308
pubis 142, 151–52
puborectalis sling 527, 528

pudendal nerves 485
pulmonary arteries 329, 330,
   368, 369, 371
pulmonary chemoreflex 456
pulmonary circulation 369,
   438, 448–449
pulmonary defence
   mechanisms 457
pulmonary embolism 407
pulmonary fibrosis 451, 458,
   459, 460
pulmonary gas
   exchange 438, 451
pulmonary oedema 451, 489
pulmonary surfactant
   438, 445
   role at birth 631
pulmonary valve 329–330,
   338–339
pulmonary veins 329, 330, 369
pulmonary ventilation
   446–448, 449–451
pulse
   arterial 372
   oximetry 397
   pressure 372
   venous 339, 340
Punnett square 657, 660, 663
pupil, reflexes of 242
purines 35, 36, 491
Purkinje fibres 335, 336
pursuit movements (of
   eyes) 248
putamen 213, 214
pyloric sphincter 506, 507,
   508, 513, 515
pylorus 506, 507
pyramidal tract (corticospinal
   tract) 208, 210
pyrexia (fever) 704–5, 750
pyridostigmine 180
pyridoxine (vitamin B_6)
   674, 676
pyrogens 704
pyruvate/pyruvic acid
   47–50, 316
   gluconeogenesis 52,
      316, 540
   glycolysis 49
   tricarboxylic acid cycle
      50, 316

Q

Q wave 338, 358
QRS complex 338, 339, 358,
   359, 360, 361, 362

QT interval 359
quadrate lobes (of liver)
   534, 535
quadriplegia
   (tetraplegia) 205, 750
quazepam 275
quinidine 351, 363
quinine 733

R

R wave 356, 358, 359
RQ (respiratory quotient) 686
R-R interval 359
radial artery 366
   of kidney 471
radial pulse 367
radioimmunoassay 285
radius 142, 150
ramelteon 275
rami communicantes
   217–19
ranitidine 505, 513, 727
raphe nuclei 274
rapid eye movement (REM)
   sleep 273–74
RDA (recommended daily
   amount) 670, 672, 674
reactive hyperaemia 378
receptive (Wernicke's)
   aphasia 268
receptive field 227, 229, 750
receptors, cellular 284
   catalytic 87–8
   drug effects 95–9
   G protein linked 88–90
   nuclear 93
   steroid 93
   see also ligand gated ion
      channel
receptor potential 227, 254
receptors, sensory 226–31,
   235, 238
   baroreceptors 379
   chemical senses 250–60
   classification 227
   in skin 229, 565–66
   stretch 200–201
   thermoreceptors 231, 700
   visceral 231–32
   see also eye, ear
reciprocal inhibition 203
recognition molecules 416
recommended daily amount,
   see RDA
rectal administration of
   drugs 727

rectum 494, 526, 527, 528–529
red blood cells (erythrocytes) 389, 391–401, 402, 409–412
  abnormalities 400–401
  acid-base balance 399
  chloride shift 399
  count 391, 392
  differentiation 391, 393
  gas carriage 396–400
  haemoglobin 394, 396–398, 399, 400, 401, 538
  high altitude and 393
  lifespan 393, 394
  pregnancy 621
  reticulocyte 392, 393
  see also anaemia
red nucleus 208, 213
red reaction 567
referred pain 236
reflex
  accommodation 242
  ankle-jerk 202
  arc 197, 750
  baroreceptor 379–380
  chemoreceptor 453–456
  conditioned (associative learning) 277
  crossed extensor 203
  dazzle 242
  defecation 528–529
  enterogastric 514
  flexion 203
  gastro-ileal 518
  Golgi tendon 203
  Hering-Breuer 456
  ileogastric 518
  knee jerk 202
  milk ejection 643
  postural 203
  pulmonary 456
  pupillary 242
  scratch 238
  sexual 608
  stretch 202
  swallowing 504–505
  vago-vagal, 511, 512
  vestibulo-ocular 257
  visceral 224
  vomiting 515
  withdrawal 203
reflex control of respiration 453–457
reflux (oesophageal) 504–505
refractory period
  heart muscle 336
  nerve 107–111

Reissners membrane 252
relative molecular mass 16, 750
relaxin 624
releasing hormones 288–89
REM (rapid eye movement) sleep 270, 273, 274
Reminyl (galantamine) 278
renal artery 467, 468
renal calyces (calyxes) 467, 468, 485
renal circulation 471–472
  autonomic regulation of 379, 471–472
renal clearance 476
renal corpuscle 467, 468, 469
renal failure 347
renal function
  fetal 617–18
  maternal 623
renal threshold 475
renal vein 467, 468
renal pelvis 467, 468
renal pyramids 467, 468,
rennin 381
renin-angiotensin-aldosterone system 381, 385, 480–481
Renshaw cells 199
reproductive system 576–605
  female 589–603
  fertilization 610–12
  gonadotrophins and ovarian steroids 594–95, 602
  male 577–588
  menarche and menopause 604
  menstrual cycle 593, 599–601
  ovarian cycle 592–595
  puberty 604–5
  sexual reflexes 608
  spermatogenesis 582–88
  testicular hormones 581–82
residual volume 441, 442
  establishment of 631–32
  measurement 443
resistance
  airways 444, 446, 458, 459, 460
  vascular 373, 375–376
resolution (inflammation) 418
respiration 431–463
  control of 452–456
  disorders of 459–463

mechanics 441–442
  in neonate 631–32
  see also respiratory system
respiratory acidosis 719, 720, 721
  chronic 722
respiratory alkalosis 720, 721, 724
respiratory epithelium 435–438, 457
respiratory exchange ratio 433, 751
  see also respiratory quotient
respiratory failure 462
respiratory function tests 441–442, 458–459
respiratory muscles 439, 442, 439, 440, 442
respiratory quotient (RQ) 433, 686
respiratory rhythm 452–53
respiratory system 431–463
  accessory muscles 439
  airways 434–438
  anatomy of 434–435
  alveolar ventilation 446–448, 449–451
  bronchial circulation 448–449
  chemoreceptors 452–455
  chest wall 439–440, 443–444
  dead space 446–448
  disorders 459–463
  fetal 630
  first breath 631
  gas exchange 438, 451
  gas laws 432–433
  hypoxia 455, 461–462
  lung volumes 441, 442, 443
  pregnancy 622–23
  pressure changes 443–445,
  pulmonary defence mechanisms 457
  pulmonary circulation 369, 438, 448–449
  reflex control 453–457
  respiratory rhythm 452–453
  see also lung
resting membrane potential 62
resting ("pill-rolling") tremor 214
rete testis 579
reticular formation 273, 515
  role in sleep 273–274
reticulocytes 392, 393

reticuloendothelial system 402
retina
  cellular organization 239–40
  image formation 243
  photoreceptors 239–40, 245
retinal ganglion cells 240
retinol (vitamin A) 673, 674
retrograde amnesia 281
reverse peristalsis 515
Reye's syndrome 237
Rhesus blood group system 410–411
rheumatoid arthritis 166, 167
Rh-factor (D antigen) 410–411, 655
rhodopsin 245
riboflavin (vitamin B_2) 674, 675
ribonucleic acid (RNA) 36, 652–53
ribosomes 44, 652–53
rickets 312, 673, 675
rifampicin 463, 539
rigor mortis 124
ringworm, treatment of 571
risedronate 162
Ritalin 275
rituximab 167
Rivastigmine (Exelon) 278
RNA 36, 652, 653
rods 240
ropivicaine 112
rosiglitazone 323
rotator cuff 165
roughage 528, 671
round window 252
rubrospinal tract 207, 208
rugae (stomach) 506, 507
Ruffini corpuscles/end organs 565–66

S

S wave 358
SA (sinoatrial) node 334, 335, 336, 354
saccades 248, 751
saccule 252, 257–58
sacrum 151
saddle joint 166
sagittal section 9, 751
salbutamol 223, 309, 440, 462, 727
salicylates 543
  excretion of 733

saliva 503, 504, 525
salivary amylase 503, 504
salivary glands 494, 502–503
salivary secretion 502–504
  regulation of 503
salmeterol 462
saltatory conduction 109, 751
Sanger F. 317
sarcolemma 120
sarcomere 120, 331, 751
sarcoplasmic reticulum 44,
  120, 134
satiety 751
satiety centre 678
saturated fats 672
saturated fatty acids 30
scabies, treatment of 571
scala media 252
scala tympani 252
scala vestibuli 252
scapula 142, 149–50
scar tissue 567, 568, 569, 570
Schild equation 99
Schwann cells 105, 106
sclera 238–39, 241, 243
scoliosis 148, 751
scopolamine 503, 572
scotopic vision 244, 751
scratch reflex 238
scrotum 577
scurvy 408, 676
seasonal affective disorder
  (SAD) 276
sebaceous glands 562–63
second messenger 86,
  88–90, 751
second class (incomplete)
  proteins 672
secondary active
  transport 59–61, 477, 751
secondary hypertension 385
secondary lymphoid
  tissue 420
secondary sexual
  characteristics
    male 582, 588, 605
    female 604
secretin 500, 537, 538, 552, 553
secretion
  bile 537, 538
  exocytosis 64–5
  gastric juice 509–513
  local mediators 91
  milk 640
  pancreatic juice 550–553
  salivary 502–504
  steroid 64

see also specific hormones,
  synaptic transmission
secretory phase (menstrual
  cycle) 600
segmentation 496–497,
  518, 751
selegiline 275
selectins 77
self/non-self
  (immunology) 416, 427
self tolerance 427
semen 578, 587, 608
  see also seminal fluid
semicircular canals 252,
  256–57
semilunar valves
  329–330, 383
seminal fluid 587
seminal vesicles 578
seminiferous tubules
  579, 580
senna 530
sensory nerves/
  neurones 105
sensory pathways 232–34
sensory receptors 226–31,
  235, 238
  adaptation 228
  baroreceptors 379
  chemoreceptors 453–54
  classification 227
  mechanoreceptors 230,
    565–560
  nociceptors 235, 238
  osmoreceptors 483, 711
  photoreceptor 239, 240,
    245–46
  proprioceptors
    200–201, 231
  of skin 565–66
  thermoreceptor 230
  touch 229–31
  visceral receptors 231–32
  volume 712
sensory transduction
  226–228
septum pellucidum 187
serosa 495
serotonin 85, 136
  role in sleep 274
  role in pain 235
serous membranes 76
serum 406
Sertoli cells 579–82, 587–88
sesamoid bones 141
sex chromosomes 582,
  647, 751

sex steroids 304, 306, 308,
  581–82, 595, 602–3, 751
sex-linked genetic
  disorders 661
sex-linked inheritance 661
  haemophilia A 662–64
  mitochondrial 664–65
sexual differentiation 636–38
sexual reflexes 608–610
SGLT 1 522
shell temperature 695–96
shivering 701
short bones 141
shoulder 142, 164–65
  bones of 149–50
  rotator cuff 165
shunts (fetal
  circulation) 628–30,
  633–34, 751
sickle cell disease 401
sigmoid colon 526, 527,
  528, 529
signal transduiction 86–8
signalling cascades
  86–93, 751
  G proteins and 88–90
signalling molecules 83–5
simple learning 277
single nucleotide
  polymorphism 653
single unit smooth
  muscle 132, 496
sino-atrial node 334, 335,
  336, 354
sinus bradycardia 360
sinusoids
  liver 535, 536, 537
  spleen 417
sinus rhythm 360
sinus tachycardia 350
skeletal muscle
  ATP 122–24
  blood supply 73
  contraction of 121–22
  disorders 180
  fatigue 124
  force summation 125–27
  force-velocity curve 127
  innervation of 124
  length-tension
    relationship 122–23, 128
  mechanical
    properties 124–29
  power of 127–28
  sliding filament
    theory 122–23
  structure 120

tendon 70, 126
tetanus 125
  twitch tension 125
  types of 125–26
skeletal muscle pump 374,
  375, 407
skeleton 140–53
  role in buffering 721
skin
  barrier to infection 415
  blood supply 564, 698–99
  carcinomas of 560
  glabrous 557, 565
  grafts 428–429
  hairy 557, 565
  pharmacology 571–72
  and temperature
    regulation 698–99
  sensory receptors 229, 230,
    565–66, 700
  structure 557–61
  thickness 556
  see also hair
skin thermoreceptors
  231, 700
skinfold thickness 679
skull 141, 143–46, 148
sleep 272–275
sleep apnoea 274
sleep deprivation 273
sleep spindles (EEG)
  270, 273
sleep stages 270, 272–273
sleep-wakefulness cycle
  272–274
  age-related changes 273
  circadian rhythm
    275–277
  regulation 273–274
sliding-filament
  theory 122–23
slow-twitch muscle
  fibres 125, 126
slow-wave sleep (ortho-
  sleep) 273, 291
small intestine 515–525
  absorption 521–525
  adaptations of
    516–518, 521
  anatomy 494, 515–516
  digestion 519–521
  emptying 518
  motility of 518
  see also
    gastrointestinal (GI)
    system
smell (olfaction) 259

smooth muscle 72–3, 131–36
  contractile response
    133–35
  distribution of 131
  excitation-contraction
    coupling 133–34
  hormonal 132–33
  innervation 132
  length-tension
    relationship 135
  multi-unit 132–33
  myogenic activity 132
  and nitric oxide 92
  pacemaker cells 134
  role of calcium ions 134
  single unit (visceral) 132, 496
  structure 131–32
  tone 92
sneeze 456
sodium
  action potential 106–8
  body fluid
    concentrations 56
  body fluid volume
    regulation 712–13
  channels 107–8, 480
  channel blockers 363
  hormonal regulation 381,
    385, 480–481
  linked transporters 60–61,
    477–78, 479–80, 522
  and local
    anaesthetics 111–12
  plasma concentration 390
  pump 59
  retention 491
  tubular transport 473–474,
    477–481
sodium channel 107–8
sodium cromoglicate 462
sodium picosulphate 530
sodium pump 59–61, 108, 751
sodium valproate 271
soft callus 160
soft palate 501, 504, 505
somatic nerves 232
somatomedins
  see IGF-1, IGF-2
somatosensory cortex
  233, 234
somatosensory
  system 229–38, 565
somatostatin 83, 85
  digestive system 512, 554
  gastric secretion 512
  growth hormone
    and 288–89, 291

islets of Langerhans 554
  plasma glucose
    regulation 554, 555
somatotopic maps (motor and
  sensory) 209, 234, 751
somatotrophin, *see growth
  hormone*
somatotrophs 288, 291
sorbitol 316, 529
sotalol 351, 363
Sotos' syndrome 295
sound
  properties of 250
  transduction 253–55
space of Disse 536
specific dynamic action (of
  protein) 688
speech 266–267, 268–269
  disorders 213, 268–269
  pathways 268–269
sperm 579–83
  development of 582–88
  fertilization 610
  structure 586
spermatocytes 582, 586
spermatogenesis 578, 582–87
  hormonal control
    of 587–88
spermatogonia 582
spermatozoa *see sperm*
spermiogenesis 586
Sperry, R.W. 266–267
sphenoid bone 143–45
sphincter 751
sphincter of Oddi 515, 516,
  534, 535, 538, 548
sphygmomanometer 367
spina bifida 625, 676
spinal anaesthesia 728
spinal cord 183, 187–90
  in motor control 199–201
spinal cord injury 205–6
  loss of bladder control 486
spinal nerves 189
spinal roots 189–90
spinal shock 205, 206, 751
spine 147–48
spinothalamic tract 232
spiral arteries 590, 599, 600
spirometer 441–442
spironolactone 95, 308,
  481, 718
splanchnic circulation 497,
  498, 751
spleen 368, 393, 394, 414,
  415, 417, 498, 516, 538
split-brain operation 266, 267

spongy bone 155, 156
squint 249
SRY gene 636, 647, 661
stapes 252
Starling forces 383, 751
  body fluid regulation 712
  capillary filtration 383, 451
  oedema 451–452, 717
Starling-Landis principle *see
  Starling forces*
Starling's Law of the Heart 344
starvation 688
steatorrhoea 523
stellate veins 471
stem cells 391, 393, 751
stercobilin 528, 538, 539
stereocilia 253–54
stereognosis 267, 751
streroscopic vision 246
stereognosis 267
sternum 148–49
steroids 31, 33, 751
steroid hormone(s) 285
  adrenal cortical
    hormones 304–9
  androgens 306, 308,
    579, 595
  oestrogens 595, 602
  progesterone 597, 602
  receptors 93, 582
  sex steroids 306, 751
  testosterone 579, 581
stomach 494, 498,
  506–515, 525
  absorption by 514
  acid secretion 509–510
  anatomy of 506–507
  autonomic
    innervation 499–500
  blood supply 497, 498, 507
  digestive function 509
  intrinsic factor 509
  motor activity 513–514
  secretory activity 508–511
  storage function 506
  wall, structure of 507–508
streptokinase 352, 407
streptomycin 255
stress 511, 513, 521
stretch receptors
  baroreceptors 379–380
  low pressure (volume)
    receptors 712
  of lung 452–453
  Golgi tendon organ 201
  muscle spindles 200
stretch reflex 202

striated muscle *see cardiac
  muscle, skeletal muscle*
stroke 195, 208, 210, 264, 265,
  406, 751
stroke volume 337, 341,
  343–347, 376
stroke work 346
ST segment (ECG) 358, 362
styloid process (stylus) 144
subclavian vein 367, 369,
  383, 384
subcutaneous tissue 561, 564
subdural haematoma 192
sublingual drug
  administration 727
sublingual glands 494,
  502–503
submandibular glands 494,
  502–503
submucosa 495
submucosal (Meissner's)
  plexus 495, 496, 499
substance P 136
substantia nigra 186, 213–14
succinylcholine 130
succus entericus 519–520
suckling 642–44
sucralphate 513
sucrose 28
sudoriferous glands, *see
  sweat glands*
sugars
  absorption 521–22
  chemical structures 28–29
  energy production 47–50
  nutrition 671, 690
  see also carbohydrates
sulcus 185, 751
sulfadiazine 570
sulfinpyrazone 167
sulphasalazine 167
sulphonylureas 323
sumatriptan 237
summation 114
superior (anatomy) 12
superior mesenteric
  artery 367, 497, 498,
  516, 527
superior mesenteric
  ganglion 500
superior vena cava 328, 329,
  330, 367, 369
supination 10, 752
supine 752
suprachiasmatic
  nucleus 275, 276
supraoptic nuclei 287, 290

surface tension 445
surfactant 438, 445
suspensory ligament 239, 243
sutures 145–46, 163, 752
swallowing 504–506
sweat (sudoriferous)
        glands 562–63, 701, 714
    role in heat loss 701
sweating 701
sympathetic ganglia 216, 217
sympathetic nervous system
    anatomical
        organization 217–19
    blood vessels and 376–377,
        378–380, 381, 537, 549
    digestive system 377,
        499–500, 503, 518, 521
    exercise 377, 537
    heart and circulation
        342–343, 345–346, 347,
        376–377, 378–380, 381, 537
    kidneys 471
    neurotransmitters 221–223
    post-ganglionic fibres 342
    preganglionic fibres 342
    respiratory system 440
sympathetic tone 221
sympathomimetic drugs 504
symphysis 151, 163
symport 58–9, 752
synapse 83, 104, 752
    excitatory 112–14
    inhibitory 112, 114
    neuromuscular
        junction 129–30
    structure of 112
synaptic signalling 83
synaptic
    transmission 112–15
    autonomic nervous
        system 221–223
    excitatory 113–14
    inhibitory 114
    mechanisms of 113–15
    neuromuscular
        junction 129–30
synarthrosis 163
synchondrosis 163
syncytiotrophoblast 613,
    615, 616
syndesmosis 163
synovial joints 163,
    165–66, 752
synovial membrane 164
systemic arterial blood
        pressure 367, 371–373,
        379–381

systemic circulation
        366–369, 752
systems of body 56
systole 337, 338, 752
systolic blood pressure 367,
    371, 372, 380

T

T cell receptor 423
T cells 393, 403, 420, 421,
    423–425, 426, 524
    classification of 423
T-lymphocytes *see T-cells*
T-tubules 120
T wave 338, 340, 358, 359
T3, T4 *see thyroid hormones*
tachycardia 341, 752
tacrolimus 429
taeniae coli 525
talus 142, 153
tarsal bones/tarsus 141, 153
taste, sense of 259
Tau protein 278
Tay-Sachs disease 44, 661
tear fluid 241
tear glands (lacrimal
        gland) 218, 241
tectoral membrane 253
tectospinal tract 208
teeth 494, 501–502
temazepam 275
temperature
    core 695–96
    fever 704–5
    measurement 696
    neonates 703
    receptors 231, 565–66
    regulation 700
    shell 695–96
temporal bone 142–44
temporal lobe 264, 268,
    280–281
temporal summation
    muscle contraction
        (tetanus) 125–26
    of synaptic potentials 114
tendon 70, 126, 168
tendon tap reflex,
    see stretch reflex
tenosynovitis 169
TENS 235, 237
teratogenic drug effects 272,
    737, 752
terbutaline 462
terminal arterioles 371
terminal bronchioles 435

terminal sulcus (of
    tongue) 502
testes
    development 636–37
    hormones 581–82
    structure of 577–81
testosterone 31, 33, 579,
    581–82, 752
    role in development
        582, 638
    metabolic effect 688
    secondary sex
        characteristics 588, 605
    secretion 581
    spermatogenesis 588
tetanus 125, 752
tetany 313
tetracycline 571, 737
tetraplegia
    (quadriplegia) 205, 752
thalamus
    anatomy 186
    sensory functions 232–33
thalassaemia 401
thalidomide 737
thecal cells 594–95
theophylline 462
therapeutic index 731
thermoreceptors 752
    central 700
    peripheral 231
thermoregulation 696–705
    in elderly 703–4
    in neonate 703
theta waves (EEG) 270,
    271, 273
thiamine (vitamin B_1) 674, 675
thiazides 490, 491
thionamides 301
thiopentone 346, 728, 731
thirst 710–11
thoracic cage 148
thoracic duct 383, 384
threadworm 530
threshold stimulus 106, 110,
    130, 752
thrombin 405, 406
thrombocytes (platelets) 391,
    392, 393, 404–406,
    407–408
thrombocytopathia 408
thrombocytopenia 407–408
thrombomodulin 406
thromboplastin (tissue
    factor) 404, 405
thrombopoietin 324, 404, 541
thrombosis 352, 375, 406, 752

thromboxane A_2 (TXA_2)
    91, 136
thrombus 406
thymus 293, 414, 415, 435
thyroglobulin 297, 298
thyroid gland 296–303
    anatomy 284, 296, 297
    blood supply 296, 297
    disorders of 301–303
    follicular cells 297–298
    histology 297–298
    hormones 284, 298–303
    parafollicular cells
        297, 298
thyroid hormones (T3,
    T4) 85, 297–303
    cardiovascular effects 300
    cold stress 298, 299–300
    disorders of secretion
        301–303
    effects on metabolism 299,
        300, 320, 687
    heat production 299–300
    regulation 286, 287, 289,
        298–299
    role in development 300,
        302–303
    secretion 298–299
    synthesis 298
    thyroxine (T4) 298
    tri-iodothyronine (T3) 298
    whole body actions 299,
        300, 302
thyroid-stimulating hormone
    (thyrotrophin, TSH) 286,
    287, 289, 298
thyrotrophin releasing
    hormone (TRH) 289
thyroxine *see thyroid
    hormones*
tibia 152–53
tidal volume 441, 442, 446
tight junction 510
timbre 250
timolol 242, 351, 363
tinnitus 256
tirofiban 350, 351
tissues
    blood and lymph 69, 383,
        388–389
    connective 70
    epithelial 74–5
    immunological 415,
        419–420, 421, 426
    muscle 72–3, 118–120
    nervous 71–2, 184–85
tissue fluid exchange 382–384

tissue plasminogen
  activator 404, 407
tissue thromboplastin
  404, 405
tissue transplantation 416,
  428–429
titratable acid 719
α-tocopherol (vitamin E) 674
tolbutamide 323
toes 153, 177
tolerance, drug 98, 737
tone
  smooth muscle 135, 496
  sound 250
  sympathetic 221
  vagal 221, 342
  vascular 376
tongue 494, 502, 505
  papillae 259–60, 502
  taste buds 260, 502
tonic activity 752
  autonomic 221, 342
  smooth muscle 135
tonicity of solutions 23, 752
tonotopic mapping 255
tonsils 415, 501
topical anaesthesia 727
topiramate 271
total body water 707–8
total lung capacity 441, 442
total peripheral
  resistance 376, 381
touch receptors 229, 230
  receptive fields 229
trabeculae (eye) 241
trabeculectomy 242
trabecular (spongy,
  cancellous) bone
  155–56
trace elements 677
trachea
  dead space 446
  epithelium 436, 437
  structure 435–436
transcellular fluid 27
transcription 652
transcutaneous electrical nerve
  stimulation see TENS
transduction
  phototransduction 245
  sensory 226–28, 253–55
  signal 86–90, 93
transfer factor (diffusing
  capacity) 451
transfer RNA (tRNA) 652
transferrin 394, 395, 541
transfusion (blood) 409–412

transient ischaemic attack
  (TIA) 407
transitional milk 640–41
translation (protein
  synthesis) 652
transplantation and the
  immune system 416,
  428–429
transport maximum
  definition of 475
  glucose 475, 476
  PAH 475, 477
transverse section
  (anatomy) 10, 752
transverse colon 494, 526, 527
trazodone 275
triamterene 490
triazolam 275
tricarboxylic acid (TCA)
  cycle 43, 316, 540
tricuspid valve 329, 337–339
tricyclic antidepressants
  237, 504
trigeminal neuralgia 236
triglycerides 31, 520, 752
trigone 485
tri-iodothyronine (T3) see
  thyroid hormones,
trimester 613, 621, 751
trimipramine 275
triple response 418, 419, 567
triprolidine 136
trisomy 21 (Down
  syndrome) 650
trophoblast 613
tropicamide 223
trunk, muscles of 170, 173,
  174, 175
trypsin 519, 550, 551
trypsinogen 519, 550, 551
TSH see thyroid stimulating
  hormone
T-tubules 120–21
tuberosities (bone) 141, 752
D-tubocurarine 130
tubular reabsorption 473,
  474, 475–476, 477–479,
  480–482, 483–485
tubular secretion 480–482
tumescence (penile
  erection) 608
tunica adventitia 370–371
tunica albuginea 577
tunica intima 370
tunica media 370, 371
turbinates 146
turbulent flow 367, 372, 446

twitch tension 125
two-point
  discrimination 229–30
tympanic membrane (ear
  drum) 251
type 1 (insulin dependent)
  diabetes 321–23
type 2 (non-insulin
  dependent)
  diabetes 321–23
tyrosine 559, 743

U

ulcers 752
  decubitus (pressure
  sores) 568, 570
  diabetic 322
  duodenal 521
  gastric 511, 513
ulcerative colitis 530
ulna 142, 150
ultradian rhythms 275, 752
ultrafiltration 473–474, 752
unipolar leads (ECG) 355
uniport 58, 61, 752
universal donor 410, 411
universal recipient 410, 411
unmyelinated axons 105,
  106, 109
unsaturated fats 672
unsaturated fatty acids 30
upper airway resistance 446
upper arm circumference 679
upper motoneuron
  lesions 205–6
urea
  cycle 543
  in liver failure 544
  in urine 487
  synthesis 542–543
ureter 467, 468, 485
urethra 467, 485, 577–78
urethral folds 637, 638
uric acid 491
  and gout 491
urinalysis (urine
  analysis) 487–489, 752
urinary incontinence 486
urinary tract 467, 485–486
  infection 489
urination (micturition)
  485–486
urine
  clarity 488
  colour in disease 487,
  489, 539

composition of 487–489
formation 473–474
pH 487, 488, 489
specific gravity 487, 488–489
urobilinogen 488, 489, 528,
  538, 539
urogenital diaphragm 485,
  486
uterine cycle 592, 599–600
uterus
  anatomy 589–91
  contraction of 620
  decidual reaction of 613
  in pregnacy 621
utricle 252, 257–58
uvula 501

V

v wave 339, 340
V/Q ratio 449–451
vaccination 424, 426
vagal reflexes 512
vagal stimulation 341–342
vagal tone 221
vagina 591
vaginal fluid 591
vaginal epithelium (mucosa)
  in ovarian cycle 601
vago-vagal reflexes 511,
  512, 552
vagus nerve 499, 500, 511,
  512, 521, 552
varices 545
vas deferens 577, 587
vasa recta 471
vascular endothelium 74,
  370, 371, 404, 752
vascular resistance 373,
  375–376
vasoactive intestinal
  polypeptide (VIP) 324,
  552
vasoconstriction 376, 404, 753
vasoconstrictor fibres 378
vasodilatation
  (vasodilation) 376,
  377–379, 380–381, 753
vasopressin (ADH) 85, 136,
  287, 289–291, 753
  in haemorrhage 379, 381
  and osmoregulation 712–13
  thirst 711
  volume regulation 712–13
  water reabsorption 483–84
  see also antidiuretic
  hormone

vecuronium 734
veins 366–369, 370, 371
  arcuate 471
  axillary 368, 369
  brachiocephalic 367, 369
  basilic 368, 369
  cardiac 332
  cephalic 368, 369
  digital 368
  femoral 368, 369
  gastric 507
  gonadal 369
  great 330
  hepatic 368, 369, 498, 534, 536
  iliac 368, 369
  jugular 297, 367–368, 369, 435
  medial cubital 368, 369
  medial vein of forearm 369
  popliteal 368, 369
  portal 498, 516
  pulmonary 329, 330, 369
  renal 368, 369, 467, 468
  saphenous 368, 369
  stellate 471
  subclavian 367, 369, 383, 384, 435
  tibial 369
venae cavae 330
  inferior 328, 329, 330, 367,369, 498, 535
  superior 328, 329, 330, 367, 369
venomotor tone 374
venous admixture 451, 753
venous pooling 374–375, 407
venous pressure 373–375
  central 371
  see also microcirculation
venous return 337, 341, 379,
ventilation (definition) 441
ventilatory capacity 458–459
ventilatory failure 462
ventilatory function tests 441–442, 458–459
ventral horns 189
ventral roots 189
ventricles
  cerebral 186, 192–93
  of heart 328, 329, 330
ventricular action potentials 335, 336
ventricular extrasystole 361
ventricular arrythmias 361

ventricular fibrillation 361, 362
ventricular myocytes 335, 336
ventricular tachycardia 361, 362, 363
venules 370, 371, 382
verapamil 350, 351, 363, 386
vertebrae, structure of 148
vertebral canal 148
vertebral column 147–48
vertigo 258
vesicles 44
  endosomes 65
  lysosomes 44, 64
  secretory 64
  synaptic 112, 113, 129
vestibular glands (of vulva) 591
vestibular system 252, 256–58
  ampulla 257
  disorders of 258
  and eye movements 257
  and posture 258
  saccule 257–58
  semicircular canals 256–57
  structure of 256
  utricle 257–58
vestibulo-ocular reflexes 257
vestibulospinal tract 207
villus/villi
  arachnoid 192
  intestinal 516, 517, 518
  intestinal, movements of 518
  of placenta 613, 615
VIP *see vasoactive intestinal polypeptide*
visceral afferents 224
visceral peritoneum 495
visceral pleura 439, 440
visceral receptors 231–32
viscerocranium 143
vision *see eye and visual pathways*
visual cortex 266
visual field 266, 267
visual illusions 246
visual pathways 266
vital capacity 441, 442
vitamin A (retinol) 626, 673
vitamin B$_1$ (thiamine) 675
vitamin B$_{12}$ (cobalamin) 401, 509, 676
  absorption of 509, 523, 524
  deficiency 401, 509, 676

vitamin B$_2$ (riboflavin) 674, 675
vitamin B$_3$, (niacin) 675–76
vitamin B$_6$ (pyridoxine) 676
vitamin B$_9$ (folic acid) 676
vitamin C 408, 676
vitamin D 312, 675
  calcium balance 312
  deficiency 312, 673–5
  intoxication 312, 673
  synthesis 312, 559–60
  toxicity 312, 673
vitamin E (α-tocopherol) 674
vitamin K 405, 407, 408, 526, 528, 675
vitamins
  absorption of 523, 524
  recommended daily allowance (RDA) 674
  see also specific vitamins
vitiligo 560
volatile acid 719
voltage-gated channels 62–3, 107–8
volume receptors 712
voluntary apnoea 452
voluntary control of respiration 452, 453
voluntary hyperpnoea 452
voluntary (goal directed) movement 208–11
vomer bone 146
vomiting 489, 509, 514–515
von Willebrand factor 408
von Willebrand's disease 408
vulva 591

W

Wada test 269
warfarin 407, 733, 737
warm receptors 231
water
  body distribution 27, 707–9
  and solutions 18–23
  tubular transport 479, 483–485
water balance 710
  diuretics 490–491
  disorders of 714–15
water channels 57, 484
  ADH regulation of 483–484
water intoxication 714
watt 683
weal/wheal formation 567
weaning 644

weight, body
  and BMI 679
weight gain 678
  in pregnancy 623, 625, 626
Wernicke's aphasia 268
Wernicke's area 268
Wernicke's encephalopathy 675
white blood cells (leukocytes) 389, 391, 392, 393, 401–403
white matter 186, 189
withdrawal (flexion) reflex 203
Wolffian ducts 636
work, muscle 127–28, 753
wound healing 567–70
wrist 151, 173

X

X chromosome 647, 662
x wave 340
xenobiotics 543, 733, 753
xerostomia 503, 504
xiphisternum 148–49

Y

Y chromosome 647
y wave 340
yolk sac 614

Z

Z line 120–21, 331
zeitgebers 275
zero-order kinetics 735
zinc 627
Zollinger-Ellison syndrome 513
zolpidem 275
zona fasciculata 304
zona glomerulosa 304
zona pellucida 594, 610
zonal reaction, *see cortical reaction*
zona reticularis 304
zonula adherens 77
zonula occludens 77
zonula of Zinn 239, *see suspensory ligament*
zygomatic bone 144, 146
zygote 576, 613, 753
zymogen granules 519, 548, 550